Pg 370
Simple Stat. de.

D1267447

Precision Aerial Delivery Systems: Modeling, Dynamics, and Control

Precision Aerial Delivery Systems: Modeling, Dynamics, and Control

EDITED BY

Oleg A. Yakimenko
Naval Postgraduate School
Monterey, California

Volume 248
Progress in Astronautics and Aeronautics

Timothy C. Lieuwen, Editor-in-Chief
Georgia Institute of Technology
Atlanta, Georgia

Published by
American Institute of Aeronautics and Astronautics, Inc.
1801 Alexander Bell Drive, Reston, VA 20191-4344

Front cover photograph shows a 10K DragonFly PADS; courtesy of Airborne Systems.

American Institute of Aeronautics and Astronautics, Inc., Reston, Virginia

1   2   3   4   5

To Tatiana

TABLE OF CONTENTS

Preface . . . . **xiii**

Conventions . . . . **xix**

Acronyms . . . . **xxi**

Common Nomenclature . . . . **xxvii**

Chapter 1 PADS and Measures of Their Effectiveness **1**

Oleg Yakimenko, *Naval Postgraduate School, Monterey, California*

1.1   Precision Aerial Delivery 2
1.2   Precision Airdrop Technology Demonstrations 16
1.3   PADS Accuracy 44

Chapter 2   Basic Analysis of Ram-Air Parachute **73**

Steven Lingard, *Vorticity Ltd., Chalgrove, Oxfordshire, United Kingdom*

2.1   General Description of Ram-Air Parachute 73
2.2   Aerodynamics of Ram-Air Wings 76
2.3   Aerodynamics of Ram-Air Parachutes 83
2.4   Flight Performance 92
2.5   Longitudinal Static Stability 93
2.6   Longitudinal Dynamics 101
2.7   Lateral Motion 115
2.8   Inflation of Ram-Air Parachutes 121

Chapter 3   Key Factors Affecting PADS Landing Precision **127**

Oleg Yakimenko, *Naval Postgraduate School, Monterey, California;*
   Horst Altmann, *Airbus Defence & Space, Manching, Germany*

3.1   Effect of Unknown Surface-Layer Winds 128
3.2   Influence of Control Rate and Actuator Dynamics 169
3.3   Utilizing Winds-Aloft Data 181

Chapter 4   Aerodynamic Characterization of Parafoils **199**

Hamid Johari, *California State University, Northridge, California;*
    Oleg Yakimenko, *Naval Postgraduate School, Monterey, California;*
    Thomas Jann, *Institute of Flight Systems of DLR, Braunschweig, Germany*

4.1   Theoretical Derivation of Aerodynamic Coefficients   200
4.2   CFD Analysis   218
4.3   Coupled FSI—CFD Analysis   245

Chapter 5   Equations of Motion **263**

Oleg Yakimenko, *Naval Postgraduate School, Monterey, California;*
    Nathan Slegers, *George Fox University, Newberg, Oregon*

5.1   Rigid-Body Models   264
5.2   Higher-Fidelity Models   296
5.3   Models of Control Efficiency   327
5.4   Linearized Models   338
5.5   Summary and Extension to Powered Systems   348

Chapter 6   Stability and Steady-State Performance **353**

Oleg Yakimenko, *Naval Postgraduate School, Monterey, California;*
    Thomas Jann, *Institute of Flight Systems of DLR, Braunschweig, Germany;*
    Gil Iosilevskii, *Technion—Israel Institute of Technology, Haifa, Israel;*
    Peter Crimi, *Andover Applied Sciences, Inc., West Boxford, Massachusetts;*
    Glen Brown, *Vorticity US, Santa Cruz, California*

6.1   Stability Analysis of the Linearized Model   354
6.2   Longitudinal Static Stability Analysis   361
6.3   Lateral-Directional Motion   369
6.4   Scaling Effects   383

Chapter 7   Guidance, Navigation, and Control **391**

Oleg Yakimenko, *Naval Postgraduate School, Monterey, California;*
    Thomas Jann, *Institute of Flight Systems of DLR, Braunschweig, Germany*

7.1   Background and General Terms   391
7.2   Navigation Solutions   395
7.3   Optimal Guidance and Practical Approaches   418
7.4   T-Approach Guidance   450
7.5   Bank of Trajectories and Band-Limited Guidance   459
7.6   Quasi-Optimal Terminal Guidance   464

7.7   Accounting for the Complex Surface-Layer Wind Models 487
7.8   PADS Control Schemes 501

Chapter 8   Glide-Angle Control 529

James Moore, *JM Technologies LLC, Tucson, Arizona*;
   Oleg Yakimenko, *Naval Postgraduate School, Monterey, California*

8.1   Drogue-Assisted Glide-Angle Control 530
8.2   Glide-Angle Control with Variable Rigging Angle 544
8.3   Glide-Angle Control Using Upper-Surface Spoilers 582

Chapter 9   Control of Non-Gliding Parachute Systems 601

Oleg Yakimenko, *Naval Postgraduate School, Monterey, California*;
   Travis Fields, *University of Missouri-Kansas City, Kansas City, Missouri*

9.1   Non-Gliding Parachute Guidance Using Canopy Distortion 601
9.2   Non-Gliding Parachute Guidance Using Reefing Control 652

Chapter 10   Flight-Test Instrumentation 685

Thomas Jann, *Institute of Flight Systems of DLR, Braunschweig, Germany*;
   Oleg Yakimenko, *Naval Postgraduate School, Monterey, California*

10.1 Onboard Instrumentation 685
10.2 Ground Equipment 714

Chapter 11   Parametrical Identification of Parachute Systems 741

Oleg Yakimenko, *Naval Postgraduate School, Monterey, California*;
   Thomas Jann, *Institute of Flight Systems of DLR, Braunschweig, Germany*

11.1 Methodology of Parameter Estimation 742
11.2 Flight-Path Reconstruction 746
11.3 Parameter Estimation Using Output-Error Method 753
11.4 Multicriteria Parametrical Identification 762
11.5 Filter-Error Method 787

Appendix A: AIAA Aerodynamic Decelerator Systems

TC Activities 829

Appendix B: Units Conversions 837

Appendix C: Canopy Data 843

Appendix D: Computation of CEP$_{DMPI}$ and CEP$_{MPI}$ for Noncircular Distribution 849

Appendix E: PADS Geometry and Mass Properties Modeling 853

Appendix F: PADS Penetration Speed 859

Appendix G: Geometry and Mass Properties of Different-Weight PADS Models 865

Appendix H: Geometry and Mass Properties of the G-12-Based AGAS 869

References 871

Contributing Authors 901

Index 905

Supporting Materials 939

PREFACE

This volume represents an effort to analyze and systematize recent developments in the area of precision aerial cargo delivery using parafoil- and parachute-based systems from the standpoint of designing guidance, navigation, and control (GNC) algorithms for them. As such, this book is primarily concerned with a variety of different issues associated with modeling of precision aerial delivery system (PADS) after its canopy is fully deployed (i.e., excluding the PADS extraction from an air carrier and canopy opening stages).

I was lucky to be involved with PADS modeling and GNC design and testing since 1999. Back then, I had a privilege of working with the Air Delivery Branch of the U.S. Army Yuma Proving Ground (YPG), Precision Aerial Delivery Team of then U.S. Army Soldier and Biological Chemical Command, and Vertigo, Inc., on the Affordable Guided Airdrop System. This system was heavily relying on predicted winds supplied by the Mission Planner, the development of the Planning Systems, Inc.. and Charles Stark Draper Laboratory, Inc., and so I got to work with their specialists as well. As a continuation of these efforts, the Aerodynamic Decelerator Systems Center (ADSC) was founded at the Naval Postgraduate School in 2001 with the goal to identify and develop open technology areas and novel approaches and/or systems with maximum potential of providing critical improvements and new capabilities to aerodynamic decelerator systems (ADS). Since then, I have had a strong relationship with YPG participating in a variety of ADS-related projects and attending all precision airdrop technology conferences and demonstrations (PATCAD) that took place in there. At about the same time, I joined the AIAA Aerodynamic Decelerator Systems Technical Committee (ADS TC) and had the honor to serve the ADS community in a variety of capacities including being on the organizing committee for 2011, 2013, and 2015 Aerodynamic Decelerator Systems Technology (ADST) conferences as a technical program cochair and a general chair.

Having some knowledge and expertise in PADS modeling and GNC design and testing, this book was not considered as an attempt to write a history of development of GNC algorithms for them. The goal was not to describe the fielded systems either, especially because information about their GNC algorithms is obviously not available. The goal, however, is to summarize the latest developments in the subject area, which involves mentioning a variety of actual fielded systems, canopies, cargo and airborne guidance unit (AGU) suspension schemes, and GNC concepts to facilitate future research efforts. This book presents the personal views of its contributors and is primarily based on the earlier open publications on different aspects of designing GNC software and other efforts supporting the design process, such as modeling and system identification. Also, this volume primarily discusses only those concepts proved in the field texts, which leaves out several recent theoretical studies on GNC design.

It took over two years to complete this book, and because of the enormous commitment it required, an earlier attempt undertaken right after the last PATCAD failed. In its current version, the outline of the book is as follows.

Chapter 1 overviews major developments on precision aerial delivery that took place in the past two decades. It starts from the overall concept of precision cargo delivery, introduces JPADS program, and specifies key performance parameters (KPPs), that is, performance attributes of PADS considered critical to the development of an effective military capability. It then talks about different-weight PADS developed by different vendors and demonstrated during major airdrop events in 2001–2009. The reminder of the chapter aims at understanding and establishing a common procedure for estimating the key measure of PADS effectiveness—touchdown accuracy.

Chapter 2 represents a slightly edited material on ram-air parachute design presented by Dr. Steven Lingard at the Precision Aerial Delivery Seminar in 1995 and serves as a starting point for the following discussions. It addresses aerodynamics of ram-air wings and parachutes, their stability and flight performance. It also addresses inflation of ram-air parachutes (even though this specific topic is out of the scope for this book because for GNC algorithms development PADS is thought of as a system with a fully deployed and stabilized main canopy).

Chapter 3 investigates the key factors affecting PADS landing precision. It is not a secret that the surface-layer winds never known precisely contribute to the touchdown error the most; hence, this chapter talks about measuring, modeling and usage of these winds, which includes computation of the so-called ballistic winds. The effect of modeling realistic controller, actuator, and atmosphere are discussed as well.

More detailed characterization of parafoil aerodynamics based on analytically derived dependences for the aerodynamic and control derivatives as well as computational fluid dynamics (CFD) analysis is presented in Chapter 4. This chapter also touches a coupled fluid-structure interaction (FSI)—CFD analysis with a goal of revealing the shape-related factors influencing parafoil aerodynamics and flight performance.

Chapter 5 is one of the main chapters from the standpoint of PADS modeling and is devoted to derivation of equations of motion. It formally introduces different systems of coordinates and PADS states, describes translational and rotation kinematics, introduces apparent masses and inertia, and presents several rigid-body models of different complexity. It then proceeds with the higher-fidelity models involving canopy-payload interaction, which depends on a specific payload suspension overviewed in Chapter 1, and suggests how the complex-shape and powered payload models could be incorporated. The chapter shows influence of control inputs on the turn rate and steady-state values of horizontal and vertical components of the airspeed vector and presents several linearized models of different complexity that could be used for simulations, stability analysis, and controllers design.

Chapter 6 addresses PADS stability and steady-state performance in longitudinal and especially in lateral-directional motion and shows how PADS are different from conventional aircraft. The chapter ends with discussing the scaling effects and their effects on PADS dynamics.

Chapter 7 is the key chapter in this book and overviews the essence and content of the GNC blocks of PADS AGU from the algorithmic standpoint. The chapter starts from showing different approaches to come up with a navigation solution, which includes estimating PADS heading, current winds, airspeed, and altitude above the surface. Next, it proceeds with a guidance portion first explaining what the optimal control would look like, what simplifications could be made to incorporate optimal guidance in the feedback control. It then focuses on the terminal guidance describing earlier maneuver-based strategies and precision placement guidance algorithms. Because most of the fielded algorithms are proprietary, this chapter is primarily based on the work accomplished by the government organizations and universities. Among those published algorithms, only algorithms that were actually tested in practice are described. As a result, the detailed discussion involves the T-approach guidance and two direct-method-based approaches, band-limited guidance, and inverse-dynamics-in-the-virtual-domain guidance. Finally, the chapter considers several published schemes to convert guidance commands to the servos control inputs to steer PADS by the corresponding deflection of its control surfaces. This includes traditional and advanced (adaptive) proportional, integral and derivative, model predictive controllers, and a nonlinear dynamic inversion controller.

Chapter 8 follows Chapter 7 with the most recent PADS control techniques under development that involve varying a glide slope angle of a descending system on the final approach. The different ways of controlling the glide angle include a discrete (one-time) drogue-assisted glide-angle control, as well as continuous glide-angle control techniques using variable canopy rigging and upper surface spoilers.

Even though PADS based on relatively high-glide ratio ram-air canopies are very attractive compared to conventional round-canopy based cargo aerial delivery systems, the latter ones are cheaper and still carry a major burden of resupplying remote locations. Chapter 9 presents two concepts of modernizing these conventional systems, so that they have some control authority to satisfy the payload delivery accuracy KPP. Even though the two discussed concepts rely on parachutes rather than parafoils and therefore require different models, the efforts on modeling and controlling them still fall under precision cargo delivery and therefore are presented in this chapter for completeness. One concept involves distortion of parachute canopy's skirt to produce some horizontal control authority without changing the rate of descent much and another, on the contrary, uses the descent rate control for the high-altitude airdrops to utilize the difference in air speed/direction at the different altitude layers to its advantage.

Developing a good control algorithm involves understanding PADS dynamics and therefore involves system identification step. To this end, Chapter 10 describes instrumentation that might be used to record and estimate the PADS states and other parameters during flight tests in the order to gather a complete set to be used in the following parameter identification efforts. It deals with both onboard instrumentation and ground instrumentation that could be used and shows different aspects of developing and testing these instrumentation capabilities.

Finally, Chapter 11 deals with parametrical identification of parachute systems itself. It starts from general methodology of parameter estimation and approaches to validate flight-test data and then proceeds with showing how parameter estimation could be accomplished using output-error method, multicriteria parametrical identification, and filter-error method. Again, as it was the case in Chapter 7, among all published approaches just a few ones, well documented, tested in practice, and proved to be able to identify of up to several dozen of unknown parameters rather just a few ones, are presented in this chapter.

Being a long-time member of the AIAA ADS TC, I took advantage of this publication to present different activities of this TC and have the most up-to-date information about 23 biennial conferences it organized since 1966, 12 biennial seminars (since 1993), biennial parachute systems short course, and awards, gathered in one place for future reference. That is why the Appendices section of the book starts from Appendix A providing this information and the links to a few valuable resources.

Throughout the text, all numerical values are given in the international system of units followed by imperial system of units (in the parentheses); however, the figures keep their original units wherever they were taken from, that is, without conversion to one specific system of units. Hence, Appendix B provides the units conversion tables so that the book could be used by those who utilize the imperial system of units and those preferring an international system of units without the need to find the conversion tables somewhere else. Mass is given in kilograms (tons) and pounds, altitude in meters and feet, vertical component of the velocity vector (descent rate) in meters/second and feet per second, and horizontal component of velocity vector in meters/second and knots.

Appendix C supplements Chapters 1, 4, and 5 and provides some representative data on different-size canopies (for modeling purposes). Appendix D includes derivation of some statistical formulas for noncircular distribution used in Chapter 1. Appendix E also supplements Chapter 5 and is devoted to PADS geometry-mass properties modeling. Appendix F supplements Chapter 6 and discusses some issues related to the maximum airspeed PADS can achieve to improve winds penetration. Appendix G provides some data on modeling of different-weight PADS geometry and mass properties to supplement investigation of the scaling effects in Chapter 6. Finally, Appendix H provides geometry and mass data on the G-12 circular parachute modeled in Chapter 9.

Quite a few authors were invited to contribute to this book. Pretty soon however, it was realized that it requires a real commitment, so not everybody could afford it. I envisioned having a couple of more chapters in this book, but even in its current version the book covers most of material related to the subject. I am thankful to those authors who agreed to work with me on this volume. I am also thankful to those individuals who could not afford spending several months on writing the chapters, but gave me a permission to use their material in this book (that is how I got involved into writing almost every chapter). To this end, my thanks go to the following:

- Dr. Yuri Mosseev for providing materials on the use of MONSTR INTE-GRATED software package for coupled FSI–CFD study of FASTWing PADS and relative research (Sec. 4.3), Dr. Simon Benolol of CIMSA Ingenieria de Sistemas S.A. for allowing using these materials as an illustrative example in this book, and Professor Oleg Lemko of National Aviation University of Ukraine for allowing the use of his material for parts of Sec. 4.3.
- Professor Ping Lu of the Iowa State University for allowing the use of the material he published with his student Branden Rademaher (Sec. 7.3), Dr. David Carter for allowing the use of his work accomplished together with his colleagues at the Draper Laboratory (in Secs. 3.3.1, 7.5, and 7.8.1), and Professor Anthony Calise of the Georgia Institute of Technology for his permission to utilize his material on the adaptive PID controllers design (in Sec. 7.8.2).
- Professor Mark Costello of the Georgia Institute of Technology for giving me a permission to use the material produced by his research group (in Secs. 3.1.5, 8.2.1, 8.2.3, 8.2.5, and 8.3).
- Dr. Robert Rogers of Rogers Engineering & Associates for providing his material on system identification of the NASA X-38 PADS (in Secs. 11.5.1–11.5.5) and others whom I communicated with on the subject of this book.

Working on this book, I had to look through a substantial body of conference and journal publications. For example, as shown in Appendix A during 23 ADST conferences covering a period of almost half a century approximately 1,500 papers dedicated to different aspects of parachute and parafoil systems development, testing, modeling, and simulations were presented. During the past decade, several dissertations devoted specifically to PADS modeling and GNC algorithms design were defended: Müller (2002), Jann and Slegers (2004), Hur (2005), Gimadieva, Om, and Prakash (2006), Rademacher and Peyada (2009), Gorman and Puranik (2011), Ward (2012), Fields (2013), and Hewgley (2014). The bibliography section of this book, composed of over 500 items, lists many of these publications that were used and mentioned throughout the book.

As just mentioned, even though almost every chapter of this book has some new material, that has not been published yet, the overall framework is based on previously published papers. Specifically, Chapter 1 is primarily based on [PATCAD 2001a; PATCAD 2001b; PATCAD 2003; PATCAD 2005; PATCAD 2007; PATCAD 2009; Yakimenko 2013]; Chapter 2 on [Lingard 1995]; Chapter 3 on [Altmann 2013; Corley and Yakimenko 2009; Benton and Yakimenko 2013; Wright et al. 2005a; Wright et al. 2005b; Yakimenko et al. 2009; Yakimenko and Slegers 2011]; Chapter 4 on [Jann 2003; Eslambolchi 2012; Mohammadi and Johari 2010; Peyada et al. 2007]; Chapter 5 on [Gorman and Slegers 2011a; Jann 2001; Mortaloni 2003; Mortaloni et al. 2003; Ochi et al. 2009; Slegers 2010; Yakimenko 2005]; Chapter 6 on [Brown 1993; Crimi 1990; Iosilevskii 1995; Jann 2004]; Chapter 7 on [Calise et al. 2007; Calise and Preston 2008; Calise and Preston 2009; Carter et al. 2007; Culpepper et al. 2013; Hattis et al. 2006; Jann 2001; Jann 2004; Jann 2005; Kaminer et al. 2005; Slegers and Costello 2005; Slegers and Yakimenko 2009; Yakimenko and Slegers 2011]; Chapter 8 on [Gavrilovski et al. 2012; Moore 2012; Moore 2013; Slegers et al. 2008; Ward et al. 2011b; Ward 2012; Ward and Costello 2013]; Chapter 9 on [Dobrokhodov et al. 2003; Fields et al. 2011; Fields et al. 2012; Fields et al. 2013; Yakimenko et al. 2004]; Chapter 10 on [Decker 2013; Jann et al. 1999; Jann 2001; Jann and Greiner-Perth 2009; Yakimenko 2005; Yakimenko et al. 2007]; Chapter 11 on [Jann and Strickert 2005; Jann 2006; Rogers 2002; Rogers 2004; Yakimenko and Statnikov 2005]. By the way, it was a real challenge to bring different notations together, but I think it was successfully accomplished throughout all chapters of the book, so that the Common Nomenclature section lists the most important ones.

Again, the main goal of this book was to overview the current status of the art in precision aerial cargo delivery from the standpoint of modeling and GNC algorithms design and as such to serve as a departing point for future research in this area. Now that the book is finished, I must say that I am quite satisfied with an outcome and think that it will serve its goal. Because it is a first book on this subject, any comments and/or corrections are very welcome (they will be incorporated in the updated edition of this book). At the 2013 ADST seminar, the 2009 Theodor W. Knacke Aerodynamic Decelerator Award winner Charles "Chuck" Lowry joked that trying to model parachute systems makes you insane. No one from the audience objected to this statement. I hope that completing this book still preserved some of my sanity.

To conclude, I would like to thank my family for encouraging me to complete this book even though it took several years from their lives watching me assembling and digesting all material related to the subject of this book and finally writing it working with all contributors. Special thanks go to my wife, who during these years mostly watched me from the back and had to cope with the fact that all horizontal surfaces in our house were covered with papers, reports, and simulation results. A long time ago our friend gifted me a rug with the following inscription: "A Pilot and A Normal Person Live Here." Well, darling, now that the book is finished, I hope to return to the normal life and at least clean the house.

Oleg Yakimenko
Seaside, California
August 2015

CONVENTIONS

To make it easier to follow the content, this textbook utilizes a set of the following conventions. Most of the variables in symbolic mathematical expressions appear in *italics*; for example, $L = mg \cos(\gamma)$. However, if the symbol represents a vector or matrix, rather than a scalar variable, **boldface type** is used, for example, $\dot{\mathbf{x}} = \mathbf{A}\mathbf{x} + \mathbf{B}u$ (in this expression $\mathbf{A}$ is a matrix, $\dot{\mathbf{x}}$, $\mathbf{x}$ and $\mathbf{B}$ are the vectors, and u is a scalar). To avoid confusion, multisymbol variables, like AR, DR, GC, and GR, are not italicized. The exception, however, is made for such commonly used variables as Re, St, and also for those comprising (a part of) the word, like *lat, lon, alt, Az, El, bias,* and *tilt*.

Most of the time variables use subscripts and/or superscripts, for example, δ_a refers to the asymmetric deflection of the trailing edge of canopy, $C_{l\delta_a}$ is the rolling-moment coefficient due to asymmetric control input, $C_{1;k}$ is an indexed variable referring to the element residing in the first row and kth column of a matrix, $\dot{\psi}^{s/p}$ is the yaw rate of store relative to parafoil, and $\chi_a^{a/c}$ is the air carrier head at release.

The rotation matrices use both subscripts and superscripts placed on the left, for example, ${}_p^b\mathbf{R}$ refers to the rotation matrix from the parafoil coordinate frame $\{p\}$ to the body coordinate frame $\{b\}$. Variables may have a bar or other symbols above them, for example, $\bar{\delta}_s$ refers to the normalized (to the range of $[0;1]$) control input, $\hat{\psi}$ denoted the estimate of the yaw angle, and $\tilde{u}$ is a component of the ground-speed vector in the canopy reference frame. The derivatives with respect to time are denoted by the dot(s), for example, $\dot{\chi}$ or $\ddot{\psi}$, the derivatives with respect to any other independent variable rather than time by the prime(s), for example, x'', λ'.

ACRONYMS

AALCT	Advanced Airdrop for Land Combat Technology
ACTD	Advanced Concept Technology Demonstration
ADP	air-data probe
ADS	aerodynamic decelerator system, also aerial delivery system
ADS TC	ADS Technical Committee
ADST	aerodynamic decelerator systems technology
AGAS	Affordable Guided Airdrop System
AGL	above ground level
AGU	aerial or airborne guidance unit
AHRS	attitude heading reference system
ALE	Arbitrary Lagrangean Eulerian (method)
ALEX	Autonomous Parafoil Landing Experiment
AoA	angle of attack
APADS	Advanced Precision Aerial Delivery System; Autonomous Precision Air Delivery System
APOS	actual position
AR	aspect ratio
ARS	advanced recovery systems
ATRK	along-track
ATW	angle to wind
BC	boundary conditions
BLG	band-limited guidance
BND	bivariate normal distribution
CADS	Controlled Aerial Delivery System
CARP	computed air release point
CDF	cumulative distribution function
CDM	canopy dynamics model
CDS	container delivery system
CEA	circular error average (aka MRE)
CEAS	Confederation of European Aerospace Societies
CEP	circular error probable (aka circular error probability or circle of equal probability)
CFD	computational fluid dynamics
c.g.	center of gravity
CI	confidence interval
CMDP	Common Mission Debrief Program
CND	circular normal distribution
COI	critical operational issues
CORS	continuously operating reference stations
c.p.	center of pressure
CRV	crew return vehicle

DFC	digital flight computer
DGPS	differential GPS
DLR	Institute of Flight Systems of German Aerospace Center (prior to 1999 known as Institute of Flight Mechanics)
DMPI	desired MPI
DNB	deployable noseboom
DNBM	DNB module
DoF	degree of freedom
DT&E	developmental and test and evaluation
DZ	drop zone
EA	error angle
EADS	European Aeronautic Defense and Space Company
ECEF	Earth-centered Earth-fixed
EGI	embedded GPS inertial navigation system
EKF	extended Kalman filter
EM	energy (altitude excess) management
EMC	energy management circle or center
EMTP	energy management turn points
ENU	East-North-Up (system of coordinates)
ERA	eigensystem realization
ESA	European Space Agency
ESC	electronic speed controller
FA	final approach
FAA	Federal Aviation Administration
FADS	flush air-data system
FAST	features from accelerated segment
FASTWing CL	foldable adaptive steerable textile wing for delivery of capital loads
FASTWing	foldable adaptive steerable textile wing
FDAS	flight data acquisition system (-L: light, -M: medium)
FEA	finite element analysis
FFT	fast Fourier transform
FMC	fully mission capable
FMS	flight management system
FPR	flight-path reconstruction
FSI	fluid-structure interaction
FTP	final turn point
GA	glide angle
GC	glide coefficient
GDGPS	global differential GPS
GDS	Generic Delivery System
GNC	guidance navigation and control
GNCS	GNC system
GNSS	Global Navigation Satellite System

GPADS	Guided Parafoil Air Delivery System
GPS	Global Positioning System
GR	glide ratio (aka lift-to-drag ratio)
GS	glide slope
GSA	glide slope angle
GSI	glide slope intercept
GUI	graphical user interface
H	heavy (-weight system)
HARP	high-altitude release point
HMMWV	high-mobility, multipurpose, wheeled vehicle
IDVD	inverse dynamics in the virtual domain
IGS	international GNSS service
IMU	inertial measurement unit
IP	impact point
IPI	intended point of impact
IQR	interquartile range
IRIG	Inter-Range Instrumentation Group
ISS	International Space Station
JMUA	joint military utility assessment
JPACD	Joint Precision Airdrop Capability Demonstration
JPADS	Joint Precision Air Drop System
KIAS	knots indicated airspeed
KPP	key performance parameter
KTM	kineto-tracking mount
L	light (-weight system)
LANC	local application control bus system
LCF	linear complementary filter
LCP	left control position
LE	leading edge
LED	light-emitting diode
LIDAR	light detecting and ranging
LSL	left-straight-left (combination of motion primitives)
LSR	left-straight-right (combination of motion primitives)
LTP	local tangent plane
LWH	length/width/height
M	medium (-weight system)
MC	maneuverable canopy
MOE	measure of effectiveness
MEMS	microelectromechanical systems
MHE	material handling equipment
ML	micro-light (-weight system)
MOVI	multicriteria optimization/vector identification
MOP	measure of performance
MPC	model predictive control

MPI	mean point of impact
MPSS	mission planner and simulation software
MRE	median radial error (aka CEA)
MSER	maximally stable extremal regions
MSL	mean sea level
NASA	National Aeronautics and Space Administration
NCF	nonlinear complementary filter
NDGPS	Nationwide Differential GPS System
NED	North-East-Down (system of coordinates)
NEF	nonlinear estimation filter
NNW	North-North-West
NRDEC	U.S. Army Research, Development, and Engineering Center
NSRDEC	Natick Soldier Research, Development and Engineering Center
OA	operating angle
OKID	observer/Kalman filter identification
OT&E	operational test and evaluation
PACD	Precision Airdrop Capability Demonstration
PADS	parafoil aerial delivery system; precision aerial delivery system
PATCAD	Precision Airdrop Technology Conference and Demonstration
PDD	power distribution device
PDPAS	payload-derived position acquisition system
PDS	parafoil dynamics simulator
PI	point of impact, also proportional-integral (controller)
PIA	Parachute Industry Association
PID	proportional, integral, and derivative (controller)
PLS	Palletized Loading System
PMA	pneumatic muscle actuator
PPG	powered paraglider
PPS	precise positioning service (military GPS service)
PSD	power spectrum density
PSI	parameter space investigation
RAD	ram-air drogue
RAeS	Royal Aeronautical Society
RAM	random-access memory
RANS	Reynolds-averaged Navier-Stokes
RCP	right control position
RF	radio frequency
RLS	recursive least squares
RMS	root mean square
RMSE	root mean squared error
RNG	random numbers generator
RP	release point
RPM	revolutions per minute
RSL	right-straight-left (combination of motion primitives)

RSR	right-straight-right (combination of motion primitives)
RT	reference trajectory
RWLS	recursive weighted least squares
S	straight (segment)
SA	selective availability
SAASM	selective availability antispoofing module
SBIR	Small Business Innovation Research
SI	International System of Units
	(from French Le Système International d'Unités)
SID	system identification
SIFT	scale-invariant feature transforms
SISO	single-input single-output
SL	straight-left (combination of motion primitives)
SMD	surface-mount device
SNCA	Système de Navigation pour Charge Accompagneè
SODAR	sonic detection and ranging
SPADeS	Smart Parafoil Autonomous Delivery System
SPS	standard positioning service (civilian GPS service)
SR	straight-right (combination of motion primitives)
SRI	Southwest Research Incorporated
SURF	speeded up robust features
TA	target area
TBL	turbulent boundary layer
TE	trailing edge
TIP	turn initiation point
TPBV	two-point boundary-value
TR	turn rate
TSPI	time-space-position information
UAV	unmanned aerial vehicle
UCL	upper confidence limit
UL	ultra-light (-weight system)
USSOCOM	U.S. Special Operations Command
VACS	variable angle of attack system
VOR/DME	VHF omnidirectional range/distance measuring equipment
WAAS	Wide-Area Augmentation System
WCA	wind correction angle
WP	waypoint
WTA	wind to track angle
XL	extra-light (-weight system)
XTRK	cross-track
YPG	U.S. Army Yuma Proving Ground
2D	two-dimensional
3D	three-dimensional

COMMON NOMENCLATURE

A, B, C, D	state, input, output, and feed-through matrices of the linear model
$a, b, c\ (\bar{c})$	canopy arc, span, and chord (mean aerodynamic chord)
b_k, l_k	width of the deflected trailing edge (relative to b) and flap arm (distance from flap's center of pressure to wing's center)
$\{b\}$	body-fixed coordinate frame
C_D	drag coefficient $[D/(QS)]$
$C_{D\alpha}$	drag-curve slope vs angle-of-attack coefficient
$C_{D\delta_s}, C_{D\delta_s^2}$	drag coefficients due to the first and second powers of a symmetric control input
C_{D0}, C_{Di}	drag coefficient at zero lift and induced drag coefficient
C_L	lift coefficient $(L/(QS))$
$C_{L0}, C_{L\alpha}$	lift coefficient at zero angle of attack and lift-curve slope coefficient versus angle of attack
$C_{L\alpha}^a, C_{L\alpha}^{a'}$	airfoil (two-dimensional) lift-curve slope coefficient and modified lift-curve slope coefficient
$C_{L\delta_s}$	lift coefficient due to a symmetric control input
C_l	rolling-moment coefficient $(L/(QS))$
$C_{l\beta}$	rolling-moment coefficient due to sideslip (aka anhedral derivative)
$C_{l\delta_a}$	rolling-moment coefficient due to asymmetric control input
C_{lp}, C_{lr}	rolling-damping coefficients due to roll rate and yaw rate
C_m	pitching-moment coefficient $[M/(QS\bar{c})]$
C_{m0}	pitching-moment coefficient at zero lift
$C_{m\alpha}$	pitching-moment coefficient due to angle of attack (aka pitch stiffness)
C_{mq}	pitch-damping coefficient due to pitch rate
$C_{m\delta_s}$	pitching-moment coefficient due to a symmetric control input
C_n	yawing-moment coefficient $[N/(QSb)]$
$C_{n\beta}$	yawing-moment coefficient due to sideslip (aka directional stability and yaw stiffness)
$C_{n\delta_a}$	yawing-moment coefficient due to asymmetric control input
C_{np}, C_{nr}	yawing-moment coefficients due to roll rate and yaw rate

C_p	pressure coefficient (P/Q)
C_Y	side-force coefficient $[Y/(QS)]$
$C_{Y\beta}$	side-force coefficient due to sideslip
$C_{Y\delta_a}$	side-force coefficient due to asymmetric control input
C_{Yp}, C_{Yr}	side-force coefficients due to roll rate and yaw rate
$\mathbf{F}$	moment vector
g	acceleration due to Earth gravity
h	altitude $(h = -z)$
$\mathbf{I}_{a.m.}$, $\mathbf{I}_{a.i.}$	apparent mass and apparent inertia matrices
$\mathbf{I}_{n \times n}$	$n \times n$ identity matrix
$I_{xx}, I_{yy}, I_{zz}, I_{xz}$	components of inertia tensor
$\mathbf{I}^*$, $\mathbf{I}_s$	inertia matrices (of canopy, suspension lines and risers, and store)
$\{I\}$	inertial coordinate frame
J	cost function
K	drag polar coefficient
K_d, K_i, K_p	derivative, integral, and proportional control gains
L, D, Y	aerodynamic lift, drag, and side force (in $\{w\}$)
L, M, N	rolling, pitching, and yawing moment (in $\{b\}$)
$\mathbf{M}$	moment vector
M_r	mass ratio $(M_r = m\rho^{-1}S^{-1.5})$
m	mass
m_e	mass of enclosed air
m_s	mass of store (payload)
m^*	mass of canopy, suspension lines, and risers
$\{n\}$	local tangent plane coordinate frame
P	pressure
$\mathbf{P}$	vector of position errors in the horizontal plane
p, q, r	roll rate, pitch rate, and yaw rate in $\{b\}$
p^*, q^*, r^*	normalized roll rate, pitch rate, and yaw rate in $\{b\}(p^* = pbV_a^{-1}, q^* = qcV_a^{-1}, r^* = rcV_a^{-1})$
$\{p\}$	canopy-fixed coordinate frame
Q	freestream dynamic pressure $(0.5\rho V_a^2)$
R	radius or length of suspension lines
$\mathbf{R}$	rotation matrix
$\mathbf{R}_\lambda$	rotation matrix by angle λ
$_n^b\mathbf{R}$	rotation matrix from one coordinate frame (lower index) to another (upper index)
Re	Reynolds number
$\mathbf{r}_{AB}$	position vector from a point $\mathbf{A}$ to a point $\mathbf{B}$
r_c	effective Earth's radius at a specific latitude
r_e	equatorial Earth's radius
$\mathbf{r}_{LOS}$	line-of-sight vector
r_p	polar Earth's radius

S	reference area
$\mathbf{S}(\boldsymbol{\omega})$	skew-symmetric matrix composed of the components of vector $\boldsymbol{\omega}$
$St = f_s d / V_\infty$	Strouhal number
$\{s\}$	store (payload)-fixed coordinate frame
T	temperature or time interval
t	time or airfoil thickness
$\mathbf{u}$	control vector (e.g., $u = \delta_s$, $u = \delta_a$, $\mathbf{u} = [\delta_s, \delta_a]^T$, or $\mathbf{u} = [\delta_T, \delta_s, \delta_a]^T$)
$\mathbf{V}$	ground-speed vector (with the components $[V_x, V_y, V_z]^T$ in $\{n\}$ and $[u, v, w]^T$ in $\{b\}$)
$\mathbf{V}_a$	airspeed vector (with magnitude V_a and components $[v_x, v_y, v_z]^T$ in $\{b\}$)
V_d	descent rate or sink rate $(-\mathrm{d}h/\mathrm{d}t)$
V_G	projection of the ground-speed vector $\mathbf{V}$ onto the local tangent plane $(V_G = \sqrt{V_x^2 + V_y^2})$
V_h	horizontal (in $\{n\}$) component of airspeed vector $\mathbf{V}_a$ (wind penetration speed)
V_v	vertical (in $\{n\}$) component of airspeed vector $\mathbf{V}_a$
V_v^*	average vertical at used in lieu of $V_v(h)$
$\mathbf{W}$	wind vector (with magnitude W and components $[w_x, w_y, w_z]^T$ in $\{n\}$)
$W^{\mathrm{bal}}(H)$	magnitude of computed ballistic winds at altitude H
$\{w\}$	wind coordinate frame
$\mathbf{x}$	state vector (e.g., $\mathbf{x} = [u, v, w, p, q, r, \phi, \theta, \psi]^T$)
x, y, z	coordinates in the inertial (navigation) coordinate plane $\{I\}$ (or $\{n\}$)
X, Y, Z	axial, normal, and side components of the aerodynamic force vector in $\{b\}$
$\mathbf{y}$	output vector
α, β	angle of attack and sideslip angle
α_i, α_g	induced and geometric angle of attack
$\alpha_{mn}, m, n = 1, \ldots, 6$	components of added (apparent) masses and inertias tensor
α_{sp}	spatial angle of attack
α_0	zero-lift angle of attack
$\Gamma, \bar{\Gamma}$	circulation (vortex strength) and dimensionless circulation
γ	Earth-referenced flight-path angle
γ_a	flight-path angle (aka air-referenced flight-path angle)
δ	planform efficiency correction factor
Δt	time-step size
δ_l, δ_r	left and right control deflection
δ_s, δ_a	symmetric and asymmetric control deflection

ε	canopy anhedral angle (varying from $-\varepsilon_b$ to ε_b)
$\varepsilon_b, \varepsilon_c$	half-span angle and chord angle
θ	pitch angle
μ	rigging angle
Ξ	vector of varied parameters
ρ	air density
ϕ	roll angle
ϕ_a	bank angle
χ	ground-track angle or course over ground or track
χ_a	heading angle
χ_W	wind direction (usually defined by the "from" angle)
$\chi_W^{\text{bal}}(H)$	direction of computed ballistic winds at altitude H
ψ	yaw angle
$\boldsymbol{\omega}_{bn}$	angular velocity (of frame $\{b\}$ with respect to frame $\{n\}$)
ω_{LOS}	angular velocity of line of sight (in the vertical plane)
$\mathbf{0}_{m \times n}$	$m \times n$ zero matrix

SUPERSCRIPTS

a/c	aircraft
$a.i.$	apparent inertia
$a.m.$	apparent mass
c	canopy
cg	center of gravity (center of mass)
e	enclosed (refers to air entrapped in ram-air canopy)
k	deflected trailing edge
l	suspension lines
p	parafoil
r	risers
s	store (payload)
s/b	store with respect to body
s/p	store with respect to canopy
$\cap$	local parafoil axis (running along the wing span)
$\sim$	value expressed in $\{p\}$
$-$	mean (or normalized) value
$\wedge$	estimated value

CHAPTER 1

PADS and Measures of Their Effectiveness

Oleg Yakimenko*
Naval Postgraduate School, Monterey, California

These days, no one argues that cargo aerial delivery systems based on the controlled gliding parachutes have a lot of advantages compared to traditional ones based on the round canopies and assuming no control capability. Since these systems were introduced in the late 1960s, almost half a century ago, dozens of different-weight precision aerial delivery systems (PADS) were developed, tested, and successfully demonstrated for a variety of traditional and some new applications. Several systems have been fielded. Compared to an aircraft, these systems are unpowered, much slower, and underactuated. And, of course, they are vulnerable to the wind. As a result, while any pilot can execute an engine-off maneuver and land "on the numbers," achieving the same accuracy for PADS poses a serious challenge. From an operational standpoint, PADS should be capable of delivering cargo within a 50-m (164-ft) diameter around a desired location; however, this objective was met by very few PADS. Heavier PADS may exhibit much larger touchdown error on the order of 300 m (984 ft). Since the invention of gliding parachutes, different canopies, rigging, and control schemes have been intensively investigated before arrival at today's kind of standard architecture. These days, a PADS usually consists of a steerable large-size ram-air parafoil controlled by electrically driven actuators. Actuators, required sensors, computers, and complementary components are housed in an airborne guidance unit (AGU) strapped in between parafoil and payload or atop payload. This section is based on the open literature and overviews the genesis and current state in the precision aerial cargo delivery. Section 1.1 introduces a concept of operations and the joint U.S. Army/U.S. Air Force program that drives the development of different-weight PADS in the United States. This section categorizes these systems and sets key performance parameters including the ones related to the terminal accuracy, which AGU is supposed to deliver. Section 1.2 shows the variety of the developed systems as demonstrated at several major airdrop events. For each system this section provides some set of

---

*Professor.

data that can be used in modeling and simulations. Section 1.3 addresses the issue of establishing the common touchdown data analysis procedure, so that PADS performance could be compared against each other and against requirements. The correct application of this procedure allows assessing of both operational effectiveness and operational suitability of PADS.

1.1 PRECISION AERIAL DELIVERY

This section introduces the parafoil-based PADS, presents the concept of their operations, and lists the challenges that need to be overcome in order to ensure high effectiveness of PADS use in a variety of applications. It also introduces the JPADS program and mentions its genesis and current status. As far as this book is concerned, this section lists key performance parameters and sets the measures of PADS effectiveness related to the accuracy of a payload delivery.

1.1.1 CONCEPT OF OPERATIONS

Accurate delivery of payload by a parachute system has always been a key requirement irrelevant of its usage domain, commercial or military, space or aerial. Up until these days, cargo delivery/recovery has been primarily accomplished using relatively inexpensive conventional unguided systems employing round or cross-type canopies or their clusters to produce drag and therefore to reduce the vertical speed at impact. (The word *parachute* itself came via French from Italian *para-*, meaning to defend, shield against, and French *para* meaning collapse/downfall.) For different reasons, such a system needs to be deployed at high altitudes. For space systems, the high-altitude deployment might be dictated by the safety requirements of a recovery item, for military applications—by the safety of an air carrier. As a result, high-altitude deployment exposes aerodynamic decelerator system (ADS) to usually unknown winds for a longer period of time, so that the actual point of impact (PI) might be very far from the intended point of impact (IPI). The high-altitude (deployment)/low-opening (HALO) technique, when a main canopy opens very close to the ground, which mitigates the effect of unknown winds, may not always be an alternative to more reliable, standard, high-altitude high-opening (HAHO) technique when a main canopy opens soon after ADS exits an aircraft. Moreover, with some knowledge about the winds aloft HAHO techniques allow keeping the airlifter away from the drop zone (DZ), which is usually required from the operational standpoint as well.

In any way, if the aircraft flies at a safe altitude the drop accuracy is compromised due to imprecision of the ADS release point, atmosphere uncertainties, and unmodeled dynamics. For recovery from space, reentry trajectories can only place ADS at the canopy deployment position with certain tolerance as well. Hence, having some control authority allowing steering in a horizontal direction while descending, that is, having some gliding capability and wind penetration to reach the designated landing site, would be really beneficial [Wailes 1998].

That is why potential advantages of the gliding parachutes were immediately recognized when they were introduced in the late 1960s. Earlier research involved the different designs, which included parasail, cloverleaf, parawing, sailwing, parafoil, and volplane (Fig. 1.1) [Eilertson 1969; Forehand and Bair 1968]. The originally required gliding performance for land recovery of space vehicles (Apollo command module) was having a targeting capability of 18 km (9.7 n miles) if deployed at 3.1 km (10,200 ft) altitude. That would require a glide ratio (GR) of 5.8:1. It was determined that neither design shown in Fig.1.1 possessed this capability. With GR of 3:1 though (during the early flight tests, ram-air parafoil demonstrated the best GR of 3.3:1), the same targeting capability could be achieved if deployment occurs at 6 km (20,000 ft) above the ground level (AGL). These days a majority of PADS use ram-air parafoils. Because of their relatively high glide capability and controllability, ram-air parafoils offer considerable scope for the delivery or recovery of payloads to IPI by utilizing the corresponding guidance, navigation, and control (GNS) algorithm linked to its control surfaces, trailing edges (TE).

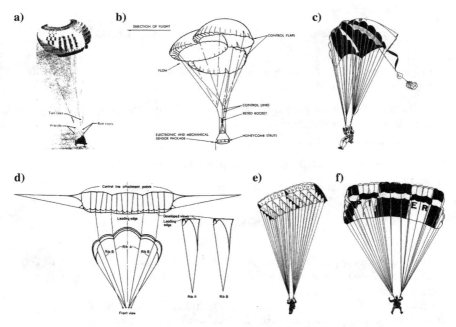

Fig. 1.1   Earlier steerable parachute concepts: a) parasail, b) cloverleaf, c) parawing, d) sailwing, e) parafoil, and f) volplane [Eilertson 1969].

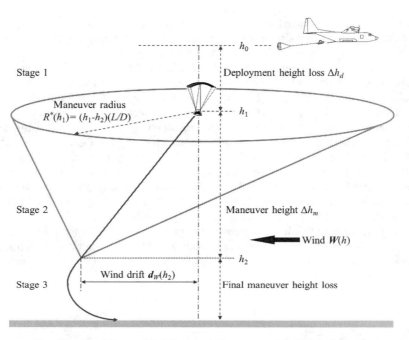

Fig. 1.2   Precision aerial delivery mission parameters [Lingard 1995].

Figure 1.2 shows a schematic diagram of a typical PADS mission. The airdrop can generally be divided into three stages [CARP 2005]:

- Release of the ram-air parachute out of an air carrier, canopy deployment, and establishment of a steady glide
- Maneuvering to the desired landing zone
- Final approach maneuver (aka terminal guidance) to land at IPI

The first stage results in some deployment altitude loss

$$\Delta h^d = h_0 - h_1 \tag{1.1}$$

This altitude loss depends on specific PADS and is usually on the order of 200–750 m (656–2,460 ft) (more specific numbers for the different-weight systems will be given in Sec. 3.3.2, Fig. 3.67).

Execution of the last stage requires a certain height (denoted in Fig. 1.3 as h_2) as well. For example, assuming the final maneuver to be comprised of the standard landing pattern starting from the base turn (with a radius R) followed by the final approach leg (of length d_f) strictly into the wind and no updrafts or downdrafts, we can write a ballpark estimate:

$$h_2 > \left(\frac{\pi R}{V_h} + \frac{d_f}{V_h - W} \right) V_v \qquad (1.2)$$

In Eq. (1.2), V_h is the horizontal projection of the (air-related) airspeed vector, V_v is the rate of descent, and W is the average magnitude of the surface-layer winds. Equation (1.2) can be further modified to

$$h_2 > \left(\frac{\pi R}{V_h} + \frac{d_f}{V_h - W} \right) \frac{V_h}{L/D} \approx \frac{\pi R + d_f(1 + \overline{W})}{L/D} \qquad (1.3)$$

where L/D is the ratio of PADS lift to drag (GR) and $\overline{W} = W/V_h$. [Derivation of Eq. (1.3) assumes $\overline{W} \ll 1$.] Typical PADS has a GR within the range of about 2:1 to 4:1, and the turn radius R varies from 15 m (50 ft) for small PADS all way up to 150 m (490 ft) for the very large ones (Sec. 5.3 discusses the turn rate of PADS in more detail).

Hence, when PADS is released at altitude h_0 the altitude available to maneuver from the release point to the desired landing zone shrinks to

$$\Delta h_m = h_0 - \Delta h_d - h_2 \qquad (1.4)$$

The volume within which PADS can maneuver to the landing zone is an inverted cone, which, in zero wind, has a half angle of $\tan^{-1}(L/D)$. The greater the L/D of a parachute, the greater the offset of realize point it can tolerate [Lingard 1995]. At any altitude $h^* \in [h_2; h_1]$, the maneuvering radius R^* from which the third stage initiation point can be reached is

$$R^* = \frac{L}{D}(h - h_2) \qquad (1.5)$$

and the effect of the winds distorting the cone can be accounted for by shifting the center of the cone by a horizontal vector $\mathbf{d}_W^*$ from the canopy deployment point

$$\mathbf{d}_W^* = \int_{h_1}^{h^*} \frac{\mathbf{W}(h)}{V_v(h)} \, dh \qquad (1.6)$$

where $\mathbf{W}(h)$ is the altitude-dependent wind vector and $V_v(h)$ is the altitude-dependent rate of descent. [Equation (1.6) will be given a more detailed consideration in Secs. 3.1.3, 3.1.4, and 7.3.4]. If PADS resides within the distorted cone, an efficient guidance system should ensure accurate payload delivery to IPI. If, because of any reason, PADS finds itself outside the cone, the PADS is likely to fail reaching the desired landing zone and will not be able to touch down at IPI. (The word "likely" is used because the reachability cone as well as the third stage depend upon actual rather than assumed winds; hence, generally speaking PADS may not reach IPI even if it stays within the cone computed using assumed winds and vice versa.)

The PADS release procedure usually employs a drogue chute to extract PADS out of an air carrier (Fig. 1.3a). Figure 1.3a also shows a fully deployed PADS consisting of the canopy, suspension lines, payload harness, and AGU. The steering lines are intended to deflect TE and provide a directional control. Different PADS designs can also include slider, risers, and other auxiliary equipment such as reefing, drogue release, and brake release cutters [Lingard 1995]. Figure 1.3b presents an example of the detailed deployment sequence for the DragonFly PADS.

Obviously, AGU serves as the brain of PADS. It includes sensors, computer board, and motors to pull the steering lines down. AGU gathers information about current PADS position with respect to IPI, estimates winds at a given altitude, and monitors overall PADS performance (navigates). Based on this information, AGU computes the best course of actions (guides) and produces and executes the servos control commands (controls). Hence, guidance, navigation, and control (GNC) software constitutes a core of AGU. Steering aerodynamically bad PADS with good GNC algorithms might be difficult, but accurately landing even aerodynamically perfect PADS with bad GNC software would simply be impossible.

The earliest PADS, like the Para-Point cargo delivery system, originally developed by SSE, Inc., and Para-Flite in the 1960s, or Parawing developed by Goodyear Aerospace for the U.S. Army Aviation Materiel Laboratories, used a pair of antenna mounted on the AGU to determine the bearing to a ground-based radio beacon [Forehand and Bair 1968]. The PADS was then simply turned toward a transmitter located at IPI. More details on such pre-GPS era guidance will be provided in Secs. 7.1 and 7.3.2, but it is clear that with a single beacon PADS does not know how far it is from IPI, cannot estimate the winds, and as a result cannot possibly achieve a desired accuracy and, in general, ensure landing into the wind. Adding more beacons and spreading them apart results in a necessity of having substantial ground infrastructure, which limits PADS autonomy. The major breakthrough in homing PADS was the introduction of satellite navigation using the global positioning system (GPS) and further cancellation of the so-called selective availability, intentionally worsening the accuracy of determining a current location. This allowed the PADS to know its current position all of the time and with much better accuracy and steer to a predetermined location without the need for a ground station.

In terms of GNC strategy and accuracy to aim at, the basic flight pattern conducted by sky divers can be looked at. Figure 1.4a shows a plan view of a typical flight, from opening to final approach. The homing maneuver might include crabbing, S turns, holding, and running to stay close to the wind line. The surface-layer wind estimates are assessed at a windcheck point about at 305 m (1,000 ft) upwind IPI. The stronger the winds, the further upwind this check must occur. The terminal maneuver suggests using half-brakes. The base

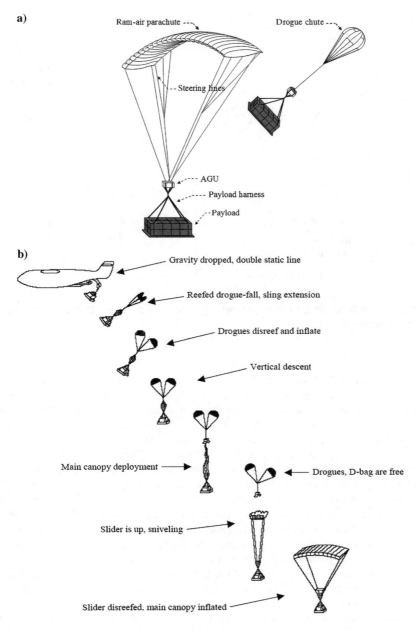

Fig. 1.3   a) PADS components and b) deployment sequence [Lingard 1995; Patel et al. 1997].

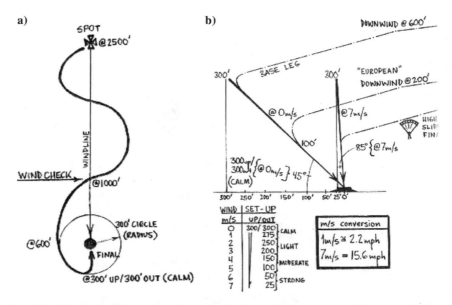

Fig. 1.4 **The basic flight pattern for sport skydiving: a) plan view and b) a vertical view [Eiff Aerodynamics 2015].**

turn initiation point is determined by the desire to end up behind IPI at the correct angle for the winds. Figure 1.4b depicts the setup points based on a starting altitude of 91 m (300 ft) for the final approach. Also shown is a shorter "European" style final approach [starting at approximately 35 m (100 ft) above IPI level]. If the winds are strong, the final approach should start well upwind of the target sliding sideways into a position only slightly behind IPI. During the final approach, brakes' adjustments are made up or down compared to the half-brakes setting depending on a relative position to the desired glide slope. Other terminal strategies (e.g., parabolic) are also available [Eiff Aerodynamics 2015]. As mentioned in [Hayhurst 1996], ram-air gliding parachutes caused a real revolution in terms of accuracy when they were introduced in skydiving in the early 1970s ending the era of round canopies. It is claimed that using a ram-air parachute and the aforementioned basic pattern, a beginner is capable of achieving a "10 jumps inside 10 m (33 ft) circle" performance in just 100 jumps [Eiff Aerodynamics 2015].

In the integrated capability-based "New Vistas Air and Space Power for 21st Century" report by the U.S Scientific Advisory Board, the global mobility in war and peace was identified as one of the six core capabilities for the Air Force in the 21st century. Specifically, the point-of-use delivery was called a truly revolutionary and highly needed capability [Macdonald et al. 1997].

The challenges in developing a point-of-use delivery capability were identified as follows:

• Cargo needs to be delivered without landing the aircraft to an accuracy of 10 to 20 m (33 to 66 ft) from altitudes up to at least 6 km (20,000 ft).

• Aircraft needs to be loaded with cargo and drop equipment at the same efficiency as for land delivery.

• Cargo should be extracted in random order.

• Drop equipment should be recovered and reused unless cost per drop unit is negligible.

Looking at the accuracy performance of the first-time jumpers (10 m), it might feel that having 10 to 20 m (33 to 66 ft) accuracy for a cargo delivery should not be a problem. The only caveat though is that a sky diver achieves the aforementioned accuracy using perfect sensors—eyes (a 20/20 vision is a requirement) and a windsock, which is always set up by IPI. In addition to that, jumpers heavily rely on a glide slope angle control that PADS do not have. (This issue is addressed in Chapter 8 in more detail.) No wonder that using the aforementioned features experienced jumpers can even achieve pin-point accuracy. According to [Eiff Aerodynamics 2015], it takes

• 300 jumps to be able to execute 10 jumps with a 5-m (16-ft) radius from IPI

• 500 jumps to be able to execute 10 jumps with a 15-cm (0.5-ft) radius from IPI

• 700 jumps to be able to execute 10 jumps with a 7-cm (2.8-in.) radius from IPI

• 900 jumps to be able to execute 10 jumps with a 4-cm (1.6-in.) radius from IPI

• 1000 jumps to be able to execute 10 jumps with a 1-cm (0.4-in.) radius from IPI

Of course, PADS developers do not have such a luxury to conduct a thousand drops to "teach" the GNC system how to achieve pin-point accuracy. Instead, PADS modeling (Chapter 5) should be used and different GNC concepts (Chapter 5) tried in order to find the ways to improve the ultimate PADS performance in computer simulations first [Hattis and Benney 1996]. Based on preliminary studies, the realistic requirements thought of back then were [Lingard 1995] as follows:

• Deployment altitude: up to 35,000 ft (10.7 km)

• Offset range: up to 20 km (10.8 n miles), meaning having GR greater than 2.5:1

• Accuracy: less than 100 m (328 ft) from IPI (threshold), less than 50 m (objective)

• Suspended mass: 225 kg to 19 tons (500 to 42,000 lb)

• Landing: soft landing into the wind

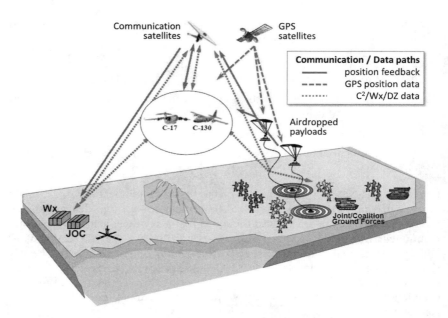

Fig. 1.5   JPADS mission support network [Benney et al. 2005a; Benney et al. 2005b; McGrath et al. 2005].

In 1997, a joint U.S. Army/U.S. Air Force program, Joint Precision Air Drop System (JPADS), was established to explore the way to cardinally improve accuracy of a cargo delivery. The Air Force was responsible for developing the mission planning computer (JPADS-MP) that would forecast wind data over DZ based on all data available and then automatically produce a computed air release point (CARP) that would minimize the landing error for the conventional unguided ADS. (This issue will be addressed in Sec. 3.3.1.) The Army was responsible for developing AGU and different-weight PADS. It was anticipated that the JPADS program pass Milestone C and reach a Full-Rate Production Decision Review in FY 2008–2009 [JPADS 2014]. Figure 1.5 shows the overall concept featuring critical components of JPADS network. The overall goal of JPADS was to meet the New Vistas report challenges and develop a new nonexisting capability ensuring delivery of required cargo to anywhere in the world within 24 hours. Military and security applications include the following:

• Providing accurate and flexible stealth supply to special force teams
• Providing navigational guide for team night insertion
• Supporting pathfinder operation

- Deploying acoustic sensing equipment into battlefield
- Deploying electronic warfare equipment
- Delivering leaflets accurately
- Deploying nuclear, biological, and chemical threat sensors
- Providing "just in time" supply of advancing troops

Of course, the list of potential users of PADS goes well beyond military applications to include the following:

- Space items recovery (as a final stage of a multistage system)
- Regular supply of remote locations
- Humanitarian aid and disaster relief deployment to inaccessible locations and unprepared DZs including field hospitals, refugee camps, and uncompounds
- All-weather equipment drop for search and rescue operations
- Equipment supply to first responders in the disaster areas
- Equipment delivery into rugged mountain areas
- Sensing equipment and video/radio uplink deployment
- Medical equipment supply
- Precision delivery of buoys and lifeboats at sea

1.1.2 JPADS PROGRAM AND KEY REQUIREMENTS

Under the long-term program plan, JPADS was envisioned as having up to four increments. Increment I objective was to handle loads up to 1 ton (2,200 lb), while Increment II would handle loads up to 4.5 tons (10,000 lb). Increment III would involve JPADS for up to 13.6 tons (30,000 lb), and Increment IV would develop a JPADS system that can handle up to 25 tons (60,000 lb). Increments III and IV were to be pursued dependent on technological success and after-action reviews associated with Increments I & II [Barber et al. 2011]. As reported in [Benney et al. 2005b; Benney et al. 2009a], JPADS self-guided PADS classification included six weight classes as follows:

- Micro-light weight (ML) $\sim$5–70 kg ($\sim$10–150 lb)
- Ultra-light weight (UL) $\sim$100–300 kg ($\sim$250–700 lb)
- Extra-light weight (XL) $\sim$300 kg–1 ton ($\sim$700–2200 lb)
- Light weight (L) $\sim$1–4.5 tons ($\sim$2200–10,000 lb)
- Medium weight (M) $\sim$4.5–13.6 tons ($\sim$10,000–30,000 lb)
- Heavy weight (H) $\sim$13.6–27 tons ($\sim$30,000–60,000 lb)

Later on, JPADS' usefulness has been skewed toward the lighter weights, so as of 2011 there were only five JPADS categories [JPADS 2014]:

- Micro-light weight (ML)   ~5–70 kg (10–150 lb)
- Ultra-light weight (UL)   ~100–300 kg (250–700 lb)
- Extra-light weight (XL)   ~300 kg–1.1 tons (700–2,400 lb)
- Light weight (L)   ~2.3–4.5 tons (5,000–10,000 lb)
- Medium weight (M)   ~4.5–19 tons (10,000–42,000 lb)

Even though JPADS was established as a joint program between the U.S. Army and U.S. Air Force, it was actually the Marine Corps Warfighting Lab (MCWL), which first purchased several MMIST's Sherpa 540 kg (1,200 lb) PADS in 2001 for evaluation. (The Sherpa PADS, as well as other PADS mentioned in this section, will be discussed in more detail in the next section.) Earlier, MMIST Sherpa versions including the body, canopy, riggings, remote control, rechargeable batteries and software, cost $68,000 vs $11,000 for a standard military cargo parachute. MCWL intended to refine the tactics, techniques, and procedures associated with GPS-guided paradrops and improve their stated requirements based on actual field use. Later on, MCWL purchased 20 more Sherpa 1200 PADS, $100,000 a piece, to be deployed at the theater [JPADS 2014].

Within JPADS and beyond, several different-weight PADS were developed, intensively tested, and demonstrated. (Many of them will be described in Sec. 1.2 in detail.) In the mid-2000s, the U.S. Army Natick Solder Center teamed with other organizations to plan and execute the JPADS Advance Concept Technology Demonstration (ACTD) with the goal of demonstrating and assessing the developed PADS (described in Sec. 1.2) capable of providing a global and in-theater delivery capability within 24 h of request the warfighter (unit/teams) worldwide [Benney et al. 2005a, 2005b]. The Joint Military Utility Assessment (JMUA) was made on the basis of three critical operational issues (COI) that were formulated as follows [Benney et al. 2005a, 2005b]:

- COI 1: Does the JPADS system-of-systems successfully support payload delivery at the target weights and standoff distances in its intended operational environment?

- COI 2: Does the JPADS system-of-systems provide the Joint Task Force Commander with an enhanced operational capability?

- COI 3: Is the JPADS system-of-systems suitable for employment in its intended environments?

Screamer 10K PADS by Strong Enterprises was one of the systems demonstrated. As a result, the U. S. Joint Forces Command initiated the development of a 900-kg (2,000-lb) variant of this PADS that was not contemplated under the original ACTD for the U.S. Special Operations Command [JPADS 2014].

As of today, Increment I and Increment II of JPADS have had a contract winner selected. Specifically, in 2011 Airborne Systems NA in Pennsauken,

New Jersey, won an $11.6 million contract for 391 JPAD 2K systems and in early 2014 the same firm received a $30 million contract for up to 110 JPADS 10K [Guided PADS 2014]. JPAD 2K and JPAD 10K systems utilize FireFly and DragonFly PADS by Airborne Systems with AGU developed by the Charles Stark Draper Laboratory. As of 2012, JPADS 2K featured circular error probable (CEP) touchdown accuracy of 150 m (490 ft) and JPADS 10K exhibited 250 m (820 ft) CEP [JPADS 2014] (CEP stands for circular error probable and will be discussed in Sec. 1.3.1). JPADS 10K can be used with Type V airdrop platforms to carry vehicles like high-mobility, multipurpose, wheeled vehicle (HMMWV), artillery pieces like an M777, or irregularly shaped items like shelters, generators, etc. By the end of 2012, Airborne Systems has sold more than 2,500 JPADS 2K/FireFly systems and more than 250 JPADS 10K/DragonFly systems to American and international customers [JPADS 2014].

A 2010 U.S. Marine Corps solicitation for the JPADS UL went the other way, aiming to deliver small payloads of 113 to 318 kg (250 to 700 lb). In 2011, Boeing subsidiary Argon ST in Fairfax, Virginia, won a $45 million contract for the procurement, testing, delivery, training, and logistical support of the JPADS-UL based on the Onyx PADS [JPADS 2014].

It might seem that the desired capability has been developed and all needs have been addressed. However, there is a lot of space for improvement. From the standpoint of landing accuracy, more sophisticated algorithms could be developed and deployed on even existing systems. That includes modeling, system identification, design of alternative GNC concepts and algorithms, utilizing new sensors, etc., and that is what this book intends to facilitate. There is also a need to make PADS more affordable. For example, in 2012 MCWL was seeking innovative, low-cost expendable alternatives to JPADS 2K because the need to recover these systems upon deployment puts detached units on the ground at greater risk. They were also seeking for a touchdown error decrease to fit a 100 by 60 m (330 by 200 ft) footprint. In 2014, the new request for information went out to find a novel way to conduct resupply of distributed forces emphasizing seeking a low-cost, expendable method to deliver 227-kg (500-lb) payload utilizing commercial of the shelve sensors and ensuring large standoff distance calling for development of systems with a much higher glide ratio than that of the existing PADS. The U.S. Marines execute about one-third of the total amount of airdrops. Their success is defined by speed, responsiveness, and flexibility. By their doctrine, responsiveness means "the right support in the right place at the right time," which is synonymous to precision airdrop [Lozar and Leaman 2009]. That is why they are keen to see the systems capable of meeting all four of the requirements set by the New Vistas report as mentioned in Sec. 1.1.1.

Now, the JMUA COI 1 formulated earlier in this section has to be supported by a set of measures of effectiveness (MOEs) that this book is concerned about. To this end, [Brown and Benney 2005] performed a system-level analysis of PADS and suggested a framework for the systematic characterization of PADS that is traceable to key performance parameters (KPP) and can assist in comparing different-type systems. Among other MOEs, they suggested using two levels of PADS accuracy:

- MOE 1.1.1: system accuracy
- MOE 1.1.2: terminal accuracy

Specifically, while 1.1.1 defines accuracy for all PADS loaded on an air carrier and intended to reach IPI, MOE 1.1.2 limits the number of PADS to consider with those that reached DZ and executed the terminal guidance maneuver as prescribed by their AGU. The difference between these two measures of effectiveness is defined by the following:

- MOE 1.1.3: PADS reliability

MOE 1.1.3 is a probability of PADS reaching the vicinity of IPI (Stage 3 initiation point in Fig. 1.2) from which a guided approach and landing can be successfully completed P_{DZ}. If $P_{DZ} = 1$, then system accuracy equals terminal accuracy. In most of the cases however, $P_{DZ} < 1$, and as a result system accuracy is worse compared to terminal accuracy. Using the language of statistics, P_{DZ} is defined by the relative number of outliers $\bar{n}_{out} = n_{out}/N$, where N is the total number of systems, so that

$$P_{DZ} = 1 - \bar{n}_{out} \tag{1.7}$$

Section 1.3 addresses these MOEs in details advocating the use of a set of specific measures of performance (MOPs) to assess the three aforementioned MOEs.

Brown and Benney [2005] also specified several failure modes affecting MOE 1.1.3. These modes include the following:

1. Air carrier missing CARP (which is especially critical for unguided ADS or PADS with a very limited control authority)
2. Parachute failure to open (can include several submodes depending on the number of PADS stages)
3. Parachute or control lines damaged
4. AGU failure to operate properly
5. Excessive actual winds aloft (compared to predicted) precluding reaching the DZ even if everything works just fine

The goal of JPADS-MP development was to mitigate Modes 1 and 5 (as explained in Sec. 3.3.1). The proper PADS design, rigging, AGU placement, and deployment procedure are supposed to take care of the remaining modes. Chapter 7 discusses different GNC concepts that are designed to minimize the probability of Mode 4. In addition to this, some recent research efforts were devoted to accommodation of partial canopy damage (i.e., making the GNC system robust enough to recognize a partial failure and adapt to it) [Cacan et al. 2005; Chiel 2015; Culpepper et al. 2013] (this subject is out of the scope of this book). Safety Fans graphical user interface (GUI) developed to assist in test planning incorporates the aforementioned modes as well [Corley and Yakimenko 2009; Mulloy et al. 2011]. Moreover, the probability of these modes to occur varies with the maturity of PADS (see Sec. 3.3.1).

To conclude this section, let us refine the JPADS KPPs as compared to those listed at the end of Sec. 1.1.1. As of today, they are thought of as follows:

- Deployment altitude: up to 25,000 ft (7.6 km)
- Offset range: up to 37 km (20 n miles) (wind dependent)
- Lift-to-drag ratio: 2.2–3.5
- Accuracy: 100 m (328 ft) threshold, less than 50-m (164-ft) objective
- CARP: large launch acceptability region
- Conops: multiple pads to the same or distributed IPI
- Suspended mass: 225 kg to 19 tons (500 to 42,000 lb)
- Landing: soft landing into the wind

From the standpoint of developing GNC algorithms, a robust JPADS AGU should be able to meet the accuracy KPP in a wide range of conditions. According to [Barber et al. 2011], AGU software must be able to navigate PADS to IPI:

- Under nominal system performance and atmospheric conditions
- Across the entire operational payload weigh range
- In winds aloft and on the surface up to and in excess of the canopy true airspeed
- In severe thermal atmospheric conditions
- From release at 7.5-km (24,500-ft) MSL
- From release at 1-km (3,500-ft) AGL
- From any location in the launch acceptability region including the edges
- Following a drogue fall phase
- Following a phase of override control
- Following the update of the programmed IPI while in flight
- With variations in canopy performance due to age or partial malfunction
- With SAASM (Selective Availability Anti-Spoofing Module, which is an enhanced GPS security architecture designed to provide over-the-air rekeying of GPS receivers in order to encrypt and decrypt the GPS signal with the so-called M-code providing antijam capabilities) GPS as the primary navigation sensor
- With failure of the primary GPS receiver
- With failure of the primary and backup GPS receivers

- With failure of the compass/inertial sensors
- With failure of a motor
- While navigating to user-specified waypoints
- While avoiding terrain or obstacles
- While avoiding other JPADS units in the air
- While landing into the wind
- While landing at a programmed bearing
- Using mission data input via a touch-screen display
- Using mission data input via the JPADS-MP

AGU software should also support multiple unique parachute sizes and designs and ensure canopy flare maneuver prior to impact for soft into-the-wind landing.

1.2 PRECISION AIRDROP TECHNOLOGY DEMONSTRATIONS

The development of the different-weight PADS within JPADS program involved a variety of demonstrations. Detailed information about the aforementioned demonstrations and PADS presented during these demonstrations can be found in [Benney et al. 2005a; Benney et al. 2007; Benney et al. 2009a; Benney et al. 2009b; Benney et al. 2009c; de Lassat de Pressigny et al. 2009; Delwarde et al 2007; McHugh et al. 2005; PATCAD 2001a; PATCAD 2001b; PATCAD 2003; PATCAD 2005; PATCAD 2007; PATCAD 2009]. In what follows in this section, these open sources and some other papers will be used to describe different PADS emphasizing their features from the standpoint of modeling. Missing geometry-mass data required for modeling can be retrieved from analysis of the photographs of these PADS (as presented in Appendix E). In addition, Appendix C lists some data on representative canopies.

To begin with, Table 1.1 lists the major demonstrations that happened in the United States and abroad in 2001–2009. Not all ADS systems demonstrated at Precision Airdrop Technology Conference and Demonstration (PATCAD), Precision Airdrop Capability Demonstration (PACD), and Joint Precision Airdrop Capability Demonstration (JPACD) events were self-guided PADS. Some systems were remotely piloted parafoil-based ADS or uncontrolled parachute-based ADS. Table 1.2 features only those systems that fit the self-guided PADS classification, and Table 1.3 lists the manufacturers of these systems. (Note, the numbers in Table 1.2 correspond to a maximum payload weight in pounds except for SPADeS PADS where they correspond to kilograms; bold font is used to denote the first demonstration of the system.) Table 1.4 shows a distribution of PADS shown at PATCAD 2001–2009 among different JPADS weight categories.

The remainder of this section briefly presents self-guided PADS listed in Tables 1.2 and 1.4 with a goal of showing their variety, configuration specifics,

TABLE 1.1   MAJOR PRECISION AIRDROP DEMONSTRATION EVENTS

Demonstration	Dates	Number of Attendees/ Nations	Number of Systems/ Airdrops
PATCAD 2001 (USA)	September 10–14, 2001	135/6	8/16
PATCAD 2003 (USA)	November 3–7, 2003	250/10	12/52
PATCAD 2005 (USA)	October 17–20, 2005	320/17	23/150
PACD 2006 (France)	July 3–6, 2005	100/10	12/25
PATCAD 2007 (USA)	October 22–26, 2007	>500/18	25/157
PACD 2008 (France)	May 26–29, 2008	>130/10	24/59
PATCAD 2009 (USA)	October 19–22, 2009	>500/15	32/>230

and overall performance. These PADS will be referred to in the following chapters when different aspects of structural and parametric identification, modeling, and GNC algorithms' development are discussed.

The Orion PADS (Fig. 1.6) demonstrated at PATCAD 2001 was the first commercially available system developed in the early 1990s by SSE, Inc., of Pennsauken, New Jersey, with high-glide parafoils supplied by Pioneer Aerospace of Melbourne, Florida [Allen 1995]. Upon exiting, a drogue chute is deployed to stabilize attitude and velocity (see Fig. 1.3). As the drogue deploys, AGU separates from the payload. After a preset time, or at a preset altitude, the main canopy is deployed. Following dereef of the main canopy AGU takes control of the system.

As described in [Allen 1995], Orion's AGU was based on a 80286 single board computer with 80287 coprocessor and utilized GPS receiver, barometric pressure altimeter, and fluxgate compass. It computed PADS current position, heading, ground speed, airspeed, wind magnitude and direction, and oscillation characteristics. The Orion Mission Planner and Simulation Software (MPSS) was written in Microsoft C and run in DOS. The same software was used to fly all payloads and canopies. The only inputs required from a user were the IPI position (geodetic coordinates), canopy size and type, and payload mass. To increase the probability of mission success, the system could also take advantage of the predefining expected winds aloft, release coordinates, waypoint coordinates, and secondary IPI coordinates. Based on a high-fidelity PADS model, MPSS would run a simulation to take all these factors into account and calculate the minimum release altitude and the altitude that the system would arrive at with planned waypoints warning the user of marginal conditions and preventing an infeasible plan from being utilized. Another important feature was that the internal MPSS model could accommodate either a load suspended by a swivel configuration as shown in Fig. 1.3 or a suspension configuration with AGU sitting directly atop payload (as many other PADS utilize). With the direct

TABLE 1.2 PADS SHOWN AT MAJOR DEMONSTRATION EVENTS

PADS	PATCAD'01	PATCAD'03	PATCAD'05	PACD'06	PATCAD'07	PACD'08	PATCAD'09
AGAS	**2000**	2000	**500**, 2000	2000	2000	2000	–
Buckeye	+	–	–	–	–	–	–
CADS	–	–	+	+	–	–	+
DragonFly	–	–	+	–	–	+	+
FireFly	–	–	+	+	+	+	+
MegaFly	–	–	–	–	+	–	+
MicroFly	–	–	–	–	+	–	+
Onyx	–	**75**	**500**, **2200**	–	**300, UL**	300, UL	**ML**, UL
Orion	+	–	–	–	–	–	–
Panther	–	–	**500**	500, **2K**	500, 2K	500	500
Para-Flite	–	**5K**	–	–	–	–	–
ParaLander	–	–	+	+	–	+	–
Pegasus	+	–	–	–	–	–	–
Screamer	–	**2K, 10K**	2K, 10K	2K	2K, 10K	+	10K
Sherpa	–	**1200**	1200, **2000**	1200	1200, 2200	+	2K, **10K**
SNCA	–	–	–	–	+	+	–
Snowbird	–	+	+	+	–	–	–
Snowflake	–	–	–	–	–	–	+
SnowGoose	+	+	CQ-10A	–	–	–	–
SPADeS	–	**400**	500, **1000**	500, 1000	300, 1000	300, 1000	300, 1000
STARA	–	**GDS**	Gnat	–	**Mosquito**	Mosquito	Mosquito
X-38	+	+	–	–	–	–	–

TABLE 1.3   DEVELOPERS OF PADS SHOWN AT MAJOR DEMONSTRATION EVENTS

PADS	Company
AGAS	Natick, Vertigo, Inc., Capewell Components Limited Liability Company
Buckeye	Southwest Research Incorporated (SRI)
CADS	Cobham Public Limited Company, UK
DragonFly	Airborne Systems
FireFly	Para-Flite Incorporated, Wamore Incorporated, Airborne Systems
MegaFly	Airborne Systems
MicroFly	Airborne Systems
Onyx	Atair Aerospace Incorporated
Orion	SSE, Inc.
Panther	Pioneer Aerospace Corporation/Aerazur
Para-Flite	Para-Flite Incorporated
ParaLander	European Aeronautic Defence and Space Company (EADS) Defence and Security Systems Division (DS)
Pegasus	FXC Corporation, Inc.
Screamer	Strong Enterprises
Sherpa	Mist Mobility Integrated Systems Technology (MMIST) Incorporated, Canada
SNCA	NAVOCAP, France
Snowbird	MMIST
Snowflake	University of Alabama in Huntsville/Naval Postgraduate School
SnowGoose	Mist Mobility Integrated Systems Technology (MMIST) Incorporated, Canada
SPADeS	Dutch Space, The Netherlands
Stara	STARA Technologies Incorporated
X-38	NASA, NASA/Pioneer Recovery System

attached load scheme, the MPSS could further accommodate the two- or four-point attachment and the geometry of the attachment (to be discussed later in this section and in Sec. 5.2.1). The direct attached load configuration involves AGU sensors capturing substantial induced horizontal oscillation of payload, which could be not easy to handle. (Section 7.8.1 provides more discussion on this subject.) Orion PADS utilized waypoint guidance towards a DZ area followed

TABLE 1.4   PADS REPRESENTATIVES BY JPADS WEIGHT CATEGORIES

Micro-Light	Ultra-Light	Extra Light	Light	Medium	Heavy
—	—	2K Screamer	10K Screamer	—	—
200 CADS	500 Panther	2K Panther	10K CADS	—	—
160 Snowbird	500 Pegasus	2K Sherpa Ranger	10K Sherpa Provider	—	—
150 Mosquito	500 MicroFly	2K FireFly	10K DragonFly	30K MegaFly	42K Gigafly
Onyx ML	Onyx UL	—	5K Para-Flite	—	—
5 Mosquito	300 SPADeS	1K SPADeS	1K SPADeS	—	—
5 Snowflake	500 AGAS	2K AGAS	5K AGAS	—	—

by an approach pattern to take account of prevailing wind conditions. In extreme wind conditions [when W in Eq. (1.2) is greater than V_h], the system would back into the target. The control module dealt with deriving the control input and timing required to execute a specific maneuver based on servo motors dynamics. [Orion' actuators had a speed of approximately 15 cm (6 in./s) so that the full control stroke took on the order of 10 s.] In the event of a GPS failure, the Orion's AGU navigated by dead reckoning using the compass and barometric altimeter data. As reported in [Allen 1995], the production standard Orion PADS was able to land within 100 m (328 ft) of IPI in 95% of the drops. As seen in the very first design, Orion PADS had all functionality of the current designs and was able to achieve a touchdown performance compatible with the today's standards.

The Orion development included 200 completely autonomous flights conducted by SSE and Pioneer plus several hundreds more by the U.S. Army Research, Development, and Engineering Center (NRDEC) with parafoils from 26.8 to 684 m^2 (288 to 7,360 ft^2), with suspended payloads of 91 kg to 12.7 tons (200 to 28,000 lb) [Allen 1995]. The

Fig. 1.6   Orion PADS [Allen 1995].

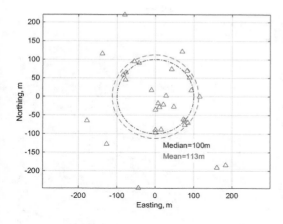

Fig. 1.7   Results of GPAS-L qualification tests [Patel et al. 1997].

Orion PADS based on GS-750 parafoil (see Table C.2) was the first Advanced Precision Air Delivery System (APADS) to be fielded to the U.S. Department of Defense. Designated by NRDEC as Guided Parafoil Air Delivery System (GPADS)–Light, it was part of a successful Advanced Airdrop for Land Combat' Technology (AALCT) demonstration undertaken in 1996. The results of successful qualification tests conducted at the U.S. Army Yuma Proving Ground (YPG) from 4 Sept. 1996, to 10 Oct. 1997, with GPADS-L rigged to standard container delivery system (CDS) bundles weighing between 318 and 500 kg (700 and 1,100 lb) delivered singly from a C-130 Hercules flying at 240 km/h (130 kt) at altitudes of 3 to 5.5 km (10,000 to 18,000 ft) MSL from standoffs of 4 to 6 km (2.5 to 3.7 miles) are shown in Fig. 1.7. (In addition to 32 successful drops shown in Fig. 1.7, there were 6 failures that account to 16% of all drops.) These airdrops demonstrated desired 100 m (328 ft) CEP accuracy and were considered representative for GPADS-L [Patel et al. 1997].

The 45-m (148-ft) wingspan version of the Orion PADS, steering a 16-metric-ton (35,000-lb) payload (back then the heaviest parafoil payload in history) to a soft, precise landing, was also a part of AALCT demonstration [Patel et al. 1997]. Based on the SSE/Pioneer Airspace design, GPADS-Heavy, with a wingspan of up to 53 m (174 ft), was under development with a goal of dropping heavier loads like HMMWV. GPADS-H was also adapted by NASA as the recovery system for the International Space Station (ISS) X-38 experimental crew return vehicle (CRV), which was also demonstrated at PATCAD 2001. The two vehicle configurations associated with the X-38 parafoil development are shown in Fig. 1.8.

The purpose of NASA's X-38 program carried out in 1998–2009 was to develop a 11.3-ton (25,000-lb) CRV with a seven-person capacity to replace the Soyuz spacecraft currently used on the International Space Station ISS [Stein et al. 2005]. Avionics was provided by the SRI and included essentially the same sensor suite as on the Orion PADS. Modified Warn truck winches, provided by Pioneer Aerospace, were used to steer the parafoil and perform the dynamic flare maneuver to assure soft landing (the dynamic flare maneuver is accomplished by pulling down parafoil TE rapidly to temporarily increase the buoyancy of parafoil that reduces the forward speed and rate of descent). The control surfaces of a parafoil were the outer 25% of TE. The control stroke setting was 80% of

a)   b)

Fig. 1.8   Vehicle configurations: a) instrumented pallet payload (System A) and b) prototype
CVR payload (System B) [Stein et al. 1999].

the stall stroke setting. The laser altimeter, built by Regal of Austria, was added to
the sensor suite as the ground proximity sensor used to trigger the flare. The AGU
software was built by Astrium Aerospace in association with the European Space
Agency (ESA). NASA modified and improved AGU software during the X-38
flight-test program. Unlike Orion's waypoint guidance approach, the X-38
AGU logics was based on generation of a curved reference trajectory to follow.
(This GNC strategy will be considered in more detail in Sec. 7.3.2 while Sec.
11.5 is devoted to X-38 system identification.) At PATCAD 2003 a smaller
390-m^2 (4,200-ft^2) canopy, also built by Pioneer Aerospace, with a standard
6-m (20-ft) Type V airdrop platform, was demonstrated to deliver 4.5 tons
(10,000 lb) of useable payload [Bennett and Fox 2005]. The 18.3-m (60-ft) ring-
slot drogue parachute had two stages of reefing (14 and 42%). The 27-cell parafoil
[with the 32.2-m (105.6-ft) wingspan and 12.2-m (40-ft) chord] was deployed in
five stages. For range safety considerations the parafoil had a built-in left-hand
turn, which was abandoned when GNC woke up [PATCAD 2003]. (If for some
reason GNC would not wake up, the parafoil remained in a built-in turn to
limit its footprint.) The 4,200-ft^2 parafoil design used the successful X-38
700-m^2 (7,500-ft^2) parafoil, also by Pioneer Aerospace, as a baseline with a
main focus to reduce the canopy cost to $3 to $6 per pound delivered, or
$30,000 to $60,000 for a 4.5-ton (10,000-lb) payload [Bennett and Fox 2003;
Bennett and Fox 2005].

The X-38 AGU logic was evaluated in the parafoil dynamics simulator (PDS) (will be addressed in Sec. 7.3.2) and on the Buckeye subscale paraglider (PPG) that was also a part of PATCAD 2001 demonstration (Fig. 1.9a). Later on, Hur and Valasek [2003] developed an observer/Kalman filter identification technique to develop an eight-degree-of-freedom (DoF) linearized model for this PADS (discussed in Secs. 5.4 and 11.5.6). Dynamics of a similar PADS was also studied by Watanabe and Ochi [2007] (see Sec. 5.4). The research group of Professor Costello of the Georgia Institute of Technology uses different-size PPGs with different (rectangular and elliptical) canopies (Fig. 1.9b) to investigate novel glide-angle control paradigms (these efforts are addressed in Secs. 8.2.3, 8.2.5, and 8.3).

Also demonstrated at PATCAD 2001 was a GPS-guided parafoil-based powered unmanned aerial vehicle (UAV) SnowGoose by MMIST (Fig. 1.10), which now became a fielded CQ-10A system used by the USSOCOM. This UAV has a payload capability of up to 261 kg (575 lb) carried in six individual cargo bays and could be dropped from a suitable cargo aircraft (C-130, C-17) or ground launched from a 365-m (1,200-ft) long unprepared level surface using HMMWV, flatbed or logistics trailer (Fig. 1.10a) [SnowGoose UAV 2014]. (These days, CQ-10B UUV is equipped with a gyro rotor head that provides near vertical takeoff and landing capability.)

One more parafoil-based PADS demonstrated at PATCAD 2001 and also belonging to GPADS-L class was the Pegasus PADS designed by FXC Corporation together with its Guardian Parachute Division over a span of five years to carry payloads of 181 to 818 kg (400 to 700 lb). Pegasus PADS featured in Fig. 1.11 utilized an eleven-cell design, 12-m (40-ft) wingspan, and 53-m^2 (575-ft^2) area canopy ram-air parafoil (2.9:1 aspect ratio) and featured 11 to 13.4 m/s (22 to 26 kt) forward speed and 2.4 to 3.7 m/s (8 to 12 ft/s) descent rate (which corresponds 3.6:1 to 4.6:1 GR) [Aguilar and Figueiredo 1995; Sego

a)   b)

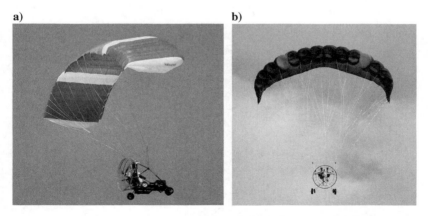

Fig. 1.9   a) Buckeye PPG and b) Georgia Tech PPG [Hur and Valasek 2003; Stein et al. 2005; Ward et al. 2012].

a)   b)

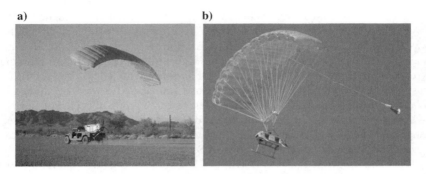

Fig. 1.10   SnowGoose a) at launch and b) in flight [PATCAD 2003].

2001]. The control surfaces of the canopy could be pulled down by as much as 0.46 m (18 in.). The sensor suite for this system included an inertial measurement unit (IMU). Even though this system had a very limited amount of drops [paraglider (PPG) primarily from a DHC-6 Twin Otter aircraft from relatively low altitudes of 1.8 km (6,000 ft) AGL], because of the complete data set available a lot of system identification and modeling efforts were undertaken. These efforts are described in Secs. 5.1.6 and 11.5. Figure 1.11 features a nice example of two different configurations of placing AGU within the same PADS as addressed earlier when the Orion PADS was discussed. Table 1.5 presents some geometry-mass properties for these two configurations when AGU was suspended at the so-called confluence point (Fig. 1.11a) compared to AGU sitting directly atop a payload (Fig. 1.11b).

The last (but not the least) system demonstrated at PATCAD 2001 was the Affordable Guided Airdrop System (AGAS), a low-cost alternative for meeting the military's requirements for precision airdrop. Designed to bridge the gap

a)   b)

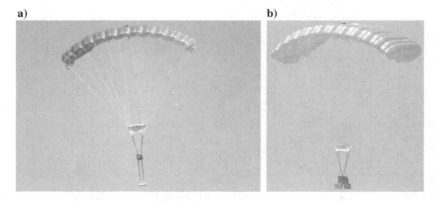

Fig. 1.11   Two configurations of Pegasus PADS [Sego 2001]. (See color insert.)

TABLE 1.5   GEOMETRY AND MASS PROPERTIES OF TWO CONFIGURATIONS OF PEGASUS PADS

	Configuration A (Fig. 1.11a)	Configuration B (Fig. 1.11b)
Top risers	1.88 m (74 in.)	1.6 m (63 in.)
Bottom risers	1.9 m (75 in.)	0
AGU, depth × width × height (D × W × H)	0.5 × 0.48 × 0.28 m (20 × 19 × 11 in.)	0.47 × 0.47 × 0.53 m (18.5 × 18.5 × 21 in.)
Payload, D × W × H	0.76 × 1.22 × 0.71 m (30 × 48 × 35 in.)	0.690 × 0.81 × 0.31 m (27 × 32 × 12 in.)
AGU mass	20 kg (44 lb)	27.8 kg (61.2 lb)
Payload mass	227 kg (500 lb)	181 kg (400 lb)

between expensive high-GR PADS and uncontrolled (ballistic) round parachutes, the AGAS concept offered the benefits of high-altitude parachute releases with a relatively large release offset range from IPI. The design goal of the AGAS development was to provide a GNC system that could be placed in line with existing fielded cargo parachute systems, like G-12, and standard delivery containers, like A-22. The accuracy requirement was the same as for high-GR PADS, that is, 100-m (328-ft) CEP threshold, with a desired goal of 50-m (164-ft) CEP. No changes to the parachute or cargo system were allowed.

The original design concept for AGAS 2000, capable of fully autonomous guidance of payloads up to 2,000 lb, included implementation of pneumatic muscle actuators (PMAs) to distort one-quarter to one-half of a round canopy by lengthening a single or two adjacent actuators. The canopy distortion was essentially equivalent to shifting the center of pressure and thus provided a drive or slip condition resulting in up to 0.8:1 GR. For the fielded implementation, PMAs were replaced with an electrical-motor-based system allowing a limited number of step lengthening of the risers (Fig. 1.12). At PATCAD 2005, another member of the AGAS family intended to deliver smaller payloads of up to 227 kg (500 lb) using a T-10 personnel parachute was demonstrated. Table 1.2 shows that AGAS proved to be quite a successful system dropped at all major demonstrations. Even though it does not necessarily fall under the category of high-GR PADS, Sec. 9.1 of this book describes its model and GNS concept as well.

At PATCAD 2003, several new PADS were introduced. That included Sherpa PADS by MMIST and Smart Parafoil Autonomous Delivery System (SPADeS) by Dutch Space of the Netherlands.

The rigging weight of Sherpa PADS originally extended from 113 to 544 kg (250 and 1,200 lb). Later on (at PATCAD 2007 and PATCAD 2009), it was extended to 907 kg (2,000 lb) and 4.5 tons (10,000 lb), respectively (Fig. 1.13). Today, MMIST offers two major configurations of Sherpa, Provider 2K, and Provider 10K, along with two smaller configurations, Provider 1200 and Provider 700.

Fig. 1.12   Two AGAS rigged to 680 and 726 kg (1,500 and 1,600 lb) [PATCAD 2005].

All four configurations feature different canopies but the same Sherpa Provider AGU.

The SPADeS PADS was originally designed to carry a payload of 22.7 to 250 kg (50 to 550 lb) under a standard G9-Galaxy parafoil by Aérazur used for a solo manned flight [limited to a maximum payload of 159 kg (350 lb)] and a tandem-parachute PBO by Aérazur with payloads up to 250 kg (550 lb) (Fig. 1.14). Over the years, SPADeS PADS has flown under a variety of different parafoils (see example in Fig. 1.15), and these days assumes three AGU configurations depending on the canopy and rigged weight (Table 1.6).

Two more new systems demonstrated at PATCAD 2003 were Para-Flite 5K and MMIST SnowBird 150. The Para-Flite 5K PADS (Fig. 1.16a) was an intermediate proof-of-concept prototype in pursuit of the development of a 4.5-ton

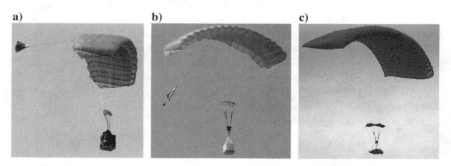

Fig. 1.13   Sherpa Provider: a, b) 2K, and c) 10K [PATCAD 2007; PATCAD 2009].
(See color insert.)

a)   b)

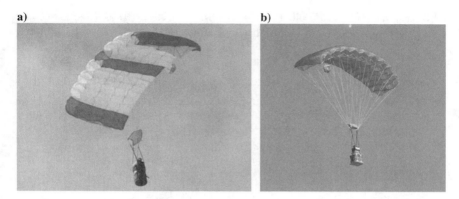

Fig. 1.14   a) SPADeS 160 kg with a G9-Galaxy canopy and b) SPADeS 250 kg with PBO canopy [Wegereef and Jentink 2003; Wegereef and Jentink 2005].

(10,000-lb) capable parafoil system. The program funded by NRDEC focused on the evaluation of low-cost manufacturing techniques and slider-controlled deployment of a large parafoil (without the use of pyrotechnic cutters) [PATCAD 2003]. At PATCAD 2003, this PADS was demonstrated under remote control. The Snow-Bird 150 PADS (Fig. 1.16b), also funded by NRDEC, explored another end of the JPADS weight spectrum. This system relied on a 14-m^2 (150-ft^2) ram-air canopy and was intended to carry up to 73 kg (160 lb) of cargo. At PATCAD 2005, a byproduct of the Sherpa system, the SnowBird family, was extended to include a larger system, SnowBird 400, capable of delivering 272 kg (600 lb) of cargo.

Even a smaller system, STARA's Generic Delivery System (GDS), capable of delivering payloads weighing between 900 g to 9 kg (2 to 20 lb), also made its debut at PATCAD 2003. The GDS was particularly designed for UAV deployment. In 2003, STARA came to PATCAD to demonstrate the concept rather than a particular system, but later on, at PATCAD 2005 and PATCAD 2007, it

a)   b)

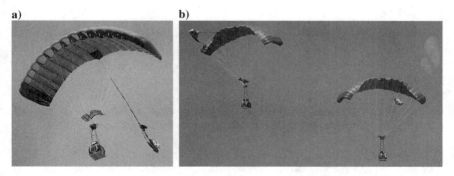

Fig. 1.15   SPADeS with Para-Flite FireFly canopies [PATCAD 2009]. (See color insert.)

TABLE 1.6   AGU CONFIGURATIONS FOR SPADES PADS WITH DIFFERENT PAYLOAD RANGE AND
CANOPY [WEGEREEF ET AL. 2007]

AGU	Parafoil	Parafoil Payload Range
Payloads up to 250 kg (550 lb)	CIMSA Range I, 37.2 m^2 (400 ft^2)*	100 to 250 kg (221 to 550 lb)
	Aérazur G9-Galaxy, 35 m^2 (377 ft^2)	90 to 190 kg (198 to 419 lb)
	Aérazur PBO, 50.2 m^2 (540 ft^2)	100 to 250 kg (221 to 550 lb)
	Para-Flite MC-5 34.4 m^2 (370 ft^2)	100 to 250 kg (221 to 550 lb)
	Performance Designs MP-360, 33.4 m^2 (360 ft^2)	100 to 250 kg (221 to 550 lb)
Payloads up to 500 kg (1100 lb)	CIMSA Range II 83.6 m^2 (900 ft^2)*	250 to 500 kg (550 to 1340 lb)
	Para-Flite FireFly, 95.2 m^2 (1025 ft^2)	300 to 1000 kg (660 to 2205 lb)
Payloads up to 1000 kg (2205 lb)	CIMSA Range III, 112 m^2 (1206 ft^2)*	300 to 1000 kg (660 to 2205 lb)
	Para-Flite FireFly, 95.2 m^2 (1025 ft^2)	300 to 1000 kg (660 to 2205 lb)

*Data on CIMSA canopies were provided by Simon Benolol.

presented the GNAT and Mosquito PADS (Figs. 1.17a and 1.17b, respectively). Again, the GNAT miniature PADS was designed to be light enough to be launched from UAV. The concept of operations included flying over IPI and dropping the 2.3- or 9-kg (5- or 20-lb) sensor packages ballistically. For these sensor packages GNAT employed 0.7- and 1.9-m^2 (8- and 20-ft^2) parafoils, respectively.

The Mosquito PADS presented at major demonstrations since 2007 was the UAV- or craft-deployable PADS capable of carrying 1.4- to 68-kg (3- to 150-lb) payload. This system was also able to release a payload midflight (as GNAT), allowing for covert placement of payloads such as unattended ground sensors or top-attack munitions. This technique allowed sensor placement with a miss distances of less than 10 m (33 ft) from IPI. These days STARA offers a family of different-size Mosquitoes with three different-size AGUs [STARA Airdrop 2014].

The two remaining PADS, first introduced at PATCAD 2003, were Screamer by Strong Enterprises and Onyx by Atair Aerospace. These systems employ a different precision payload delivery paradigm and rely on two stages as opposed to just one for all aforementioned systems. Both the 907-kg (2,000-lb)

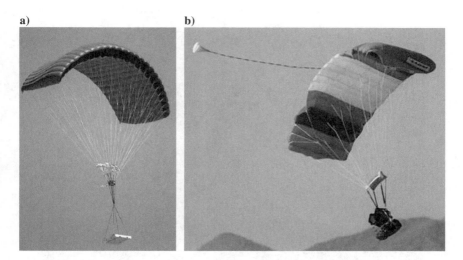

Fig. 1.16   a) Para-Flite 5K and b) Snowbird PADS [PATCAD 2003; PATCAD 2005].
(See color insert.)

Screamer and 4.5-ton (10,000-lb) Screamer utilize a ram-air parachute as the drogue (first stage) with a typical wing loading in excess of 49 kg^2/m (10 lb/ft^2). Screamer 2K utilizes 20 m^2 (220 ft^2) ram-air drogue (RAD) and Screamer 10K – 79 m^2 (850 ft^2) RAD [PATCAD 2005]. This allows penetrating most upper-level wind conditions at a forward speed of about 45 m/s (87.5 kt) and descending from a 7.6 km (25,000 ft) altitude in just 7 min. The onboard AGU, developed by

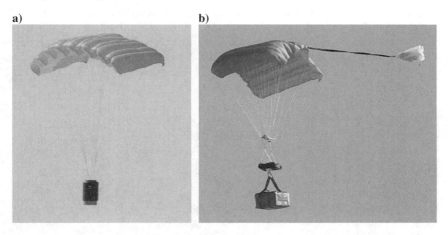

Fig. 1.17   a) STARA's 1 stage GNAT and b) 18 kg (40 lb) Mosquito PADS [PATCAD 2005;
PATCAD 2007]. (See color insert.)

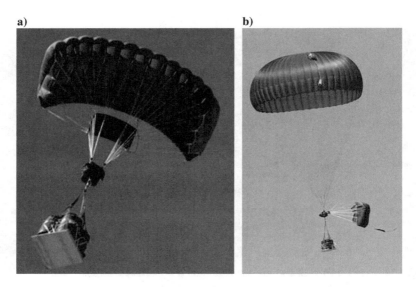

**Fig. 1.18   The a) first and b) second stages of the Screamer 2K PADS [PATCAD 2005].
(See color insert.)**

Robotek, guided PADS to a point 335 m (1,100 ft) above IPI at which time one
Strong Pocket G-12 parachute (for Screamer 2K) or a pair of round G-11 cargo
recovery parachutes (for Screamer 10K) is deployed (second stage). In a latter
case, G-11 parachutes are deployed simultaneously by the apex using a 5-m
(17-ft) drogue parachute, which is deployed by an AGU software command.
This effectively arrests a forward speed causing the payload to descend at 5.5 to
9 m/s (18 to 26 ft/s) sink rate. Screamer 2K utilizes a modified A-22 configured
payload container while Screamer 10K is compatible with a variety of standard
airdrop pallets, including the ECDS, Type V, and 463L. These systems are
shown in Figs. 1.18 and 1.19, respectively. A completely rigged Screamer 10K

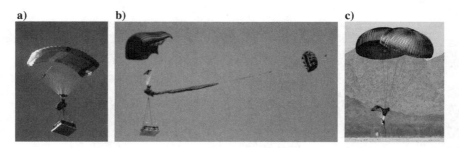

Fig. 1.19   The a) first and b) second stages of the Screamer 10K PADS [PATCAD 2005].

Fig. 1.20   Onyx 500 with a rigged weight 62 kg (136 lb) deploying the second stage
[PATACD 2005].

decelerator components weigh approximately 363 kg (800 lb) [McGrath et al. 2005; Strong et al. 2005].

PATCAD 2003 also featured the two drops of the 34-kg (75-lb) Onyx PADS developed via the Small Business Innovation Research (SBIR) program with NRDEC [PATCAD 2003]. Similar to Screamer, the Onyx PADS utilizes ram-air parachute for guidance and dumb round parachute that opens at the lowest possible altitude of about 107 m (350 ft) above IPI for soft landing. The Onyx system has been developed in three payload categories including 0–9 kg (0–20 lb) Micro Onyx, 0–227 kg (0–500 lb) Onyx 500, and 227–998 kg (500–2,200 lb) Onyx 2200. The latter two systems were demonstrated at PATCAD 2005 (Fig. 1.20). The two-stage Onyx was reintroduced at PATCAD 2007 as Onyx 300. The Onyx 300 used a 7-m^2 (75-ft^2) elliptical ram-air canopy to rapidly autopilot cargo to just overhead of a target where it transitions to the landing 28-m^2 (299-ft^2) round parachute for a final descent.

PATCAD 2007 also introduced the UL version of Onyx that followed the one-stage PADS paradigm (Fig. 1.21). The Onyx UL uses an autonomous GPS guided AGU that was capable of interfacing with multiple military parachutes

a)   b)

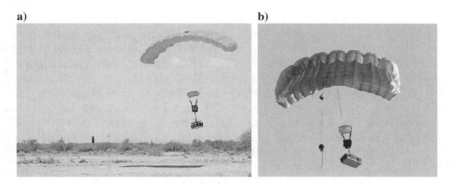

Fig. 1.21   Onyx UL [PATCAD 2007; PATCAD 2009].

a)   b)

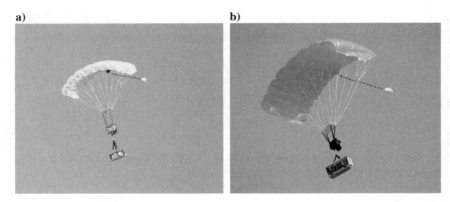

Fig. 1.22   Onyx ML under a) 4.7 m^2 (50 ft^2) and b) 11 m^2 (120 ft^2) canopies [PATCAD 2009]. (See color insert.)

for cargo delivery of 90 to 318 kg (200 to 700 lb) payload. Parafoils that can be used with this system include the Maneuverable Canopy MC-5, MP-360, TP-400, and Atair C350. The system can utilize surplus or decommissioned military parachutes as well. A smaller version, the one-stage Onyx ML, was shown at PATCAD 2009 (Fig. 1.22).

Compared to most other PADS, Onyx AGU included an integrated GPS/INS IMU and featured a variety of novel capabilities such as flocking/swarming (formation flying), active collision avoidance, and adaptive control (self-learning functions for varying cargo weights) [Calise and Preston 2006; Calise et al. 2007; Calise and Preston 2008; Calise and Preston 2009; Onyx PADS 2014]. Some aspects of the Onyx GNC design are presented in Sec. 7.83.

The Controlled Aerial Delivery System (CADS) that entered service with the U.K. Armed Forces in 1991 was among several new systems presented at PATCAD 2005. Not fully autonomous, the concept of operations for this PADS assumed controlling it over a radio frequency (RF) link using either a ground-based or airborne transmitter. The ground transmitter had two modes, auto and manual. Any unit on the same channel steers towards the ground station in the auto mode. Once PADS reaches DZ and is acquired visually, the transmitter is switched to manual for a controlled, flared landing. The airborne transmitter can be used by a parachutist who follows PADS and controls it to DZ [PATCAD 2005]. Figure 1.23 shows CADS rigged with a BT-80 canopy. The French Système de Navigation pour Charge Accompagneè (SNCA) PADS presented at PATCAD 2007 relied on a jumper escort as well.

Two more traditional self-guided PADS introduced at PATCAD 2005 were Panther, by Pioneer Airspace/Aerazur and ParaLander by EADS DS. The Panther 500 with Aerazur G9-Galaxy parafoil (Fig. 1.24) had a payload capability of 91–227 kg (200–500 lb) and was intended to provide service for the military

a) b)

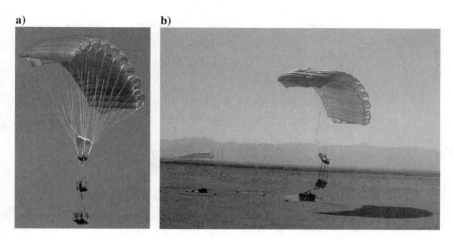

Fig. 1.23   CADS PADS [PATCAD 2005]. (See color insert.)

A7 class of CDS. Panther's 21-kg (46-lb) AGU provided accurate guidance to a landing point, ensured landing into the wind, and enabled flaring right before touchdown. Similar to other PADS, the major elements of the mission were to deploy, navigate to a holding pattern, perform altitude loss as required in a holding pattern, and then execute a final approach to a landing. The holding pattern maneuver keeps the unit relatively close to IPI until it is time to begin the

a) b)

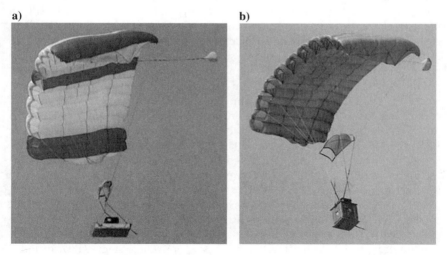

Fig. 1.24   Panther 500 under a 34-m^2, 5.4-kg (365-ft^2, 12-lb) parafoil [PATCAD 2005]. (See color insert.)

a)   b)   c)

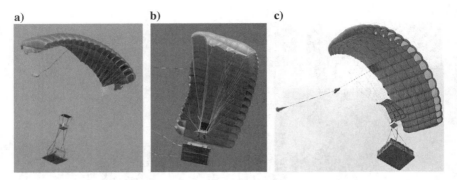

Fig. 1.25   ParaLander PADS [Altmann 2011; PATCAD 2005].

final approach. If the winds are too strong for a normal mission, the system holds upwind until the wind decreases, or until it is time to turn directly to IPI and land into the wind. The Panther 2K, introduced at PACD 2006, was compatible with the military A22 container and had a 1.1-ton (2,500-lb) payload capacity. The Panther 2K AGU weighed 43 kg (94 lb) and had an optional ground unit interface that allowed navigating in an automatic or manual control mode. The system flew under a 102-m² (1,100-ft²), 25-kg (55-lb) parafoil [PATCAD 2007].

The ParaLander PADS (Fig. 1.25) had a capacity of up to 500 kg (1,100 lb) and was essentially the cargo version of the manned ParaFinder system also presented at PATCAD 2005. In the case of ParaFinder, upon deployment from an aircraft, the system acquired GPS lock and provided directional vectors to the jumper (wearing a helmet) to the preprogrammed waypoints and ultimately to IPI. In the case of ParaLander, a navigation computer compared preprogrammed flight altitude with the actual flight altitude. Then, any variation was converted to a corresponding symmetrical control signal to lengthen or shorten the flight to compensate for the time-related variance in altitude. The range-optimized programmed course was calculated from the current position. The autopilot converted the variance and preprogrammed an actual course into an asymmetrical control input into motor control electronics. The landing sequence consisted of downwind leg, crosswind leg, and final approach against the forecast wind. The ParaLander AGU had a bidirectional data link with the ground station allowing a control to be taken over to land PADS manually [PATCAD 2005]. (A human-in-the-loop control concept aimed at controlling descending PADS from the ground was recently looked at on a smaller scale in [Cacan et al. 2015].)

The two PADS that later became the fielded JPADS 2K and JPADS 10K, FireFly and DragonFly, were first introduced at PATCAD 2005 as well. The 10K DragonFly was 3- to 4.5-ton (6,500- to 10,000-lb) payload range PADS developed under the U.S. Army JPADS ACTD program. The system, which included a 4:1 GR 325-m² (3,500-ft²) innovative main canopy and hardened AGU, had been developed under a multicontractor effort and managed by NRDEC. The decelerator system was provided by Para-Flite, Inc.; the AGU has been developed by Wamore, Inc.; the avionics suite was produced by Robotek Engineering; and the GNC software was developed by the Draper Laboratory, Inc. The system is gravity dropped on an ECDS, 2.4-m (8-ft) Type V, or 463L platform. A 15-m (48-ft) ringslot drogue parachute is used to stabilize the system for 10 s allowing the payload to achieve a nearly vertical orientation at a dynamic pressure of 49 kg^2/m (10 lb/ft^2) safe for main canopy deployment [PATCAD 2005]. A timed drogue release mechanism was used to cut a set of Kevlar® release straps, allowing the deployment bag to be pulled from the AGU and the main canopy to be freed from the bag. The deployment bag is retained with the drogues and descends to the ground where it is recovered for reuse. The spanwise inflation of the canopy forces the slider down the rigging lines. The main canopy deployment sequence is shown in the first three photos of Fig. 1.26 [George et al. 2005]. Total altitude loss prior to initiation of autonomous flight is approximately 914 m (3,000 ft).

Once the canopy fully inflates, the cutters are fired, releasing the TE lines, and the canopy reaches its steady-state flight condition. While in flight, the 80-kg (175-lb) AGU is suspended between the parafoil and payload. The AGU contained a pair of small servo motors with worm gear reducers powered by a 24-V DC battery, a dual-channel motor controller, and a 12-V DC battery to power the avionics, as well as an avionics suite to generate trajectory information for the flight software. The avionics for DragonFly consists of a 44.2-MHz 8-bit microprocessor with 1024 Kb static RAM and 8-Mb serial flash memory, a commercial dual-antenna GPS receiver, and a 900-MHz spread spectrum RF modem. The dual-antenna GPS provided heading and heading rate information, in addition to position and the velocity vector with respect to the ground provided by standard single-antenna GPS receivers. The GNC algorithms have been developed to accommodate an aforementioned low-cost processor by utilizing very simple command logic and a lookup table driven trajectory profile for final-descent maneuvers [George et al. 2005; PATCAD 2005]. The details of the GNC approach will be addressed in Secs. 7.5 and 7.8.1.

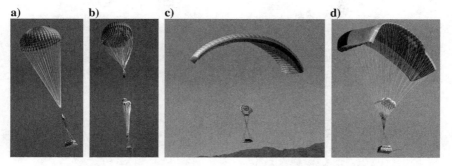

Fig. 1.26   DragonFly PADS deployment, glide, and flaring [George et al. 2005; PATCAD 2005]. (See color insert.)

a)   b)   c)

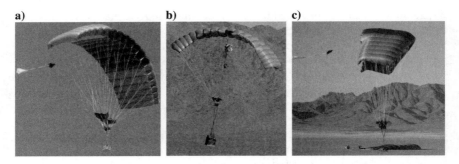

Fig. 1.27   FireFly PADS [Guided PADS 2014; PATCAD 2005]. (See color insert.)

The 2K FireFly (Fig. 1.27) for a payload range of 430 to 1,000 kg (950 to 2,200 lb) was also a collaborative development between Para-Flite, Inc., for the decelerator and GNC and Wamore, Inc., for the AGU, and as such it had maximum component commonality with the DragonFly PADS. A 6-s drogue fall was used to stabilize the system. After transition to a main high-performance canopy with a 4:1 GR, the drogue is retained to parafoil's TE. During drogue descent, the FireFly PADS undergoes an altitude loss of approximately 610 m (2,000 ft). The FireFly's low-cost AGU weighs approximately 20 kg (45 lb) [Guided PADS 2014; PATCAD 2005].

At PATCAD 2007, Airborne Systems introduced two more PADS, MicroFly and MegaFly, extending the FireFly and DragonFly line of products at both ends of the weight- class spectrum. The MicroFly PADS (Fig. 1.28a) was capable of carrying payloads from 45 to 318 kg (100 to 700 lb). It can be used with any 21- to

a)   b)

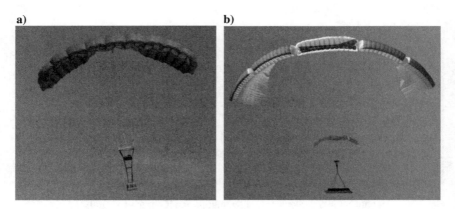

Fig. 1.28   a) MicroFly PADS and b) MegaFly PADS [PATCAD 2009].

40-m^2 (230- to 430-ft^2) canopy manufactured by Airborne Systems including the MT-1X, MC-4, MC-5, MT-1Z, Intruder, and HG-380 (with an elliptical HG-380 ram-air parachute MicroFly can archive a 6:1 GR) [Guided PADS 2014]. By selecting the appropriate canopy, MicroFly PADS can match the speed and rate of descent of a jumper under canopy and therefore can be used to lead jumper teams to IPI during insertion. The MicroFly AGU weighed 13.6 kg (30 lb) and had a battery life of five airdrops from approximately 5.5 km (18,000 ft) AGL [PATCAD 2005].

The MegaFly PADS (Fig. 1.28b) was capable of carrying payloads from 9 to 13.6 tons (20,000 to 30,000 lb) and had an 836-m^2 (9,000-ft^2), 52-m (170-ft) long, 408-kg (900-lb) modular canopy that was made of five separate segments for ease of recovery. The recovery procedure for the MegaFly consisted of separating the canopy segments, slider segments, and AGU interface frame segments from one another. Each component weighed less than 113 kg (250 lb). The weight of the AGU, suspended between the parachute and payload, was 170 kg (375 lb) [PATCAD 2009]. Being one of JPADS platforms developed and manufactured by Airborne Systems, the MegaFly PADS operated with common GNC algorithm and user interface. The packing methodology for all different-weight PADS was identical as were the AGU interfaces, so little additional training was required to qualify riggers on different systems. MegaFly was extracted from an aircraft using a standard airdrop equipment and is deployed as a conventional ram air canopy without pyrotechnic cutters or timers.

In September of 2008, Airborne Systems successfully deployed the world's largest self-guided PADS capable of carrying 19 tons (42,000 lb) under a single ram-air parachute. The new system, called the GigaFly (Fig. 1.29), utilized a 970-m^2 (10,400-ft^2) ram-air parachute with a wingspan of 60 m (195 ft), nearly as wide as the wingspan of a Boeing 747-400 of 64.4 m (211 ft). A 15-ton (33,000-lb) load was deployed from a Lockheed C-130 Hercules aircraft from 4.5 km (15,000 ft) mean sea level (msl). The system landed fully autonomously at a gentle 4.3-m/s (14-ft/s) rate of descent less than 275 m (900 ft) from IPI. In November of 2008, GigaFly surpassed this

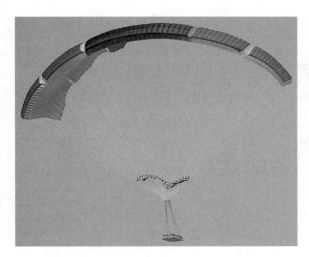

Fig. 1.29 GigaFly PADS [JPADS 2014].

a)   b)

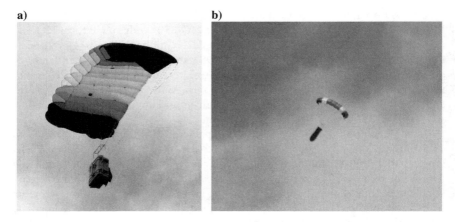

Fig. 1.30   Snowflake PADS.

payload record delivering 18-ton (40,000-lb) payload 100 m (328 ft) short of IPI [Guided PADS 2014].

Ironically, at about the same time, in October of 2008, the thorough testing of one of the smallest PADS was conducted at the same place [Yakimenko et al. 2009]. The miniature 1.4-kg (3-lb) Snowflake PADS, developed by the University of Alabama in Huntsville and Naval Postgraduate School, featured a 0.85-m^2 (9-ft^2), 1.37-m (4.4-ft) span low-performance ram-air canopy with only 2:1 GR, exhibiting the forward speed of 7.2 m/s (14 kt), descent rate of 3.6 m/s (12 ft/s), and minimum turning radius of about 15 m (50 ft) (Fig. 1.30). Even though this experimental PADS capable of carrying about 2-kg (4.5-lb) payload was developed to study novel GNC concepts and algorithms rather than compete with other systems, its superb performance of landing within 10 m (33 ft) from the stationary or moving target and extremely low-cost found potential users and was demonstrated at the last PATCAD deployed from UAV [PATCAD 2009; Yakimenko et al. 2011]. Throughout the book modeling and flight-test data available for this PADS are used quite intensively, especially in Chapters 3, 5, and 7. A few novel applications for a small PADS-like Snowflake are mentioned at the end of Sec. 7.7.

The Snowflake PADS was not the first and only experimental PADS developed exclusively for research. In the 1990s, the extensive research involving system identification, modeling, and GNC concept proving was undertaken by the Institute of Flight Systems of the German Aerospace Center (DLR) using Small Autonomous Parafoil Landing Experiment (ALEX) PADS [Doherr and Jann 1997b; Gockel 1998; Gockel and Jann 1998; Jann et al. 1999]. To develop and validate mathematical models of different complexities, two ALEX PADS having a mass of about 100 kg (220 lb) were developed and equipped with an extensive flight-test instrumentation package including GPS, inertial and air-data sensors,

Fig. 1.31   ALEX PADS flight testing [Jann 2006].

magnetometer, and actuator position transducers (Fig. 1.31). A video camera looking upward towards the canopy captured the relative motion between parafoil and payload [Jann 2006]. Both remote/autonomously controlled drops were conducted from a helicopter from altitudes 600 to 2,000 m (2,000 to 6,660 ft) AGL. Sections 4.1, 5.1.3, 6.1, 7.4, 10.1, 11.2, and 11.3 provide more details about modeling, instrumentation package, and system identification technique used in conjunction with the experimental PADS.

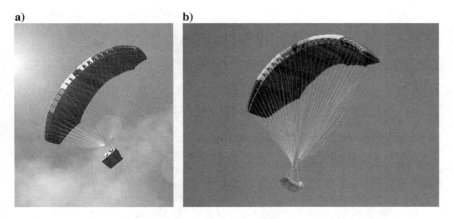

Fig. 1.32   a) FASTWing PADS and b) FASTWing CL PADS [Jann and Greiner-Per 2009; Wegereef et al. 2011].

TABLE 1.7   SOME REPORTED AND COMPUTED PROPERTIES OF
DIFFERENT-WEIGHT PADS

	ASU Suspended Between Canopy and Payload		AGU Atop Payload	
Swivel (one-point attachment)	CADS (Fig. 1.23)	Onyx (Figs. 1.20–1.22)	—	
	DragonFly (Fig. 1.26)	Orion (Fig. 1.6)		
	FireFly (Fig. 1.27)	Para-Flite (Fig. 1.16a)		
	MegaFly (Fig. 1.28b)	Screamer (Figs. 1.18, 1.9)		
	Mosquito (Fig. 1.17b)	Sherpa 2K (Fig. 1.13b)		
Two-point attachment	GigaFly (Fig. 1.29)	Pegasus (Fig. 1.11a)	ALEX (Fig. 1.31)	Snowbird (Fig. 1.16b)
	MicroFly (Fig. 1.28a)	Sherpa 2K (Fig. 1.13a)	Buckeye (Fig. 1.9)	SnowGoose (Fig. 1.10)
	ParaLander (Fig. 1.25)	Sherpa 10K (Fig. 1.13c)	FASTWing (Fig. 1.32)	SPADeS (Figs. 1.14, 1.15)
			Panther (Fig. 1.24)	X-38 (Fig. 1.8)
			Pegasus (Fig. 1.11b)	
Four-point attachment	—		AGAS (Fig. 1.12)	Snowflake (Fig. 1.30)
			Gnat (Fig. 1.17a)	

The main parachute system of ALEX was a 23.4-m^2 (252-ft^2) Parafoil 252-7 Lite that was activated by a static line. For emergency cases a stable cross-type parachute initiated by a standard Cypres device was available. As an impact attenuation system, a sandwich construction was used, which consisted of two plywood panels and a flat block of plastic foam containing holes to insert empty soft drink cans in between. A tail fin was applied to stabilize the vehicle and prevent yawing during the transfer flight beneath the helicopter towards the air release point. For steering the parafoil via the control lines, the vehicle contained two direct-current motors with gearboxes that were connected to an analog actuator control unit. This unit contained a command converter

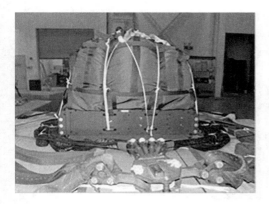

Fig. 1.33   DragonFly AGU attached to the deployment bag with a packed canopy [George et al. 2005].

that allowed switching between manual remote control and autonomous control, the last with commands coming from the onboard computer. In case the vehicle gets out of reach for the remote control, a fail-safe mode has been included that pulled the control lines to an asymmetrical deflection. Batteries were combined in a powerpack providing the actuators and electronic devices with the required power [Jann 2006].

The experience gained working on the ALEX PADS was later used for system identification of the Folding, Adaptive, Steerable Textile Wing Structure (FAST-Wing) PADS developed by Autoflug, CIMSA Ingeniería de Sistemas, Dutch Space, National Aerospace Laboratory NLR, DLR, EADS-CESA, CFDn, and the Technion University with a goal of investigating the application of a 300-m^2 (3,230-ft^2) high-performance parafoil with GR > 5:1 for delivering payloads up to 6,000 kg (13,230 lb) [Wegereef et al. 2011]. Figure 1.32a shows a FASTWing PADS with a 160-m^2 (1,720-ft^2) parafoil carrying 1.1-ton (2,425-lb) payload. Figure 1.32b features a flight of FASTWing project successor, now fielded FASTWing CL (Capital Loads) PADS with a 300-m^2 (3,230-ft^2), 35-m (115-ft) wingspan parafoil carrying 3-ton (6,700-lb) 463L pallet and featuring the forward speed of 15.5 m/s (30 kt) and descent rate of 3.7 m/s (12 ft/s) [Wegereef et al. 2011].

a)   b)

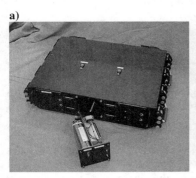

Fig. 1.34   DragonFly AGU a) with avionics tray and b) avionics tray by itself [George et al. 2005].

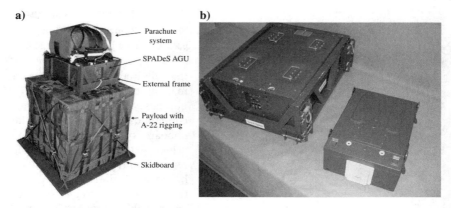

Fig. 1.35   a) Rigged SPADeS system and b) SPADeS-1000 (1 ton) and SPADeS-UL (350 kg) AGUs [Wegereef and Jentink 2007].

An overview of the major PADS developed and demonstrated for the past two decades reveals that they differ by payload weight, canopy size, and apparently by GNC algorithms. From the standpoint of modeling, they differ by the way the AGU and payload are suspended under PADS canopy. To this end, Table 1.7 combines different PADS into several groups by the way AGU is suspended (at the confluence point or atop payload) and the way payload is suspended (one-point, two-point, or four-point suspension). While the way payload is suspended defines the number of DoF in the high-fidelity models (addressed in Sec. 5.2.1), the AGU suspension scheme affects PADS control algorithms (addressed in Secs. 7.8.1 and 7.8.2). Having AGU atop payload seems to be a simpler solution whereas AGU suspended between canopy and payload provides a very flexible attachment point for a variety of payload sizes and shapes and, what is probably even more important, simplifies interface between AGU and canopy and simplifies GNC algorithms for sensor data processing.

To this end, Fig. 1.33 shows an example of DragonFly PADS's canopy being packed into the deployment bag and rigged to AGU. The AGU (Fig. 1.34a) is constructed of heavy-gauge aluminum to ensure survivability during landings and has the same area footprint as the main canopy deployment bag. The control line interface with the canopy's TE rigging lines is designed to make detaching AGU from the canopy simple and quick, even in severe landing conditions. The AGU's control line reels are readily accessible from the exterior aft face of the unit. When the system is completely packed, the control lines exit the parafoil deployment bag and connect directly to the AGU reels. The rigger can use toggle switches on the AGU to adjust the motor reel position as necessary for the proper control line geometry, and it is not necessary to power up the whole AGU to do this, nor is special equipment necessary. Figure 1.34a shows AGU with a forward-facing avionics tray in the foreground removed. By design, removal and reinstallation of the avionics tray (Fig. 1.34b) can be done even when the parafoil is rigged on top of the AGU. The front panel connector of the avionics tray is used to load flight software and mission information prior to flight and to download telemetry data, which have been logged to flash memory during a flight test afterwards [George et al. 2005]. Figure 1.35a shows an example of AGU installed atop payload. In any case, the different-size PADS require different-size AGU (Fig. 1.35b) because a smaller payload mass can be steered with the smaller steering line forces meaning smaller actuators, smaller power consumption, and lower-capacity batteries.

Table 1.8 generalizes available information about different PADS to show a variety of canopies, their aspect ratio (AR), GR, maximum turn rate (TR), and components of airspeed. Appendix C has information on some other representative canopies that can be used in modeling.

TABLE 1.8   SOME REPORTED AND COMPUTED PROPERTIES OF REPRESENTATIVE DIFFERENT-WEIGHT PADS

		SPADeS	FASTWing	DragonFly	X-38	MegaFly
Weight	kg	160	6,000	4,536	4,536	11,793
	lbs	353	13,228	10,000	10,000	26,000
Area	m²	34	300	325	390	836
	ft²	366	3,229	3,500	4,200	9,000
Chord	m	3.9	8.6	10.7	12.2	16.1
	ft	12.8	28.2	35.1	40	52.8
Span	m	8.6	35	30	32.3	52
	ft	28.2	114.8	98.4	106	170.6
Control authority	m	–	2.5	2.5	4	–
	ft	–	8.2	8.2	13.1	–
Control speed	m/s	–	–	0.61	0.46	–
	ft/s	–	–	2	1.5	–
Weight-to-area ratio	kg/m²	4.7	20	13.9	11.6	14.1
	lb/ft²	1.0	4.1	2.8	2.4	2.9
GR		3.25	3.83	3.9	3.34	3.6
TR	°/s	47	9	10	12	5
V_h	m/s	11.7	15.7	18	22.7	21
	ft/s	38.4	51.5	59.1	74.5	68.9
V_v	m/s	3.6	4.1	4.6	6.8	5.9
	ft/s	11.8	13.5	15.1	22.3	19.4

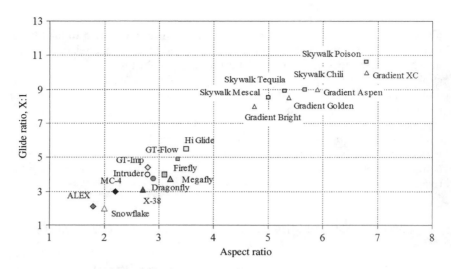

Fig. 1.36   Reported glide ratio vs aspect ratio for several parafoil systems [Ward 2012].

Figure 1.36 shows a variety of PADS in terms of reported and estimated GR and AR graphically. For comparison, it also shows these data for paragliding canopies designed for soaring flight [Ward 2012]. One observation is that PADS form their own group featuring twice as low AS and GR as compared to those of paragliding canopies. Another observation is that GR ~ 1.3 AR. In Fig. 1.37, data on PADS considered in this section are represented vs the rate of descent (for the nominal wing loading at the sea level). Finally, Fig. 1.38 shows PADS rate of descent V_v vs their forward speed V_h to visualize the glide slope angle. (Figure 1.38b zooms in by excluding AGAS PADS and the two-stage Screamer and Onyx PADS.) This latter figure will serve as a reference while discussing different ways to control a glide slope angle in Chapter 8.

Now, when we learned about a wide spectrum of developed different-weight PADS, let us address an issue of correctly assessing one of their main MOEs—touchdown performance.

1.3   PADS ACCURACY

The MOEs addressing the issue of PADS accuracy and introduced in Sec. 1.1.2 [which includes Eq. (1.7)] rely on several MOPs. Test methodology and requirements' interpretation of the advanced tactical parachute system as presented in [Mortaloni 2001] includes the proper flight-test instrumentation and data collection. The actual touchdown coordinates compared to those of IPI along with the release conditions data, PADS data, and wind data are used to assess these

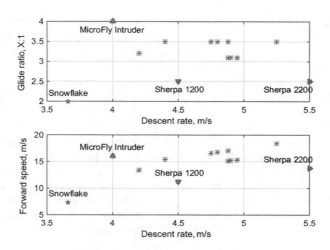

Fig. 1.37   PADS performance envelope.

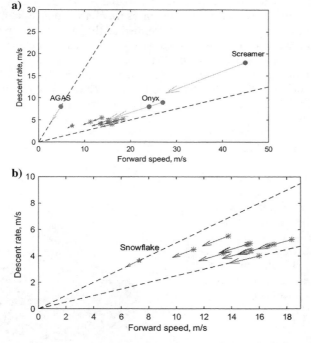

Fig. 1.38   Descent rate vs forward speed a) for most of PADS presented in Table 1.4 and b) for only one-stage parafoil-based PADS.

MOPs and, in turn, the aforementioned MOEs. When estimating PADS touch-down accuracy, most of the time a very simple approach involving easily obtained and robust median radial error (MRE) is used. This approach, however, assumes certain conditions for providing the true estimates that do not necessarily hold for self-guided PADS. Compared to the circular normal distribution (CND), PADS touchdown error distributions feature heavy tails so that the chance of PADS landing beyond a three MRE circle around IPI is about two orders of magnitude higher. As a result, MRE might only serve as a lower bound of a circular-error-probable (CEP) estimate while analysis of touchdown error distribution could provide much deeper insight into performance of a particular PADS (and address all three MOEs laid down in Sec. 1.1.2). The section advocates for a unified methodology for evaluating touchdown error performance of self-guided PADS adapted from that of the science of ballistics and overviews touchdown-error distribution for a variety of different-weight PADS presented in Sec. 1.2. (Their guidance strategies will be discussed in Chapter 7.)

1.3.1 CIRCULAR ERROR PROBABILITY

In the military science of ballistics, CEP (circular error probability or circle of equal probability) is used as an intuitive measure of a weapon system's precision [Driels 2013]. For guided systems, it is defined as the radius of a circle, centered about the IPI (aim point), whose boundary is expected to include 50% of the population within it. The key word in the definition is *population*, not sample. The CEP might not contain exactly 50% of the sample data points, but it is expected to include 50% of the "true" population. That is why it is important to employ and properly use the appropriate statistical tools.

The original concept of CEP was based on CND with CEP as a parameter of the CND just as the mean μ and standard deviation σ are parameters of the normal distribution. The derivation of CEP for CND starts with the general bivariate normal distribution (BND), the statistical elliptical distribution with a probability density function expressed in planar Cartesian coordinates x (range) and y (deflection) (Fig. 1.39)

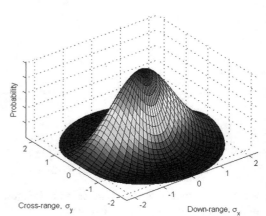

Fig. 1.39   Bivariate normal distribution.

$$p(x, y) = \frac{1}{2\pi\sigma_x\sigma_y\sqrt{1-\rho^2}}e^{-\frac{z}{2(1-\rho^2)}} \tag{1.8}$$

where z is the so-called z-score

$$z = \frac{(x-\mu_x)^2}{\sigma_x^2} - \frac{2\rho(x-\mu_x)(y-\mu_y)}{\sigma_x\sigma_y} + \frac{(y-\mu_y)^2}{\sigma_y^2} \tag{1.9}$$

and ρ is the correlation coefficient between x and y

$$\rho = cor(x, y) = \frac{cov(x, y)}{\sigma_x\sigma_y} = \frac{\sum_{i=1}^{n}(x_i-\mu_x)(y_i-\mu_y)}{\sqrt{\sum_{i=1}^{n}(x_i-\mu_x)^2\sum_{i=1}^{n}(y_i-\mu_y)^2}} \tag{1.10}$$

If $\sigma_x \approx \sigma_y = \sigma$, there is no bias ($\mu_x = \mu_y = 0$), and $x - y$ data are not correlated ($\rho = 0$), and then BND becomes CND or Rayleigh distribution with a probability density function

$$p(r) = p(x, y) = \frac{1}{2\pi\sigma^2}e^{-\frac{x^2+y^2}{2\sigma^2}} = \frac{1}{2\pi\sigma^2}e^{-\frac{r^2}{2\sigma^2}} \tag{1.11}$$

Using this distribution, the probability of radius r being less than some value a is computed as

$$P(r \le a) = \int_0^a 2\pi r p(r)dr = \frac{1}{2\pi\sigma^2}\int_0^a 2\pi r e^{-\frac{r^2}{2\sigma^2}}dr = 1 - e^{-\frac{a^2}{2\sigma^2}} \tag{1.12}$$

The CEP is the radius obtained using the preceding equation for a probability P of 0.5

$$P(r \le CEP) = 0.5 = 1 - e^{-\frac{CEP^2}{2\sigma^2}} \tag{1.13}$$

or solving for CEP

$$CEP = \sqrt{2\ln(2)}\sigma \approx 1.1774\sigma \tag{1.14}$$

If CEP is n meters, 50% of rounds (if we are talking about ballistics) land within n meters of IPI, 43% between n and $2n$, and 7% between $2n$ and $3n$ meters, and the proportion of rounds that lands farther than three times the CEP from IPI is less than 0.2%.

To determine if we are sampling from a CND, the following assumptions must be tested:

- The x and y components of the miss distance are statistically independent ($\rho = 0$).
- The distributions of x and y are both normal.

- The distribution is circular ($\sigma_x = \sigma_y$).
- The mean point of impact (MPI) is at IPI ($\mu_x = \mu_y = 0$).

These assumptions are to be verified with a series of tests as shown in Table 1.9.

However in practice, these assumptions are seldom tested because if they were, the use of CEP would be found inappropriate. Specifically, precision-guided munitions generally have more "close misses" and therefore are not normally distributed. Munitions may also have larger standard deviation of range errors than the standard deviation of deflection errors, resulting in an elliptical confidence region. Finally, munition samples might not be exactly on target, that is, the mean vector is not a zero vector, but rather biased ($\mu_x \neq 0$, $\mu_y \neq 0$). For the latter two cases, approximate formulas are available to convert the distributions along the two axes into the equivalent circle radius for the specified percentage.

Let us now lay out a detailed procedure for statistically correct estimation of CEP helping to reveal some features associated with PADS as opposed to ballistic systems.

1.3.2 EXAMPLE OF PADS' TOUCHDOWN ACCURACY ANALYSIS

Formally, the procedure for CEP estimation looks as follows:

- Gather x and y data pairs. (By gathering just the values of radial miss distance, we may be losing a lot of useful information.)
- Eliminate obvious outliers.

TABLE 1.9   PARAMETRIC AND NONPARAMETRIC TESTS FOR CEP COMPUTATION

	Parametric Test (Normal Distribution)	Nonparametric Test (Distribution Free)
Assumption 1: The x and y components are statistically independent.	t-test [Hayter 2012]	Spearman's rho test [Spearman 1904]
Assumption 2: The distributions of x and y are both normal.	Chi-square test [Hayter 2012]	Lilliefors test [Lilliefors 1967]*
Assumption 3: The distribution is circular.	F-test [Hayter 2012]	Siegel–Tukey test [Spearman 1904]
Assumption 4: The MPI is at IPI.	t-test [Hayter 2012]	Wilcoxon signed-rank test [Hayter 2012; Wilcoxon 1945]

*Lilliefors' test is used to test whether data come from a normally distributed population without specifying the expected value and variance of the distribution.

- Test the x and y data sets for normality.
- Test for $x - y$ data sets' correlation, and rotate original data if necessary.
- Test whether MPI is at IPI, that is, compute a bias, and check whether this bias is statistically significant.
- Test whether the distribution is circular, and compute CEP_{MPI} with respect to MPI.
- If the bias happens to be statistically significant, calculate CEP_{DMPI}, that is, CEP with respect to the desired MPI (DMPI).
- Calculate the upper confidence level for CEP_{MPI} (CEP_{DMPI}).

The concept of CEP_{MPI} vs CEP_{DMPI} is illustrated in Figs. 1.40a and 1.40b. CEP_{DMPI} has such a value that it encompasses 50% of data (Fig. 1.40c).

There is no analytical solution for CEP_{DMPI}, so its computation is based on a simple expression

$$CEP_{DMPI} = CEP_{MPI}k_{Bias} \qquad (1.15)$$

where the coefficient k_{Bias} represents a numerical approximation. The discussion on how to find this coefficient is given in Appendix D, but for self-guided PADS featuring $V = Bias/CEP_{MPI} < 0.5$ (meaning that actual PI spread/precision larger than the bias/accuracy), the following quadratic regression can be used:

$$k_{Bias} = 0.3525V^2 - 0.00146V + 1 \qquad (1.16)$$

As seen from Fig. 1.41, even in the case of a statistically significant bias, accounting for it does not change the computed CEP much—at most it can increase it by about 8%, but for mature PADS $k_{Bias} \sim 1$ and $CEP_{DMPI} \sim CEP_{MPI}$.

To demonstrate how the CEP estimation routine outlined in the beginning of this section works, Fig. 1.42 features some real drop test data and their preliminary analysis.

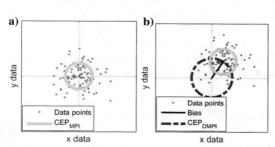

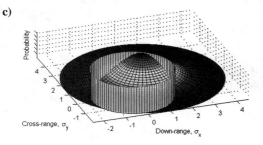

Fig. 1.40   a) CEP_{MPI} vs b) CEP_{DMPI} and c) graphical illustration of CEP_{DMPI} corresponding to Fig. 1.40b.

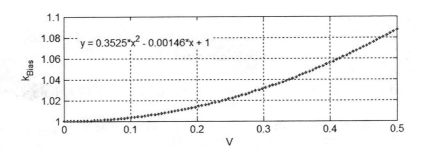

Fig. 1.41   Coefficient $k_{Bias} = k_{Bias}(V)$ to determine CEP$_{DMPI}$ from CEP$_{MPI}$.

Specifically, we start from producing a nonparametric box plot (upper-right portion of the figure) revealing several outliers out of the total of 19 data points.

By construction, a box plot (or a box-and-whisker diagram or plot) in Fig. 1.42 graphically depicts numerical data without making any assumptions of the underlying statistical distribution through their five-number summaries: the box itself is composed of the lower quartile (Q1), median (Q2), and upper quartile (Q3) while the ends of the whiskers in this case represent ± 1.5 interquartile range (IQR = Q3 − Q1) above Q3 and below Q1. What appears beyond the whiskers maybe treated as outliers.

As seen from Fig. 1.42, in our specific case we have three obvious outliers, probably caused by structural malfunction (ripped off or twisted control lines) or GNC algorithms not working properly (which might also include an effect of improper mitigation of unknown winds or unmodeled dynamics). Eliminating these three outliers as shown in the bottom-left portion of the figure (leaving $n = 16$ data points to work with) leads to the box plots covering (explaining statistically) all of the remaining data. (Compare the upper-right and bottom-right plots in Fig. 1.42 corresponding to and taking into account all 19 and just 16 data points.)

In other cases eliminating outliers might not be so straightforward. Consider Fig. 1.43, which represents the relationship between the box plot constructed as just described for the standard normal distribution with a well-known bell-shape probability density function for this distribution. As seen for the normal distribution, the box plot would cover $[-2.698\sigma; 2.698\sigma]$ range or about 99.3% of the set (compared to about 99.7% for the $[-3\sigma; 3\sigma]$ range). That means that even for the normal distribution we can expect to see about 1% of data points beyond the whiskers that are not necessarily outliers (i.e., cannot be eliminated from the further analysis). For the data points not necessarily distributed normally, this amount might be much larger.

According to Table 1.9, the following procedures rely on population distribution; therefore, the normality test needs to be conducted. For example, the Kolmogorov–Smirnov normality test (Fig. 1.44) compares the empirical cumulative

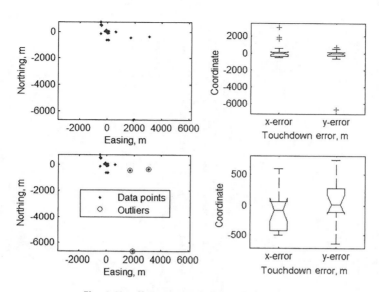

Fig. 1.42   Nonparametric data conditioning.

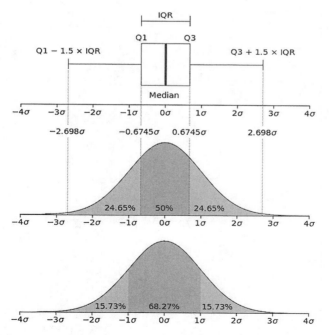

Fig. 1.43   Explanation of the box plot concept as applied to the normal distribution.

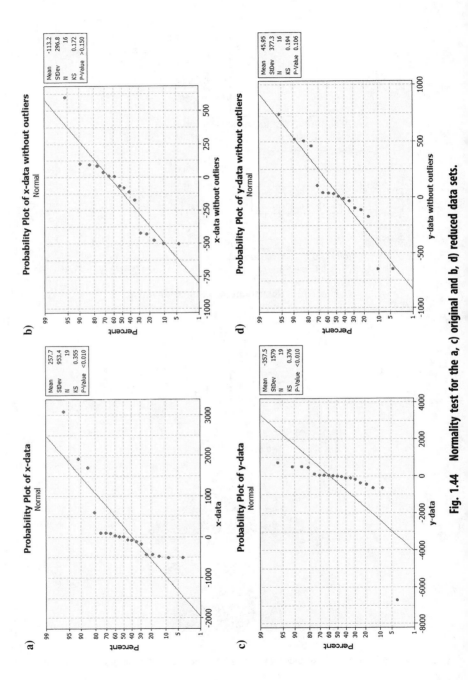

Fig. 1.44 Normality test for the a, c) original and b, d) reduced data sets.

distribution function of sample data with the distribution expected if the data were normal [Hayter 2012]. The outcome of this test is the p-value—a small p-value is an indication that the null hypothesis (that the observations follow a normal distribution) is false (that is, what $p < 0.01$ in Fig. 1.44 stands for). As seen from Fig. 1.44, elimination of three outliers at the previous step allows increasing the p-value for both x and y data sets to the level where the null hypothesis cannot be rejected (to $p > 0.15$ and $p > 0.106$, respectively), so we can assume that the data points are distributed normally.

Next, the correlation analysis to reveal a possible correlation between x (Northing) and y (Easting) data takes place. The correlation coefficient ρ defined in Eq. (1.10) is used in the t test with the t-statistic computed as

$$t_s = \frac{|\rho|\sqrt{n-2}}{\sqrt{1-\rho^2}} \tag{1.17}$$

This value is then compared with the critical value of t-statistic (a ratio of the departure of an estimated parameter from its notional value and its standard error). In our case (for $n = 16$) $\rho = -0.56$, which leads to $t_s = 2.54$. This value happens to be larger than the critical value $t_{0.02,14} = 2.26$. [Here 0.02 stands for $98\% = 100\,(1 - 0.02)\%$ confidence level and $14 = n - 2$ defines the number of DoF.] This means that in our specific case the correlation happens to be significant [which, for instance, might be caused by the northwest (NW) direction of the prevailing surface winds during the tests]; hence, the data set needs to be rotated by some angle computed as

$$\theta = \frac{1}{2}\tan_{-1}\left(\frac{2\rho\sigma_x\sigma_y}{\sigma_x^2 - \sigma_y^2}\right) \tag{1.18}$$

so that the data point coordinates in the new coordinate frame become

$$\begin{bmatrix} u \\ v \end{bmatrix} = \mathbf{R}_\theta \begin{bmatrix} x \\ y \end{bmatrix} = \begin{bmatrix} \cos(\theta) & \sin(\theta) \\ -\sin(\theta) & \cos(\theta) \end{bmatrix}\begin{bmatrix} x \\ y \end{bmatrix} \tag{1.19}$$

In our case this rotation happens to be $\theta = 33$ deg (Fig. 1.45a).

The next step is to evaluate a statistical significance of the bias [which in our case happens to be 122 m (400 ft)]. We do it by comparing the critical value of $t_{0.02,15} = 2.25$ with the values of corresponding u and v statistics

$$t_u = \frac{\bar{u}}{s_u/\sqrt{n}} \quad \text{and} \quad t_v = \frac{\bar{v}}{s_v/\sqrt{n}} \tag{1.20}$$

[which evaluates the distance from the (0,0) point using the data-spread "yardstick"]. As mentioned in the beginning of this section, for PADS, because of the large data spread, it is usually insignificant (see Fig. 1.45a), but, even if it were the low, V values (0.3 in this specific case) would increase CEP by only 3% (see Fig. 1.41).

a)

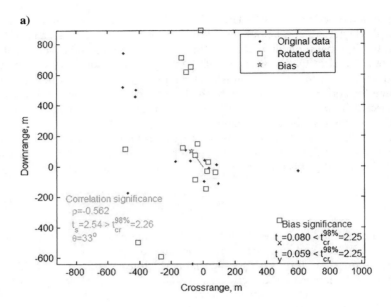

b)

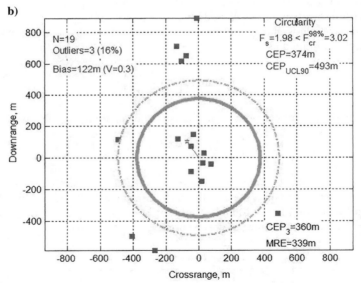

Fig. 1.45   a) Data rotation and b) CEP$_{MPI}$ calculation.

Now, we can check whether the distribution of our data is circular. We do it by calculating the F-statistic

$$F_s = \frac{\sigma_L^2}{\sigma_S^2} \tag{1.21}$$

[where σ_L is the larger standard deviation (range) and σ_S is the smaller standard deviation (deflection)] and comparing it to the critical value. In our case, despite an obvious difference in the values of standard deviation for the u and v data sets (after rotation) the Fisher's F test reveals that this difference (σ_L is $1.4 \approx \sqrt{1.98}$ times larger than σ_S) is insignificant (compared to the critical value of $F_{0.02,15,15} = 2.86$, where $15 = n-1$ defines the DoF for numerator and denominator), that is, in our case we can be 98% confident that distribution is circular. Hence, we can apply a generic formula for CEP applicable to CND

$$CEP = 1.1774 \frac{\sigma_S + \sigma_L}{2} = 0.5887\sigma_S + 0.5887\sigma_L \tag{1.22}$$

Obviously, CEP by itself is not the entire story; a confidence interval (CI) should be calculated and reported along with CEP. Specifically, the upper confidence limit (UCL) for the one-sided CI can be calculated using

$$CEP_{UCL_{(1-\alpha)\%}} = CEP\sqrt{\frac{(1 + K^2)(n - 1)}{\chi^2_{\{1-\alpha,\text{int}[(1+K^2)(n-1)]\}}}} \tag{1.23}$$

where $K = \sigma_S/\sigma_L < 1$, n is the number of data points, and the integer part of $(1 + K^2)(n - 1)$ represents the DoF to compute the χ^2 (chi-square) value for the appropriate confidence level $(1 - \alpha)$. In the case of two-sided CI, the lower and upper limits are computed as

$$CEP_{LCL_{(1-\alpha)\%}} = CEP\sqrt{\frac{(1 + K^2)(n - 1)}{\chi^2_{\{\alpha/2,\text{int}[(1+K^2)(n-1)]\}}}}$$

$$\tag{1.24}$$

$$CEP_{UCL_{(1-\alpha)\%}} = CEP\sqrt{\frac{(1 + K^2)(n - 1)}{\chi^2_{\{1-\alpha/2,\text{int}[(1+K^2)(n-1)]\}}}}$$

In our case CEP $= 374$ m (1,227 ft) and $CEP_{UCL_{90}} = 493$ m (1,618 ft) as shown in Fig. 1.45b.

What if we do not have the $x - y$ data set to begin with, but are rather given the radial miss distance values? Well, in this case the circular error average (CEA) (or MRE), still applicable for CND in the form of Eq. (1.11), can be computed as

$$\text{CEA} = \int_0^\infty \frac{r^2}{\sigma^2} e^{-r^2/2\sigma^2}\, dr = \sigma\sqrt{\frac{\pi}{2}} \tag{1.25}$$

Because CEP $=\sqrt{2\ln(2)}\sigma$ [see Eq. (1.14)], it can be related to CEA as

$$\text{CEP} = \frac{\sqrt{2\ln(2)}\text{CEA}}{\sqrt{\pi/2}} = 0.9394\text{CEA} \tag{1.26}$$

Hence, the maximum likelihood unbiased estimate of CEP becomes

$$\text{CEP}_3 = 0.9394\frac{1}{n}\sum_{i=1}^{n} r_i \tag{1.27}$$

If you cannot assume normality, you can still estimate CEP using the median, which is also a measure of location, so that 50% of the data values are on each side of it. The median radial error (MRE) is computed as

$$\text{MRE} = \text{median}(r) \tag{1.28}$$

Both estimates (CEP$_3$ and MRE) are also shown in the bottom-right corner of Fig. 1.45b. The fact that they are relatively close to the CEP estimate shown in the upper-right corner indicates that the actual distribution is really close to CND as we have formally proven it using the Kolmogorov–Smirnov test already.

Figure 1.46 shows a relationship between different estimates of CEP obtained for some other PADS test data. Such a behavior is quite typical, and all accuracy estimates are usually characterized by the following inequality:

$$\text{MRE} < \text{CEP}_3 < \text{CEP}_{\text{MPI}} < \text{CEP}_{\text{DMPI}} < \text{CEP}_{\text{UCL}(1-\alpha)\%} \tag{1.29}$$

While decreasing the size of the sample by eliminating outliers, the CEP$_{\text{MPI}}$ estimate converges to that of MRE. (Note that the MRE estimate decreases with an

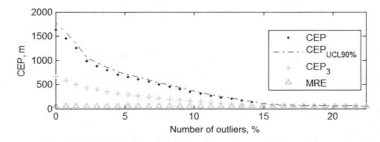

Fig. 1.46   Convergence of all CEP metrics vs the number of eliminated outliers.

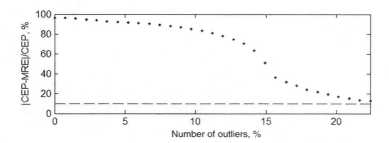

Fig. 1.47   Relative error between CEP$_{MPI}$ and MRE.

increase of the number of outliers as well; it is just the scale of Fig. 1.46 that prevents us from seeing it). Hence, MRE can serve as a lower bound of a CEP estimate. Figure 1.47 shows this specific (relative) difference between CEP$_{MPI}$ and MRE that correlates with data in Fig. 1.46. This difference can probably serve as an implicit indicator of the number of outliers in test data. In this specific case, when the number of outliers reaches about 21%, the relative difference between two estimates becomes less than 15%. Obviously, further increase of the number of outliers leads to even closer match between CEP$_{MPI}$ and MRE; however, there is a danger of increasing the system's operational effectiveness (speaking the Operational Test and Evaluation language [Giadrosich 1995]) at the expense of operational suitability. To this end, Fig. 1.48 shows a weighted sum of a normalized CEP$_{MPI}$ and the number of outliers this CEP estimate was computed without. For this specific case, it is clear that going beyond ~21% of outliers is not rational.

To conclude this section, let us address one more issue. What if the F statistic [Eq. (1.21)] turns out to be larger than the critical value indicating that we do not have a CND? In this case (BND), we can still utilize the CEP concept, which means exactly the same—the radius of a circle encompassing 50% of

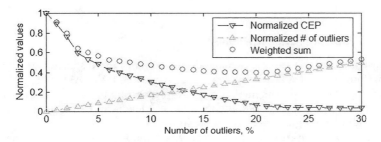

Fig. 1.48   Effectiveness vs suitability.

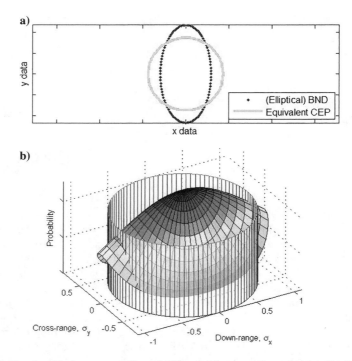

Fig. 1.49   Graphical representation of CEP$_{MPI}$ in the case of noncircular distribution.

data (Fig. 1.49). The only difference is that, instead of Eq. (1.22), we would need to use

$$CEP = 0.6183\sigma_S + 0.5619\sigma_L \qquad (1.30)$$

(Derivation of this equation is presented in Appendix D.)

Now that the approach to estimate an accuracy of self-guided PADS has been formalized, let us try to apply it to establish the overall trends in precision airdrop.

1.3.3   TRENDS IN PRECISION AIRDROP

Figures 1.50 and 1.51 represent statistical analysis of miss-distance data obtained during four of five PATCAD events [PATCAD 2003; PATCAD 2005; PATCAD 2007; PATCAD 2009]. The methodology for processing test data was the same as established in Sec. 1.3.2. (Note, the axis names, down- and cross-range, in these and the following figures are used quite loosely.)

As seen from Figs. 1.50 and 1.51, from the standpoint of statistics there is a lot of similarities between test data obtained at all four events. Elimination of the

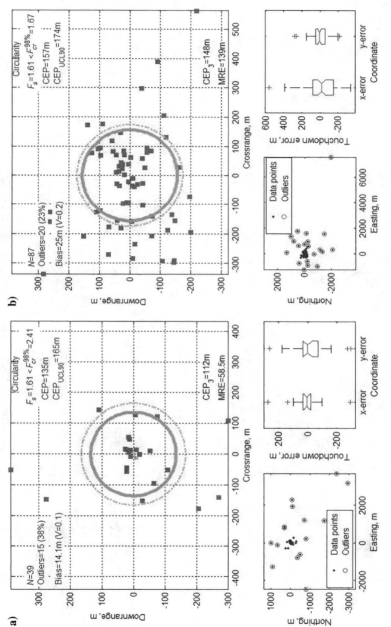

Fig. 1.50   Overall performance at a) PATCAD 2003 and b) PATCAD 2005.

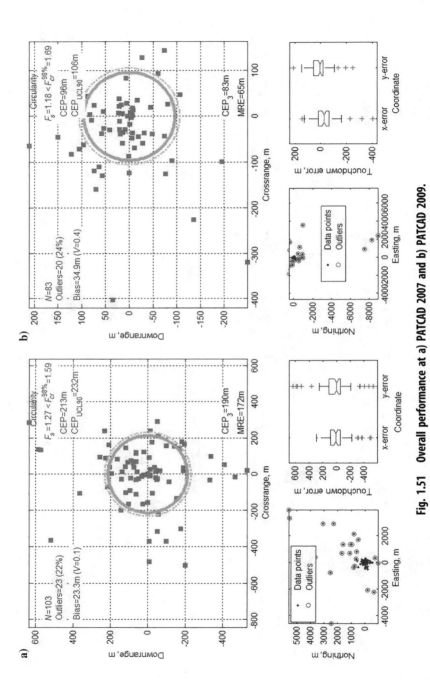

Fig. 1.51   Overall performance at a) PATCAD 2007 and b) PATCAD 2009.

essential number of outliers brings miss distance distributions close to normal. (The box plots characterize reduced data sets after exclusion of most obvious outliers.) The overall reliability of the systems becomes better, so that the number of outliers decreases from 38% down to about 25% (among all PADS demonstrated at each event as indicated in Table 1.2). The V value happens to be small for the first three events and relatively large (still below 0.5) for the fourth event. The latter was caused partly because of reduced CEP but also because of the strong winds (on some occasions of the order of the forward speed of PADS). However, the bias happens to be not statistically significant for any data set. All distributions are close to circular, which is probably caused by the specific features of GNC algorithms typical for most of PADS and the fact that data from different PADS are considered together.

Figure 1.52 features the MRE estimates obtained by consequent elimination of more and more worse data points for each of the five PATCAD events computed according to Eq. (1.28). (Again, MRE represents a lower bound of a CEP estimate as explained in Sec. 1.3.2.) The horizontal axis shows the percentage of test data that was considered for MRE computation. This figure clearly indicates the overall improvement of aerial payload delivery accuracy in the course of eight years no matter how many points one decides to eliminate due to the different reasons.

Figure 1.53 represents miss distances normalized by the corresponding MRE, so that all dependences intersect a 50%, (1-MRE) point. If data were distributed according to CND, 99.8% of all data points would reside within 3 MRE (CEP). However, it is not the case. On average, about 25% of data fall outside the 3-MRE limit. Interestingly enough, this percentage is about the same as the percentage of accepted outliers (see Figs. 1.50 and 1.51). Another observation is that for PATCAD 2001 the overall distribution happens to be much closer to that of CND. This suggests that most likely it is basic PADS flight performance and GNC paradigm that affect touchdown accuracy the most because 30% of the data points for PATCAD 2001 belonged to parachute-based PADS with a GNC scheme cardinally different from those of parafoil-based PADS.

In fact, it happens to be a well-known fact in a guided weapon design. To this end, Fig. 1.54 shows three sets of miss-distance distribution data normalized

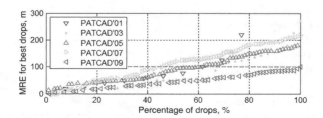

Fig. 1.52   PATCAD 2001–2009 performance.

Fig. 1.53 Miss distance distribution (in the units of MRE).

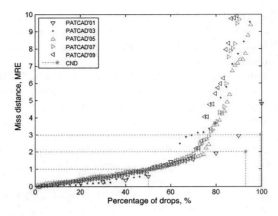

by the corresponding MRE belonging to three similar weapon systems differing by the added control authority. Even though these data are not related to PADS, they show a general tendency. Although the miss-distance distribution for unguided weapon fits CND very well, adding some (system 2) and more (system 1) control authority for a guided version of the same weapon disturbs the distribution causing the heavy tails. Comparison of Fig. 1.53 data and Fig. 1.54 data reveals a lot of similarities—adding more control authority (which, for example, includes a higher glide ratio) leads to having larger percentage of "beyond 3-MRE (CEP)" tail.

Do not be confused; having more control authority serves its goal, that is, CEP becomes smaller and smaller, but it is not clearly seen in Fig. 1.53 because everything is normalized by the corresponding CEP. Another comment to make though is that in terms of accuracy, the miss-distance distribution for all PATCAD events shown in Fig. 1.53 is centered on IPI. This indicates that GNC algorithms for all PADS work properly, introducing no bias. It is not the case for unguided weapon data of Fig. 1.54 though. As seen, there is some systematic error preventing distribution to be centered on IPI. (The distribution does not start from a zero MRE.) This would be typical for unguided ADS as well; if multiple ADS were to be released from the same CARP, they would feature the same distribution, exhibiting a nonzero bias. Adding even a minimum control authority (for guided weapon 2) serves its goal and brings the bias down to zero.

To elaborate on this topic, let us consider the issue of

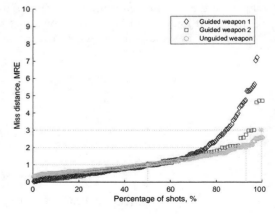

Fig. 1.54 Typical miss-distance distribution for guided and unguided weapons.

having heavy tails for several different PADS separately as opposed to the data sets composed of multiple PADS as discussed in this section. This way we will try to determine the effect of PADS's size and GNC concept.

To this end, Figs. 1.55 and 1.56 demonstrate statistical analysis of miss distances for some typical PADS in different weight categories. Even though in this case the data points characterize a single system as opposed to several systems at once as it was the case in Sec. 1.3.3, all major observations are still the same. The number of outliers needed to be removed varies from 36% for more complex M PADS all the way down to 13% for the lighter ones attesting to their maturity (operational suitability). The biases are statistically insignificant with the small values of V even for the systems with low CEP (Fig. 1.52); the distributions are close to CND (for XL PADS $\sigma_L = 1.07\sigma_S$).

For comparison, Fig. 1.57 shows statistical data for two more UL PADS. All observations just presented are generally valid for these two PADS as well (with a slightly larger number of outliers, but smaller values of the CEP estimates).

Figure 1.58 exhibits the same type of dependencies as that of Fig. 1.52, but for representative M/L/XL/UL PADS. Again, it shows the overall trends based on the lower bound of the CEP estimate. Clearly, nothing else but PADS's weight affects the overall performance. Weight of PADS though may have an indirect influence via a turn radius, which happens to be proportional to the value of a CEP estimate for each system.

As just mentioned, if data were to follow CND, the actual PIs would tend to cluster around IPI, with most reasonably close, progressively fewer and fewer further away, and very few at long distance. To this end, Fig. 1.59a shows actual touchdown error distributions for all systems presented in Figs. 1.55 and 1.56a, while Fig. 1.59b presents distributions for three different UL PADS shown in Figs. 1.56b and 1.57. As seen, in the case of self-guided PADS it is about 20% (on average) of the drops that are farther than 3 MRE (CEP) as opposed to just 0.2% for the classical CND. What is interesting is that these three different-weight-category PADS in Fig. 1.59a exploit about the same guidance strategy whereas three same-weight-category PADS in Fig. 1.59b exercise different guidance strategies! As a result, three distributions of Fig. 1.59a are much closer to each other compared to those three in Fig. 1.59b. Hence, as it was hypothesized when observing the plots in Fig. 1.53, the guidance strategy not only affects the overall touchdown accuracy by itself, but also causes changes in the tails of distributions making them either heavy, like $PADS_2$ in Fig. 1.59, or closer to those of a classical CND ($PADS_3$). Finally, let us address the nature and distribution of these outliers.

1.3.4 ANALYSIS OF OUTLIERS

Although the preceding sections concentrated on a proper procedure to remove outliers in order to estimate PADS's performance in terms of CEP, from the

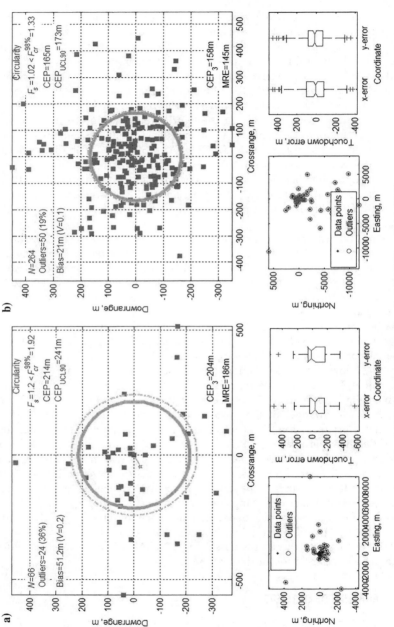

Fig. 1.55   Typical performance of a) M PADS and b) L PADS.

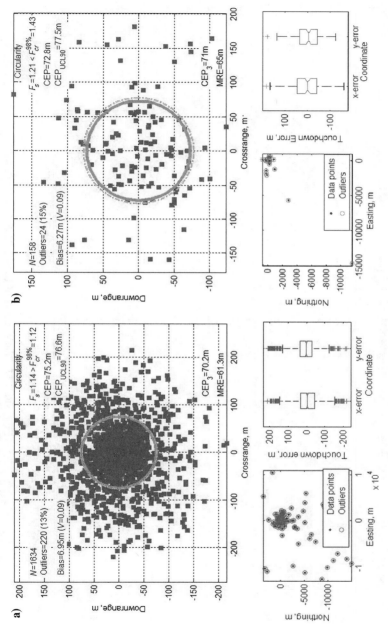

Fig. 1.56 Typical performance of a) XL PADS and b) UL PADS.

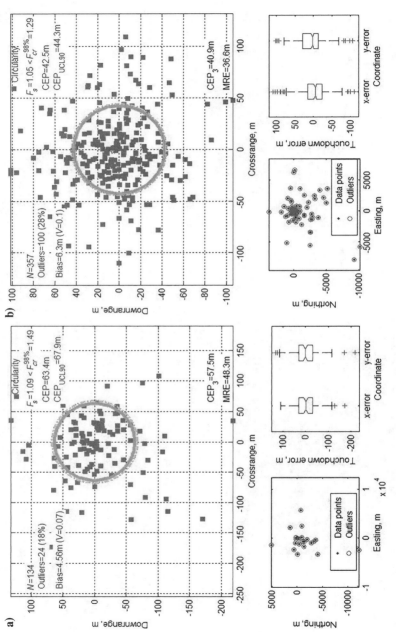

Fig. 1.57   Performance of two UL PADS.

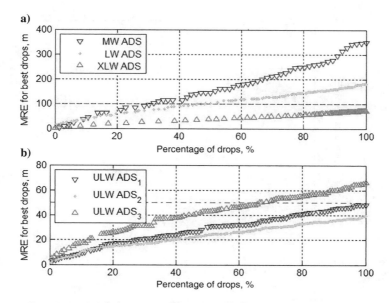

Fig. 1.58   a) M/L/XL PADS performance and b) UL PADS performance.

standpoint of a test planning officer these outliers represent a major interest/concern. The test planning officer needs to know how far these outliers might be off the 3-CEP circle around IPI, what a probability of such an event is, what distribution these outliers exhibit, and what the worst-case scenario would be (estimating the largest miss distance). Toward this goal, Figs. 1.60–1.64 represent the analysis of outliers that were removed to estimate CEP as it was shown in Figs. 1.50, 1.51, 1.55–1.57.

The first observation is that in most of the cases there is a strong correlation between x and y sets of data (revealed in this case by the Spearman's rho test as indicated in Table 1.9) resulting in the necessity to rotate the data sets. (Again, the axis names, down- and cross-range, are used loosely.) It is probably safe to attribute it to the direction of the prevailing winds (which by the way also defined the position of a release point with respect to IPI).

The second observation is that some data sets contain obvious outliers that are most likely caused by the complete or partial failure to deploy a canopy (see Figs. 1.60b, 1.61b, 1.62a, 1.63b, and 1.64b). Another reason might be that the IPI coordinates were not entered into the system correctly, so it tries to steer to the point that is miles away from the actual IPI. Both events happen quite rarely (especially for a mature PADS), and the only way to account for them is to compute a full-fledged safety fan based on parameters of a release point and PADS's characteristics (glide ratio), as will be shown in Sec. 3.3.2. If these extreme outliers are

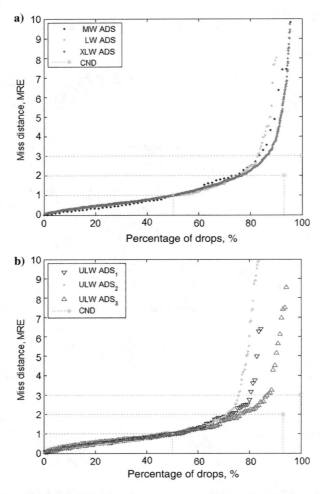

Fig. 1.59 Miss distance distribution (in the units of MRE) for a) different-weight-category PADS and b) the same-weight-category PADS.

eliminated from the further analysis, then we usually have a distribution featuring a slightly larger dispersion of data along one axis of the correlation compared to that on another axis. If characterized by the standard deviation (which might not be exactly the right thing to use in this case), on the average we have about 25% difference (which, as explained in the preceding section, is most likely caused by the essence of GNC algorithms). Running a formal nonparametric Siegel–Turkey test (see Table 1.9) does not necessarily treat these differences as being statistically

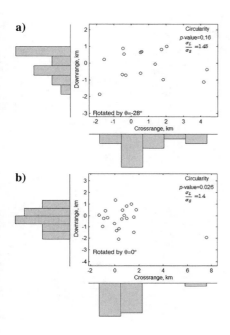

Fig. 1.60   Outliers corresponding to data shown in Fig. 1.50.

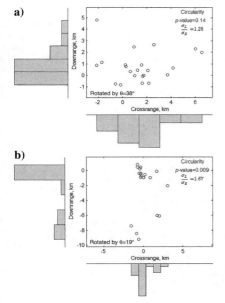

Fig. 1.61   Outliers corresponding to data shown in Fig. 1.51.

Fig. 1.62 Outliers corresponding to data shown in Fig. 1.55.

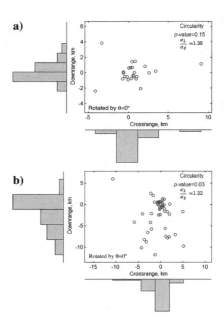

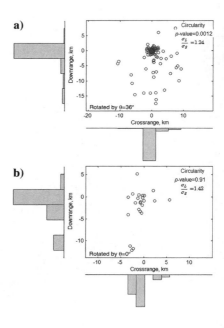

Fig. 1.63 Outliers corresponding to data shown in Fig. 1.56.

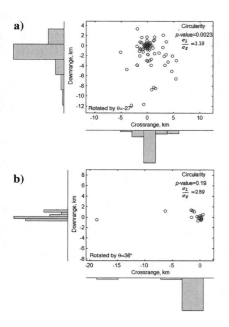

Fig. 1.64   Outliers corresponding to data shown in Fig. 1.57.

significant (see Figs. 1.60a, 1.61a, 1.62a, 1.63b, and 1.64b). On the contrary, the outlier distributions can be thought of being as close to circular.

Section 3.3.2 also uses the analysis of outliers' distribution for computation of the safety fans during flight tests.

CHAPTER 2

Basic Analysis of Ram-Air Parachute

Steven Lingard*
Vorticity Ltd., Chalgrove, Oxfordshire, United Kingdom

This chapter features slightly edited material on ram-air parachute design presented at the Precision Aerial Delivery Seminar in 1995 (see Table A2 in Appendix A) [Lingard 1995]. That was about the time when the first attempts to design practical guidance, navigation, and control algorithms for autonomous payload delivery systems had commenced. The purpose of this particular presentation was to discuss the performance and design of ram-air parachutes with particular reference to the current requirements of precision aerial delivery systems. To meet these requirements, the parachute must have predictable high glide ratio; predictable flight speeds compatible with wind penetration; predictable, stable, dynamic performance in response to control inputs and wind gusts; predictable turn characteristics; reliable deployment and inflation characteristics; and the ability to flare to reduce landing speed. This material summarized all prior knowledge on modeling of ram-air parachutes and, as such, served as a starting point for many modeling efforts undertaken in the 2000s. Compared with other chapters in the book, all derivations in this chapter assume still air (all three component of the wind are considered zero). As a result, this chapter uses notation V to denote a magnitude of airspeed (later on, we will distinguish between the groundspeed vector V, as measured by GPS, and airspeed vector V_a). Another feature is that the rigging angle μ is defined as a positive quantity while later on we might need to account for its (negative) sign. Lastly, this chapter utilizes symbol γ to denote the glide angle rather than the flight-path angle (negative to glide angle) utilized in all other chapters.

2.1 GENERAL DESCRIPTION OF RAM-AIR PARACHUTE

A significant advance in parachute design occurred in the early 1960s when the airfoil or ram-air parachute (Fig. 2.1) was invented by [Jalbert 1966].

---

*Technical Director.

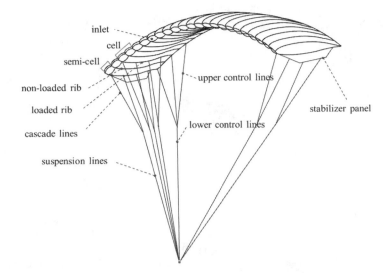

Fig. 2.1   Ram-air parachute.

The ram-air parachute, when inflated, resembles a low-aspect-ratio wing. It is entirely constructed from fabric with no rigid members, which allows it to be packed and deployed in a manner similar to a conventional parachute canopy. The wing has upper and lower membrane surfaces, an airfoil cross section, and a rectangular planform. The airfoil section is formed by airfoil-shaped ribs sewn chordwise between the upper and lower membrane surfaces at a number of spanwise intervals forming a series of cells. The leading edge of the wing is open over its entire length so that ram-air pressure maintains the wing shape. The ribs usually have apertures cut in them. This allows the transmission of pressure from cell to cell during inflation and pressure equalization after. The fabric used in the manufacture of ram-air parachutes is as imporous as possible to obviate pressure loss.

Suspension lines are generally attached to alternate ribs at multiple chordwise positions with typically 1-m spacing. Although this results in a large number of suspension lines, it is necessary in order to maintain the chordwise profile of the lower surface. Lines are often cascaded to reduce drag. Some early ram-air parachute designs have triangular fabric panels, "flares," distributed along the lower surface to which the suspension lines are attached. In addition to distributing evenly the aerodynamic load to the suspension lines and thus helping to maintain the lower surface profile, these panels partially channel the flow into a two-dimensional pattern, reducing tip losses and also aiding in directional stability. While flares produce a truer airfoil shape with fewer distortions, improving aerodynamic efficiency, there is a penalty to be paid in weight, bulk, and construction complexity. Indeed, the drag of the flares themselves possibly out-weighs their advantage. Most modern designs dispense with the flares, the aero-dynamic loads being distributed by tapes sewn to the ribs. Stabilizer panels are often used to provide partial end-plating and to enhance directional stability. On current designs, rigging lines are typically 0.6...1 spans in length, with the wing crown rigged, that is, the lines in a given spanwise bank are equal in length. The wing therefore flies with arc anhedral.

Several airfoil sections have been used on ram-air parachutes: most early wings used the Clark Y section with a section depth of typically 18% chord (Fig. 2.2a); however, recent designs have benefited from glider technology and use a range of low-speed sections (e.g., NASA LS1-0417, Fig. 2.2b). Various nose aperture shapes have also been investigated as nose shape affects inflation and airfoil drag. There is also a trend to reduce section depth to reduce drag, but this has proceeded slowly because inflation performance can be adversely affected. Means for lateral-directional and longitudinal control are provided by steering lines attached to the trailing edge (TE) of the canopy. These lines form a crow's-foot pattern such that pulling down on one line causes TE on one side of the canopy to deflect. Turn control is affected by an asymmetric deflection of the steering lines, and angle of attack control and flare-out are accomplished by a symmetric deflection.

The ram-air parachute has basically been refined within sports parachut-ing, with much of the early work experimental in nature. However, because of the perceived military and space applications and the desire to improve glide ratios, increasingly theoretical studies of ram-air performance were undertaken [Goodrick 1975; Goodrick 1979b; Lingard 1981]. Goodrick [1975] presents a comprehensive bibliography of ram-air literature up to 1980. Development of

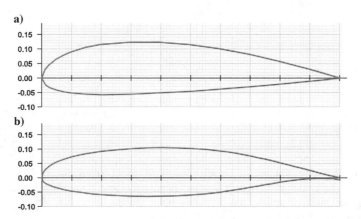

Fig. 2.2   a) 18% Clark YM-18 and b) 17% NASA LS1-0417 airfoils [AID 2014; McGhee and Beasley 1973; Patel et al. 1997].

the ram-air since that time has comprised continued refinement of the basic para-chute in the sports parachuting world, work to increase glide ratio [Lingard 1986], and development of the ram-air parachute at small scale for submunition use, up to very large scale for space applications. Mayer et al. [1986] and Puskas [1989] document the development of large Para-Flite canopies up to $279\,\mathrm{m}^2$ ($3{,}000\,\mathrm{ft}^2$). Wailes [1993] reports on Pioneer Systems/NASA work on the design of a 47-cell, 454-kg (1,000-lb), $1{,}022\text{-}\mathrm{m}^2$ ($11{,}000\text{-}\mathrm{ft}^2$) ram-air for the recov-ery of 27.2-ton (60,000-lb) loads. Goodrick [1984] has also published a useful paper examining scaling effects on ram-air parachutes.

Chatzikonstantinou [1989, 1993] presents a sophisticated and elegant numeri-cal analysis to predict the behavior of an elastic membrane ram-air wing under aerodynamic loading, solving coupled aerodynamic and structural equations. This type of analysis is essential in the extension of ram-air technology to high wing loading where the deformation of the wing is extremely significant. The equations of motion and apparent (or added) masses related to ram-air para-chutes is discussed in [Lingard 1981; Lissaman and Brown 1993]. Brown [1993] analyzes turn control. The aerodynamic characteristics of ram-air wings are inves-tigated by Gonzalez [1993] who applies panel methods to determine lift-curve slope and lift distribution. Ross [1993] investigates the application of compu-tational fluid dynamics (CFD) to the problem and demonstrates the potential for improved lift-to-drag ratio by restricting the airfoil inlet size to the range of movement of the stagnation point.

2.2   AERODYNAMICS OF RAM-AIR WINGS

Once inflated, a ram-air parachute is essentially a low-aspect-ratio wing, and thus conventional wing theory is applicable. Although CFD Navier–Stokes and vortex panel methods can give accurate predictions of the performance of ram-air wings, a simpler approach is adopted here in order to bring out clearly the influence of the various parameters on wing performance.

2.2.1   LIFT COEFFICIENT

Before embarking on the analysis of the ram-air wing, it is useful to consider the two-dimensional flow around the wing section at various angles of attack. The flow pattern can be determined using a vortex singularity method [Lingard 1981]. The canopy is modeled by a vortex sheet placed to coincide with the phys-ical location of the canopy. The strength of the sheet is determined by the appli-cation of velocity boundary conditions. The sheet strength is then used to determine the pressure distribution and circulation and hence lift coefficient. By varying the angle of attack, the two-dimensional lift-curve slope $C_{L\alpha}^a = \mathrm{d}C_L/\mathrm{d}\alpha$, and the angle of attack for zero lift α_0 are found.

Figure 2.3 shows lift coefficient as a function of angle of attack. The value of $C_{L\alpha}^a = 6.89\,\mathrm{rad}^{-1}$ is greater than the theoretical value of $2\pi\,\mathrm{rad}^{-1}$ for a thin airfoil,

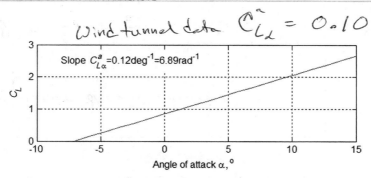

Fig. 2.3   Theoretical section lift coefficient for an 18% Clark Y ram-air parachute airfoil.

but is slightly less than the value of 7.18 rad^{-1} for an 18% thick airfoil. The zero lift angle α_0 is found from Fig. 2.2 to be -7 deg.

Although lifting-line theory represents well those wings with aspect ratios A above 5, in the case of low-aspect-ratio wings TE of the wing is exposed to a flow that has been deflected by the leading edge; thus, the lift cannot be assumed to act on a line and must be spread over the surface of the airfoil. Also, the influence of lateral edges (a minor effect for larger aspect ratios) becomes increasingly significant as aspect ratio is reduced.

Several attempts have been made to extend lifting-line theory into the range of small aspect ratios, or to replace it by lifting-surface theory, which accounts for the effect of wing chord on lift characteristics. Lifting-line theory (e.g., [Pope 1951]) gives the lift curve for a wing as

$$C_{L\alpha} = \frac{\pi C_{L\alpha}^a AR}{\pi AR + C_{L\alpha}^a (1 + \tau)} \tag{2.1}$$

where $C_{L\alpha}^a$ is the two-dimensional lift-curve slope, AR is the aspect ratio, and τ is a small positive factor that increases the induced angle of attack over that for the minimum case of elliptic loading. The parameter τ is plotted for rectangular planform wings in Fig. 2.4.

Hoerner and Borst [1985] suggest that for small-aspect-ratio wings the two-dimensional lift-curve slope $C_{L\alpha}^a$ is reduced by a factor k; that is,

$$C_{L\alpha}^{a\,\prime} = C_{L\alpha}^a k \tag{2.2}$$

where

$$k = \frac{2\pi AR}{C_{L\alpha}^a} \tanh\left(\frac{C_{L\alpha}^a}{2\pi AR}\right) \tag{2.3}$$

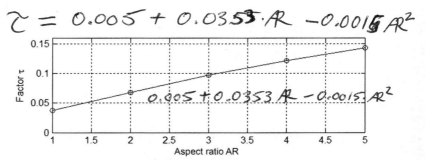

$$\tau = 0.005 + 0.0353 \cdot AR - 0.0015 \, AR^2$$

Fig. 2.4   Parameter τ for rectangular planform wings vs aspect ratio.

and hence

$$C_{L\alpha} = \frac{\pi AR C_{L\alpha}^{a}{}'}{\pi AR + C_{L\alpha}^{a}{}'(1 + \tau)} \, \text{rad}^{-1} = \frac{\pi^2 AR C_{L\alpha}^{a}{}'}{180(\pi AR + C_{L\alpha}^{a}{}'(1 + \tau))} \, \text{deg}^{-1} \qquad (2.4)$$

Lift-curve slope derived from this function with $C_{L\alpha}^{a} = 6.89 \, \text{rad}^{-1}$ is given in Fig. 2.5.

The lift-curve slope of a low-aspect-ratio wing increases with angle of attack over and above the basic slope determined from Eq. (2.4). The increase is a function of aspect ratio, the shape of the wing's lateral edges, and the component of velocity normal to the wing. This nonlinear component has been the subject of several investigations but is still not well understood. In some ways, however, it appears to be caused by the drag based on the normal velocity component, and thus tentatively the lift increment ΔC_L can be written [Hoerner and Borst 1975]

$$\Delta C_L = k_1 \sin^2 (\alpha - \alpha_0) \cos(\alpha - \alpha_0) \qquad (2.5)$$

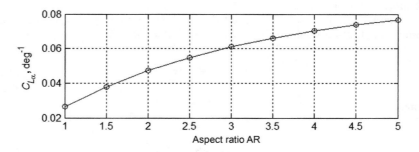

Fig. 2.5   Slope of lift curve for a low-aspect-ratio wing vs aspect ratio.

where k_1 is a function of aspect ratio and the shape of the wing's lateral edges. Hoerner and Borst [1975] suggest that experimental data are best fitted as

$$k_1 = \begin{cases} 3.33 - 1.33\text{AR}, & 1 < \text{AR} < 2.5 \\ 0, & \text{AR} \geq 2.5 \end{cases} \qquad (2.6)$$

The total lift for a low-aspect-ratio rectangular wing before the stall can therefore be written as

$$C_L = C_{L\alpha}(\alpha - \alpha_0) + k_1 \sin^2(\alpha - \alpha_0)\cos(\alpha - \alpha_0) \qquad (2.7)$$

where $C_{L\alpha}$ is determined from Eq. (2.4) and k_1 from Eq. (2.6).

Figure 2.6 shows theoretical lift coefficient curves for a ram-air wing of AR = 3.0 compared with experimental data [Nicolaides 1971; Ware and Hassell 1969]. The zero-lift angle α_0 is taken to be -7 deg (see Fig. 2.6). Up to the stall, the match is acceptable. A small decrease in α_0 would further enhance the fit. Below $\alpha = 0$ deg, the NASA data show a deviation from the nearly constant value of $C_{L\alpha}$ predicted theoretically. This appears to be caused by wing distortion, and below $\alpha = 5$ deg the NASA data are suspect. Both sets of experimental data show that ram-air canopies stall at lower angles of attack than rigid wings of corresponding section and aspect ratio [Zimmerman 1932]. Ware and Hassell [1969] and Ross [1993] propose that the reason for this early stall is that the sharp leading edge of the upper lip of the open nose causes leading-edge separation at relatively

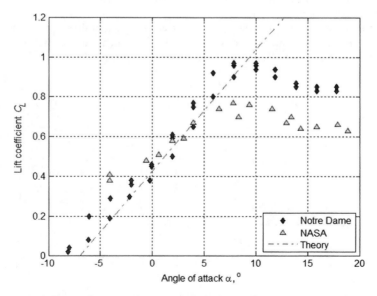

Fig. 2.6   Experimental and theoretical lift coefficients for a ram-air wing with AR = 2.5.

low lift coefficients. Conversely, [Speelman et al. 1972] state that separation takes place progressively from TE. The shapes of the lift curves found by Ware and Hassell are however reminiscent of long bubble leading-edge separation described by [Hoerner and Borst 1975]. Also, the large values of suction near the upper lip shown in [Lingard 1981] would in practice promote leading-edge separation. The NASA results show earlier stall than the Notre Dame data, stall occurring at approximately $C_L = 0.7$ compared with $C_L = 0.85$. This could be attributed to the fact that the NASA models were flexible whereas the Notre Dame models were semirigid. Speelman et al. [1972] emphasize the importance of maintaining a smooth, undistorted airfoil section under load, with particular attention being paid to the inlet shaping, if the aerodynamic characteristics of the wing are not to be adversely affected.

Figure 2.7 shows C_L data for aspect ratios AR = 2.0 ... 4.0. The lift-curve slope increases with increasing aspect ratio.

2.2.2 DRAG COEFFICIENT

For a rectangular wing lifting-line theory gives

$$C_D = C_{D0} + \frac{C_L^2(1 + \delta)}{\pi \text{AR}} \quad (2.8)$$

where C_{D0} is the profile drag and the second term represents the induced drag. The parameter δ is a small factor to allow for nonelliptic loading. Theoretical values for δ are plotted in Fig. 2.8.

For small-aspect-ratio wings a further drag component exists, corresponding to the nonlinear lift component described earlier. Again, assuming that this results from drag due to normal velocity, we can write [Hoerner 1965]

$$\Delta C_D = k_1 \sin^3 (\alpha - \alpha_0) \quad (2.9)$$

where k_1 is identical to that for the lift component and can be found from Eq. (2.6).

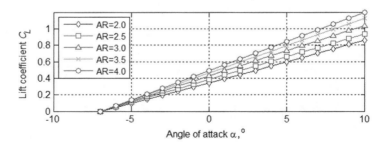

Fig. 2.7   Theoretical lift coefficient for a ram-air wing for different aspect ratios.

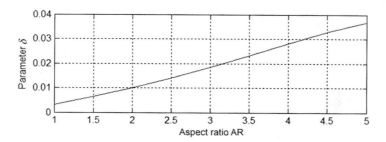

Fig. 2.8   Parameter δ for rectangular planform wings vs aspect ratio.

The total drag coefficient for a low-aspect-ratio rectangular wing before the stall can thus be written as

$$C_D = C_{D0} + \frac{C_{LC}^2(1 + \delta)}{\pi \mathrm{AR}} + k_1 \sin^3(\alpha - \alpha_0) \tag{2.10}$$

In this equation $C_{LC} = C_{L\alpha}(\alpha - \alpha_0)$ is the circulation lift coefficient not the total lift coefficient given by Eq. (2.7), which includes the nonlinear lift increment. Clearly, for a given lift coefficient the second and the third terms in Eq. (2.10) decrease with increasing aspect ratio, and, hence with C_{D0} remaining constant, the total drag coefficient decreases.

Ware and Hassell [1969] estimate the profile drag of ram-air wings by summing the following contributory elements:

1. Basic airfoil drag—for an airfoil of typical section—$C_D = 0.015$

2. Surface irregularities and fabric roughness—$C_D = 0.004$

3. Open airfoil nose—$C_D = 0.5 \, h/c$
    where h is the inlet height and c is the chord length

4. Drag of pennants and stabilizer panels—for pennants and stabilizers that do not flap, this is very small (0.0001). If flapping occurs

$$C_D = 0.5 \, S_p/S \tag{2.11}$$

where S_p is the area of pennants and stabilizer panels and $S = bc$ is the constructed canopy with b being the constructed wing span.

Theoretical and experimental coefficients for a ram-air wing are presented in Fig. 2.9 for AR = 3 [Nicolaides 1971; Ware and Hassell 1969]. The agreement lends some credibility to the scheme of calculation proposed. Figure 2.10 shows drag coefficient data for a range of aspect ratios. Drag coefficient at a given angle of attack varies little with aspect ratio.

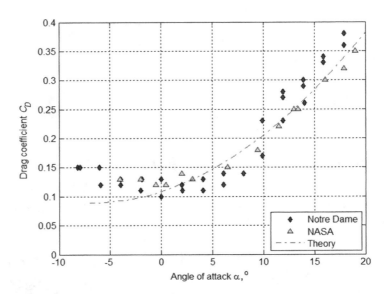

Fig. 2.9   Experimental and theoretical drag coefficient for a ram-air wing with AR = 3.0.

2.2.3   LIFT-TO-DRAG RATIO

Theoretical and experimental values of L/D for the wing alone are shown in Fig. 2.11.

Up to the stall, the agreement with the NASA data is quite good, particularly in respect to the maximum value of L/D of a little over 5. Notice that the maximum theoretical value of L/D occurs close to the stall (see Fig. 2.6).

Figure 2.12 shows L/D ratio for a range of aspect ratios AR = 2 . . . 4. The expected improvement of L/D ratio with increasing aspect ratio is shown. Theory predicts that for the wing alone maximum L/D increases from 4.15 at AR = 2 to 5.86 at AR = 4.

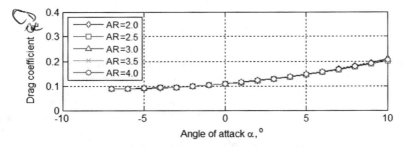

Fig. 2.10   Theoretical drag coefficient for ram-air wings with AR = 2 . . . 4.

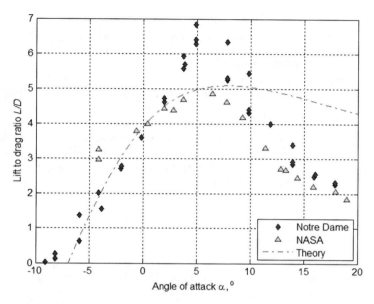

Fig. 2.11 Experimental and theoretical L/D ratio for a ram-air wing with AR = 3.0.

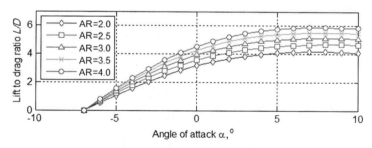

Fig. 2.12 Theoretical L/D for ram air wings with different aspect ratios.

2.3   AERODYNAMICS OF RAM-AIR PARACHUTES

The incorporation of a ram-air wing into a parachute system necessitates the addition of suspension lines. Some early ram-air parachutes were rigged such that the wing flew flat. Lingard [1981] shows that this necessitates very long suspension lines (more than 2.25 spans) to give spanwise stability of the wing. Lines of this length create substantial drag, overwhelming any gains in lift achieved. Ram-air wings are thus rigged with the lines in any chordwise bank of equal length, giving the wing arc anhedral. This is called crown rigging. Arc anhedral is slightly detrimental to the wings' lifting properties while the drag coefficient for a given angle of attack remains as that for a flat wing. The suspension lines themselves add considerably to the total drag.

The amount of arc anhedral is a function of the ratio of line length R to span b. From the preceding, it is apparent that the wing-alone performance improves with increasing aspect ratio but, for a given value of R/b, the larger the aspect ratio the greater the line length. In addition, the number of lines tends to increase with aspect ratio. Thus, increasing aspect ratio gives markedly higher line drag. Reducing R/b for a given aspect ratio reduces line drag but yields an increasingly inefficient wing. It is clear, therefore, that the performance of the complete parachute system is significantly different to that for the wing alone. The result of introducing arc anhedral and line drag is therefore considered here.

2.3.1 LIFT COEFFICIENT

With the wing incorporated into the ram-air parachute, arc anhedral is introduced. For a conventional wing with dihedral angle defined as shown in Fig. 2.13, [Hoerner and Borst 1975] indicate that

$$C_L = C_{L;\varepsilon=0} \cos^2(\varepsilon) \tag{2.12}$$

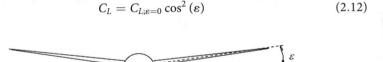

Dihedral angle for a conventional wing

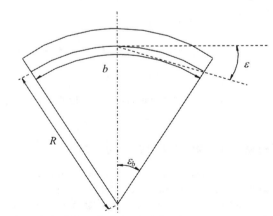

Anhedral angle for a ram-air wing

Fig. 2.13   Definition of anhedral angle for a ram-air wing.

This equation can be shown to remain valid for a wing with arc anhedral if ε is defined as in Fig. 2.13

$$\varepsilon = \varepsilon_b/2 \qquad (2.13)$$

Referring to Fig. 2.13,

$$\varepsilon_b = \frac{b}{2R} \qquad (2.14)$$

Hence,

$$\varepsilon = \frac{b}{4R}\,\text{rad} = 45\,\frac{b}{\pi R}\,\text{deg} \qquad (2.15)$$

Equation (2.7) can now be rewritten for a ram-air wing with arc anhedral

$$C_L = C_{L\alpha}(\alpha - \alpha_0)\cos^2(\varepsilon) + k_1 \sin^2(\alpha - \alpha_0)\cos(\alpha - \alpha_0) \qquad (2.16)$$

where in the absence of any information on the effect of arc anhedral on k_1 it is assumed that k_1 remains unchanged.

The effect of line length on wing lift coefficient is shown in Fig. 2.14.

2.3.2 DRAG COEFFICIENT

The overall drag of a gliding parachute system comprises the sum of wing drag, line drag, and store drag; line drag and store drag contribute significantly.

To simplify the estimation of line drag, it is assumed that all lines are the same length and are subject to the same normal velocity $V\cos(\alpha)$ where V is the system velocity. For typical Reynolds numbers the drag coefficient of a suspension line would be approximately 1. The contribution of line drag to the total system drag can therefore be estimated from

$$C_D^l = \frac{nRd\cos^3(\alpha)}{S} \qquad (2.17)$$

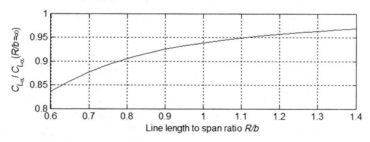

Fig. 2.14   Influence of line length on the lift coefficient of a ram-air wing AR = 3.

where n is the number of lines, R is mean line length, d is line diameter, and S is the canopy area. Store drag coefficient can be written

$$C_D^s = \frac{(C_D S)_s}{S} \tag{2.18}$$

where $(C_D S)_s$ is the store drag area.

Induced drag and profile drag for a wing with anhedral remain approximately constant [Hoerner 1965]; therefore, the drag coefficient for a ram-air parachute system with arc anhedral can finally be written from Eq. (2.10)

$$C_D = C_{D0} + C_D^l + C_D^s + \frac{C_{L\alpha}^2 (\alpha - \alpha_0)^2}{\pi \text{AR}} (1 + \delta) + k_1 \sin^3 (\alpha - \alpha_0) \tag{2.19}$$

where AR is the aspect ratio based on constructed span b.

2.3.3 LIFT-TO-DRAG RATIO

Typically, the maximum theoretical value of lift-to-drag ratio occurs close to or beyond the stall for a conventional ram-air parachute. The maximum L/D is therefore not a useful measure to use for the performance of a ram-air parachute because the wing cannot operate effectively close to the stall. Current ram-air parachutes are generally rigged to fly with a lift coefficient of around 0.5. To compare the performance of various configurations, we will therefore consider L/D ratio at this practical value of C_L.

The parameters of a basic parachute design used in this section are presented in Table 2.1.

The value of L/D ratio for varying aspect ratios and line lengths is shown in Fig. 2.15, for a ram-air parachute of 300 m^2 (3,230 ft^2) area with 270 lines. It is clear that for Large ram-air parachutes performance is considerably worse than for the wing alone. The best performance is achieved with the largest aspect ratio and the lower line-length-to-span ratios of 0.5–0.6. For all aspect ratios performance decreases with increasing line length. At line-length-to-span ratio

TABLE 2.1 BASIC PARACHUTE DESIGN PARAMETERS

Parameter	Value
Aspect ratio	AR $= 3.0$
Inlet height	$h = 0.14c$
Line length	$R = 0.8b$
Line diameter	$d = 2.5$ mm (0.1 in.)
Number of lines	1 line per 1.11 m^2 (12 ft^2)
Payload drag coefficient	$C_D^s = 0.006$

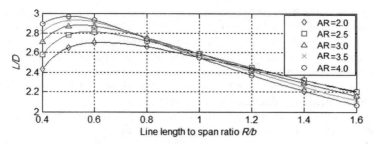

Fig. 2.15   Theoretical L/D for a 300-m^2 (3,230-ft^2) ram-air parachute with various aspect ratios and line-length-to-span ratios.

0.8, there is little improvement in performance with increasing aspect ratio. At large line-length-to-span ratios, the smaller-aspect-ratio wings yield the best performance. These trends are explained by the influence of line drag. At low values of R/b, the reduction in induced drag with increasing aspect ratio exceeds the rise in drag due to the longer lines. At higher values of R/b, the increasing line drag with aspect ratio overwhelms the decreases in induced drag.

Figure 2.16 shows L/D ratio for varying aspect ratios and line lengths for a 36-m^2 (388-ft^2) ram-air parachute with 32 lines. Different trends to those for the large canopy are apparent. Again, performance is significantly poorer than for the wing alone, but not as bad as for the large ram-air. At all values of R/b, performance increases with aspect ratio, but gains become less significant at the longer line lengths. That implies that at this scale line drag is less significant. The reduction of induced drag with increasing aspect ratio is always greater than the increase in line drag. For AR $<$ 3.5, there is an optimum value of R/b of 0.8. Below $R/b = 0.8$, the falling wing efficiency is more significant than the decrease in line drag. Above $R/b = 0.8$, wing efficiency gains are less significant, and line drag increases predominate.

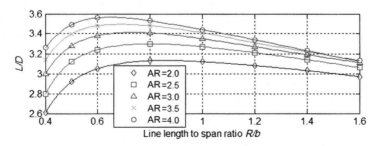

Fig. 2.16   Theoretical L/D for a 36-m^2 (388-ft^2) ram-air parachute with various aspect ratios and line-length-to-span ratios.

The difference in performance of the two sizes of ram-air parachute is of most concern for PADS. An examination of Eqs. (2.17) and (2.19) reveals the problem. All contributions to total drag are proportional to wing area with the exception of line drag. Typically, ram-air parachutes are designed with the number of suspension lines proportional to the area of the parachute. Moreover, if the canopy is scaled with constant wing loading, and a constant inflation g load is assumed, line diameter is also constant.

Hence, nd/S is constant, and from Eq. (2.17) it follows that C_D^l increases with line length R increase. Therefore, simply scaling up a ram-air wing reduces its performance. For the basic design with inlet height $0.14c$ and $C_L = 0.5$, the L/D ratio varies with wing area as shown in Fig. 2.17.

For gliding parachutes to have acceptable performance at large scale, it is therefore necessary to considerably improve the wing design.

2.3.4 GLIDING PARACHUTE PERFORMANCE IMPROVEMENT

Improvements in system glide performance can be achieved by reducing system drag, which comprises the profile drag of the store, suspension lines and wing, and the induced drag of the wing, or by increasing working lift coefficient (C_{Lmax}). To achieve acceptable performance, it is necessary to consider both approaches. Generally, effort aimed at reducing drag, as with all airborne devices, yields the most benefit.

It is useful to examine the drag contributions to a typical ram-air parachute to give an indication of where effort at drag reduction will be most effective. Drag contributions for two different-canopy-size ram-air parachutes of 300 and 39 m² (3,230 and 420 ft²) (hereinafter referred to as large and small parachutes, respectively) are given in Tables 2.2 and 2.3. The same data are also shown in Fig. 2.18 to emphasize the substantially larger contribution of line drag for Large canopy (in both Figs. 2.18a and 2.18b the two offset pie slices correspond to the basic and roughness drag, counted counterclockwise).

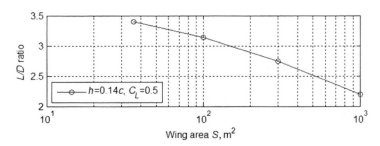

Fig. 2.17   Lift-to-drag ratio vs wing area for the basic design with inlet height 0.14c and $C_L = 0.5$.

TABLE 2.2   TYPICAL DRAG CONTRIBUTIONS FOR LARGE RAM-AIR PARACHUTE

Parameter	Value
Basic airfoil drag	0.015 (8.2%)
Roughness drag	0.004 (2.2%)
Inlet drag 0.5 h/c	0.070 (38.5%)
Induced drag at $C_L = 0.5$	0.033 (18.1%)
Line drag	0.054 (29.7%)
Store drag	0.006 (3.3%)
Total	0.182

For the large parachute canopy, basic airfoil drag and roughness drag only contribute 10.4% of total drag and are clearly difficult to improve. Nonetheless, more modern wing sections, such as the LS-1 series, have lower profile drag than the Clark Y from which these data were derived and could yield benefits. Maintenance of section by correct rigging and load distribution is essential to keep these factors low. Store drag is not within the control of the parachute designer. Reduction of the three remaining drag-producing elements is where most gains can be made.

For large ram-air wings, low values of R/b should be used. There is extensive practical experience at $R/b = 0.6$, and therefore this value is recommended because, although $R/b = 0.5$ is regularly used for sports parachutes, the gains in using shorter lines are not substantial and we are here at the limits of the theory. For a given line length, line drag can be decreased by reducing the

TABLE 2.3   TYPICAL DRAG CONTRIBUTIONS FOR SMALL RAM-AIR PARACHUTE

Parameter	Value
Basic airfoil drag	0.015 (10.2%)
Roughness drag	0.004 (2.7%)
Inlet drag 0.5 h/c	0.070 (47.6%)
Induced drag at $C_L = 0.5$	0.033 (22.5%)
Line drag	0.019 (12.9%)
Store drag	0.006 (4.1%)
Total	0.147

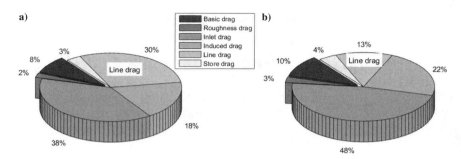

Fig. 2.18   Typical drag contributions for a) large and b) small ram-air parachutes.

number of lines provided the wing remains undistorted under load. Cascading of lines is standard practice on sports ram-air parachutes and is now being used on Larger ram-air parachutes. Cascading would reduce line drag by approximately 20%; thus, L/D for the Large parachute would rise from 2.75 to 2.9.

Together, inlet drag and induced drag contribute 56.6% of the total drag of the large system. Clearly, before any significant improvement in ram-air wing performance can be realized, one or both of these elements must be considerably reduced. Figure 2.15 shows that increasing AR beyond 3 without other improvements is ineffective because reduced induced drag is simply traded with increased line drag. To yield significant improvements, inlet drag must be reduced.

The basic 18% Clark Y airfoil has an 11% inlet height at the ribs with 15% between ribs. This gives an average inlet height of 14%. Increasing scale will reduce the amount of bulge between ribs because the cell width typically remains constant in scaling up the wing—due to constant line spacing—while the height of the cell increases. For a 36-m^2 (388-ft^2) wing with unloaded ribs, the cell depth to width is typically 0.8, but for the Large parachute with only loaded ribs this becomes 1.6. Bulge will therefore be approximately halved giving a smoother wing with height typically 12.5%. Applying this correction increases L/D for the large wing from 2.75 to 2.86.

Once the parachute is gliding, it is strictly only necessary to have an inlet height that covers the range of movement of the stagnation streamline. Ross [1993] shows that a 4% inlet is feasible; however, because inflation must not be compromised, reductions in inlet height should be conservative. Wings with 8.4% inlet height at the ribs are flying; therefore, 8% overall height is realistic. With inlet height 0.08c and $C_L = 0.5$, performance is significantly enhanced as shown in Fig. 2.19.

An additional benefit of reduced inlet height is reduced disruption of the flow over the upper leading edge of the wing and consequently delayed stall. Ross [1993] shows that for an 8.4% inlet stall does not occur until $C_L = 0.85$ and with a 4% inlet stall is further delayed to $C_L = 1.55$. It is therefore possible with

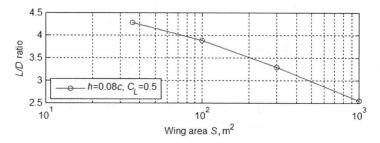

Fig. 2.19   Lift-to-drag ratio vs wing area for the basic design with inlet height 0.08c and $C_L = 0.5$.

reduced inlet height to increase the design lift coefficient. Figure 2.20 demonstrates the increased benefits of this strategy. A further improvement in performance comes out of increasing design lift coefficient in that if flight speed is maintained then wing area can be reduced giving further L/D gains.

Clearly for large-scale ram-air parachutes, detailed attention to drag reduction is essential:

- Greater attention must be paid to inlet design. Inlet height should be minimized subject to maintaining inflation reliability. Vortex panel methods can be used to determine to stagnation point range.
- The number of lines should be minimized subject to maintaining the airfoil shape without undue distortion.
- Cascading of lines should be employed.
- Line length should be optimized with $R/b = 0.6$ being a good design point.
- Aspect ratio AR = 3 is a good design point.

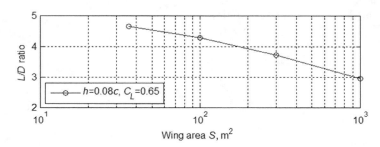

Fig. 2.20   Lift-to-drag ratio vs wing area for the basic design with inlet height 0.08c and $C_L = 0.65$.

2.4   FLIGHT PERFORMANCE

Figure 2.21 shows a gliding parachute in steady descent in still air with an airspeed V and glide angle γ. (Note, in other chapters, when winds are considered, we will distinguish between the airspeed vector $\mathbf{V}_a$ and groundspeed vector $\mathbf{V}$; also, in this chapter γ denotes the glide angle whereas in all other chapters this notation will be used for the flight-path angle, which is negative to the glide angle.) All angles shown in Fig. 2.21 are positive.

Resolving the forces acting on the system horizontally and vertically,

$$(L_c + L_l + L_s)\sin(\gamma) - (D_c + D_l + D_s)\cos(\gamma) = 0 \qquad (2.20)$$

$$(m_s + m_c)g - (L_c + L_l + L_s)\cos(\gamma) - (D_c + D_l + D_s)\sin(\gamma) = 0 \qquad (2.21)$$

where L_c is the lift force acting on the canopy, L_l is the lift force acting on the suspension lines, L_s is the lift force acting on the payload, D_c is the drag force acting on the canopy, D_l is the drag force acting on the suspension lines, D_s is the drag force acting on the payload, m_c is the mass of the canopy, m_s is the mass of the store, and g is the acceleration due to gravity. From Eq. (2.20)

$$L/D = C_L/C_D = 1/\tan(\gamma) \qquad (2.22)$$

where $L = L_c + L_l + L_s$ is the total system lift, $D = D_c + D_l + D_s$ is the total system drag, $C_L = L/(QS)$ is the lift coefficient, $C_D = D/(QS)$ is the drag coefficient, $Q = 0.5\rho V^2$ is the dynamic pressure, S is the reference area, and ρ is the air density.

Equation (2.22) is the standard equation of aerodynamic efficiency in gliding flight. The smaller the glide angle the higher becomes the lift-drag ratio and the greater the gliding range for a given height loss. Write Eq. (2.21) in coefficient form, and transform

$$W = QS[C_D\cos(\gamma) + C_L\sin(\gamma)] \qquad (2.23)$$

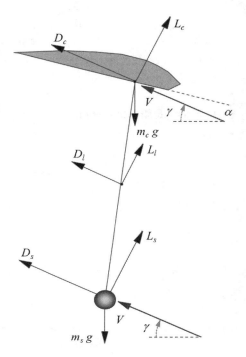

Fig. 2.21   Ram-air parachute in steady gliding flight.

where $W = (m_s + m_c)g$ is the system weight. Substituting from Eq. (2.22) and simplifying give

$$W = QS(C_L^2 + C_D^2)^{0.5} = 0.5\rho V^2 SC_T \tag{2.24}$$

where

$$C_T = (C_L^2 + C_D^2)^{0.5} = C_L(1 + (L/D)^{-2})^{0.5} \tag{2.25}$$

is the tangent coefficient.

The velocity of the parachute system in steady gliding flight can therefore be written as

$$V = \left(\frac{2}{\rho}\frac{W}{S}\frac{1}{C_T}\right)^{0.5} = \left[\frac{2g}{\rho}\frac{(m_s + m_c)}{S}\frac{1}{C_T}\right]^{0.5} \tag{2.26}$$

The horizontal u and vertical w velocity components of the system can be calculated from

$$V_h = V\cos(\gamma), \quad V_v = V\sin(\gamma) \tag{2.27}$$

Equation (2.22) shows that in still-air conditions glide angle, and therefore glide distance for a given height loss, is a function only of lift-to-drag ratio. However, Eq. (2.26) indicates that velocity V is dependent on wing loading W/S, air density, and the aerodynamic characteristics of the parachute. Consequently, in still-air conditions, a given parachute will travel the same distance for a given height loss whatever the altitude or wing loading, but velocity down the glide path will increase with increasing altitude and wing loading.

It is obvious that penetration of the wind is essential for effective PADS operation. Using a design lift coefficient of 0.5 and a range of L/D ratios, typical vertical and horizontal velocities for ram-air parachutes for varying wing loading at sea level are shown in Fig. 2.22. For PADS systems to give some margin over possible winds, flight velocity is usually chosen to be greater than 20 m/s (39 kt), that is, a wing loading in the range of 15–20 kg/m² (3.1–4.1 lb/ft²).

2.5 LONGITUDINAL STATIC STABILITY

2.5.1 CANOPY RIGGING

So far in this discussion, it has been assumed that the system can be induced to fly at the angle of attack corresponding to the optimum L/D. The practical means of achieving this is by rigging the canopy; that is, positioning the

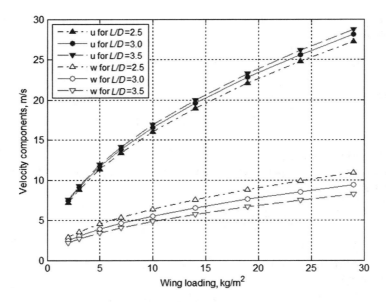

Fig. 2.22   Ram-air parachute flight velocities vs wing loading.

payload, and hence the c.g. of the system, such that the equilibrium attitude is at the required angle of attack. The parachute will be in stable equilibrium when the sum of moments acting on the system is zero and when the slope of the pitching-moment curve $C_{m\alpha} = dC_m/d\alpha$ is negative, that is, when any small disturbance from equilibrium results in a restoring couple. Assuming that the system is rigid and that the canopy mass is lumped at the quarter-chord point, the moment M about the wing quarter-chord of the system can be written with reference to Fig. 2.23 as

$$M = M_{c/4} + R[L_s \sin(\alpha + \mu) - D_s \cos(\alpha + \mu)]$$

$$+ \frac{R}{2}[L_l \sin(\alpha + \mu) - D_l \cos(\alpha + \mu)] - m_s gR \sin(\theta) \qquad (2.28)$$

where $M_{c/4}$ is the pitching moment of the canopy about 25% chord point, L_l is the lift force acting on the suspension lines, L_s is the lift force acting on the payload, D_l is the drag force acting on the suspension lines, D_s is the drag of the payload, m_s is the mass of the payload, g is the acceleration due to gravity, R is the distance from the quarter-chord point of the canopy to the payload, and μ is the so-called rigging angle as defined in Fig. 2.23. Rigging angle for a parafoil is the same as an incidence angle for an aircraft, that is, a small angle formed by the chord line and longitudinal axis. As shown in Fig. 2.23, this angle is negative, but hereinafter in this chapter its sign will be neglected.

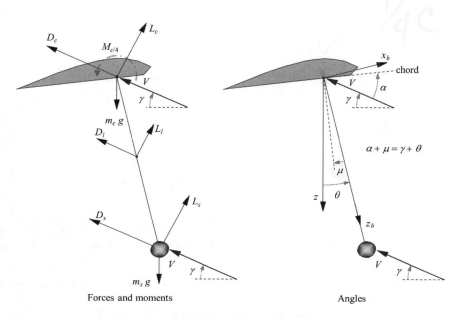

Forces and moments   Angles

Fig. 2.23   Definitions for static stability analysis.

Line force is assumed to act normal to and at the midpoint of the line joining the 25% chord point to the payload. Therefore, line drag and lift can be written as

$$C_D^l = \frac{ndR\cos^3(\alpha + \mu)}{S} \tag{2.29}$$

$$C_L^l = -\frac{ndR\cos^2(\alpha + \mu)\sin(\alpha + \mu)}{S} \tag{2.30}$$

where d is the line's diameter.

Writing Eq. (2.28) in the coefficient form results in

$$C_m = C_{m;c/4} + \frac{R}{c}\left[C_L^s\sin(\alpha + \mu) - C_D^s\cos(\alpha + \mu)\right]$$

$$+ \frac{R}{2c}\left[C_L^l\sin(\alpha + \mu) - C_D^l\cos(\alpha + \mu)\right] - \frac{m_sgR\sin(\theta)}{0.5\rho V^2 Sc} \tag{2.31}$$

or simplifying

$$C_m \doteq C_{m;c/4} + \frac{R}{c}\left[C_L^s\sin(\alpha + \mu) - C_D^s\cos(\alpha + \mu)\right]$$

$$- \frac{R}{2c}\frac{nRd\cos^2(\alpha + \mu)}{S} - \frac{m_sgR\sin(\theta)}{0.5\rho V^2 Sc} \tag{2.32}$$

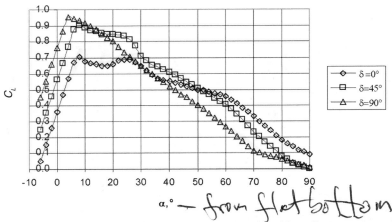

Fig. 2.24   Lift coefficient for a ram-air wing without lines.

Lift, drag, and pitching-moment data for a ram-air wing of AR = 2.5 are shown in Figs. 2.24–2.26.

The variation of pitching moment with rigging angle derived from Eq. (2.33) is illustrated in Fig. 2.27. System parameters are $S = 300$ m^2 (3,230 ft), AR = 2.5, $R/b = 0.6$, $m_s = 5{,}229$ kg (11,528 lb), $n = 270$, $d = 0.0025$, $C_D = 0.006$, and $C_L^s = 0$.

It is clear that variation of μ is an effective means of selecting the stable angle of attack α; a wide range of stable values is available for fairly small changes in μ. In this case the system will fly at $\alpha = 2.8$ deg resulting in $C_L = 0.5$ if rigged with

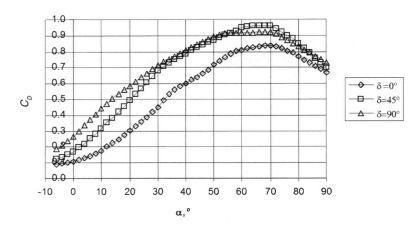

Fig. 2.25   Drag coefficient for a ram-air wing without lines.

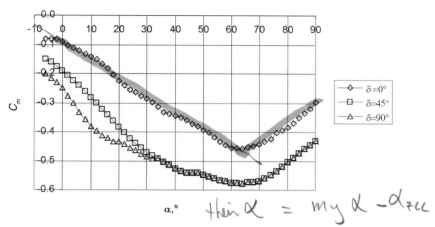

Fig. 2.26   Pitching-moment coefficient for a ram-air wing without lines.

$\mu = 5$ deg. In this configuration $C_D = 0.165$, giving an L/D ratio of 3.03. Rigging the canopy with μ too high results in a lower trim angle of attack and a fall off in performance; setting μ too low can lead to the trim angle of attack being beyond the stall, resulting in a considerable drop in performance. This sensitivity to rigging emphasizes the need for accurate rigging using suspension lines with very low elasticity.

The effect of increasing line length to $R/b = 1.0$ is shown in Fig. 2.28.

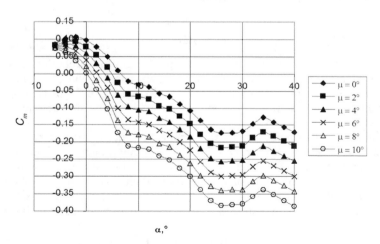

Fig. 2.27   Pitching-moment coefficient for a ram-air parachute with various rigging angles for $R/b = 0.6$.

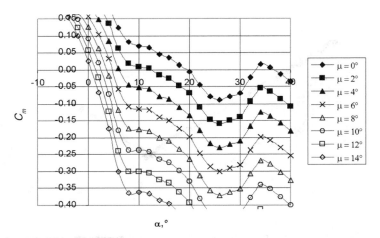

Fig. 2.28   Pitching-moment coefficient for a ram-air parachute with various rigging angles for $R/b = 1.0$.

With longer lines, it is necessary to increase the rigging angle to $\mu = 9.1$ deg to attain the desired trim lift coefficient of 0.5. The parachute will fly at $\alpha = 2.9$ deg, $C_D = 0.187$, giving an L/D ratio of 2.67. The magnitude of $C_{m\alpha}$ at $C_m = 0$ becomes greater with longer lines implying increased static stability.

Increasing aspect ratio also results in improved static stability with the value of $C_{m\alpha}$ again increasing due to the greater line length.

2.5.2   EFFECT OF TRAILING-EDGE DEFLECTION

On a conventional wing, deflecting plain flaps changes the camber of the section and results in increases of lift coefficient at a given angle of attack, maximum lift coefficient, profile drag, and induced drag. Deflecting TE of a ram-air wing with the parameters given in the preceding section produces similar effects as is demonstrated in Figs. 2.24–2.26.

The effect of flap deflection (we will denote it δ_s where s stands for symmetric) to half- and full-brakes position on the system pitching moment about the wing quarter-chord point is shown in Figs. 2.29 and 2.30 for the standard system. The influence of flap deflection on the various flight parameters is shown in Figs. 2.31–2.33.

Deflection of TE results in only a small changes in trim angle of attack. The values of C_L and C_D increase almost in proportion resulting in L/D reducing (γ increasing) only slightly from 3.03 to 2.52 with δ_s changing from 0 to 90 deg. The significant increases in lift and drag coefficients result only in reduced flight velocity. Thus, deflecting TE does not strongly influence glide angle for

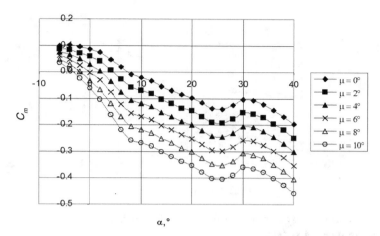

Fig. 2.29 Pitching-moment coefficient for a ram-air parachute with various rigging angles for $R/b = 0.6$ and $\delta_s = 45$ deg.

the standard system; γ changes only from 18.3 to 21.6 deg. Therefore, in this configuration controlling glide slope in zero wind conditions is not feasible.

This is not true for all rigging configurations. With increased line length, TE deflection can result in a significant change in trim angle as shown in Figs. 2.34 and 2.35. With the system rigged off-design with $\mu = 6$ deg, fully deflecting TE results in a shift of trim angle of attack from 3.8 to 36.3 deg.

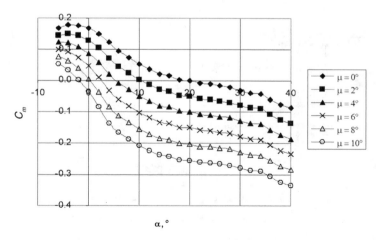

Fig. 2.30 Pitching-moment coefficient for a ram-air parachute with various rigging angles for $R/b = 0.6$ and $\delta_s = 90$ deg.

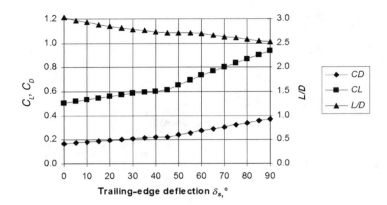

Fig. 2.31   Influence of symmetric TE deflection on lift and drag coefficients and lift-to-drag ratio for the standard system.

If the system is flying into the wind, as it ideally would be for landing, deflection of TE can cause a more significant change in glide angle for the standard system. This occurs because forward velocity decreases as δ_s increases and the headwind has a proportionally greater effect. This is illustrated in Fig. 2.36. With a 10-m/s (19.4-kt) headwind, the glide angle can be adjusted by TE deflection from 31 to 48 deg. The possibility to adjust glide slope is even greater for man-carrying systems. This adjustment is used by jumpers in accuracy competitions.

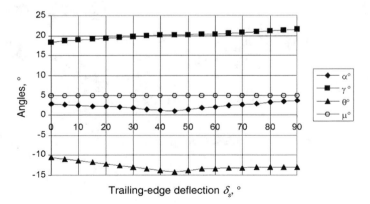

Fig. 2.32   Influence of symmetric TE deflection on trim values of α, γ, and θ for the standard system.

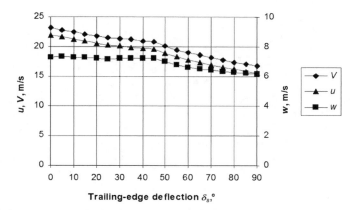

Fig. 2.33   Influence of symmetric TE deflection on flight velocities for the standard system.

2.6   LONGITUDINAL DYNAMICS

In this section, consideration is given to the dynamics of ram-air parachute flight. Dynamical analysis is important in that it can indicate conditions under which steady glide is rapidly achieved and conditions where dynamic instability can occur or transient motions are only lightly damped. Of particular importance is the ability of the system to move smoothly from one state to another. This includes release of TE of the parachute following inflation, stall recovery, response to dynamic TE deflection—glide path modification and the landing flare maneuver—and gust response. Here motions are limited to the pitch plane. In the following section, lateral control and stability will be discussed.

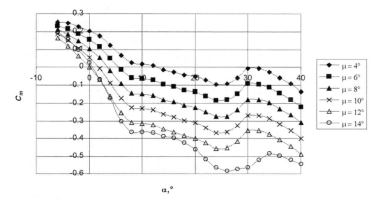

Fig. 2.34   Pitching-moment coefficient for a ram-air parachute with various rigging angles, angles for $R/b = 1$ and $\delta_s = 45$ deg.

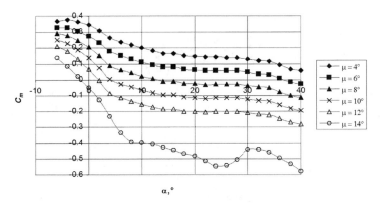

Fig. 2.35   Pitching-moment coefficient for a ram-air parachute with various rigging angles for $R/b = 1$ and $\delta_s = 90$ deg.

The dynamic stability model used to derive the results described in this section is detailed in [Lingard 1981]. To obtain correct predictions, this model includes the so-called apparent or added masses calculated according to [Lingard 1981; Lissaman and Brown 1993]. The concept of apparent masses and added mass matrix will be introduced in Sec. 2.7. The model is capable of simulating the effects on flight dynamics of changes in system parameters such as canopy size, line length, rigging angle, payload mass and aerodynamic characteristics, and environmental parameters. In this section, it is applied to review the response of a ram-air parachute system to control inputs and gusts and to evaluate the

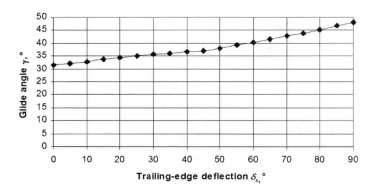

Fig. 2.36   Influence of symmetric TE deflection on glide angle for the standard system with a 10-m/s (19.4-kt) headwind.

impact of variations in parachute size, payload mass, line length, and altitude on system stability. The major findings of this section are as follows:

- Acceptable longitudinal stability is achievable for large parachutes appropriate to the PADS role, but scaling must be on the basis of mass ratio. Mass ratios in the range of 0.8 are recommended, but severely adverse dynamics does not appear to occur above $M_r = 0.4$.

- Increasing altitude increases mass ratio and therefore stability.

- Increasing mass for a given canopy size increases stability.

- Increasing line length is destabilizing.

- Effective reduction in vertical velocity by the flare maneuver is achievable at the recommended mass ratio, but residual horizontal velocities increase with system size: 10.7 m/s (20.8 kt) for a 36-m^2 (388-ft^2) wing, 16 m/s (31 kt) for a 300-m^2 (3,230-ft^2) wing, and more than 20 m/s (38.9 kt) for a 1000-m^2 (10,764-ft^2) wing.

2.6.1 LONGITUDINAL DYNAMICS OF SMALL RAM-AIR PARACHUTES

Consider initially a small ram-air parachute system with the parameters given in Table 2.4.

It is well known that small ram-air parachutes can perform an effective flare maneuver to reduce landing velocity. This is shown in Fig. 2.37a. Flight is commenced at 400 m (1,312 ft). Initially, the parachute is in steady glide for 3 s; subsequently, TE is deflected to the full-brakes position over 1 s. During the first 3 s, the system maintains a horizontal velocity of 13.3 m/s (25.9 kt) and a vertical velocity of 3.7 m/s (7.2 kt). The L/D ratio is 3.6. The stable angle of attack is 2.9 deg, the payload is swung backward with $\theta = -9.3$ deg, and the glide angle is 15.7 deg.

The short period during and immediately after control deflection simulates the flare maneuver. The objective of a flared landing is to simultaneously reduce both horizontal and vertical velocity by increasing lift and drag. Referring

TABLE 2.4   DESIGN PARAMETERS FOR SMALL RAM-AIR PARACHUTE

Parameter	Value
Reference canopy area	$S = 36$ m^2 (388 ft^2)
Aspect ratio	AR $= 2.5$
Line-length-to-span ratio	$R/b = 0.6$
Rigging angle	$\mu = 3.5$ deg
System's mass	$m_s = 217$ kg (478 lb)
Store drag coefficient	$C_D^s = 0.006$

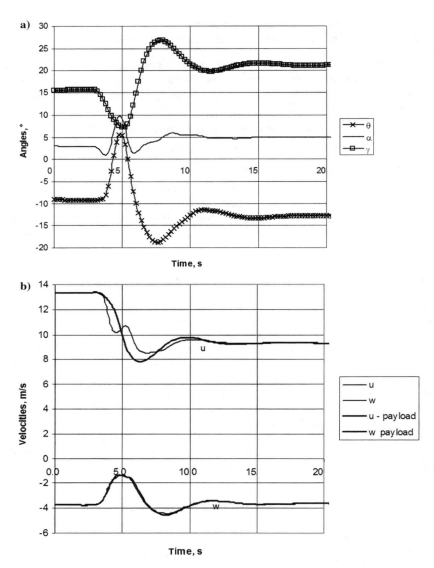

Fig. 2.37   Response of small ram-air parachute to symmetric TE deflection at $h = 400$ m (1,312 ft).

to Fig. 2.37a, it is noticed that immediately following initiation of control deflection the canopy horizontal velocity falls rapidly with the payload deceleration lagging. The system therefore pitches up, with θ eventually reaching 5.6 deg forward 1.8 s from control initiation. Glide angle reduces, attaining 7.2 deg coincident with maximum pitch forward. Following an initial slight reduction (predictable from static analysis for δ_s up to 45 deg), α rises to 10 deg at maximum pitch forward of the payload. The vertical velocities of the canopy and payload, which are almost identical throughout the maneuver, initially fall to 1.35 m/s (4.4 ft/s) before rising again to more than 4.0 m/s (13 ft/s). The minimum vertical velocity occurs at maximum pitch forward, which is the time at which the payload would ideally land. The horizontal velocity at this time is 10.7 m/s (20.8 kt) and has not reached its minimum value, which occurs at the bottom of the backswing of the payload. When minimum horizontal velocity occurs, vertical velocity has risen again to 3.0 m/s (9.8 ft/s). The time between control initiation and ideal touchdown is 1.8 s with a height loss of 4.6 m (15 ft).

It is clear that the flare is essentially a dynamic maneuver. If control initiation is too early, vertical velocity will be high; if too late, horizontal velocity will be high. A headwind can significantly improve the quality of the landing. In this case, with a wind of 7.5 m/s (14.6 kt), the ideal landing would result in a vertical velocity of 1.35 m/s (4.4 ft/s) and a horizontal velocity of only 3.2 m/s (6.2 kt).

Subsequent to the simulated flare, a highly damped phugoid motion occurs. All disturbance is damped in two cycles, 12 s after control initiation. Steady gliding flight at full brakes follows with $\alpha = 4.7$ deg, $u = 9.3$ m/s (18 kt), $w = 3.6$ m/s (11.8 ft/s), and $L/D = 2.6$.

2.6.2 SCALING EFFECTS

The small ram-air configuration clearly has good dynamic characteristics, and in scaling up to large parachutes, appropriate to the PADS role, it is essential that these are retained. It is therefore useful here to digress in order to discuss scaling parameters. At first sight, one can assume that a good option would be to maintain wing loading in order to match flight speed. However, there is another option. In considering general dynamic parachute motion, three nondimensional parameters that influence parachute behavior in unsteady motion regularly arise: mass ratio $M_r = m_s/\rho D_0^3$, Froude number $F_r = V^2/gD_0$, and dimensionless time $\tau = Vt/D_0$. In these ratios, D_0 is the representative length of a canopy. For a circular canopy, D_0 is its diameter; for ram-air parachutes we can assume $D_0 = S^{0.5}$.

Mass ratio M_r represents the ratio of payload mass to an air mass associated with the canopy. This parameter influences many unsteady parachute phenomena including inflation and wake recontact.

In classical hydrodynamics, Froude number F_r represents the ratio of fluid inertial forces to gravitational forces. For a parachute in steady descent,

$$k\rho V^2 D_0^2 = m_s g \qquad (2.33)$$

where k is a constant $(\pi C_T/8)$. Hence, substituting in the Froude number term for V^2, we obtain

$$F_r = \frac{1}{k}\frac{m_s g}{\rho D_0^2}\frac{1}{D_0 g} = \frac{1}{k}\frac{m_s}{\rho D_0^2} \tag{2.34}$$

Thus, for a parachute in steady descent Froude number is equivalent to mass ratio, and we can use either parameter.

Finally, for comparing response times of systems of different scales, it is useful to dimensionalize time by multiplying it by a velocity and dividing by a characteristic length.

Consider the small ram-air parachute system defined in Table 2.4. The wing loading for the system m_s/S is 6.02 kg/m^2 (1.2 lb/ft^2). Mass ratio, defined for ram-air parachutes as $m_s/\rho S^{1.5}$, is 0.82. Scaling up to a 300-m^2 (3,230-ft^2) wing on the basis of wing loading would give a payload mass of 1,808 kg (3,986 lb), and matching mass ratio gives 5,229 kg (11,528 lb). Other parameters for the scaled-up system are given in Table 2.5.

The time to deflect TE is also scaled on the dimensionless time basis. For the small system,

- The steady-state glide velocity is 13.9 m/s (27 kt).
- The representative length is $S^{0.5} = 6.0$ m (19.7 ft).
- The deflection time is $t_p = 1.0$ s.

Thus, dimensionless time of pull

$$\tau_p = \frac{V t_p}{S^{0.5}} = \frac{13.9 \times 1.0}{6.0} = 2.32 \tag{2.35}$$

For the payload scaled on mass ratio, steady glide velocity is 23.4 m/s (45.5 kt), and the representative length is 17.3 m (56.8 ft). Deflection time should therefore be 1.7 s. That is, time is scaled by a factor of 1.7 between the two systems.

TABLE 2.5   DESIGN PARAMETERS FOR A LARGE RAM-AIR PARACHUTE

Parameter	Value
Reference canopy area	$S = 300$ m^2 (3,230 ft^2)
Aspect ratio	AR = 2.5
Line-length-to-span ratio	$R/b = 0.6$
Rigging angle	$\mu = 5$ deg
System's mass	$m_s = 1,808$ or 5,229 kg (5,986 or 11,528 lb)
Store drag coefficient	$C_D^s = 0.006$

To evaluate the two scaling parameters, the test flight used for the small system is repeated for the scaled-up system. Figure 2.38 shows the response of the system scaled on the basis of mass ratio. The response is qualitatively very similar to the small system. An effective stall is achieved with the subsequent motion well damped.

During the first 3 s, the system flies at a horizontal velocity of 22.3 m/s (43.4 kt) and a vertical velocity of 7.3 m/s (24 ft/s). The L/D ratio is 3.0. The stable angle of attack is 2.9 deg, the payload is swung backward with $\theta = -10.3$ deg, and the glide angle is 18.2 deg. During the flare, the parachute pitches up to a maximum θ of 3.2 deg, slightly after which vertical velocity is reduced to 2.8 m/s (9.2 ft/s) and horizontal velocity to 16 m/s (31.1 kt). Minimum vertical velocity is achieved 3.8 s after control initiation. Height loss during the maneuver is 18 m. The ratio of horizontal velocity at flare to initial horizontal velocity is 72%, and the ratio of vertical velocities is 38%, compared with 79 and 36% for the small system. System response to maximum pitch up was 1.77 times that for the small system—close to the predicted ratio. Motion after flare is favorable with all disturbance damped within two cycles. Phugoid cycle time is 10.8 s compared to 6.5 s for the small system, a ratio of 1.7. The performance of the large system scaled on the basis of mass ratio is thus very close to the small prototype.

Figure 2.39 show the performance for the system scaled on the basis of wing loading. The response to control input is somewhat different. The flare maneuver is acceptable but weaker with less pitch-up of the payload. The subsequent motion is a lightly damped oscillation of pitch angle and angle of attack of period 6.3 s.

For the first 3 s of steady flight, the system glides at a horizontal velocity of 13.3 m/s (25.9 kt) and a vertical velocity of 4.4 m/s (14.4 ft/s), and L/D is 3.0. The stable angle of attack is 2.7 deg, the payload is swung backward with $\theta = -10.8$ deg, and the glide angle is 18.2 deg. In the flare, the parachute only pitches to θ of -2 deg at which time vertical velocity is reduced to 2.2 m/s (7.2 ft/s) and horizontal velocity to 10 m/s (19.4 kt). Minimum vertical velocity is achieved 3.6 s after control initiation with height loss 10 m (32.8 ft). Vertical velocity is only reduced to 50% of the initial value, and horizontal velocity is reduced to 75% of the steady glide value. The poorly damped pitching motion after simulated flare would be of concern for a PADS operation because it would indicate that this configuration is sensitive to control inputs and gusts and would give unpredictable performance.

It can be demonstrated that the effect identified is related to mass ratio, and not some other feature of the configuration, if the simulation of the system scaled on the basis of wing loading is repeated with the parachute flying at high altitude such that mass ratio is also matched. This occurs at an altitude of 10,000 m (32,808 ft). Figure 2.40 shows that the performance is now identical to the system originally scaled on mass ratio.

Thus, to match dynamic performance of ram-air parachutes, scaling must be on the basis of mass ratio.

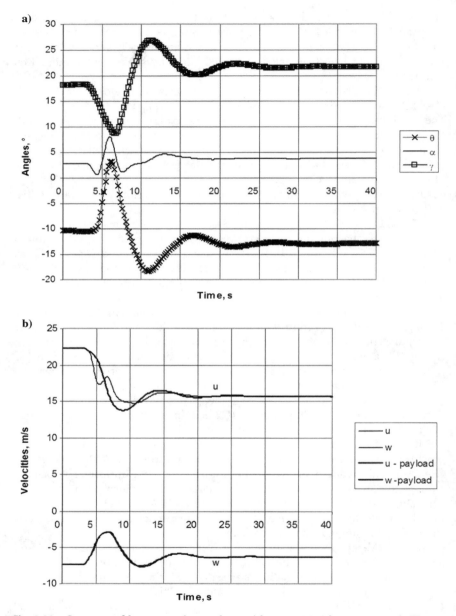

Fig. 2.38   Response of large ram-air parachute with $m_s = 5{,}229$ kg to symmetric TE deflection at $h = 400$ m (1,312 ft).

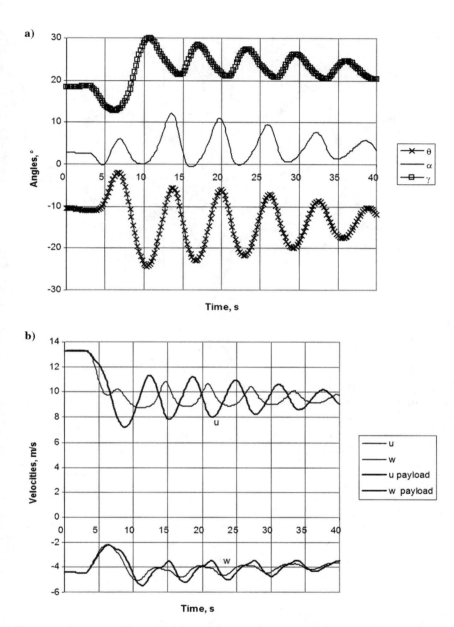

Fig. 2.39 **Response of large ram-air parachute with $m_s = 1,808$ kg (5,229 lb) to symmetric TE deflection at $h = 400$ m (1,312 ft).**

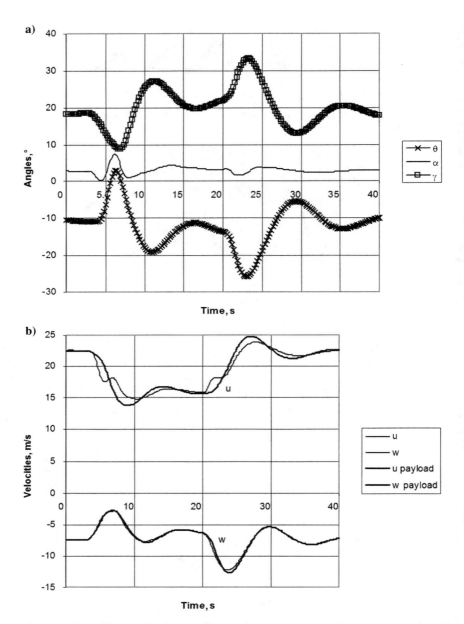

Fig. 2.40   Response of large ram-air parachute with $m_s = 1{,}808$ kg (3,986 lb) to symmetric TE deflection at $h = 10{,}000$ m (32,808 ft).

2.6.3   EFFECTS OF TRAILING-EDGE RELEASE

During inflation, TE of ram-air parachutes are retained in the full-brake position to prevent surge during inflation. The ability to smoothly transition from this state to full glide is essential to PADS. The standard large system was subjected to a simulated flight with brakes fully applied over 1.7 s after 3 s of gliding flight and released over 1.7 s after 20 s of flight. The results are shown in Fig. 2.41.

As the brakes are released, the canopy accelerates rapidly with the load lagging. The system pitches down, eventually reaching $\theta = -25$ deg. Perturbation of angle of attack is very small. As the system dives and accelerates, both vertical and horizontal velocity increase. When the payload swings forward again, vertical velocity rapidly recovers to its steady glide value. Subsequent oscillations are highly damped. The ability to effect a good transition is thus demonstrated.

2.6.4   EFFECTS OF GUSTS

The effect of a 7.5-m/s (14.6-kt) step tail gust on the standard large system is shown in Fig. 2.42.

As the gust hits the parachute at time 3 s, the response shown is similar to the release of brakes. The system pitches down rapidly and accelerates to recover the lost lift. Rate of descent initially rises but recovers as the payload pitches forward and the parachute glides out of the dive. Again, α variations are only very transient. Following the gust, classical damped phugoid motion occurs with α constant. As the gust ceases at 25 s, the effect is the equivalent of a head gust. The canopy responds by decelerating rapidly. The payload lags, and the system pitches up. Rate of descent transiently falls to almost zero with the additional lift before the system pitches down and steady glide is reestablished. With the standard system, gusts are accommodated with the resulting perturbations rapidly damped.

2.6.5   EFFECT OF LINE LENGTH ON LONGITUDINAL PERFORMANCE

To investigate the effect of line length on dynamic performance, the standard large system was modeled with line length increased to $R/b = 1.0$. The results are shown in Figs. 2.43.

The system is initially established in a steady glide with $u = 21.8$ m/s (42.4 kt), $w = 8.1$ m/s (26.6 ft/s), and $L/D = 2.7$. The flare is poor with minimum vertical velocity occurring 1.8 s before maximum pitch-up. Vertical velocity is only reduced to 58% of the initial glide velocity with horizontal velocity at this time still 90% of the steady glide figure. At maximum pitch-up, horizontal velocity is 15 m/s (29.2 kt) (69% of the steady glide value), but rate of descent has risen to 6.2 m/s (20.3 ft/s) (77% of the steady glide value). The perturbation in angle of attack is large with α reaching 25 deg, well beyond the stall. Recovery from the maneuver is also poor with a sustained pitching motion ensuing. The poor performance is related to the increased lag between canopy and payload response.

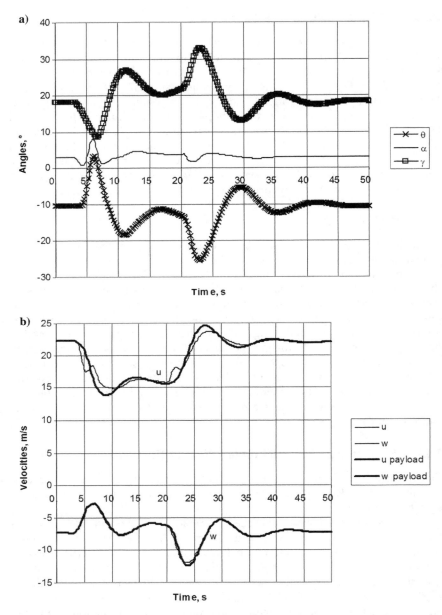

Fig. 2.41   Response of large ram-air parachute with $m_s = 5,229$ kg (11,528 lb) to symmetric TE deflection at $h = 500$ m (1,640 ft).

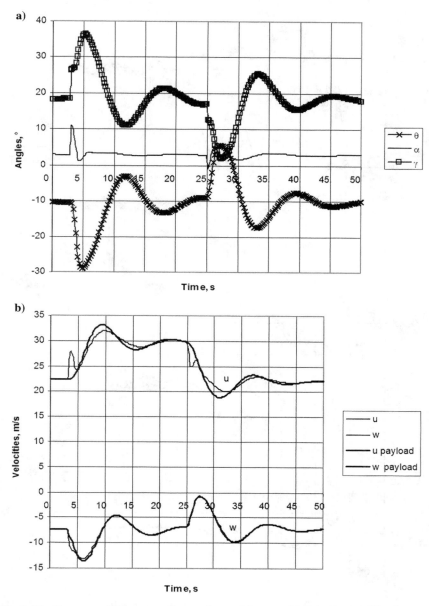

Fig. 2.42 Response of large ram-air parachute with $m_s = 5,229$ kg (11,528 lb) to a step gust at $h = 500$ m (1,640 ft).

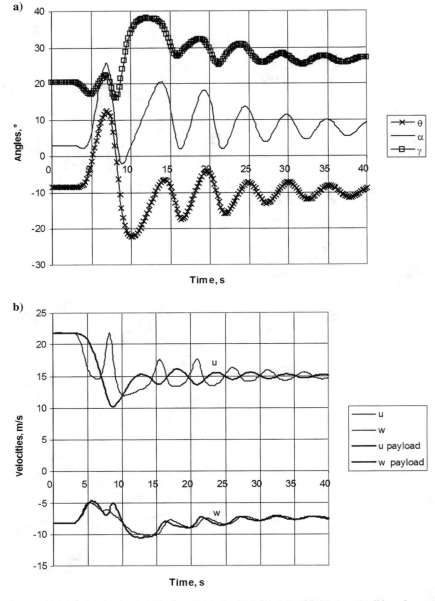

Fig. 2.43   Response of large ram-air parachute with $m_s = 5,229$ kg (11,528 lb) and $R/b = 1.0$ to symmetric TE deflection at $h = 400$ m (1,312 ft).

Clearly then, increased line length not only reduces glide performance but also compromises longitudinal stability.

2.6.6   EFFECT OF RATE OF RETRACTION OF TRAILING EDGE ON FLARE

It is obviously easier to design retraction devices that do not have to deflect TE very rapidly. The impact of much slower TE deflection on the effectiveness of the flare of the standard large system is therefore simulated. Figure 2.44 shows the results of control actuation over 5 s.

Again, glide is established with $u = 22.3$ m/s (43.4 kt) and $w = 7.3$ m/s (24 ft/s). Response of the system to control deflection is slower, and pitch-up of the system is less pronounced, with θ only reaching -1.7 deg, 5.6 s after the start of control movement. At this time, vertical velocity is at a minimum of 3.3 m/s (10.8 ft/s), 45% of the initial value, and horizontal velocity is 17.2 m/s (33.4 kt), 77% of the initial value. Flare is therefore seen to be less effective with slower application of brakes but might still be acceptable. A low vertical velocity is still achieved, and horizontal velocity is only 1.2 m/s (2.3 kt) greater than with rapid TE deflection, a small percentage of the overall forward speed.

2.7   LATERAL MOTION

The most important aspects of lateral motion for PADS are predictable turns and lateral dynamic stability. In analyzing lateral motion, a model based on the equations set out by [Doherr and Saliaris 1987] has been used. Aerodynamic data were derived from [Lingard 1981; Brown 1993]. The axes x_b, y_b, and z_b representing the coordinate frame associated with the PADS as a whole and definitions of a six-degree-of-freedom (DoF) model (three translational and three rotational DoFs) are shown in Fig. 2.45.

As mention in the preceding section, a correct model should include the so-called added (apparent) masses. In fluid mechanics, added mass or virtual mass is the inertia added to a system because an accelerating or decelerating body must move (or deflect) some volume of surrounding fluid as it moves through it. Combining components of PADS speed vector and angular velocity vector into a single six-element vector $\mathbf{x} = [u, v, w, p, q, r]^T$, and also combining three components of force and three components of moment imposed on PADS into a generalized vector $\mathbf{F} = [X, Y, Z, L, M, N]^T$, the second Newton's law can be written in the form

$$\mathbf{F}^* = \mathbf{A}\dot{\mathbf{x}} \qquad (2.37)$$

It is matrix $\mathbf{A}$ that has to include all added masses. For the potential flow this matrix is symmetric. If the center of mass of the body is chosen as the origin of the coordinate frame, then this matrix has only seven unique nonzero elements. If this is not the case, the number of nonzero elements is greater. For PADS,

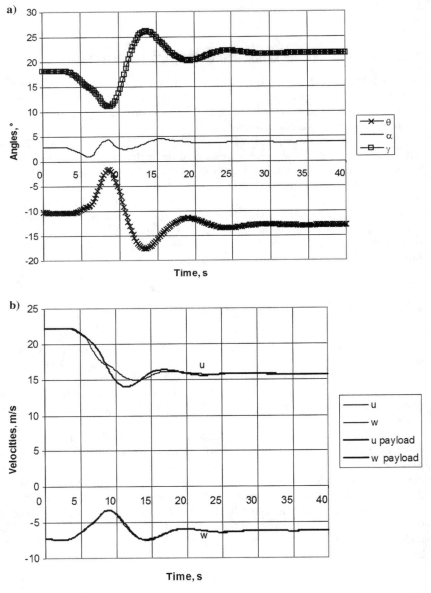

Fig. 2.44 Response of large ram-air parachute with $m_s = 5,229$ kg (11,528 lb) to slow TE deflection at $h = 400$ m (1,312 ft).

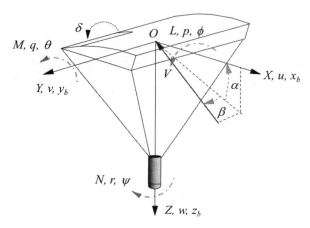

Fig. 2.45   Definition of ram-air parachute axes.

having a single plane of symmetry, the number of unique nonzero elements is reduced to 12 [Brennen 1982]

$$
\mathbf{A} =
\begin{bmatrix}
m + \alpha_{11} & 0 & \alpha_{13} & 0 & \alpha_{15} + mz_s & 0 \\
0 & m + \alpha_{22} & 0 & \alpha_{24} - mz_s & 0 & \alpha_{26} + mx_s \\
\alpha_{13} & 0 & m + \alpha_{33} & 0 & \alpha_{35} - mx_s & 0 \\
0 & \alpha_{24} - mz_s & 0 & I_{xx} + \alpha_{44} & 0 & I_{xz} + \alpha_{46} \\
\alpha_{15} + mz_s & 0 & \alpha_{35} - mx_s & 0 & I_{yy} + \alpha_{55} & 0 \\
0 & \alpha_{26} + mx_s & 0 & I_{xz} + \alpha_{46} & 0 & I_{zz} + \alpha_{66}
\end{bmatrix}
$$

$$(2.37)$$

In this matrix α_{ij} are the added masses referred to the axes defined in Fig. 2.45, m is the PADS mass, I_{ij} are the elements of PADS inertia tensor, x_s and z_s are the x and z positions of payload center of gravity (c.g.) with respect to the canopy's c.g. Chapter 5 presents a detailed derivation of equations of motions and also discusses how to compute added masses α_{ij}. The coefficients α_{ij} are also called apparent masses and inertias [Yavus and Cockrell 1981].

2.7.1   TURN MANEUVER

Turns are initiated on ram-air parachutes by deflection of all or part of the canopy TE on one side of the canopy. As discussed in Sec. 2.4, deflection of TE of the wing increases the camber of the section and results in increases of lift coefficient at a given angle of attack, maximum lift coefficient, profile drag, and induced drag. In response, in a TE deflection δ_a ("a" stands for asymmetric) the wing initially yaws in the positive direction. The lift caused by the deflection creates an adverse rolling moment. Centrifugal forces on the payload roll the wing in the correct direction and sustain the turn.

Of interest for PADS are achievable rates of turn, so first consider steady turning flight. The equations of motion of the system in body axes for about the yaw and roll axes in steady turning flight are

$$QSb\left(C_{n\delta_a}\delta_a + C_{n\beta}\beta + C_{nr}\frac{rb}{2V}\right) = (a_{22} - a_{11})uv - a_{13}vw + a_{26}ur \quad (2.38)$$

$$QSb\left(C_{l\delta_a}\delta_a + C_{l\beta}\beta + C_{lr}\frac{rb}{2V}\right) - m_sgR\sin(\phi)\cos(\theta)$$

$$= -(m_sR + a_{15})ur + a_{13}uv + (a_{33} - a_{22})vw - (a_{26} + a_{35})wr \quad (2.39)$$

where

$C_{n\delta_a}$ = yawing-moment coefficient due to control deflection

$C_{n\beta}$ = yawing-moment coefficient due to sideslip

C_{nr} = yaw-damping coefficient due yaw rate

$C_{l\delta_a}$ = rolling-moment coefficient due to control deflection

$C_{l\beta}$ = rolling-moment coefficient due to sideslip

C_{lr} = roll-damping-moment coefficient due to yaw rate

r = yaw rate, rad/s

Considering Eq. (2.38), the terms a_{13} and a_{26} are small, and $a_{11} \approx a_{22}$ are proportional to the mass of air enclosed in the canopy. Therefore, the right-hand side of the equation is approximately zero. On the left-hand side of the equation, for the wings with zero sweep, $C_{n\beta} = 0$ [Dommasch et al. 1951]. Therefore, Eq. (2.38) reduces to

$$C_{n\delta_a}\delta_a + C_{nr}\frac{rb}{2V} = 0 \quad (2.40)$$

or

$$r = -\frac{C_{n\delta_a}}{C_{nr}}\frac{2V}{b}\delta_a \quad (2.41)$$

Rate of turn, for a given asymmetric TE deflection, is increased by velocity. The larger the parachute is, the lower the turn rate for a given flight velocity and deflection. With values for $C_{n\delta_a}$ and C_{nr}, a good approximation for rate of turn r as a function of asymmetric TE deflection is available. Dommasch et al. [1951] propose that

$$C_{nr} = -C_D/6 \quad (2.42)$$

For a typical ram-air parachute operating at a design lift coefficient of 0.5 with glide ratio 3:1, then $C_D = 0.167$, and $C_{nr} = -0.028$.

Assuming that the yawing moment produced by asymmetric TE deflection is entirely due to the difference in drag on each side of the wing, then from Fig. 2.46 we can write

$$C_{n\delta_a} = \frac{b_k \bar{y}}{b^2} \frac{dC_D}{d\delta_a} \tag{2.43}$$

where $dC_D/d\delta_a$ is the rate of change of drag with asymmetric deflection of the complete TE at the design angle of attack.

Further, assume the control lines act on the outer half of each side of the wing, and then

$$C_{n\delta_a} = \frac{3}{32} \frac{dC_D}{d\delta_a} \tag{2.44}$$

From Fig. 2.25,

$$\frac{dC_D}{d\delta_a} \approx 0.11 \, \text{rad}^{-1} \tag{2.45}$$

resulting in $C_{n\delta_a} = 0.01$. Substituting the values for C_{nr} and $C_{n\delta_a}$ into Eq. (2.41), we finally have

$$r = 0.71 \frac{V}{b} \delta_a \tag{2.46}$$

Thus, for a typical large PADS system with $b = 30$ m (98 ft) and $V = 20$ m/s (38.9 kt), only 21-deg (0.366-rad) asymmetric TE deflection is needed to produce a turn rate of 10 deg/s (0.175 rad/s). The radius of turn R_c for this

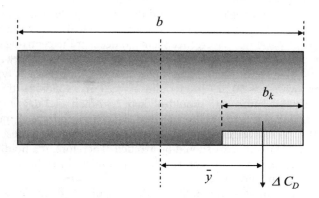

Fig. 2.46   Yawing moment on ram-air wing.

case is therefore

$$R_c = \frac{u}{r} = \frac{V\cos(\gamma)}{r} = \frac{19.0}{0.175} = 108\text{ m (354 ft)} \qquad (2.47)$$

Now consider Eq. (2.39). The apparent (added) mass terms are negligible with the exception of $(a_{33} - a_{22})vw$. This term, however, has a value in the absence of acceleration and would be included in wind-tunnel measured aerodynamic data as a part of $C_{l\beta}$. It therefore can also be neglected. On the right-hand side of the equation, the aerodynamic terms are an order of magnitude smaller than the gravitational term and thus to a first approximation

$$m_s g R \sin(\phi)\cos(\theta) = (m_s R)ur \qquad (2.48)$$

or because θ is small,

$$\phi = \sin^{-1}(ur/g) \qquad (2.49)$$

Hence, the motion of a ram-air parachute system simplifies to that of a pendulum undergoing curved motion. The sine of the angle of bank is approximately the ratio of the centrifugal acceleration and gravitational acceleration. For the preceding case

roll / bank angle $\qquad \phi = \sin^{-1}\dfrac{19.0 \times 0.175}{9.81} = 19.8\text{ deg} \qquad (2.50)$

2.7.2 LATERAL INSTABILITIES

Doherr et al. [1994] show that large ram-air parachutes are undamped in spiral divergence and only lightly damped in Dutch roll.

Spiral divergence occurs when a turn in one direction is sustained. Rate of turn, velocity, and bank angle increase, and the path helix steepens. Mathematical modeling indicates that slow turns can be sustained, but above a certain turn rate spiral divergence develops. Divergence seems to occur if the centrifugal acceleration ur exceeds a critical value. Modeling suggests that the boundary is approximately 4 m/s^2 (7.8 kt) or when bank angle exceeds 24 deg. For a typical large PADS system with $b = 30$ m (98.4 ft) and $V = 20$ m/s (38.9 kt), this criterion means that sustained turn rate should be kept less than 11.5 deg/s (0.2 rad/s). Moreover, because the turn rate for a given deflection and speed increase, the problem of the spiral divergence would seem to be exacerbated. Because the analysis is tentative, sustained turns should be avoided. Therefore, if a holding pattern close to the target is part of the control strategy, a figure of eight is better than a sustained turn in one direction.

The Dutch roll mode seems similar to the phugoid mode in that shorter suspension lines and increased mass ratio are stabilizing.

2.8   INFLATION OF RAM-AIR PARACHUTES

Analysis of films showing the inflation of ram-air parachutes reveals that the canopy initially inflates in a manner similar to a conventional parachute, until it first reaches normal flying size, with little cell inflation. Subsequently, it starts to collapse and pitches forward. As the pitching motion occurs, the cells inflate, and the parachute begins to fly.

Without inflation, control ram-air parachutes develop high inflation loads. This is not surprising because the initial inflation phase is extremely rapid. Some trials data have indicated dimensionless inflation times $\tau_0 = V_s t_i / D_0$ of approximately 2...5, where V_s is the snatch velocity, t_i is the inflation time, and D_0 is the diameter of a circle equal in area to the planform area of the parachute. These times should be compared to values of 10 for conventional, man-carrying, circular parachutes.

In addition to the opening load, a second force peak can occur because of the lift and drag forces generated during the rapid pitch forward phase of inflation.

Because of these high loads, a reefing device is employed to reduce the initial opening load, and the second peak load during the rapid pitch forward is prevented by initial TE deflection.

2.8.1   REEFING TECHNIQUES

Early ram-air parachutes used a variety of reefing techniques including the "ropes and rings technique." The pilot chute was attached to a bridle that was routed down to the center of the canopy and then passed outboard through a series of rings along the canopy surface. The bridle thus acts as a drawstring, resisting the spanwise inflation of the canopy. Upper- and lower-surface versions of this technique were used. In the latter, the bridle passed through grommets in the upper and lower surfaces of the canopy to rings on the lower surface. This method is more effective than the upper-surface version because lower forces are needed for positive control of inflation.

Modern canopies for man-carrying purposes, however, usually use slider reefing. The slider comprises four rings, each of which passes around one of the four sets of suspension lines from the four risers. The lines are located at the corners of a square of fabric. The slider area is typically 2% of the canopy area. At the start of the inflation, the slider is positioned at the top of the lines adjacent to the canopy, effectively reefing the canopy. During inflation, the tension in the lines forces the slider down toward the risers. This motion is resisted by air drag on the slider, and thus inflation rate is controlled. On some systems, the pilot chute is attached to a webbing cross, and downward motion of the slider is resisted by the pilot chute drag.

With the increasing size of ram-air parachutes for cargo and space missions, other reefing techniques have been developed. Puskas [1989] describes the use of a sail slider for a 279-m^2 (3,000-ft^2) system. The advanced recovery system (ARS) uses a Pioneer proprietary system called Mid-Span Reefing. The effect of this system is to completely remove sets of reefed cells from the airflow by compressing them spanwise. A fringe benefit of this system is that it enables the reefed cells and their rigging lines to be fabricated from lighter materials than the first stage cells because they never see the high loads associated with the deployment dynamic pressure. Weight savings in excess of 30% are claimed. Reefing ratios used on the ARS were 25.9 and 48.1%.

2.8.2 INFLATION ANALYSIS

The complex inflation characteristics of a ram-air parachute with reefing make analysis of this phase difficult. Engineering calculations can be made using semi-empirical methods. The following approach is typical and works reasonably well.

Unsteady parachute inflation force data obtained from experiments are found to be well-correlated when plotted in the form of force coefficient $C_F = F/qS_0$ vs dimensionless time.

This implies the following:

- Inflation force coefficient is independent of mass ratio and Froude number and is a function only of dimensionless time, that is, a given parachute type has a dimensionless inflation force–time signature.
- Inflation time occurs in a fixed dimensionless time $\tau_0 = V_s t_i / D_0$.

The dimensionless inflation time and inflation force time signature can be extracted from a limited number of tests.

The majority of simple semi-empirical methods such as those of [Knacke 1992; Ludtke 1973; Lingard 1979] rely, with some individual refinements, on these assumptions.

The equations of motion of the system shown in Fig. 2.47 can be written as

$$m_s \frac{dV}{dt} = m_s g \cos(\theta) - \frac{1}{2}\rho V^2 C_F(t) S_0 \tag{2.51}$$

$$\frac{d\theta}{dt} = -g\,\frac{\sin(\theta)}{V} \tag{2.52}$$

where $C_F(t)$ is the varying force coefficient. For slider reefed parachutes, the force coefficient can be approximated as

$$C_F = \begin{cases} 1.5\left(\dfrac{t}{t_i}\right)^3, & t \le t_i \\[2mm] 1, & t > t_i \end{cases} \tag{2.53}$$

where

$$t_i = \frac{D_0}{V_s}\tau_0 \tag{2.54}$$

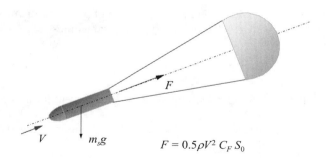

$$F = 0.5\rho V^2\, C_F\, S_0$$

Fig. 2.47   Schematic diagram of inflating parachute.

Dimensionless inflation time τ_0 is in the range 12–18 depending on the slider design, with 14 a good starting point. Equations (2.51) and (2.52) yield a set of equations that can be easily solved for V and θ allowing inflation force F to be calculated.

The preceding inflation model does not explicitly take apparent mass into account. Indeed, in using experimentally measured dimensionless force–time curves as a basis for the model, it is assumed that apparent mass is automatically accounted for. As mentioned before, apparent masses for parachutes are significant, often being much larger than the suspended mass. The forces due to apparent mass effects depend on the rate of change of apparent mass and on the magnitude of the acceleration. During the inflation process, large rates of change of apparent mass occur. This is particularly true for rapidly inflating parachutes. If midspan reefing is used, inflation between stages can be rapid because when the parachute is not reefed free, unrestrained inflation takes place. To allow for this, a different type of model, in which apparent mass is explicitly treated, is useful.

Assume that the drag coefficient of the parachute is constant and the profile drag is simply a function of the parachute's instantaneous projected diameter. Additionally, assume that there is an apparent mass a_{11} associated with the parachute, which is also a simple function of the instantaneous projected diameter. Equation (2.51) then can be represented by

$$m_s \frac{\mathrm{d}V}{\mathrm{d}t} = m_s g \cos(\theta) - \frac{1}{2}\rho V^2 C_D S' - a_{11}\frac{\mathrm{d}V}{\mathrm{d}t} - V\frac{\mathrm{d}a_{11}}{\mathrm{d}t} \qquad (2.55)$$

where $S' = 0.25\pi D^2$. For inviscid flow, the apparent mass of a disc in the direction of its axis of symmetry is

$$a_{11} = \rho\frac{D^3}{3} \qquad (2.56)$$

To complete the model, a function for the variation of projected diameter against time and a value for C_D are required. For ram-air parachutes,

$$D = D_0 \left(\frac{t}{t_i}\right)^{1.5} \tag{2.57}$$

where $D_0 = (4S/\pi)^{0.5}$. A typical steady-state force coefficient for a ram-air parachute with full brakes is 1.

Finally, the inflation time for these parachutes can be approximated as follows:

For free inflation:

$$t_i = \frac{3.5D_0}{V_s} \tag{2.58}$$

For slider reefing:

$$t_i = \frac{14D_0}{V_s} \tag{2.59}$$

With this type of model, the inflation factor is automatically generated. This model was applied to the slider reefed MC-4 parachute with 163-kg (359-lb) suspended mass deployed at 72 m/s (140 kt). The results are compared with test in Fig. 2.48. The match obtained is reasonable.

The model was also used to simulate drop test 10 of the ARS [Wailes 1993]. The 334 m² (3,595 ft²) is two-stage reefed: 25.9 and 48.1%. The parachute was

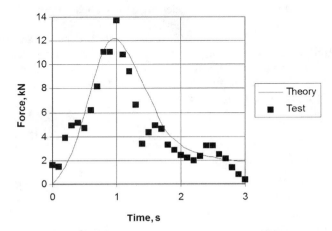

Fig. 2.48   Simulation of the inflation load of a slider reefed ram-air parachute.

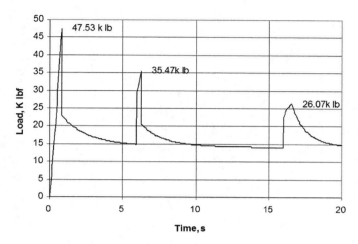

Fig. 2.49   Simulation of the inflation load for a midspan reefed parachute, ARS drop test 10.

dropped with a load of 6,350 kg (14,000 lb). The simulated data are shown in Fig. 2.49. In the test, the three inflation peaks were 21.56 tons (47,530 lb), 16 tons (35,470 lb), and 11.83 tons (26,070 lb). Again, the match is satisfactory for a design calculation.

In conclusion, although complete analysis of the inflation of ram-air parachutes is almost intractable, engineering methods that give adequate predictions for design calculations are available.

CHAPTER 3

Key Factors Affecting PADS Landing Precision

Oleg Yakimenko*

Naval Postgraduate School, Monterey, California

Horst Altmann[†]

Airbus Defence & Space, Manching, Germany

As shown in Sec. 1.3, achieving a required touchdown precision of air delivery is a challenging problem. The factors contributing to touchdown error are limited control authority inherent in commonly used unpowered PADS; unknown and constantly changing surface-layer winds; nonstandard and also changing atmosphere resulting in erroneous estimates of height above IPI, descent rate, and time to touchdown; canopy asymmetry causing biased control inputs, controls rate and dynamics; and imperfection in GNC algorithms. This chapter is primarily based on [Altmann 2013; Corley and Yakimenko 2009; Benton and Yakimenko 2013; Ward et al. 2010a; Wright et al. 2005a; Yakimenko et al. 2009; Yakimenko and Slegers 2011] and investigates an effect of most of the aforementioned factors on landing precision based on real flight data and simulations involving a generic PADS with the fixed guidance strategy (to exclude the effect of GNC algorithms, which will be considered in more detail later, in Chapter 7). Obviously, the major contributors to a touchdown error are unknown winds. Unknown or erroneous winds aloft can cause PADS not to reach the intended touchdown area at all whereas the lack of knowledge about the surface-layer winds results in messing up terminal guidance. As just mentioned, other important contributors to the overall touchdown error are nonstandard atmosphere and inaccuracies in estimating the altitude above IPI. These factors and the ways to mitigate them are addressed in Sec. 3.1. The influence of other real-world effects like controls rate and dynamics are considered in Sec. 3.2. This section also shows how the effect of many factors would be mitigated if PADS would have an extended control authority allowing glide angle control. (This topic will be considered in more detail in Chapter 8.) Finally, Sec. 3.3 introduces two software tools making a good use of the winds aloft to compute a desirable release area, subject to airlifter altitude, speed and ground track, and to estimate a touchdown footprint in the case of PADS partial or total malfunction.

---

*Professor.

[†]Chief Engineer, Parafoil Systems.

3.1   EFFECT OF UNKNOWN SURFACE-LAYER WINDS

Let us start by addressing the issue of effect of the varying surface winds acting on PADS during its terminal descent onto IPI. Even though all PADS attempt to estimate winds on their way down to the surface, the winds below PADS's current altitude are either unknown at all or only known with a certain probability. This section starts with a representative example showing how a misjudgment on these winds can easily result in a huge touchdown error even for the best and most robust PADS, followed by a discussion on how such a negative effect can possibly be mitigated.

3.1.1   ASSUMED VS REAL SURFACE-LAYER PROFILES

To investigate an effect of the surface-layer winds on landing performance of a ULW PADS, which in ideal conditions would exhibit high touchdown precision, a series of tests was conducted at U.S. Army Yuma Proving Ground (YPG). Figure 3.1a presents a setup of the test site featuring the wind tower that provided the updates of ground winds every minute. Figure 3.1b shows another view of the solar-powered wind tower transmitting wind information to the YPG ground station residing on a truck shown on the left.

The UH-1A helicopter used for the drops (seen at the background in Fig. 3.1a) carried Snowflake PADS along with several YPG Windpack dropsondes strapped in the helicopter's bay. (Four of these Windpack systems are shown in Fig. 3.2a.) The drops were conducted from 762 to 914 m (2,500 to 3,000 ft) above ground level (AGL). About 1 min prior to each PADS drop, the Windpack system was released to measure the actual winds. After the tests, these actual winds were compared to the wind estimates produced onboard Snowflake PADS while it was descending. Two different IPIs were assigned in the vicinity of the wind tower and were used depending on the direction of the ground winds. Snowflake canopy deployment occurred about 152 m (500 ft) lower than that of the corresponding

Fig. 3.1   Snowflake test site at a) YPG and b) wind tower.

Fig. 3.2   a) Windpack parachutes and b) geodetic survey of a touchdown point.

Windpack. The entire trajectory and other PADS state parameters were recorded by onboard AGU. The impact points were also surveyed with a subcentimeter accuracy using a mobile DGPS receiver as shown in Fig. 3.2b.

Although specifics of the most current Snowflake GNC paradigm will be discussed later in Chapter 7, here we will briefly mention that all Snowflake trajectories look alike and are similar to the ones shown in Fig. 3.3. Although most of the drops land pretty close to IPI as shown in Fig. 3.3b, it is a trajectory shown in Fig. 3.3a that will be a focus of discussion. Upon release from a helicopter (phase 0), the canopy deploys, and PADS transitions to a loiter area depicted in Fig. 3.3 with the four waypoints (phase 1). After making several passes within a loiter area to estimate the current magnitude of the winds along the predetermined final approach direction (phase 2), PADS AGU exits loiter pattern towards the target (phase 3) and proceeds along the downwind leg (phase 4), continuing estimating the wind, until a base turn followed by a final approach (phase 5). The base-turn initiation point is based entirely on the surface winds below the current altitude as they are perceived by PADS AGU.

As seen in Fig. 3.3, both impact points lie slightly to the right of the intended landing direction, which is due to the crosswind component unaccounted for. For these specific tests, it was done intentionally by setting the intended landing direction slightly off the current ground wind direction. These offsets for the two drops presented in Fig. 3.3 are shown in Table 3.1 along with the crosswind component at impact.

With perfect knowledge of the winds, any PADS would land almost exactly at IPI. Unfortunately, the only winds available onboard are the wind estimates, provided by the GNC algorithm in real time. As mentioned already for these specific tests with an earlier version of AGU algorithms, these estimates were obtained during two phases, loitering and downwind descent. Of course, these estimates are not perfect, but that does not constitute a major problem. The last estimate on the wind magnitude comes at the end of the downwind leg, that is, the construction of the final turn trajectory is based on this last update, for Snowflake

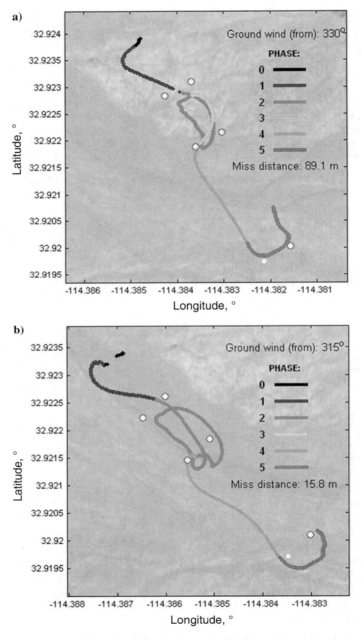

Fig. 3.3   Two representative drops of Snowflake PADS: a) drop A and b) drop B.

TABLE 3.1   GROUND WIND OFFSETS AND CROSSWIND COMPONENT AT IMPACT

Drop Number	A	B
Ground wind offset	45 deg	20 deg
Crosswind component	0.73 m/s (1.4 kt)	4.35 m/s (8.5 kt)

PADS obtained at an altitude of about 100 m (328 ft) AGL. This altitude is defined by a simple equation. If the winds below 100 m (328 ft) are the same as estimated at 100 m (328 ft), then the PADS would land in a close vicinity of IPI, where the only source of error then would be an error in estimating winds at 100 m (328 ft). Of course, having the same winds throughout the last 100 m (328 ft) of the drop is rarely the case (if ever at all).

Let us now consider the two trajectories in Fig. 3.3 from exactly this point of view—how accurate the estimates of the winds were at 100 m (328 ft) and how much of a wind change the ADSs had to face on their final 100 m (328 ft) descent to the ground. Analysis of the data collected during the drops by Wind-packs and wind tower allowed the revealing of the cause of the difference in touch-down performance for drop A and drop B, with only slightly over an hour time difference between them!

To begin, Figs. 3.4 and 3.5 show the complete sets of recorded data for both drops. Figure 3.6 presents the data collected by the wind tower. First, one can observe the drift in the barometric pressure that the estimates of the PADS alti-tude are based upon. As seen, the onboard barometric pressure sensor settings have to be changed constantly to accommodate these changes in the ground pressure. Failure to do that results in the PADS thinking that it is higher above the ground level than it actually is (because the barometric pressure drops down as the sun heats the surface). Second, during the four sets of drops shown in Fig. 3.6, the winds did change their direction and speed. Starting with the light winds in the morning (below 1 m/s), the ground winds got stronger up to about 4 m/s for the last sets of drops. The wind direction changed drastically before and during the first set of drops settling up after about 0830–0900 hrs to NNW winds. On the time scale of Fig. 3.6, drop A was the fourth drop, and drop B was the sixth drop of the day.

Next, Fig. 3.7 presents data collected by two Windpacks several minutes prior to drop A and drop B. In this figure, they are presented as recorded by Windpack, where altitude is provided with respect to mean sea level (MSL). It is seen that during the last 100 m (328 ft) the winds do change, and that is what causes the touchdown error. Figure 3.8 addresses this issue in more detail, providing the comparison of Windpack data with that of the wind tower and Snowflake esti-mates. (In this figure, Windpack altitude data are converted from MSL to AGL, and the Snowflake PADS altitude data are corrected to accommodate errors in altitude estimates, that is, all data are altitude-synchronized.)

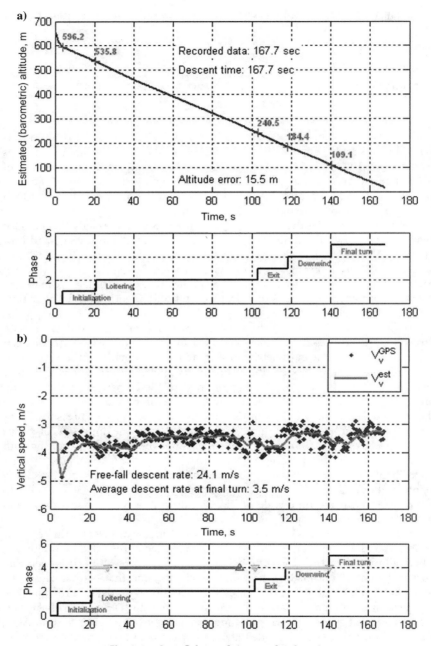

Fig. 3.4   Snowflake performance for drop A.

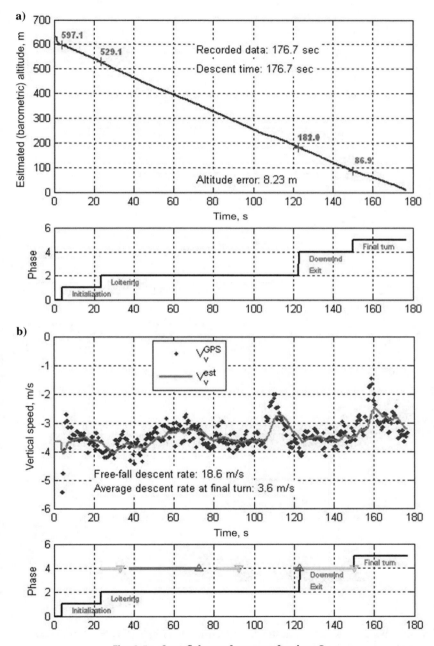

Fig. 3.5   Snowflake performance for drop B.

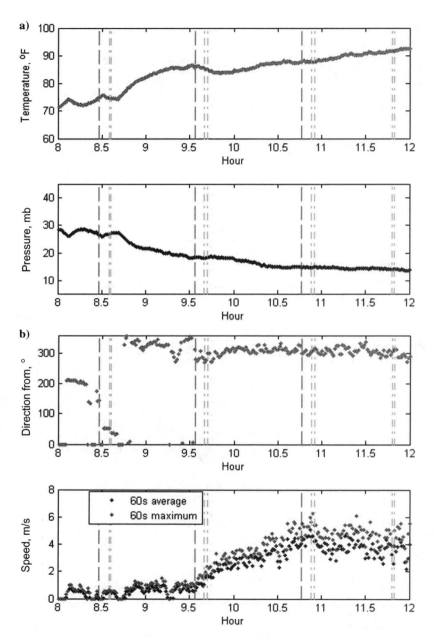

Fig. 3.6 Wind tower data with the dashed vertical lines indicating Windpack release followed by two dashed-dotted lines for two Snowflake PADS releases.

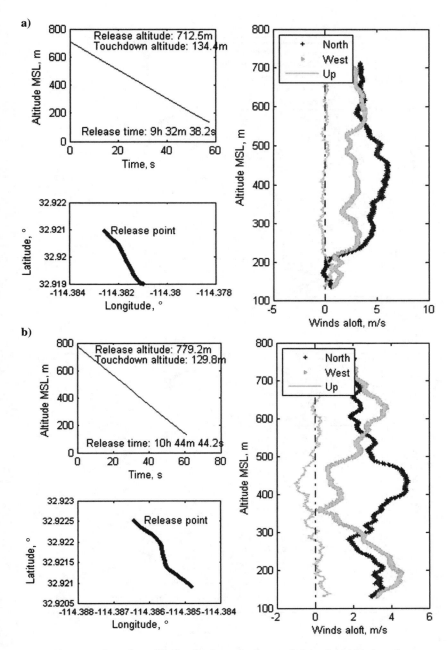

Fig. 3.7   Data from Windpack released prior to a) drop A and b) drop B.

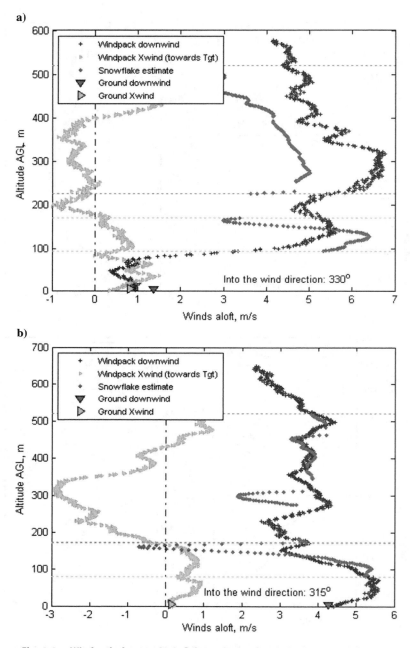

Fig. 3.8   Windpack data vs Snowflake estimate for a) drop A and b) drop B.

Figures 3.9a and 3.9b present a more detailed analysis of the data shown in Figs. 3.7a and 3.7b, respectively. Specifically, the last phase of the guided descent, final turn to the target, is considered. The left portion of each plot shows the difference between the tailwinds predicted by the Snowflake PADS during its downwind leg and the actual winds measured by the Windpack, and the crosswind component that the PADS had to overcome on its way down. On the right portion of each plot, the vertical speed as recorded by the Snowflake PADS barometric altimeter is presented. Several calculated parameters shown on these plots represent a rough estimate of contribution that unknown winds in the downwind direction might have on overshooting IPI.

Consider Fig. 3.7a (drop A). As seen from Fig. 3.4, the last estimate of the along-the-course wind component by the Snowflake autopilot occurred at about 93.6 m (307 ft) corrected altitude [109.1 m (358 ft) to 15.5 m (51 ft)] AGL and provided the value of 5.3 m/s. However, the real winds suddenly die and amount to only about 1 m/s all the way down to the ground. Hence, there is a 4-m/s difference between the last Snowflake estimate and the actual winds (Fig. 3.8a) all of the way down. These unaccounted-for tailwinds that acted during the last 27 s of the guided descent contribute to a large system overshoot of about 116 m. (The actual overshoot happened to be smaller than that because of the error in altitude estimate that in this particular case was of the opposite sign.)

Now, consider Fig. 3.7b (drop B). Compared to the preceding case, the wind estimate at the corrected 79 m (259 ft) [86.9 m (285 ft)–8.23 m (27 ft)] AGL (Fig. 3.5) is almost perfect (5.2 m/s), and the winds do not change much during the final turn (Fig. 3.8b). This would result in landing only 5 m (16.4 ft) short of IPI (again, due to the error in altitude estimate the actual error happened to be slightly different).

The data for these two Snowflake drops are consolidated in Table 3.2. This table shows upwind and crosswind components of the miss distance and then analyzes the possible contribution of the two sources of error, which are an altitude estimation error ($GR\Delta h$) and winds that are unaccounted for during the last phase of the flight, the final turn to the target ($\int_0^{h_{turn}} \Delta W dh$). The last column shows the values of a miss distance that could be achieved if both errors were eliminated.

In their paper, Carter and Rasmussen [2015] considered terminal guidance as a differential game against unknown winds. At each time increment during a descent the winds [defined by two components, $w_x(t)$ and $w_y(t)$] are doing their best not to allow PADS to land at IPI; in turn, PADS implements the best possible guidance [by varying its turn rate $\dot{\psi}(t)$] to mitigate a negative effect of these winds. Using a simple kinematic model, Carter and Rasmussen showed that the landing accuracy depends on the initial distance and relative heading to IPI, remaining time to impact, PADS airspeed and maximum turn rate, and, of course, maximum wind speed W_{max}. The best wind strategy is to use the maximum magnitude W_{max} all the time blowing PADS from the target. The best PADS guidance is to fly straight or execute a maximum turn rate maneuver depending on its

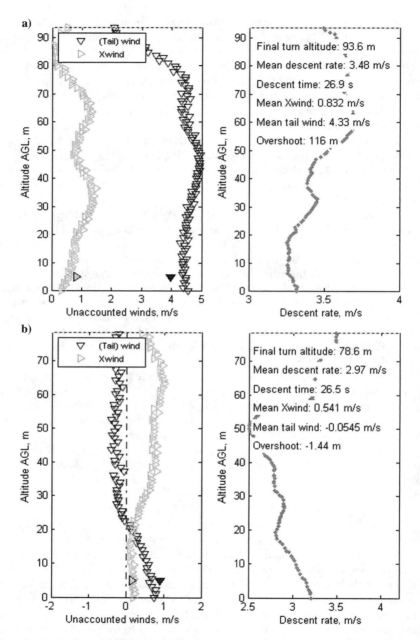

Fig. 3.9   Downwind and crosswind components of unaccounted winds for a) drop A and b) drop B.

TABLE 3.2   ESTIMATES OF POSSIBLE CAUSES FOR OVER- /UNDERSHOOTING THE TARGET

Drop ID	Miss-distance Components		Altitude Error	Correction due to the Altitude Error	Correction due to Unknown Winds	Estimated Upwind Miss Distance
	Upwind	Crosswind				
Drop A	102.1 m	34.8 m	15.5 m	31 m	−116 m	17.1 m
	(335 ft)	(114 ft)	(51 ft)	(102 ft)	(381 ft)	(56 ft)
Drop B	−5.1 m	15.2 m	8.23 m	16.5 m	−1.4 m	9.9 m
	(17 ft)	(50 ft)	(27 ft)	(54 ft)	(4.7 ft)	(33 ft)

position with respect to IPI [$\dot{\psi}(t) \in \{-\dot{\psi}_{max}, 0, \dot{\psi}_{max}\}$]. Surely, no matter what PADS does, the winds always win. Specifically, for the final turn of the Snowflake PADS [characterized by the final turn altitude of about 90 m (295 ft) AGL, the airspeed of 7.2 m/s (14 kt), descent rate of 3.5 m/s (11.5 ft/s kt), and $\dot{\psi}_{max}$ of 27 deg/s], Carter and Rasmussen established a lower bound on the miss distance being 0, 9, and 23 m (0, 30, 75 ft) when the unknown winds magnitude is 0, 25, and 50% of a PADS airspeed, respectively. In the case of drop of course, actual winds might not be an intelligent adversary, but the actual guidance strategy is not set to fight these winds either. Obviously, for the larger PADS with a slower dynamics these lower bounds will be higher.

The bottom line from the preceding discussion is that the varying surface-layer winds play a major role in touchdown precision for any PADS. The better PADS AGU knows these winds, a lesser chance they have to mess up a precise delivery. If a current value of the ground winds (along with a current barometric pressure) were available and uplinked to the descending PADS, its AGU suite might be able to take advantage of it. Hence, let us consider what means can be used to obtain these data.

3.1.2   SURFACE-LAYER WIND MEASURING

The primary source for measuring atmospheric pressure, temperature, humidity and wind speed are weather balloons (or sounding balloons). They can carry appropriate instrumentation to transmit data back to the ground or be tracked by radar (to measure just wind data). An alternative (or supplemental) source is another weather reconnaissance device, a dropsonde, dropped from an aircraft at altitude and utilizing a small stabilizing parachute to slow its descent (e.g., see [Hock and Franklin 1999]). Both devices are routinely used in conjunction with airdrop testing. For example, Fig. 3.10a shows a Radiosonde Wind Sounding (RAWIN) balloon, which is considered a standard of wind estimation used throughout the test community, WindPack radiosonde (see it packaged in Fig. 3.2a), and a smaller dropsonde by QinetiQ North America (QNA), used at YPG. For the in-house developed WindPack system that employs 3-m (9.82-ft)

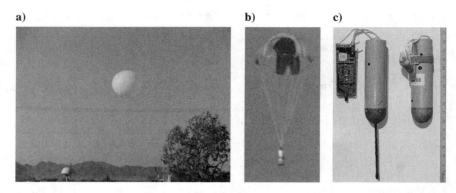

Fig. 3.10   a) RAWIN balloon right after launch, b) WindPack dropsonde under a trilobe canopy in flight, and c) QNA dropsonde assembly [Rogers 2009; Fraser 2011].

and 0.84-m (2.75-ft) trilobe parachutes with the aerodynamic drag coefficients of 0.56 and 0.48, respectively, the descent speed varies within [4.6;10.7] m/s ([15;35] ft/s) and [18.3;24/4] m/s ([60;80] ft/s). [The actual value depends on the weight of the system, which for the larger canopy ranges from 23 kg (50 lb) to 35 kg (76 lb), and for the smaller one ranges from 14 kg (30 lb) to 16 kg (35.5 lb).] The atmospheric data are sampled at 10-Hz rate and stored on a flash card [Kelly and Pena 2001].

Technology improvement and miniaturization allows the lowering of the cost of both weather balloons and dropsondes, so that they can also be used in the field conditions. For example, the Tactical Atmospheric Sounding Kit (TASK) (Fig. 3.11), also developed by QNA, consists of the 6-ft^3(32-in. diam) weather balloon and less than 40-g ultra high frequency (UHF) radiosonde continuously measuring and broadcasting atmospheric data. (The complete kit shown in Fig. 3.11b consists of the four sets, transceiver, and high-pressure helium bottle.) The 110-g (3.5-oz) QNA's Micro Air-Launched Expendable Meteorological Sensor (MAXMS) dropsonde (Fig. 3.12) was specifically developed to support

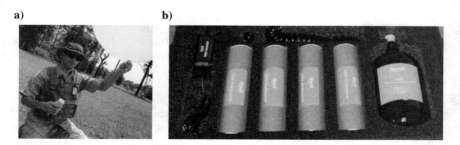

Fig. 3.11   Tactical atmospheric sounding kit by QNA [TASK Sensors 2014].

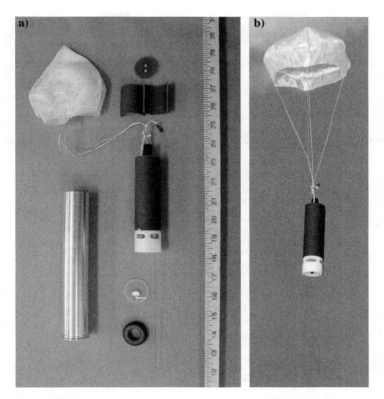

Fig. 3.12 a) MAXMS assembly within a MJU-38/B form factor and b) MAXMS dropsonde in flight [MAX Sensors 2014].

precision air delivery operations. It can be packaged for a hand launch from an aircraft or deployed from a standard airborne countermeasures dispenser system (CMDS) using an explosive squib (note that AN/ALE-47 CMDS can hold up to four MAXMS). The fall rate at the sea level is about 7.6 m/s (25 ft/s).

Both TASK and MAXMS are PADS/JPADS compatible, enabling the unit requiring resupply to send highly localized and current data to an incoming airlifter. Both systems utilize the same sensor suite that allow the measuring of the wind speed, wind direction, atmospheric pressure, air temperature, and relative humidity with ± 0.5 m/s (1 kt), ± 1 deg, ± 2 mb, $\pm 1°$C, $\pm 2.5\%$ accuracy, respectively. TASK and MAXMS allow outputting measured data at 4-Hz sample rate with a vertical resolution of less than 1 and 2.5 m (3.3 and 8.2 ft), respectively.

Figures 3.13–3.17 show sample formats of atmospheric data collected by different systems. Wind data are usually provided in the heading-magnitude

| Time | Press | Temp | RH | Wdir | Wspd | dz | Lon | Lat | GPS Alt | Sa |
sec	mb	°C	%	deg	m/s	m/s	deg	deg	m	#s
0.0	932.5	11.9	64.7	310.1	4.29	-15.9	-120.773	35.717	692.8	15
0.2	932.9	12.0	64.6	311.6	4.15	-16.1	-120.773	35.717	689.1	15
0.5	933.3	12.0	64.6	310.2	4.30	-15.8	-120.773	35.717	685.3	15
1.5	934.9	12.1	64.4	299.9	3.04	-15.3	-120.773	35.717	670.6	15
2.0	935.7	12.2	64.3	311.7	2.34	-15.3	-120.773	35.717	663.3	15
2.2	936.1	12.2	64.1	327.3	2.36	-15.5	-120.773	35.717	659.8	15
2.5	936.5	12.3	64.2	311.5	2.69	-15.2	-120.773	35.717	656.2	15

Fig. 3.13   Example of output data for MAXMS dropsonde and TASK radiosonde.

Line	AGL(m)	T(°C)	RH(%)	WSPD (m/s)	WDD(deg)	P(mb)
1	2.0	11.76	50.93	9.33	182.93	968.04
2	15.2	11.64	51.16	10.97	182.91	966.53
3	56.2	11.26	52.30	13.50	182.37	961.80
4	111.2	10.73	54.07	14.80	181.77	955.47
5	173.0	10.15	56.25	15.57	180.91	948.40
6	242.0	9.53	58.50	16.02	179.44	940.56
7	319.0	8.85	60.97	16.48	178.28	931.86

Fig. 3.14   Example of a winds sounding file.

Altitude (ft MSL)	Press (mb)	Temp (°F)	RH (%)	Dew Pt Temp (°F)	Air Density (g/m^3)	Wind Direct (deg)	Wind Speed (kn)
1322	968.0	45.6	44	24.9	1197.94	124.0	9.0
1400	965.2	45.4	41	23.0	1195.07	127.0	8.9
1500	961.6	45.5	41	23.3	1190.34	129.5	10.7
1600	958.0	45.4	42	23.5	1186.13	132.4	11.4
1700	954.5	45.8	43	24.8	1180.75	136.8	11.2
1800	951.0	46.5	44	26.0	1174.59	143.9	10.9
1900	947.4	46.7	44	26.0	1169.75	152.2	11.1

Fig. 3.15   Example of a RAWIN winds file.

Time(UTC)	Lat(deg)	Long(deg)	Altitude(m HAE)	Q	StdDev	VE(m/s)	VN(m/s)	VUp(m/s)
53728.0	33.282676544	-114.383564006	5613.796	4	0.697	-0.970	91.957	-0.030
53728.1	33.282759262	-114.383565067	5613.817	4	0.697	-0.967	91.811	0.142
53728.2	33.282841830	-114.383566246	5613.828	4	0.697	-1.097	91.665	0.102
53728.3	33.282924537	-114.383567418	5613.869	4	0.697	-1.090	91.842	0.316
53728.4	33.283007193	-114.383568702	5613.910	4	0.697	-1.181	91.788	0.355
53728.5	33.283089876	-114.383570015	5613.939	4	0.697	-1.227	91.822	0.265
53728.6	33.283172538	-114.383571381	5613.972	4	0.697	-1.255	91.774	0.318

Fig. 3.16   Example of a Windpack winds file.

Height(press) (kft AGL)(mb)	Dir (deg)	Speed (kn)	Temp (°C)	Press alt (kft MSL)	D-value ft)	Ballistic winds (deg)	(kn)
SFC(959)	93	10	20	1.61	+0198	93	10
1 (926)	101	36	22	2.47	+0245	96	18
2 (894)	119	25	25	3.45	+0294	108	18
3 (863)	129	21	24	4.36	+0355	108	18
4 (834)	135	19	23	5.30	+0418	117	17
5 (805)	139	17	21	6.23	+0485	117	17
6 (777)	145	16	18	7.17	+0545	120	16

Fig. 3.17   Example of a JAAWIN winds forecast file.

format (Figs. 3.13–3.15, 3.17); vertical component of the winds can be provided explicitly (Fig. 3.16) or computed implicitly using altitude vs time data (Figs. 3.13 and 3.16). Another important piece of information is air density, affecting the vertical speed of the descending ADS. Again, it is either provided explicitly (Fig. 3.15) or can be computed implicitly using temperature, pressure, and humidity data (Figs. 3.13–3.15) as will be discussed in Sec. 3.1.4. An example of graphical representation of raw data on wind speed and direction along with the ascent rate collected by TASK and MAXMS during Snowflake PADS tests is shown in Figs. 3.18a and 3.18b, respectively.

Obviously, the closeness of proximity and time of the wind estimate and airdrop plays a crucial role in PADS landing performance. To this end, Fig. 3.19 shows an experiment when the winds were measured by two weather balloons, launched three hours apart, both on the way up and then on the way down after the balloon has burst. Note that Fig. 3.19a only presents altitudes relevant to that specific day PADS drop altitude, but the balloon actually flew to about 1,524 m (5,000 ft) MSL. The second balloon (Fig. 3.19b) did not go very high (because of a helium leak), so that data shown in Fig. 3.19b represent all continuous recording.

As seen from Fig. 3.19a, at the higher altitudes [between 152 and 427 m (500 and 1,400 ft) AGL] radiosonde measured about the same wind profile on the way up and down while at the lower altitudes light winds exhibited fairly large variations in wind direction. In 3 h, the ground winds became stronger, and there were fewer variations in direction. The observed differences might be attributed to slight variations of location (which would be especially true for the mountainous terrain) and/or time. Even this simple test demonstrates that winds do change, and they change very fast, so that the descending ADS can never possess the exact winds. More studies on this subject were conducted by [Kelly and Pena 2001; Brocato 2003].

Figures 3.20 and 3.21 show an example when the WindPack data were collected at the same location hourly, starting from 0600 hrs. By looking at these data, there is no doubt that atmospheric data change throughout a day drastically, so that even a 1-h difference could result in a completely different wind profile (Fig. 3.21). Also shown in Fig. 3.20 for air temperature, pressure and density are parameters of the International Standard Atmosphere (ISA) often

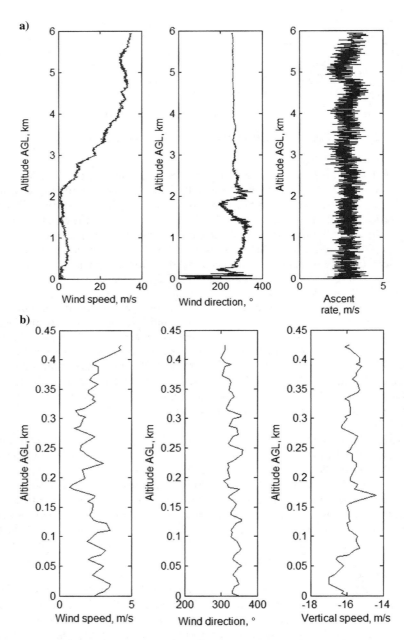

Fig. 3.18   Examples of the wind profiles collected by a) TASK balloon and b) MAXM dropsonde.

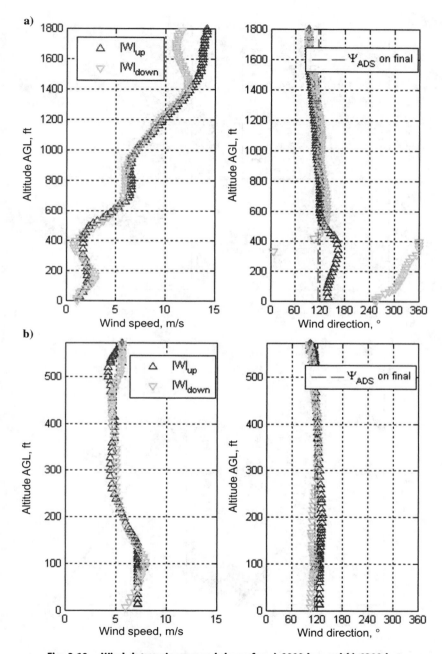

Fig. 3.19   Wind data going up and down for a) 0900 hrs and b) 1200 hrs.

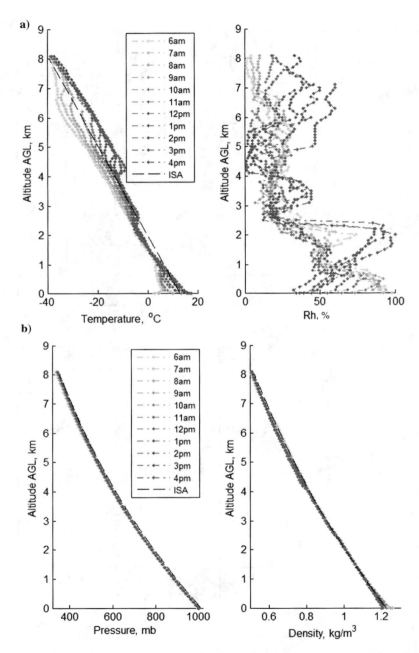

Fig. 3.20   Sounding balloon data taken hourly at the same location throughout a day.

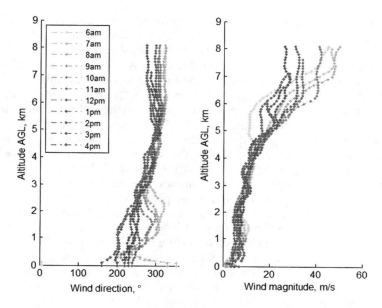

Fig. 3.21 Wind data taken by the same sounding balloon as in Fig. 3.20.

used in simulation. As seen, actual atmospheric data might differ quite a bit from that of ISA.

Usually, some atmospheric data are uploaded into PADS AGU way before the drop during a mission planning step, but with introduction of JPADS Mission Planner (discussed in Sec. 3.3.1) it can be done immediately prior a drop. All available most recent atmospheric data are then used to come up with a short-term prognosis. This forecast is then used primarily to compute a release point, so that ADS reaches the intended impact area. Depending on a GNC paradigm (discussed in Chapter 7), it can also be used to choose/correct a guidance (especially terminal guidance) strategy.

If PADS GNC algorithms allow accommodating surface-wind data, then it might be worth providing the descending system with the most current winds update rather than rely on a forecast. This idea was explored in experiments with Snowflake PADS.

The simplest solution was to set up a miniature portable weather station like the one shown in Fig. 3.22a. In this particular case, it was based on the Kestrel 4500 pocket weather tracker that measures wind speed, wind direction (when mounted on a vane as shown in Fig. 3.22b), atmospheric pressure, air temperature, and relative humidity with ± 0.1 m/s (0.2 kt), ± 5 deg, ± 1 mb, $\pm 0.5°C$, $\pm 3\%$ accuracy, respectively. This allows the computing of current height above MSL with a 7.2-m accuracy. Note that in this case wind speed and direction are measured explicitly rather than via GPS data. Bluetooth interface allows communicating wirelessly and transmitting ground atmospheric data at the IPI to the descending PADS in real time. In Fig. 3.22a, Kestrel tracker is paired with a Blackberry cellphone, which then broadcasts winds and barometric altimeter setting data using GSM network. (In this case PADS avionics also includes a cell phone talking to its autopilot.) If GSM network is not available, a common RF-based station is used (Fig. 3.22b). In Fig. 3.22a, this weather station is shown on a tripod so that in this case it measures headwind/tailwind in the predetermined (desired) landing direction. If mounted on a vane (Fig. 3.22b), it broadcasts a current ground wind direction as well, allowing PADS landing exactly into the wind. Introduction of this weather station alone allowed the drastically decreasing of the Snowflake's CEP and eventually avoiding situations like those discussed in Sec. 3.1.1 when the winds estimated before the final base turn to the IPI are way too different as compared to the current ground winds.

Another, even more advanced, solution explored in the Snowflake PADS tests was to drop a MAXMS dropsonde right before the drop, so that not only the ground winds but the complete wind vs altitude and density vs altitude profiles are uplinked to PADS upon MAXMS touchdown. Figure 3.23a shows how the dropsonde was accommodated under one wing of an unmanned aerial vehicle (UAV) while the actual PADS under another. For operational use on a fielded Tier I (Silver Fox)/Tier II (Predator, Reaper) UAVs, QNA developed dispensers requiring very little direct integration with UAVs (Fig. 3.23b). The Tier I dispenser

Fig. 3.22 Miniature weather station measuring ground winds and uplinking them either via a) GSM network or b) RF-based station.

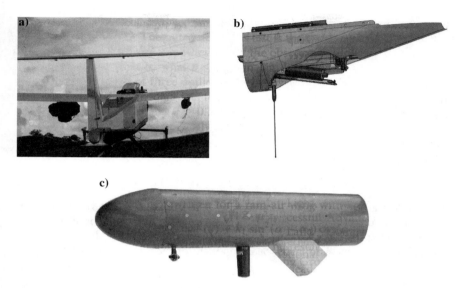

Fig. 3.23   Examples of accommodating MAXMS dropsonde on a UAV: a) attached instead of an external pod and b) fielded Tier I/Tier II dispensers [MAX Sensors 2014].

system is packaged as a modular payload that can be integrated into the payload cavity of a UAV. The Tier II dispenser system is packaged as an add-on standalone pod that is typically mounted on one of the available external attachment points under the wing. When needed, MAXMS dropsondes are deployed using a nonexplosive springloaded expulsion device for Tier I UAVs or an explosive squib in the case of Tier II UAVs.

The question of how the winds provided by either balloon or dropsonde can possibly be used by AGU algorithms is discussed next.

3.1.3   BALLISTIC WINDS

To compute an ADS trajectory, subject to winds $w_x(h)$, $w_y(h)$, and $w_z(h)$, descending with the vertical component of an airspeed vector $V_v(h)$, we can write

$$\dot{x} = w_x(h)$$
$$\dot{y} = w_y(h) \tag{3.1}$$
$$\dot{h} = -[V_v(h) + w_z(h)]$$

We can rewrite the third expression in Eq. (3.1) as

$$\mathrm{d}t = -[V_v(h) + w_z(h)]^{-1}\mathrm{d}h \tag{3.2}$$

and substitute it into the first two equations of Eq. (3.1), which yields

$$dx = -[V_v(h) + w_z(h)]^{-1} w_x(h)\, dh$$
$$dy = -[V_v(h) + w_z(h)]^{-1} w_y(h)\, dh \tag{3.3}$$

In Eqs. (3.1–3.3), the horizontal components of the wind $w_x(h)$, $w_y(h)$ are assumed to be provided by a weather balloon or dropsonde (Fig. 3.16). To be more precise, most often (see Figs. 3.13–3.15, 3.17) these data are provided as triplets h_k, W_k, and $\chi_{W;k}$, where $k = 1, \ldots, M$, which calls for a conversion to triplets h_k, $w_{x;k}$, and $w_{y;k}$

$$w_{x;k} = W_k \cos(\chi_{W;k}) \quad \text{and} \quad w_{y;k} = W_k \sin(\chi_{W;k}) \tag{3.4}$$

The vertical component $w_z(h)$ might also be available (Fig. 3.16). Quite often however, the unknown overall effect of the vertical winds throughout the entire descent is chosen to be neglected. Hence, we can assume that the descent rate V_d is about the same as the vertical component of the airspeed vector V_v

$$V_d(h) = V_v(h) + w_z(h) \approx V_v(h) \tag{3.5}$$

The vertical component of an airspeed vector V_v varies with altitude, and this issue will be explored in the next section, but in this section, for the sake of deriving simple analytical equations to allow the quick estimation of the overall effect of the winds aloft onto the horizontal shift of the landing location, we will use some average constant value of the descent rate V_d^*.

For operational use, to avoid integrating Eq. (3.3), the wind profiles measured by weather balloons or dropsondes can be reduced to the so-called ballistic winds. The idea is that if at some altitude H we have a ballistic wind of magnitude $W^{bal}(H)$ and direction $\chi_W^{bal}(H)$, then the effect of variable winds $w_x(h)$ and $w_y(h)$ acting on ADS descending with the descent rate V_d^* on its way down from altitude H to the surface can be reduced to simple formulas:

$$x(H) = \frac{H}{V_d^*} W^{bal}(H) \cos\left[\chi_W^{bal}(H)\right], \quad y(h) = \frac{H}{V_d^*} W^{bal}(H) \sin\left[\chi_W^{bal}(H)\right] \tag{3.6}$$

In other words,

$$\int_0^H \frac{1}{V_d^*} w_x(h)\, dh = \frac{H}{V_d^*} W^{bal}(H) \cos\left[\chi_W^{bal}(H)\right]$$

$$\int_0^H \frac{1}{V_d^*} w_y(h)\, dh = \frac{H}{V_d^*} W^{bal}(H) \sin\left[\chi_W^{bal}(H)\right] \tag{3.7}$$

Note, hereinafter the minus sign in the left-hand side of Eq. (3.7) [compare to Eq. (3.3)] disappeared due to inverse order of the limits of integration.

Substituting definite integrals in Eq. (3.7) with the finite sum of trapezoids based on the discrete values of h_k, $w_{x;k}$, and $w_{y;k}$, where $k = 1, \ldots, M$, we get

$$\sum_{k=2}^{M} (h_k - h_{k-1}) \frac{w_{x;k} + w_{x,k-1}}{2} = h_M W_M(h_M) \cos[\chi_W^{bal}(h_M)]$$

$$\sum_{k=2}^{M} (h_k - h_{k-1}) \frac{w_{y;k} + w_{y,k-1}}{2} = h_M W_M(h_M) \sin[\chi_W^{bal}(h_M)]$$

(3.8)

The index starts from 2 because by definition the winds' measurements at the lowest altitude are considered ballistic winds at this altitude.

From Eqs. (3.8), it further follows that

$$\chi_W^{bal}(h_M) = \tan^{-1} \frac{\sum_{k=2}^{M} (h_k - h_{k-1})(w_{y;k} + w_{y;k-1})}{\sum_{k=2}^{M} (h_k - h_{k-1})(w_{x;k} + w_{x;k-1})}$$

$$W^{bal}(h_M) = \frac{1}{2h_M} \sqrt{\left[\sum_{k=2}^{M} (h_k - h_{k-1})(w_{xk} + w_{x,k-1})\right]^2 + \left[\sum_{k=2}^{M} (h_k - h_{k-1})(w_{yk} + w_{y,k-1})\right]^2}$$

(3.9)

For the specific case when $h_k - h_{k-1} = \Delta h = \text{const}$, where $k = 2, \ldots, M$, Eqs. (3.9) can be further reduced to

$$\chi_W^{bal}(h_M) = \tan^{-1} \frac{\sum_{k=2}^{M} (w_{yk} + w_{y,k-1})}{\sum_{k=2}^{M} (w_{xk} + w_{x,k-1})}$$

(3.10)

$$W^{bal}(h_M) = \frac{\Delta h}{2h_M} \sqrt{\left[\sum_{k=2}^{M} (w_{x;k} + w_{x;k-1})\right]^2 + \left[\sum_{k=2}^{M} (w_{y;k} + w_{y;k-1})\right]^2}$$

Consider a simple example. Suppose that the winds aloft were measured as presented in Table 3.3.

Then, converting these winds to the ballistic winds [using Eq. (3.10)] yields a profile given in Table 3.4.

Now, if, for instance, we want to find a horizontal shift (due to winds) of a point of impact of an uncontrolled ADS with $V_d^* = 6\,\text{m/s}$ (20 ft/s) if it were deployed at 914 m (3,000 ft) AGL, then instead of integrating Eq. (3.3) with the

TABLE 3.3   SAMPLE WINDS ALOFT DATA FILE

Altitude, ft	Direction, deg	Magnitude, m/s	Magnitude, kt
0	93	5.14	10
1,000	101	18.52	36
2,000	119	12.86	25
3,000	129	10.80	21
4,000	135	9.77	19
5,000	139	8.75	17
6,000	145	8.23	16
7,000	150	8.23	16
8,000	152	8.23	16
9,000	151	8.23	16
10,000	144	8.23	16

varying winds of Table 3.3 (which would result in a trajectory depicted in Fig. 3.24 with a solid line) we could simply use a single (third) line from a ballistic winds of Table 3.4 to accurately compute the forecast shift of a point of impact (shown in Fig. 3.24 as a solid circle).

TABLE 3.4   BALLISTIC WINDS FOR TABLE 3.3 WIND DATA

Altitude, ft	Direction, deg	Magnitude, m/s	Magnitude, kt
0	93.0	5.14	10
1,000	99.3	11.83	23
2,000	104.4	13.63	26.5
3,000	110.2	12.86	25
4,000	114.7	12.09	23.5
5,000	118.2	11.37	22.1
6,000	121.2	10.80	21
7,000	124.1	10.34	20.1
8,000	126.8	9.98	19.4
9,000	129.1	9.67	18.8
10,000	130.6	9.52	18.5

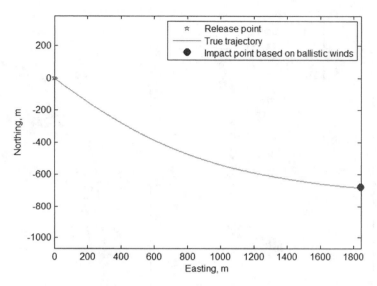

Fig. 3.24   Calculation of a predicated impact point based on ballistic winds.

Figure 3.25 shows the ballistic winds computed for the hourly wind collection presented in Fig. 3.21. By construction, they are much smoother than the original raw winds and clearly show a general tendency. Using these winds, it is quite easy to see what error might the usage of the dated winds introduce.

For weather balloons, the winds are measured upwards from the surface, so that the ballistic winds can be computed during ascent and be available immediately (as shown in the last two columns of Fig. 3.17). Figure 3.26 shows an example of the ballistic winds computed by the Snowflake PADS ground station based on the raw data provided online by QNA balloon and dropsonde, which was presented in Fig. 3.18. These ballistic winds are then uplinked and used onboard to guide Snowflake PADS to IPI. Specifically, it is the surface portion of these ballistic winds (Fig. 3.27) that is used during a final turn maneuver, which is a crucial phase of a guided descent.

Now let us get back to the question how bad the assumption $V_d(h) = V_d^* = $ const is.

3.1.4   RATE OF DESCENT AND TIME TO TOUCHDOWN

Consider a quasi-steady-state descent of ADS

$$\frac{1}{2}C_D\rho V_v^2 S \approx mg \tag{3.11}$$

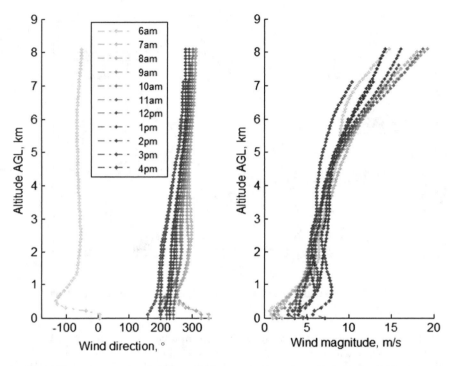

Fig. 3.25   Hourly wind data of Fig. 3.21 converted to the ballistic winds.

where C_D is the aerodynamic drag coefficient. (For gliding PADS, C_D should be replaced with the lift coefficient C_L.) For the tropospheric drops (from an aircraft), let us assume that the atmosphere holds properties of the standard atmosphere model below 11 km (36,000 ft)

$$\rho(H) = 1.225(1 - 0.00002256H[\text{m}])^{4.2559}[\text{kg/m}^3] \qquad (3.12)$$

where a geopotential altitude H is measured with respect to the mean sea level. For example, according to Eq. (3.12), at 300-m (984-ft) elevation (drop zone elevation) $\rho_{0.3} = 0.97\rho_0$, and at the release altitude of 3 km MSL $\rho_3 = 0.74\rho_0$. How does this affect the vertical component of an airspeed vector V_v?

Note that gravitational acceleration in Eq. (3.11) g decreases with altitude because moving up means moving away from the Earth's center

$$g(h) = g_0 \left(\frac{r_c}{r_c + h} \right)^2 \qquad (3.13)$$

where r_c is the effective Earth's radius at a specific latitude and g_0 is the gravitational acceleration at this latitude at the sea level. This problem of decreasing g

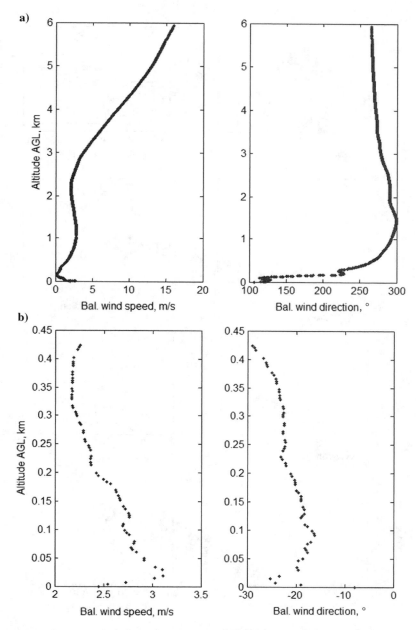

Fig. 3.26   Ballistic winds computed for the wind profiles of Figs. 3.18a and 3.18b, respectively.

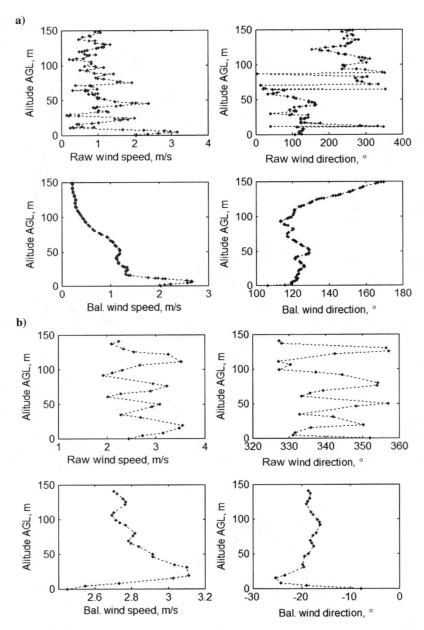

Fig. 3.27   The raw vs ballistic winds 150 m above the IPI altitude for the wind profiles of Figs. 3.18a and 3.18b, respectively.

is dealt with by defining a transformation from a real geometric altitude h to an abstraction called "geopotential altitude" H, used in Eq. (3.12)

$$H = \frac{r_c h}{r_c + h} \qquad (3.14)$$

The idea is that the amount of work done lifting a test mass m to height h through an atmosphere where gravity decreases with altitude is the same as the amount of work done lifting that same mass to a height H through an atmosphere where g remains equal to g_0. That is why the hydrostatic equation [that Eq. (3.12) was derived from] uses this geopotential altitude H instead of geometric altitude h. For our application however, the difference between these two is very small. To demonstrate this, Fig. 3.28 shows the difference $\Delta h = h - H$ computed for some mean Earth's radius defined by the International Union of Geodesy and Geophysics (IUGG) via equatorial radius r_e and polar radius r_p as $R_1 = \frac{1}{3}(2r_e + r_p) = 6{,}371$ km. Surely, for the typical ADS drops from below 6.1 km (20,000 ft), this difference is quite small, and so hereinafter we will keep using a geometric altitude.

Suppose we know that with a nominal mass m the vertical speed of some ADS at the sea level ($H = 0$) is V_{v0}. Then, rewriting Eq. (3.11) as

$$\rho V_v^2 = \frac{2\,mg}{C_D S} = \text{const} \qquad (3.15)$$

yields

$$\rho_0 V_{v0}^2 = \rho_H V_v^2 \qquad (3.16)$$

or

$$V_v = \sqrt{\frac{\rho_0}{\rho_H}} V_{v0} \qquad (3.17)$$

Hence, $V_{v0.3} = 1.02 V_{v0}$, and at the release altitude of 3 km MSL $V_{v3} = 1.16 V_{v0}$. What about the time of descent from the altitude of canopy deployment H_d to $H = 0$ (both altitudes are with respect to MSL)? Integrating Eq. (3.2) and

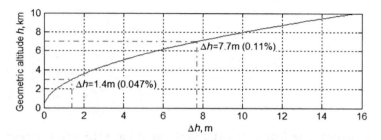

Fig. 3.28   Difference between the geometric and geopotential altitudes.

accounting for Eq. (3.17) (and also assuming the vertical component of the wind to be zero), we arrive at

$$T = \frac{1}{V_{v0}} \int_0^{H_d} \sqrt{\frac{\rho_H}{\rho_0}} \, dh \qquad (3.18)$$

For the low-altitude drops [assuming a standard atmosphere described by Eq. (3.12)], Eq. (3.18) reduces to

$$T = \frac{1}{V_{v0}} \int_0^{H_d} (1 - 0.00002256h)^{2.128} \, dh$$

$$= -14{,}171 \frac{1}{V_{v0}} \left[(1 - 0.00002256 H_d)^{3.128} - 1 \right] [s] \qquad (3.19)$$

The time to descend to some nonzero altitude H_{DZ} (also in MSL) could then be computed as

$$T = -14{,}171 \frac{1}{V_{v0}} \left[(1 - 0.00002256 H_d)^{3.128} - (1 - 0.00002256 H_{DZ})^{3.128} \right] [s]$$

$$(3.20)$$

Figure 3.29 represents Eq. (3.19) graphically for a nominal descent rate of 1 m/s (3.3 ft/s). As seen, for the low-altitude drops, this dependence happens to be very close to linear. Also shown in this figure is the descent time based on a simple linear model

$$T = \frac{H_d}{V_{v0}} \qquad (3.21)$$

For Eq. (3.20), it would be

$$T = \frac{H_d - H_{DZ}}{V_{v0}} \qquad (3.22)$$

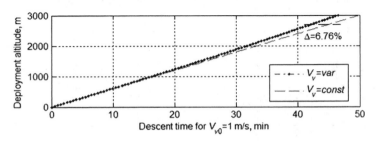

Fig. 3.29 Descent time T for a nominal descent rate of 1 m/s (3.3 ft/s) vs a deployment altitude (in the troposphere).

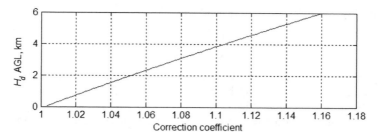

Fig. 3.30   Correction coefficient for calculating a descent time *T*.

Obviously, for these altitudes the difference between estimates provided by Eqs. (3.19) and (3.21) might be neglected. For instance, for our example of a canopy deployment at 3 km and drop zone altitude of 300 m (984 ft), the difference between the models described by Eqs. (3.19) and (3.21) is only about 6.5%. [Because of linearity, this value was taken at $3{,}000 - 300 = 2{,}700$ m (8,860 ft) AGL altitude.] The truth of matter is that the error in knowing V_{v0} (and C_D) might be much bigger. Besides, there will always be some contribution from the vertical winds. Hence, the assumption made in Sec. 3.1.3 that for the sake of computing the ballistic winds for the relatively low-altitude drops we can assume $V_v(h) = V_v^* = $ const does not result in large errors.

Therefore, when using the ballistic winds to estimate the horizontal shift of ADS landing location due to the winds, the value of

$$V_v^* = V_{vDZ} = \sqrt{\frac{\rho_0}{\rho_{DZ}}} V_{v0} \tag{3.23}$$

could be used. To account for increase of V_v with altitude, one can choose to correct it as

$$V_v^* = \kappa V_{vDZ} \tag{3.24}$$

where the coefficient κ depends on the canopy deployment altitude H_d and can be computed by equating Eq. (3.19) to Eq. (3.21) with κV_{v0} instead of V_{v0}. The dependence $\kappa = \kappa(H_d)$ is shown in Fig. 3.30. [Accounting for H_{DZ}, that is, doing the same procedure with Eqs. (3.20) and (3.22), introduces a very small change that can be neglected.]

Rewriting Eq. (3.11) as

$$V_v = \sqrt{\frac{2\,mg}{C_D \rho S}} = \rho^{-0.5} \text{ const} \tag{3.25}$$

and taking its variation yields

$$\delta V_v = -0.5 \delta \rho \rho^{-1.5} \text{ const} \tag{3.26}$$

Dividing Eq. (3.26) by Eq. (3.25) results in

$$\frac{\delta V_v}{V_v} = -0.5 \frac{\delta \rho}{\rho} \qquad (3.27)$$

Therefore, a 10% decrease in air density leads to a 5% increase of the descent speed. The same observations hold for the drag coefficient C_D and reference area S (the property to be utilized in Sec. 9.2).

Performing similar analysis of the effect of the varying mass leads to the following:

$$V_d = m^{0.5} \text{ const} \qquad (3.28)$$

$$\frac{\delta V_d}{V_d} = 0.5 \frac{\delta m}{m} \qquad (3.29)$$

Hence, the 10% increase in mass leads to a 5% increase of the descent rate (and 5% decrease of the descent time T).

For the stratospheric drops (from a balloon), we should account for two additional density models. Along with Eq. (3.12), which is valid for $H \le 11 \text{ km}$ (36,000 ft), we need to add a model for the tropopause $H \in [11;20] \text{ m}$ ([36,000;65,617] ft)

$$\rho(H) = 0.364 \exp(1.7346 - 0.0001577 H[\text{m}])[\text{kg/m}^3] \qquad (3.30)$$

and the lower stratosphere $H \in [20;32] \text{m}$ ([65,617;105,000] ft)

$$\rho(H) = 0.0880345(0.9077 + 0.000004616 H[\text{m}])^{-35.163}[\text{kg/m}^3] \qquad (3.31)$$

In practice, to avoid having three piecewise approximations for the air density [Eqs. (3.12), (3.30), (3.31)], a single exponential approximation is often used in one of the forms:

$$\rho(H) = \rho_* \exp(-H/H_*) \qquad (3.32)$$

or

$$\rho(H) = 1.225 \exp(-H/H_*) \qquad (3.33)$$

Figure 3.31a shows the optimum values of parameters ρ_* and H_* of these models depending on the altitude range the regression was produced for. For several selected altitudes, these parameters are also presented in Table 3.5. Figure 3.31b shows the maximum and average errors of both models, again depending on the altitude range the regression model was produced for. Figures 3.32a and 3.32b show examples of models (3.32) and (3.33) produced for the altitude range of [0;15] and [0;25] km ([0;49,200] and [0;82,000] ft), respectively.

Figure 3.33 shows an example of using a single exponential regression like Eq. (3.32) to approximate the descent rate DR. [In view of Eq. (3.5) by DR, we will actually mean $V_v(h)$.] Because DR is proportional to $\rho(h)^{-0.5}$, we could expect the regression power multiplier to be of the order of $0.06 = 1/8.32/2$

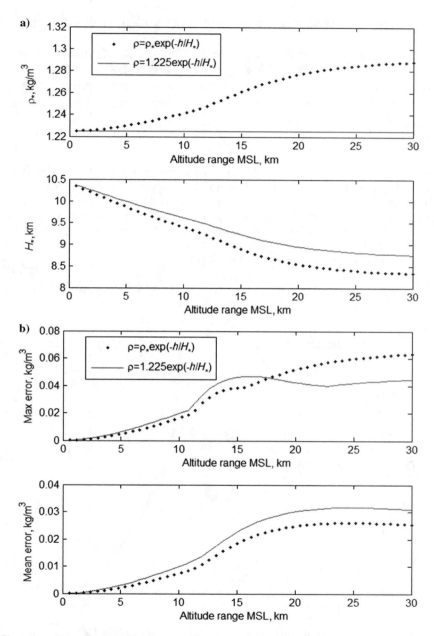

Fig. 3.31   Parameters of the a) air density regression models and b) regression errors depending on the altitude range.

TABLE 3.5   PARAMETERS OF EXPONENTIAL APPROXIMATIONS OF AIR DENSITY

		[0;10] km	[0;15] km	[0;20] km	[0;25] km	[0;30] km	[0;35] km	[0;40] km
Model (3.32)	ρ_*, kg/m^3	1.2414	1.2609	1.2770	1.2849	1.2881	1.2890	1.2889
	H_*, km	9.39869	8.88938	8.55144	8.40087	8.33973	8.31798	8.31240
Model (3.33)	H_*, km	9.62071	9.22529	8.94745	8.81391	8.75502	8.73054	8.72074

(using the value of H_* from the [0;35] km column of Table 3.5). As seen from Fig. 3.33, in this particular drop this number happened to be slightly higher than that, 0.073, which should not be a surprise; while Table 3.5 data were computed for the standard atmosphere, real atmospheric conditions were used to produce the DR regression (dashed-dotted line). As seen, this regression still matches the ISA data (solid line) fairly well.

With this model, integration of Eq. (3.2) yields

$$T = \frac{1}{V_{v0}} \int_0^{H_d} e^{-0.000073h}\, \mathrm{d}h = -13{,}700\,\frac{1}{V_{v0}}\left(e^{-0.000073 H_d} - 1\right) \tag{3.34}$$

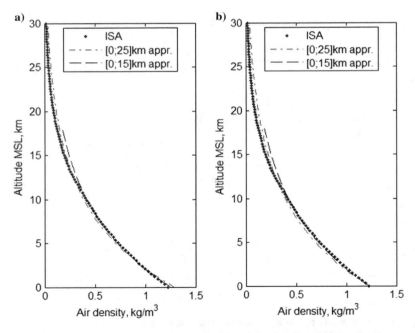

Fig. 3.32   Air density models described by a) Eq. (3.32) and b) Eq. (3.33).

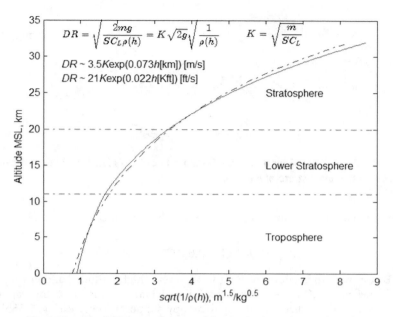

Fig. 3.33 **Standard atmosphere model vs a single exponential model.**

If we want to account for a DZ altitude, Eq. (3.34) changes to

$$T = -13,700 \frac{1}{V_{v0}} \left(e^{-0.000073 H_d} - e^{-0.000073 H_{DZ}} \right) \qquad (3.35)$$

Figure 3.34 represents Eq. (3.34) graphically for a nominal descent speed of 1 m/s similar to that of Fig. 3.29. Clearly, the linear model of Eq. (3.21) for the stratosphere drops yields a huge error.

So far, we were assuming standard atmosphere parameters, as if the actual density data were not available. If, however, these data are available (Fig. 3.15), then integration of Eqs. (3.2) and (3.3) should involve accounting for a variable speed of descent

$$V_{v;k} = \sqrt{\frac{\rho_0}{\rho_k}} V_{v0} \qquad (3.36)$$

Actual air density ρ_k can also be calculated from local atmospheric conditions, temperature T_k, humidity Rh_k, and barometric pressure P_k (as available in Figs. 3.13–3.15)

$$\rho_k [\text{kg/m}^3] = \frac{P_k [\text{Pa}]}{287.05 T_k [\text{K}]} \left(1 - \frac{0.378 P_{v;k} [\text{Pa}]}{P_k [\text{Pa}]} \right) \qquad (3.37)$$

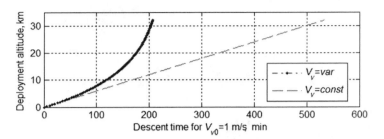

Fig. 3.34   Descent time T for a nominal descent rate of 1 m/s (3.28 ft/s) vs a deployment altitude (in the lower stratosphere).

where the partial pressure of water vapor can be computed as

$$P_{v;k}[\text{Pa}] = Rh_k[\%] \times 0.061078 \times 10^{\frac{7.5T_k[°C]}{237.3+T_k[°C]}} \tag{3.38}$$

To estimate the validity of Eq. (3.34), several high-altitude balloon drops of the Snowflake PADS were conducted (Fig. 3.35a). To maintain the canopy's shape, Snowflake utilized a semirigid canopy support structure (Fig. 3.35b), so that the ram-air parafoil was inflated right after cutdown from the balloon at 50,000 ft (15,240 m) altitude (Fig. 3.36).

Figure 3.37 demonstrates a typical altitude profile as recorded by two units of the Automatic Packet Reporting System (APRS) residing within PADS payload

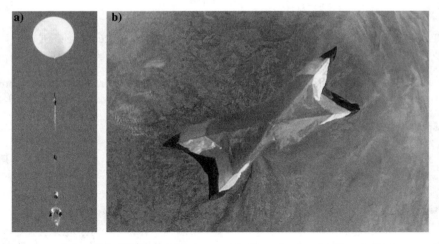

Fig. 3.35   a) Snowflake PADS carried away on a balloon shortly after launch; b) downward view from the fish-eye camera from another payload above Snowflake.

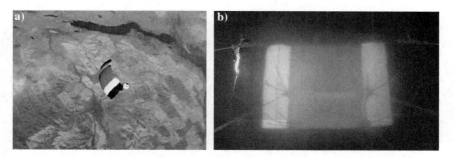

Fig. 3.36   a) PADS cutdown and b) looking-up camera snapshot upon successful canopy inflation at 50,000-ft AGL altitude.

and the rest of the balloon train. After releasing Snowflake, the balloon continued to ascend until it burst. The remains of the system (with other balloon train payloads) were being carried down by a circular uncontrolled parachute. Figure 3.38 features time history of the vertical speed of Snowflake corresponding to data of Fig. 3.37. Figures 3.39 and 3.40a represent the ascent/descent profiles for the balloon and Snowflake for the same drop.

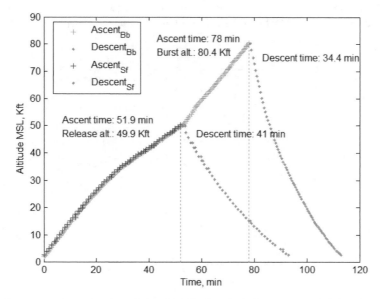

Fig. 3.37   Time histories of the altitudes for the balloon and Snowflake PADS.

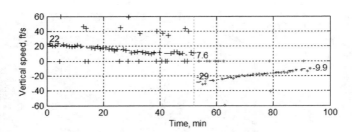

Fig. 3.38   Time histories of the vertical speed of Snowflake PADS.

Figure 3.40b shows an example of GPS data for Snowflake ADS for another drop. As seen, the regression curves for a descent rate in Fig. 3.40 coincide with that of Fig. 3.33, and therefore the actual descent time proved to be very close to that predicted by Eq. (3.34).

3.1.5   SURFACE-LAYER WINDS MODELING

Modeling of the surface-layer winds is important both for simulations and for a real-time implementation of GNC algorithms within PADS AGU. With no knowledge of the surface-layer winds, the assumption made at the terminal guidance phase (final approach) explicitly affects PADS landing precision.

Modelling the near-surface winds within AFU software can employ different approaches. Figure 3.41 shows a variety of models tested within the Snowflake

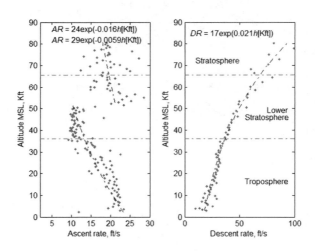

Fig. 3.39   Vertical speed vs altitude for a balloon.

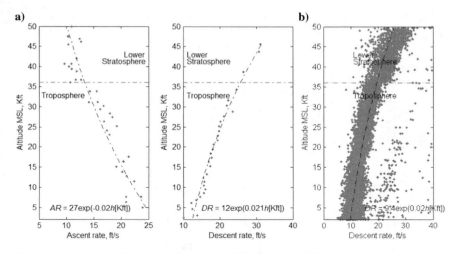

Fig. 3.40   Examples of the vertical speed vs altitude for Snowflake as recorded by a) APRS and b) IMU.

PADS development program. (The last available estimate of wind magnitude at a certain altitude before the final maneuver is of course varied, so that actual values in Fig. 3.41 are given just to understand the order of magnitude of these parameters.) With absolutely no information about the surface winds, the linear model

$$W(h) = ah + b \qquad (3.39)$$

would have $b = 0$; otherwise, it might use whatever measurement is available at the ground. The logarithmic model

$$W(h) = a\ln(h) + b \qquad (3.40)$$

is supposedly the most accurate model theoretically, but in practice it might not provide a good match with the actual winds at all. Section 7.3.4 addresses these models that can be used within PADS AGU in more details, but it is clear that whatever model is used it does not preclude from having totally unexpected situations like discussed for the two air drops in Sec. 3.1.1.

For the modeling purposes, a surface-layer wind modeling is important from the standpoint of obtaining more or less realistic results. If the wind model is too simple and does not allow realistic variations the real PADS face, the Monte Carlo simulation can result in too optimistic CEP values.

Ward et al. [2010a] studied the sensitivity of a simulated PADS landing precision to the assumed wind model considering three cases of

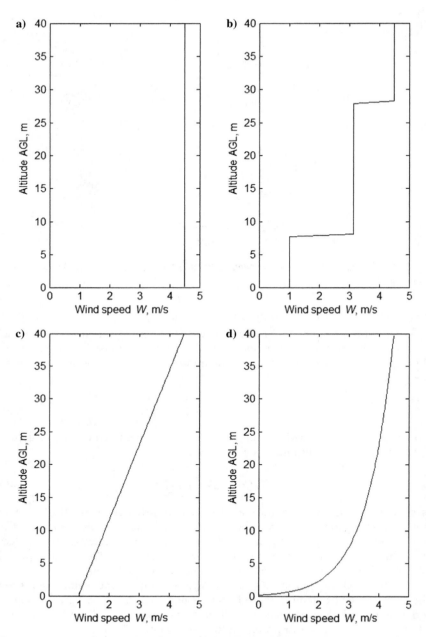

Fig. 3.41   Near-surface wind models: a) constant, b) piecewise constant, c) linear, and d) logarithmic.

increasing complexity. The first case assumed only a horizontal wind of a constant magnitude and direction with Gaussian white noise added to simulate turbulence. The second case used still constant horizontal wind components but with turbulence in all three axes added according to the Dryden turbulence spectrum. The third case used a three-dimensional, time-varying model of the atmosphere around the drop zone, supplemented with Dryden turbulence. The most complicated, third case relied on the Weather Research and Forecasting Model (WRF) [WRF Model 2014], developed by the NCAR, the National Oceanic and Atmospheric Administration (NOAA), the National Centers for Environmental Prediction and the Forecast Systems Laboratory (FSL), the Air Force Weather Agency (AFWA), the Naval Research Laboratory, the University of Oklahoma, and the Federal Aviation Administration (FAA).

Figure 3.42 presents a snapshot of a four-dimensional simulation grid spanning 1 km in each direction with 25-m (82-ft) resolution. This simulation is capable of generating realistic wind fields incorporating effects such as gravity waves, thermals, and wind shear and uses fully compressible nonhydrostatic equations to propagate the wind field forward in time. It requires some wind profile to begin with, and then it produces 30 min worth of simulated data to allow large-scale atmospheric features to develop. After this initialization period, some additional data for, say a 10-min time span, can be generated to be used in the airdrop simulation. This WRF model outputs data at a user-defined grid size at sampling intervals as low as 1 s.   •

An example of wind profiles utilized in simulations generated using three different approaches just mentioned is shown in Fig. 3.43. (In this figure the w_x component has the largest magnitude, and the WRF winds exhibit a substantial updraft.) Clearly, the usage of these winds should result in different landing performance. Figure 3.44 proves it clearly indicating that the usage of the most complex four-dimensional winds model results in poorer, yet more realistic simulated PADS performance and might catch some of the phenomena addressed in Sec. 3.1.1. In the particular set of simulations shown in Fig. 3.44, it is not the spread by itself but rather the increase of the touchdown points spread that is important. To this end, the CEP for two more complicated cases is related to the simplest case as 1.2:1 and 1.7:1, respectively.

3.2 INFLUENCE OF CONTROL RATE AND ACTUATOR DYNAMICS

Now that the effect of the winds and nonstandard atmosphere on the landing precision are understood, this section aims at looking at the influence of other factors. The analysis is based on computer simulations for generic PADS featuring some fixed guidance strategy (slightly different from that presented in Sec. 3.1.1 but also based on tracking some predetermined landing pattern). It also shows that implementing a control in a vertical channel could mitigate the effect of various contributors to the touchdown error all together.

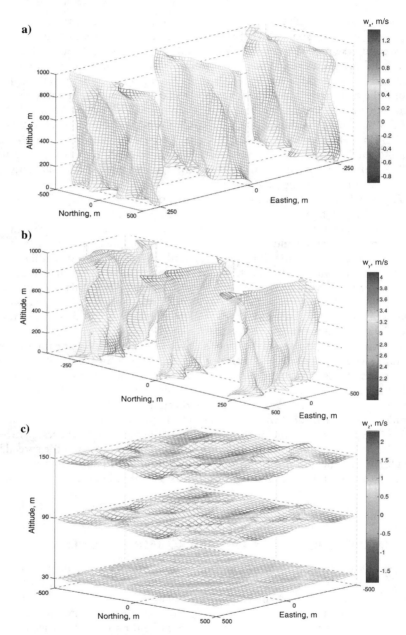

Fig. 3.42   The *x*, *y*, and *z* components of the modeled surface-layer winds [Ward et al. 2010a].

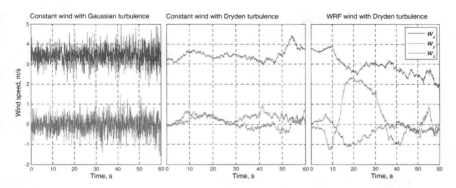

Fig. 3.43   Comparison of wind models used in computer simulations [Ward et al. 2010a].

3.2.1   SIMULATION SETUP AND REFERENCE SIMULATIONS

Let us consider some generic GNC strategy combining a parachutist-like s-turn loitering with a base turn (turnaround) maneuver transforming into a final glide against the wind towards the target as shown in Fig. 3.45.

All simulations use a fixed wind direction blowing from the South with a constant or variable magnitude $w_x(h)$. Starting from some initial point, PADS

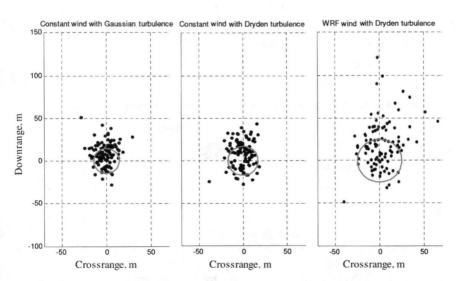

Fig. 3.44   Landing precision for Monte Carlo simulations utilizing different wind models [Ward et al. 2010a].

**Fig. 3.45   Illustration of guidance strategy
adopted in computer simulations.**

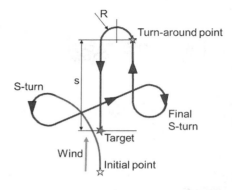

performs an s-turn maneuver to dissipate an excess of potential energy following by the downwind leg towards
a turnaround initiation point as
shown in Fig. 3.45. The decision to
abort the series of s-turns is based on
comparing current glide capability
hGR against the needed glide distance
regarding transfer flight to the other
side and a full landing maneuver.

As was the case in the algorithm presented in Sec. 3.1.1, the decision to initiate
the turnaround maneuver is based on the estimates of current altitude above IPI h,
descent rate V_d, constant prevailing winds w_x (in light of Sec. 3.1.3, this estimate
can be considered as a projection of ballistic wind onto the landing direction), turn
radius R, and PADS glide ratio GR and is triggered when

$$h\text{GR} < \pi R + s + w_x \frac{h}{V_d} \qquad (3.41)$$

Simulation involves a three-DoF model of PADS and in the ideal case assumes
the following:

- Wind profile with constant speed and direction
- No crosswinds
- Constant air density
- Control commands rate set equal to overall simulation rate of 0.1 s
- Infinite actuator speed/immediate dynamic system response
- No bias in controls
- No glide path control

Two simulations using the preceding ideal conditions with a constant wind
speed of $w_x = 0$ and $w_x = 5$ m/s are shown in Fig. 3.46. The parameters entering
Eq. (3.41) were GR = 3.33 (3.33:1), $R = 57$ m (187 ft), and $V_d = 4.5$ m/s (15 ft/
s). [The last two values correspond to the sea level with $\rho = 1.225$ kg/m^3
(0.0765 lb/ft^3).] The initial starting point was set at $x = -250$ m (820 ft), $y = 0$,
and $h = 280$ m (917 ft). The miss distances for these two drops were identified
as 8 and 16 m (26 and 52 ft), respectively. Because the implemented control functionality cannot support geometrically perfect circle trajectories, a nonzero precision error even in zero wind will prevail causing PADS to land slightly too

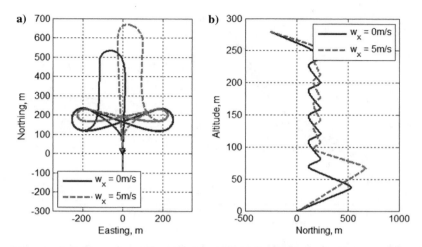

Fig. 3.46 a) Bird's-eye and b) vertical views of two "ideal simulations" with $w_x = 0$ and $w_x = 5$ m/s.

short. As the wind increases, these characteristics become even more effective (Fig. 3.46a).

For the purpose of this study, this increase of a touchdown error with the stronger winds does not really matter as long as the GNC algorithm remains the same because we want to explore an effect of other factors, not the winds themselves we know about from Sec. 1.1 already. To establish some nominal performance, a broader set of Monte Carlo simulations involving 100 different winds rather than just the two already considered was conducted. The result in terms of landing precision is presented in Fig. 3.47. The spread of errors for about the same winds speed is caused by variations in the initial starting point ($x \in [-250; -500]$ m ($[-820; -1,640]$ ft)) and the 0.1-s GNC update rate producing random delays with respect to turnaround initiations.

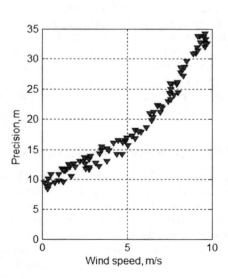

Fig. 3.47 Monte Carlo simulation in the ideal case varying a constant value of wind w_x.

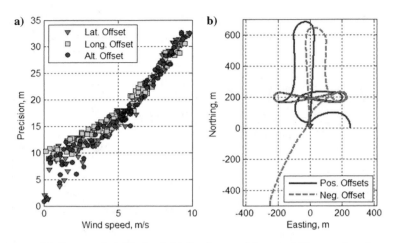

Fig. 3.48 **a) Influence of simulation initialization conditions and b) examples of two trajectories with extreme positive and negative offsets.**

Now that we have some nominal performance presented in Fig. 3.47, we can start exploring the effect of other factors. However, before we do that, let us exclude the influence of variation of the initial conditions for simulation. As mentioned, for a setup shown in Fig. 3.46a, a position at $h = 280$ m (919 ft) AGL, 250 m South of IPI with northern heading, was chosen as the nominal starting position. Figure 3.48a presents the results of Monte Carlo simulations when this initial position was varied within $250 \times 250 \times 250$ m ($820 \times 820 \times 820$ ft) cube centered around the nominal starting position. The trajectories for two cases with all offsets being set to positive and negative values are illustrated in Fig. 3.48b. As seen, variations of starting position do not have a significant effect on the touchdown performance as compared to the nominal case of Fig. 3.48 meaning that the GNC algorithm used in simulation handles these variations relatively well (compared to those in winds). Therefore, all further simulations will utilize the nominal starting position, but dropping one by one the assumptions formulated for the ideal case just discussed.

3.2.2 EFFECT OF MODELING REALISTIC CONTROLLER, ACTUATOR, AND ATMOSPHERE

First, let us consider the influence of a discrete controller. According to the GNC strategy adopted in these simulations, the guidance functionality basically compares the actual position against a reference position and commands the actuators accordingly to adjust PADS path. For example, during the final approach PADS steers towards the IP. In practice, control command rate is tied to GPS sensor update rate, which is usually set to 1 Hz. Figure 3.49a presents the Monte Carlo simulation with everything set as in the ideal case but with a discrete controls rate set to 1 Hz (zero-order hold with a sample time of 1 s). As seen, this leads to a significant degradation of landing precision. This degradation seems to be caused by two effects:

• Delay in initialization of a turning maneuver (by up to 1 s) so that PADS has to fly farther downwind
• Overall retardation in flight path control adding to short landings

The first effect contributes by a ±15-m (49-ft) scatter in the data, and the second one leads to a more or less constant shift in the data by approximately 35 m (115 ft) compared to the reference performance of Fig. 3.47. Figure 3.49b shows an example of a trajectory with a discrete controller.

Next, let us introduce a finite actuator speed, that is, account for controller dynamics. Usually, to pull the control lines PADS utilize electrically driven motors. However, the rotational speed of the motors is limited, and thus the dynamic response of the system is lagged. As inertial resistance of the system will lead to a similar result, both effects can be combined and represented by limiting the actuator speed. This set of Monte Carlo simulations was based on the reference set but with actuator speed limited to 20%/s (compared to the instantaneous unlimited deflection of the reference simulation). Figure 3.50 presents the simulation results. As expected, a limited actuator speed contributes to short landings by approximately 13 m across all values of the winds.

So far, simulations did not account for the realistic atmosphere with a varying air density. Air density increases during PADS descent, and thus horizontal as well as vertical components of airspeed decrease. As explained in Sec. 3.1.4, this

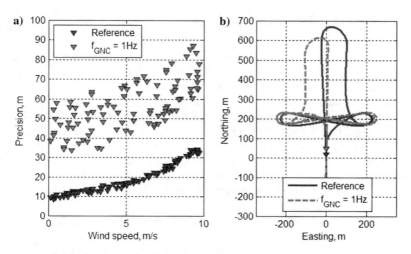

Fig. 3.49 a) Landing precision with GNC execution rate set to 1 Hz and b) an example of more realistic trajectory.

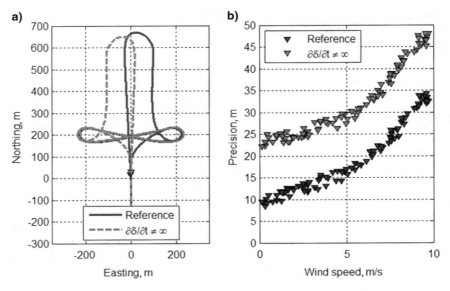

Fig. 3.50   a) An example of a trajectory and b) landing precision with a finite actuator speed of 20 %/s.

effect depends on the release altitude and can be quite significant. Because initiation of the turnaround maneuver [Eq. (3.41)] is based on some constant value of V_d, simulation of nonconstant air density leads to PADS being affected by a headwind longer than predicted by Eq. (3.41). Figure 3.51 demonstrates this effect and shows that in the current implementation the stronger winds result in larger touchdown errors.

Combination of all three real-world effects introduced in this section results in simulations characterized in Fig. 3.52. The expected landing precision degrades significantly and is on the average fivefold of that for the ideal case.

Finally, to make simulations more realistic, crosswinds were introduced. This was done by randomly rotating the originally southerly winds by 0, 15, 30, and 45 deg and adding these crosswinds contributions to the real-world simulations shown in Fig. 3.52. The results presented in Fig. 3.53 seem to have a little difference compared to those of Fig. 3.52 thus suggesting that other real-world effects simply absorb those lesser effects of crosswinds. Besides, there is one more major effect to consider, and that is varying surface-layer winds.

3.2.3   INFLUENCE OF CHANGING SURFACE-LAYER WINDS AND A WAY TO MITIGATE THEM

As discussed in Sec. 3.1.1, changing surface-layer winds might have a dramatic effect on PADS touchdown performance. The ideal case assumed a constant

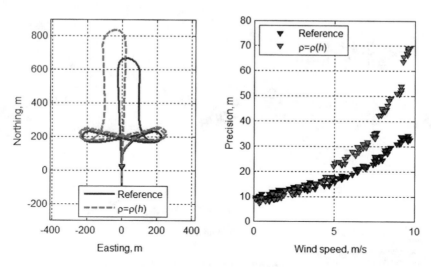

Fig. 3.51 Landing precision with influence of atmosphere.

wind profile all the way down to the ground, but, surely, it is rarely the case. The worst-case scenario is when, due to friction within the boundary layer, these winds die.

To estimate an effect of these winds usually unaccounted for, a set of Monte Carlo simulations with the winds decreasing their magnitude at a certain altitude

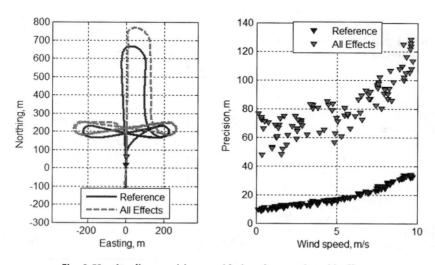

Fig. 3.52 Landing precision considering three real-world effects.

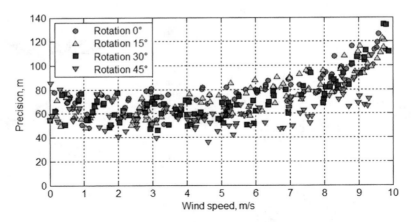

Fig. 3.53   Landing precision involving a crosswind component.

was conducted. The following four wind profiles shown in Fig. 3.54 were investigated (compare Fig. 3.41):

- Constant speed (the reference case)
- Linear decrease from a full speed to half-speed at the surface
- Linear decrease from a full speed to zero speed at the surface
- Linear decrease from a full speed to half-speed in an opposite direction

A height of 250 m (820 ft) where the nonconstant wind profile starts to deviate from its higher altitude constant value was chosen.

The reference case for these simulations was the simulation shown in Fig. 3.52, that is, when three real-world effects were included. Figure 3.55 presents the results of these realistic wind profiles simulations in terms of a touchdown error. Obviously, dying winds introduce a huge overshoot error, which is much larger than all real-world effects. Again, that seems to be a major contribution to the landing performance as presented in Sec. 1.2.

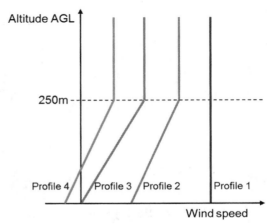

Fig. 3.54   Dying wind profiles.

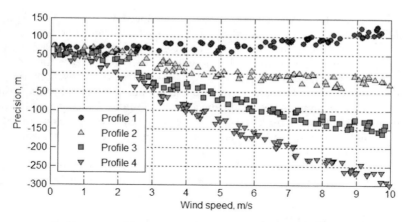

Fig. 3.55   Landing precision considering change in wind speed.

While different GNC strategies discussed in Chapter 7 aim at dealing with the effects introduced by real hardware, variable winds, and other factors, there is an interest of exploring a possibility of a solution that solves the problems all together by adding to the current PADS what human jumpers have, and that is a glide slope angle control. As discussed in Sec. 1.1, experienced human jumpers can usually land much closer to the IPI than cargo PADS, and that is what cargo PADS are supposed to be able to do.

A simple aerodynamic model to simulate application of brakes (symmetric deformation of the trailing edge) suggested by PADS behavior observed in real flight tests might be based upon

$$C_L = 0.40 + 0.50\delta_s, \quad C_D = 0.12 + 0.15\delta_s + C_{D\delta_s^2}\delta_s^2 \qquad (3.42)$$

In these equations δ is the brakes deflection, and the coefficient $C_{D\delta_s^2}$ will be a varied parameter assuming three different values 0, 0.0375, or 0.075. Depending on the value of this coefficient, the aerodynamic lift and drag, lift-to-drag ratio, and components of the airspeed can have a linear or quadratic dependency on δ_s as shown in Fig. 3.56.

Applying brakes at higher altitudes (as opposed to doing it only for a flare maneuver when close to the ground) is thought to do the following:

- Reduce a rate of descent allowing PADS to be exposed to the head surface-layer winds longer mitigating a possible overshoot
- Decrease a glide ratio resulting in a steeper glide path

The additional control authority employed in the vertical channel on a final approach (only) will then be based on trying to apply brakes in response to a difference between the actual GPS speed-based glide slope angle (given by a ratio of vertical to horizontal components of a speed vector) to the required position-based glide path angle (defined by the actual height over the ground and distance to IPI). Using a basic PID controller, any difference is fed back to modulate the brake line command. The entire idea is to increase the slope of descent on a final approach, so that PADS should be set to fly high (higher than usual) because the suggested control could not possibly be held in decreasing the glide slope angle if PADS falls below a required glide slope angle. Therefore, the original simulation setup described in the preceding chapter was modified to include some lead time with respect to turnaround maneuver initialization to avoid any short landings.

Thus, based on the aerodynamic model of Eq. (3.42) with three different values of $C_{D\delta^2}$, the Monte Carlo simulation runs were performed with following conditions:

- Control rate set at 1 Hz, limited actuator speed, and air density varying with height
- Linear wind profile decreasing from full speed at 250 m (820 ft) AGL down to half-speed at the ground
- Initialization of the turnaround maneuver with a lead time of 4 s

For the sake of comparison, no-glide slope angle control cases were included into these simulation runs as well. These results are shown in Fig. 3.57. As seen, for the chosen range of the surface-layer winds simulations without a glide slope angle control result in as much as around 90-m (295-ft) overshoots. Introduction of the glide slope angle controller allows reducing it. It is clear that even with a $C_{D\delta^2} = 0$ case when glide ratio remains constant (see Fig. 3.33) scatter can be reduced and landing performance improves by approximately 40 m (131 ft), down to 50-m (164-ft) overshoot with a wind speed of 10 m/s

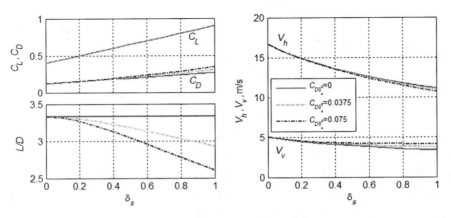

Fig. 3.56   PADS aerodynamics modeling.

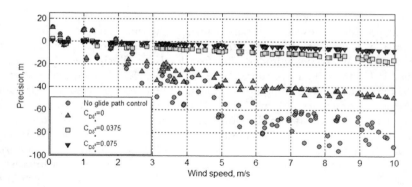

Fig. 3.57   Landing precision with a glide slope angle control.

(19.4 kt). Increasing the value of coefficient $C_{D\delta^2}$ leads to further improvement of landing performance all way down to about less than 10 m for $C_{D\delta^2} = 0.075$. Even though these simulations show a good potential of utilizing brakes to control a glide slope angle, in practice PADS do not exhibit such high values of $C_{D\delta^2}$ basically maintaining the same glide slope angle when the brakes are applied and only changing airspeed magnitude. Chapter 8 explores other ways of utilizing the same idea but using different means to change the glide-slope angle (reducing PADS lift or increasing aerodynamic drag or changing an incidence angle).

3.3   UTILIZING WINDS-ALOFT DATA

As discussed throughout this chapter, these are mainly the surface-layer winds that ultimately define a touchdown precision of aerial payload delivery. However, on a broader scale, for all nonpowered PADS, the winds aloft are also needed to be known. Without this knowledge, it would be impossible to specify an area (at a given altitude) where PADS needs to be released from an aircraft so that it reaches the designated touchdown area. This section describes two major applications intended to utilize these data. The first one is a mission planner that takes advantage of all available winds-aloft data to produce four-dimensional winds forecasts to be used to compute a desired release point. The second application, Safety Fans GUI (graphical user interface), pursues an exactly opposite goal: to predict a footprint of possible touchdown area in the case of failure of different PDS components.

3.3.1   CARP AND MISSION PLANNER

For the high-altitude drops of uncontrolled ADS, the issue of computing a desired release point is crucial; uncontrolled ADS totally rely on knowing current winds aloft. Such winds, however, are rarely available. The National Weather Service of NOAA provides winds aloft, officially known as the winds and temperatures aloft forecast, as a forecast of specific atmospheric conditions in terms of wind and temperature at certain altitudes. The first level for which a wind forecast is issued is 457 m (1,500 ft) or more above the station elevation, otherwise starting at 914 m (3,000 ft) and going up to 3,658 m (12,000 ft) with a 914-m (3,000-ft) increment [sometimes 457-m (1,500-ft) increment] and a 1,829-m (1,500-ft) increment or higher afterwards. However, weather stations are spread spatially (e.g., out of 160 stations in the continental United States, 12 are located in California, but only four in Nevada, two in Utah, etc.) and do not necessarily provide information in the specific area of interest. Moreover, this forecast is made four times a day, so it might be 6 hours old. Finally, a wind direction in this forecast is rounded to the nearest 10 deg.

That is why data from more recent and accurate dropsonde or weather balloons launched near the intended point of impact site are always preferable. However, what if dropsonde or weather balloon data are outdated? Clearly (as is shown in the example presented in Fig. 3.21), winds-aloft data age fairly fast resulting in unreliable computed air release point (CARP). Figure 3.58 shows a drift of CARP location based on Fig. 3.25 data for ADS with $V_v = 10\,\text{m/s}$ (32.8 ft/s) released at 3 km (9,843 ft) AGL. The only possible solution is to use the sophisticated meteorological models in attempt to predict

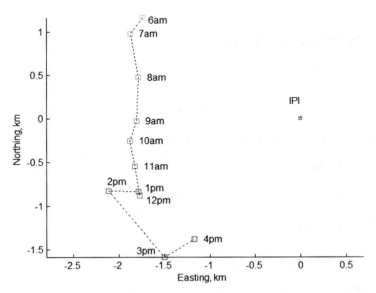

Fig. 3.58   CARP locations for Fig. 3.25 wind data based on a 5-min descent.

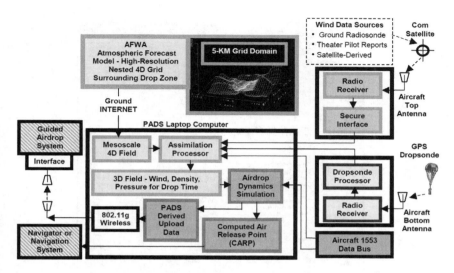

Fig. 3.59   PADS architecture and top-level functions [Wright et al. 2005a].

the change of winds aloft on a short time horizon based on all meteorological data available up to the current point in time.

This idea was realized in what now is known as a common mission planning system for precision airdrop operations JPADS Mission Planner (JPADS-MP). The basic architecture of MP is shown in Fig. 3.59. The two major components of MP software are WindPADS (stands for the Wind Profile Aerial Delivery System) and PAPS (Precision Airdrop Planning System). The top-level GUI is shown in Fig. 3.60; the GUI that displays the PAPS simulation resulting in the ballistic CARP for given mission planning and weather data is shown in Fig. 3.60b.

The top-level GUI (Fig. 3.60a) enables the operator to activate GUIs for mission planning data entry (aircraft type, drop zone, payload weight, load station, decelerator type, guided system selection, release and performance data, aircraft airdrop parameters, altimeter setting); for weather data acquisition and assimilation; for calculating the ballistic payload CARP; for calculating the guided payload CARP and allowable CARP range (earliest/latest CARPs along the run-in course); for aircraft navigation data monitoring en route to the CARP; and for upload of mission files (winds and IPI) to the AGUs of guided systems before payload release. At the top level, the operator enters the current aircraft altimeter setting. Right below it, the calculated ballistic winds at the current altitude (see Sec. 3.1.3) produced by the WindPADS component of MP show up.

The WindPADS component, developed by Planning Systems, Inc., produces a high-fidelity, high-resolution three-dimensional grid of winds, pressure, and

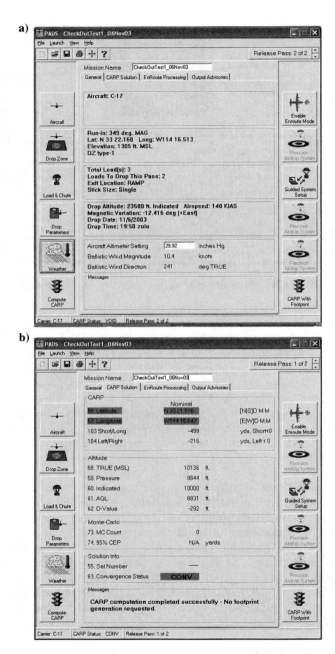

Fig. 3.60   a) Top-level GUI and b) CARP solution tab [Wright et al. 2005a].

density valid at the intended drop time. WindPADS represents an atmospheric data acquisition, assimilation, and short-range forecasting system that parallels the functions of the NOAA's national operational weather forecast centers. At these centers the Local Analysis and Prediction System (LAPS), developed by the NOAA FSL, utilizes dynamic/physical forecast models executed on supercomputer platforms to produces the forecast fields, valid at the next analysis time and used as the first estimate of the atmospheric fields at that time [Albers 1995; Snook et al. 1995; Kim et al. 2007]. Received observations of global atmospheric data valid within approximately 2 h of the analysis time are assimilated with the forecast fields, as influenced by the underlying topography, to produce the best analysis estimate. From this analysis, the dynamic/physical forecast models are numerically integrated to produce three-dimensional grid forecasts of the atmosphere in the future. Global analysis and forecast models use grids with a horizontal resolution on the order of 90 km (56 miles). Because of the large computational load, even with the use of supercomputers, higher-resolution horizontal forecast grid domains are produced using a complex grid nesting numerical computation procedure, with the horizontal distance between grid points being one-third that of the parent coarser grid.

Similarly, using a lighter MP-tailored version of LAPS, MP downloads and can assimilate AFWA high-resolution forecast fields at 5-, 15-, and 45-km (3.1-, 9.3-, 28-miles) resolution, depending on availability and on the required forecast lead time. It also utilizes on-scene in situ wind and weather data observations (from hand-launched GPS dropsonde and at-altitude wind data derived by the airdrop aircraft's navigation system). The algorithm enables assimilation of observed data over a larger time of about 4 h, executing a number of assimilations with a short-range extrapolation serving as the analysis for the next extrapolation until reaching the planned time of drop. The assimilation is done using a topographic database with 1-km resolution residing on a MP laptop. The topographic database is very important as topography drives the atmospheric flow in the lower layers of the atmosphere, especially in complex, rugged terrain.

The time difference between the observed wind data and the planned time of drop is on the order of 15 to 30 min. Surely, for a better precision this time difference must be as short as the mission timeline will allow. LAPS produces the short-range forecast valid at the planned time of drop by linear extrapolation of the rate-of-change of wind, pressure, and density (determined by the time difference between three-dimensional forecast fields), from the analysis time to the intended time of drop (Fig. 3.61).

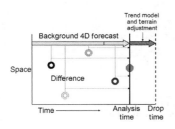

Fig. 3.61 LAPS data assimilation process [Wright et al. 2005a].

The WindPADS main GUI is shown in Fig. 3.62. As seen, it has two main panels. The Weather Acquisition panel show weather data inventory with valid times relative to (before) the planned drop. The Wind File Production panel serves to control assimilation of acquired weather data, by selecting various assimilation options. (The order of the assimilation buttons represents the order of decreasing accuracy and fidelity.)

The PAPS component of PADS, developed by Draper Laboratory, produces a high-fidelity, three-DoF simulation of all of the components of ballistic ADS from initiation of the release through touchdown. Centering this simulated touchdown point with IPI produces a desired CARP (Fig. 3.60b). For ballistic payload mission planning, the simulation is based on aircraft type, altitude above IPI, airspeed, run-in course, payload weight, decelerator type, and load

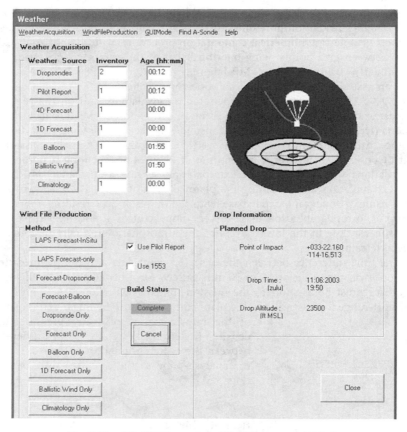

Fig. 3.62   WindPADS weather GUI [Wright et al. 2005a].

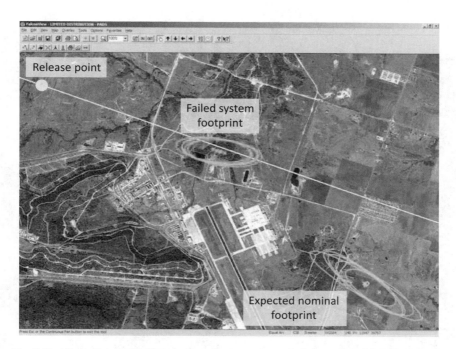

Fig. 3.63   Example of a bird's-eye view of expected nominal descent and failed canopy dispersion footprints [Campbell et al. 2005].

station. A Monte Carlo analysis tool accommodates random variations in the key model parameters such as aircraft release time, deceleration characteristics during canopy deployment, ADS drag coefficient, wind and air density, to compute expected precision footprints for a given probability. This information can be computed for both nominal descent and descent following canopy failure (Fig. 3.63). Some of the models also used in another application will be described in the next section.

For guided systems, based on their glide ratio, MP produces the reachability sets from where PADS could be deployed to reach IPI in a self-guided descent. The entire forecast wind profile can be wirelessly transmitted to PADS AGU before deployment to assist in most control-effective descent and landing into the winds (Fig. 3.64).

3.3.2   SAFETY FANS TOOL

Whereas the preceding section's concern was to use available wind vs altitude profile to compute an inverted cone with its apex centered at IPI and base defining

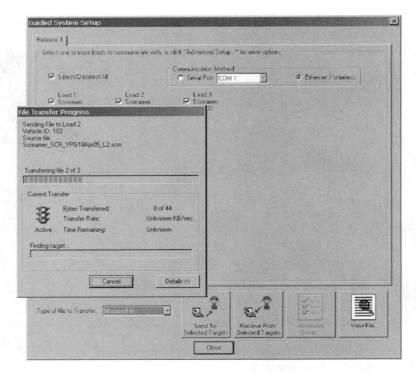

Fig. 3.64   PADS wireless communication GUI [Wright et al. 2005a].

the reachability area at a given altitude, this section addresses an opposite question of computing a cone centered at the release point and its base representing an area where PADS might land in the case of hardware or software failures/ malfunctions. For example, what would happen if a canopy does not open, or it opens but GNC algorithm steers it in one direction all the time because the IPI coordinates were entered with an error?

Let us consider a generic three-stage system. (Relations for the one-stage and two-stage systems can then be derived by eliminating extra stages.) To begin with, let us assume that upon release from an aircraft PADS goes down ballistically. The three-DoF equations driving a PADS motion in this case can be written as [compare Eq. (3.1)]

$$\dot{x} = W(h) \cos\left[\chi_W(h)\right] + V_0 \cos(\Psi_0)$$
$$\dot{y} = W(h) \sin\left[\chi_W(h)\right] + V_0 \sin(\Psi_0) \qquad (3.43)$$
$$\dot{h} = gt$$

Here, the vertical motion for the altitude h (the last equation) is driven by the gravity force g (aerodynamic drag is neglected), while the horizontal motion (coordinated x and y) is caused by the initial horizontal velocity $V_h^{a/c}$ as released from an aircraft flying $\chi_a^{a/c}$ heading and variable winds [with the magnitude $W(h)$ and direction $\chi_W(h)$]. Equations (3.43) can be rewritten as

$$x'_h \dot{h} = \frac{dx}{dh}\frac{dh}{dt} = W(h)\cos[\chi_W(h)] + V_h^{a/c}\cos(\chi_a^{a/c})$$

$$y'_h \dot{h} = \frac{dy}{dh}\frac{dh}{dt} = W(h)\sin[\chi_W(h)] + V_h^{a/c}\sin(\chi_a^{a/c}) \qquad (3.44)$$

$$\dot{h} = \frac{dh}{dt} = gt$$

where x'_h and y'_h denote derivatives with respect to altitude.
From the last equation of Eq. (3.44), it follows that

$$t = \sqrt{\frac{2}{g}(H_0 - h)} \qquad (3.45)$$

where H_0 is the release altitude. With this relationship between the elapsed time and altitude, the first two equations in Eq. (3.44) can be further rewritten as

$$\frac{dx}{dh} = \frac{W(h)\cos[\chi_W(h)] + V_h^{a/c}\cos(\chi_a^{a/c})}{\sqrt{2g(H_0 - h)}}$$

$$\frac{dy}{dh} = \frac{W(h)\sin[\chi_W(h)] + V_h^{a/c}\sin(\chi_a^{a/c})}{\sqrt{2g(H_0 - h)}} \qquad (3.46)$$

Now, the x and y coordinates at any arbitrary altitude h can be obtained via integrating Eq. (3.46):

$$x(h) = x_0 + \int_h^{H_0} \frac{W(h)\cos[\chi_W(h)] + V_h^{a/c}\cos(\chi_a^{a/c})}{\sqrt{2g(H_0 - h)}}\, dh$$

$$y(h) = y_0 + \int_h^{H_0} \frac{W(h)\sin[\chi_W(h)] + V_h^{a/c}\sin(\chi_a^{a/c})}{\sqrt{2g(H_0 - h)}}\, dh \qquad (3.47)$$

The integrals in these equations can be further split into two separate components, so that

$$x(h) = x_0 + \int_h^{H_0} \frac{W(h)\cos[\chi_W(h)]}{\sqrt{2g(H_0 - h)}}\,dh + \sqrt{2\frac{H_0 - h}{g}}V_h^{a/c}\cos(\chi_a^{a/c})$$

$$y(h) = y_0 + \int_h^{H_0} \frac{W(h)\sin[\chi_W(h)]}{\sqrt{2g(H_0 - h)}}\,dh + \sqrt{2\frac{H_0 - h}{g}}V_h^{a/c}\sin(\chi_a^{a/c})$$

(3.48)

If the parameters $W(h)$ and $\chi_W(h)$ are given as a look-up table with the elements h_k, W_k, and $\chi_{W;k}$, where $k = 1, \ldots, M$, $(h_k > h_{k-1})$, then the effect of the winds can be reduced to the following finite sums:

$$\int_h^{H_0} \frac{W(h)\cos[\chi_W(h)]}{\sqrt{2g(H_0 - h)}}\,dh \approx \sum_{k=m}^{M-1} \frac{W_{k+1}\cos(\chi_{W;k+1})(h_{k+1} - h_k)}{\sqrt{2g(H_0 - h_k)}}$$

$$\int_h^{H_0} \frac{W(h)\sin[\chi_W(h)]}{\sqrt{2g(H_0 - h)}}\,dh \approx \sum_{k=m}^{M-1} \frac{W_{k+1}\sin(\chi_{W;k+1})(h_{k+1} - h_k)}{\sqrt{2g(H_0 - h_k)}}$$

(3.49)

(where H_0 corresponds to h_M and h corresponds to h_m).

Applying Eq. (3.48) to the range $h \in [H_0 - \Delta H; H_0]$, characterized by the altitude loss before canopy opening ΔH, yields

$$x_1 = x_0 + \int_{H_0 - \Delta H}^{H_0} \frac{W(h)\cos[\chi_W(h)]}{\sqrt{2g(H_0 - h)}}\,dh + \sqrt{2\frac{\Delta H}{g}}V_h^{a/c}\cos(\chi_a^{a/c})$$

$$y_1 = y_0 + \int_{H_0 - \Delta H}^{H_0} \frac{W(h)\sin[\chi_W(h)]}{\sqrt{2g(H_0 - h)}}\,dh + \sqrt{2\frac{\Delta H}{g}}V_h^{a/c}\sin(\chi_a^{a/c})$$

(3.50)

For the sake of computing the safety fans, we can further assume that by the time the canopy opens the initial horizontal speed $V_h^{a/c}$ is phased out completely [Jann 2011]. Thus, after canopy deployment the trajectory obeys the following set of equations:

$$\dot{x} = w(h)\cos[\chi_W(h)] + V_{v1}\mathrm{GR}_1\cos(\chi_a)$$

$$\dot{y} = w(h)\sin[\chi_W(h)] + V_{v1}\mathrm{GR}_1\sin(\chi_a)$$

$$\dot{h} = V_{v1}$$

(3.51)

Here, GR_1 is a glide ratio, V_{v1} is the descent rate, and subindex 1 indicates the first stage. (We will derive formulas for a generic three-stage system.) In the light of discussion in Sec. 3.1.4, Eq. (3.51) assumes that the descent rate V_{v1} is about the same throughout a descent, that is, it does not depend on the altitude (air density) and that PADS experiences a control failure so that GR_1 and χ_a are constant throughout the entire descent as well. Using the third equation of Eq. (3.51) as demonstrated earlier, the first two equations in Eq. (3.51) can be reduced to

$$\frac{dx}{dh} = V_{v1}^{-1} W(h) \cos[\chi_W(h)] + GR_1 \cos(\chi_a)$$

$$\frac{dy}{dh} = V_{v1}^{-1} W(h) \sin[\chi_W(h)] + GR_1 \sin(\chi_a)$$

(3.52)

Now, we can integrate Eq. (3.52) with respect to an altitude, so that

$$x(h) = x_1 + \int_h^{H_0 - \Delta H} \left\{ V_{v1}^{-1} W(h) \cos[\chi_W(h)] + GR_1 \cos(\chi_a) \right\} dh$$

$$y(h) = y_1 + \int_h^{H_0 - \Delta H} \left\{ V_{v1}^{-1} W(h) \sin[\chi_W(h)] + GR_1 \sin(\chi_a) \right\} dh$$

(3.53)

Again, the integrals can be split into two parts:

$$x(h) = x_1 + \int_h^{H_0 - \Delta H} V_{v1}^{-1} W(h) \cos[\chi_W(h)] dh + (H_0 - \Delta H - h) GR_1 \cos(\chi_a)$$

$$y(h) = y_1 + \int_h^{H_0 - \Delta H} V_{v1}^{-1} W(h) \sin[\chi_W(h)] dh + (H_0 - \Delta H - h) GR_1 \sin(\chi_a)$$

(3.54)

where the third terms represent the contribution from the stuck controls and the second terms provide the contribution of the variable winds.

By the altitude of deployment of the second stage, we will have

$$x_2 = x_1 + \int_{H_2}^{H_0 - \Delta H} V_{v1}^{-1} W(h) \cos[\chi_W(h)] dh + (H_0 - \Delta H - H_2) GR_1 \cos(\chi_a)$$

$$y_2 = y_1 + \int_{H_2}^{H_0 - \Delta H} V_{v1}^{-1} W(h) \sin[\chi_W(h)] dh + (H_0 - \Delta H - H_2) GR_1 \sin(\chi_a)$$

(3.55)

If stage 2 is present, then the equations of motion will be similar to those of Eqs. (3.53–3.55). Specifically, we can write

$$x_3 = x_2 + \int_{H_3}^{H_2} V_{v1}^{-1} W(h) \cos[\chi_W(h)] dh + (H_2 - H_3)GR_2 \cos(\chi_a)$$

$$\qquad (3.56)$$

$$y_3 = y_2 + \int_{H_3}^{H_2} V_{v1}^{-1} W(h) \sin[\chi_W(h)] dh + (H_2 - H_3)GR_2 \sin(\chi_a)$$

respectively.

Finally, for stage 3 (if present) we will have

$$x(0) = x_3 + \int_0^{H_3} V_{v1}^{-1} W(h) \cos[\chi_W(h)] dh + H_3 GR_3 \cos(\chi_a)$$

$$\qquad (3.57)$$

$$y(0) = y_3 + \int_0^{H_3} V_{v1}^{-1} W(h) \sin[\chi_W(h)] dh + H_3 GR_3 \sin(\chi_a)$$

Let us consider an example of how the developed equations apply to the two-stage system having parameters presented in Table 3.6.

Assume we release this PADS from $H_0 = 1,500$ m (4,920 ft) at $V_h^{a/c} = 60$ m/s (117 kt) due North-West. The winds are given in a look-up table. Depending on the data collection method, the wind measurements might be distributed unevenly in altitude. Hence, before computing definite integrals in Eqs. (3.50), (3.54), (3.56), or (3.118) numerically (e.g., using simple trapezoidal rule), the original look-up table h_k, W_k, $\chi_{W;k}$ should be approximated with, for example, piecewise cubic Hermite interpolating polynomials $W(h)$ and $\chi_W(h)$. Figure 3.65 shows the results of simulation.

Upon release from an aircraft, PADS flies ballistically. [Figure 3.65 shows the contribution of the second and third terms in Eq. (3.50), that is, with and without winds.] Both trajectories are extended all of the way down to the ground to show what would happen if the PADS canopy does not deploy at all. [In this case, the lower limit in the integrals of Eq. (3.50) are set to 0, that is, in this case $\Delta H = H_0$.]

TABLE 3.6   PARAMETERS OF A GENERIC TWO-STAGE PADS

	$V_{v;i}$, m/s	GR_i	$\Delta H(H_2)$, m
Stage 1	18	2.5	666
Stage 2	6	0	333

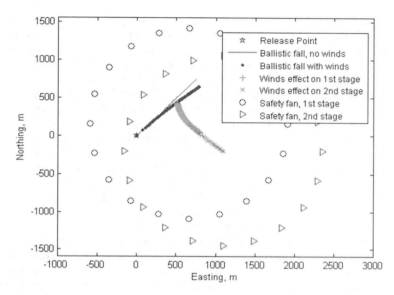

Fig. 3.65 Safety fans in case the controls are stuck in a fixed position.

As seen, the winds are pushing the trajectory to the South-East, but because of fast descent cause only about 100-m shift of the impact point in this direction.

Successful deployment of the first-stage canopy means that PADS slows down, and the wind affecting PADS longer produces a larger shift. (Again, this trajectory is propagated all of the way down to the surface rather than to H_2.) A trajectory only shows a contribution of the second terms in Eq. (3.54) while contribution of the third terms is depicted with a circumference of radius $(H_0 - \Delta H)GR_1$ centered at the point of impact in case the second stage does not open. This circumference illustrates what the boundaries of reachability are in case PADS is directed to fly (or flies unintentionally because of some malfunctions) with a constant arbitrary heading. Of course, this safety fan represents a worst-case scenario because in practice the failed GNC algorithm still forces PADS to change heading during its descent, so that PADS lands somewhere within this safety fan.

Figure 3.65 shows a successful deployment of the second stage as well. Because the descent rate for the second stage is three times smaller than that of the first stage (see Table 3.6), the winds play an even greater role shifting the point of impact further South-East. In this particular case, the second stage assumes no control authority (the glide ratio is equal to zero, i.e., a round parachute is used to simply slow the descent before a touchdown), so the safety fan computed for the first stage is simply shifted South-East to be centered around the second stage point of impact.

Figure 3.66 presents a screen shot of the developed safety fans GUI that requires a minimum amount of PADS data stored in the form of a table shown in Fig. 3.67 to produce safety fans assisting testing officers in airdrop planning. For operational use, these fans can also be treated as the reachability sets, similar to those produced by the mission planner (Fig. 3.63).

The winds panel of this GUI (the top right corner of Fig. 3.66) allows an operator to browse for the wind files, preview them, and choose the most appropriate one (i.e., most recent, closest to IPI, covering enough altitude range). The wind source files are sounding, Windpack, balloon, and JAAWIN winds as presented in Figs. 3.13–3.17. The program automatically recognizes what type of wind data file was chosen and how to read it properly populating all of the fields in the winds panel with relevant data taken from the chosen file.

The drop zone panel (below the winds panel) allows the operator to choose one of the drop zones with the coordinates stored in an ASCII file. Once chosen, the corresponding map pops up in the bird's-eye view window. By default, the center of the chosen DZ becomes IPI (which can be changed by selecting a new point on the map or typing new values in the windows). The user can zoom the map in and out, move it, measure distances between selected points, get the coordinates of any point selected with the cursor, and read the coordinates of any point in one of the chosen formats (UTM or lat/lon). That is what the map

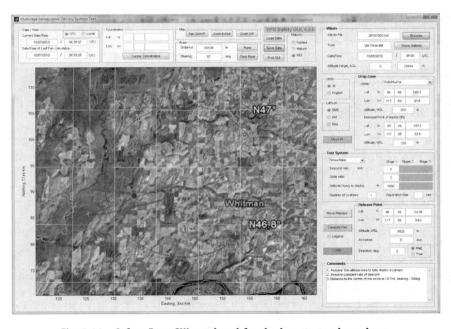

Fig. 3.66 Safety Fans GUI employed for the low-stratosphere drops.

System ID	Number of stages	Weight, lb	Stage 1			Stage 2			Stage 3		
			Descent rate, m/s	Glide ratio	Altitude loss to deploy, ft	Descent rate, m/s	Glide ratio	Altitude loss to deploy, ft	Descent rate, m/s	Glide ratio	Altitude loss to deploy, ft
System 1	1		22	0	2,000						
System 2	1		8	0.6	2,000						
System 3	3		55	0	12,000	8.5	3	3,000	4	2	1,000
System 4	1		8.5	3	100						
System 5	2		58	0	-	7	0	3,300			
System 6	1		5.25	3.5	2,000						
System 7	1		4.88	3.1	1,500						
System 8	1		5.25	3.5	2,500						
System 9	1		8.53	0	2,000						
System 10	2		55	0	12,000	8.5	2	3,000			
System 11	1		4.8	0.8	315						
System 12	1		4	4	1,000						
System 13	1		4.75	3.5	1,000						
System 14	1		9	3	1,000						
System 15	2		8	3	1,000	5	0	1,000			

Fig. 3.67   Examples of PADS input data.

and coordinates panels appearing to the left of the winds panel (Fig. 3.66) were designed for.

The test system panel allows the choosing of one of the ADSs stored in an Excel file in a format shown in Fig. 3.67. This system can have up to three stages. (The corresponding windows for unused stages are shadowed.) The operator can edit any of ADS parameters in the corresponding fields manually or create an entirely new system. Altitude, airspeed, and aircraft heading at the release point are entered in the release point panel. The correctness of the entered data can be verified graphically as shown in Fig. 3.68. Additionally, in the case of multiple drops during a single aircraft pass, the number of systems to be released and time separation between them have to be entered. These data along with the winds aloft and release point information are used to compute the safety fans.

The idea behind the release point panel is that based on the system being tested/deployed and available winds data, the operator enters various release points and computes safety fans. If the safety requirements are not met, the user moves the release point, and a new safety fan can be generated. The release point can be moved by clicking on a new location in the map, in which case the new coordinates immediately appear in the corresponding fields of the release point panel, or by entering a new latitude and longitude manually. The maturity panel allows the shrinking of the size of the safety fans based on maturity of PADS because it is less likely that the fielded PADS will glide with a constant heading all of the way down to the surface. Usually, in the case of detected malfunction, the fielded PADS would be entered into a spiral mode. This can also be treated as varying the threshold of probability of landing away from a nominal landing area or, in other words, varying the level of risk.

The Compute Fan button allows computing of CARP automatically, so that the center of the safety fan is centered at IPI. (Essentially, it is the same

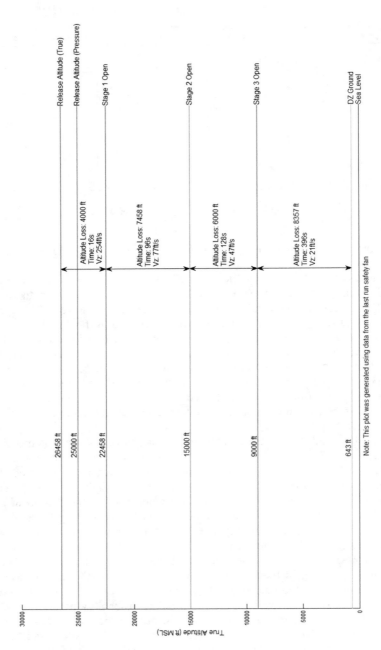

Fig. 3.68   Entered ADS data correctness check [Mulloy et al. 2011].

capability of the mission planner GUI as shown in Fig. 3.60b.) If an airlifter crew releases uncontrolled ADS like System 1 or 9 shown in Fig. 3.67 at CARP (computed for this ADS), there will be no touchdown error. (Of course, in practice, because of changing winds. it will never be the case.)

Other operational functions in the GUI include units, timing, loading and saving data, and commenting. The start clock button asks the operator for the computer's time zone setting and starts a continuously updating clock in Coordinated Universal Time (UTC). The time of the most recent calculation appears under the running clock. The GUI can be saved with the save data button. Once a GUI has been saved, it can be loaded with the load data button. When reloaded, the GUI is still interactive. This means that though it can be used as a simple figure capturing one state in time, it can also be used as the starting point for continuing operations.

CHAPTER 4

Aerodynamic Characterization of Parafoils

Hamid Johari*
California State University, Northridge, California

Oleg Yakimenko†
Naval Postgraduate School, Monterey, California

Thomas Jann‡
Institute of Flight Systems of DLR, Braunschweig, Germany

For simulation and dynamic stability analysis of a parafoil-load system, a mathematical model that accounts for the kinematics and aerodynamics of the system is required. In contrast to the well-known kinematic relationships, there is still some uncertainty in the determination of aerodynamic coefficients of ram-air parachute canopies. Past investigations have either evaluated the aerodynamic coefficients from wind-tunnel tests [Burk and Ware 1967; Nicolaides et al. 1971; Tribot et al. 1997; Ware and Hassell 1969], have taken airplane aerodynamics as a model [Crimi et al. 1990], or have used numerical methods to compute them for a rigid wing configuration [Brown 1993; Müller 2002; Ross 1993; Tribot et al. 1997]. This approach was utilized in Chapter 2 where analytical expressions for a low-aspect-ratio wing was modified to include the additional drag produced by the inlet, suspension lines, stabilizers, and payload of a parafoil-payload system. Strictly speaking, this type of approach is applicable to flat wings and not to inflated parafoil canopies with an anhedral arc. Hence, simple analytical expressions to account for the influence of dihedral angle on lift were included; however, for some parafoil canopies an anhedral arc cannot be characterized by a single anhedral angle. Gonzales [1993] and Jann [2003] used Prandtl's lifting-line theory and Weissinger's method to derive analytical forms of C_L and C_D for wings with arc anhedral as a function of arc–to–span ratio. Iosilevskii [1993] conducted a theoretical study of the aerodynamics of a paraglider. An advantage of the analytical approach is that it provides a simple way of estimating the parafoil aerodynamic characteristics. However, it is necessary to recognize that some assumptions in the analytical approach are too stringent. For example, it is usually

---

*Professor.
†Professor.
‡Scientist.

supposed that the parafoil canopy has identical geometry in the spanwise direction coinciding with the cut pattern of the rib. However, it is easily observed that not only the rib gets deformed, but also the lower and upper surfaces are wavy in the spanwise direction. Thus, it is necessary to identify some "average" section profile that is difficult to predict in advance. Furthermore, it is assumed that the planform of the deformed canopy is known in advance, or even coincides with the design form. But a real parafoil canopy in flight is always wavy in the spanwise direction and contracts noticeably in the chordwise and spanwise directions from the cut pattern. This chapter starts from Sec. 4.1, which describes an analytical approach as a convenient tool for a quick estimation of aerodynamic characteristics. Then, Sec. 4.2 shows an example of what can be achieved using computational-fluid-dynamics (CFD) analysis. Finally, Sec. 4.3 briefly addresses some computational tools developed specifically for coupled fluid-structure interaction (FSI)–CFD simulations of PADS and the outcome of this analysis.

4.1 THEORETICAL DERIVATION OF AERODYNAMIC COEFFICIENTS

This section, based on a slightly edited work by [Jann 2003], presents a theoretical approach for the calculation of aerodynamic coefficients based on the extended lifting-line theory for an elliptical rigid wing with arc anhedral. As a result, algebraic expressions that allow the computation of aerodynamic derivatives (including lateral and dynamic derivatives) for wings with different aspect ratios (AR) and curvatures are derived. The aerodynamic effects responsible for the typical behavior of an arched wing are also discussed. Clearly, as the flexibility of parafoil canopy is neglected in this approach, the computed coefficients should only be used as estimates rather than precise values. More studies are needed to account for the position and shift of center of pressure on a flexible canopy, flow separation, and forces as result of the flexible trailing-edge (TE) deflection, the role of derivatives C_{Yp}, C_{np}, C_{Yr} and C_{lr}, and unsteady effects. Nevertheless, the suggested approach provides a set of consistent aerodynamic parameters that can be used as a starting point for simulations and system identification.

4.1.1 ARC-ANHEDRAL GEOMETRY AND FORCE DISTRIBUTION

These days, dozens of different canopies are used for personnel and cargo delivery. Figure 4.1 features some of the rectangular low-AR parafoils originally designed to carry jumpers and then adopted to carry small-weight [up to 227-kg (500-lb)] cargo. Modern constructions include tapered, semi-elliptical or elliptical planforms. Some manufacturer and estimated data on these and other relatively small canopies are given in Appendix C.

One main characteristic of gliding parachutes is the arc anhedral, that is, the spanwise curvature of the wing, produced by the suspension lines joined together at the confluence points. The anhedral geometry creates a side force that is

Fig. 4.1   MT-4 canopy carrying a) a jumper and b) cargo, c) BT-80, d) Set-366, and e) PD500 canopies.

required to spread the flexible wing and produce a stable shape. Equation (2.15) has already shown that the anhedral angle is a function of the ratio of wing span b and suspension line length R. If the line length R is constant along the span, the spanwise geometry of the canopy can be approximated by a circular arc (Fig. 4.2).

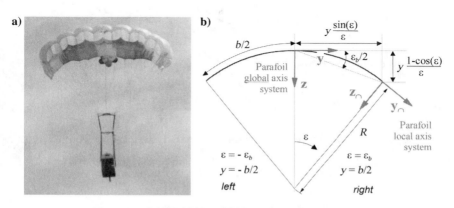

Fig. 4.2   a) ALEX PADS and b) its canopy shape model.

In the following discussion, the flexibility of the canopy is (mostly) ignored. Instead, the canopy is assumed to be rigid so that the lifting-line theory can be applied. The flexibility is only taken into account in the context of the semi-analytical estimation of the control derivatives due to TE deflection. Other simplifications include the following:

- Influence of the fins
- Flow separation
- Apparent mass effects are neglected

Because of these limitations, the validity of the results has to be scrutinized. Nevertheless, the presented analytical approach might help to improve the understanding of certain aerodynamic phenomena associated with ram-air parachutes. In this sense, this section should not be understood as the final answer to all aerodynamic questions but rather a contribution to diminish the still remaining gaps in the understanding of parafoil aerodynamics.

The forces and moments on the canopy are related to the reference point at the center of the wing at $c/2$. In the following procedure, they are computed by integrating the local force and moment distributions over the canopy surface. To calculate the local forces and moments, it is appropriate to compute the local airspeed vector first. In polar coordinates, this vector is described by the angle of attack α, the angle of sideslip β, and the magnitude of the airspeed V. The local distribution of these quantities along the wing depends on the canopy anhedral ε, the rotational motion, and the "global" airspeed, measured at the center of the wing in the wing-fixed coordinate system. Approximating $\alpha \ll 1$, $\beta \ll 1$, and assuming moderate rotational speed (i.e., neglecting quadratic terms of p, q, r), the local airspeed in the parafoil local axis system (index $\cap$) can be written as

$$\alpha_\cap \approx \alpha \cos(\varepsilon) - \beta \sin(\varepsilon) + \frac{py}{V}\frac{\sin(\varepsilon)}{\varepsilon} - \frac{qx}{V}\cos(\varepsilon) - \frac{rx}{V}\sin(\varepsilon) \qquad (4.1)$$

$$\beta_\cap \approx \alpha \sin(\varepsilon) + \beta \cos(\varepsilon) + \frac{py}{V}\frac{[1 - \cos(\varepsilon)]}{\varepsilon} - \frac{qx}{V}\sin(\varepsilon) + \frac{rx}{V}\cos(\varepsilon) \qquad (4.2)$$

$$V_\cap \approx V - ry\frac{\sin(\varepsilon)}{\varepsilon} + qy\frac{[1 - \cos(\varepsilon)]}{\varepsilon} \qquad (4.3)$$

The force distribution in the parafoil local axis system can be computed as

$$dX_\cap = dL_\cap \sin(\alpha_\cap) - dD_\cap \cos(\alpha_\cap)$$
$$dY_\cap = dY_\cap \qquad (4.4)$$
$$dZ_\cap = -[dL_\cap \cos(\alpha_\cap) + dD_\cap \sin(\alpha_\cap)]$$

Next, these local forces are converted into the global parafoil axis system (at the wing center)

$$dX = dX_\cap$$
$$dY = dY_\cap \cos(\varepsilon) - dZ_\cap \sin(\varepsilon) \qquad (4.5)$$
$$dZ = dY_\cap \sin(\varepsilon) + dZ_\cap \cos(\varepsilon)$$

Then, the lift, drag, and side-force distributions can be determined using the global angle of attack (AoA)

$$dL = dX \sin(\alpha) - dZ \cos(\alpha)$$
$$dY = dY \qquad (4.6)$$
$$dD = -[dX \cos(\alpha) + dZ \sin(\alpha)]$$

The resultant forces and moments at the canopy center are computed by integrating the corresponding distributions over the reference area S. For example, for canopy lift we have

$$L = \iint_S dL(x, y) \qquad (4.7)$$

Using the local lift, drag, and side-force distributions obtained from lifting-line theory, the resultant forces and the rolling moment can be integrated over the span as follows:

$$L \approx \int_{-b/2}^{b/2} \cos(\varepsilon)dL_\cap - \int_{-b/2}^{b/2} \sin(\varepsilon)dY_\cap \qquad (4.8)$$

$$Y \approx \int_{-b/2}^{b/2} \sin(\varepsilon)dL_\cap + \int_{-b/2}^{b/2} \cos(\varepsilon)dY_\cap \qquad (4.9)$$

$$D \approx \int_{-b/2}^{b/2} dD_\cap \qquad (4.10)$$

The forces acting at a distance from the canopy center produce the following moments (see Fig. 4.2b):

$$dL^a = y\frac{\sin(\varepsilon)}{\varepsilon}dZ - y\frac{1 - \cos(\varepsilon)}{\varepsilon}dY$$

$$dM^a = y\frac{1 - \cos(\varepsilon)}{\varepsilon}dX - x\,dZ \qquad (4.11)$$

$$dN^a = -y\frac{\sin(\varepsilon)}{\varepsilon}dX + x\,dY$$

For pitching and yawing moments, the variable distribution of the forces along the chord has to be taken into account as well. Although the forces depend on the chordwise pressure distribution, a simpler approach is used here to estimate the contributions to these moments. First, the center of pressure (c.p.) is assumed to be at $c/2$ (see following discussion). Second, for normal flight conditions (β, p, q, $r = 0$), it is assumed that the local lift $dL_\cap$ and side force $dY_\cap$ are equally distributed in chordwise direction, so that the amounts of lift and sideforce ahead of and behind c.p. are the same. Third, a rectangular planform is assumed for ease of integration. With these simplifications, one arrives at the following expressions for the moments on the canopy:

$$L^a \approx - \int_{-b/2}^{b/2} y \frac{\sin(\varepsilon)}{\varepsilon} dL_\cap + \int_{-b/2}^{b/2} y \frac{1 - \cos(\varepsilon)}{\varepsilon} dY_\cap \qquad (4.12)$$

$$M^a \approx - \int_{-b/2}^{b/2} y \frac{1 - \cos(\varepsilon)}{\varepsilon} dD_\cap + \int_{-b/2}^{b/2} \int_{-c/2}^{c/2} x \cos(\varepsilon) dL_\cap \qquad (4.13)$$

$$N^a \approx \int_{-b/2}^{b/2} y \frac{\sin(\varepsilon)}{\varepsilon} (-\alpha_\cap dL_\cap + dD_\cap)$$

$$+ \int_{-b/2}^{b/2} \int_{-c/2}^{c/2} x \underbrace{[\sin(\varepsilon) dL_\cap + \cos(\varepsilon) dY_\cap]}_{dY} \qquad (4.14)$$

Now, the distribution of the local lift $dL_\cap$ and induced drag $dD_{i\cap}$ can be calculated from the local aerodynamic circulation [vortex strength $\Gamma(y)$], the local airspeed $V(y)$, and the induced AoA $\alpha_i(y)$ using the Kutta–Joukowski theorem [Schlichting and Truckenbrodt 1959]

$$dL_\cap(y) = \rho V(y) \Gamma(y) dy \qquad (4.15)$$

$$dD_{i\cap}(y) = \rho V(y)\alpha_i(y)\Gamma(y) dy \qquad (4.16)$$

Using notations $\overline{\Gamma} = \Gamma/(bV)$ for the dimensionless circulation and $\eta = 2y/b \in [-1;1]$, Eqs. (4.15) and (4.16) can be rewritten as

$$dL_\cap(\eta) = \underbrace{QS\,AR \left[\frac{V(\eta)}{V}\right]^2 \overline{\Gamma}(\eta) \, d\eta}_{dC_L} \qquad (4.17)$$

$$dD_{in}(\eta) = QS\,AR\underbrace{\left[\frac{V(\eta)}{V}\right]^2\alpha_i(\eta)\overline{\Gamma}(\eta)\,d\eta}_{dC_{Di}} \tag{4.18}$$

where $Q = 0.5\rho V^2$ is the dynamic pressure. If the pitch and yaw rates are zero, then $V(\eta)V^{-1} = 1$.

The dimensionless circulation $\overline{\Gamma}(\eta)$ and the induced AoA can be approximated by Fourier series using $\eta = \cos(\vartheta)$ [Schlichting and Truckenbrodt 1959]

$$\overline{\Gamma}(\vartheta) = 2\sum_{n=1}^{M} a_n \sin(n\vartheta) \tag{4.19}$$

$$\alpha_i(\vartheta) = \sum_{n=1}^{M} na_n \frac{\sin(n\vartheta)}{\sin(\vartheta)} \tag{4.20}$$

Substituting $d\eta = -d\vartheta\,\sin(\vartheta)$ into Eq. (4.17) results in the aerodynamic lift coefficient [with $V(\eta)V^{-1} = 1$]

$$C_L = AR\int_{-1}^{1}\overline{\Gamma}(\eta)\,d\eta = AR\int_{0}^{\pi}\overline{\Gamma}(\vartheta)\sin(\vartheta)\,d\vartheta \tag{4.21}$$

The Fourier coefficients a_n, where $n = 1, \ldots, M$ in Eqs. (4.19) and (4.20) can be determined by solving a set of linear equations defined by the local AoA $\alpha_n(\eta)$ and the planform geometry $f(\eta)$. However, according to the extended lifting-line theory (Weissinger's 3/4-chord method), the Fourier coefficients for an elliptical wing can also be computed from the following integral equation:

$$a_n = \frac{1}{\sqrt{k^2 + n^2} + n}\frac{2}{\pi}\int_{0}^{\pi}\alpha_g(\vartheta)\sin(\vartheta)\sin(n\vartheta)d\vartheta \tag{4.22}$$

with $k = AR\,\pi C_{L\alpha}^{a-1}$ [see Eq. (2.3)]. In Eq. (4.22), α_g is the geometric AoA measured against the zero-lift angle α_0, that is,

$$\alpha_g = \alpha_n - \alpha_0 \tag{4.23}$$

For nonelliptical planforms, the Fourier coefficients will be slightly different. Nevertheless, in such cases the results for an elliptical wing can be normally taken as a good approximation. Then, the influence on the induced drag can be taken into account by applying the Oswald efficiency factor e (empirical planform correction factor).

The total drag is the result of the sum of profile drag and induced drag:

$$dD_\cap = dD_{i\cap} + dD_{0\cap} \qquad (4.24)$$

with the assumption that the profile drag distribution is constant along the span

$$dD_{0\cap}(y) = QS\frac{C_{D0}}{b}\left[\frac{V(y)}{V}\right]^2 dy \qquad (4.25)$$

The local side-force distribution is a result from the profile drag for transverse flow over the wing, which is proportional to $\sin(\beta_\cap)$

$$dY_\cap(y) = -QS\frac{C_{D0}}{b}\left[\frac{V(y)}{V}\right]^2 \sin(\beta_\cap)\,dy \qquad (4.26)$$

The numerical solution of Eqs. (4.15) and (4.26) followed by transformations (4.4–4.11) gives the lift and side-force distribution. Figure 4.3 shows an example of the normalized distribution for the right half-span of an elliptical wing with $\varepsilon_b = 40$ deg anhedral and a zero-lift AoA $\alpha_0 = -7$ deg.

As shown by Gonzalez [1993], the lift decreases as result of the anhedral. Compared to a straight wing with the same AR, the lift is reduced to approximately 92%. At the same time, side forces of the magnitude of 15% of the total lift appear on each side.

4.1.2 AERODYNAMIC AND CONTROL COEFFICIENTS

Using the derivations from the preceding section, one can compute the aerodynamic coefficients for a specific wing. The analytical solution of the integrals (4.8–4.10) and (4.12–4.14) allows deriving simple expressions for the dimensionless aerodynamic forces and moments C_L, C_D, C_Y, C_l, C_m, and C_n (defined in the wind coordinate frame $\{w\}$). From these the corresponding aerodynamic forces (lift L, drag D, side force Y) and moments (rolling moment L^a, pitching moment M^a, and yawing moment N^a) can

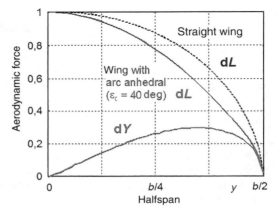

Fig. 4.3 Normalized lift and side-force distribution for a wing with arc anhedral at $\alpha = 5$ deg.

be computed as

$$L = QSC_L = QS[C_L(\alpha) + C_{L\delta_s}\delta_s] \tag{4.27}$$

$$D = QSC_D = QS[C_{D0} + C_{Di}(\alpha) + C_{D\delta_s}\delta_s] \tag{4.28}$$

$$Y = QSC_Y = QS\left(C_{Y\beta}\beta + C_{Yp}\frac{b}{2V}p + C_{Yr}\frac{b}{2V}r + C_{Y\delta_a}\delta_a\right) \tag{4.29}$$

$$L^a = QSbC_l = QSb\left(C_{l\beta}\beta + C_{lp}\frac{b}{2V}p + C_{lr}\frac{b}{2V}r + C_{l\delta_a}\delta_a\right) \tag{4.30}$$

$$M^a = QS\bar{c}C_m = QS\bar{c}\left(C_{m\alpha}\alpha + C_{mq}\frac{c}{V}q + C_{m\delta_s}\delta_s\right) \tag{4.31}$$

$$N^a = QSbC_n = QSb\left(C_{n\beta}\beta + C_{np}\frac{b}{2V}p + C_{nr}\frac{b}{2V}r + C_{n\delta_a}\delta_a\right) \tag{4.32}$$

Considering longitudinal aerodynamics, the contribution from the local side forces is on the order of 1% of the local lift and will be neglected hereafter. Hence, one can write

$$L \approx \int_{-b/2}^{b/2} \cos(\varepsilon)\,dL_n \approx \underbrace{QS\,AR \int_0^\pi \cos(\varepsilon)\overline{\Gamma}(\vartheta)\sin(\vartheta)\,d\vartheta}_{C_L} \tag{4.33}$$

According to Fig. 4.2b, the local anhedral angle ε is a function of y, respectively η and ϑ

$$y = \varepsilon R \implies \varepsilon = \varepsilon_b \cos(\vartheta) \tag{4.34}$$

To evaluate the integral in Eq. (4.33) analytically, it is appropriate to write the expression $\cos(\varepsilon)$ as a series and neglect the higher-order terms

$$\cos(\varepsilon) \approx 1 - \varepsilon^2/2 = 1 - 0.5\varepsilon_b^2\cos^2(\vartheta) \tag{4.35}$$

Using this assumption,

$$C_L(\alpha) \approx \frac{\pi AR}{\sqrt{k^2+1}+1}\left[\left(1 - \frac{\varepsilon_b^2}{4} + \frac{\varepsilon_b^4}{64}\right)\alpha - \left(1 - \frac{\varepsilon_b^2}{8}\right)\alpha_0\right] \tag{4.36}$$

By suitable trigonometric functions, to replace the series expression one can write

$$C_L(\alpha) \approx C_{L\alpha}[\alpha\cos(\varepsilon_b/2) - \alpha_0]\cos(\varepsilon_b/2) \tag{4.37}$$

where

$$C_{L\alpha} \approx \frac{\pi AR}{\sqrt{k^2+1}+1} = \frac{\pi ARC_{L\alpha}^a}{\sqrt{(\pi AR)^2 + C_{L\alpha}^a{}^2} + C_{L\alpha}^a} \tag{4.38}$$

is the Helmbold equation for low-AR nonswept wings [compare Eq. (2.1)] [McCormick 1979]. This principle is applied in the derivation of all the following "approximate" equations to compute the coefficients.

Applying the same procedure to the induced drag yields

$$C_{Di}(\alpha) \approx \frac{C_{L\alpha}^2}{e\pi AR}[\alpha \cos(\varepsilon_b/2) - \alpha_0]^2 \qquad (4.39)$$

[compare to Eq. (2.8), which is equivalent to Eq. (4.39) with Oswald efficiency factor $e = (1 + \delta)^{-1}$ that is assumed here with a value of 0.8].

The pitching moment depends on the distance of c.p. from the wing's reference point. For airfoils with positive curvature, c.p. position is behind the quarter-chord point and changes with AoA. In agreement with wind-tunnel tests and observations, it seems likely that c.p. is situated between $c/3$ and $c/2$ (for ram-air parachutes at normal flight conditions) [Ware and Hassell 1969].

For the sake of simplicity, the c.p. position is assumed to be stationary and coincident with the reference point at $c/2$. For this reason, the coefficients C_{m0} and $C_{m\alpha}$ become zero, and the pitch damping derivative yields

$$C_{mq} \approx -\frac{C_{L\alpha}}{12}\cos^2(\varepsilon_b/2) \qquad (4.40)$$

Solving the integrals in Eqs. (4.9), (4.12), and (4.14) for $\beta \neq 0$, one obtains the dependency of the lateral aerodynamic coefficients on the sideslip angle

$$C_{Y\beta} \approx -C_{L\alpha}k_1 \frac{\varepsilon_b \sin(\varepsilon_b)}{4} - C_{D0}\frac{1 + 2\cos(\varepsilon_b)}{3} \qquad (4.41)$$

$$C_{l\beta} \approx C_{L\alpha}k_1 \frac{\sin(\varepsilon_b)}{8} \qquad (4.42)$$

$$C_{n\beta} \approx -\frac{1}{8}C_{L\alpha}k_1 k_2 [\sin(\varepsilon_b)\alpha_0 - 2\sin(3\varepsilon_b/2)\alpha] \qquad (4.43)$$

In Eqs. (4.41–4.45),

$$k_1 = \frac{\sqrt{k^2 + 1} + 1}{\sqrt{k^2 + 4} + 2} \quad \text{and} \quad k_2 = \frac{\sqrt{k^2 + 4} - 1}{\sqrt{k^2 + 1} + 1}$$

These factors depend on the wing AR and have values between $1/2$ and 1. As expected, the results show a negative side-force derivative and a positive rolling-moment derivative. According to the preceding results, a positive yawing moment is produced by the sideslip, that is, the wing itself is directionally stable (for the reference point at $c/2$). However, because the magnitude of this derivative is relatively small, the sign might change due to other effects, which have been neglected here (e.g., fins or flexibility).

For rotation, the local airspeed vector changes depending on the distance to the reference point. Lift, drag, and pitching moment are almost unaffected by the roll rate. For the lateral aerodynamic derivatives, one obtains

$$C_{Yp} \approx C_{L\alpha}k_1\frac{\sin(\varepsilon_b)}{4} \tag{4.44}$$

$$C_{lp} \approx -C_{L\alpha}k_1\frac{\sin(\varepsilon_b)}{8\varepsilon_b} \tag{4.45}$$

$$C_{np} \approx \frac{1}{8\varepsilon_b}C_{L\alpha}k_1k_2[\sin(\varepsilon_b)\alpha_0 - 2\sin(3\varepsilon_b/2)\alpha] \tag{4.46}$$

In contrast to wings with positive dihedral, here a rolling motion produces a side force in the same direction. A negative yawing-moment derivative means that the side rolling downwards is pulled forward and vice versa. A positive yaw rate increases the airspeed on the left side and reduces the airspeed on the right side of the wing. Because of the anhedral, the side forces on the left and right sides of the wing become asymmetric. This effect particularly increases the yaw damping.

Similarly, the derivatives with respect to yaw rate can be written as

$$C_{Yr} \approx 0.5C_{L\alpha}[\sin(\varepsilon_b)\alpha_0 - 2\sin(\varepsilon_b/2)\cos^2(\varepsilon_b/2)\alpha] \tag{4.47}$$

$$C_{lr} \approx -\frac{1}{4\varepsilon_b}C_{L\alpha}[\sin(\varepsilon_b)\alpha_0 - 2\sin(\varepsilon_b/2)\cos^2(\varepsilon_b/2)\alpha] \tag{4.48}$$

The yaw damping is composed of four different factors (asymmetric profile drag, induced drag, force vector inclination, and side forces) that make it more complicated to estimate the resultant derivatives. The following equation can be used to approximately compute the derivative:

$$C_{nr} \approx -\frac{C_{D0}}{3}\left(1 - \frac{1}{5}\varepsilon_b^2\right) + \frac{C_{L\alpha}^2}{\pi AR}\left(\frac{1}{2} - \frac{\varepsilon_b^2}{24}\right)\alpha_0^2 + C_{L\alpha}\frac{k_1}{AR^2}\frac{\varepsilon_b^2}{24}$$

$$ - \frac{1}{4}\left[\frac{C_{L\alpha}^2}{\pi AR}\left(4 - \frac{7}{12}\varepsilon_b^2\right) - C_{L\alpha}\left(1 - \frac{5}{12}\varepsilon_b^2\right)\frac{\alpha_0}{4}\right]\alpha \tag{4.49}$$

[dependence on α^2 in Eq. (4.49) is neglected]. Given the negative side-force derivative and the positive rolling derivative, the part of the wing moving forward is pulled sideways and upwards by the aerodynamic forces. The opposite happens to the side moving backwards.

Although rotation of the wing does not necessarily deform the canopy, pulling the control lines certainly does. In addition, the flow starts to separate at the deflected TE making the theoretical description even more difficult. Because of these issues, a rather simple semi-analytical approach is proposed assuming the additional forces are linearly dependent on the control line deflection.

From theory, we can approximately estimate the shift of the zero-lift angle for full control line stroke [Schlichting and Truckenbrodt 1959]. For the following calculations, a value of $\Delta\alpha_0(\delta_{max}) \approx -11$ deg is assumed. Also, the ratio of the deflected surface width and the span b_k/b (see Fig. 2.46), the mean anhedral ε_k of the deflected edge, and the distance to the center of the wing l_k have to be taken into account as follows:

$$\varepsilon_k = \varepsilon_b \left(1 - \frac{b_k}{b} \right) \tag{4.50}$$

$$l_k = \frac{b}{2} \left(1 - \frac{b_k}{b} \right) \frac{\sin(\varepsilon_k)}{\varepsilon_k} \tag{4.51}$$

The pitching moment depends on the change of c.p. with respect to TE deflection. In agreement with wind-tunnel test results, it seems likely that the c.p. position moves backward producing a negative pitching moment [Burke and Ware 1967; Ware and Hassell 1969; Chapter 2 of this book]. This effect is taken into account in a very simplistic manner by assuming a lever of 0.25 c for the deflected TE behind the reference point at $c/2$. At a certain point the flow starts to separate near the TE, increasing the profile drag and decreasing the lift. According to [Babinsky 1999], the flow separation does not occur suddenly but rather in a continuous way. Because of the linear approach, an additional profile drag of $\Delta C_{D0\delta_s} \approx 0.2$ should be added. Assembling all of these elements together, we obtain estimations of the derivatives for a symmetric control line deflection

$$C_{L\delta_s} \approx C_{L\alpha} \Delta\alpha_0(\delta_{max}) \frac{2b_k}{b} \cos(\varepsilon_k) \tag{4.52}$$

$$C_{D\delta_s} \approx \frac{2b_k}{b} \left[\frac{C_{L\alpha}^2}{e\pi AR} \Delta\alpha_0(\delta_{max}) \left(\Delta\alpha_0(\delta_{max}) + 2\alpha_0 + 2\alpha \right) + \Delta C_{D0\delta_s} \right] \tag{4.53}$$

$$C_{m\delta_s} \approx -0.25 C_{L\delta_s} \tag{4.54}$$

where

$$\delta_s = 0.5(\delta_l + \delta_r) \tag{4.55}$$

The left δ_l and right δ_r control line deflections are normalized to fit in the interval [0;1], so that $\delta_s \in [0;1]$. Note, because $\Delta\alpha_0(\delta_{max})$ is negative, the result for the control derivative in Eq. (4.52) is positive.

For asymmetric control line deflection,

$$C_{Y\delta_a} \approx \frac{C_{L\delta_s}}{2} \frac{\sin(\varepsilon_k)}{\cos(\varepsilon_k)} \tag{4.56}$$

$$C_{l\delta_a} \approx -\frac{C_{D\delta_s}}{2} \frac{\cos^2(\varepsilon_k/2)}{\cos(\varepsilon_k)} \frac{l_k}{b} \tag{4.57}$$

$$C_{n\delta_a} \approx \frac{C_{D\delta_s}}{2} \frac{l_k}{b} \tag{4.58}$$

where

$$\delta_a = \delta_r - \delta_l, \quad \delta_a \in [-1;1] \tag{4.59}$$

Reviewing Eqs. (4.27–4.32), it should be noted that coefficients $C_{n\beta}$, C_{np}, C_{Yr}, C_{lr}, C_{nr}, and $C_{D\delta_s}$, as derived in Eqs. (4.43), (4.46–4.49), and (4.53), respectively, include an explicit dependence on the AoA. However, analytical expressions for the aerodynamic coefficients refer to the parafoil canopy reference point CP, which is located in a distance from the system's center of mass CM and payload's center of mass CL as denoted in Fig. 4.4. This distance is essentially defined by the suspension line and harness lengths. The distance acts like a lever arm and creates additional portions of airflow on the canopy for rotations of the system around CM, or better, center of mass. The suspension line geometry also defines the rigging angle μ that inclines the canopy and the associated axis system with respect to the body-fixed coordinate frame. (Note that here we keep the notation introduced in Fig. 2.23 of Chapter 2, so that the rigging angle μ as shown in Fig. 4.4 is positive.) Hence, compared with the AoA at payload, the AoA at canopy is smaller by μ in steady flight.

Let us consider an example of assessing aerodynamic derivatives for ALEX PADS (Table 4.1). Table 4.2 summarizes the approximate values for the aerodynamic coefficients for ALEX PADS that are partly based on Chapter 2 and derived from the analytical approach just described for a wing with arc anhedral and zero-lift AoA $a_0 = -0.122$ rad (-7 deg). As mentioned earlier, $c/2$ is used as canopy reference point CP. The dependence of several coefficients on AoA was eliminated by setting $\alpha = 5$ deg.

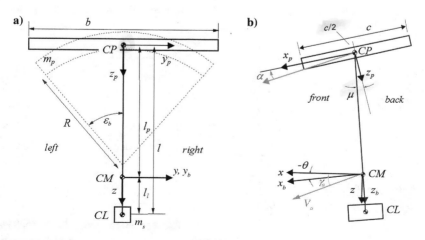

Fig. 4.4   Parafoil-payload model in a) y–z and b) x–z planes.

TABLE 4.1   GEOMETRY AND MASS PROPERTIES OF ALEX PADS

Parameter	Notation	Value	Parameter	Notation	Value
Parafoil span	b	5.8 m (19 ft)	Rigging angle [at $c/2$ (CP)]	μ	18 deg
Parafoil chord	c	3.4 m (11.2 ft)	Parafoil mass (including air)	m_p	13.7 kg (30.2 lb)
Parafoil aspect ratio	AR	1.7	Payload mass	m_s	97.6 kg (215.2 lb)
Suspension line length	R	4.2 m (13.8 ft)	Distance CP-CM	l_p	4.8 m (15.7 ft)
Anhedral angle	ε_b	40 deg	Distance CL-CM	l_l	0.7 m (2.3 ft)

Comparison with other approaches like vortex lattice and panel methods showed that most of the coefficients have the same order of magnitude and the same sign. Differences seem to appear mainly because of the extremely simplified treatment of the chordwise force distribution leading to different

TABLE 4.2   STABILITY AND CONTROL DERIVATIVES FOR ALEX PADS RESULTING FROM THE LIFTING LINE THEORY

Force Derivative	Value	Moment Derivative	Value
$C_{L\alpha}$	2.36	C_{m0}	0
$C_{L\alpha}$ ($\alpha = 5$ deg)	2.088	$C_{m\alpha}$	0
$C_{L\delta_s}$	0.292	C_{mq}	−0.174
C_{D0}	0.084	$C_{m\delta_s}$	−0.073
$C_{D\alpha}$ ($\alpha = 5$ deg)	0.500	$C_{l\beta}$	0.104
$C_{D\delta_s}$ ($\alpha = 5$ deg)	0.252	C_{lp}	−0.149
$C_{Y\beta}$	−0.216	C_{lr} ($\alpha = 5$ deg)	0.096
C_{Yp}	0.208	$C_{l\delta_a}$	−0.048
C_{Yr} ($\alpha = 5$ deg)	−0.155	$C_{n\beta}$ ($\alpha = 5$ deg)	0.019
$C_{Y\delta_a}$	0.070	C_{np} ($\alpha = 5$ deg)	−0.027
		C_{nr} ($\alpha = 5$ deg)	−0.047
		$C_{n\delta_a}$	0.039

assumptions for the c.p. position, which, in fact, is not constant. For this reason, coefficients C_{m0} and $C_{m\alpha}$ in Table 4.2 are zero although they are actually not zero for a geometrically fixed reference point. In reality, lift and drag, namely, the coefficients C_{Lq} and C_{Dq}, are also influenced by the pitch rate that is not considered here. The bottom line, however, is that the analytical approach presented in this section is the least expensive and does produce feasible results that could be used in modeling and simulation efforts and as a starting point for system identification.

4.1.3   DEPENDENCE OF AERODYNAMIC COEFFICIENTS ON ANGLE OF ATTACK

Another approach suggested by [Peyada et al. 2007] uses even simpler canopy geometry and also provides the dependences of the aerodynamic and control derivatives on the AoA. This approach employs two basic characteristics, lift and drag dependence on the AoA, and then derives the lateral aerodynamic coefficients from there. Specifically, the experimental lift and drag curves, obtained in wind-tunnel tests by [Ware and Hassell 1969; Nicolaides et al. 1970] (Figs. 2.6 and 2.9 or Figs. 2.24 and 2.25, involving different flaps settings) can be used. As a reminder, the same data as in Figs. 2.6, 2.9, and 2.11 are also shown in Fig. 4.5, along with the matching computational-fluid-dynamics (CFD) simulations by [Cao and Zhu 2013]. (A CFD approach will be addressed in the next section.)

From $C_L(\alpha)$ and $C_D(\alpha)$ [and the associated gradients $C_{L\alpha}(\alpha)$ and $C_{D\alpha}(\alpha)$], the coefficients $C_{Y\beta}(\alpha)$, $C_{l\beta}(\alpha)$, $C_{n\beta}(\alpha)$, $C_{lp}(\alpha)$, $C_{np}(\alpha)$, $C_{lr}(\alpha)$, and $C_{nr}(\alpha)$ can be derived based on simplified models breaking the parafoil canopy onto left and right consoles as shown in Figs. 4.6–4.9. Flying with a nonzero sideslip angle or banking or yawing, two consoles appear to be at the different AoA producing different lift and drag, and that is what causes the side force and rolling and yawing moments. In addition to that, when flying with a nonzero sideslip angle the inlet drag D_{Ram}, or in other words the force on cells ribs, adds an additional side force and yawing moment.

Following Peyada et al. [2007], when flying with a nonzero sideslip angle $\beta \ll 1$, the total side force can be computed as (see Fig. 4.6)

$$Y = \left(L_{\text{left}} \cos\left(\alpha_{\text{left}}\right) + D_{\text{left}} \sin\left(\alpha_{\text{left}}\right)\right) \sin(\eta)$$
$$- \left(L_{\text{right}} \cos\left(\alpha_{\text{right}}\right) + D_{\text{right}} \sin\left(\alpha_{\text{right}}\right)\right) \sin(\eta) - D_{\text{inlet}} \sin(\beta) \quad (4.60)$$

In this equation,

$$L_{\text{left}} = 0.5QS(C_L - C_{L\alpha}\Delta\alpha) \quad D_{\text{left}} = 0.5QS(C_D - C_{D\alpha}\Delta\alpha) \quad \alpha_{\text{left}} = \alpha - \Delta\alpha$$
$$L_{\text{right}} = 0.5QS(C_L + C_{L\alpha}\Delta\alpha) \quad D_{\text{right}} = 0.5QS(C_D + C_{D\alpha}\Delta\alpha) \quad \alpha_{\text{right}} = \alpha + \Delta\alpha$$
$$(4.61)$$

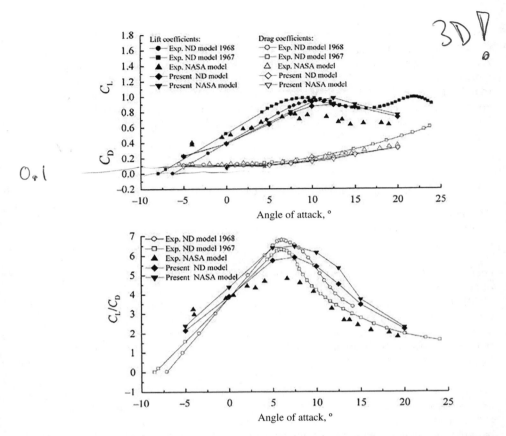

Fig. 4.5 Comparison of lift, drag coefficients and lift-drag ratios between wind-tunnel test data and results of numerical simulation [Cao and Zhu 2013].

Using trigonometric equalities $\cos(a + b) = \cos(a)\ \cos(b) - \sin(a)\ \sin(b)$ and $\sin(a + b) = \cos(a)\ \sin(b) + \sin(a)\ \cos(b)$, one can write

$$
\begin{aligned}
\cos(\alpha_{\text{left}}) &= \cos(\alpha - \Delta\alpha) \approx \cos(\alpha) + \sin(\alpha)\Delta\alpha \\
\sin(\alpha_{\text{left}}) &= \sin(\alpha - \Delta\alpha) \approx -\cos(\alpha)\Delta\alpha + \sin(\alpha) \\
\cos(\alpha_{\text{right}}) &= \cos(\alpha + \Delta\alpha) \approx \cos(\alpha) - \sin(\alpha)\Delta\alpha \\
\sin(\alpha_{\text{right}}) &= \sin(\alpha + \Delta\alpha) \approx \cos(\alpha)\Delta\alpha + \sin(\alpha)
\end{aligned}
\tag{4.62}
$$

Accounting for the fact that there is a small additional AoA due to asymmetry

$$
\Delta\alpha \approx \beta\eta
\tag{4.63}
$$

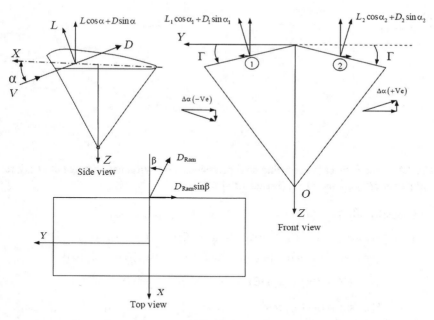

Fig. 4.6 Schematic of forces acting on a parafoil-payload system for estimation of $C_{Y\beta}$ and $C_{l\beta}$ [Peyada et al. 2007].

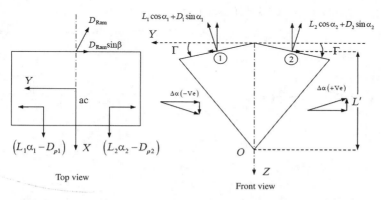

Fig. 4.7 Schematic of forces acting on a parafoil-payload system at an angle of attack for estimation of $C_{n\beta}$ [Peyada et al. 2007].

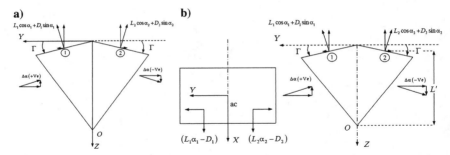

Fig. 4.8 **Schematic of forces acting on a parafoil-payload system at an angle of attack for estimation of a) C_{lp} and b) C_{np} [Peyada et al. 2007].**

and substituting Eqs. (4.61) and (4.62) into Eq. (4.60) yields

$$
\begin{aligned}
Y &\approx 0.5QS\{(C_L - C_{L\alpha}\beta\eta)[\cos(\alpha) + \sin(\alpha)\beta\eta] - (C_L + C_{L\alpha}\beta\eta) \\
&\quad \times [\cos(\alpha) - \sin(\alpha)\beta\eta] + (C_D - C_{D\alpha}\beta\eta)[-\cos(\alpha)\beta\eta + \sin(\alpha)] \\
&\quad - (C_D + C_{D\alpha}\beta\eta)[\cos(\alpha)\beta\eta + \sin(\alpha)]\}\eta - C_D^{\text{inlet}}QS\beta \\
&\approx QS[C_L \sin(\alpha) - C_{L\alpha} \cos(\alpha) - C_D \cos(\alpha) - C_{D\alpha} \sin(\alpha)]\beta\eta^2 - C_D^{\text{inlet}}QS\beta
\end{aligned}
$$

$$(4.64)$$

As a result, we obtain

$$
C_{Y\beta}(\alpha) \approx [C_L \sin(\alpha) - C_{L\alpha} \cos(\alpha) - C_D \cos(\alpha) - C_{D\alpha} \sin(\alpha)]\eta^2 - C_D^{\text{inlet}} \quad (4.65)
$$

Similarly, the total rolling moment acting about the c.g. of parafoil due to sideslip angle can be derived from the relation (see Fig. 4.5)

$$
\begin{aligned}
QSbC_{l\beta} = 0.25b\big\{ &\left[L_{\text{right}} \cos\left(\alpha_{\text{right}}\right) + D_{\text{right}} \sin\left(\alpha_{\text{right}}\right)\right] \\
&- \left[L_{\text{left}} \cos\left(\alpha_{\text{left}}\right) + D_{\text{left}} \sin\left(\alpha_{\text{left}}\right)\right]\big\}
\end{aligned}
$$

$$(4.66)$$

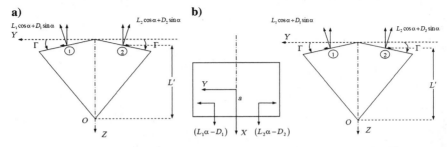

Fig. 4.9 **Schematic of forces acting on a parafoil-payload system for estimation of a) C_{lr} and b) C_{nr} [Peyada et al. 2007].**

Substituting the parameters from Eqs. (4.61–4.63) yields

$$C_{l\beta}(\alpha) \approx 0.25[-C_L \sin(\alpha) + C_{L\alpha} \cos(\alpha) + C_D \cos(\alpha) + C_{D\alpha} \sin(\alpha)]\eta \quad (4.67)$$

Following the same approach, the total yawing moment acting about the c.g. of parafoil due to sideslip angle is derived from (see Fig. 4.7)

$$
\begin{aligned}
QSbC_{n\beta} = {}& 0.25b\{[L_{\text{right}} \sin(\alpha_{\text{right}}) - D_{\text{right}} \cos(\alpha_{\text{right}})] - [L_{\text{left}} \sin(\alpha_{\text{left}}) - D_{\text{left}} \cos(\alpha_{\text{left}})]\} \\
& - 0.25c\{[L_{\text{right}} \cos(\alpha_{\text{right}}) + D_{\text{right}} \sin(\alpha_{\text{right}})] - [L_{\text{left}} \cos(\alpha_{\text{left}}) + D_{\text{left}} \sin(\alpha_{\text{left}})]\} \\
& \times \sin(\eta) + 0.5cD_{\text{inlet}} \sin(\beta)
\end{aligned}
\quad (4.68)
$$

yielding

$$
\begin{aligned}
C_{n\beta}(\alpha) \approx {}& 0.25[C_L \cos(\alpha) + C_{L\alpha} \sin(\alpha) + C_D \sin(\alpha) - C_{D\alpha} \cos(\alpha)]\eta \\
& + 0.25cb^{-1}[C_L \sin(\alpha) - C_{L\alpha} \cos(\alpha) - C_D \cos(\alpha) - C_{D\alpha} \sin(\alpha)]\eta^2 \\
& + 0.5cb^{-1}C_D^{\text{inlet}}
\end{aligned}
\quad (4.69)
$$

Derivation of the rolling and yawing moments about c.g. due to the roll rate follows the same paradigm with the only difference being a small additional AoA in this case:

$$\Delta\alpha \approx pb/(4V) \quad (4.70)$$

Using notations of Fig. 4.8, one can write

$$
\begin{aligned}
QSb\frac{pb}{2V}C_{lp} = {}& 0.25b\{[L_{\text{right}} \cos(\alpha_{\text{right}}) + D_{\text{right}} \sin(\alpha_{\text{right}})] \\
& - [L_{\text{left}} \cos(\alpha_{\text{left}}) + D_{\text{left}} \sin(\alpha_{\text{left}})]\}
\end{aligned}
\quad (4.71)
$$

and

$$
\begin{aligned}
QSb\frac{pb}{2V}C_{np} = {}& 0.25b\{(L_{\text{right}} \sin(\alpha_{\text{right}}) - D_{\text{right}} \cos(\alpha_{\text{right}})) - (L_{\text{left}} \sin(\alpha_{\text{left}}) - D_{\text{left}} \cos(\alpha_{\text{left}})) \\
& + [(L_{\text{left}} \cos(\alpha_{\text{left}}) + D_{\text{left}} \sin(\alpha_{\text{left}})) \\
& - (L_{\text{right}} \cos(\alpha_{\text{right}}) + D_{\text{right}} \sin(\alpha_{\text{right}}))] \sin(\eta)\}
\end{aligned}
\quad (4.72)
$$

Equations (4.70) and (4.71) result in

$$C_{lp}(\alpha) \approx 0.125[C_L \sin(\alpha) - C_{L\alpha} \cos(\alpha) - C_D \cos(\alpha) - C_{D\alpha} \sin(\alpha)] \quad (4.73)$$

$$
\begin{aligned}
C_{np}(\alpha) \approx {}& 0.125cb^{-1}[-C_L \sin(\alpha) + C_{L\alpha} \cos(\alpha) + C_D \cos(\alpha) + C_{D\alpha} \sin(\alpha)]\eta \\
& - 0.125[C_L \cos(\alpha) + C_{L\alpha} \sin(\alpha) + C_D \sin(\alpha) - C_{D\alpha} \sin(\alpha)]
\end{aligned}
\quad (4.74)
$$

Derivation of the rolling and yawing moments about c.g. due to yaw relies on notations of Fig. 4.9 where

$$L_{\text{left}} = 0.5\rho S(V - \Delta V)^2 0.5C_L \quad D_{\text{left}} = 0.5\rho S(V - \Delta V)^2 0.5C_D$$
$$L_{\text{right}} = 0.5\rho S(V + \Delta V)^2 0.5C_L \quad D_{\text{right}} = 0.5\rho S(V + \Delta V)^2 0.5C_D$$

(4.75)

and

$$\Delta V \approx 0.25rb \tag{4.76}$$

For the rolling moment (see Fig. 4.9a), one can write

$$QSb\frac{pb}{2V}C_{lr} = 0.25b\{[L_{\text{right}}\cos(\alpha) + D_{\text{right}}\sin(\alpha)]$$
$$- [L_{\text{left}}\cos(\alpha) + D_{\text{left}}\sin(\alpha)]\}\cos(\eta)$$

(4.77)

resulting in

$$C_{lr}(\alpha) \approx 0.25[C_L\cos(\alpha) + C_D\sin(\alpha)] \tag{4.78}$$

For the yawing moment (see Fig. 4.9b),

$$QSb\frac{pc}{2V}C_{nr} = 0.25b\{(L_{\text{right}}\sin(\alpha) - D_{\text{right}}\cos(\alpha)) - (L_{\text{left}}\sin(\alpha) - D_{\text{left}}\cos(\alpha))$$
$$+ 0.25c[(L_{\text{left}}\cos(\alpha) + D_{\text{left}}\sin(\alpha)) - (L_{\text{right}}\cos(\alpha) + D_{\text{right}}\sin(\alpha))]\sin(\eta)\}$$

(4.79)

and therefore

$$C_{nr}(\alpha) \approx 0.25[C_L\sin(\alpha) - C_D\cos(\alpha)] - 0.25[C_L\cos(\alpha) + C_D\sin(\alpha)]\eta cb^{-1} \quad (4.80)$$

Following this approach, analytical expressions for the side-force and moment coefficients and the relevant derivatives are found from the canopy lift and drag curves.

4.2 CFD ANALYSIS

As shown in the preceding section, the lifting-line theory can be used to estimate the aerodynamic characteristics of a canopy. This involves a series of assumptions to account for the effects of the opening at the canopy leading edge and the arc-anhedral geometry of the canopy. Obviously, the spanwise variation of the canopy thickness, due to the ribs in between the cells, is difficult to incorporate into analytical models. Hence, CFD can be used to compute the flowfield and the resulting forces on a model canopy. Subsequently, the fidelity of lifting-line theory in predicting the aerodynamic forces on the ram-air canopy can be assessed. Ram-air canopies have typically much higher drag when compared to a rigid wing with the same AR and based on the same airfoil profile. Two- and three-dimensional potential flow computations of [Ross 1993] show that the cell opening cut at the leading edge has a significant effect on the performance, and the lift-to-drag ratio can be increased by reducing the cell opening and modifying the opening position with respect to the airfoil stagnation point. Moreover, [Chatzikonstantinou 1993, 1999] has carried out coupled finite element and potential (vortex lattice) analysis of a ram-air personnel canopy. Several two-dimensional Reynolds-averaged Navier–Stokes computations of ram-air canopy airfoil profiles also have been carried out by [Balaji et al. 2005; Mittal et al. 2001; Mohammadi and Johari 2010]. The former two studies used a finite element solver and the Baldwin–Lomax turbulence model whereas the latter used a finite volume solver along with the Spalart–Allmaras turbulence model. This section, based on a slightly edited work by [Eslambolchi and Johari 2013 as well as Mkrtchyan and Johari 2011], presents an example of full-scale, viscous, three-dimensional computations of ram-air canopy.

4.2.1 CANOPY MODELING

If not known a priori, the specifications of inflated canopy geometry such as the spanwise curvature, relative inflation, and orientation of each cell can be extracted from close-up images of canopy in flight, such as the one in Fig. 4.10 (or Fig. 4.1a). This information along with the canopy cut pattern can be employed to create a full-scale solid model of the canopy (without the suspension lines) using any solid modeling software package. To generate the three-dimensional (3D) model of the canopy, a two-dimensional (2D) rib should be created from the parachute drawings first. The angle between the freestream and the chord line of the rib at the canopy centerline constitutes AoA.

Once the rib is formed, information extracted from the canopy photographs along with the canopy dimensions can be used to construct one of the two half-cells

Fig. 4.10 Close-up image of the MC-4 parachute in flight.

in the center of the canopy. The process of creating this half-cell is carried out in a way that performing a mirror function produced the center cell of the canopy. Subsequently, the mirror function can be used to generate the full 3D canopy. Lastly, the stabilizers should be added to the two ends of the canopy. All surfaces are combined together into a single solid model with surface lofts and a finite thickness in preparation for the surface and volume mesh production. Figure 4.11 shows several views of the canopy solid model. This model is then imported into the CFD preprocessor to generate the volume mesh.

The specific canopy shown in Figs. 4.10 and 4.12, MC-4, has a constructed chord length of $c = 3.96$ m (13 ft) and a span of $b = 8.69$ m (28.5 ft), and it is made from low-permeability PIA-C-44378, Type IV fabric. The permeability of this fabric is in the range of 0.5 to 3.0 ft^3/min/ft^2 (0.15 to 0.9 m^3/min/m^2), which translates to an average flow through the fabric of less than 1 cm/s at pressure differentials comparable in typical flight conditions. This value is quite small when compared with the freestream velocity of 12.2 m/s (23.7 kt), and therefore the effect of flow through the canopy surface can be neglected. The computational domain for the CFD analysis should be chosen such that there is about one canopy span on either side of the canopy, at least several canopy chord lengths ahead and downstream, and several chord lengths above and below the canopy. These suggested minimum values tend to minimize the effects of the boundaries on the flow in the vicinity of the canopy. Specifically, for the MC-4 canopy the 3D computational domain was chosen as a rectangular volume having a cross section of 28×23 m (92×75.5 ft) and a length of 27 m (88.6 ft), corresponding to $7.1 \times 5.8 \times 6.8$ canopy chord lengths with the relative placement of the canopy within the computational domain is shown schematically in Fig. 4.12. The canopy occupied less than 0.1% of the computational volume, and hence blockage effects were negligible. In the spanwise direction, the computational domain is

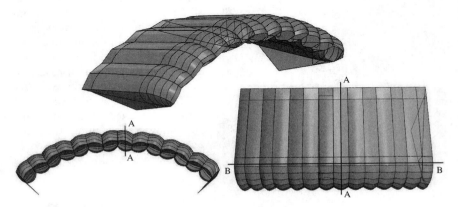

Fig. 4.11   Several views of the model canopy.

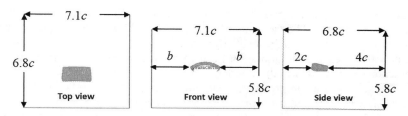

Fig. 4.12 An example of a computational domain.

over three times the canopy constructed span. There are 2c of clearance in front of the canopy and 4c of clearance behind the canopy. There are about 3c of clearance at the top and bottom of the canopy. The choice of this domain was based on a study where the flowfield around a 2D slice of a ram-air canopy was simulated [Mohammadi and Johari 2010]. In that study, the domain size around an airfoil was systematically varied, and the computed lift and drag were compared to the experimental values. The results showed that the smallest domain size that provided a domain-independent solution was $7c \times 4c$. Thus, a domain consisting of at least 2c above, below, upstream, and 4c downstream of the canopy is expected to provide a domain-independent solution. This was verified in a set of runs discussed in the next section. Also, allowing for one span length of clearance on either side of the canopy ensures that the domain boundary effects will not interfere with the canopy flow near its tips.

A robust (Octree) tetrahedral/mixed unstructured grid can be used to discretize the entire domain, inside and outside of the canopy. The Octree is a top-down approach that generates a tetrahedral mesh. An existing surface mesh is not required because one gets created by the Octree process. It is based on a spatial subdivision algorithm that ensures refinement of the mesh where necessary, but maintains larger elements where possible, allowing for faster computation. Once the "root" tetrahedron, which encloses the entire geometry, has been initialized, the meshing algorithm subdivides the root tetrahedron until all element size requirements are met. The mesh generator balances the mesh so that elements sharing an edge or face do not differ in size by more than a factor of two. The mesh is smoothed by moving nodes, merging nodes, swapping edges, and in some cases, deleting poor elements. Further details can be found in [Eslambolchi 2012].

The maximum size of surface elements on the canopy was set at $0.003c$, which is equivalent to 12 mm (0.5 in.). This value was found by trial and error and the limitation of 40 million mesh elements stemming from the available memory. The mesh was concentrated around the canopy surface, and the size of elements increased with the distance from the canopy surface. This created a very dense mesh close to the canopy surface. The initial mesh for each case was ~30 million elements, and then multiple dynamic refinements were performed during the computation to allow placement of more refined elements in areas with steep velocity gradients. The refinements were performed until the available memory was reached. The y^+ values varied for different conditions and across the canopy surface, but the majority of elements had values of 5 or less for the densest mesh used for each orientation. For most cases, there were more than 35 million elements and about 8.5 million nodes.

The steady, incompressible Reynolds-averaged Navier–Stokes (RANS) equations with constant fluid properties were solved over a rectangular volume containing the canopy. The one-equation Spalart–Allmaras turbulence model [Spalart and Allmaras 1992], which is appropriate for aerodynamic flows at higher Reynolds numbers, was used to account for the turbulent stresses. The equations were solved by ANSYS Fluent, which is a well-established commercial finite volume flow solver. The canopy fabric was assumed to be impermeable and rigid, and the no-slip boundary condition was applied to the interior and exterior surfaces of the canopy. The boundary condition on the computational domain consisted of uniform velocity of $V = 12.2$ m/s (23.7 kt) (freestream value) at the inflow and vanishing viscous stress (traction-free) at the outflow. The normal velocity component on, as well as all gradients normal to, the top and bottom plus the right and left boundaries of the computational domain was set to zero (free slip), which makes these boundaries into stream surfaces. A sufficient number of iterations were carried out for each case such that fluctuations of the forces were within 4% of the mean value. Moreover, typically between 2,000 and 3,000 iterations were performed for each case, and the fluctuations were about 1% of the mean value.

For angles of attack other than zero, the canopy and coordinate system were fixed in space, and the computational domain was rotated so that the chordline of the canopy center rib was at the desired angle with the freestream velocity. For the cases with nonzero sideslip angles, the computational domain was rotated sideways. For each orientation, a new mesh was generated with the procedure described earlier. Even though the number of mesh elements was different for each orientation, the mesh resolution and density were comparable for all the cases examined.

To examine the dependence of the forces on the mesh density, the case of canopy at AoA of $\alpha = 6$ deg was created with mesh densities ranging from 7 to 38 million elements. Then, CFD simulations were performed for a sufficient number of iterations until the force fluctuations had reached within 1%. A comparison of the forces and pitching moment on the canopy indicated that for mesh densities greater than 15 million elements, the lift and drag remained within 0.5% and the pitching moment within 1.1%. For mesh densities as small as 7 million elements, the lift, drag, and pitching moment were only 0.8, 1.7, and 1.5% different from the case with the densest mesh. As the data reported here are from the densest mesh, we conclude that the solutions are mesh-independent.

To quantify the discretization uncertainty of the present computations, the procedure outlined in [Celik et al. 2008], which is based on the Richardson extrapolation, was followed. Three different cases with mesh densities of 7, 15, and 38 million elements at $\alpha = 6$ deg were created, and the computed lift and drag coefficients were chosen as the key variables. The mesh refinement factor between the mesh pairs was 1.3. The approximate relative error was 0.5 and 0.04% for the lift and drag, respectively. The extrapolated relative error was 3.8 and 0.04% for the lift and drag. The fine-grid convergence index was 4.9 and 0.009% for the lift and drag, respectively. These values for pitching moment are within the ones reported for the lift and drag. Thus, a relative error of less than 4% and a fine-grid convergence index of less than 5% are reported for the computations carried out in this study.

4.2.2 CANOPY AERODYNAMIC COEFFICIENTS

Following CFD simulations, flowfield characteristics as well as the aerodynamic force and moment coefficients can be computed. In this section, the results are divided into two categories for the cases of steady glide and with sideslip. The AoA was varied from $\alpha = -4$ to 14 deg, and the freestream velocity was $V = 12.2$ m/s corresponding to a Reynolds number of $Re = 3.2 \times 10^6$. The flowfield is presented in terms of the pressure coefficient C_p and velocity magnitude contours. The forces and moments extracted from the CFD dataset were scaled by the freestream dynamic pressure of $Q = 91$ Pa and the canopy constructed area of $S = 370$ ft^2 (34.4 m^2). For the moment coefficient, the canopy chord length was used as the length scale.

As an example, the following shows a representative case for a flow simulation at $\alpha = 10$ deg without sideslip. Two specific slices of the flowfield were examined: a streamwise plane through the middle of the central half-cell and a spanwise plane at the quarter-chord of the canopy. These slices correspond to the A-A and B-B cuts in Fig. 4.11. The canopy upper and lower surfaces in the A-A plane correspond to the most inflated section of the cell.

The pressure coefficient distribution within the streamwise plane through the middle of the central half-cell at $\alpha = 10$ deg is shown in Fig. 4.13 where the darkest shade is indicative of the stagnation pressure of $C_p = 1$, and the lightest shade the lowest pressure coefficient of $C_p \approx -1.9$. The maximum pressure appears on the canopy lip (leading edge) where the stagnation point is located. The pressure within this plane starts at $C_p \approx 0.6$ just outside the opening and increases to $C_p \approx 0.95$ inside the canopy. There are two low-pressure regions, one on the top of the canopy (as expected) and the other below the lower lip. The latter is associated with a separated region, which was also observed in the earlier 2D simulations [Mohammadi and Johari 2010]. The pressure on the upper exterior surface of the canopy increases continuously indicating an attached boundary layer.

The velocity magnitude distribution within the streamwise plane through the middle of the first cell at $\alpha = 10$ deg is shown in Fig. 4.14. Again, the darkest shade represents the stagnation region (zero velocity). The flow barely enters the opening, which results in minor low-speed motions near the canopy opening as indicated by the lighter shaded region, but the majority of the flow

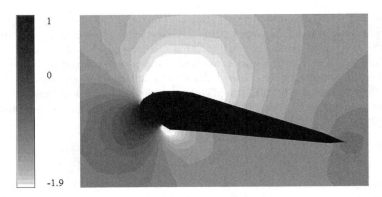

Fig. 4.13   Pressure coefficient distribution in a streamwise plane at α = 10 deg.

follows the opening to the lower lip where the flow has to turn sharply resulting in the separation region. The stagnation point at the upper lip is followed by acceleration over the top of the canopy and a slow recovery toward TE. The boundary layer on the upper surface is thin indicating attached flow. On the other hand, the flow on the lower surface starts out with a separated region on the lower lip, and a thick boundary layer follows all the way to TE. This separation region is the cause of the low pressure on the lower lip.

The velocity magnitude distribution within the spanwise plane at the quarter-chord of the canopy at α = 10 deg is also shown at the bottom of Fig. 4.14. The maximum velocity in this plane is 1.5 times the freestream velocity. The higher velocity regions over the central section on the top of the canopy are expected as a result of the flow acceleration over the curved surface. The flow within the canopy is at very low velocity or stagnant in this plane. The surprising aspect of the flow is the nonuniform low-velocity regions below the canopy that corre-

spond to the separation bubbles. The separation bubbles are absent from the concavities on the lower surface, beneath each rib. The elimination of the low-speed regions beneath the ribs

Fig. 4.14   Velocity magnitude distribution in the streamwise and spanwise planes at α = 10 deg.

is most likely due to the spanwise variation of adverse pressure gradients seen by the oncoming flow.

The other plane through the data that was explored was a spanwise plane perpendicular to the freestream at the quarter-chord of the canopy, corresponding to the B-B cut in Fig. 4.11. The pressure coefficient distribution in this spanwise plane at the quarter-chord of the canopy at $\alpha = 10$ deg is shown in Fig. 4.15. The lowest pressure in this plane corresponds to $C_p \approx -0.95$. Pressure within the canopy is not uniform, and it decreases from the central portion toward the canopy edges. Furthermore, the low pressure across the top of the canopy also decreases towards the canopy far edges. This is expected considering the load distribution on a finite wing. The lowest pressure on top of the canopy is in between the cells in the central section of the canopy.

The computed lift and drag coefficients as well as the lift-to-drag ratio and the pitching moment are plotted in Fig. 4.16. The lift coefficient increases linearly with AoA up to $\alpha = 12$ deg; beyond this AoA, the canopy approaches stall. By fitting several lines to the linear segments of the data, the best-fit line results in a lift slope $C_{L\alpha}$ of 0.03 deg^{-1}. The best-fit line was based on the data in the range of $\alpha = -2$ to 12 deg. The zero-lift AoA is $\alpha_0 = -4$ deg. The computed lift characteristics are typical for a finite wing at comparable Reynolds numbers. The drag coefficient plotted in Fig. 4.16b has a minimum at angle $\alpha = 2$ deg. For large angles of attack, the drag increases rapidly as flow separation extends over the upper surface of the canopy. For the AoA of $\alpha = 14$ deg with the canopy approaching stall, drag coefficient is about twice the minimum value. The lift-to-drag ratio L/D increases with AoA, and the largest lift-to-drag ratio of 5.7 for the canopy is at $\alpha = 10$ deg. Beyond this angle, the lift-to-drag ratio decreases as the canopy approaches stall.

The pitching-moment coefficient C_m about the quarter-chord, and halfway between the upper and lower surfaces, is plotted in Fig. 4.16d; it increases with AoA in the range of $\alpha = -4$ to 2 deg, and then it decreases nearly linearly with AoA from $\alpha = 2$ to 12 deg. For the AoA of $\alpha = 14$ deg, the pitching moment increases suddenly because of stall. The pitching-moment trend appears to

Fig. 4.15   Pressure coefficient distribution in the spanwise plane at $\alpha = 10$ deg.

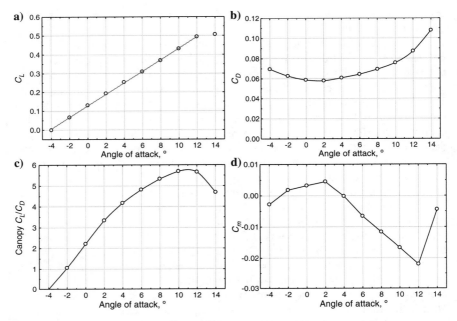

Fig. 4.16   a) Lift coefficient, b) drag coefficient, c) lift-to-drag ratio, and d) pitching-moment coefficient.

reveal that the canopy is statically stable in pitch in the AoA range of 2 deg $\leq \alpha \leq$ 12 deg.

The effect of varying the sideslip angle between 0 deg $\leq \beta \leq$ 20 deg at the fixed AoA of $\alpha = 10$ deg on the canopy forces and moments are presented next. The choice of $\alpha = 10$ deg was based on the fact that the maximum lift-to-drag ratio is observed at this angle. The flow characteristics at the sideslip angle of $\beta = 10$ deg and AoA of $\alpha = 10$ deg are presented next as a representative case. As in the preceding section, the streamwise plane through the middle of the central half-cell (A-A) and the spanwise plane at quarter-chord of the canopy (B-B) are examined in detail.

The pressure distribution in the streamwise plane through the middle of the first cell at $\beta = 10$ deg and $\alpha = 10$ deg is shown in Fig. 4.17. The maximum pressure appears on the canopy lip at the leading-edge stagnation point. Furthermore, the pressure inside the canopy within this plane is $C_p \approx 0.9$ behind the opening and the upper lip stagnation point. Further inward, there is a region with $C_p \approx 0.8$ as a result of the minor flow through the cell opening. Toward TE of the canopy, pressure increases to $C_p \approx 0.9$ again. There are two low-pressure regions, one on the top of canopy (as expected) and the other below the lower lip. Similar to the zero sideslip angle case, the low-pressure region

Fig. 4.17   **Pressure coefficient distribution in the streamwise plane at $\beta = 10$ deg.**

below the canopy lower lip is associated with a separated region. Interestingly, the separated region is smaller than the case without the sideslip angle. The pressure on the rest of the canopy exterior surface increases continuously indicating attached boundary layers.

The velocity magnitude distribution within the streamwise plane through the middle of the central half-cell at $\beta = 10$ deg and $\alpha = 10$ deg is shown in Fig. 4.18. There is an area near the cell opening with minor low-speed motions as indicated by the lighter shades. These low-speed motions have velocities greater than the case without sideslip. It is suspected that the flow inside the canopy is caused by the spanwise pressure gradient. Also, these low-speed motions penetrate to about halfway within the canopy.

The stagnation point on the upper lip is followed by acceleration over the top of the canopy and the slow recovery toward TE. The boundary layer on the upper surface is quite thin indicating attached flow although there appears to be a small separation region at about 50% of the chord length. Subsequently, the flow reattaches prior to complete separation near TE. On the other hand, the flow on the lower surface starts out with a

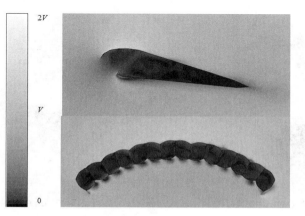

Fig. 4.18   **Velocity magnitude distribution within the streamwise and spanwise planes at $\beta = 10$ deg.**

separated region on the lower lip, and a thick boundary layer follows all of the way to TE. This separation region is similar to that observed for the case without sideslip and is the cause of the low pressure on the lower lip.

The velocity magnitude distribution within the spanwise plane at the quarter-chord of the canopy at $\beta = 10$ deg and $\alpha = 10$ deg is also shown at the bottom of Fig. 4.18. The higher-velocity regions over the central section on the top of canopy are expected as a result of the flow acceleration on this curved surface. The flow within the canopy is at very low velocity or stagnant at this location. The interesting aspect of the flow is the asymmetric low-velocity regions below each half-cell that correspond to the separation bubbles. These separation bubbles are larger on the left side of the canopy compared to the right side. Also, these separation bubbles are generally smaller when compared to the no-sideslip case. The concavities on the lower surface appear to energize a portion of the boundary layer and eliminate the low-speed separated regions in between adjacent half cells. Another aspect that was not present in the case without sideslip is the low-velocity regions in the concavities on the upper surface of the right side of the canopy that correspond to a separated boundary layer.

The other cut-plane considered is the spanwise plane at the quarter-chord location. The asymmetry of the flowfield is clearly visible in the datasets for this plane as the left side of the canopy is upstream of the right side. The pressure distribution in this plane at $\beta = 10$ deg and $\alpha = 10$ deg is shown in Fig. 4.19 where the lowest pressure regions correspond to $C_p \approx -1.2$. Pressure within the canopy is not uniform or symmetric as a result of the sideslip. Pressure increases from the left end of the canopy toward the right end of the canopy. At the left end of the canopy, $C_p \approx 0.4$; in the central portion of the canopy, the pressure increases to $C_p \approx 0.8$, and it reaches the maximum of $C_p \approx 0.99$ at the right end of the canopy. Although the low-pressure across the top of the canopy is not uniform or symmetric, it generally decreases towards the canopy edges, and the lowest pressure is in between the cells in the middle section of the canopy. As a result

Fig. 4.19 Pressure coefficient distribution in the spanwise plane at $\beta = 10$ deg.

of the sideslip, Fig. 4.19 features low-pressure regions to the right side of the stabilizer.

The lift, drag, and side-force coefficients on the canopy extracted from the simulations are presented in Fig. 4.20 for the four sideslip angles of $\beta = 5$ deg, 10, 15, and 20 deg at AoA of $\alpha = 10$ deg. As before, the forces were scaled with the freestream dynamic pressure Q and the canopy constructed surface area S. The lift coefficient varies very slightly (by less than $\pm 3\%$ of the average) for the range of sideslip angles considered. However, there appears to be a trend of decreasing C_L with β for sideslip angles greater than $\beta = 5$ deg. In contrast to

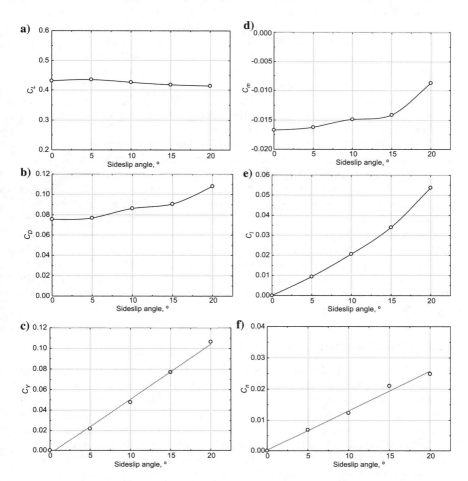

Fig. 4.20   a) Lift, b) drag, c) side-force coefficients along with d) pitching-, e) rolling-, and f) yawing-moment coefficients at $\alpha = 10$ deg.

the lift, the drag is strongly dependent on the sideslip angle. Drag coefficient increases minimally at the 5-deg sideslip angle but subsequently increases rapidly with sideslip angle. At $\beta = 20$ deg, the drag coefficient is 1.4 times the value for the no-sideslip case.

The side-force coefficient C_Y at the 10-deg AoA increases linearly with the sideslip angle, and the rate of increase is $C_{Y\beta} = 0.0054$ deg^{-1}. The side force is much smaller than the lift at low sideslip angles; however, it reaches a value of about one-fourth of the lift at the highest sideslip angle considered. Similarly, the side force is less than one-third of the drag at low sideslip angles, but C_Y increases to a value equal to the drag at the highest sideslip angle of $\beta = 20$ deg.

The pitching-, rolling-, and yawing-moment coefficients about the quarter-chord location, halfway between the upper and lower surfaces, are presented in Figs. 4.20b, 4.20d, and 4.20f. The chord length was used as the scaling length in computing the moment coefficients. The pitching-moment coefficient plotted in Fig. 4.20b increases with sideslip angle. The increase of pitching moment with sideslip angle is nonlinear and can be fitted approximately by the second-order polynomial $C_m = (3 \times 10^{-5})\beta^2 - (2 \times 10^{-4})\beta - 0.0164$. At the largest sideslip angle of $\beta = 20$ deg, the pitching-moment coefficient is about 48% greater than the value at zero sideslip.

The variation of rolling-moment coefficient with the sideslip angle is analogous to the pitching-moment trend. The rolling-moment coefficient plotted in Fig. 4.20d increases nonlinearly with the sideslip angle, and the variation can be described by the second-order polynomial $C_l = (7 \times 10^{-5})\beta^2 + 0.0013\beta + 0.0004$, where β is in degrees. The rolling-moment coefficient at $\beta = 20$ deg is 5.7 times the value at $\beta = 5$ deg.

In contrast to the pitching and rolling moments, the yawing-moment coefficient in Fig. 4.20f increases linearly with sideslip angle. The rate of increase of yawing moment with the sideslip angle is $C_{n\beta} = 0.0013$ deg^{-1}. The trends of rolling- and yawing-moment coefficients indicate that as the canopy starts to turn, the rolling moment increases much faster than the yawing moment, at least for the $\alpha = 10$ deg AoA.

4.2.3 COMPARISON WITH LIFTING-LINE THEORY

As mentioned already, the lifting-line theory has been proven to be successful in predicting the aerodynamic characteristics of finite wings for large Reynolds numbers and AR greater than about 3. For the low-AR wings, TE of the wing is exposed to a flow that has been affected by the tip flow, and lift cannot be assumed to act on a line. In other words, the effects of the lateral edges become significant when AR is small. Several efforts have been made to extend the lifting-line theory for small AR and swept wings (e.g., Helmbold's equation). Here we use the lifting-line theory as outlined in Chapter 2 to verify its applicability to ram-air canopies with arc anhedral, complex surface geometry, and leading-edge opening. Specifically, the lift and drag were calculated from a 2D computation of the flow around a slice of the MC-4 canopy. The 2D data were extended to the 3D for AR corresponding to the MC-4 using the lifting-line theory, and the results were compared with the lift and drag computed for the 3D MC-4 canopy model. Allowances were made for the rectangular planform and arc anhedral of the canopy. The key expressions used to extend the 2D load coefficients are presented in Chapter 2.

The lift and drag coefficients for a rectangular wing before stall can be described by Eqs. (2.7) and (2.10), respectively. For the wings with the arc-anhedral angle, the lift coefficient is corrected according to Eq. (2.16) while the drag coefficient is not affected by the arc-anhedral angle.

The nominal AR for the MC-4 canopy is AR $= 2.19 = 8.7$ m/5 m (28.5 ft/ 13 ft). The 2D lift-curve slope was found by fitting a line to the 2D lift coefficient data up to 8 deg; the slope was $C_{L\alpha}^{a} = 0.1024 \deg^{-1}$, and the zero-lift AoA was $\alpha_0 = -1.6$ deg. The resulting finite-wing lift-curve slope was calculated to be $C_{L\alpha} = 0.0534 \deg^{-1}$. Using these values and the preceding lifting-line expressions, the lift and drag coefficients for the MC-4 canopy were estimated.

The lifting-line predicted lift coefficient along with that from the 3D CFD model are plotted in Fig. 4.21. Clearly, the lift-curve slope of the 3D model is much less than the lifting-line prediction; the latter is 78% greater than the 3D CFD value. Moreover, the 3D model zero-lift AoA of $\alpha_0 = -4$ deg is smaller than the $\alpha_0 = -1.6$ deg value from the 2D data. Interestingly, the two curves in Fig. 4.16 cross at about $\alpha = 4$ deg. For angles of attack smaller than $\alpha = 4$ deg, the lifting line underpredicts the lift coefficient for the MC-4 canopy compared to the 3D results. On the other hand, for angles of attack larger than $\alpha = 4$ deg, the lifting line overpredicts compared to the 3D results. For the 2 deg $\leq \alpha \leq 10$ deg AoA range, the lift coefficients differ from about -15 to $+26$%. An alternative approach would employ the

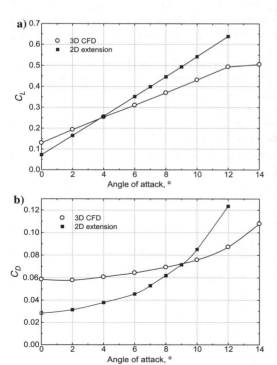

Fig. 4.21 Comparison of a) lift and b) drag coefficients from the 3D model with the lifting-line prediction.

Helmbold equation [Eq. (4.38)]. Applying this equation results in a lift-curve slope of 0.0473 deg^{-1}. Even though this slope is still 57% greater than the 3D CFD value, the use of Eq. (4.38) is recommended for canopies with AR less than 3.

Figure 4.21 compares the drag coefficient from the extension of 2D data with that from the 3D CFD model. When compared to the data from the lifting-line extension, the drag coefficient for the 3D model increases much more gradually with increasing AoA. The two curves cross over at the AoA of 9 deg. For angles of attack less than 9 deg, the 2D extension underpredicts the drag coefficient compared to the 3D CFD results. On the other hand, for angles of attack larger than 9 deg, the 2D extension overpredicts the drag coefficient for the MC-4 canopy compared to the 3D CFD results.

The computed lift-to-drag ratio of the MC-4 canopy is compared with that obtained from the lifting-line extension of the 2D data in Fig. 4.22. The lift-to-drag ratio from the 3D CFD simulations increases with AoA, and the largest lift-to-drag ratio of 5.7 is achieved at $\alpha = 10$ deg; after this AoA, the lift-to-drag ratio decreases as the canopy approaches stall. On the other hand, the lift-to-drag ratio from the extensions of 2D data achieves its largest value of 7.7 at $\alpha = 6$ deg. There is a 4-deg difference between maximum lift-to-drag ratio AoA found from the 3D CFD simulations and the extensions of the 2D data. For nearly the entire AoA range, the lifting-line theory overpredicts the lift-to-drag ratio, by as much as 65%, compared to the 3D CFD simulations. This lift-to-drag ratio refers to the canopy only and not the entire system; the latter is expected to be significantly smaller.

There are multiple reasons for the observed differences between the computed aerodynamic coefficients of the 3D model of the MC-4 and those obtained from the lifting-line extension of the 2D data. The two most important factors are the low AR, where the influence of lateral edges becomes increasingly significant, and the geometry of the upper and lower surfaces. The local curvatures and concavities on the upper and lower surfaces cannot be accounted for in the 2D data or the lifting-line theory. The 3D geometry of the canopy not only affects the slope of the lift curve, but also the induced drag, which, in turn, modifies the drag coefficient. Based on the data examined in this study, it does not appear that any 2D cut of the canopy, whether from the rib or any other section, could be used to accurately predict the

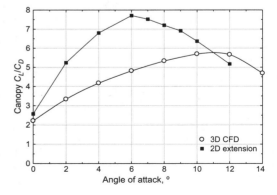

Fig. 4.22 Comparison of lift-to-drag ratio for the 3D model with the lifting-line prediction.

aerodynamic coefficients of the canopy. Interestingly, the extension of the 2D data does produce reasonable values for the lift and drag over certain, limited AoA ranges. For the case considered here, the canopy is expected to operate at 8 deg $\leq$ $\alpha \leq 10$ deg, and the drag coefficient is estimated reasonably (within $\pm 10\%$) from the 2D data over this range. On the other hand, the lift coefficient is overpredicted by more than 25% over the same range. Hence, even though the lifting-line extension of the 2D data does not result in fundamentally correct predictions, this approach can be used for arriving at first-order engineering estimates without the resources needed to carry out a full-scale CFD simulation.

4.2.4   AERODYNAMICS OF SUSPENSION LINES, PAYLOAD, AND SLIDER

With canopy being a major contributor of aerodynamic forces and moments, other components of PADS include the suspension lines, risers, payload, AGU (attached atop the payload or at a confluence point), and slider (which controls canopy opening). All of these components are clearly seen in Fig. 4.23 featuring two examples of PADS with AGU atop the payload connected to the main canopy by short risers (Fig. 4.23a) and more a complex configuration with elongated risers (Fig. 4.23b). The following attempts to estimate the contribution of these other components to the aerodynamic loads.

Both circular and noncircular lines constructed from several materials are used as suspension lines in ram-air parachutes. The Reynolds number of suspension lines in flight is typically on the order of thousands. Drag of rigid circular cylinders in this Reynolds number range is approximately one [Hoerner 1965]. Traditionally, the total length of the lines along with the drag coefficient of one is used to estimate the suspension line drag, as in Chapter 2. However, suspension lines can vibrate depending on the flight conditions and the tension in the lines. Drag of vibrating lines could be significantly greater than stationary lines. To determine the drag of suspension lines, first the stationary drag of lines is

a)   b)

Fig. 4.23   PADS examples: a) Sherpa 2200 and b) SPADeS 1000 [PATCAD 2005, PATCAD 2009].

computed, and then the additional drag due to vibration is added. Blevins [1990] provides a procedure to determine whether a circular cylinder vibrates and the amount of the associated drag increase. The procedure for the determination of the line drag is outlined next.

As the orientation of the lines with respect to the oncoming airstream determines the drag of the lines, the angle θ formed by each line segment and the freestream flow has to be determined using the CAD model of the parachute system. Hoerner [1965] shows that the forces on a rigid cylinder depend on the angle θ as follows:

$$C_D^l = C_{D0} \sin^3(\theta), \quad C_L^l = C_{D0} \sin^2(\theta) \cos(\theta) \tag{4.81}$$

When the line is perpendicular to the flow, $\theta = 90$ deg, the drag coefficient is C_{D0}, the baseline value, and the lift coefficient is zero as expected. At the other limit of $\theta = 0$ deg, the line is aligned with the flow, and lift and drag coefficients both vanish. Of course, the viscous drag is neglected in this limit. For ram-air parachutes, the baseline line-drag coefficient is expected to be $C_{D0} = 1$. Knowing the orientation of each line segment, the drag of all suspension lines under stationary conditions can be calculated.

The ratio of natural frequency to vortex shedding frequency of the suspension lines is an important parameter in determining whether the line segments of the suspension line system fall in a lock-in regime where vibration of the lines becomes significant. To assess the vibration of lines, the procedure outlined in [Blevins 1990] is followed where it is necessary to know the tension, damping characteristics, and natural and shedding frequencies of each line segment. For each line segment, the length l, diameter d, mass density m (mass per unit length), tension T, line angle θ, elastic modulus E, and the Poisson's ratio is required. The flow velocity and fluid properties allow the line Reynolds number $Re = \rho V d / \mu$ to be calculated, where ρ and μ are the air density and dynamic viscosity, respectively. Then, the vortex shedding frequency f_s for a stationary line is found from the Strouhal number $S = f_s d / V$. The Strouhal number is a function of Reynolds number, and for the range of parameters encountered by ram-air parachutes it is approximately constant at a value of 0.22. The vortex shedding frequency is then given by $f_s = SV/d$.

The next step is to compute the fluid damping term

$$\xi = \frac{\pi \rho d^2}{2 \; m} \left(\frac{\mu/\rho}{\pi f_s d^2} \right)^{1/2} \tag{4.82}$$

and the modified damping term $2\pi\xi(2m)/(\rho d^2)$. Both of these terms are dimensionless. The natural frequency of each line segment is computed assuming that the ends of each line segment are rigidly held

$$f_n = \frac{n}{2L} \sqrt{\frac{T}{m}} \tag{4.83}$$

Here, n is the vibration mode, and under most circumstances the fundamental $n = 1$ mode results in the largest amplitudes. Thus, only the fundamental mode is considered.

To calculate the natural frequency of each line segment, the tension in that segment is needed. Tension in each line segment can be determined by using the solid CAD model of lines and a finite element based analysis software package. In our work, we have used SolidWorks, a 3D CAD software suite, to create the model and to carry out the analysis of the lines. The ends of lines connecting to the canopy, at a given AoA, are assumed to be held rigidly, and a force equal to the weight of the payload is applied to the lowest section of the lines connected to the confluence point. In this manner, the von Mises stress in each line segment can be computed using the structural analysis tool in the SolidWorks Simulation suite. The computed stresses in each line segment allow the calculation of tension and, subsequently, the natural frequency of that line segment.

The procedure for computing the average drag of vibrating line segments is outlined in Fig. 4.24. Two main questions need to be answered. First, is the modified damping parameter $2\pi\xi(2m)/(\rho d^2)$ less than 64? Next, does the ratio of natural frequency over the shedding frequency fall in the lock-in (synchronization) regime? To answer the second question, the amplitude of vibration A_y/d is needed. Figure 4.25 provides the vibration amplitude as a function of a damping parameter and the line l/d ratio.

As the line diameters are very small compared to the length and the damping term in the abscissa is typically less than one, the vibration amplitude is taken to be $A_y/d = 1$. Knowing the amplitude, the lock-in range in the map of Fig. 4.26 could be used to determine whether each line segment falls within or outside of the synchronization regime. However, the lock-in map in [Blevins 1990] only extends up to $A_y/d \sim 0.5$. This map can be extended by fitting a parabola to the data. The plot in Fig. 4.27 shows that the line segments that fall within the lock-in regime have natural frequencies between 0.4 and 1.6 times the shedding frequency. Within this lock-in regime, drag coefficient increases to 3.1 due to the line vibration [$C_D' = 1 + 2.1(A_y/d)$ with $A_y/d = 1$]. Outside this region, vortex shedding does not cause significant vibration of the lines, and the drag coefficient of the lines is expected to remain the same as the stationary value of one. As the canopy loading changes, the tension in the lines varies, and, in turn, the natural frequency of the line segments changes. At the same time, the parachute flight velocity depends on the system loading, and thus the shedding frequency of the lines changes with the loading as well. Because the ratio of the natural to shedding frequency is the key parameter in determining whether each line segment falls within or outside the lock-in regime, general statements about the status of the lines and their contribution to the total drag of the system cannot be made. For each parachute system and operating condition, the analysis has to be performed to establish the drag of each line segment and the total line drag.

For cargo-carrying PADS, the payload can be modeled as a parallelepiped container (Fig. 4.28a) or low-AR platform with cargo on top of it (Fig. 4.28b).

Required Input

- Suspension line tension T and mass density m
- Suspension line diameter d
- Flow velocity V and line angle θ
- Fluid density ρ and kinematic viscosity v ($\mu = \rho v$)

Vortex Shedding Frequency

Compute the Reynolds and Strouhal numbers
Compute the vortex shedding frequency, $f_s = SV / d$

Non-Dimensional Parameters

Compute the modified damping parameter, $2m(2\pi\xi)\rho^{-1}d^{-2}$
Compute the ratio of natural to shedding frequency, f_n / f_s

---

Is there Significant Vibration Amplitudes?

- Is $2m(2\pi\xi)\rho^{-1}d^{-2} < 64$?
- Is f_n/f_s in the synchronization range (see Figs. 4.26 and 4.27)?
If the answer to both questions is yes, proceed

Peak Resonant Amplitude

Compute A_y (amplitude of vibration) using the data in Fig. 4.25

Line Drag

Compute the drag coefficient, $C_D^l = 1 + 2.1 A_y d^{-1}$
Compute the line drag, $D^l = 0.5\rho V^2 dl C_D^l$

Fig. 4.24   Outline of procedure for determination of line vibration.

For example, the A-22 Container Delivery System (CDS) (Fig. 4.29a) has almost a cubic geometry with a dimension of 1.22 m (4 ft); other CDS can have different dimensions. A-22 CDS is an adjustable nylon cloth and webbing container used to deliver food, medicine, ammunition, supply type loads, and disassembled, or ready-to-use equipment in a convenient 1-ton (2200-lb) bundle. It can also be used to drop folded Zodiac-type boats, motorcycles, snowmobiles, and arctic sleds. A typical CDS consists of a sling assembly, cargo cover, and four suspension webs connecting the container to the parachute (Fig. 4.29b). The entire bundle is attached to a plywood skidboard and uses paper honeycomb as energy-dissipating material.

For these type of payloads, [White 2003] provided a drag coefficient of 1.07 (for a cube aligned with the flow) at Reynolds numbers above 10^4. Cargo containers used in ram-air systems are in the turbulent regime, and flow separation exists on the edges and corners causing large drag forces. Estimation of

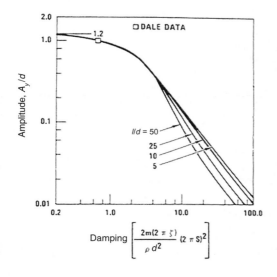

Fig. 4.25   Vibration amplitude as a function of the damping parameter [Blevins 1990].

drag coefficients for parallelepiped geometries other than cube can be made by examining two-dimensional rectangular sections. Hoerner [1965] provided values of drag coefficients from 0.9 to 2 depending on the section AR.

Figure 4.30 shows an example of CFD simulations of payload rotating in a homogenous airflow. The payload is modeled as a cubic solid with the dimensions $2.31 \times 1.89 \times 1.05$ m $(7.6 \times 6.2 \times 3.4$ ft) (L × W × H), which were adopted from a 364L pallet with approximately 20-cm (7.9-in.) space from the edges. The aerodynamics are described by the lift, drag, and pitching-moment coefficients as functions of the AoA (Fig. 4.30b). The largest drag coefficient of 1.1 appears at $\alpha = 90$ deg and $\alpha = 270$ deg, that is, when the airflow encounters the bottom or top surface of 4.37 m^2 (47 ft^2) that is also used as reference area. The lowest drag coefficient of approximately 0.4 appears at $\alpha = 0$ deg and $\alpha = 180$ deg, when the flow encounters the front or back surface of 1.98 m^2 (21.3 ft^2). Because of the simulated rotation, the coefficients are not exactly symmetric as

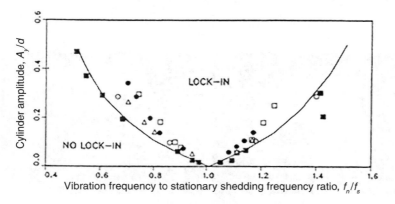

Fig. 4.26   Map of lock-in regime [Blevins 1990].

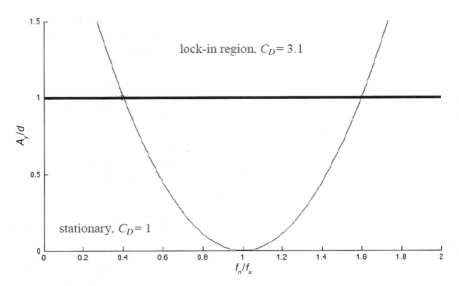

Fig. 4.27   Classification of line segments that fall in or out of the lock-in regime.

they also contain unsteady aerodynamic effects. Based on this analysis, for a cuboid shaped payload in steady-state flight (with a small AoA) a drag coefficient of approximately one can be assumed with a reference area equal to the frontal surface of payload.

A similar study was conducted for a $4.88 \times 2.75 \times 0.15$ m ($16 \times 9 \times 6$ in.) type-V Low Velocity Airdrop Delivery System (LVADS) platform with a nose bumper and $4.86 \times 2.06 \times 0.74$ m ($16 \times 6.8 \times 2.4$ ft) cargo atop it as shown in Fig. 4.31 [Alexander et al. 2007; McQuilling et al. 2011].

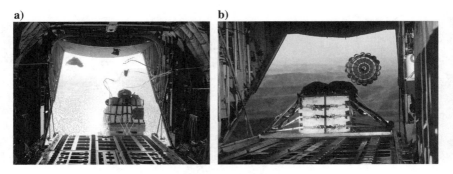

Fig. 4.28   Payload extraction.

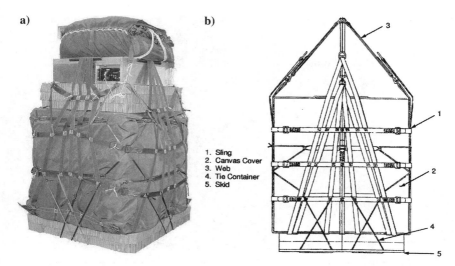

a)   b)

1. Sling
2. Canvas Cover
3. Web
4. Tie Container
5. Skid

Fig. 4.29   a) A-22 cargo bag with AGU and canopy on the top of it and b) schematics of a cargo bag.

In that study, the authors used the shear stress transport k-ω turbulence model, and all CFD simulations were carried out using a finite volume flow solver on a 10-node computer cluster composed of the total of 80 2.6-GHz processors. The solver employed an unstructured tetrahedral mesh using prism elements. For this study 32 to 61 million elements were used to cover a 3D computational domain spanned 15 m (49.2) away from the center of the geometry in each direction and 30 m (98.4) sidewise and downstream. The typical run time for each simulation (angle of attack) for two forward velocities of 9.1 m/s (17.7 kt) and 30.5 m/s (60 kt) was on the order of 12–14 h and 3 days, respectively [McQuilling et al. 2011].

Figure 4.32 shows steady-state lift coefficient, drag coefficient, and pitch-moment coefficient as computed for a variety of angles of attack with 13-deg increments along with other CFD analysis data obtained for the same payload by [Potvin et al. 2007] and experimental study for a 0.56 AR flat plate at a Reynolds numbers 3.8×10^4–5.7×10^4 by [Desabrais 2005].

For comparison, Fig. 4.32 also shows the same dependences for a 2.74 $\times$ 2.24 $\times$ 1.6 m (9 $\times$ 7.4 $\times$ 5.2 ft) 10K JPADS CDS. As seen, they are quite similar to those shown in Fig. 4.30b.

Similar to the cargo containers, AGU can be also modeled as a slender rectangular box, and therefore the drag coefficient of one can be used for computation of the guidance unit drag values.

Besides cargo packages, the payload for ram-air can be human beings as well. For completeness, let us provide the estimates of the drag coefficient in this case as

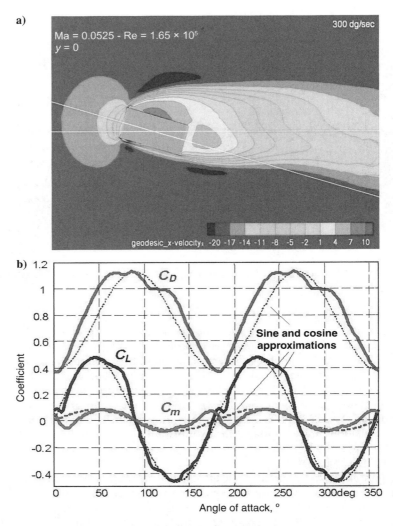

Fig. 4.30   The a) x-velocity distribution and b) aerodynamic coefficients of a cubic solid payload [Jann 2011].

well. A first estimate is provided by [Hoerner 1965] where the drag area is listed as 0.84 m^2 (9 ft^2) for an "average person" standing against the wind. It is understood that the "average person" probably does not represent a paratrooper, and the latter will likely have a drag area larger than the value provided by Hoerner. An alternative method is to follow the procedure outlined in the work by [Penwarden et al.

Fig. 4.31   Type-V platform with a nose bumper [McQuilling et al. 2011].

1978]. The surface area of a standing human is referred to as the DuBois area A_{Du} as follows:

$$A_{Du} = 0.0769 W^{0.425} H^{0.725} \tag{4.84}$$

where W is the weight in Newtons and H is the height in meters. Penwarden et al. [1978] suggest a frontal area A based on the DuBois area, $A \approx 0.35 \, A_{Du}$.

The drag coefficient based on frontal area A was measured for a number of subjects in a wind tunnel, and the data are listed in Table 4 of [Penwarden et al. 1978]. The table suggests a value of $C_D = 1.17$ for a male subject with coat and trousers. Thus, the drag area would be

$$AC_D = A_{Du}(A/A_{Du})C_D = A_{Du}(0.35)1.17 \tag{4.85}$$

Given height and weight, the drag area could be estimated using this expression. Let's assume that the average man has a height of 1.8 m (5.9 ft) and weighs 800 N (180 lb). The DuBois area is 2.02 m^2 (21.7 ft^2), and the drag area is 0.83 m^2 (8.9 ft^2). This value is quite close to the value provided by Hoerner of 9 ft^2. Thus, the expressions in Eqs. (4.84) and (4.85) can be used to arrive at a drag area, and subsequently the drag of a parachutist/paratrooper could be estimated.

Besides the canopy, lines, payload drag, one might also want to consider the drag produced by a slider that is used to control the opening of ram-air canopies and gets fixed above the confluence point after full inflation is achieved. Sliders consist of rectangular fabric elements retained horizontally at the four corners. To estimate the slider drag, one might consider the drag coefficient of flags aligned with flow. It is understood that there are differences between a slider and a flag fluttering in the wind; however, the drag coefficients are expected to be in the same range. Drag of a flag depends on AR and the fabric area. Hoerner [1965] provides a plot of the drag coefficient as a function of flag AR; see Fig. 4.33. The chord length is the dimension along the flow, and the span is in the transverse direction. Ram-air parachute sliders have AR typically less than one. Examination of several ram-air systems shows sliders have AR from about one to 0.36. The drag coefficient of flags with AR in this range is approximately 0.05. This value is suggested as the slider drag coefficient, and the slider drag is computed using the slider area and the flight dynamic pressure.

The drag of the entire parachute system can now be computed by adding the values for the various components, and the relative contributions of each component can be assessed. Moreover, as the lift is almost entirely created by the canopy, the system lift-to-drag ratio (glide ratio) during steady glide can be found.

Fig. 4.32   Coefficients of a) lift, b) drag, and c) pitch-moment [McQuilling et al. 2011].

4.2.5   AERODYNAMICS OF COMPLETE PADS

This section provides an example of estimating performance of a model ram-air parachute system as a whole based on contributions from its individual components, that is, canopy, lines, and payload, as outlined in the preceding sections. In this example, the canopy was assumed to have a rectangular planform with an area of 93 m^2 (1,000 ft^2). The AR was assumed to be 3.2 and the arc-anhedral angle to be 26 deg. The payload was taken to be an A-22 CDS weighing 998 kg (2,200 lb). To compute the canopy aerodynamic coefficients, the two-dimensional slice discussed in [Mohammadi and Johari 2010] was chosen as the mean canopy cross section. The CFD simulations provided the lift and drag coefficients of the slice as a function of the AoA. The computed lift slope was $C_{L\alpha}^{a} = 5.67$ rad^{-1}, and the zero-lift AoA was $\alpha_0 = -0.8$ deg. Using Eq. (2.1), the lift slope of the canopy in the linear regime was computed and plotted in Fig. 4.34a. The lift coefficient varies linearly up to the stall angle $C_L = 0.041 + 0.049\alpha$. The AoA α in this expression is in degrees. The AoA at which the canopy generated lift was equal to the system weight was taken as the equilibrium flight

a)

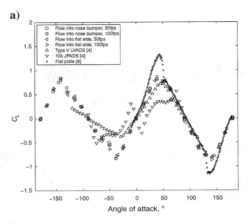

Fig. 4.32   Coefficients of a) lift, b) drag, and c) pitch-moment [McQuilling et al. 2011].

b)

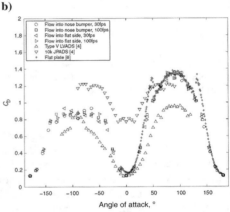

c)

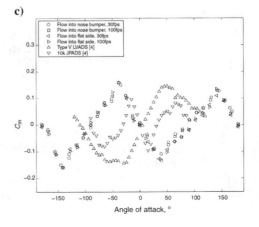

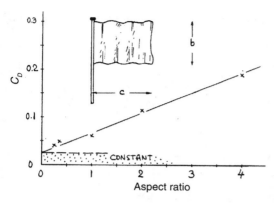

AoA, and for this model parachute system it was $\alpha = 7$ deg.

The profile drag of the slice computed in [Mohammadi and Johari 2010] along with the computed lift coefficient allowed the calculation of the canopy drag coefficient, which is plotted in Fig. 4.34b. The profile drag coefficient at the flight AoA is 0.04, and the canopy drag coefficient estimated through Eq. (2.10) is 0.062. About 35% of the canopy drag is due to the induced drag at this AoA.

The canopy lift-to-drag (L/D) ratio was then computed by taking the ratio of the data in Fig. 4.34. The resulting canopy lift-to-drag ratio is presented in Fig. 4.35. The data in this plot indicate that the canopy L/D varies with AoA, and it reaches its maximum value at $\alpha \approx 6.5$ deg. The canopy lift-to-drag ratio of approximately 6 appears reasonable for this model canopy at $\alpha \approx 7$ deg.

The next step was to compute the drag of suspension lines. A model of the suspension lines allowed the determination of the orientation of the line segments with respect to the oncoming flow. The lines were assumed to be 2,000-lb Spectra, which is round with a diameter of 4 mm (0.16 in.). At first, the lines were assumed to be stationary (nonvibrating), and the total drag of lines was computed using a drag coefficient of one. The line drag under the stationary assumption was 463 N (104 lb). Then, the procedure for the assessment of line vibration outlined in Sec. 4.2.4 was followed. The map demarcating the vibrating lines from the stationary lines is shown in Fig. 4.36. Each suspension line segment is denoted individually on this map as an open symbol. The majority of lines fall in the lock-in regime resulting in a large C_D. Once the vibration of lines was taken into account, the total line drag increased to 1,118 N (251 lb). The vibration of lines increased the line drag by a factor of 2.4, which is a major contribution to the total drag of the system.

To illustrate the line segments that fall in the lock-in regime, Fig. 4.37 shows a schematic of the suspension lines from one side of the parachute. The line pattern is symmetric with respect to the canopy centerline. The majority of the lines are vibrating. The line segments connected to the outer edge of the canopy as well as several others were found to be stationary. The reason for the lines on the outer edges behaving as stationary lines is that the tension on the outer edges is expected to be less due to the lower loads on the canopy edge. Lower tension leads to lower natural frequencies and thus further to the left of the lock-in regime boundary.

Fig. 4.34   Model canopy a) lift and b) drag coefficients.

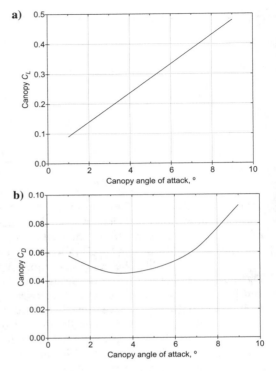

The model parachute was assumed to have a slider with a surface area of 5.3 m² (57 ft²), resulting in a drag contribution of 70 N (15.7 lb). The payload was assumed to be an A-22 CDS bundle, a cube of 1.2 m (4 ft) on each side. The payload produced a drag of 418 N (94 lb) corresponding to a drag coefficient of 1.07. Summing the drag of the canopy, suspension lines, payload and the slider results in a system drag of 3,235 N (727 lb). The system glide ratio was computed to be just over 3. This value appears to be quite comparable to the values commonly quoted for cargo ram-air parachutes. The variation of the parachute system glide as a function of the canopy AoA is shown in Fig. 4.38. The glide ratio increases continuously until $\alpha = 7$ deg, and subsequently levels out. Thus, it appears that the equilibrium canopy AoA results in the maximum glide ratio for this model ram-air parachute system.

The relative contribution of each component to the total drag of the system is listed in Table 4.3 (compare with data of Table 2.2 and 2.3). Clearly, the canopy contributes the largest share of the total drag; however, the suspension lines also contribute a substantial share (35%) of the total drag. Had the vibration of the suspension lines been ignored, the drag of the lines would be less, and the relative contribution of the lines would be reduced to 18%. Moreover, the computed system glide ratio would be 3.9 if the line vibration were not taken into account. Thus, it is important to consider the vibration of suspension lines when estimations of the system drag and glide ratio are made.

To examine the relative importance of the suspension line vibration on the drag distribution, two other model ram-air parachute systems were examined. One canopy was assumed to have a canopy area smaller and the other larger than the model considered this section. The smaller canopy parachute was assumed to have a loading of 4.9 kg/m² (1 lb/ft²) while the larger canopy was assumed to have loading of 14.6 kg/m² (3 lb/ft²). The model canopy just

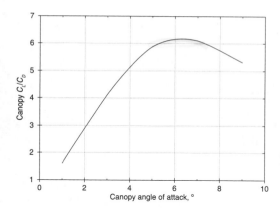

Fig. 4.35 Model canopy lift-to-drag ratio.

examined had a loading of 10.7 kg/m² (2.2 lb/ft²). The loading changes the tension in the lines, and consequently the natural frequency of each line segment changes. It is found that the larger canopy with the higher loading would result in the shift of the line segments to the right in a plot analogous to that in Fig. 4.36. On the other hand, the opposite would be true for the case of the smaller canopy with reduced loading. Also, the smaller canopy would have a smaller number of lines compared to the other two models. Computation of the line drag for all cases indicated that accounting for line vibration increases the line drag. However, the increase is not uniform and does not correlate directly with the canopy loading. For the larger canopy, accounting for the vibration of lines in the lock-in regime only increased the line drag by slightly over 5%. On the other hand, the small canopy parachute model had its line drag increase by about 41% due to the line vibration. These values are to be compared against the 141% increase computed for the model described in Fig. 4.34b.

The relative contribution of the line drag to the system was also examined for the three cases considered here. As listed in Table 4.3 and shown graphically in Fig. 4.39, the suspension line drag contributes 35% of the total drag on the model parachute system evaluated in Fig. 4.34b (in Fig. 4.39 the offset pie slice corresponds to the canopy drag). For the larger canopy model, lines contribute 23% of the drag whereas for the smaller model with the lower loading only 17% of the drag comes from the suspension lines. These estimates indicate that the relative contribution of the line drag does not scale with the canopy size or the loading directly even though the latter parameters are key elements of the parachute system. These few models appear to suggest that each parachute design needs to be evaluated separately, and generalizations regarding the effects of line vibration or the relative contribution of the line drag to the system drag and performance cannot be made based on the ram-air parachute scale or loading. However, the canopy generates at least 50% of the total drag of the system in all three cases considered.

4.3 COUPLED FSI—CFD ANALYSIS

The inflated shape of a parafoil canopy depends not only on its design parameters, but also on the aerodynamic load. In turn, this load is determined

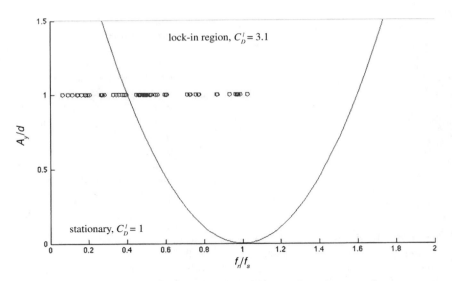

Fig. 4.36   Map of line segments in the lock-in and stationary regimes.

by the form of a parachute, and therefore aerodynamic characteristics should be predicted using FSI analysis. This approach was developed by [Chatzikonstantinou 1999] in the FSI code PARA3D [structural finite element analysis (FEA) + 3D vortex lattice method] and later in the PRA2G (second generation) code for PC; by [Kalro et al. 1997] for parallel computing on CRAY2 and M5 supercomputers (structural FEA + aerodynamic 3D Navier-Stokes + analytical turbulent model); by [Mosseev 2001b] for PC (structural FEA and aerodynamic ALE); by [Zhu et al. 2001] (structural FEA for prescribed pressure on the parafoil surfaces); by [Taylor 2003; Tutt and Taylor 2005; Tutt et al. 2005] demonstrated capabilities of LS-DYNA code (structural FEA + aerodynamic ALE); and by [Altmann and Windl 2005] in code AIRPAC (vortex

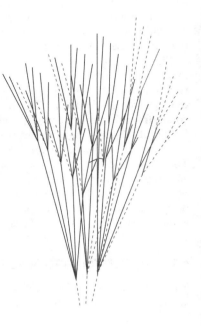

Fig. 4.37   Schematic of stationary and vibrating suspension lines (which are to the left of stationary ones).

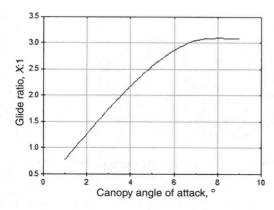

Fig. 4.38 Glide ratio of model ram-air parachute system.

panel method aerodynamic code). Results of the FSI analysis in these articles testify to the promising capabilities of detailed research on characteristics of parafoils; however, these codes demand significant computational resources and time, and the researchers using them need to be competent in the fields of aerodynamics and structural dynamics. Specifically, the aerodynamic analysis is the most time-consuming aspect of these codes (>95% of total FSI analysis time). This section introduces one specific toolkit that enables serial calculations and optimization of parafoil design based on some simplifications that drastically accelerate the attainment of a solution. Some findings resulting from this analysis as applied to the ram-air parachutes are presented as well.

4.3.1 SIMULATION TOOLS FOR SHAPE AND AERODYNAMICS ANALYSIS

In general, the FSI analysis can have two different objectives. First, it might concentrate on predicting just the parafoil shape without conducting an aerodynamic analysis. This is a well-known approach in parachute canopy analysis. Although in reality the pressure distribution over the parafoil is not constant, it can be assumed constant everywhere on the surface as a rough approximation. This assumption barely affects the predicted geometry of the inflated canopy. The same approach was used for parafoil analysis by [Zhu et al. 2001; Altmann and Windl 2005; Moseev et al. 1987]. In practice, this is confirmed by a more or less stable canopy shape in flight (or in a wind tunnel) for various AoA and in light turbulence. This approach is a simple way of predicting the canopy shape and

TABLE 4.3 RELATIVE CONTRIBUTION OF EACH COMPONENT OF THE MODEL RAM-AIR PARACHUTE

Component	Drag Contribution
Canopy	50%
Suspension lines	35%
Payload	13%
Slider	2%

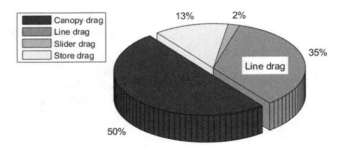

Fig. 4.39   Estimate of drag contribution of ADS components (compare to Fig. 2.18).

optimizing the structural design (load-bearing rib design, suspension system, etc.) to avoid defects like bulges or wrinkles on the surface that deteriorate aerodynamic performance. For example, the reliable outcomes can be obtained if the pressures on the upper C_p^{upper} and lower C_p^{lower} skins are assumed to be constant and positive, additionallly enforcing $C_p^{upper} > C_p^{lower}$ to simulate the lift force. This method can be applied to "well-designed" parafoils without stability problems like end cells closure stemming from a local negative pressure. Evidently, this approach is not suited for full-scale structural analysis because the tension in fabric elements depends on the pressure distribution.

Second, the FSI analysis might have an objective of performing an aerodynamic analysis of parafoil without a detailed examination of the structural analysis. This is also a well-known approach in the parachute design and analysis practice. Both flight experiments and numerical simulations revealed that the canopy shape response to the global pressure changes is often minor. As a result, the canopy surface might be considered as "frozen." For example, the drag coefficient dependence on the number and length of suspension lines in this approximation is in a very good agreement with experiments and enhanced simulations. Of course, with the frozen canopy (when only aerodynamic analysis is needed) FSI analysis can be accomplished much faster. This approach, however, should be applied with caution to avoid critical situations when the aerodynamic loads change the canopy shape dramatically. In a full-scale FSI analysis, it has been noted that this approach is acceptable if a pressure everywhere on a canopy surface is positive with some margin (e.g., $C_p > 0.2$).

The MONSTR INTEGRATED software package designed specifically for parachutes and parafoils includes both structural and several aerodynamic codes communicating through the appropriate interfaces to run a complete FSI analysis [Mosseev 2001b]. It consists of the MONSTR, PARAD, and DVM tools.

MONSTR is a structural code applicable to arbitrary textile/film structures and is based on a dynamic finite element method with the solver modified for explicit nonmatrix equations in the vector form. This tool allows importing pressure from aerodynamic code (PARAD and/or DVM) and exporting the current surface geometry back. A static problem can be solved as a dynamic one by adding the dampening forces.

PARAD is a 3D solver of unsteady aerodynamics for the explicit Arbitrary Lagrangean Eulerian (ALE) template of Eulerian equations in a structured mesh for compressible fluid (Mach number from 0.1 through 6) and is applicable to the arbitrary solid body or movable permeable surfaces. The steady problems can be solved dynamically for bodies appearing instantly in an undisturbed flow. In the case the flow over a body/surface is unsteady, an average value (in time) is computed and exported to structural code (MONSTR).

DVM is aerodynamic code based on the 2D discrete vortex method for standard and ram-air airfoils as well as 3D wings. Either Prandtl lifting-line theory (planar wings having rectangular, elliptical, and trapezoidal planforms) or Weissinger method can be used. (The Weissinger method is preferable because it works with arbitrary wing planforms having varying degrees of twist, anhedral arc, sweep angle, etc.) The extra drag due to skin friction, suspension lines, payload, flares and stabilizers is added from analytical expressions. Because the vortex panel methods cannot define a flow separation point, it can be assigned directly from visualization experiments. For example, such experiments for ram-air airfoils have been conducted in water and wind tunnels by [Lemko et al. 1983; Lemko et al. 1986]. After the separation point, a tangential vortex sheet from the upper wing surface is implemented downstream. A resulting surface-pressure distribution can be exported back to a structural code (MONSTR).

The results discussed in the following two sections were produced using primarily DVM code. For traditional (rectangular, trapezoidal, or elliptical) planforms wings with AR between 3 and 5, this FSI code produces about the same results as Prandtl and Weissinger methods presented in Sec. 4.1. To this end, Fig. 4.40 presents an example of a close match between the results of the DVM code and a wind-tunnel test dataset for the Clark-YH-17% airfoil at $\alpha = 3.5$ deg ($C_L = 0.6$). Of course, in this case airfoil geometry was maintained the same in the spanwise direction; hence, the local flow could be considered planar. For some low-AR parafoils however, "wavy" when inflated, DVM would not be used by itself but rather needs to be complemented with PARAD code.

The static results can be obtained in FSI analysis in several iterations owing to efficient interfaces (import and export of the shape and loads between appropriate parts of the package). This iterative analysis usually exhibits a very good convergence requiring only a couple of iterations. In the case of stability loss (e.g., leading-edge collapse), more iterations might be needed. The relative time budget for obtaining an FSI solution using MONSTR INTEGRATED package as of 2006 was as follows (these estimates were obtained on a 2.4-GHz Pentium 4 PC with 1-MB RAM):

- Parafoil mesh generation: 10–15 min
- Mesh import to aerodynamic codes: seconds
- Structural FEA analysis: 10 min (mesh of 100,000 FE)

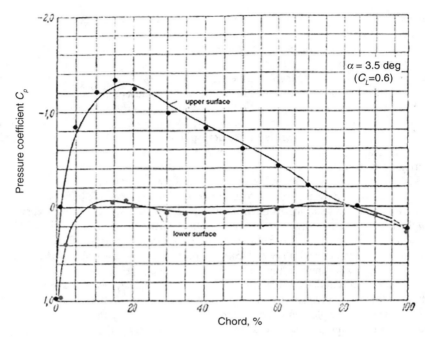

Fig. 4.40 Chordwise pressure coefficient distribution over the top and bottom surfaces of a Clark-YH-17% airfoil [Mosseev 2007]: — experiment in wind tunnel, ••• and pressure coefficient predicted in DVM code.

- Aerodynamic 3D mesh generation in PARAD
  preprocessor: minutes
- Aerodynamic 2D mesh generation in DVM
  preprocessor: minutes
- Pressure export to structural code: seconds (DVM), minutes (PARAD)

After that, the aerodynamic solution was obtained in

- 3 s, using DVM for a mesh of 300 panels
- 10 min, using the 2D PARAD solver for a mesh of 200×100 panels
- 2–4 h, using the 3D PARAD solver for a mesh of $200 \times 200 \times 200$ panels

4.3.2 AERODYNAMICS OF RAM-AIR AIRFOILS AND WINGS

Surely, wing performance depends explicitly on airfoil characteristics. (An airfoil can be considered as an infinite-AR wing.) Traditionally, ram-air airfoil contours are chosen from the standard airfoil geometry with the properly defined inlet;

however, aerodynamic properties of an airfoil and the associated ram-air airfoil are different.

The typical objectives in the common airfoil design include maximum lift and minimum drag to get a larger lift/drag ratio, lift range, prescribed pressure distribution over airfoil, etc. There is an enormous database of 1,000+ airfoils tested in wind tunnels, and there are well-defined recommendations on how to select proper airfoil geometry for definite flight conditions. However, most of these airfoils are not explicitly applicable to ram-air parafoils because of the following:

- There is a rather large opening at the leading edge contaminating the local flow.
- Airfoil must be thick enough (usually above 15–17%) to ensure stability under the aerodynamic loads.
- Parafoil sections are deformed in flight so that they become more bluff and thicker than the original airfoil selected as a load-bearing rib.
- Flow separation on the leading edges of inlet, fabric roughness, and surface waviness induce thick turbulent boundary layers and early stall.
- Even though modern parafoils are made of nonporous fabrics, after several deployments the fabric becomes slightly porous, which worsens airfoil performance.

As mentioned earlier, an experimental study for many typical ram-air airfoils in a wind/water tunnel was conducted by [Lemko et al. 1983; Lemko et al. 1986]. Along with determining lift and drag coefficients, surface pressure was visualized (both on the inner and outer surfaces), and the flow over an airfoil was visualized (using ink in a water tunnel or smoke in a wind tunnel). The very same airfoils were also studied theoretically in [Davydov and Moseev 1990; Dribnoy et al. 1989; Moseev 2001b; Rysev et al. 1987; Rysev et al. 1996], exhibiting a good agreement with the experimental results. The LS(1)-0417 and CIM-2016 ram-air airfoils (see Fig. 4.41) were studied by [Ross 1993; Benolol and Zapirain 2005]. In short, the results of these studies can be generalized as follows:

- The pressure distribution over the surface of ram-air and standard airfoils are qualitatively are close to each other except at the leading-edge area.
- For all airfoils, a separation bubble is formed in tests on both inlet edges, resulting in a turbulent boundary layer (TBL) that is especially thick on

a)   b)

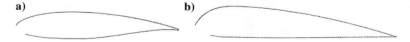

Fig. 4.41   a) LS(1)-0417 and b) CIM-2016 ram-air airfoils [AID 2014; Benolol and Zapirain 2005].

the upper surface. The TBL separation is observed at moderate AoA of $\alpha > 5$ ... -10 deg on the upper surface, and global separation from the leading edge is observed at large AoA of $\alpha > 15$ deg. A central line of the separated mixing layer has inclination at about -12 deg with respect to a flow at infinity (Fig. 4.42b shows three specific points: S_1 – the leading-edge separation point; i.e. the origin of a turbulent bubble; S_2 – global separation on a top surface; A – the reattachment point, i.e. the end of a bubble).

•   Inside an airfoil, a deep stagnation is observed. There is almost constant pressure at $C_p = 1$ (theoretical), $C_p = 0.97 ... 0.98$ (measured for a nonporous model), and $C_p = 0.95$ (measured for a low permeable fabric model), except for regions very close to the leading edge.

•   An open inlet is the main cause for large C_D, much higher (12 times) than that expected for a standard airfoil. The drag is almost directly proportional to the projection of inlet size in flow direction.

•   The variation of C_D with α is almost parabolic for small and moderate AoA.

a)

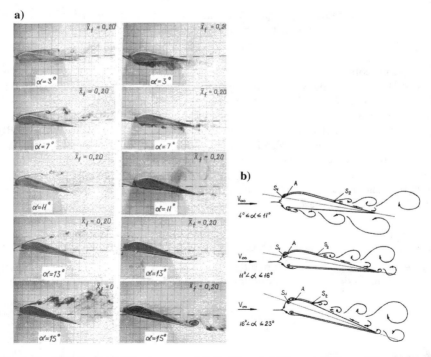

Fig. 4.42   Visualization of flow over an airfoil with an inlet in a water tunnel [Lemko et al. 1983].

- The lift force coefficient C_L is almost linear for moderate AoA; its maximum value is significantly lower than that of the same standard airfoil. Figure 4.43 demonstrates that for all semirigid parafoils (parafoils with rigid ribs and low porous fabric skin as in Fig. 4.44), composed of traditional and exotic airfoils with an inlet the coefficient $C_{L\alpha}$ is about the same.

- Roughness of the surface has no significant influence on aerodynamics presumably because of a thick TBL on both sides of the airfoil.

- Porous fabric on the upper surface causes TBL growth and early separation for smaller AoA. As a result, C_D enlarges, and C_L diminishes considerably with α.

Obviously, parafoils in flight are not rigid. Aerodynamic load or control line action deforms them as well as the underlying airfoil. Figure 4.45 shows an example of parafoil cross section deformation without and with TE deflected (as predicted in MONSTR + DVM code) [Moseev 2001b]. Depending on whether it is a load-bearing rib, a nonloaded rib, or a section in between the ribs, the pressure distribution and, as a result, all aerodynamic loads are quite different from those of the original airfoil. Deformation of the shape of the load-bearing rib compared to original airfoil geometry and deformation of the entire canopy shape obviously worsens the overall performance of a parafoil. Another issue is what to accept as a parafoil AoA in a situation when the wing and airfoil are deformed. A traditional way is to measure AoA as an angle between chord (straight line from the leading-edge to TE tip) and flow direction at infinity. However, for the deformed canopy this would not work because various sections of parafoil canopy (load-bearing rib, not loaded rib, section in between ribs) have a different AoA and in addition to that both the leading edge and TE could be deflected in flight.

4.3.3 SHAPE-RELATED FACTORS INFLUENCING PARAFOIL AERODYNAMICS

Compared to a traditional planar wing, a parafoil canopy is arched, has an inlet at the leading edge, exhibits more complex "wavy" distorted shape due to inflation, and features a deflected TE. These shape-related factors obviously influence the parafoil's aerodynamics as well. That is why the comprehensive studies of parafoil aerodynamics (flow visualization, pressure distribution, and integral coefficients C_L, C_D, and C_m computation) have been performed (in wind tunnels). For example, [Lemko et al. 1983; Lemko et al. 1986] studied the rigid and semirigid models (rigid ribs, fabric top and bottom surface), [Nicolaides 1971] investigated performance of soft and semirigid models, and [Ware and Hassel 1969; Bashkina et al. 1989] analyzed performance of full-scale parafoils.

One of the major 3D effects consists of the vortex sheet leaving TE and forming the tip vortices trailing back from the wing. A vortex pattern over a wing explains why the flow above and below the wing surface is not parallel to the freestream direction (see Fig. 4.46). The streamlines converge slightly on the top and diverge on the bottom surface. This effect is weak for the large-span wings, but quite significant for the small ones. Hence, for the small-span wings,

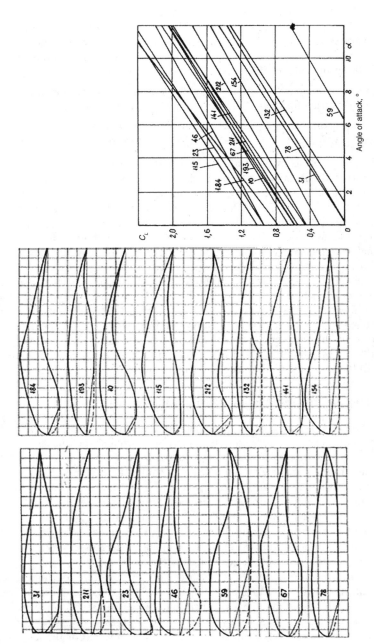

Fig. 4.43   Wind-tunnel testing of semirigid parafoil models [Lemko et al. 1986].

Fig. 4.44   Example of a semirigid parafoil model [Mosseev 2007].

the problem of airfoil section cannot be treated as a 2D problem. The flow over the inflated shape of a parafoil does not go exactly along the ribs, resulting in a normal velocity component "across" the wing as well. In other words, flow goes not along airfoils, but along some "wavy" surface, which decreases the effective airspeed.

An effect of the anhedral-arc shape of a parafoil was addressed in Chapter 2 by using an analytical approach for low-AR parafoils inherited from evaluation valid for dihedral angle of planar semiwings and also in Sec. 4.1 by applying the lifting-line theory extended to consider a curvature of this line. However, heavy-weight PADS require parafoils with AR greater than 2.5–2.7, and two or more connection points to a payload with a relatively large spread between them (comparable with a length of suspension lines). As a result, a front view of the inflated canopy might have an anhedral bi-arc shape (one arc for each semiwing as shown in Fig. 4.47) or arc—straight line—arc shape as opposed to just a single anhedral-arc shape [Benolol and Zapirain 2005; Smith et al. 1999]. Hence, a single anhedral-arc assumption utilized in the aforementioned analytical approaches might be violated, which requires using a more flexible lifting-line theory like the Weissinger method, which doesn't depend on arc geometry, or again calls for the CFD and FSI analysis. The use of end-cell stabilizers that can be considered as winglets or wing endplates diminishes a tip vortice's influence on the wing surface and thus enlarges the lift force and L/D ratio. Surely, it can only be accounted for in CFD and FSI models.

Perhaps the major contributor that can only be accounted for in a FSI model is parafoil deformation, both spanwise and chordwise. It affects both the

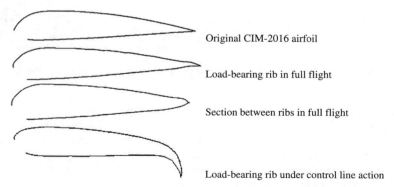

Original CIM-2016 airfoil

Load-bearing rib in full flight

Section between ribs in full flight

Load-bearing rib under control line action

Fig. 4.45   Various sections of FASTWing parafoil in flight [Mosseev 2007].

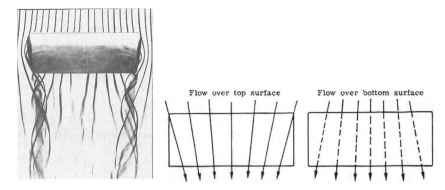

Fig. 4.46   Flow direction over the top and bottom surfaces of a typical wing load-bearing rib [Head 1982].

Fig. 4.47   Wind-tunnel test of a FASTWing parafoil [Benolol and Zapirain 2005].

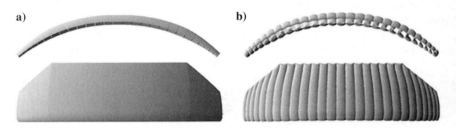

Fig. 4.48   a) An "ideal" parafoil shape vs b) an inflated one [Mosseev 2007].

anhedral-arc radius and projected planform in flight. Figure 4.48 shows an example of how an "ideal" parafoil shape (before inflation or having an infinite number of cells) shrinks its effective span (and therefore projected area and AR) because of inflation of its 26 cells. That is because the spanwise section looks like usual inflatable pneumo-panel made of cylinders (Fig. 4.49). Obviously, the upper and lower surfaces are smoother for larger number of ribs (sections) or in other words for a smaller rib-rib distance.

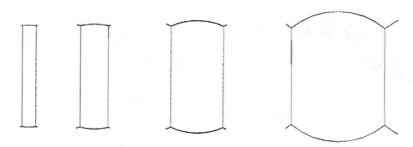

Fig. 4.49   Cell shape for rib-rib distance 12, 25, 50, and 100% of a rib height [Mosseev 2007].

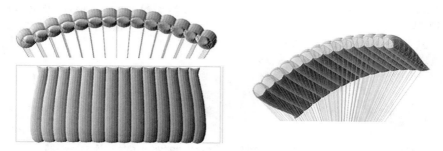

Fig. 4.50   Predicted shape for a 14-cell rectangular parafoil [Mosseev 2007].

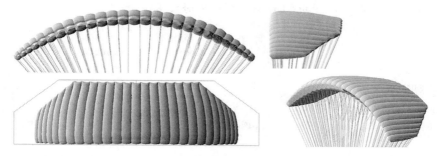

Fig. 4.51   Predicted shape for a FASTWing tapered parafoil [Mosseev 2007].

The shapes predicted by the FSI analysis (using MONSTR and PARAD) for various parafoils composed of CIM-2016 ram-air airfoil are shown in Figs. 4.50–4.53. Figures 4.50 and 4.51 show examples of canopy shrinkage for a relatively low-AR rectangular parafoil and FASTWing tapered parafoil. The top views feature an actual planform as compared to the nominal planform. Tables

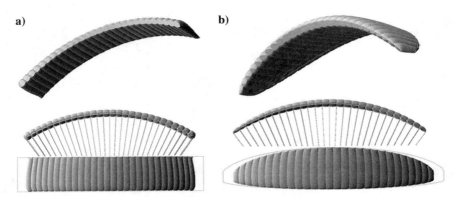

Fig. 4.52   Predicted shape for a) large-span rectangular and b) elliptical parafoils [Mosseev 2007].

Fig. 4.53   Predicted shape a 10-triple-cells rectangular parafoil [Mosseev 2007].

TABLE 4.4   RELATIVE SHRINKAGE FOR THE LOW-ASPECT-RATIO RECTANGULAR PARAFOIL

	Span, *l*	Root Chord, *b*	Projected Planform Area, *S*	Aspect Ratio
Design parameters	2.5	1	2.5	2.5
Inflated canopy parameters	2.153	0.973	2.03	2.28
Relative shrinkage	14%	3%	19%	9%

4.4 and 4.5 show parameters of the inflated canopy as related to design parameters. Obviously, the platform area and appropriate AR change significantly. For high-AR parafoils, this effect is not significant (Fig. 4.52 and Tables 4.6 and 4.7). The double- and triple-cell designs can introduce even more changes, especially for the high-AR wings. For example, utilizing a triple-cell technology for the parafoil shown in Fig. 4.52a resulting in a model shown in Fig. 4.53 leads to a relative change of parameters as shown in Table 4.8 (compare with data in Table 4.6).

TABLE 4.5   RELATIVE SHRINKAGE FOR A TAPERED FASTWING PARAFOIL

	Span, l	Root Chord, b	Area, S	Aspect Ratio
Design parameters	3.66	1	3.31	4.04
Inflated canopy parameters	3.04	0.941	2.6	3.55
Relative shrinkage	17%	6%	21%	12%

TABLE 4.6   RELATIVE SHRINKAGE FOR A LARGE-ASPECT-RATIO RECTANGULAR PARAFOIL

	Span, l	Root Chord, b	Area, S	Aspect Ratio
Design parameters	5	1	5	5
Inflated canopy parameters	4.68	0.985	4.42	4.68
Relative shrinkage	6%	2%	12%	6%

TABLE 4.7   RELATIVE SHRINKAGE FOR A LARGE-ASPECT-RATIO ELLIPTICAL PARAFOIL

	Span, l	Root Chord, b	Tip Chord	Area, S	Aspect Ratio
Design parameters	5	1	0.25	4	6.23
Inflated canopy parameters	4.54	0.985	0.23	3.6	5.72
Relative shrinkage	9%	2%	8%	10%	8%

TABLE 4.8   RELATIVE SHRINKAGE FOR A 10-TRIPLE-CELLS RECTANGULAR PARAFOIL

	Span, l	Root Chord, b	Area, S	Aspect Ratio
Design parameters	5	1	5	5
Inflated canopy parameters	4.42	0.974	4.29	4.566
Relative shrinkage	12%	3%	14%	9%

a)   b)

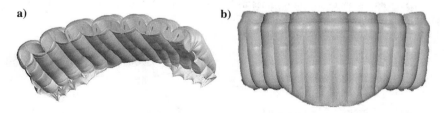

Fig. 4.54   Parawing shape in a brake regime as compared to the nominal constructed shape [Mosseev 2007].

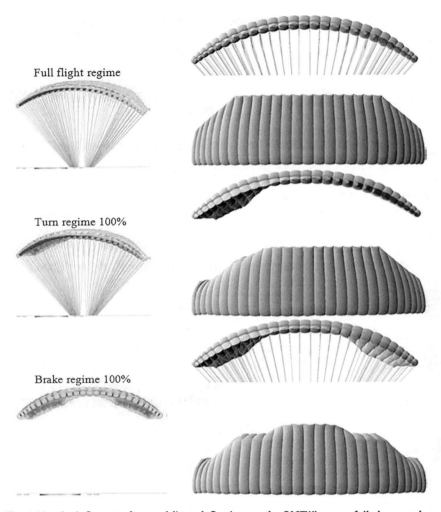

Full flight regime

Turn regime 100%

Brake regime 100%

Fig. 4.55   An influence of control lines deflection on the FASTWing parafoil shape and pressure distribution over wing [Mosseev 2007].

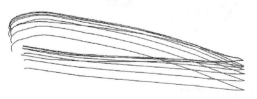

Fig. 4.56   Contours of load-bearing ribs for the rectangular parafoil depicted in Fig. 4.48 [Mosseev 2007].

To summarize, if a nominal planform is assumed instead of a real, deformed one, the error in aerodynamic performance (the values of coefficients C_L and C_D) can vary between about 10% for an elliptical wing with an AR of 5 to 20% for a rectangular 7-cell low-AR canopy with AR $= 2$. However, the L/D ratio remains almost the same for both nominal and deformed planforms.

The TE deflection causes even further changes of a planform. As an example, Fig. 4.54 shows what happens to a low-AR double-cell rectangular planform parafoil in a brake regime, and Fig. 4.55 shows the predicted FASTWing shapes for the straight flight, left turn (only one side of TE is fully deflected) and brake (both sides of TE are fully deflected). Although deformations change drag and lift coefficients, the L/D ratio remains unaffected again.

Finally, FSI analysis also reveals a negative wing twisting (wash-in) effect caused by the fact that the suspension lines are more loaded in a wing quarter-line area and close to a center line and in the case of low modulus of material stiffness are elongated more. For example, Fig. 4.56 shows a negative -5-deg spanwise twist for a rectangular parafoil shown in Fig. 4.48.

CHAPTER 5

Equations of Motion

Oleg Yakimenko[*]
Naval Postgraduate School, Monterey, California

Nathan Slegers[†]
George Fox University, Newberg, Oregon

This chapter attempts to develop several different-complexity models of PADS that can be used for studying PADS stability and flight performance as well as design and validation of GNC algorithms. In the recent decade, several high-fidelity models were developed and published worldwide (e.g., [Mooij et al. 2003; Prakash and Ananthkrishnan 2006; Redelinghuys 2007; Toglia and Vendittelli 2010; Watanabe and Ochi 2007; Wise 2006]); however, this chapter will primarily be based on [Gorman and Slegers 2011a; Jann 2001; Mortaloni 2003; Mortaloni et al. 2003; Ochi et al. 2009; Slegers 2010; Yakimenko 2005] to develop a whole range of models, starting from a simple three-degree-of-freedom (DoF) model of PADS and advancing to seven-/eight-/nine-DoF models accounting for relative dynamics of a payload-parafoil system, in a consecutive manner. The chapter starts from Sec. 5.1 formally introducing the major systems of coordinates and Euler angles to transfer quantities between these systems. It also presents translational and rotational kinematics and dynamics, followed by the development of the simple one-DoF model of the vertical descent motion and three- and four-DoF models for lateral dynamics. It proceeds with introducing apparent masses and inertia (based on [Lissaman and Brown 1998; Barrows 2002]) and develops a complete six-DoF model showing some computer implementations of this model. Next, Sec. 5.2 deals with a step-by-step development of the higher-order models for different rigging schemes defining relative parafoil-payload motion. Major differences caused by having more degrees of freedom are also discussed. Section 5.3 presents the results of real-flight identification of controls efficiency and attempts to build a universal model describing PADS reaction to the symmetric and asymmetric trailing-edge deflection. The chapter concludes with Sec. 5.4 introducing linearized and reduced-order models and Sec. 5.5 providing a brief overview of all developed models and showing how to incorporate complex shapes of a payload.

---

[*]Professor.
[†]Associate Professor.

5.1 RIGID-BODY MODELS

This section presents some basic models that consider canopy and payload to be one (rigid) body. The section starts with some basic definitions and then proceeds with the simplest three-DoF and four-DoF models that are usually used for trajectory planning (PADS guidance). It further discusses apparent mass and inertia models used in higher-fidelity models and concludes with a complete six-Dof model of a parafoil-payload system and its computer implementations.

5.1.1 BASIC DEFINITIONS AND ROTATION MATRICES

Let us start from formally introducing three right-handed, Cartesian coordinate systems to be used for derivation of equations of motion for parafoil aerial delivery systems (PADS) deployed from an aircraft (in troposphere). The first coordinate system, the Earth frame $\{I\}$, is established as follows:

- Origin: arbitrary, fixed relative to the surface of the Earth [most often it is associated with the intended point of impact (IPI)]
- x axis: positive in the direction of true North
- y axis: positive in the direction of East
- z axis: positive towards the center of the Earth

This coordinate frame is also referred to as the North-East-Down (NED) frame. Another system of coordinates frequently used to describe PADS position is the East-North-Up (ENU) coordinate frame. Because of a short duration of a controlled descent, relatively slow speed and a limited range, both NED and ENU are considered to be an inertial frame.

The other two reference frames are body-fixed, with their origins moving along with the parafoil system, typically at the center of gravity. The body-fixed $\{b\}$ frame is defined as follows:

- Origin: PADS center of gravity
- x_b axis: positive out the longitudinal axis of PADS in the plane of symmetry of PADS
- z_b axis: perpendicular to the x_b axis, in the plane of PADS symmetry, positive pointing down (from canopy towards payload)
- y_b axis: perpendicular to the x_b–z_b plane, positive determined by the right-hand rule (out to the right)

The wind frame $\{w\}$ is defined as follows:

- Origin: PADS center of gravity
- x_w axis: positive in the direction of the velocity vector relative to the air, airspeed vector $\mathbf{V}_a$

- z_w axis: perpendicular to the x_w axis, in the plane of symmetry, pointing down
- y_w axis: perpendicular to the x_w–z_w plane, positive determined by the right-hand rule (out to the right).

The Earth frame is a convenient frame to express PADS translational kinematics. The body frame is often of interest because the origin and the axes remain fixed relative to PADS. This means that the relative orientation of the Earth and body frames describes the PADS attitude. The wind frame is a convenient frame to express the aerodynamic forces and moments acting on PADS components. In particular, the net aerodynamic force can be divided into components along the wind frame axes, with the drag force D in the $-x_w$ direction and the lift force L in the direction $-z_w$.

In addition to these three coordinate systems, let us define an auxiliary one on the navigational frame $\{n\}$, which is parallel to $\{I\}$ with the origin at PADS center of gravity.

The relative orientation of these three coordinate frames with respect to each other is defined as follows:

- $\{b\}$ to $\{I\}$ ($\{n\}$): by three Euler angles, the roll angle ϕ, pitch angle θ, and yaw angle ψ
- $\{b\}$ to $\{w\}$: by two Euler angles, the angle of attack α and sideslip angle β
- $\{w\}$ to $\{I\}$ ($\{n\}$): by three Euler angles, the bank angle ϕ_a, flight path angle γ_a, and heading angle χ_a

The formal definition of these angles is as follows (Fig. 5.1):

- The yaw angle ψ is the angle from North to the longitudinal axis x_b.

- The pitch angle θ is the angle from a horizontal plane to the longitudinal axis x_b.

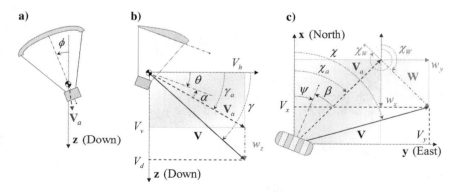

Fig. 5.1 Views of PADS: a) front, b) side, and c) bird's-eye.

- The formal definition of the roll angle ϕ is too cumbersome because it refers to a not explicitly defined intermediate coordinate plane after two rotations by the yaw angle ψ and pitch angle θ, so let us just say that the rotation by this angle completes transformation from $\{n\}$ to $\{b\}$.
- The angle of attack α is the angle from the projection of the airspeed vector $\mathbf{V}_a$ on to the x_b–z_b plane to the longitudinal axis x_b.
- The sideslip angle β is the angle from the airspeed vector $\mathbf{V}_a$ to its projection on to the x_b–z_b plane.
- The bank angle ϕ_a represents a rotation of the lift force $(-z_w)$ around the airspeed vector $\mathbf{V}_a$ from the vertical plane including the vector $\mathbf{V}_a$.
- The flight path angle (also referred to as the air-referenced flight path) γ_a is the angle from a horizontal plane to the airspeed vector $\mathbf{V}_a$.
- The heading angle χ_a is the angle from North to the horizontal component of the airspeed vector $\mathbf{V}_a$.

Figures 5.1b and 5.1c show a situation with no banking; otherwise, the angle of attack α and sideslip angle β would be depicted incorrectly. According to the right-hand rule, the angles ϕ, α, ψ, χ_a, χ shown in Fig. 5.1 are positive whereas the angles θ, γ_a, γ, and β have a negative value.

Figure 5.1 also features the wind vector $\mathbf{W}$ with its components $\left[w_x, w_y, w_z\right]^T$ in $\{I\}$. Unlike aircraft, the magnitude of winds can be of the order of $|\mathbf{V}_a| = V_a$ or even greater. As a result, a groundspeed vector $\mathbf{V}$

$$\mathbf{V} = \mathbf{V}_a + \mathbf{W} \tag{5.1}$$

with its components $\left[V_x, V_y, V_z\right]^T$ in $\{I\}$ can be quite different compared to that of $\mathbf{V}_a$.

While the vertical component of the airspeed vector V_v defines the flight path angle γ_a, so that

$$\gamma_a = -\sin^{-1}(V_v/V_a) \tag{5.2}$$

the descent rate or sink rate

$$V_d = V_z = V_v + w_z \tag{5.3}$$

defines the Earth-referenced flight path angle γ, which is the angle in a vertical plane between the groundspeed vector $\mathbf{V}$ and its x–y projection V_G

$$V_G = \sqrt{V_x^2 + V_y^2} \tag{5.4}$$

so that

$$\gamma = -\sin^{-1}(V_d/V) \tag{5.5}$$

Most of the time the vertical component of the wind is neglected, which makes

$$\gamma \approx \gamma_a \tag{5.6}$$

The V_x and V_y components of the horizontal projection of the groundspeed vector $\mathbf{V}$ define the ground track angle or course over ground or simply track χ. Crab angle is the amount of correction parafoil that must be turned into the wind in order to maintain the desired track ($|\gamma - \gamma_a|$), resembling the sideways motion of a crab. The angle between the PADS track and the longitudinal axis of PADS is called the crab angle. Wind direction is usually defined not by the "to" angle χ_W, but by the "from" angle χ_W^*. The positive descent rate causes coordinate z to increase and altitude h, which is

$$h = -z \tag{5.7}$$

to decrease.

A rotation of any vector $\boldsymbol{\xi} = \left[\xi_x, \xi_y, \xi_z\right]^T$ by some angle λ means premultiplying it from the left by the corresponding rotation matrix $\mathbf{R}_\lambda$. For example, if vector $\boldsymbol{\xi}^n$, expressed in $\{n\}$, needs to be expressed in $\{b\}$, it needs to undergo three consecutive rotations

$$\boldsymbol{\xi}^b = \mathbf{R}_\phi \mathbf{R}_\theta \mathbf{R}_\psi \boldsymbol{\xi}^n \tag{5.8}$$

where three rotation matrices are

$$
\mathbf{R}_\psi = \begin{bmatrix} \cos(\psi) & \sin(\psi) & 0 \\ -\sin(\psi) & \cos(\psi) & 0 \\ 0 & 0 & 1 \end{bmatrix}
$$

$$
\mathbf{R}_\theta = \begin{bmatrix} \cos(\theta) & 0 & -\sin(\theta) \\ 0 & 1 & 0 \\ \sin(\theta) & 0 & \cos(\theta) \end{bmatrix} \tag{5.9}
$$

$$
\mathbf{R}_\phi = \begin{bmatrix} 1 & 0 & 0 \\ 0 & \cos(\phi) & \sin(\phi) \\ 0 & -\sin(\phi) & \cos(\phi) \end{bmatrix}
$$

Hence, the product of three elementary rotation matrices in Eq. (5.8) constitutes a rotation matrix from $\{n\}$ to $\{b\}$

$$
{}^b_n\mathbf{R} = \mathbf{R}_\phi \mathbf{R}_\theta \mathbf{R}_\psi
$$
$$
= \begin{bmatrix} \cos(\psi)\cos(\theta) & \sin(\psi)\cos(\theta) & -\sin(\theta) \\ \cos(\psi)\sin(\theta)\sin(\phi) - \sin(\psi)\cos(\phi) & \sin(\psi)\sin(\theta)\sin(\phi) + \cos(\psi)\cos(\phi) & \cos(\theta)\sin(\phi) \\ \cos(\psi)\sin(\theta)\cos(\phi) + \sin(\psi)\sin(\phi) & \sin(\psi)\sin(\theta)\cos(\phi) - \cos(\psi)\sin(\phi) & \cos(\theta)\cos(\phi) \end{bmatrix}
$$
$$\tag{5.10}$$

Using matrices of Eq. (5.9), the rotation matrix between NED and ENU can be developed as follows:

$$\mathbf{_{NED}^{ENU}R} = \mathbf{R}_{\phi=180°}\,\mathbf{R}_{\psi=90°} = \begin{bmatrix} 1 & 0 & 0 \\ 0 & -1 & 0 \\ 0 & 0 & -1 \end{bmatrix}\begin{bmatrix} 0 & 1 & 0 \\ -1 & 0 & 0 \\ 0 & 0 & 1 \end{bmatrix}$$

$$= \begin{bmatrix} 0 & 1 & 0 \\ 1 & 0 & 0 \\ 0 & 0 & -1 \end{bmatrix} \tag{5.11}$$

If PADS rotates with an angular rate $\boldsymbol{\omega}_{bn}$ (frame $\{b\}$ rotates with respect to frame $\{n\}$), its rotation can be described in $\{n\}$ via roll rate $\dot\phi$, pitch rate $\dot\theta$, and yaw rate $\dot\psi$, or in the vector form

$$\boldsymbol{\omega}_{bn} = \dot{\boldsymbol{\psi}} + \dot{\boldsymbol{\theta}} + \dot{\boldsymbol{\phi}} \tag{5.12}$$

It can also be described by the corresponding angular rates in $\{b\}$. These latter rates are denoted as $\boldsymbol{\omega}_{bn}^{b} = \boldsymbol{\omega}^{b} = [p,\,q,\,r]^{T}$ (for simplicity, hereinafter superscript b will be omitted). To express the vector $\boldsymbol{\omega}$ via $\dot\phi$, $\dot\theta$, and $\dot\psi$, the following rotations should take place:

$$\boldsymbol{\omega} = \begin{bmatrix} p \\ q \\ r \end{bmatrix} = \mathbf{R}_{\phi}\begin{bmatrix} \dot\phi \\ 0 \\ 0 \end{bmatrix} + \mathbf{R}_{\phi}\mathbf{R}_{\theta}\begin{bmatrix} 0 \\ \dot\theta \\ 0 \end{bmatrix} + \mathbf{R}_{\phi}\mathbf{R}_{\theta}\mathbf{R}_{\psi}\begin{bmatrix} 0 \\ 0 \\ \dot\psi \end{bmatrix} \tag{5.13}$$

The latter expression reduces to

$$\begin{bmatrix} p \\ q \\ r \end{bmatrix} = \begin{bmatrix} 1 & 0 & -\sin(\theta) \\ 0 & \cos(\phi) & \cos(\theta)\sin(\phi) \\ 0 & -\sin(\phi) & \cos(\theta)\cos(\phi) \end{bmatrix}\begin{bmatrix} \dot\phi \\ \dot\theta \\ \dot\psi \end{bmatrix} \tag{5.14}$$

and, if inverted, takes the following form:

$$\begin{bmatrix} \dot\phi \\ \dot\theta \\ \dot\psi \end{bmatrix} = \begin{bmatrix} 1 & \sin(\phi)\dfrac{\sin(\theta)}{\cos(\theta)} & \cos(\phi)\dfrac{\sin(\theta)}{\cos(\theta)} \\ 0 & \cos(\phi) & -\sin(\phi) \\ 0 & \sin(\phi)\dfrac{1}{\cos(\theta)} & \cos(\phi)\dfrac{1}{\cos(\theta)} \end{bmatrix}\begin{bmatrix} p \\ q \\ r \end{bmatrix} \tag{5.15}$$

Accounting for PADS specifics, namely, assuming $\theta \approx 0$ and $\dot\theta \approx 0$, Eq. (5.14) can be simplified to

$$\begin{bmatrix} p \\ q \\ r \end{bmatrix} \approx \begin{bmatrix} \dot\phi - \dot\psi\theta \\ \dot\psi\sin(\phi) \\ \dot\psi\cos(\phi) \end{bmatrix} \approx \begin{bmatrix} \dot\phi \\ \dot\psi\sin(\phi) \\ \dot\psi\cos(\phi) \end{bmatrix} \tag{5.16}$$

Rate gyros of the onboard inertial measurement unit do not measure $\boldsymbol{\omega}$, but rather measure PADS' angular rate components with respect to the true inertial frame, but for PADS the contribution of Earth's sidereal rotation (about $\sim 7.3 \times 10^{-5}\,\mathrm{s}^{-1}$) and transport rate (defined by negligibly slow changes of PADS latitude and longitude).

Similar to Eq. (5.10), the rotation from the inertial frame $\{I\}$ to the wind frame $\{w\}$ can be developed as

$$
{}^{w}_{n}\mathbf{R} = \mathbf{R}_{\gamma_a}\mathbf{R}_{\chi_a} =
\begin{bmatrix}
\cos(\gamma_a) & 0 & -\sin(\gamma_a) \\
0 & 1 & 0 \\
\sin(\gamma_a) & 0 & \cos(\gamma_a)
\end{bmatrix}
\begin{bmatrix}
\cos(\chi_a) & \sin(\chi_a) & 0 \\
-\sin(\chi_a) & \cos(\chi_a) & 0 \\
0 & 0 & 1
\end{bmatrix}
$$

$$
=
\begin{bmatrix}
\cos(\gamma_a)\cos(\chi_a) & \cos(\gamma_a)\sin(\chi_a) & -\sin(\gamma_a) \\
-\sin(\chi_a) & \cos(\chi_a) & 0 \\
\sin(\gamma_a)\cos(\chi_a) & \sin(\gamma_a)\sin(\chi_a) & \cos(\gamma_a)
\end{bmatrix}
\tag{5.17}
$$

5.1.2 TRANSLATIONAL AND ROTATIONAL DYNAMICS

As known, the Newton's second law in the (nonrotating) inertial coordinate frame $\{I\}$ can be written in the following form:

$$
m\dot{\mathbf{V}} = m
\begin{bmatrix}
\dot{V}_x \\
\dot{V}_y \\
\dot{V}_z
\end{bmatrix}
= \mathbf{F}^I
\tag{5.18}
$$

Here, m is the mass of a dynamic object, and $\mathbf{F}^I$ is the force vector expressed in $\{I\}$. For the aerial vehicles producing lift L and drag D, this expression is usually written in the wind coordinate frame $\{w\}$ for the magnitude of the airspeed vector and two angles defining its orientation, bank angle ϕ_a and heading angle χ_a (refer to Figs. 5.1b and 5.1c). For unpowered PADS, the Newton's second law in $\{w\}$ then takes the form [Kaminer and Yakimenko 2003]

$$
\dot{V}_a = -\frac{1}{m}[D + mg\sin(\gamma_a)]
$$

$$
\dot{\gamma}_a = \frac{1}{mV_a}[L\cos(\phi_a) - mg\cos(\gamma_a)]
\tag{5.19}
$$

$$
\dot{\chi}_a = \frac{L\sin(\phi_a)}{mV_a\cos(\gamma_a)}
$$

In the noninertial (rotating) body frame $\{b\}$, the Newton's second law of motion for a rigid body takes the form

$$
m\dot{\mathbf{v}} + \boldsymbol{\omega} \times m\mathbf{v} = m
\begin{bmatrix}
\dot{u} \\
\dot{v} \\
\dot{w}
\end{bmatrix}
+ m
\begin{bmatrix}
p \\
q \\
r
\end{bmatrix}
\times
\begin{bmatrix}
u \\
v \\
w
\end{bmatrix}
= \mathbf{F}
\tag{5.20}
$$

where $\mathbf{v} = [u, v, w]^T$ is the groundspeed vector $\mathbf{V}$ expressed in the body frame $\{b\}$ and $\mathbf{F}$ is the force vector expressed in $\{b\}$. The cross-product terms in Eq. (5.20) arise along any given axis due to angular velocities about the remaining two axes. The vector product in Eq. (5.20) can be replaced with multiplication by a skew-symmetric matrix $\mathbf{S}(\boldsymbol{\omega})$, which is formed from the components of vector $\boldsymbol{\omega}$ as follows:

$$\mathbf{S}(\boldsymbol{\omega}) = \begin{bmatrix} 0 & -r & q \\ r & 0 & -p \\ -q & p & 0 \end{bmatrix} \tag{5.21}$$

Using Eq. (5.21), Eq. (5.20) becomes

$$m \begin{bmatrix} \dot{u} \\ \dot{v} \\ \dot{w} \end{bmatrix} + m\mathbf{S}(\boldsymbol{\omega}) \begin{bmatrix} u \\ v \\ w \end{bmatrix} = \mathbf{F} \tag{5.22}$$

or

$$\begin{bmatrix} \dot{u} \\ \dot{v} \\ \dot{w} \end{bmatrix} = \frac{1}{m}\mathbf{F} - \mathbf{S}(\boldsymbol{\omega}) \begin{bmatrix} u \\ v \\ w \end{bmatrix} \tag{5.23}$$

Similarly, for the rotational dynamics expressed in the body frame $\{b\}$ becomes

$$\mathbf{I} \begin{bmatrix} \dot{p} \\ \dot{q} \\ \dot{r} \end{bmatrix} + \begin{bmatrix} p \\ q \\ r \end{bmatrix} \times \mathbf{I} \begin{bmatrix} p \\ q \\ r \end{bmatrix} = \mathbf{M} \tag{5.24}$$

or

$$\mathbf{I} \begin{bmatrix} \dot{p} \\ \dot{q} \\ \dot{r} \end{bmatrix} = \mathbf{M} - \mathbf{S}(\boldsymbol{\omega})\mathbf{I} \begin{bmatrix} p \\ q \\ r \end{bmatrix} \tag{5.25}$$

where $\mathbf{I}$ is the inertia matrix or inertia tensor and $\mathbf{M}$ is the moment vector expressed in $\{b\}$. For PADS, characterized by having an x_b–z_b symmetry plane, the inertia matrix has only four unique components

$$\mathbf{I} = \begin{bmatrix} I_{xx} & 0 & I_{xz} \\ 0 & I_{yy} & 0 \\ I_{xz} & 0 & I_{zz} \end{bmatrix} \tag{5.26}$$

where $I_{xx} = \int (y_b^2 + z_b^2)dm$ is the moment of inertia about the x_b axis (an integral is taken over infinitesimal mass elements), $I_{yy} = \int (x_b^2 + z_b^2)dm$ is the moment of inertia about the y_b axis, $I_{zz} = \int (x_b^2 + y_b^2)dm$ is the moment of inertia about the z_b axis, and $I_{xz} = -\int x_b z_b \, dm$ is the $x_b z_b$ product of inertia.

5.1.3 THREE-AND FOUR-DoF MODELS

Depending on the availability of the components of the velocity vector, a rigid-body translational kinematics can be expressed in one of the following forms:

$$\begin{bmatrix} \dot{x} \\ \dot{y} \\ \dot{z} \end{bmatrix} = \begin{bmatrix} V_x \\ V_y \\ V_z \end{bmatrix} \quad \text{or} \quad \dot{\mathbf{p}} = \mathbf{V} \tag{5.27}$$

$$\begin{bmatrix} \dot{x} \\ \dot{y} \\ \dot{z} \end{bmatrix} = {}^b_n\mathbf{R}^T \begin{bmatrix} u \\ v \\ w \end{bmatrix} \quad \text{or} \quad \dot{\mathbf{p}} = {}^b_n\mathbf{R}^T\mathbf{v} \tag{5.28}$$

$$\begin{bmatrix} \dot{x} \\ \dot{y} \\ \dot{z} \end{bmatrix} = V_a \begin{bmatrix} \cos(\gamma_a)\cos(\chi_a) \\ \cos(\gamma_a)\sin(\chi_a) \\ -\sin(\gamma_a) \end{bmatrix} + \begin{bmatrix} w_x \\ w_y \\ w_z \end{bmatrix} \quad \text{or} \quad \dot{\mathbf{p}} = {}^w_n\mathbf{R}^T \begin{bmatrix} V_a \\ 0 \\ 0 \end{bmatrix} + \mathbf{W} \tag{5.29}$$

$$\begin{bmatrix} \dot{x} \\ \dot{y} \\ \dot{z} \end{bmatrix} = \begin{bmatrix} V_h\cos(\chi_a) \\ V_h\sin(\chi_a) \\ V_v \end{bmatrix} + \begin{bmatrix} w_x \\ w_y \\ w_z \end{bmatrix} \tag{5.30}$$

Equation (5.27) assumes availability of the groundspeed vector $\mathbf{V}$ expressed in $\{I\}$, Eq. (5.28) projects the components of the groundspeed vector $\mathbf{V}$ expressed in $\{b\}$ onto the axes of inertial frame using the transpose of the rotation matrix (5.10), Eq. (5.29) relies on knowing the airspeed vector $\mathbf{V}_a$ then rotated by the corresponding Euler angles, Eq. (5.30) takes advantage of the horizontal

$$V_h = V_a\cos(\gamma_a) \tag{5.31}$$

and vertical V_v components of vector $\mathbf{V}_a$, and Equations (5.29) and (5.30) include the components of wind vector explicitly.

The magnitude of a horizontal component of a groundspeed vector [Eq. (5.4)] can then be defined as

$$V_G = \sqrt{V_x^2 + V_y^2} = \sqrt{[V_h\cos(\chi_a) + w_x]^2 + [V_h\sin(\chi_a) + w_y]^2} \tag{5.32}$$

and then track μ as

$$\chi = \tan^{-1}\left(\frac{V_h\sin(\chi_a) + w_y}{V_h\cos(\chi_a) + w_x}\right) \tag{5.33}$$

Using V_G, the Earth-referenced flight path angle is defined as [compare (5.5)]

$$\gamma = -\tan^{-1}\left(\frac{V_v + w_z}{V_G}\right) = -\tan^{-1}\left(\frac{V_d}{V_G}\right) \tag{5.34}$$

The simplest three-DoF model of PADS as a controlled object usually assumes that horizontal and vertical motions are decoupled. In this case the model for uncontrolled vertical motion reduces to the third row of Eq. (5.29)

$$\dot{z} = V_v + w_z \tag{5.35}$$

while a controlled horizontal motion can be described by the first two kinematic equations complemented with an equation describing yaw dynamics. Hence, the simplest model trajectory optimization (guidance) algorithms most often are based off of the following form:

$$\begin{bmatrix} \dot{x} \\ \dot{y} \end{bmatrix} = \begin{bmatrix} V_h \cos(\chi_a) \\ V_h \sin(\chi_a) \end{bmatrix} + \begin{bmatrix} w_x \\ w_y \end{bmatrix} \tag{5.36}$$

$$\dot{\psi} = u_c \tag{5.37}$$

Here u_c is a controlled yaw rate. In this model β is considered to be small, so that it could be neglected, which makes $\chi_a \approx \psi$.

A more accurate three-DoF model was developed by Jann [2001] and includes a first-order dynamics of the yaw rate

$$T_\psi \ddot{\psi} + \dot{\psi} = K_\psi \delta_a \tag{5.38}$$

In the last equation, T_ψ and K_ψ are constants, and δ_a is the differential (asymmetric) trailing-edge (TE) deflection. The yaw dynamics [Eq. (5.38)] can be represented in the state-space form as

$$\begin{bmatrix} \dot{\psi} \\ \ddot{\psi} \end{bmatrix} = \begin{bmatrix} 0 & 1 \\ 0 & T_\psi^{-1} \end{bmatrix} \begin{bmatrix} \psi \\ \dot{\psi} \end{bmatrix} + T_\psi^{-1} \begin{bmatrix} 0 \\ K_\psi \end{bmatrix} \delta_a \tag{5.39}$$

As an example, Table 5.1 presents parameters of models (5.35), (5.36), and (5.38) as identified for the ALEX PADS. Figure 5.2a shows an example of how this model fits real flight data.

To take into account possible asymmetries, for example, caused by different left $\bar{\delta}_l$ and right $\bar{\delta}_r$ control line offsets ALEX PADS model identification also utilized actuator parameters $\Delta \delta$ and K_δ, so that the symmetric and asymmetric

TABLE 5.1   IDENTIFIED PARAMETERS OF A THREE-DoF MODEL OF ALEX PADS [JANN 2001]

Parameter	Value	Description
V_h	9.89 m/s (19.2 kt)	horizontal component of $\mathbf{V}_a$
V_v	4.72 m/s (15.5 ft/s)	vertical component of $\mathbf{V}_a$
K_ψ	0.341 rad/s	ψ-model gain
T_ψ	0.43 s	ψ-model time constant
$\Delta \delta$	−0.221	asymmetric actuator offset
K_δ	2.821	right actuator coefficient

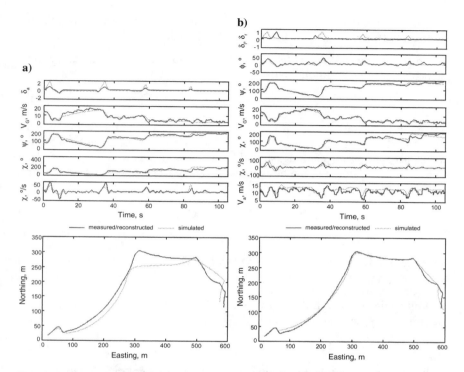

Fig. 5.2   Match of the a) three-DoF and b) four-DoF models of ALEX PADS to flight-test data [Jann 2001].

control inputs become

$$\overline{\delta}_s = 0.5(K_\delta \overline{\delta}_r + \overline{\delta}_l) \qquad (5.40)$$

$$\overline{\delta}_a = K_\delta \overline{\delta}_r - \overline{\delta}_l + \Delta\delta \qquad (5.41)$$

Within the parameter identification of the three-DoF model of ALEX PADS, actuator parameters of model (5.40) and (5.41) were identified as presented in Table 5.1. Figure 5.2a shows an example of how this model fits real flight data.

As seen from Fig. 5.2a, the three-DoF model is capable of representing some of the most important vehicle characteristics. It can also be used to principally check the functionality of GNC algorithms. However, many aspects are not modeled. Among those the most important ones are as follows:

- In turns, the roll angle ϕ of a vehicle can change significantly; it depends on the size of PAGS but for small systems can exceed 60 deg.
- In turns, the inclination of the lift vector causes the sink rate to increase.

- Horizontal and vertical components of an airspeed vector, along with lift-to-drag (L/D) ratio, change with symmetric TE deflection (flaring, braking).

To account for just a latter effect, the vertical motion [Eq. (5.35)] can be complemented with an additional term

$$\dot{z} = V_v + w_z + k_{v\dot{\psi}}|\dot{\psi}| \tag{5.42}$$

where the weighting coefficient $k_{v\dot{\psi}}$ is to be determined from the flight data (see Sec. 5.3). However, to address all aforementioned issues, a more accurate model, accounting for banking, should be developed. Following [Jann 2001], in such a model lift and drag can be assumed to be linearly dependent on symmetric TE deflection

$$D = 0.5\rho V_a^2 S(C_{D0} + C_{D\delta s}\delta_s) \tag{5.43}$$

$$L = 0.5\rho V_a^2 S(C_{L0} + C_{L\delta s}\delta_s) \tag{5.44}$$

while the aerodynamic side force Y along with pitch and sideslip angles is assumed to be negligibly small resulting in $Y \approx 0$, $\theta \approx 0$, $\beta \approx 0$, $v \approx 0$, $V_a \approx \sqrt{u^2 + w^2}$. (We consider nonwind conditions.)

Roll angle changes due to asymmetric TE deflection can be modeled with a first-order delay similar to that of Eq. (5.38)

$$T_\phi \dot{\phi} + \phi = K_\phi \delta_a \tag{5.45}$$

{In this model, for simplicity aerodynamic forces are thought to be independent of angle of attack, and aerodynamic moments and damping are taken into account implicitly by the roll angle delay model [Eq. (5.45)]}.

Next, using Eq. (5.23) and neglecting apparent masses bring us to

$$\begin{bmatrix} \dot{u} \\ 0 \\ \dot{w} \end{bmatrix} = \frac{1}{m}\mathbf{F} - \begin{bmatrix} qw \\ ru - pw \\ -qu \end{bmatrix} \tag{5.46}$$

where m is PADS mass. Accounting for Eq. (5.16) yields

$$\begin{bmatrix} \dot{u} \\ 0 \\ \dot{w} \end{bmatrix} = \frac{1}{m}\mathbf{F} - \begin{bmatrix} \dot{\psi}\sin(\phi)w \\ \dot{\psi}\cos(\phi)u - \dot{\phi}w \\ -\dot{\psi}\sin(\phi)u \end{bmatrix} \tag{5.47}$$

Transforming aerodynamic and gravitational force into the body-fixed coordinate system (with $\theta \approx 0$ and $\beta \approx 0$) results in

$$\mathbf{F} = \begin{bmatrix} L\sin(\alpha) - D\cos(\alpha) \\ 0 \\ -L\sin(\alpha) - D\cos(\alpha) \end{bmatrix} + m\begin{bmatrix} 0 \\ g\sin(\phi) \\ g\cos(\phi) \end{bmatrix} \tag{5.48}$$

(here g is the acceleration due to gravity). Finally, substituting Eq. (5.48) into Eq. (5.47) and resolving the second equation with respect to $\dot{\psi}$ results in the four-DoF model [Jann 2001]:

$$
\begin{bmatrix} \dot{u} \\ \dot{\psi} \\ \dot{w} \end{bmatrix} = \frac{1}{m} \begin{bmatrix} L\sin(\alpha) - D\cos(\alpha) \\ 0 \\ -L\sin(\alpha) - D\cos(\alpha) \end{bmatrix} + \begin{bmatrix} 0 \\ \dfrac{g\tan(\phi)}{u} \\ g\cos(\phi) \end{bmatrix} - \begin{bmatrix} w\dot{\psi}\sin(\phi) \\ -\dfrac{\dot{\phi}w}{u\cos(\phi)} \\ -u\dot{\psi}\sin(\phi) \end{bmatrix}
$$

$$
= \begin{bmatrix} \dfrac{L\sin(\alpha) - D\cos(\alpha)}{m} - w\dot{\psi}\sin(\phi) \\[2ex] \dfrac{g}{u}\tan(\phi) + \dfrac{w\dot{\phi}}{u\cos(\phi)} \\[2ex] \dfrac{-L\cos(\alpha) - D\sin(\alpha)}{m} + g\cos(\phi) + u\dot{\psi}\sin(\phi) \end{bmatrix} \tag{5.49}
$$

[The steady-state part of the second equation in Eq. (5.37) describes approximately the pendulum motion of a parafoil-load system in turn.]

Using the same flight data for ALEX PADS as in Fig. 5.2a, Table 5.2 presents identified parameters of the four-DoF model (5.43–5.45) and (5.49) [with $m = 122$ kg (269 lb), $S = 23.36$ m^2 (251.5 ft^2), and angle of attack determined as $\alpha = \tan^{-1}(w/u)$]. Reconstructed and simulated time histories using this four-DoF model along with the actuator parameters (5.40) and (5.41) are shown for comparison in Fig. 5.2b.

Obviously, the four-DoF model ensures a better match compared to the three-DoF model. As seen from Fig. 5.2b, the four-DoF model is able to simulate the increasing sink rate during turns and also reproduces the steady effects of symmetric TE deflection on the velocities and L/D ratio. In contrast to the three-DoF

TABLE 5.2   IDENTIFIED PARAMETERS OF A FOUR-DoF MODEL OF ALEX PADS [JANN 2001]

Parameter	Value	Description
C_{L0}	0.502	zero-lift coefficient
$C_{L\delta_s}$	0.892	$\partial C_L/\partial \delta_s$ derivative
C_{D0}	0.173	zero-drag coefficient
$C_{D\delta_s}$	1.086	$\partial C_D/\partial \delta_s$ derivative
K_ϕ	0.504 rad	ϕ-model gain
T_ϕ	0.994 s	ϕ-model time constant
$\Delta\delta$	−0.141	asymmetric actuator offset
K_δ	1.647	right actuator coefficient

model, the four-DoF model better reproduces the reduction of forward velocity during turns. Also, its actuator parameters are probably more reliable.

A simpler three-DoF model was also developed and used for the ParaLander PADS. Lift and drag coefficients were identified as [Altmann 2011]

$$C_L = 0.58 + 0.8(\bar{\delta}_s - \delta_0) \tag{5.50}$$

$$C_D = 0.21 + 0.27(\bar{\delta}_s - \delta_0) \tag{5.51}$$

where

$$\bar{\delta}_s = 0.5(\bar{\delta}_r + \bar{\delta}_l) \tag{5.52}$$

[compare to Eq. (5.40)] and $\delta_0 = 0.3$ refers to the design point. The roll dynamics were neglected resulting in

$$\phi = 3\bar{\delta}_a \tag{5.53}$$

[which is equivalent to Eq. (5.45) with $T_\phi = 0$ and $K_\phi = 3$] with

$$\bar{\delta}_a = 0.5(\bar{\delta}_r - \bar{\delta}_l) \tag{5.54}$$

[compare to Eq. (5.41)].

Although for GNC algorithm development the simple three- and four-DoF models just presented are sufficient, further development is needed if there is a need to address the following:

- Dynamic flare, caused by the load swinging forward and increasing dynamically the angle of attack

- Spiral divergence

- Aerodynamic damping in the longitudinal motion and the side force contribution on the lateral motion

- Nonlinear effect of the TE deflection (e.g., dependency on angle of attack), etc.

To address the aforementioned dynamic effects, the following sections present the higher-fidelity models.

To conclude this section, let us derive a couple of equations to be further used in Chapter 7 for trajectory optimization. To be more specific, let us exclude time from Eq. (5.30). To accomplish that, the first two equations should be divided by the third one, which, accounting for Eq. (5.7), become

$$\begin{bmatrix} x'_h \\ y'_h \end{bmatrix} = \frac{-V_h}{V_v + w_z} \begin{bmatrix} \cos(\chi_a) \\ \sin(\chi_a) \end{bmatrix} - \frac{1}{V_v + w_z} \begin{bmatrix} w_x \\ w_y \end{bmatrix} \tag{5.55}$$

The "prime" and subindex "h" in this equation mean a derivative with respect to h, for example, $x'_h = dx/dh = -dx/dz = -(dx/dt)/(dz/dt) = -\dot{x}/\dot{z}$. Similarly, time can be excluded from the model represented by Eq. (5.37)

$$\psi'_h = \frac{-1}{V_v + w_z} u_c \tag{5.56}$$

or Eq. (5.39)

$$\begin{bmatrix} \psi'_h \\ \dot{\psi}'_h \end{bmatrix} = \frac{-1}{V_v + w_z} \begin{bmatrix} 0 & 1 \\ 0 & T_\psi^{-1} \end{bmatrix} \begin{bmatrix} \psi \\ \dot{\psi} \end{bmatrix} - \frac{T_\psi^{-1}}{V_v + w_z} \begin{bmatrix} 0 \\ K_\psi \end{bmatrix} \delta_a \tag{5.57}$$

5.1.4 APPARENT MASS AND INERTIA TENSORS

Equation (5.48) utilized a simple force model acting on PADS. This force $\mathbf{F}$ consisting of the aerodynamic and gravity components $\mathbf{F}_a$ and $\mathbf{F}_g$ appears on the right-hand side of Eq. (5.22) and drives a translational motion. Similarly, Eq. (5.24) describing rotational dynamics depends upon the moments acting on a body. In turn, these moments are caused by aerodynamic and gravitational forces.

However, there is one more force we should take into account. When a body moves in the air (or fluid), it sets the air around it into a motion itself. In turn, this motion introduces pressure forces on the body, which is called the *apparent mass pressures*. The magnitude of these apparent mass pressures is inversely proportional to the mass ratio, representing the ratio of a mass of vehicle m to an air mass displaced or associated with this vehicle (see Sec. 2.6.2). For a standard aircraft, this ratio is huge, and consequently the apparent mass terms are negligibly small compared to the mass effects. However, for the lighter-than-air vehicles and parachute systems, the apparent mass terms are not small at all. For parafoils, the mass ratio (introduced in Sec. 2.6.2) can be represented as

$$M_r = \frac{m}{\rho S^{3/2}} \tag{5.58}$$

Based on data of Tables 1.6, C.1–C.3 and assuming $\rho \approx 1 \, \text{kg/m}^3$ (0.062 lb/ft^3), parameter M_r is usually of the order of 0.8. It might be as low as 0.5 for large PADSs (MegaFly) and as large as $\sim$6 for the smallest PADS (Snowflake) (or even larger for the two-stage penetrator systems like Screamer and Onyx). Not only are parafoils lightly loaded compared to aircraft, but they also feature a significant displacement of the canopy, a major contributor of aerodynamic forces, and the payload, where most of the mass is concentrated.

Forces and moments from apparent mass and inertia for a body with three orthogonal planes of symmetry, acting at the center of this body, can be found by considering the fluid's kinetic energy

$$2T^{\text{air}} = A\tilde{u}^2 + B\tilde{v}^2 + C\tilde{w}^2 + I_A\tilde{p}^2 + I_B\tilde{q}^2 + I_C\tilde{r}^2 \tag{5.59}$$

In Eq. (5.59) A, B, and C are the principal apparent masses while I_A, I_B, and I_C are principal apparent moments of inertia. Apparent mass and inertia contributions expressed in the rotating parafoil coordinate frame can be written as

$$\tilde{\mathbf{F}}_{a.m.} = -\left(\mathbf{I}_{a.m.} \begin{bmatrix} \dot{\tilde{v}}_x \\ \dot{\tilde{v}}_y \\ \dot{\tilde{v}}_z \end{bmatrix} + \mathbf{S}(\tilde{\boldsymbol{\omega}})\mathbf{I}_{a.m.} \begin{bmatrix} \tilde{v}_x \\ \tilde{v}_y \\ \tilde{v}_z \end{bmatrix} \right) \tag{5.60}$$

$$\tilde{\mathbf{M}}_{a.i.} = -\left(\mathbf{I}_{a.i.} \begin{bmatrix} \dot{\tilde{p}} \\ \dot{\tilde{q}} \\ \dot{\tilde{r}} \end{bmatrix} + \mathbf{S}(\tilde{\boldsymbol{\omega}})\mathbf{I}_{a.i.} \begin{bmatrix} \tilde{p} \\ \tilde{q} \\ \tilde{r} \end{bmatrix} + \mathbf{S}(\tilde{\mathbf{V}}_a)\mathbf{I}_{a.m.} \begin{bmatrix} \tilde{v}_x \\ \tilde{v}_y \\ \tilde{v}_z \end{bmatrix} \right) \tag{5.61}$$

In Eqs. (5.59–5.61), $[\tilde{v}_x, \tilde{v}_y, \tilde{v}_z]^T$ and $[\tilde{p}, \tilde{q}, \tilde{r}]^T$ are the components of the canopy airspeed vector $\tilde{\mathbf{V}}_a$ and canopy angular velocity vector $\tilde{\boldsymbol{\omega}}$. These components are expressed in the parafoil canopy frame $\{p\}$ (which is what "$\sim$" stands for), which is different from the body frame $\{b\}$ because of nonzero rigging angle. The apparent mass force [Eq. (5.60)] and apparent inertia moment [Eq. (5.61)] act at the centroid of apparent mass of the body. This centroid depends on the body's geometry and for ellipsoidal canopy shape resides at approximately volumetric canopy centroid. Apparent mass and inertia matrices in Eqs. (5.60) and (5.61), $\mathbf{I}_{a.m.}$ and $\mathbf{I}_{a.i.}$, have the diagonal form

$$\mathbf{I}_{a.m.} = \begin{bmatrix} A & 0 & 0 \\ 0 & B & 0 \\ 0 & 0 & C \end{bmatrix}, \quad \mathbf{I}_{a.i.} = \begin{bmatrix} I_A & 0 & 0 \\ 0 & I_B & 0 \\ 0 & 0 & I_C \end{bmatrix} \tag{5.62}$$

For a planar wing with an elliptical, noncambered cross section in potential flow with t, c, and b being the thickness, chord, and span of the inflated parafoil canopy (Fig. 5.3), the simplified expressions for six apparent mass terms in Eq. (5.62) can be related to a volume of the corresponding cylinder as shown in Fig. 5.4 [Lissaman and Brown 1998]

$$A = k_A \rho \frac{\pi}{4} t^2 b, \quad B = k_B \rho \frac{\pi}{4} t^2 c, \quad C = k_C \rho \frac{\pi}{4} c^2 b \tag{5.63}$$

$$I_A = k_A^* \rho \frac{\pi}{48} c^2 b^3, \quad I_B = k_B^* \rho \frac{4}{48\pi} c^4 b, \quad I_C = k_C^* \rho \frac{\pi}{48} t^2 b^3 \tag{5.64}$$

In these equations ρ is the air density and coefficients are supposed to account for the tip effects related to aspect ratio, and the nonplanar effects related to the arc-to-span ratio. (See Sec. 4.3.3 explaining the difference between planforms of inflated and flat parafoils.) Note that the

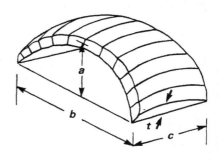

Fig. 5.3   Parafoil geometry [Lissaman and Brown 1998].

a)

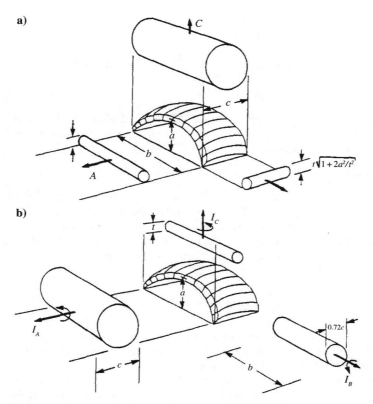

b)

Fig. 5.4 Volumetric representation of a) apparent masses and b) moments [Lissaman and Brown 1998].

expression for I_B assumes the effective cylinder radius of $[8/(3\pi^2)]^{0.25}c \approx 0.72c$ (Fig. 5.4b) and the effective cylinder radius for B for the arched wing will be adjusted as shown in Fig. 5.4a later as well. Following [Adkins 1982], Lissaman and Brown derived more accurate expressions based on the actual three-dimensional shape of inflated canopy and presented them using aspect ratio $AR = bc^{-1}$, arc-to-span ratio $a^* = ab^{-1}$, and relative thickness $t^* = tc^{-1} = tb^{-1}AR$

$$A = 0.666\rho\left(1 + \frac{8}{3}a^{*2}\right)t^2b$$

$$B = 0.267\rho\left[1 + 2\frac{a^{*2}}{t^{*2}}AR^2(1 - t^{*2})\right]t^2c \qquad (5.65)$$

$$C = 0.785\rho\sqrt{1 + 2a^{*2}(1 - t^{*2})}\frac{AR}{1 + AR}c^2b$$

$$I_A = 0.055\rho \frac{AR}{1 + AR} c^2 b^3$$

$$I_B = 0.0308\rho \frac{AR}{1 + AR} \left[1 + \frac{\pi}{6}(1 + AR)ARa^{*2}t^{*2} \right] c^4 b \qquad (5.66)$$

$$I_C = 0.0555\rho(1 + 8a^{*2})t^2 b^3$$

If we assume $a^* = 0$ and relate Eqs. (5.65) and (5.66) back to Eqs. (5.63) and (5.64), for a flat wing with $AR = 3$ we will obtain the values of coefficients in Eqs. (5.63) and (5.64) as follows:

$$k_A = 0.848, \quad k_B = 0.34, \quad k_C = 0.75, \quad k_A^* = 0.63, \quad k_B^* = 0.871, \quad k_C^* = 0.848 \quad (5.67)$$

When we account for an arched wing, we should probably expect these coefficients to be slightly larger. The exception is the coefficient k_B, which would be about 20 times larger. The reason is that for the arched wing a major contribution in the B term in Eq. (5.63) comes not from the volume of air $0.25\pi t^2 c$, but rather from the volume $0.5\pi a^2 c$, and so the expression for B in this case should look like

$$B = k_B \rho \frac{\pi}{4}(t^2 + 2a^2)c = k_B \rho \frac{\pi}{4} \left(1 + 2\frac{a^{*2}}{t^{*2}} AR^2 \right) t^2 c \qquad (5.68)$$

[As seen, for the arched wing with $a^* \sim t^{*2}$, the second term in Eq. (5.68) is $2AR^2$ times larger than the first term.] Hence, making comparison to the second equation in Eq. (5.63) makes no sense.

Let us now relate Eqs. (5.65) and (5.66) to Eqs. (5.63) and (5.64) [the equation for B in Eq. (5.65) will be related to Eq. (5.68)] and obtain the three-dimensional correction coefficients for a typical PADS with $AR = 3$, and $a^* = t^* = 0.15$:

$$k_A = 0.899, \quad k_B = 0.34, \quad k_C = 0.766, \quad k_A^* = 0.63, \quad k_B^* = 0.874, \quad k_C^* = 1 \quad (5.69)$$

As seen, coefficients in Eq. (5.69) are about the same as those used for a planar nonarched wing of Eq. (5.67) with the major effect in increasing the terms A (by 6%) and I_C (by 18%) as expected.

To confirm the values of coefficients in Eqs. (5.67) and (5.69), [Barrows 2002] utilized VSAERO panel code with 1,600 panels for the main portion of the wing plus 160 panels for each tip (Fig. 5.5). The values for the three-dimensional correction coefficients obtained for the same parafoil ($AR = 3$, $a^* = t^* = 0.15$) happened to be about the same as those of Eqs. (5.67) and (5.69). Specifically, utilizing Eqs. (5.63) and (5.64) along with data by Barrows for the flat wing (Fig. 5.5a), they were found to be

$$k_A = 0.879, \quad k_B = 0.996, \quad k_C = 0.823, \quad k_A^* = 0.589, \quad k_B^* = 0.812, \quad k_C^* = 0.646 \quad (5.70)$$

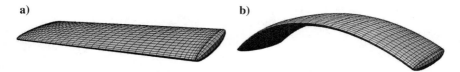

Fig. 5.5   a) Flat wing vs b) arched wing [Barrows 2002].

Utilizing the same equations but Eq. (5.68) and arched wing data (Fig. 5.5b), the coefficients were found to be

$$k_A = 0.945, \quad k_B = 0.614, \quad k_C = 0.806, \quad k_A^* = 0.589, \quad k_B^* = 0.784, \quad k_C^* = 0.953$$

$$(5.71)$$

Although the absolute values of coefficients in Eqs. (5.70) and (5.71) are slightly different from those presented in Eqs. (5.67) and (5.69), respectively, the two effects observed earlier are about the same: the A term in Eq. (5.71) compared to that of Eq. (5.70) increases by 8%, and the I_C term increases by 48%. With regard to the B term, Barrows [2002] found that it is very sensitive to the tip shape and for the flat wing can vary by the factor of four. Specifically, for the flat wing represented by an ellipsoid with the semi-principal axes related as 3:1:0.15, Barrows reported the same value of $k_B = 0.34$ as in Eq. (5.67). For the wing with ellipsoidal caps shown in Fig. 5.5a, the k_B value was three times larger, as shown in Eq. (5.70). The largest value of k_B was found for the wing with flat end caps. Obviously, for the arched wing (Fig. 5.5b) the effect of the tip shape is smaller because of a major contribution of the second term in Eq. (5.68). Yet, the value of $k_B = 0.614$ obtained in this case [Eq. (5.71)] is still 1.8 times larger than that of $k_B = 0.34$ in Eq. (5.69).

A vast majority of the known PADS models use Eqs. (5.63), (5.64), and (5.68) with coefficients given in Eq. (5.69) (e.g., [Gupta et al. 2011a; Mooij et al. 2003; Prakash and Ananthkrishnan 2006; Toglia and Vendittelli 2010; Wise 2006]). The coefficients given in Eq. (5.71) in their explicit form have never been published, but can now be used as well. Sections 11.4.3 and 11.5.3 address the issue of identifying these coefficients from flight data. As shown in there, they can be quite different from those of Eq. (5.68) or Eq. (5.71).

5.1.5   SIX-DoF MODEL

Let us now develop a complete six-DoF model of PADS as a rigid body. Figure 5.6 shows a schematic of a parafoil-payload system. With the exception of movable parafoil brakes, the parafoil canopy is considered to be a fixed shape once

Fig. 5.6   Generic six-DoF parafoil-payload system.

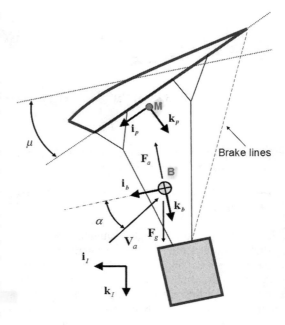

it has completely inflated. A body frame $\{b\}$ is fixed at the system mass center with the unit vectors $\mathbf{i}_b$ and $\mathbf{k}_b$ oriented forward and down. The canopy-fixed coordinate frame $\{p\}$ is rotated with respect to $\{b\}$ by some negative rigging angle μ. (Compared to Sec. 2.5.1, which only considered a magnitude of angle μ, hereinafter we will account for the sign of μ as well.) Note that Fig. 5.6 features the origin of $\{p\}$ being at the apparent mass center M, which is not necessarily the case. For other high-fidelity models developed later in this chapter, the apparent mass center will not coincide with the origin of $\{p\}$ (see Sec. 5.2.1).

Dynamic equations are formed by summing forces and moments about the system mass center, both in the body reference frame $\{b\}$ and equating them to the time derivatives of linear and angular momentum. Hence, we start by combining translational and rotational dynamics described by Eqs. (5.23) and (5.25) into a single matrix equation

$$
\begin{bmatrix}
(m+m_e)\mathbf{I}_{3\times3} & \vdots & \mathbf{0}_{3\times3} \\
\cdots & \cdots & \cdots \\
\mathbf{0}_{3\times3} & \vdots & \mathbf{I}
\end{bmatrix}
\begin{bmatrix}
\dot{u} \\ \dot{v} \\ \dot{w} \\ \cdots \\ \dot{p} \\ \dot{q} \\ \dot{r}
\end{bmatrix}
=
\begin{bmatrix}
\mathbf{F}-(m+m_e)\mathbf{S}(\boldsymbol{\omega})\begin{bmatrix} u \\ v \\ w \end{bmatrix} \\
\cdots \\
\mathbf{M}-\mathbf{S}(\boldsymbol{\omega})\mathbf{I}\begin{bmatrix} p \\ q \\ r \end{bmatrix}
\end{bmatrix}
\tag{5.72}
$$

In this equation, m represents PADS mass, m_e denotes the mass of air entrapped in ram-air canopy (also referred to as enclosed or included air mass), $\mathbf{0}_{3\times3}$ and $\mathbf{I}_{3\times3}$ represent 3×3 zero and identity matrices, respectively.

The total force $\mathbf{F}$ and moment $\mathbf{M}$ acting on the parafoil and payload have contributions from weight, aerodynamic loads, and apparent mass of the canopy.

Weight contribution of the system is expressed in the body frame $\{b\}$ as

$$\mathbf{F}_g = mg \begin{bmatrix} -\sin(\theta) \\ \cos(\theta)\sin(\phi) \\ \cos(\theta)\cos(\phi) \end{bmatrix} \tag{5.73}$$

[compare it with the second term on the right-hand side of Eq. (5.48)]. Aerodynamic forces and moments have contributions from both the canopy and payload. For the larger PADS, a canopy contributes the most, so that the forces and moments on a payload might be neglected. For the sake of simplicity, both canopy and payload contributions are to be combined into a single aerodynamic model using standard aerodynamic derivatives. The aerodynamic force in $\{b\}$ is modeled as

$$\mathbf{F}_a = QS_w^b\mathbf{R} \begin{bmatrix} C_{D0} + C_{D\alpha^2}\alpha^2 + C_{D\delta_s}\bar{\delta}_s \\ C_{Y\beta}\beta \\ C_{L0} + C_{L\alpha}\alpha + C_{L\delta_s}\bar{\delta}_s \end{bmatrix} \tag{5.74}$$

[compare to Eqs. (5.43) and (5.44)] where S is the parafoil canopy area, $\bar{\delta}_s = \delta_s/\delta_{s\,max} \in [0; 1]$ is the symmetric TE deflection, and $_w^b\mathbf{R}$ is the transformation from the aerodynamic (wind) coordinate system $\{w\}$ to the body frame $\{b\}$

$$_w^b\mathbf{R} = \mathbf{R}_\alpha\mathbf{R}_\beta = \begin{bmatrix} \cos(\alpha) & 0 & -\sin(\alpha) \\ 0 & 1 & 0 \\ \sin(\alpha) & 0 & \cos(\alpha) \end{bmatrix} \begin{bmatrix} \cos(\beta) & \sin(\beta) & 0 \\ -\sin(\beta) & \cos(\beta) & 0 \\ 0 & 0 & 1 \end{bmatrix}$$
$$= \begin{bmatrix} \cos(\alpha)\cos(\beta) & \cos(\alpha)\sin(\beta) & -\sin(\alpha) \\ -\sin(\beta) & \cos(\beta) & 0 \\ \sin(\alpha)\cos(\beta) & \sin(\alpha)\sin(\beta) & \cos(\alpha) \end{bmatrix} \tag{5.75}$$

If the angles of attack and sideslip are small (less than 10 deg), the rotation matrix of Eq. (5.75) reduces to

$$_w^b\mathbf{R} \approx \begin{bmatrix} 1 & \beta & -\alpha \\ -\beta & 1 & 0 \\ \alpha & 0 & 1 \end{bmatrix} \tag{5.76}$$

The angles of attack and sideslip are determined by the components of the airspeed vector $\mathbf{V}_a$ in the $\{b\}$ frame, $[v_x, v_y, v_z]^T$

$$\alpha = \tan^{-1}\left(\frac{v_z}{v_x}\right), \quad \beta = \tan^{-1}\left(\frac{v_y}{\sqrt{v_x^2 + v_z^2}}\right) \tag{5.77}$$

These components are determined using Eq. (5.1) as

$$\begin{bmatrix} v_x \\ v_y \\ v_z \end{bmatrix} = \begin{bmatrix} u \\ v \\ w \end{bmatrix} - {}^b_n\mathbf{R} \begin{bmatrix} w_x \\ w_y \\ w_z \end{bmatrix} \tag{5.78}$$

Similarly to Eq. (5.74), the aerodynamic moment $\mathbf{M}$ in $\{b\}$ can be written as

$$\mathbf{M}_a = \frac{\rho V_a^2 S}{2} \begin{bmatrix} b\left(C_{l\beta}\beta + \dfrac{b}{2V_a}C_{lp}p + \dfrac{b}{2V_a}C_{lr}r + C_{l\delta_a}\bar{\delta}_a\right) \\ \bar{c}\left(C_{m0} + C_{m\alpha}\alpha + \dfrac{c}{2V_a}C_{mq}q\right) + C \, m \, \delta_s \cdot \hat{\delta}_s \\ b\left(C_{n\beta}\beta + \dfrac{b}{2V_a}C_{np}p + \dfrac{b}{2V_a}C_{nr}r + C_{n\delta_a}\bar{\delta}_a\right) \end{bmatrix} \tag{5.79}$$

where $\bar{c} = \frac{2}{S}\int_0^{b/2} c(y)\,dy$ is a mean aerodynamic chord and $\bar{\delta}_a = \delta_a/\delta_{s\max} \in [-1; 1]$ is the asymmetric TE deflection.

The apparent mass terms expressed by Eqs. (5.60) and (5.61) and acting at the apparent mass center M should be projected into $\{b\}$

$$\mathbf{F}_{a.m.} = {}^p_b\mathbf{R}^T\tilde{\mathbf{F}}_{a.m.} = -{}^p_b\mathbf{R}^T\left(\mathbf{I}_{a.m.}\begin{bmatrix} \dot{\tilde{v}}_x \\ \dot{\tilde{v}}_y \\ \dot{\tilde{v}}_z \end{bmatrix} + \mathbf{S}(\tilde{\boldsymbol{\omega}})\mathbf{I}_{a.m.}\begin{bmatrix} \tilde{v}_x \\ \tilde{v}_y \\ \tilde{v}_z \end{bmatrix}\right) \tag{5.80}$$

$$\mathbf{M}_{a.i.} = {}^p_b\mathbf{R}^T\tilde{\mathbf{M}}_{a.i.} + \mathbf{S}(\mathbf{r}_{BM})\mathbf{F}_{a.m.}$$

$$= -{}^p_b\mathbf{R}^T\left(\mathbf{I}_{a.i.}\begin{bmatrix} \dot{\tilde{p}} \\ \dot{\tilde{q}} \\ \dot{\tilde{r}} \end{bmatrix} + \mathbf{S}(\tilde{\boldsymbol{\omega}})\mathbf{I}_{a.i.}\begin{bmatrix} \tilde{p} \\ \tilde{q} \\ \tilde{r} \end{bmatrix}\right) + \mathbf{S}(\mathbf{r}_{BM})\mathbf{F}_{a.m.} \tag{5.81}$$

where $\mathbf{r}_{BM} = [x_{BM}, y_{BM}, z_{BM}]^T$ is the vector from the system mass center (origin of $\{b\}$) to the apparent mass center (in this simplified case—the origin of $\{p\}$), so that the second term in Eq. (5.81) is due to the translation from the origin of $\{p\}$ to the origin of $\{b\}$. The rotation matrix ${}^p_b\mathbf{R}$ is determined by a single-axis transformation from the body to canopy reference frame by the rigging angle μ

$$ {}^p_b\mathbf{R} = \mathbf{R}_\mu = \begin{bmatrix} \cos(\mu) & 0 & -\sin(\mu) \\ 0 & 1 & 0 \\ \sin(\mu) & 0 & \cos(\mu) \end{bmatrix} \tag{5.82}$$

Following [Slegers 2010], the expression for $\mathbf{M}_{a.i.}$ in Eq. (5.81) has one term less than that of Eq. (5.61) assuming that aerodynamic coefficients of Eq. (5.79) related to steady or quasi-steady motions cover the corresponding effects already.

The final step before substituting forces [Eqs. (5.73), (5.74), and (5.80)] and moments [Eqs. (5.79) and (5.81)] into Eq. (5.72) is to specify vectors $[\tilde{u}, \tilde{v}, \tilde{w}]^T$ and $[\tilde{p}, \tilde{q}, \tilde{r}]^T$ expressed in $\{p\}$ via their components in $\{b\}$. The angular velocity and acceleration expressed in the canopy frame $\{p\}$ become

$$\tilde{\boldsymbol{\omega}} = \begin{bmatrix} \tilde{p} \\ \tilde{q} \\ \tilde{r} \end{bmatrix} = {}^p_b\mathbf{R}\begin{bmatrix} p \\ q \\ r \end{bmatrix}, \quad \begin{bmatrix} \dot{\tilde{p}} \\ \dot{\tilde{q}} \\ \dot{\tilde{r}} \end{bmatrix} = {}^p_b\mathbf{R}\begin{bmatrix} \dot{p} \\ \dot{q} \\ \dot{r} \end{bmatrix} \tag{5.83}$$

The airspeed vector of the canopy at the apparent mass center M can be expressed in $\{p\}$ as

$$\begin{bmatrix} \tilde{v}_x \\ \tilde{v}_y \\ \tilde{v}_z \end{bmatrix} = {}^p_b\mathbf{R}\left(\begin{bmatrix} u \\ v \\ w \end{bmatrix} + \mathbf{S}(\boldsymbol{\omega})\begin{bmatrix} x_{BM} \\ y_{BM} \\ z_{BM} \end{bmatrix} - {}^b_n\mathbf{RW} \right)$$

$$= {}^p_b\mathbf{R}\left(\begin{bmatrix} u \\ v \\ w \end{bmatrix} - \mathbf{S}(\mathbf{r}_{BM})\begin{bmatrix} p \\ q \\ r \end{bmatrix} - {}^b_n\mathbf{RW} \right) \tag{5.84}$$

[Note, Eq. (5.84) utilizes an equality $\mathbf{S}(\mathbf{a})\mathbf{b} = -\mathbf{S}(\mathbf{b})\mathbf{a}$.] Similar to Eq. (5.84), with account of Eq. (5.78) the acceleration vector of the canopy at the apparent mass center M can be expressed as

$$\begin{bmatrix} \dot{\tilde{v}}_x \\ \dot{\tilde{v}}_y \\ \dot{\tilde{v}}_z \end{bmatrix} = {}^p_b\mathbf{R}\left(\begin{bmatrix} \dot{u} \\ \dot{v} \\ \dot{w} \end{bmatrix} + \mathbf{S}(\dot{\boldsymbol{\omega}})\begin{bmatrix} x_{BM} \\ y_{BM} \\ z_{BM} \end{bmatrix} \right) = {}^p_b\mathbf{R}\left(\begin{bmatrix} \dot{u} \\ \dot{v} \\ \dot{w} \end{bmatrix} - \mathbf{S}(\mathbf{r}_{BM})\begin{bmatrix} \dot{p} \\ \dot{q} \\ \dot{r} \end{bmatrix} \right) \tag{5.85}$$

Substituting Eqs. (5.83–5.85) into Eqs. (5.80) and (5.81) yields

$$\mathbf{F}_{a.m.} = -{}^p_b\mathbf{R}^T\left[\mathbf{I}_{a.m.}\,{}^p_b\mathbf{R}\left(\begin{bmatrix} \dot{u} \\ \dot{v} \\ \dot{w} \end{bmatrix} - \mathbf{S}(\mathbf{r}_{BM})\begin{bmatrix} \dot{p} \\ \dot{q} \\ \dot{r} \end{bmatrix} \right)\right.$$

$$\left. + {}^p_b\mathbf{R}\mathbf{S}(\boldsymbol{\omega}){}^p_b\mathbf{R}^T\mathbf{I}_{a.m.}\,{}^p_b\mathbf{R}\left(\begin{bmatrix} u \\ v \\ w \end{bmatrix} - \mathbf{S}(\mathbf{r}_{BM})\begin{bmatrix} p \\ q \\ r \end{bmatrix} - {}^b_n\mathbf{RW} \right) \right] \tag{5.86}$$

$$\mathbf{M}_{a.i.} = -{}^p_b\mathbf{R}^T\left(\mathbf{I}_{a.i.}\,{}^p_b\mathbf{R}\begin{bmatrix} \dot{p} \\ \dot{q} \\ \dot{r} \end{bmatrix} + \mathbf{S}(\boldsymbol{\omega}_p)\mathbf{I}_{a.i.}\,{}^p_b\mathbf{R}\begin{bmatrix} p \\ q \\ r \end{bmatrix} \right) + \mathbf{S}(\mathbf{r}_{BM})\mathbf{F}_{a.m.} \tag{5.87}$$

Substituting forces [Eqs. (5.73), (5.74), and (5.86)] and moments [Eqs. (5.79) and (5.87)] into Eq. (5.72), employing the equality $\mathbf{S}(\tilde{\boldsymbol{\omega}}) = \mathbf{S}({}^p_b\mathbf{R}\boldsymbol{\omega}) = {}^p_b\mathbf{R}\mathbf{S}(\boldsymbol{\omega}){}^p_b\mathbf{R}^T$ (which was not used in the [Gorman 2011; Gorman and Slegers 2011a]), and rearranging the terms yield final dynamic equations of motion expressed compactly in the matrix form

$$
\begin{bmatrix}
(m + m_e)\mathbf{I}_{3\times3} + \mathbf{I}'_{a.m.} & \vdots & -\mathbf{I}'_{a.m.}\mathbf{S}(\mathbf{r}_{BM}) \\
\cdots & \cdots & \cdots \\
\mathbf{S}(\mathbf{r}_{BM})\mathbf{I}'_{a.m.} & \vdots & \mathbf{I} + \mathbf{I}'_{a.i.} - \mathbf{S}(\mathbf{r}_{BM})\mathbf{I}'_{a.m.}\mathbf{S}(\mathbf{r}_{BM})
\end{bmatrix}
\begin{bmatrix}
\dot{u} \\ \dot{v} \\ \dot{w} \\ \cdots \\ \dot{p} \\ \dot{q} \\ \dot{r}
\end{bmatrix}
$$

$$
= \begin{bmatrix}
\mathbf{B}_1 \\ \cdots \\ \mathbf{B}_2
\end{bmatrix} \tag{5.88}
$$

where

$$
\mathbf{B}_1 = \mathbf{F}_a + \mathbf{F}_g - (m + m_e)\mathbf{S}(\boldsymbol{\omega})\begin{bmatrix} u \\ v \\ w \end{bmatrix} - \mathbf{S}(\boldsymbol{\omega})\mathbf{I}'_{a.m.}\left(\begin{bmatrix} u \\ v \\ w \end{bmatrix} - \mathbf{S}(\mathbf{r}_{BM})\begin{bmatrix} p \\ q \\ r \end{bmatrix} \right)
$$

$$
+ \mathbf{S}(\boldsymbol{\omega})\mathbf{I}'_{a.m.}{}^{b}_{n}\mathbf{RW} = \mathbf{F}_a + \mathbf{F}_g - \mathbf{S}(\boldsymbol{\omega})\left[(m + m_e)\mathbf{I}_{3\times3} + \mathbf{I}'_{a.m.} \right]\begin{bmatrix} u \\ v \\ w \end{bmatrix}
$$

$$
+ \mathbf{S}(\boldsymbol{\omega})\mathbf{I}'_{a.m.}\mathbf{S}(\mathbf{r}_{BM})\begin{bmatrix} p \\ q \\ r \end{bmatrix} + \mathbf{S}(\boldsymbol{\omega})\mathbf{I}'_{a.m.}{}^{b}_{n}\mathbf{RW} \tag{5.89}
$$

$$
\mathbf{B}_2 = \mathbf{M}_a - \mathbf{S}(\boldsymbol{\omega})\mathbf{I}\begin{bmatrix} p \\ q \\ r \end{bmatrix} - \mathbf{S}(\boldsymbol{\omega})\mathbf{I}'_{a.i.}\begin{bmatrix} p \\ q \\ r \end{bmatrix}
$$

$$
- \mathbf{S}(\mathbf{r}_{BM})\mathbf{S}(\boldsymbol{\omega})\mathbf{I}'_{a.m.}\left(\begin{bmatrix} u \\ v \\ w \end{bmatrix} - \mathbf{S}(\mathbf{r}_{BM})\begin{bmatrix} p \\ q \\ r \end{bmatrix} - {}^{b}_{n}\mathbf{RW} \right)
$$

$$
= \mathbf{M}_a - \left[\mathbf{S}(\boldsymbol{\omega})(\mathbf{I} + \mathbf{I}'_{a.i.}) - \mathbf{S}(\mathbf{r}_{BM})\mathbf{S}(\boldsymbol{\omega})\mathbf{I}'_{a.m.}\mathbf{S}(\mathbf{r}_{BM}) \right]\begin{bmatrix} p \\ q \\ r \end{bmatrix}
$$

$$
- \mathbf{S}(\mathbf{r}_{BM})\mathbf{S}(\boldsymbol{\omega})\mathbf{I}'_{a.m.}\begin{bmatrix} u \\ v \\ w \end{bmatrix} + \mathbf{S}(\mathbf{r}_{BM})\mathbf{S}(\boldsymbol{\omega})\mathbf{I}'_{a.m.}{}^{b}_{n}\mathbf{RW} \tag{5.90}
$$

Equation (5.88) also benefits from utilizing the standard notation used for tensors of second rank such that $I'_\xi = {}^P_b R^T I_\xi {}^P_b R$. This model can be further reduced to the following compact form:

$$\dot{V}^* = (A_U + A_r A_L)^{-1} \left\{ \begin{bmatrix} F_a + F_g \\ M_a \end{bmatrix} - [A_\omega A_U + A_r A_\omega A_L - A_r S(\omega) I'_{a.m.} {}^b_n RW] V^* \right\}$$

(5.91)

with $V^* = [u, v, w, p, q, r]^T$ and matrices A_U, A_L, A_r, A_ω defined as

$$A_U = \begin{bmatrix} (m + m_e) I_{3\times3} + I'_{a.m.} & I'_{a.m.} S(r_{BM}) \\ 0_{3\times3} & I + I'_{a.i.} \end{bmatrix}, \quad A_r = \begin{bmatrix} I_{3\times3} & 0_{3\times3} \\ 0_{3\times3} & S(r_{BM}) \end{bmatrix}$$

$$A_L = \begin{bmatrix} 0_{3\times3} & 0_{3\times3} \\ I'_{a.m.} & I'_{a.m.} S(r_{BM}) \end{bmatrix}, \quad A_\omega = \begin{bmatrix} S(\omega) & 0_{3\times3} \\ 0_{3\times3} & S(\omega) \end{bmatrix}$$

(5.92)

The translational and rotational kinematics of the $\{b\}$ frame is described by Eqs. (5.28) and (5.15), respectively.

Figure 5.7 shows a modification of the PADS model of Fig. 5.2 in respect to how canopy aerodynamics is modeled [Slegers 2004].

In this case system's dynamics will be described by the same set of equations with the only difference that the aerodynamic force F_a [Eq. (5.89)] will be modeled not by Eq. (5.74), but be rather composed of the aerodynamic force on the payload and lift and drag contributions of individual canopy panels, modeled as

$$C_{L,i} = C_{L0;i} + C_{L\alpha;i}\alpha_i + k(C_{L\delta i}\delta_i + C_{L\delta_i^3}\bar{\delta}_i^3)$$

$$C_{D,i} = C_{D0;i} + C_{D\alpha^2;i}\alpha_i^2 + k(C_{D\delta i}\delta_i + C_{D\delta_i^3}\bar{\delta}_i^3)$$

(5.93)

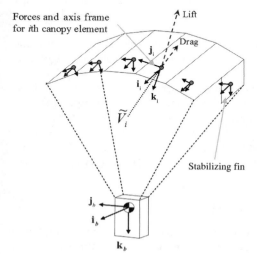

Forces and axis frame for ith canopy element

Lift

Drag

Stabilizing fin

These individual terms (Fig. 5.8) are functions of the angle of attack of each element, $\alpha_i = \tan^{-1}(\tilde{v}_z / \tilde{v}_x)$, and for the outboard panels are amended by two additional terms due to individual control deflections δ_i. [The coefficient k in Eq. (5.93) in this case changes its value from 0 to 1.] In addition to the linear term, as in model (5.74), a term proportional to the third power of

Fig. 5.7 Representing a canopy composed of several panels.

Fig. 5.8 Lift and drag coefficients for various brake deflections.

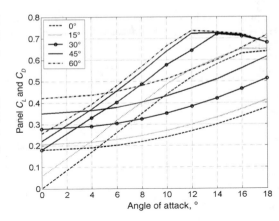

control deflection is added to account for the rapid changes in lift and drag coefficient observed for large brake deflections.

In this case,

$$\mathbf{F}_a = \mathbf{F}_a^s + \sum_i \mathbf{F}_a^{p_i}$$

$$= -\frac{1}{2}\rho S_s C_D^s \sqrt{v_x^{s\,2} + v_y^{s\,2} + v_z^{s\,2}} \begin{bmatrix} v_x^s \\ v_y^s \\ v_z^s \end{bmatrix}$$

$$-\frac{1}{2}\rho \sum_i S_{i_b}^{p_i} \mathbf{R}^T \left(C_{L,i}\sqrt{\tilde{v}_{xi}^2 + \tilde{v}_{zi}^2} \begin{bmatrix} \tilde{v}_{zi} \\ 0 \\ -\tilde{v}_{xi} \end{bmatrix} + C_{D,i}\sqrt{\tilde{v}_{xi}^2 + \tilde{v}_{yi}^2 + \tilde{v}_{zi}^2} \begin{bmatrix} \tilde{v}_{xi} \\ \tilde{v}_{yi} \\ \tilde{v}_{zi} \end{bmatrix} \right) \quad (5.94)$$

$$\mathbf{M}_a = \mathbf{S}(\mathbf{r}_{BS})\mathbf{F}_a^s + \sum_i \mathbf{S}(\mathbf{r}_{B p_i})\mathbf{F}_a^{p_i} \quad (5.95)$$

where the aerodynamic force on the payload $\mathbf{F}_a^s$ consists entirely of profile drag and $\sum_i \mathbf{F}_a^{p_i}$ is the sum of lift and drag contributions of individual canopy panels converted from the individual canopy element frame $\{p_i\}$ to $\{b\}$ using the corresponding rotation matrices $_b^{p_i}\mathbf{R}$.

Model (5.93) assumes minimal knowledge of parafoil aerodynamics and might be useful, for example, for studying dynamics and control of damaged parafoils [Culpepper et al. 2013].

Equation (5.91) represents a set of coupled, nonlinear differential equations. The matrix on the left-hand side of Eq. (5.91), composed of three simpler matrices $\mathbf{A}_U$, $\mathbf{A}_r$, and $\mathbf{A}_L$, is a function of the mass and geometry properties of the parafoil. The geometry of the parafoil is assumed to be fixed, and so this matrix is constant and only needs to be inverted once at the beginning of the simulation. With specified initial conditions, the states can be numerically integrated forward in time.

5.1.6 IMPLEMENTATIONS OF SIX-DoF MODEL

This section presents some of the documented models that were proven and used on multiple occasions. The first model is the model of the ultra-light-weight

TABLE 5.3 IDENTIFIED PARAMETERS OF A SIX-DOF MODEL OF SNOWFLAKE PADS

Parameter	Value
System mass m	2.4 kg (5.2 lb)
Canopy span b, chord c, thickness t	1.35 m (4.45 ft), 0.75 m (2.47 ft), 0.075 m (0.25 ft)
Canopy reference area S	1 m² (11.0 ft²)
Rigging angle μ	−12 deg
Steady-state aerodynamic velocity V_a	7.3 m/s (14.2 kt)
Inertia matrix $\mathbf{I}$ elements	$\mathbf{I} = \begin{bmatrix} 0.42 & 0 & 0.03 \\ 0 & 0.40 & 0 \\ 0.03 & 0 & 0.053 \end{bmatrix}$ kg·m², $\mathbf{I} = \begin{bmatrix} 9.97 & 0 & 0.71 \\ 0 & 9.5 & 0 \\ 0.71 & 0 & 1.26 \end{bmatrix}$ lb·ft²
Apparent mass matrix	$\mathbf{I}_{a.m.} = \mathrm{diag}([0.012, 0.032, 0.42])$ kg, $\mathbf{I}_{a.m.} = \mathrm{diag}([0.027, 0.071, 0.93])$ lb
Apparent inertia matrix	$\mathbf{I}_{a.i.} = \mathrm{diag}([0.054, 0.14, 0.0024])$ kg·m², $\mathbf{I}_{a.i.} = \mathrm{diag}([1.29, 3.34, 0.06])$ lb·ft²
Vector from the mass center to the apparent mass center	$\mathbf{r}_{BM} = [0.046, 0, -1.11]^T$ m, $\mathbf{r}_{BM} = [0.15, 0, -3.64]^T$ ft
Maximum brake deflection $\delta_{s\,max}$	0.13 m (0.42 ft or 5 in.)
Aerodynamic coefficients	$C_{D0} = 0.25$, $C_{D\alpha^2} = 0.12$ $C_{Y\beta} = -0.23$ $C_{L0} = 0.091$, $C_{L\alpha} = 0.90$ $C_{m0} = 0.35$, $C_{m\alpha} = -0.72$, $C_{mq} = -1.49$ $C_{l\beta} = -0.036$, $C_{lp} = -0.84$, $C_{lr} = -0.082$, $C_{l\delta_a} = -0.0035$ $C_{n\beta} = -0.0015$, $C_{np} = -0.082$, $C_{nr} = -0.27$, $C_{n\delta_a} = 0.0115$

system Snowflake (Fig. 1.30). Table 5.3 shows all parameters necessary to build a six-DoF model [Eq. (5.91)].

Note that the aerodynamic and control coefficients of the model presented in Table 5.3 do not depend on the angle of attack. Another model, discussed next, attempts to incorporate this dependence from general data obtained in wind-tunnel testing of small-aspect-ratio parafoils. Figure 5.9 shows a Simulink implementation of a six-DoF model [Eq. (5.91)] for a Pegasus PADS (Fig. 1.8).

The only available data on the aerodynamic coefficients for this PADS at the time the modeling was carried out were the wind-tunnel data from [Entchev and Rubenstein 2001; Ware and Hassel 1969] and computational fluid dynamics data obtained for the steady-state turn in [Brown 1993]. However, it was impossible to implement these data directly. While data in [Brown 1993; Entchev and Rubenstein 2001] contain just one single point in terms of dependence on the angle of attack, data in [Ware and Hassel 1969], despite the richness of different dependences for three aspect ratios, were found to be somewhat incomplete and contradictory. Figures 5.10–5.12 give an example of lateral and longitudinal coefficient dependencies on the angle of attack after some necessary corrections. As seen, these data are quite nonlinear and nonconsistent and obviously use the different scale for the angle of attack. After thorough analysis of data, it was assumed that a setup for the wind-tunnel experiment included about a 10-deg rigging angle so that the angle-of-attack axis for the longitudinal channel (Figs. 5.10b and 5.11) should be shifted left by this angle.

Instead of modeling the aerodynamic coefficients along the full range of angles of attack ([−10; 80] deg), it was suggested to replace them with the linear dependences within the operable range of angles of attack ([0; 20] deg) as shown by the

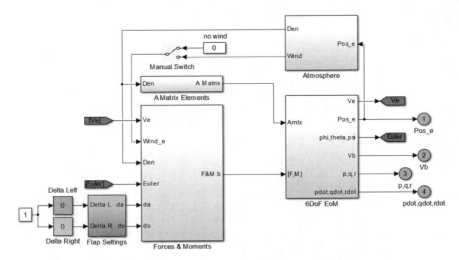

Fig. 5.9   Six-DoF model implemented in the Simulink developmental environment.

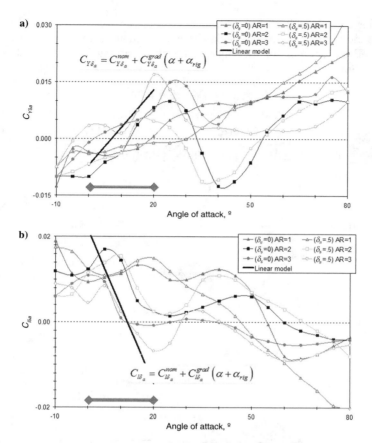

Fig. 5.10   Modeling of the coefficient a) $C_{Y\delta_a}(\alpha)$ and b) $C_{l\delta_a}(\alpha)$.

corresponding regressions in Figs. 5.10–5.12. Not only did that ensure a reasonable behavior and give an acceptable accuracy, but it also enabled the use of a simple identification technique where only a few parameters for each dependence have to be optimized. Aerodynamics modeling also included the usage of LinAir software [Kroo 1987]. To this end, Figs. 5.13 and 5.14 show examples of two geometric models, one with no TE deflection and one with asymmetric TE deflection, and the angle of attack and sideslip angle sweeps for the corresponding model. In this model the symmetric and asymmetric (differential) control inputs were modeled based on individual left and right TE deflections as follows:

$$\delta_s = \min(\delta_l, \delta_r) \in [0; 1]$$
$$\delta_a = \delta_l - \delta_r$$

[compare Eqs. (4.55) and (4.59) or Eqs. (5.40) and (5.41)].

negative of (4.59)! (handwritten annotation)

After careful analysis and preliminary simulation runs, the applied aerodynamics model was developed as a database for all major aerodynamic coefficients expressed as functions of the angle of attack α, sideslip angle β, and control inputs follows:

$$C_D = 0.14 + (0.25 + 0.2\delta_s)C_L^2 \tag{5.96}$$

$$C_Y = (-0.005 - 0.0001\alpha)\beta + (-0.007 + 0.0012\alpha)\delta_a \tag{5.97}$$

$$C_L = 0.0375(\alpha + 10°) + 0.2(\delta_s + 0.5|\delta_a|) \tag{5.98}$$

$$C_l = -(0.0014 + 0.001\alpha)\beta - \frac{b}{2V_a}(0.15p - 0.0775r)$$

$$-(0.0063 + 0.001\alpha)\delta_a \tag{5.99}$$

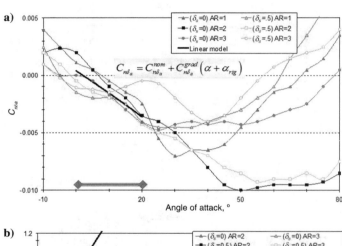

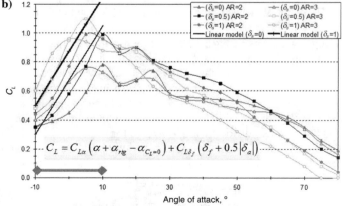

Fig. 5.11 Modeling of the coefficient a) $C_{n\delta_a}(\alpha)$ and b) $C_L(\alpha)$.

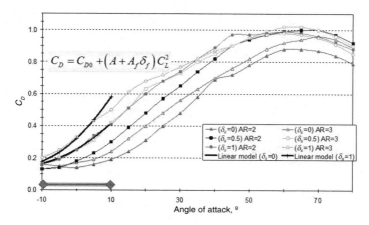

Fig. 5.12   Modeling of the coefficient $C_D(\alpha)$.

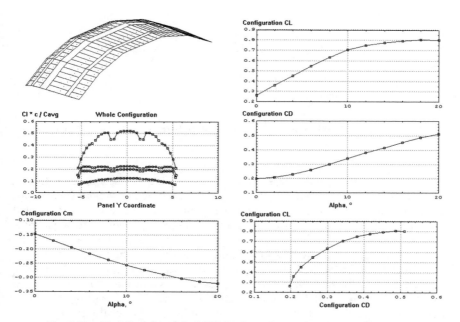

Fig. 5.13   Modeling the plain PADS configuration with a linear panel code.

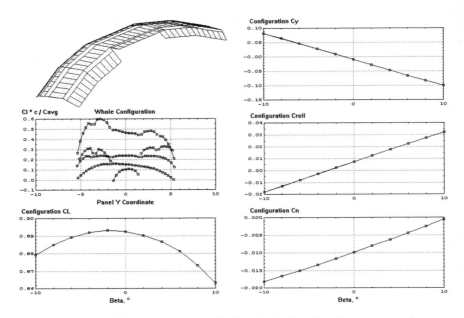

Fig. 5.14   Modeling asymmetric deflection of the TE with a linear panel code.

$$C_m = -0.33 - 6.39 \frac{c}{V_a} q + \frac{d m_s}{m_s + m_c} \left(\frac{1}{c} + \frac{\rho C_{B0}}{2W/S} \right) [C_D \quad C_L]$$

$$\times \begin{bmatrix} \cos(\alpha + 8°) \\ -\sin(\alpha + 8°) \end{bmatrix} \quad\quad (5.100)$$

$$C_n = (0.007 - 0.0003\alpha)\beta + \frac{b}{2V_a}(0.023p - 0.0936r) + (0.019 - 0.001\alpha)\delta_a$$

$$(5.101)$$

In this model, C_L, C_D, and C_Y are the lift, drag, and side-force coefficients; C_l, C_m, and C_n are the rolling-moment, pitch-moment, and yawing-moment coefficients; p, q, and r are the rolling, pitch, and yawing moments; b and c are the inflated span and chord; d is the distance from a canopy ($c/4$ point) to a payload; $C_{B0} = 2UgS^{-1}c^{-1}$ is the buoyancy constant (U is a canopy volume); W is the canopy loading; V is the total velocity; and m_s and m_c are the masses of payload and canopy, respectively. The rigging angle of 8 deg and $\alpha_{C_L=0} = -10$ deg were applied. Essentially, Eqs. (5.96–5.101) represent the same aerodynamic model as in Eqs. (5.74) and (5.79), but with the most aerodynamic and control derivatives depending on angle of attack. The derivation of Eq. (5.100) for the coefficient C_m was based on the analysis presented in [Goodrick 1975].

Figure 5.15 shows the longitudinal response to a pulse input of $\delta_s = 0.5$ after the parafoil was trimmed in flight at an altitude of 3,000 m (9,840 ft). [The steady-state values with no flaps are $\alpha = 9$ deg, $\theta = -12$ deg, $\gamma_a = 21$ deg, $V_G = 11.25$ m/s (22 kN), $V_d = 4.25$ m/s (14 ft/s), and $L/D \approx 2.65$.] With a symmetric TE deflection, the angle of attack increases, pitch angle increases, and flight-path angle is steeper, as is expected. A short-period type of dynamic response is also noted, which is heavily damped. Figure 5.16 demonstrates the roll-rate and yaw-rate responses due to the asymmetric control input. The steady-state bank angle is about 4 deg, and the steady yaw rate is 2.8 deg/s. Figure 5.17 shows the trajectory due to this asymmetric TE deflection.

This model was validated against flight data collected by airborne sensors suite that included the following:

- The IMU provided roll rate, pitch rate, yaw rate, x acceleration, y acceleration, and z acceleration sampled at 100 Hz.

- The compass provided attitude data measured in frame $\{b\}$ and sampled at 4 Hz.

- The GPS unit provided latitude, longitude, altitude, ground speed, track angle, East, North, Up components on the speed vector sampled at 10 Hz.

The WindPack dropsonde, dropped at the same time as the descending PADS, provided the real-time wind profile update with the frequency of 10 Hz.

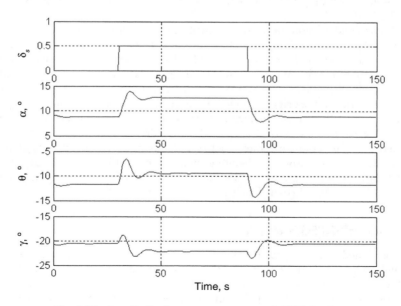

Fig. 5.15   Longitudinal response to symmetric TE deflection.

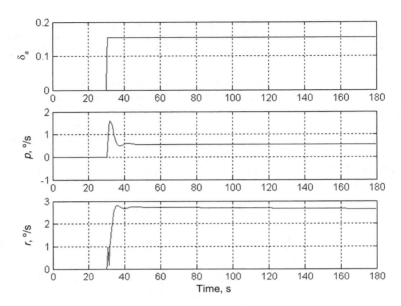

Fig. 5.16   Roll and yaw rate responses to asymmetric TE deflection.

A sample of vertical profiles for both parafoil and WindPack dropsonde is shown on Fig. 5.18.

A nonstandard nonlinear system identification technique including employment of the multicriteria optimization to find reasonable ranges of the optimization parameters and zero-order Hooke–Jeeves method to actually define the best solutions for optimization parameters was applied to tune the initial aerodynamic dependences and apparent mass terms as well. Figure 5.19a features the discrepancy of the original model in the lateral channel. Although the original model matches the integral behavior of the system in terms of glide ratio (GR) of 2.5 and vertical velocity of 3.7 m/s (12 ft/s) fairly well, some correction was needed to match the turn rate 6 deg/s demonstrated during the flight tests. Figure 5.19b shows how the changing of the only coefficient $C_{n\delta_a}^{nom}$ from 0.019 to 0.032 improved the model. Further improvement involved changing more coefficients (Fig. 5.20). (Further details are provided in Chapter 11.)

5.2   HIGHER-FIDELITY MODELS

As mentioned in Sec. 5.1 for the development of guidance strategies, it is usually enough to have a three-DoF or four-DoF model. To develop control architecture and test developed GNC algorithms usually requires only a six-DoF model where parafoil and payload are treated as one rigid body. Such a model was presented in Sec. 5.1.5. Further enhancement of PADS model fidelity, which might be

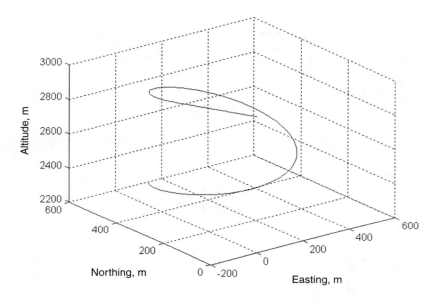

Fig. 5.17   Trajectory due to asymmetric TE deflection.

required for GNC algorithms validation in a more realistic environment, involves accounting for a rigging geometry and riser connection to the payload as discussed in this section.

5.2.1   RIGGING SCHEMES

Chapter 1 presented several PADS featuring different rigging geometries. Consequently, parafoil-payload models can be represented with a variety of different rigging geometries as well, resulting in different DoF systems.

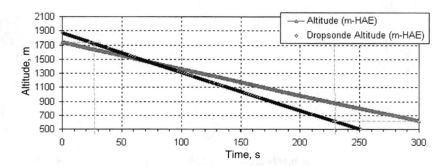

Fig. 5.18   Example of vertical profiles vs time.

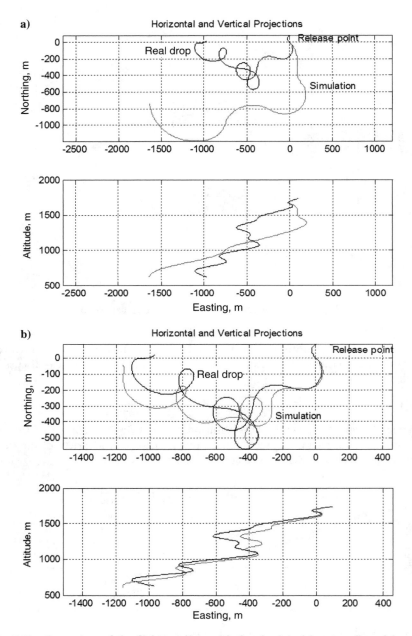

Fig. 5.19 Comparison of the flight-test data with the a) original (not-tuned) model and b) $C_{n\delta_a}^{nom}$-tuned model.

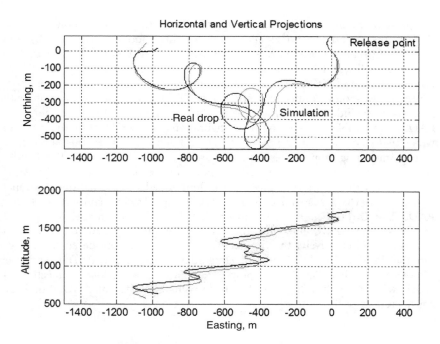

Fig. 5.20   Example of the further tuning.

As usual, six degrees of freedom include three inertial position components of some point in the body frame $\{b\}$ (e.g., it may be a connection point C shown in Figs. 5.21–5.23) and three Euler orientation angles of $\{b\}$. The seven-/eight-/nine-DoF models include additional Euler orientation angles for the payload (coordinate frame $\{s\}$). The number of DoF depends on rigging geometry and riser connection to the payload (Fig. 5.24).

Compared to Fig. 5.6, representing a simple six-DoF model, we will now separate the apparent mass center and origin of $\{p\}$ fixed to the canopy aerodynamic center. As a result, the airspeed vectors computed at the apparent mass center and canopy aerodynamic center will be slightly different. Computing both vectors will still rely on Eqs. (5.84) and (5.85) but be based on r_{BM} and r_{BP}, respectively. We will also need to introduce one more coordinate frame $\{s\}$, associated with the payload center of gravity, where s stands for the store.

As seen from Figs. 5.21 and 5.24, the seven-DoF model assumes a four-point rigging harness connection to the payload (see examples in Table 1.7). This connection prohibits roll and pitch relative motion but allows for yawing motion requiring introduction of one more state, an Euler yaw angle ψ_s. For the eight-DoF (Fig. 5.22), a two-point connection at the sore permits yawing, described by ψ_s, and pitching, described by the Euler pitch angle θ_s, but prevents rolling motion (see examples in Table 1.7). A single-point connection of the nine-DoF model allows for all three motions (ϕ_s, θ_s, ψ_s) and requires introducing one more state described by the Euler roll angle ϕ_s (see examples in Table 1.7). Looking at Table 1.7, it occurs that only two PADS (Gnat and Snowflake) utilize a four-point connection, 15 PADS use a more common two-point connection, and 10 PADS use a swivel connection with long risers extended down from the confluence point. On the other hand, 12 PADS from Table 1.7 employ a configuration with AGU atop the payload while 16 PADS have AGU placed at the confluence point. In this latter case, several modeling approaches can be used. Some PADS, like ParaLander (Fig. 1.25), might utilize the seven-DoF model of Fig. 5.21 for the canopy-AGU interaction with two more degrees of freedom added to describe the rotational motion of payload suspended under AGU using a swivel connection. Other PADS, like Onyx (Fig. 1.19), would employ the eight-DoF model of Fig. 5.22 for the canopy—AGU link plus another two for pitching and rolling motion of payload. Alternatively, the canopy-AGU link can be considered fixed [e.g., for the FireFly (Figs. 1.27) and DragonFly (Figs. 1.26) PADS] and the nine-DoF model of Fig. 5.23 employed for the canopy-payload connection as usual (with the only difference that the weight and drag of AGU should be considered as well).

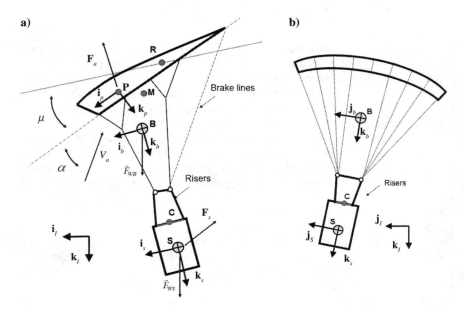

Fig. 5.21   Views of the seven-DoF model: a) side and b) front.

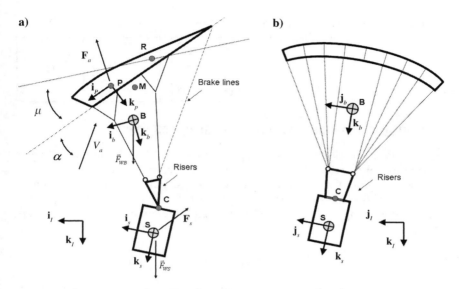

Fig. 5.22   Views of the eight-DoF model: a) side and b) front.

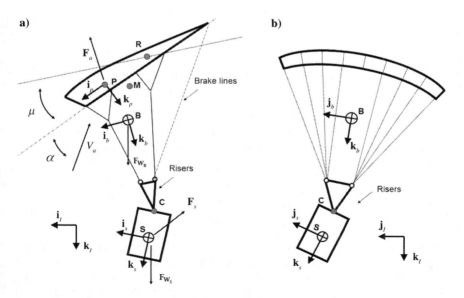

Fig. 5.23   Views of the nine-DoF model: a) side and b) front.

Fig. 5.24   Payload top view showing connection points.

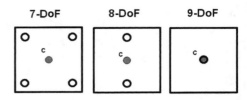

5.2.2   SIX-DoF MODEL REWRITTEN FOR THE CONNECTION POINT

Let us start developing seven-/eight-/nine-DoF models by rewriting a six-DoF model described by Eqs. (5.88–5.90) to incorporate a velocity vector at a connection point C ($[u_c, v_c, w_c]^T$) shown in Figs. 5.21–5.23 rather than that of the origin of the coordinate frame $\{b\}$. We will use some reference point R fixed at the canopy (Figs. 5.21–5.24) to simplify computation of vectors $\mathbf{r}_{CP}$ and $\mathbf{r}_{CM}$

$$\mathbf{r}_{CP} = \mathbf{r}_{CR} + {}_{b}^{P}\mathbf{R}^T \mathbf{r}_{RP}^{p}, \quad \mathbf{r}_{CM} = \mathbf{r}_{CR} + {}_{b}^{P}\mathbf{R}^T \mathbf{r}_{RM}^{p} \qquad (5.102)$$

As before, we can start from a general matrix equation, similar to Eq. (5.72) and written in $\{b\}$, with the only difference that because our speed vector $[u_c, v_c, w_c]^T$ is now defined not at the origin of frame $\{b\}$ we need to add two more terms to the right-hand side of the first (force) equation

$$
\begin{bmatrix} (m+m_e)\mathbf{I}_{3\times3} & \vdots & \mathbf{0}_{3\times3} \\ \cdots & \cdots & \cdots \\ \mathbf{0}_{3\times3} & \vdots & \mathbf{I} \end{bmatrix}
\begin{bmatrix} \dot{u}_c \\ \dot{v}_c \\ \dot{w}_c \\ \cdots \\ \dot{p} \\ \dot{q} \\ \dot{r} \end{bmatrix}
$$

$$
= \begin{bmatrix} \mathbf{F} - (m+m_e)\mathbf{S}(\boldsymbol{\omega})\begin{bmatrix} u_c \\ v_c \\ w_c \end{bmatrix} + (m+m_e)\mathbf{S}(\mathbf{r}_{CB})\begin{bmatrix} \dot{p} \\ \dot{q} \\ \dot{r} \end{bmatrix} - (m+m_e)\mathbf{S}(\boldsymbol{\omega})\mathbf{S}(\boldsymbol{\omega})\mathbf{r}_{CB} \\ \cdots \\ \mathbf{M} - \mathbf{S}(\boldsymbol{\omega})\mathbf{I}\begin{bmatrix} p \\ q \\ r \end{bmatrix} \end{bmatrix} \qquad (5.103)
$$

These two terms are tangential components of acceleration due to the angular acceleration $\dot{\boldsymbol{\omega}}$ [the sign is changed because of the change of the order in the cross product, i.e., $\mathbf{S}(\mathbf{a})\mathbf{b} = -\mathbf{S}(\mathbf{b})\mathbf{a}$] and centrifugal acceleration. Note that $\boldsymbol{\omega} = [p, q, r]^T$ will still pertain to the angular velocity vector expressed in $\{b\}$.

The apparent mass terms have to be related to the new point C as well, so instead of using the vector $\mathbf{r}_{BM}$ in Eqs. (5.84–5.86) we will use the vector $\mathbf{r}_{CM}$ resulting in the following version of Eq. (5.86):

$$
\mathbf{F}_{a.m.} = -\mathbf{I}'_{a.m.} \left(\begin{bmatrix} \dot{u}_c \\ \dot{v}_c \\ \dot{w}_c \end{bmatrix} - \mathbf{S}(\mathbf{r}_{CM}) \begin{bmatrix} \dot{p} \\ \dot{q} \\ \dot{r} \end{bmatrix} \right)
$$
$$
+ \mathbf{S}(\boldsymbol{\omega})\mathbf{I}'_{a.m.} \left(\begin{bmatrix} u_c \\ v_c \\ w_c \end{bmatrix} - \mathbf{S}(\mathbf{r}_{CM}) \begin{bmatrix} p \\ q \\ r \end{bmatrix} - {}^b_n\mathbf{RW} \right) \qquad (5.104)
$$

[Note that Eq. (5.104) still relies on airspeed and acceleration components computed at the apparent mass center M.] Consequently, even though Eq. (5.87) has exactly the same form

$$
\mathbf{M}_{a.i.} = -\mathbf{I}'_{a.i.} \begin{bmatrix} \dot{p} \\ \dot{q} \\ \dot{r} \end{bmatrix} - \mathbf{S}(\boldsymbol{\omega})\mathbf{I}'_{a.i.} \begin{bmatrix} p \\ q \\ r \end{bmatrix} + \mathbf{S}(\mathbf{r}_{BM})\mathbf{F}_{a.m.} \qquad (5.105)
$$

its last term will now use $\mathbf{F}_{a.m.}$ defined by Eq. (5.104). Adding the right-hand side of Eqs. (5.103) and (5.104) to the corresponding right-hand sides of Eq. (5.103) and rearranging the terms result in a new version of the six-DoF model developed for the state vector $[u_c, v_c, w_c, p, q, r]^T$ as opposed to $[u, v, w, p, q, r]^T$ as in Eq. (5.72)

$$
\begin{bmatrix} (m + m_e)\mathbf{I}_{3\times3} + \mathbf{I}'_{a.m.} & \vdots & -\mathbf{I}'_{a.m.}\mathbf{S}(\mathbf{r}_{CM}) - (m + m_e)\mathbf{S}(\mathbf{r}_{CB}) \\ \cdots & \cdots & \cdots \\ \mathbf{S}(\mathbf{r}_{BM})\mathbf{I}'_{a.m.} & \vdots & \mathbf{I} + \mathbf{I}'_{a.i.} - \mathbf{S}(\mathbf{r}_{BM})\mathbf{I}'_{a.m.}\mathbf{S}(\mathbf{r}_{CM}) \end{bmatrix} \begin{bmatrix} \dot{u}_c \\ \dot{v}_c \\ \dot{w}_c \\ \cdots \\ \dot{p} \\ \dot{q} \\ \dot{r} \end{bmatrix} = \begin{bmatrix} \mathbf{B}_1 \\ \cdots \\ \mathbf{B}_2 \end{bmatrix}
$$

$$(5.106)$$

featuring the extra terms in the second column of the state matrix caused by the term $(m + m_e)\mathbf{S}(\mathbf{r}_{CB})\dot{\boldsymbol{\omega}}$ in Eq. (5.103). The term $(m + m_e)\mathbf{S}(\boldsymbol{\omega})\mathbf{S}(\boldsymbol{\omega})\mathbf{r}_{CB}$ in Eq. (5.103) remains on the right-hand side and appears as an additional term in $\mathbf{B}_1$

$$\mathbf{B}_1 = \mathbf{F}_a^p + \mathbf{F}_a^s + \mathbf{F}_g - (m + m_e)\mathbf{S}(\boldsymbol{\omega}) \begin{bmatrix} u_c \\ v_c \\ w_c \end{bmatrix}$$

$$- \mathbf{S}(\boldsymbol{\omega})\mathbf{I}'_{a.m.} \left(\begin{bmatrix} u_c \\ v_c \\ w_c \end{bmatrix} - \mathbf{S}(\mathbf{r}_{CM}) \begin{bmatrix} p \\ q \\ r \end{bmatrix} - {}_n^b\mathbf{R}\mathbf{W} \right) - (m + m_e)\mathbf{S}(\boldsymbol{\omega})\mathbf{S}(\boldsymbol{\omega})\mathbf{r}_{CB}$$

$$= \mathbf{F}_a^p + \mathbf{F}_a^s + \mathbf{F}_g - \mathbf{S}(\boldsymbol{\omega})\left[(m + m_e)\mathbf{I}_{3\times3} + \mathbf{I}'_{a.m.} \right] \begin{bmatrix} u_c \\ v_c \\ w_c \end{bmatrix}$$

$$+ \mathbf{S}(\boldsymbol{\omega})\mathbf{I}'_{a.m.}\mathbf{S}(\mathbf{r}_{CM}) \begin{bmatrix} p \\ q \\ r \end{bmatrix} + \mathbf{S}(\boldsymbol{\omega})\mathbf{I}'_{a.m.}{}_n^b\mathbf{R}\mathbf{W} - (m + m_e)\mathbf{S}(\boldsymbol{\omega})\mathbf{S}(\boldsymbol{\omega})\mathbf{r}_{CB} \quad (5.107)$$

$$\mathbf{B}_2 = \mathbf{M}_a^p + \mathbf{S}(\mathbf{r}_{CP})\mathbf{F}_a^p + \mathbf{M}_a^s + \mathbf{S}(\mathbf{r}_{CS})\mathbf{F}_a^s - \mathbf{S}(\boldsymbol{\omega})\mathbf{I} \begin{bmatrix} p \\ q \\ r \end{bmatrix}$$

$$- \mathbf{S}(\boldsymbol{\omega})\mathbf{I}'_{a.i.} \begin{bmatrix} p \\ q \\ r \end{bmatrix} - \mathbf{S}(\mathbf{r}_{BM})\mathbf{S}(\boldsymbol{\omega})\mathbf{I}'_{a.m.} \left(\begin{bmatrix} u_c \\ v_c \\ w_c \end{bmatrix} - \mathbf{S}(\mathbf{r}_{CM}) \begin{bmatrix} p \\ q \\ r \end{bmatrix} - {}_n^b\mathbf{R}\mathbf{W} \right)$$

$$= \mathbf{M}_a^p + \mathbf{S}(\mathbf{r}_{CP})\mathbf{F}_a^p + \mathbf{M}_a^s + \mathbf{S}(\mathbf{r}_{CS})\mathbf{F}_a^s$$

$$- \left[\mathbf{S}(\boldsymbol{\omega})(\mathbf{I} + \mathbf{I}'_{a.i.}) - \mathbf{S}(\mathbf{r}_{BM})\mathbf{S}(\boldsymbol{\omega})\mathbf{I}'_{a.m.}\mathbf{S}(\mathbf{r}_{CM}) \right] \begin{bmatrix} p \\ q \\ r \end{bmatrix}$$

$$- \mathbf{S}(\mathbf{r}_{BM})\mathbf{S}(\boldsymbol{\omega})\mathbf{I}'_{a.m.} \begin{bmatrix} u_c \\ v_c \\ w_c \end{bmatrix} + \mathbf{S}(\mathbf{r}_{BM})\mathbf{S}(\boldsymbol{\omega})\mathbf{I}'_{a.m.}{}_n^b\mathbf{R}\mathbf{W} \quad (5.108)$$

Another difference in Eqs. (5.107) and (5.108) compared to Eqs. (5.89) and (5.90) is that aerodynamic forces and moments—$\mathbf{F}_a^p$, $\mathbf{F}_a^s$, $\mathbf{M}_a^p$, $\mathbf{M}_a^s$, $\mathbf{S}(\mathbf{r}_{CP})\mathbf{F}_a^p$, $\mathbf{S}(\mathbf{r}_{CS})\mathbf{F}_a^s$—are computed separately for parafoil and payload.

The model (5.106–5.108) is complemented with translation kinematics

$$\begin{bmatrix} \dot{x}_c \\ \dot{y}_c \\ \dot{z}_c \end{bmatrix} = {}_b^n\mathbf{R} \begin{bmatrix} u_c \\ v_c \\ w_c \end{bmatrix} \quad (5.109)$$

[analog of Eq. (5.28)] and rotational kinematics of Eq. (5.15), which constitutes yet another representation of the six-DoF model of a parafoil-payload system.

5.2.3   NINE-DoF MODEL

Now that we refined a six-DoF model to include six states u_c, v_c, w_c, p, q, and r, we may proceed with developing the higher-fidelity models. As mentioned in Sec. 5.2.1, to develop the seven-/eight-/nine-DoF models, we will have to relax the rigid-body assumptions and consider two subsystems: a parafoil with lines, risers and harness (with the corresponding coordinate frame $\{b\}$ now pertaining just to this part of PADS), and a payload (with the coordinate frame $\{s\}$). The two systems are connected at the point C (see Figs. 5.21–5.23). As before, the aerodynamic forces acting on a canopy are computed in the coordinate frame $\{p\}$ with its origin at the point P, which is rotated with respect to $\{b\}$ around the point R by the rigging angle μ [with a rotation matrix described by Eq. (5.82)]. In Figs. 5.21–5.23, the point M is where the apparent mass forces and moments of an arched canopy are applied.

Compared to the six-DoF model, the origin of the coordinate frame $\{b\}$ now resides higher, at the center of gravity of a PADS's subsystem composed of parafoil plus lines, risers, and harness only. The origin of the coordinate frame $\{s\}$ resides at the center of gravity of payload. Each subsystem is described by its own inertia matrix, $\mathbf{I}^*$ and $\mathbf{I}_s$, and mass, m^* and m_s (so that $m = m^* + m_s$). Inertia matrices $\mathbf{I}^*$ and $\mathbf{I}_s$ can be approximated with simple formulas for solid rectangular cuboids of appropriate dimensions

$$
\mathbf{I}^* \approx \frac{m^*}{12} \begin{bmatrix} b^2 + t^2 & 0 & 0 \\ 0 & c^2 + t^2 & 0 \\ 0 & 0 & b^2 + c^2 \end{bmatrix}
$$

$$
\mathbf{I}_s = \frac{m_s}{12} \begin{bmatrix} W^2 + H^2 & 0 & 0 \\ 0 & D^2 + H^2 & 0 \\ 0 & 0 & D^2 + W^2 \end{bmatrix} \tag{5.110}
$$

where D, W, and H are payload depth (along x_s axis), width (along y_s axis), and height (along z_s axis) (Fig. 5.25). For the (solid) cylindrical payload with a radius R like in Figs. 1.14a, 1.17a, and 1.30b,

$$
\mathbf{I}^* = \frac{m^*}{12} \begin{bmatrix} 3R^2 + H^2 & 0 & 0 \\ 0 & 3R^2 + H^2 & 0 \\ 0 & 0 & 6R^2 \end{bmatrix} \tag{5.111}
$$

For large PADS the inertia of suspension lines and AGU should also be properly accounted for. More accurate formulas are given in Appendix E.

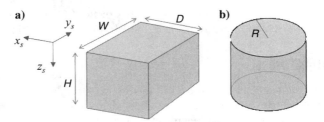

Fig. 5.25   a) Solid rectangular cuboid and b) solid cylinder.

Although the six aforementioned states correspond to the motion of the frame $\{b\}$, a few more states are needed to describe the motion of frame $\{s\}$ relative to $\{b\}$. Specifically, for the nine-DoF model these three additional states are the components of the angular velocity of $\{s\}$ with respect to the inertial frame $\{I\}$ (expressed in $\{s\}$), $\boldsymbol{\omega}_s^s = [p_s, q_s, r_s]^T$. Virtually "separating" two subsystems at the connection point C, we have to also introduce the components of the constraint force at the connection point (expressed in $\{b\}$) $\mathbf{F}_c = [F_{cx}, F_{cy}, F_{cz}]^T$. For the nine-DoF model, we will assume a zero-moment connection, that is, a connection that transmits no moments across it (i.e., $\mathbf{M}_c = \mathbf{0}_{3\times1}$). This assumption is usually true (see swivel connections in Figs. 1.13b, 1.16a, 1.18a, 1.21, 1.22, etc.); however, in some cases (e.g., Fig. 1.11a) line twisting might allow some relatively small twisting resistance.

Subsystem 1 (parafoil) dynamics will be described with the same set of equations (5.106–5.108) substituting m with m^*, $\mathbf{I}$ with $\mathbf{I}^*$, $\mathbf{F}_g$ with $\mathbf{F}_g^*$ and adding forces and moments caused by subsystem 2 (payload), with $\mathbf{F}_c$ and $\mathbf{S}(\mathbf{r}_{BC})\mathbf{F}_c$, respectively

$$
\begin{bmatrix}
(m^* + m_e)\mathbf{I}_{3\times3} + \mathbf{I}'_{a.m.} & \vdots & -\mathbf{I}'_{a.m.}\mathbf{S}(\mathbf{r}_{CM}) - (m^* + m_e)\mathbf{S}(\mathbf{r}_{CB}) \\
\cdots & \cdots & \cdots \\
\mathbf{S}(\mathbf{r}_{BM})\mathbf{I}'_{a.m.} & \vdots & \mathbf{I}^* + \mathbf{I}'_{a.i.} - \mathbf{S}(\mathbf{r}_{BM})\mathbf{I}'_{a.m.}\mathbf{S}(\mathbf{r}_{CM})
\end{bmatrix}
\begin{bmatrix}
\dot{u}_c \\
\dot{v}_c \\
\dot{w}_c \\
\cdots \\
\dot{p} \\
\dot{q} \\
\dot{r}
\end{bmatrix}
$$

$$
= \begin{bmatrix}
\mathbf{B}_1 \\
\cdots \\
\mathbf{B}_2
\end{bmatrix}
\tag{5.112}
$$

$$\mathbf{B}_1 = \mathbf{F}_a^* + \mathbf{F}_c + \mathbf{F}_g^* - \mathbf{S}(\boldsymbol{\omega})\left[(m^* + m_e)\mathbf{I}_{3\times3} + \mathbf{I}_{a.m.}^*\right]\begin{bmatrix} u_c \\ v_c \\ w_c \end{bmatrix}$$

$$+ \mathbf{S}(\boldsymbol{\omega})\mathbf{I}_{a.m.}' \mathbf{S}(\mathbf{r}_{CM})\begin{bmatrix} p \\ q \\ r \end{bmatrix} + \mathbf{S}(\boldsymbol{\omega})\mathbf{I}_{a.m.n}'{}^b\mathbf{RW}$$

$$- (m^* + m_e)\mathbf{S}(\boldsymbol{\omega})\mathbf{S}(\boldsymbol{\omega})\mathbf{r}_{CB} \tag{5.113}$$

$$\mathbf{B}_2 = \mathbf{M}_a^* + \mathbf{S}(\mathbf{r}_{BP})\mathbf{F}_a^* + \mathbf{S}(\mathbf{r}_{BC})\mathbf{F}_c$$

$$- \left[\mathbf{S}(\boldsymbol{\omega})(\mathbf{I}^* + \mathbf{I}_{a.i.}') - \mathbf{S}(\mathbf{r}_{BM})\mathbf{S}(\boldsymbol{\omega})\mathbf{I}_{a.m.}'\mathbf{S}(\mathbf{r}_{CM})\right]\begin{bmatrix} p \\ q \\ r \end{bmatrix}$$

$$- \mathbf{S}(\mathbf{r}_{BM})\mathbf{S}(\boldsymbol{\omega})\mathbf{I}_{a.m.}'\begin{bmatrix} u_c \\ v_c \\ w_c \end{bmatrix} + \mathbf{S}(\mathbf{r}_{BM})\mathbf{S}(\boldsymbol{\omega})\mathbf{I}_{a.m.n}'{}^b\mathbf{RW} \tag{5.114}$$

Subsystem 2 (payload) dynamics can be described by a similar set of equations. Specifically, we can start from a generic equation (5.103) written for a payload

$$\begin{bmatrix} m_{s_b}^s\mathbf{R} & \vdots & \mathbf{0}_{3\times3} \\ \cdots & \cdots & \cdots \\ \mathbf{0}_{3\times3} & \vdots & \mathbf{I}_s \end{bmatrix}\begin{bmatrix} \dot{u}_c \\ \dot{v}_c \\ \dot{w}_c \\ \cdots \\ \dot{p}_s \\ \dot{q}_s \\ \dot{r}_s \end{bmatrix}$$

$$= \begin{bmatrix} \mathbf{F}_s - m_{s_b}^s\mathbf{RS}(\boldsymbol{\omega})\begin{bmatrix} u_c \\ v_c \\ w_c \end{bmatrix} + m_s\mathbf{S}(\mathbf{r}_{CS}^s)\begin{bmatrix} \dot{p}_s \\ \dot{q}_s \\ \dot{r}_s \end{bmatrix} - m_s\mathbf{S}(\boldsymbol{\omega}_s^s)\mathbf{S}(\boldsymbol{\omega}_s^s)\mathbf{r}_{CS}^s \\ \cdots \\ \mathbf{M}_s - \mathbf{S}(\boldsymbol{\omega}_s^s)\mathbf{I}_s\begin{bmatrix} p_s \\ q_s \\ r_s \end{bmatrix} \end{bmatrix} \tag{5.115}$$

Equation (5.115) benefits from the fact that the derivative of a speed vector at the connection point C has been defined in (the rotating coordinate frame) $\{b\}$ already. As such, we can simply multiply the left-hand side of Eq. (5.22) (we started the derivation of the dynamic equations from) by the rotation matrix ${}^s_b\mathbf{R}$. As a result, the first row in a matrix equation (5.115) simply represents the following expression:

$$m_s {}^s_b\mathbf{R}(\dot{\mathbf{V}}_c + \boldsymbol{\omega} \times \mathbf{V}_c) + m_s \dot{\boldsymbol{\omega}}^s_s \times \mathbf{r}^s_{CS} + m_s \boldsymbol{\omega}^s_s \times \boldsymbol{\omega}^s_s \times \mathbf{r}^s_{CS} = \mathbf{F}_s \qquad (5.116)$$

The rotation matrix ${}^s_b\mathbf{R}$ has the same form as the one of Eq. (5.10) and is obtained using three consecutive rotations

$$
\begin{aligned}
{}^s_b\mathbf{R} &= \mathbf{R}_{\phi_s}\mathbf{R}_{\theta_s}\mathbf{R}_{\psi_s} \\
&= \begin{bmatrix}
\cos(\psi_s)\cos(\theta_s) & \sin(\psi_s)\cos(\theta_s) & -\sin(\theta_s) \\
\cos(\psi_s)\sin(\theta_s)\sin(\phi_s) - \sin(\psi_s)\cos(\phi_s) & \sin(\psi_s)\sin(\theta_s)\sin(\phi_s) + \cos(\psi_s)\cos(\phi_s) & \cos(\theta_s)\sin(\phi_s) \\
\cos(\psi_s)\sin(\theta_s)\cos(\phi_s) + \sin(\psi_s)\sin(\phi_s) & \sin(\psi_s)\sin(\theta_s)\cos(\phi_s) - \cos(\psi_s)\sin(\phi_s) & \cos(\theta_s)\cos(\phi_s)
\end{bmatrix}
\end{aligned}
$$
$$(5.117)$$

Bringing Eq. (5.115) to the form of Eq. (5.112) (and excluding all apparent mass terms) results in

$$
\begin{bmatrix}
m_s {}^s_b\mathbf{R} & \vdots & -m_s\mathbf{S}(\mathbf{r}^s_{CS}) \\
\cdots & \cdots & \cdots \\
\mathbf{0}_{3\times 3} & \vdots & \mathbf{I}_s
\end{bmatrix}
\begin{bmatrix}
\dot{u}_c \\ \dot{v}_c \\ \dot{w}_c \\ \cdots \\ \dot{p}_s \\ \dot{q}_s \\ \dot{r}_s
\end{bmatrix}
=
\begin{bmatrix}
\mathbf{B}_3 \\ \cdots \\ \mathbf{B}_4
\end{bmatrix}
\qquad (5.118)
$$

$$\mathbf{B}_1 = \mathbf{F}^s_a - {}^s_b\mathbf{R}\mathbf{F}_c + \mathbf{F}^s_g - m_s{}^s_b\mathbf{R}\mathbf{S}(\boldsymbol{\omega})\begin{bmatrix} u_c \\ v_c \\ w_c \end{bmatrix} - m_s\mathbf{S}(\boldsymbol{\omega}^s_s)\mathbf{S}(\boldsymbol{\omega}^s_s)\mathbf{r}^s_{CS} \qquad (5.119)$$

$$\mathbf{B}_2 = \mathbf{M}^s_a - \mathbf{S}(\mathbf{r}^s_{SC}){}^s_b\mathbf{R}\mathbf{F}_c - \mathbf{S}(\boldsymbol{\omega}^s_s)\mathbf{I}_s\begin{bmatrix} p_s \\ q_s \\ r_s \end{bmatrix} \qquad (5.120)$$

Note, in Eqs. (5.119) and (5.120) the connecting point force $\mathbf{F}_c$ acting at the payload subsystem has a negative sign compared to the force acting on the parafoil subsystem [in Eqs. (5.113) and (5.114)].

Now we can combine Eq. (5.112) and Eq. (5.118) into a single matrix equation

$$
\begin{bmatrix}
(m^* + m_e)\mathbf{I}_{3\times3} + \mathbf{I}'_{a.m.} & -\mathbf{I}'_{a.m.}\mathbf{S}(\mathbf{r}_{CM}) - (m^* + m_e)\mathbf{S}(\mathbf{r}_{CB}) & \mathbf{0}_{3\times3} & -\mathbf{I}_{3\times3} \\
\mathbf{S}(\mathbf{r}_{BM})\mathbf{I}'_{a.m.} & \mathbf{I}^* + \mathbf{I}'_{a.i.} - \mathbf{S}(\mathbf{r}_{BM})\mathbf{I}'_{a.m.}\mathbf{S}(\mathbf{r}_{CM}) & \mathbf{0}_{3\times3} & \mathbf{S}(\mathbf{r}_{CB}) \\
m_s{}^s_b\mathbf{R} & \mathbf{0}_{3\times3} & -m_s\mathbf{S}(\mathbf{r}^s_{CS}) & {}^s_b\mathbf{R} \\
\mathbf{0}_{3\times3} & \mathbf{0}_{3\times3} & \mathbf{I}_s & -\mathbf{S}(\mathbf{r}^s_{CS}){}^s_b\mathbf{R}
\end{bmatrix}
$$

$$
\times
\begin{bmatrix}
\dot{u}_c \\
\dot{v}_c \\
\dot{w}_c \\
\cdots \\
\dot{p} \\
\dot{q} \\
\dot{r} \\
\cdots \\
\dot{p}_s \\
\dot{q}_s \\
\dot{r}_s \\
\cdots \\
F_{cx} \\
F_{cy} \\
F_{cz}
\end{bmatrix}
=
\begin{bmatrix}
\mathbf{B}_1 \\
\cdots \\
\mathbf{B}_2 \\
\cdots \\
\mathbf{B}_3 \\
\cdots \\
\mathbf{B}_4
\end{bmatrix}
\tag{5.121}
$$

where

$$
\mathbf{B}_1 = \mathbf{F}^*_a + \mathbf{F}^*_g - \mathbf{S}(\boldsymbol{\omega})[(m^* + m_e)\mathbf{I}_{3\times3} + \mathbf{I}'_{a.m.}]\begin{bmatrix} u_c \\ v_c \\ w_c \end{bmatrix} + \mathbf{S}(\boldsymbol{\omega})\mathbf{I}'_{a.m.}\mathbf{S}(\mathbf{r}_{CM})\begin{bmatrix} p \\ q \\ r \end{bmatrix}
$$

$$
+ \mathbf{S}(\boldsymbol{\omega})\mathbf{I}'_{a.m.}{}^b_n\mathbf{RW} - (m^* + m_e)\mathbf{S}(\boldsymbol{\omega})\mathbf{S}(\boldsymbol{\omega})\mathbf{r}_{CB}
\tag{5.122}
$$

$$
\mathbf{B}_2 = \mathbf{M}^*_a + \mathbf{S}(\mathbf{r}_{BP})\mathbf{F}^*_a - \left[\mathbf{S}(\boldsymbol{\omega})(\mathbf{I}^* + \mathbf{I}'_{a.i.}) - \mathbf{S}(\mathbf{r}_{BM})\mathbf{S}(\boldsymbol{\omega})\mathbf{I}'_{a.m.}\mathbf{S}(\mathbf{r}_{CM})\right]\begin{bmatrix} p \\ q \\ r \end{bmatrix}
$$

$$
- \mathbf{S}(\mathbf{r}_{BM})\mathbf{S}(\boldsymbol{\omega})\mathbf{I}'_{a.m.}\begin{bmatrix} u_c \\ v_c \\ w_c \end{bmatrix} + \mathbf{S}(\mathbf{r}_{BM})\mathbf{S}(\boldsymbol{\omega})\mathbf{I}'_{a.m.}{}^b_n\mathbf{RW}
\tag{5.123}
$$

$$
\mathbf{B}_3 = \mathbf{F}^s_a + \mathbf{F}^s_g - m_s{}^s_b\mathbf{RS}(\boldsymbol{\omega})\begin{bmatrix} u_c \\ v_c \\ w_c \end{bmatrix} - m_s\mathbf{S}(\boldsymbol{\omega}^s_s)\mathbf{S}(\boldsymbol{\omega}^s_s)\mathbf{r}^s_{CS}
\tag{5.124}
$$

$$
\mathbf{B}_4 = \mathbf{M}^s_a - \mathbf{S}(\boldsymbol{\omega}^s_s)\mathbf{I}_s\begin{bmatrix} p_s \\ q_s \\ r_s \end{bmatrix}
\tag{5.125}
$$

When combining these two equations together, we brought the terms with the force at the connection point C to the left-hand side of a resulting equation making the three components of $\mathbf{F}_c$ additional states. The reason for this is that these components are not known, and rather than eliminating them from highly coupled translational and rotational dynamics algebraically it is easier to find them at each step of a numerical solution along with the other model states. In addition, the solutions to $\mathbf{F}_c$ can be useful for design.

In Eqs. (5.121) and (5.122),

- $\mathbf{F}_a^*$ and $\mathbf{M}_a^*$ are the aerodynamic force and moment acting on subsystem 1 (parafoil) and expressed in $\{b\}$ [compare Eqs. (5.73) and (5.79)]:

$$\mathbf{F}_a = QS_{Pb}^{\,P}\mathbf{R}^{T\,p}_w\mathbf{R} \begin{bmatrix} C_{D0} + C_{D\alpha^2}\alpha^2 + C_{D\delta_s}\bar{\delta}_s \\ C_{Y\beta}\beta \\ C_{L0} + C_{L\alpha}\alpha + C_{L\delta_s}\bar{\delta}_s \end{bmatrix} \tag{5.126}$$

$$\mathbf{M}_a = QS_{Pb}^{\,P}\mathbf{R}^{T} \begin{bmatrix} b\left(C_{l\beta} + \dfrac{b}{2V_a}C_{l\tilde{p}} + \dfrac{b}{2V_a}C_{lr}\tilde{r} + C_{l\delta_a}\bar{\delta}_a \right) \\ c\left(C_{m0} + C_{m\alpha}\alpha + \dfrac{c}{2V_a}C_{mq}\tilde{q} \right) \\ b\left(C_{n\beta} + \dfrac{b}{2V_a}C_{np}\tilde{p} + \dfrac{b}{2V_a}C_{nr}\tilde{r} + C_{n\delta_a}\bar{\delta}_a \right) \end{bmatrix} \tag{5.127}$$

- $\mathbf{F}_a^s$ and $\mathbf{M}_a^s \approx 0$ are the aerodynamic force and moment acting on subsystem 1 (payload) [compare the first term in Eq. (5.94)]

$$\mathbf{F}_a^s = -\frac{1}{2}\rho S_s C_D^s \sqrt{v_x^{s2} + v_y^{s2} + v_z^{s2}} \begin{bmatrix} v_x^s \\ v_y^s \\ v_z^s \end{bmatrix} \tag{5.128}$$

- $\mathbf{F}_g^*$ and $\mathbf{F}_g^s$ are the corresponding weight contributions expressed in their coordinate frames, $\{b\}$ and $\{s\}$, respectively [compare Eq. (5.73)]:

$$\mathbf{F}_g^* = m^*g \begin{bmatrix} -\sin(\theta) \\ \cos(\theta)\sin(\phi) \\ \cos(\theta)\cos(\phi) \end{bmatrix}, \quad \mathbf{F}_g^s = m_s g\,_b^s\mathbf{R} \begin{bmatrix} -\sin(\theta) \\ \cos(\theta)\sin(\phi) \\ \cos(\theta)\cos(\phi) \end{bmatrix} \tag{5.129}$$

Along with dynamics equations (5.121), the complete nine-DoF model includes translational kinematics described by Eq. (5.109), rotational kinematics of the parafoil subsystem described by Eq. (5.15), and its analogous representing rotational kinematics for the payload subsystem

$$
\begin{bmatrix} \dot{\phi}_s \\ \dot{\theta}_s \\ \dot{\psi}_s \end{bmatrix} = \begin{bmatrix} 1 & \sin(\phi_s)\dfrac{\sin(\theta_s)}{\cos(\theta_s)} & \cos(\phi_s)\dfrac{\sin(\theta_s)}{\cos(\theta_s)} \\ 0 & \cos(\phi_s) & -\sin(\phi_s) \\ 0 & \sin(\phi_s)\dfrac{1}{\cos(\theta_s)} & \cos(\phi_s)\dfrac{1}{\cos(\theta_s)} \end{bmatrix} \begin{bmatrix} p_s \\ q_s \\ r_s \end{bmatrix} \tag{5.130}
$$

[The Euler angles ϕ_s, θ_s, and ψ_s defined in Eq. (5.130) are used in Eq. (5.117).]

5.2.4 PAYLOAD CONSTRAINED ROTATIONAL KINEMATICS

In the nine-DoF rigging scheme of Fig. 5.23, a payload is rotating freely with respect to the connecting point C, and that is why rotational kinematics of a payload is not related to that of a parafoil subsystem. For the seven- and eight-DoF rigging schemes (models) however, relative rotational kinematics of two subsystems is constrained. To be more specific, for the eight-DoF rigging of Fig. 5.22, it is a roll motion that is related (i.e., relative roll is constrained), and for the seven-DoF rigging of Fig. 5.21 they are both roll and pitch motions that are related. (Relative roll and pitch are constrained.)

As a result, when developing the seven- and eight-DoF models, we will not be able to simply combine two matrix equations for two subsystems as we did for the nine-DoF model in Eq. (5.121). That is because we could not use independent components of $\boldsymbol{\omega}_s^s$ as three model states anymore: for the seven-DoF model we will only be able to use r_s and for the eight-DoF model only r_s and q_s. In turn, this results in having nonzero matrix blocks in both moment equations. To be more specific, $\mathbf{0}_{3 \times 3}$ elements at intersection of the second row and the third column, and the fourth row and the second column will need to be replaced with specific kinematic dependence vectors $\boldsymbol{\omega}$ and $\boldsymbol{\omega}_s^s$. The goal of this section is to establish these dependences for seven- and eight-DoF models.

The angular velocity of payload with respect to parafoil (expressed in $\{s\}$) $\boldsymbol{\omega}_{s/b}^s$ can be written in two ways. First, it can be represented via $\boldsymbol{\omega}_s^s$ and $\boldsymbol{\omega}$

$$
\boldsymbol{\omega}_{s/b}^s = \boldsymbol{\omega}_s^s - {}_b^s\mathbf{R}\boldsymbol{\omega} \tag{5.131}
$$

Second, it can be written using explicit rotations similar to Eq. (5.13). For the eight-DoF model, assuming $\phi_s = 0$ ($\mathbf{R}_{\phi_s} = \mathbf{I}_{3 \times 3}$), we can write

$$
\boldsymbol{\omega}_{s/b}^s = \mathbf{R}_{\theta_s}\begin{bmatrix} 0 \\ \dot{\theta}_s \\ 0 \end{bmatrix} + \mathbf{R}_{\theta_s}\mathbf{R}_{\psi_s}\begin{bmatrix} 0 \\ 0 \\ \dot{\psi}_s \end{bmatrix} \tag{5.132}
$$

and utilizing two other matrices of Eq. (5.9)

$$
\boldsymbol{\omega}^s_{s/b} =
\begin{bmatrix} \cos(\theta_s) & 0 & -\sin(\theta_s) \\ 0 & 1 & 0 \\ \sin(\theta_s) & 0 & \cos(\theta_s) \end{bmatrix}
\begin{bmatrix} 0 \\ \dot{\theta}_s \\ 0 \end{bmatrix}
+
\begin{bmatrix} \cos(\theta_s) & 0 & -\sin(\theta_s) \\ 0 & 1 & 0 \\ \sin(\theta_s) & 0 & \cos(\theta_s) \end{bmatrix}
$$

$$
\times
\begin{bmatrix} \cos(\psi_s) & \sin(\psi_s) & 0 \\ -\sin(\psi_s) & \cos(\psi_s) & 0 \\ 0 & 0 & 1 \end{bmatrix}
\begin{bmatrix} 0 \\ 0 \\ \dot{\psi}_s \end{bmatrix}
$$

$$
=
\begin{bmatrix} -\dot{\psi}_s \sin(\theta_s) \\ \dot{\theta}_s \\ \dot{\psi}_s \cos(\theta_s) \end{bmatrix}
\tag{5.133}
$$

For the seven-DoF model, additionally assuming $\theta_s = 0$ ($\mathbf{R}_{\theta_s} = \mathbf{I}_{3\times3}$) leads to

$$
\boldsymbol{\omega}^s_{s/b} = \mathbf{R}_{\psi_s}
\begin{bmatrix} 0 \\ 0 \\ \dot{\psi}_s \end{bmatrix}
=
\begin{bmatrix} \cos(\psi_s) & \sin(\psi_s) & 0 \\ -\sin(\psi_s) & \cos(\psi_s) & 0 \\ 0 & 0 & 1 \end{bmatrix}
\begin{bmatrix} 0 \\ 0 \\ \dot{\psi}_s \end{bmatrix}
=
\begin{bmatrix} 0 \\ 0 \\ \dot{\psi}_s \end{bmatrix}
\tag{5.134}
$$

Let us now proceed with the seven-DoF model. (We will repeat everything for the eight-DoF model as well.) Equating the right-hand sides of Eqs. (5.131) and (5.134) results in the following:

$$
\begin{bmatrix} p_s \\ q_s \\ r_s \end{bmatrix}
- {}^s_b\mathbf{R}
\begin{bmatrix} p \\ q \\ r \end{bmatrix}
=
\begin{bmatrix} 0 & 0 & 0 \\ 0 & 0 & 0 \\ 0 & 0 & 1 \end{bmatrix}
\begin{bmatrix} 0 \\ 0 \\ \dot{\psi}_s \end{bmatrix}
\tag{5.135}
$$

The three unknowns are the yaw rate $\dot{\psi}_s$ [appearing on the right-hand side of the Eq. (5.135)], along with the two constraint components of the angular rate, p_s and q_s, appearing on the left-hand side of Eq. (5.135). Rearranging the terms with the goal of having all unknowns on the right-hand side results in

$$
\begin{bmatrix} 0 \\ 0 \\ r_s \end{bmatrix}
- {}^s_b\mathbf{R}
\begin{bmatrix} p \\ q \\ r \end{bmatrix}
=
\begin{bmatrix} -1 & 0 & 0 \\ 0 & -1 & 0 \\ 0 & 0 & 1 \end{bmatrix}
\begin{bmatrix} p_s \\ q_s \\ \dot{\psi}_s \end{bmatrix}
\tag{5.136}
$$

Using matrix inversion and the fact that for seven-DoF ${}^s_b\mathbf{R} = \mathbf{R}_{\psi_s}$, Eq. (5.136) reduces to

$$
\begin{bmatrix} p_s \\ q_s \\ \dot{\psi}_s \end{bmatrix} = \begin{bmatrix} -1 & 0 & 0 \\ 0 & -1 & 0 \\ 0 & 0 & 1 \end{bmatrix}^{-1} \left(\begin{bmatrix} 0 \\ 0 \\ r_s \end{bmatrix} - {}^s_b\mathbf{R} \begin{bmatrix} p \\ q \\ r \end{bmatrix} \right)
$$

$$
= \begin{bmatrix} 0 \\ 0 \\ r_s \end{bmatrix} + \begin{bmatrix} \cos(\psi_s) & \sin(\psi_s) & 0 \\ -\sin(\psi_s) & \cos(\psi_s) & 0 \\ 0 & 0 & -1 \end{bmatrix} \begin{bmatrix} p \\ q \\ r \end{bmatrix} \tag{5.137}
$$

The first two rows of Eq. (5.137) are kinematic constraints

$$
\begin{bmatrix} p_s \\ q_s \end{bmatrix} = \begin{bmatrix} \cos(\psi_s) & \sin(\psi_s) & 0 \\ -\sin(\psi_s) & \cos(\psi_s) & 0 \end{bmatrix} \begin{bmatrix} p \\ q \\ r \end{bmatrix} \tag{5.138}
$$

The last row

$$
\dot{\psi}_s = r_s - r \tag{5.139}
$$

describes kinematics of a payload yaw angle. Differentiating Eq. (5.138) yields

$$
\begin{bmatrix} \dot{p}_s \\ \dot{q}_s \end{bmatrix} = \dot{\psi}_s \begin{bmatrix} -\sin(\psi_s) & \cos(\psi_s) & 0 \\ -\cos(\psi_s) & -\sin(\psi_s) & 0 \end{bmatrix} \begin{bmatrix} p \\ q \\ r \end{bmatrix}
$$

$$
+ \begin{bmatrix} \cos(\psi_s) & \sin(\psi_s) & 0 \\ -\sin(\psi_s) & \cos(\psi_s) & 0 \end{bmatrix} \begin{bmatrix} \dot{p} \\ \dot{q} \\ \dot{r} \end{bmatrix} \tag{5.140}
$$

Adding one more row $\dot{r}_s = \dot{r}_s$ and utilizing Eq. (5.139) result in a relation we were looking for:

$$
\begin{bmatrix} \dot{p}_s \\ \dot{q}_s \\ \dot{r}_s \end{bmatrix} = \mathbf{G}^7 + \mathbf{K}_1^7 \dot{r}_s + \mathbf{K}_2^7 \begin{bmatrix} \dot{p} \\ \dot{q} \\ \dot{r} \end{bmatrix} \tag{5.141}
$$

where

$$
\mathbf{G}^7 = (r_s - r) \begin{bmatrix} -\sin(\psi_s) & \cos(\psi_s) & 0 \\ -\cos(\psi_s) & -\sin(\psi_s) & 0 \\ 0 & 0 & 0 \end{bmatrix} \begin{bmatrix} p \\ q \\ r \end{bmatrix}
$$

$$
\mathbf{K}_1^7 = \begin{bmatrix} 0 \\ 0 \\ 1 \end{bmatrix}, \quad \mathbf{K}_2^7 = \begin{bmatrix} \cos(\psi_s) & \sin(\psi_s) & 0 \\ -\sin(\psi_s) & \cos(\psi_s) & 0 \end{bmatrix}
$$

(5.142)

An angular acceleration of a parafoil subsystem $\dot{\boldsymbol{\omega}}$ defines angular acceleration of a payload subsystem $\dot{\boldsymbol{\omega}}_s^s$, so we should use Eq. (5.115) for the seven-DoF model.

Let us repeat the derivation of relative angular kinematics for the eight-DoF model. Equating Eqs. (5.131) and (5.133) results in the following:

$$
\begin{bmatrix} p_s \\ q_s \\ r_s \end{bmatrix} - {}_b^s \mathbf{R} \begin{bmatrix} p \\ q \\ r \end{bmatrix} = \begin{bmatrix} 0 & 0 & -\sin(\theta_s) \\ 0 & 1 & 0 \\ 0 & 0 & \cos(\theta_s) \end{bmatrix} \begin{bmatrix} 0 \\ \dot{\theta}_s \\ \dot{\psi}_s \end{bmatrix}
$$

(5.143)

Rearranging the terms in the first row of Eq. (5.143), so that three unknowns $\dot{\theta}_s$, $\dot{\psi}_s$, and p_s are isolated with the flowing usage of matrix inversion and the fact that for eight-DoF ${}_b^s \mathbf{R} = \mathbf{R}_{\theta_s} \mathbf{R}_{\psi_s}$, results in

$$
\begin{bmatrix} p_s \\ \dot{\theta}_s \\ \dot{\psi}_s \end{bmatrix} = \begin{bmatrix} -1 & 0 & -\sin(\theta_s) \\ 0 & 1 & 0 \\ 0 & 0 & \cos(\theta_s) \end{bmatrix}^{-1} \left(\begin{bmatrix} 0 \\ q_s \\ r_s \end{bmatrix} - \mathbf{R}_{\theta_s} \mathbf{R}_{\psi_s} \begin{bmatrix} p \\ q \\ r \end{bmatrix} \right) = \begin{bmatrix} 0 & -\tan(\theta_s) \\ 1 & 0 \\ 0 & 1/\cos(\theta_s) \end{bmatrix} \begin{bmatrix} q_s \\ r_s \end{bmatrix}
$$

$$
+ \begin{bmatrix} \cos(\psi_s)/\cos(\theta_s) & \sin(\psi_s)/\cos(\theta_s) & 0 \\ \sin(\psi_s) & -\cos(\psi_s) & 0 \\ -\cos(\psi_s)\tan(\theta_s) & -\sin(\psi_s)\tan(\theta_s) & -1 \end{bmatrix} \begin{bmatrix} p \\ q \\ r \end{bmatrix}
$$

(5.144)

It is just the first row that represents the kinematic constraint for the eight-DoF model being multiplied by $\cos\theta_s$; it takes the form

$$
\cos(\theta_s) p_s = \begin{bmatrix} 0 & -\sin(\theta_s) \end{bmatrix} \begin{bmatrix} q_s \\ r_s \end{bmatrix} + \begin{bmatrix} \cos(\psi_s) & \sin(\psi_s) & 0 \end{bmatrix} \begin{bmatrix} p \\ q \\ r \end{bmatrix}
$$

(5.145)

Differentiating Eq. (5.140) results in

$$
-\dot{\theta}_s \sin(\theta_s) p_s + \cos(\theta_s)\dot{p}_s = \dot{\theta}_s [0 \quad -\cos(\theta_s)] \begin{bmatrix} q_s \\ r_s \end{bmatrix} + [0 \quad -\sin(\theta_s)] \begin{bmatrix} \dot{q}_s \\ \dot{r}_s \end{bmatrix}
$$

$$
+ \dot{\psi}_s [-\sin(\psi_s) \quad \cos(\psi_s) \quad 0] \begin{bmatrix} p \\ q \\ r \end{bmatrix}
$$

$$
+ [\cos(\psi_s) \quad \sin(\psi_s) \quad 0] \begin{bmatrix} \dot{p} \\ \dot{q} \\ \dot{r} \end{bmatrix} \tag{5.146}
$$

and rearranging the terms yields

$$
\dot{p}_s = \begin{bmatrix} p_s \tan(\theta_s) - r_s & -p\dfrac{\sin(\psi_s)}{\cos(\theta_s)} + q\dfrac{\cos(\psi_s)}{\cos(\theta_s)} \end{bmatrix} \begin{bmatrix} \dot{\theta}_s \\ \dot{\psi}_s \end{bmatrix}
$$

$$
+ [0 \quad -\tan(\theta_s)] \begin{bmatrix} \dot{q}_s \\ \dot{r}_s \end{bmatrix} + \begin{bmatrix} \dfrac{\cos(\psi_s)}{\cos(\theta_s)} & \dfrac{\sin(\psi_s)}{\cos(\theta_s)} & 0 \end{bmatrix} \begin{bmatrix} \dot{p} \\ \dot{q} \\ \dot{r} \end{bmatrix} \tag{5.147}
$$

Utilizing the two last rows in Eq. (5.144) for $\begin{bmatrix} \dot{\theta}_s \\ \dot{\psi}_s \end{bmatrix}$ and adding two rows, $\dot{q}_s = \dot{q}_s$ and $\dot{r}_s = \dot{r}_s$, finally result in a relation similar to Eq. (5.141). We were looking for

$$
\begin{bmatrix} \dot{p}_s \\ \dot{q}_s \\ \dot{r}_s \end{bmatrix} = \mathbf{G}^8 + \mathbf{K}_1^8 \begin{bmatrix} \dot{q}_s \\ \dot{r}_s \end{bmatrix} + \mathbf{K}_2^8 \begin{bmatrix} \dot{p} \\ \dot{q} \\ \dot{r} \end{bmatrix} \tag{5.148}
$$

In Eq. (5.148),

$$
\mathbf{G}^8 = \begin{bmatrix} p_s \tan(\theta_s) - r_s, & -p\dfrac{\sin(\psi_s)}{\cos(\theta_s)} + q\dfrac{\cos(\psi_s)}{\cos(\theta_s)} \end{bmatrix}
$$

$$
\times \left(\begin{bmatrix} q_s \\ r_s \\ \cos(\theta_s) \end{bmatrix} + \begin{bmatrix} \sin(\psi_s) & -\cos(\psi_s) & 0 \\ -\cos(\psi_s)\tan(\theta_s) & -\sin(\psi_s)\tan(\theta_s) & -1 \end{bmatrix} \begin{bmatrix} p \\ q \\ r \end{bmatrix} \right)
$$

$$
\tag{5.149}
$$

$$
\mathbf{K}_1^8 = \begin{bmatrix} 0 & \tan(\theta_s) \\ 1 & 0 \\ 0 & 1 \end{bmatrix}, \quad \mathbf{K}_2^8 = \begin{bmatrix} \cos(\psi_s)/\cos(\theta_s) & \sin(\psi_s)/\cos(\theta_s) & 0 \\ 0 & 0 & 0 \\ 0 & 0 & 0 \end{bmatrix} \tag{5.150}
$$

Again, it is Eq. (5.148); we should use Eq. (5.115) for the eight-DoF model.

5.2.5   SEVEN- AND EIGHT-DoF MODELS

Now that we have relations (5.141) and (5.148) to be used in Eq. (5.115) for the seven- and eight-DoF rotational dynamics, the only thing we need to start composing seven- and eight-DoF equations is the moment acting at the connecting point. (If you recall, for the nine-DoF model $\mathbf{M}_c = \mathbf{0}_{3\times1}$.)

The only known moment is the twisting moment depending on the actual parafoil-payload mass and geometry configuration. In the general case the line twist can be modeled as a nonlinear spring and damper

$$M_{cz} = K_\psi(\psi_s)\psi_s + K_r(\psi_s)\dot{\psi}_s \qquad (5.151)$$

where the stiffness K_ψ and damping K_r coefficients can be functions of the rotation angle ψ_s. The other two components of the moment $\mathbf{M}_c$—M_{cx} and M_{cy} for the seven-DoF model and just M_{cy} for the eight-DoF model—are unknown. Having in mind to cast these unknown components as the additional states [the way we did it for the components of $\mathbf{F}_c$ in Eq. (5.121)], we can express $\mathbf{M}_c$ (in $\{b\}$) as

$$\mathbf{M}_c = \begin{bmatrix} 0 \\ 0 \\ M_{cz} \end{bmatrix} + {}_b^s\mathbf{R}\begin{bmatrix} M_{cx} \\ M_{cy} \\ 0 \end{bmatrix} = \begin{bmatrix} 0 \\ 0 \\ M_{cz} \end{bmatrix} + {}_b^s\mathbf{R}\begin{bmatrix} 1 & 0 \\ 0 & 1 \\ 0 & 0 \end{bmatrix}\begin{bmatrix} M_{cx} \\ M_{cy} \end{bmatrix}$$

$$= \begin{bmatrix} 0 \\ 0 \\ M_{cz} \end{bmatrix} + {}_b^s\mathbf{R}\mathbf{E}^7\begin{bmatrix} M_{cx} \\ M_{cy} \end{bmatrix} \qquad (5.152)$$

and

$$\mathbf{M}_c = \begin{bmatrix} 0 \\ 0 \\ M_{cz} \end{bmatrix} + {}_b^s\mathbf{R}\begin{bmatrix} M_{cx} \\ 0 \\ 0 \end{bmatrix} = \begin{bmatrix} 0 \\ 0 \\ M_{cz} \end{bmatrix} + {}_b^s\mathbf{R}\begin{bmatrix} 1 \\ 0 \\ 0 \end{bmatrix}M_{cx}$$

$$= \begin{bmatrix} 0 \\ 0 \\ M_{cz} \end{bmatrix} + {}_b^s\mathbf{R}\mathbf{E}^8 M_{cx} \qquad (5.153)$$

for the seven- and eight-DoF models, respectively.

Now we can proceed with modifying Eqs. (5.121) and (5.122) for the seven- and eight-DoF models using relations (5.141), (5.148) and (5.152), (5.153) in Eq. (5.115), respectively. Compared to Eq. (5.121), the seven-DoF will lack two states, p_s and q_s (because they are dependent on $\boldsymbol{\omega}$) but will include new equations for M_{cx} and M_{cy}. The eight-DoF model will lack one state p_s but will include a new equation for M_{cx}.

Thus, the seven-DoF model can be represented by the following set of equations modified from those of Eq. (5.121):

$$
\begin{bmatrix}
(m^* + m_e)\mathbf{I}_{3\times3} + \mathbf{I}'_{a.m.} & -\mathbf{I}'_{a.m.}\mathbf{S}(\mathbf{r}_{CM}) - (m^* + m_e)\mathbf{S}(\mathbf{r}_{CB}) & \mathbf{0}_{3\times1} & -\mathbf{I}_{3\times3} & \mathbf{0}_{3\times2} \\
\mathbf{S}(\mathbf{r}_{BM})\mathbf{I}'_{a.m.} & \mathbf{I}^* + \mathbf{I}'_{a.i.} - \mathbf{S}(\mathbf{r}_{BM})\mathbf{I}'_{a.m.}\mathbf{S}(\mathbf{r}_{CM}) & \mathbf{0}_{3\times1} & \mathbf{S}(\mathbf{r}_{CB}) & \mathbf{E}^7 \\
m_s{}^s_b\mathbf{R} & -m_s\mathbf{S}(\mathbf{r}^s_{CS})\mathbf{K}^7_2 & -m_s\mathbf{S}(\mathbf{r}^s_{CS})\mathbf{K}^7_1 & {}^s_b\mathbf{R} & \mathbf{0}_{3\times2} \\
\mathbf{0}_{3\times3} & \mathbf{I}_s\mathbf{K}^7_2 & \mathbf{I}_s\mathbf{K}^7_1 & -\mathbf{S}(\mathbf{r}^s_{CS})^s_b\mathbf{R} & -{}^s_b\mathbf{R}\mathbf{E}^7
\end{bmatrix}
$$

$$
\times
\begin{bmatrix}
\dot{u}_c \\
\dot{v}_c \\
\dot{w}_c \\
\cdots \\
\dot{p} \\
\dot{q} \\
\dot{r} \\
\cdots \\
\dot{r}_s \\
\cdots \\
F_{cx} \\
F_{cy} \\
F_{cz} \\
\cdots \\
M_{cx} \\
M_{cy}
\end{bmatrix}
=
\begin{bmatrix}
\mathbf{B}_1 \\
\cdots \\
\mathbf{B}_2 \\
\cdots \\
\mathbf{B}_3 \\
\cdots \\
\mathbf{B}_4
\end{bmatrix}
\tag{5.154}
$$

Compared to Eq. (5.121), the state matrix of Eq. (5.154) has one more column with nonzero blocks for both moment equations (second and fourth rows). It also has the new elements in the third column and two last rows of the second column. The three last rows at the right-hand side of Eq. (5.154) will also have a few new terms (third term in the second and third rows and the first two terms in the fourth row):

$$
\mathbf{B}_1 = \mathbf{F}^*_a + \mathbf{F}^*_g - \mathbf{S}(\boldsymbol{\omega})[(m^* + m_e)\mathbf{I}_{3\times3} + \mathbf{I}'_{a.m.}]
\begin{bmatrix}
u_c \\
v_c \\
w_c
\end{bmatrix}
$$

$$
+ \mathbf{S}(\boldsymbol{\omega})\mathbf{I}'_{a.m.}\mathbf{S}(\mathbf{r}_{CM})
\begin{bmatrix}
p \\
q \\
r
\end{bmatrix}
+ \mathbf{S}(\boldsymbol{\omega})\mathbf{I}'_{a.m.}{}^b_n\mathbf{R}\mathbf{W}
$$

$$
- (m^* + m_e)\mathbf{S}(\boldsymbol{\omega})\mathbf{S}(\boldsymbol{\omega})\mathbf{r}_{CB}
\tag{5.155}
$$

$$\mathbf{B}_2 = \mathbf{M}_a^* + \mathbf{S}(\mathbf{r}_{BP})\mathbf{F}_a^* - \begin{bmatrix} 0 \\ 0 \\ M_{cz} \end{bmatrix} - \left[\mathbf{S}(\boldsymbol{\omega})(\mathbf{I}^* + \mathbf{I}_{a.i.}') - \mathbf{S}(\mathbf{r}_{BM})\mathbf{S}(\boldsymbol{\omega})\mathbf{I}_{a.m.}'\mathbf{S}(\mathbf{r}_{CM}) \right] \begin{bmatrix} p \\ q \\ r \end{bmatrix}$$

$$- \mathbf{S}(\mathbf{r}_{BM})\mathbf{S}(\boldsymbol{\omega})\mathbf{I}_{a.m.}' \begin{bmatrix} u_c \\ v_c \\ w_c \end{bmatrix} + \mathbf{S}(\mathbf{r}_{BM})\mathbf{S}(\boldsymbol{\omega})\mathbf{I}_{a.m.}' {}_n^b\mathbf{R}\mathbf{W} \qquad (5.156)$$

$$\mathbf{B}_3 = \mathbf{F}_a^s + \mathbf{F}_g^s + m_s\mathbf{S}(\mathbf{r}_{CS}^s)\mathbf{G}^7 - m_{sb}^s\mathbf{R}\mathbf{S}(\boldsymbol{\omega}) \begin{bmatrix} u_c \\ v_c \\ w_c \end{bmatrix} - m_s\mathbf{S}(\boldsymbol{\omega}_s^s)\mathbf{S}(\boldsymbol{\omega}_s^s)\mathbf{r}_{CS}^s \qquad (5.157)$$

$$\mathbf{B}_4 = -\mathbf{I}_s\mathbf{G}^7 + {}_b^s\mathbf{R} \begin{bmatrix} 0 \\ 0 \\ M_{cz} \end{bmatrix} - \mathbf{S}(\boldsymbol{\omega}_s^s)\mathbf{I}_s \begin{bmatrix} p_s \\ q_s \\ r_s \end{bmatrix} \qquad (5.158)$$

Equation (5.154) along with the translational kinematic equation (5.109) and rotational kinematics described by Eqs. (5.15) and (5.137) constitute the seven-DoF model.

Similarly, the eight-DoF model can be represented by the following set of equations, modified from those of Eq. (5.154):

$$\begin{bmatrix} (m^* + m_e)\mathbf{I}_{3\times3} + \mathbf{I}_{a.m.}' & -\mathbf{I}_{a.m.}'\mathbf{S}(\mathbf{r}_{CM}) - (m^* + m_e)\mathbf{S}(\mathbf{r}_{CB}) & \mathbf{0}_{3\times2} & -\mathbf{I}_{3\times3} & \mathbf{0}_{3\times1} \\ \mathbf{S}(\mathbf{r}_{BM})\mathbf{I}_{a.m.}' & \mathbf{I}^* + \mathbf{I}_{a.i.}' - \mathbf{S}(\mathbf{r}_{BM})\mathbf{I}_{a.m.}'\mathbf{S}(\mathbf{r}_{CM}) & \mathbf{0}_{3\times2} & \mathbf{S}(\mathbf{r}_{CB}) & \mathbf{E}^8 \\ m_{sb}^s\mathbf{R} & -m_s\mathbf{S}(\mathbf{r}_{CS}^s)\mathbf{K}_2^8 & -m_s\mathbf{S}(\mathbf{r}_{CS}^s)\mathbf{K}_1^8 & {}_b^s\mathbf{R} & \mathbf{0}_{3\times1} \\ \mathbf{0}_{3\times3} & \mathbf{I}_s\mathbf{K}_2^8 & \mathbf{I}_s\mathbf{K}_1^8 & -\mathbf{S}(\mathbf{r}_{CS}^s){}_b^s\mathbf{R} & -{}_b^s\mathbf{R}\mathbf{E}^8 \end{bmatrix}$$

$$\times \begin{bmatrix} \dot{u}_c \\ \dot{v}_c \\ \dot{w}_c \\ \cdots \\ \dot{p} \\ \dot{q} \\ \dot{r} \\ \cdots \\ \dot{q}_s \\ \dot{r}_s \\ \cdots \\ F_{cx} \\ F_{cy} \\ F_{cz} \\ \cdots \\ M_{cx} \end{bmatrix} = \begin{bmatrix} \mathbf{B}_1 \\ \cdots \\ \mathbf{B}_2 \\ \cdots \\ \mathbf{B}_3 \\ \cdots \\ \mathbf{B}_4 \end{bmatrix} \qquad (5.159)$$

Compared to Eq. (5.121), the state matrix of Eq. (5.154) has one more column with nonzero blocks for both moment equations (second and fourth rows). It also has the new elements in the third column and two last rows of the second column. The three last rows at the right-hand side of Eq. (5.154) will also have one new term (third in second and third rows and second in fourth row):

$$\mathbf{B}_1 = \mathbf{F}_a^* + \mathbf{F}_g^* - \mathbf{S}(\boldsymbol{\omega})[(m^* + m_e)\mathbf{I}_{3\times3} + \mathbf{I}_{a.m.}'] \begin{bmatrix} u_c \\ v_c \\ w_c \end{bmatrix}$$

$$+ \mathbf{S}(\boldsymbol{\omega})\mathbf{I}_{a.m.}' \mathbf{S}(\mathbf{r}_{CM}) \begin{bmatrix} p \\ q \\ r \end{bmatrix} + \mathbf{S}(\boldsymbol{\omega})\mathbf{I}_{a.m.}' {}_n^b \mathbf{RW}$$

$$- (m^* + m_e)\mathbf{S}(\boldsymbol{\omega})\mathbf{S}(\boldsymbol{\omega})\mathbf{r}_{CB} \tag{5.160}$$

$$\mathbf{B}_2 = \mathbf{M}_a^* + \mathbf{S}(\mathbf{r}_{BP})\mathbf{F}_a^* - \begin{bmatrix} 0 \\ 0 \\ M_{cz} \end{bmatrix}$$

$$- \left[\mathbf{S}(\boldsymbol{\omega})(\mathbf{I}^* + \mathbf{I}_{a.i.}') - \mathbf{S}(\mathbf{r}_{BM})\mathbf{S}(\boldsymbol{\omega})\mathbf{I}_{a.m.}' \mathbf{S}(\mathbf{r}_{CM})\right] \begin{bmatrix} p \\ q \\ r \end{bmatrix}$$

$$- \mathbf{S}(\mathbf{r}_{BM})\mathbf{S}(\boldsymbol{\omega})\mathbf{I}_{a.m.}' \begin{bmatrix} u_c \\ v_c \\ w_c \end{bmatrix} + \mathbf{S}(\mathbf{r}_{BM})\mathbf{S}(\boldsymbol{\omega})\mathbf{I}_{a.m.}' {}_n^b \mathbf{RW} \tag{5.161}$$

$$\mathbf{B}_3 = \mathbf{F}_a^s + \mathbf{F}_g^s + m_s\mathbf{S}(\mathbf{r}_{CS}^s)\mathbf{G}^8 - m_{sb}^s\mathbf{RS}(\boldsymbol{\omega}) \begin{bmatrix} u_c \\ v_c \\ w_c \end{bmatrix} - m_s\mathbf{S}(\boldsymbol{\omega}_s^s)\mathbf{S}(\boldsymbol{\omega}_s^s)\mathbf{r}_{CS}^s \tag{5.162}$$

$$\mathbf{B}_4 = -\mathbf{I}_s\mathbf{G}^8 + {}_b^s\mathbf{R} \begin{bmatrix} 0 \\ 0 \\ M_{cz} \end{bmatrix} - \mathbf{S}(\boldsymbol{\omega}_s^s)\mathbf{I}_s \begin{bmatrix} p_s \\ q_s \\ r_s \end{bmatrix} \tag{5.163}$$

Equation (5.159) along with the translational kinematic described by Eq. (5.109) and rotational kinematics described by Eqs. (5.15) and (5.144) constitutes the eight-DoF model.

5.2.6 MODEL COMPARISON

Mass-geometry properties, apparent mass terms, and aerodynamic coefficients for the Snowflake PADS used in simulations are provided in Table 5.4.

TABLE 5.4   IDENTIFIED PARAMETERS OF THE SEVEN-/EIGHT-/NINE-DoF MODELS OF SNOWFLAKE PADS (FIG. 1.30A)

Parameter	Value
Subsystem 1 (payload and suspension lines) mass m^*	0.45 kg (1 lb)
Subsystem 2 (payload) mass m_s	1.9 kg (4.23 lb)
Canopy span b, chord c, thickness t	1.3 m (4.25 ft), 0.7 m (2.27 ft), 0.07 m (0.23 ft)
Canopy reference area S	1 m² (11.0 ft²)
Payload dimensions, D × W × H	0.123 × 0.246 × 0.27 m (0.4 × 0.81 × 0.89 ft)
Payload reference area S_s	0.042 m² (0.45 ft²)
Rigging angle μ	−12 deg
Steady-state aerodynamic velocity V_a	9.6 m/s (18 kt)
Inertia matrices	$\mathbf{I}^* = \begin{bmatrix} 0.042 & 0 & -0.0068 \\ 0 & 0.027 & 0 \\ -0.0068 & 0 & 0.054 \end{bmatrix}$ kg·m², $\quad \mathbf{I}_s = \begin{bmatrix} 0.0102 & 0 & 0 \\ 0 & 0.0092 & 0 \\ 0 & 0 & 0.0069 \end{bmatrix}$ kg·m²,
	$\mathbf{I}^* = \begin{bmatrix} 1 & 0 & -0.16 \\ 0 & 0.64 & 0 \\ -0.16 & 0 & 1.28 \end{bmatrix}$ lb·ft², $\quad \mathbf{I}_s = \begin{bmatrix} 0.24 & 0 & 0 \\ 0 & 0.22 & 0 \\ 0 & 0 & 0.16 \end{bmatrix}$ lb·ft²
Apparent mass matrix	$\mathbf{I}_{a.m.} = \mathrm{diag}([0.012, 0.032, 0.42])$ kg, $\quad \mathbf{I}_{a.m.} = \mathrm{diag}([0.027, 0.071, 0.93])$ lb
Apparent inertia matrix	$\mathbf{I}_{a.i.} = \mathrm{diag}([0.054, 0.14, 0.0024])$ kg·m², $\quad \mathbf{I}_{a.i.} = \mathrm{diag}([1.29, 3.34, 0.06])$ lb·ft²
Vector from the mass center to the apparent mass center	$\mathbf{r}_{BM} = [0.046, 0, -1.11]^T$ m, $\quad \mathbf{r}_{BM} = [0.15, 0, -3.64]^T$ ft
Vector from the point C to the canopy	$\mathbf{r}_{CR} = [-0.15, 0, -0.82]^T$ m, $\quad \mathbf{r}_{CR} = [-0.5, 0, -2.7]^T$ ft

Vector from the point R to the parafoil aerodynamic center P	$\mathbf{r}_{RP}^p = [0.19, 0, 0]^T$ m,	$\mathbf{r}_{RP}^p = [0.63, 0, 0]^T$ ft
Vector from the point R to the apparent mass center M	$\mathbf{r}_{RM}^p = [0.18, 0, 0.061]^T$ m,	$\mathbf{r}_{RM}^p = [0.59, 0, 0.2]^T$ ft
Vector from the point C to parafoil mass center B	$\mathbf{r}_{CB} = [0.15, 0, -0.69]^T$ m,	$\mathbf{r}_{CB} = [0.5, 0, -2.25]^T$ ft
Vector from the point C to the payload mass center S	$\mathbf{r}_{CS}^s = [0, 0, 0.31]^T$ m,	$\mathbf{r}_{CS}^s = [0, 0, 1]^T$ ft
Maximum brake deflection $\delta_{s\,max}$	0.25 m (0.75 ft or 9 in.)	

Aerodynamic coefficients

$$C_{D0} = 0.15, \quad C_{D\alpha^2} = 0.90$$
$$C_{Y\beta} = -0.05$$
$$C_{L0} = 0.25, \quad C_{L\alpha} = 0.68$$
$$C_{m0} = 0.0, \quad C_{m\alpha} = 0, \quad C_{mq} = -0.265$$
$$C_{l\beta} = -0.036, \quad C_{lp} = -0.355, \quad C_{lr} = -0, \quad C_{l\delta_a} = -0.00032$$
$$C_{n\beta} = -0.0015, \quad C_{np} = -0, \quad C_{nr} = -0.09, \quad C_{n\delta_a} = 0.003$$

Payload drag coefficient	$C_D^s = 0.40$
Rotational stiffness and damping from risers	$K_\psi = 0.07$ kg · m/rad, $\quad K_r = 0.005$ kg · m · s/rad

The six-/seven-/eight-/nine-DoF models presented in the preceding sections are numerically integrated using a fourth-order Runge–Kutta method with a time step of 0.01 s. The simulation starts at zero cross range and downrange, from 400 m (1,300 ft) above sea level, with payload and parafoil pitch angles of -1.8 and -1 deg, respectively. The velocity vector components are $u_c = 8.6$ m/s (16.7 kt) and $v_c = 4.2$ m/s (13.9 ft/s), with all other states equal to zero.

An example of simulation similar to the real situation experienced by the parafoil/payload system in the flight test with a series of 44% left and right brake deflection is shown in Fig. 5.26. The resulting trajectory is a maneuver shown in Fig. 5.27. The ground track illustrates the difference between the six- and the seven-/eight-/nine-DoF model trajectory. The differences show that payload twist has an effect on trajectory while similarity between the seven-/eight-/nine-DoF models demonstrates relative payload pitch and roll have little effect on the trajectory.

Figures 5.28–5.30 show angular rates of the six-/seven-/eight-/nine-DoF system for the parafoil and payload. In Fig. 5.28, the parafoil yaw rate oscillates at a significantly lower magnitude than the payload yaw rate, similar to experimental results. The parafoil steady-state yaw rate is equivalent; however, the yaw rate for the six-DoF, which does not take into account relative payload yaw rate, has substantially lower motion. For parafoil and payload pitch rate in Fig. 5.29, the seven-/eight-/nine-DoF models show only small differences when compared to the six-DoF model. Because the seven-DoF model, which does not include relative payload pitch, produces similar results to the eight-/nine-DoF models, the differences are due to yaw roll coupling excited by the relative payload yawing. Similar to the pitch rate, payload roll rate in Fig. 5.30 is very similar for the seven-/eight-/nine-DoF models. The difference between six- and seven-/eight-/nine-DoF models is small and generated mainly from the relative yawing motion of the payload.

Figures 5.31–5.33 illustrate simulated Euler angles of the parafoil and payload. In Fig. 5.31, the differences between six- and seven-/eight-/nine-DoF model yaw angles are slight but generate changes in the turn trajectory. Minimal variations in the seven-/eight-/nine-DoF models indicate that all can accurately represent relative parafoil-payload twist and that relative pitch and roll are not significantly coupled to payload yaw.

The payload yaw has oscillations that are about four times the magnitude of the parafoil oscillations. Figure 5.32 shows that the differences between pitch in the

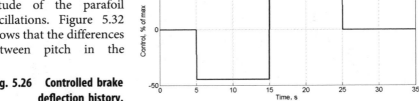

Fig. 5.26   Controlled brake deflection history.

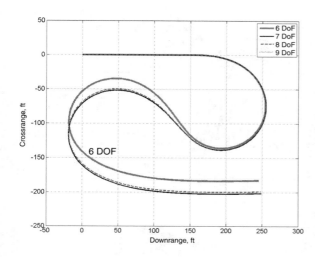

Fig. 5.27 Ground track comparison.

models are small. Ability of the payload to pitch respect to the canopy, however, allows a slight variation in the steady-state pitch of the canopy in the eight-/ nine-DoF models. Roll in Fig. 5.33 similarly shows negligible variations in the parafoil while allowing the payload to remain more vertical during the turns.

When comparing the simulated results with experimental data, a difference that is the observed is the appearance of persistent oscillation in the experimental data while the simulation decays to a steady state. The reason for this is not model error but rather the pristine atmosphere in the simulation. During experiments, the atmosphere experience changes in wind magnitude, wind direction,

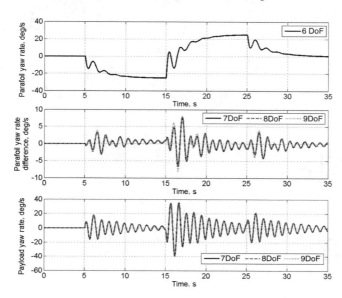

Fig. 5.28 Parafoil and payload yaw rate.

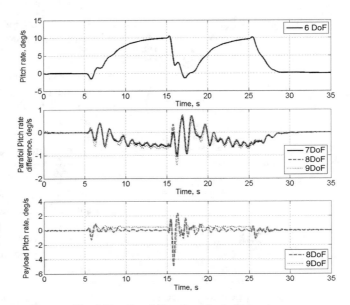

Fig. 5.29   Parafoil and payload pitch rate.

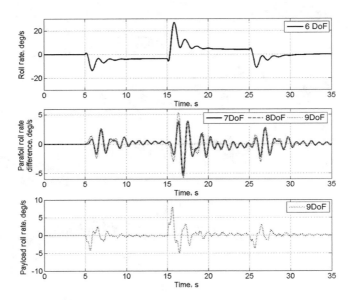

Fig. 5.30   Parafoil and payload roll rate.

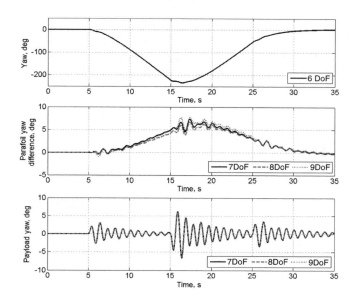

Fig. 5.31   Parafoil and payload yaw.

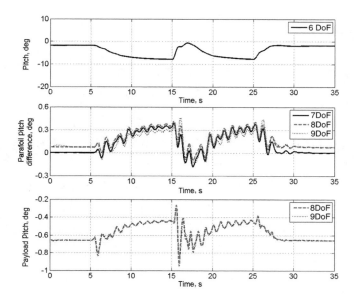

Fig. 5.32   Parafoil and payload pitch.

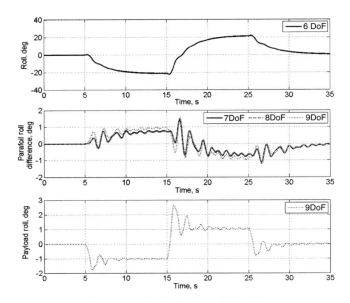

Fig. 5.33   Parafoil and payload roll.

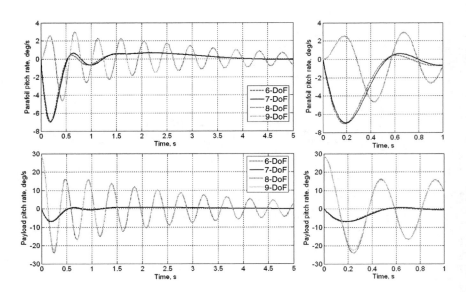

Fig. 5.34   Parafoil and payload pitch rate.

updrafts, and turbulence, all exciting motion continually. When the payload motion is excited in the simulation, the natural oscillations of the parafoil-payload can be seen in Figs. 5.34–5.36. For 5 s, the payload is given a payload a rate of $p_s = q_s = r_s = 30$ deg/s, which is similar to that experienced in the flight test. Both the parafoil and payload yaw rate have a frequency of 1 Hz from the twist of the lines as also seen in the flight data. The coupling between parafoil yaw and roll rate can be seen in the first second of Fig. 5.34 before it is quickly damped out. Similar to Figs. 5.29 and 5.30, Fig. 5.35 and 5.36 demonstrate the 2.2-Hz frequency of the pitch rate and roll rate of the payload and the 1-Hz frequency of the parafoil roll and pitch rate.

To conclude this section, let us provide data on larger PADS as reported by [Toglia and Vendittelli 2010] (see Table 5.5) and powered paraglider (PPG) used by [Hur 2005] (Table 5.6). These data are consistent with those of Table 5.4 and can also be used in simulations.

5.3 MODELS OF CONTROL EFFICIENCY

Throughout this chapter, different assumptions have been made with regard to effect of symmetric and asymmetric TE deflection on PADS lift, drag, and turn rate. For example, Eqs. (5.35) and (5.36) assumed constant descent rate and horizontal component of airspeed regardless of control inputs; turn dynamics of Eq. (5.38) assumed a symmetric linear dependence of a steady-state yaw rate, and vertical motion described by Eq. (5.42) accounted for effect of the yaw

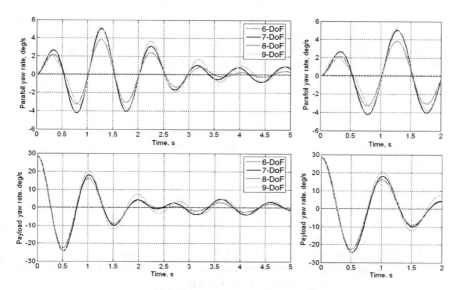

Fig. 5.35   Parafoil and payload yaw rate.

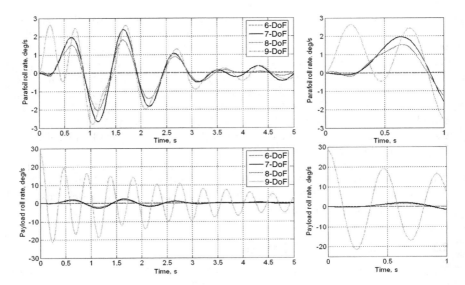

Fig. 5.36   Parafoil and payload roll rate.

dynamics. This section elaborates on this issue and uses real flight-test data available for different PADSs to establish some common grounds.

The issue of introducing a nonlinear control efficiency helping to cope with the effects of canopy deformation due to control surface deflections was explicitly addressed in [Jann and Strickert 2005; Jann 2006]. The semi-empiric approach based on *sine*-functions allowed transforming the normalized deflections of the left and right flaps according to a suitable characteristic

$$\overline{\delta}_{l(n)} = \overline{\delta}_l + K_1 \sin(\pi\overline{\delta}_l) + K_2 \sin(2\pi\overline{\delta}_l) \tag{5.164}$$

$$\overline{\delta}_{r(n)} = \overline{\delta}_r + K_1 \sin(\pi\overline{\delta}_r) + K_2 \sin(2\pi\overline{\delta}_r) \tag{5.165}$$

Based on these corrected deflections $\overline{\delta}_{l(n)}$ and $\overline{\delta}_{r(n)}$, the normalized symmetric and asymmetric control inputs into the corresponding force and moment equations are defined as

$$\overline{\delta}_s = 0.5(\overline{\delta}_{r(n)} + \overline{\delta}_{l(n)}) \tag{5.166}$$

$$\overline{\delta}_a = \overline{\delta}_{r(n)} - \overline{\delta}_{l(n)} - \Delta\delta \tag{5.167}$$

where $\Delta\delta$ is the asymmetric actuator offset.

The results of identification using Eqs. (5.164–5.167) for the six-DoF model of ALEX PADS (as discussed in detail in Sec. 11.3) are shown in Fig. 5.37. They confirm that small control line deflections are apparently absorbed by the flexibility of the canopy until larger deflections led to a significant increase of the local drag. However, the exact values of K_1, K_2, $\Delta\delta$ and their corresponding

TABLE 5.5 PARAMETERS OF A NINE-DoF MODEL OF PADS BY THE UNIVERSITÀ DI ROMA

Parameter	Value
Subsystem 1 (canopy, suspension lines) mass m^*	13 kg (29 lb)
Subsystem 2 (payload) mass m_s	135 kg (298 lb)
Canopy span b, chord c, thickness t	7 m (23 ft), 3 m (9.8 ft), 0.3 m (1 ft)
Canopy reference area S	21 m² (226 ft²)
Payload dimensions, D × W × H	0.5 × 0.5 × 0.5 m (1.6 × 1.6 × 1.6 ft)
Payload reference area S_s	0.25 m² (2.7 ft²)
Vector from the point C to parafoil mass center	$\mathbf{r}_{CB} = [0, 0, 7.5]^T$ m ($[0, 0, 24.6]^T$ ft)
Vector from the point C to the payload mass center	$\mathbf{r}_{CS}^s = [0, 0, 0.5]^T$ m ($[0, 0, 1.6]^T$ ft)
Aerodynamic coefficients	$C_{D0} = 0.15$, $\quad C_{D\alpha} = 1$, $\quad C_{D\delta_s} = 0.3$ $\quad C_{D\delta_a} = 0.0001$ $C_{L0} = 0.4$, $\quad C_{L\alpha} = 2$, $\quad C_{L\delta_s} = 0.21$ $\quad C_{L\delta_a} = 0.0001$ $C_{m0} = 0.018$, $\quad C_{m\alpha} = -0.2$, $\quad C_{mq} = -2$ $C_{l\phi} = -0.05$, $\quad C_{lp} = -0.1$, $\quad C_{lr} = 0$, $\quad C_{l\delta_a} = 0.0021$ $C_{n\beta} = 0$, $\quad C_{np} = 0$, $\quad C_{nr} = -0.07$, $\quad C_{n\delta_a} = 0.004$
Payload drag coefficient	$C_D^s = 0.4$
Rotational stiffness and damping from risers	$K_\psi = 0.07$ N·m/rad (0.05 lb·ft/rad) $\quad K_r = 0.005$ N·m·s/rad (0.037 lb·ft·s/rad)

TABLE 5.6 PARAMETERS OF AN EIGHT-DoF MODEL OF BUCKEYE PPG BY THE TEXAS A&M UNIVERSITY (FIG. 1.9A)

Parameter	Value
Subsystem 1 (canopy, enclosed air, suspension lines) mass	25.3 kg (55.8 lb)
Subsystem 2 mass m_s [including 13.6 kg (5 gal, 6 lb/gal) of fuel]	280 kg (617.3 lb)
Canopy span b and chord c	13.3 m (43.5 ft), 3.5 m (11.5 ft)
Canopy reference area S	46.5 m² (500 ft²)
Rigging angle μ	−7 deg
Steady-state glide conditions	$V_a = 12$ m/s (23 kt)[$V_h = 11.2$ m/s (21.8 kt), $V_v = 3.3$ m/s (10.8 ft/s)], $\theta = -0.8$ deg
Inertia matrices	$I^* = \begin{bmatrix} 348 & 0 & 0 \\ 0 & 28 & 0 \\ 0 & 0 & 342 \end{bmatrix}$ kg·m², $\quad I_s = \begin{bmatrix} 41 & 0 & 0 \\ 0 & 167 & 0 \\ 0 & 0 & 167 \end{bmatrix}$ kg·m² $I^* = \begin{bmatrix} 8270 & 0 & 0 \\ 0 & 900 & 0 \\ 0 & 0 & 8124 \end{bmatrix}$ lb·ft², $\quad I_s = \begin{bmatrix} 975 & 0 & 0 \\ 0 & 3951 & 0 \\ 0 & 0 & 3951 \end{bmatrix}$ lb·ft²
Distance from the confluence point C to canopy c.g.	5 m (19.6 ft)
Distance from the confluence point C to payload c.g.	0.7 m (2.3 ft)
Aerodynamic coefficients:	$C_{D0} = 0.23$, $C_{D\alpha} = 1$, $C_{D\delta_s} = 0.02$/in. $C_{L0} = 0.72$, $C_{L\alpha} = 3.6$, $C_{L\delta_s} = 0.04$/in. $C_{Y\beta} = -0.26$, $C_{Y\delta_a} = -0.0008$/in. $C_{m0} = -0.032$, $C_{m\alpha} = -0.077$, $C_{mq} = -0.3$, $C_{m\delta_a} = -0.004$/in. $C_{l\beta} = -0.037$, $C_{lp} = -0.15$, $C_{lr} = 0.005$, $C_{l\delta_a} = 0.00053$/in. $C_{n\beta} = 0.0057$, $C_{np} = 0.023$, $C_{nr} = -0.094$, $C_{n\delta_a} = -0.000038$/in.
Payload drag coefficient	$C_D^s = 0.36$

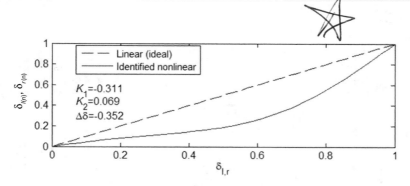

Fig. 5.37   Nonlinear control efficiency.

characteristics depend not only on the canopy design but also on the trim of the control lines that may again be subject to change between different configurations of the same parafoil. Another interpretation of data in Fig. 5.37 is that if the control lines are trimmed in a 20 to 40% position (as opposed to a 0% position) the parafoil system would react to flap deflection much faster.

As has already been discussed in Sec. 5.1.3, in order to take into account possible asymmetries Eqs. (5.166) and (5.167) were modified to those of Eqs. (5.40) and (5.41). Also, neither Eqs. (5.164–5.167) nor Eqs. (5.166) and (5.167) account for actuator dynamics that should be modeled separately. The longer it takes for an actuator to move the control lines to the desired position, the bigger influence the actuator has on PADS flight dynamics. For ALEX PADS, the time to perform a full stroke, both pull and release, was about 2.5 s. In practice, because of the tension in the control lines, pulling takes a little longer than releasing the control lines (the actual difference depends on the motor power).

More experimental data collected for different systems are presented in Figs. 5.38–5.42. These figures show the influence of symmetric and asymmetric TE deflection on the GR (Fig. 5.38), horizontal (Fig. 5.39), and vertical (Fig. 5.40) components of an airspeed vector, as well as the effect of the asymmetric TE deflection on the turn rate (Figs. 5.39) and (5.40). Data shown in Figs. 5.38–5.41 were taken from [Stein et al. 2005; Carter et al. 2005; Carter et al. 2007; Wegereef and Jentink 2005; Wegereef et al. 2011], converted to the same units and normalized (with respect to the flap deflection). The most comprehensive analysis of longitudinal and lateral-directional aerodynamics of large parafoils in flight, performed within the X-38 program with 335, 508, and 700 m² (3,600; 5,468; and 7,536 ft²) canopies, can be found in [Iacomini and Cerimele 1999a; Iacomini and Cerimele 1999b; Madsen and Cerimele 2000; Madsen and Cerimele 2003]. The major trends were then generalized in [Stein et al. 2005], and that is what Figs. 5.38a, 5.39a, and 5.40a refer to as X-38.

Regression analysis reveals that data presented in Figs. 5.38–5.40 can be accurately described with the same quadratic equation

$$\xi = \xi_0(1 - p_\delta \bar{\delta}_{s;a}^2) \tag{5.168}$$

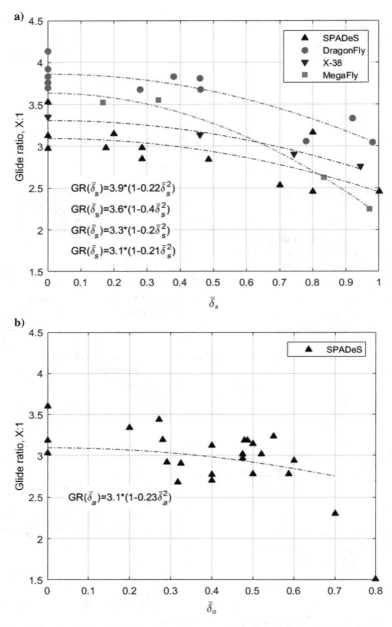

Fig. 5.38   Variation of GR for the a) symmetric and b) asymmetric TE deflection.

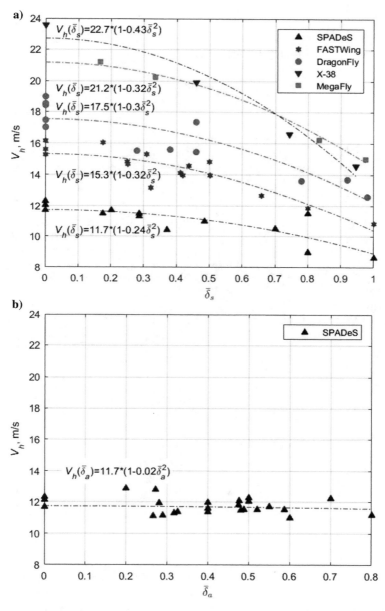

Fig. 5.39 Variation of a horizontal speed for the a) symmetric and b) asymmetric TE deflection.

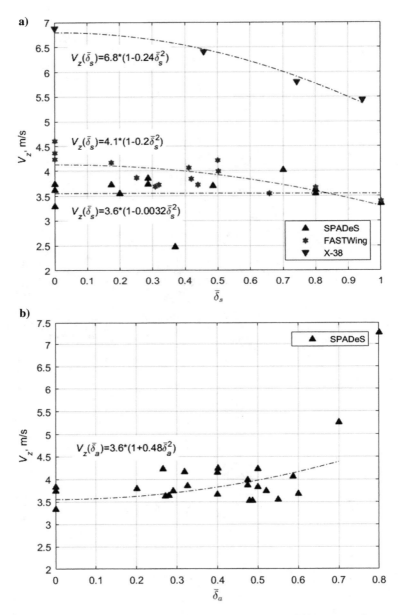

Fig. 5.40 Variation of a vertical speed for the a) symmetric and b) asymmetric TE deflection.

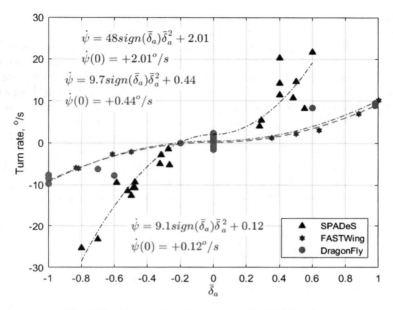

Fig. 5.41   Turn rate against asymmetric TE deflection.

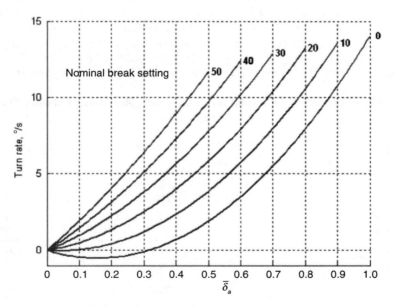

Fig. 5.42   The 7,500-ft² X-38 turn rate against asymmetric TE deflection [Soppa et al. 2005].

(Other regressions were tested as well, but this one happens to give the best fit overall.) In Eq. (5.168), ξ_0 is the value at $\overline{\delta}_{s;a} = 0$, and parameter p_δ shows the decrease in the value at full actuator deflection ($\overline{\delta} = 1$). The values of parameters ξ_0 and p_δ for the GR (Fig. 5.38a), horizontal speed (Fig. 5.39a), and descent rate (Fig. 5.40a) in response to $\overline{\delta}_s$ shown in Figs. 5.38a, 5.39a, and 5.40a are repeated in Table 5.7. For the SPADeS PADS with a G9-Galaxy canopy, the GR, horizontal speed, and descent in response to $\overline{\delta}_a$ are also available. These data are shown in Figs. 5.38b, 5.39b, and 5.40b, respectively, using the same y scale as for the symmetric TE deflection plots for the sake of fair comparison. The regression analysis for these data also used Eq. (5.168), but the values of $\xi_0(\overline{\delta}_a = 0)$ were considered fixed and equal to those at $\xi_0(\overline{\delta}_s = 0)$. The three regression coefficients p_δ are shown in Table 5.7 in the parenthesis in the first row.

To check for the consistency of data, we can use the following self-explanatory formula:

$$\xi^{GR} = \frac{\xi^{Vh}}{\xi^{Vz}} = \frac{\xi^{Vh}}{\xi^{Vz}} \frac{1 - p_\delta^{Vh} \overline{\delta}_{s;a}^2}{1 - p_\delta^{Vz} \overline{\delta}_{s;a}^2} \approx \frac{\xi^{Vh}}{\xi^{Vz}} (1 - p_\delta^{Vh} \overline{\delta}_{s;a}^2)(1 + p_\delta^{Vz} \overline{\delta}_{s;a}^2)$$

$$\approx \frac{\xi^{Vh}}{\xi^{Vz}} [1 - (p_\delta^{Vh} - p_\delta^{Vz}) \overline{\delta}_{s;a}^2] \qquad (5.169)$$

For example, for the X-38 PADS this formula yields $\xi^{Vh}/\xi^{Vz} = 3.34$ and $p_\delta^{Vh} - p_\delta^{Vz} = 0.19$, which are about the same as the ones obtained explicitly (3.3 and 0.2, respectively). For the SPADeS PADS, $\xi^{Vh}/\xi^{Vz} = 3.25$ and $p_\delta^{Vh} - p_\delta^{Vz} = 0.24$ vs 3.1 and 0.21, respectively. For asymmetric TE deflection $\left| p_\delta^{Vh} - p_\delta^{Vz} \right| = 0.48$ vs 0.23. This error might be caused by not accounting for the winds correctly, so that instead of V_h data actually features V_G. Regressions in Figs. 5.38b, 5.39b, and 5.40b do not involve the very last data point obtained at 80% (asymmetric) TE

TABLE 5.7   SYMMETRIC TE DEFLECTION EFFECT ON LONGITUDINAL MOTION PARAMETERS FOR DIFFERENT PADS

PADS	Glide Ratio C_L/C_D		Horizontal Speed V_h		Descent Rate V_z	
	ξ_0	p_δ	$\xi_0,$ m/s	p_δ	$\xi_0,$ m/s	p_δ
SPADeS with a G9-Galaxy canopy	3.1	0.21 (0.23)	11.7	0.24 ($\sim$0)	3.6	$\sim$0 (-0.48)
FASTWing	—	—	15.3	0.32	4.1	0.20
DragonFly	3.9	0.22	17.5	0.30	—	—
X-38	3.3	0.20	22.7	0.43	6.8	0.24
MegaFly	3.6	0.40	21.2	0.32	—	—

deflection. With these data points the corresponding coefficients in Table 5.7 (in the second row for the SPADeS PADS) would be 0.38, ~ 0, and -0.79.

As seen, a full symmetric TE deflection causes about 20% decrease of the descent rate and about a doubled decrease of the horizontal component of the airspeed. As a result, full symmetric TE deflection usually leads to about 20% decrease of GR (40% for MegaFly). This decrease is gradual so that at $\bar{\delta}_s = 0.33$ it is only one-tenth of it, meaning that for the relatively small $\bar{\delta}_s$ the effect can be considered negligible. Asymmetric TE deflection happens to have a much stronger effect. While having almost no effect on a horizontal component of the airspeed, because of banking it has a strong effect on the descent rate causing it to increase. This increase can be quite substantial (up to 80% for $\bar{\delta}_a = 0.8$ for the SPADeS PADS with a G9-Galaxy canopy).

It turned out that the effect of asymmetric TE deflection on the turn rate (Figs. 5.41 and 5.42) can essentially be modeled with the same regression described by Eq. (5.152). Slightly modified to allow asymmetry and nonzero turn rate at $\bar{\delta}_a = 0$, it takes the form

$$\dot{\psi} = K_\psi \, \mathrm{sign}(\bar{\delta}_a) \bar{\delta}_a^2 + \dot{\psi}_0 \qquad (5.170)$$

Compared to Eq. (5.38), this dependence is not linear and resembles that presented in Fig. 5.37 for the ALEX PADS. Table 5.8 shows the values of parameters K_ψ and $\dot{\psi}_0$ for three systems. Interestingly, for about the same-area canopies (Fastwing and DragonFly), the turn rate happens to be about the same. The nine-time smaller canopy system (G9-Galaxy) is (about five times) more agile.

Equation (5.170) for the turn rate and Eq. (5.168) for the descent rate at asymmetric TE deflections allow establishing the following dependence:

$$V_v = V_{v0}(1 - p_{\delta V v} K_\psi^{-1} |\dot{\psi}|) = V_{v0} + k_{v\dot{\psi}} |\dot{\psi}| \qquad (5.171)$$

This dependence intuitively suggested in Eq. (5.42) now allows the estimation of the order of coefficient $k_{v\dot{\psi}} = -V_{v0} p_{\delta V v} K_\psi^{-1}$. For the G9-Galaxy SPADeS PADS, $k_{v\dot{\psi}} = 3.6 \cdot 0.79/48 \cdot 180/\pi = 3.4 \; \frac{\mathrm{m}}{\mathrm{s}} \frac{\mathrm{s}}{\mathrm{rad}}$. For the larger systems V_{v0} can be twice as much (Fig. 5.40a) and the turn rate five time smaller (Table 5.8). Hence, for the large PADS the coefficient $k_{v\dot{\psi}}$ is probably one order of magnitude larger. Equation (5.171) can now be used in the design of GNC algorithms tying maneuvering in the horizontal plane with the total altitude loss.

TABLE 5.8 ASYMMETRIC TE DEFLECTION EFFECT ON THE TURN RATE FOR DIFFERENT PADS

PADS	K_ψ, deg/s	$\dot{\psi}_0$, deg/s
SPADeS with a G9-Galaxy canopy	48	2
FASTWing	9.1	0.12
DragonFly	9.7	0.44

5.4 LINEARIZED MODELS

To study PADS stability and develop controllers, the nonlinear models developed in Secs. 5.2.2–5.2.5 need to be linearized. For example, the six-DoF PADS model of Eq. (5.91) can be represented as

$$\dot{\mathbf{x}} = \mathbf{f}(\mathbf{x}) + \mathbf{g}(\mathbf{x})\mathbf{u} \tag{5.172}$$

In this equation $\mathbf{x}$ is the state vector, $\mathbf{u}$ is the input or control vector, and $\mathbf{f}$ and $\mathbf{g}$ are the vectors of nonlinear functions of the state vector. The higher-fidelity models of Secs. 5.2.3 and 5.2.5 can be reduced to this form as well.

For the equilibrium,

$$\mathbf{0} = \mathbf{f}(\mathbf{x}_0) + \mathbf{g}(\mathbf{x}_0)\mathbf{u}_0 \tag{5.173}$$

Linearizing Eq. (5.172) with respect to equilibrium of Eq. (5.173) using state perturbations yields

$$\delta\dot{\mathbf{x}} = (\mathbf{f}'_\mathbf{x} - \mathbf{f}\mathbf{g}^{-1}\mathbf{g}'_\mathbf{x})\big|_{\mathbf{x}=\mathbf{x}_0}\,\delta\mathbf{x} + \mathbf{g}_0\delta\mathbf{u} = \mathbf{A}\delta\mathbf{x} + \mathbf{B}\delta\mathbf{u} \tag{5.174}$$

In this equation, $\mathbf{A}$ and $\mathbf{B}$ are the state and input matrices, respectively. The `linearize(sys,io)` function of MATLAB allows linearization of Eq. (5.172) around the operating (equilibrium) point `io` numerically.

The most general state-space representation of a linear system compliments the state equation of Eq. (5.172) with the output equation

$$\delta\mathbf{y} = \mathbf{C}\delta\mathbf{x} + \mathbf{D}\delta\mathbf{u} \tag{5.175}$$

where $\delta\mathbf{y}$ is the output vector and matrices $\mathbf{C}$ and $\mathbf{D}$ are the output and feedthrough matrices.

Depending on application, the state vector can include a different set of parameters. For Eq. (5.91), the state and the control vector would be represented by

$$\mathbf{x} = [\delta u,\, \delta v,\, \delta w,\, \delta p,\, \delta q,\, \delta r]^T \quad \text{and} \quad \mathbf{u} = [\delta\bar{\delta}_s,\, \delta\bar{\delta}_a]^T \tag{5.176}$$

In practice however, the kinematic equation (5.15) is also included, extending the state vector to

$$\mathbf{x} = [\delta u,\, \delta v,\, \delta w,\, \delta p,\, \delta q,\, \delta r,\, \delta\phi,\, \delta\theta,\, \delta\psi]^T \tag{5.177}$$

Although these quantities are pertinent to the six-DoF model, extending the model beyond that will involve adding more states describing the motion of payload. An example of analyzing a full linearized six-DoF model to investigate different motion modes will be given in Sec. 6.1, and in this section we will proceed with examples of the reduced-order models.

Quite often to analyze PADS behavior and/or develop a controller for the longitudinal motion, the state vector is reduced to

$$\mathbf{x}^{\text{lon}} = [u,\, w,\, q,\, \theta]^T \tag{5.178}$$

with

$$\mathbf{u} = \delta_s \tag{5.179}$$

and for the lateral-directional motion, to

$$\mathbf{x}^{lat} = [v, p, r, \phi, \psi]^T \tag{5.180}$$

with

$$\mathbf{u} = \delta_a \tag{5.181}$$

(for simplicity we dropped the deltas). For example, Gorman and Slegers [2011b] developed such linearized models for the six-DoF model of the Snowflake PADS, described by Eq. (5.106) and parameters defined in Table 5.4, for a nominal 8-deg pitch angle glide 762 m (2,500 ft) above sea level as

$$\begin{bmatrix} \dot{u}_c \\ \dot{w}_c \\ \dot{q} \\ \dot{\theta} \end{bmatrix} = \begin{bmatrix} -0.844 & 1.28 & -8.9 & -31.1 \\ -0.869 & -2.22 & 28.8 & 2.78 \\ 0.644 & -1.5 & -3.42 & -0.8 \\ 0 & 0 & 1 & 0 \end{bmatrix} \begin{bmatrix} u_c \\ w_c \\ q \\ \theta \end{bmatrix} \tag{5.182}$$

and

$$\begin{bmatrix} \dot{v}_c \\ \dot{p} \\ \dot{r} \\ \dot{\phi} \\ \dot{\psi} \end{bmatrix} = \begin{bmatrix} -0.714 & 10.1 & -31.7 & 31.1 & 0 \\ -0.606 & -4.21 & 0.087 & -0.652 & 0 \\ 0.781 & -0.928 & -1.89 & 0.598 & 0 \\ 0 & 1 & -0.141 & 0 & 0 \\ 0 & 0 & 1.01 & 0 & 0 \end{bmatrix} \begin{bmatrix} v_c \\ p \\ r \\ \phi \\ \psi \end{bmatrix} + \begin{bmatrix} -0.106 \\ -0.111 \\ 0.654 \\ 0 \\ 0 \end{bmatrix} \delta_a \tag{5.183}$$

[TE deflection inputs are in radians, and engine throttle input is in Newtons with no TE deflection and about 190 N (43 lb) input required for a straight and level flight.]

To capture a payload motion in the high-fidelity models of Secs. 5.2.3 and 5.2.5, the state vectors of Eqs. (5.178) and (5.180) can be extended all of the way up to

$$\mathbf{x}^{lon} = [u, w, q, \theta, q_s, \theta_s]^T \quad \text{and} \quad \mathbf{x}^{lat} = [v, p, r, \phi, \psi, p_s, r_s, \phi_s, \psi_s]^T \tag{5.184}$$

(based on the nine-DoF model), respectively. For example, applying Eq. (5.174) for the seven-DoF model of the Snowflake PADS and linearizing it about the same steady nominal glide as above results in [Gorman 2011;

Gorman and Slegers 2011b]

$$
\begin{bmatrix} \dot{u}_c \\ \dot{w}_c \\ \dot{q} \\ \dot{\theta} \end{bmatrix} = \begin{bmatrix} -0.793 & 1.16 & -9.17 & -31.1 \\ -0.954 & -2.02 & 29.3 & 2.88 \\ 0.594 & -1.38 & -3.16 & -0.742 \\ 0 & 0 & 1 & 0 \end{bmatrix} \begin{bmatrix} u_c \\ w_c \\ q \\ \theta \end{bmatrix}
$$
(5.185)

and

$$
\begin{bmatrix} \dot{v}_c \\ \dot{p} \\ \dot{r} \\ \dot{\phi} \\ \dot{\psi} \\ \dot{r}_s \\ \dot{\psi}_s \end{bmatrix} = \begin{bmatrix} -0.665 & 10.5 & -31.5 & 31.2 & 0 & 0 & -0.61 \\ -0.563 & -3.82 & 0.147 & -0.602 & 0 & 0 & -0.538 \\ 0.812 & -1.43 & -2.19 & 0.608 & 0 & 0 & 4.49 \\ 0 & 1 & -0.141 & 0 & 0 & 0 & 0 \\ 0 & 0 & 1.01 & 0 & 0 & 0 & 0 \\ 0 & 0 & 0 & 0 & 0 & 0 & -39.2 \\ 0 & 0 & -1 & 0 & 0 & 1 & 0 \end{bmatrix}
$$

$$
\times \begin{bmatrix} v_c \\ p \\ r \\ \phi \\ \psi \\ r_s \\ \psi_s \end{bmatrix} + \begin{bmatrix} -0.106 \\ -0.111 \\ 0.72 \\ 0 \\ 0 \\ 0 \\ 0 \end{bmatrix} \delta_a
$$
(5.186)

As seen, the corresponding elements in these matrices are fairly close to those of Eqs. (5.182) and (5.183).

For the linearized eight-DoF model of Buckeye PPG, [Hur 2005] used

$$
\mathbf{x}^{\text{lon}} = [u, w, q, \theta, q_s, \theta^{s/p}]^T
$$
(5.187)

and

$$
\mathbf{x}^{\text{lat}} = [v, p, r, \phi, r_s, \psi^{s/p}]^T
$$
(5.188)

(Superscript s/p means the relative motion of the store s about canopy p.) For PPG characterized in Table 5.6 and shown in Fig. 5.43a, [Hur 2009] came up with

$$
\begin{bmatrix} \dot{u} \\ \dot{w} \\ \dot{q} \\ \dot{\theta} \\ \dot{q}_s \\ \dot{\theta}^{s/p} \end{bmatrix} =
\begin{bmatrix}
-10.4 & 29.5 & -8.58 & -32.2 & -0.0005 & 1.16 \\
-0.466 & -15.0 & 36.7 & 0.45 & -0.0004 & 2.08 \\
0.539 & -1.52 & -0.138 & 0 & -0.0005 & 1.20 \\
0 & 0 & 1 & 0 & 0 & 0 \\
-0.0187 & 0.0618 & 0.0725 & 0 & 0 & -11.3 \\
0 & 0 & -1 & 0 & 1 & 0
\end{bmatrix}
$$

$$
\times \begin{bmatrix} u \\ w \\ q \\ \theta \\ q_s \\ \theta^{s/p} \end{bmatrix} +
\begin{bmatrix} -2.07 \\ -0.954 \\ 0.107 \\ 0 \\ -0.0028 \\ 0 \end{bmatrix} \delta_s
\tag{5.189}
$$

and

$$
\begin{bmatrix} \dot{v} \\ \dot{p} \\ \dot{r} \\ \dot{\phi} \\ \dot{r}_s \\ \dot{\psi}^{s/p} \end{bmatrix} =
\begin{bmatrix}
-9.08 & 8.65 & -37.1 & 32.2 & 0 & -0.243 \\
-0.415 & -0.121 & -0.0279 & 0 & 0 & -0.0334 \\
-0.097 & -0.153 & -0.195 & 0 & 0 & -0.707 \\
0 & 1 & -0.014 & 0 & 0 & 0 \\
-0.0078 & 0.0017 & -0.0004 & 0 & 0 & 1.45 \\
0 & -0.0719 & -1 & 0 & 0 & 0
\end{bmatrix}
$$

$$
\times \begin{bmatrix} v \\ p \\ r \\ \phi \\ r_s \\ \psi^{s/p} \end{bmatrix} +
\begin{bmatrix} -0.194 \\ -0.0088 \\ -0.002 \\ 0 \\ -0.0002 \\ 0 \end{bmatrix} \delta_a
\tag{5.190}
$$

To develop a nonlinear model of two-body PPG dynamics, Hur used the Kane's equation introduced in [Kane and Levinson 1985]. He then conducted an analytical linearization deriving the elements of matrices **A** and **B** in Eqs. (5.189) and (5.190) analytically. The values of aerodynamic derivatives used to compute the values of the elements of matrices **A** and **B** were scaled down from those of the Parafoil Dynamic Simulator created for the X-38/CRV program by NASA (Fig. 1.8) [Jessick 1990b]. As a consequence, the linearized models of Eqs. (5.189) and (5.190) produce dynamics, which is not realistic for the small PPG

a)   b)

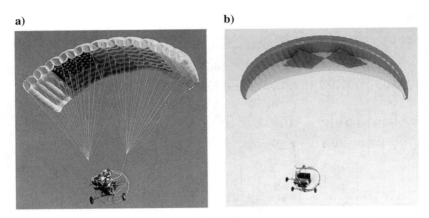

Fig. 5.43   The a) 46.5 m^2 (500 ft^2) and b) 29 m^2 (312 ft^2) powered paragliders [Hur 2005; Watanabe and Ochi 2007].

[Hur 2005]. [In fact, Eq. (5.190) represents an unstable system.] The values of matrices **A** and **B**, however, were not used for simulations but rather as initial guesses for a system identification procedure as described in Sec. 11.5.6.

Note, Eq. (5.190) features a single control input while, in general, the PPG control vector for the longitudinal cannel should also include an engine throttle setting

$$\mathbf{u} = [\delta_T, \delta_s]^T \tag{5.191}$$

Compared to Eqs. (5.187) and (5.188), [Ochi et al. 2009] used the slightly different state vectors

$$\mathbf{x}^{\text{lon}} = [u, w, q, \theta, q^{s/P}, \theta^{s/P}]^T \tag{5.192}$$

and

$$\mathbf{x}^{\text{lat}} = [v, p, r, \phi, \psi, r^{s/P}, \psi^{s/P}]^T \tag{5.193}$$

with the control vector of the form of Eq. (5.191). For their PPG characterized by the wing span of 10 m (33 ft), mean aerodynamic chord of 3 m (9.8 ft), wing area 29 m^2 (312 ft^2), the length of suspension lines of 6.5 m (21.3 ft), canopy mass 6.4 kg (14 lb), and mass of payload of 93 kg (205 lb) (Fig. 5.43b), [Ochi et al.

2009] came up with

$$
\begin{bmatrix} \dot{u} \\ \dot{w} \\ \dot{q} \\ \dot{\theta} \\ \dot{q}^{s/p} \\ \dot{\theta}^{s/p} \end{bmatrix}
=
\begin{bmatrix}
-0.537 & 0.071 & 2.48 & -2.88 & -0.238 & 0.610 \\
-0.231 & -1.55 & 1.16 & -0.862 & 0.122 & 0.387 \\
0.0293 & 0.192 & -0.99 & -1.06 & 0.380 & 0.670 \\
0 & 0 & 1 & 0 & 0 & 0 \\
0.296 & -1.43 & 2.84 & 1.05 & -6.73 & -15.8 \\
0 & 0 & 0 & 0 & 1 & 0
\end{bmatrix}
$$

$$
\times
\begin{bmatrix} u \\ w \\ q \\ \theta \\ q^{s/p} \\ \theta^{s/p} \end{bmatrix}
+
\begin{bmatrix}
1.30e-3 & -32.7 \\
8.22e-4 & -18.7 \\
1.42e-3 & 5.2 \\
0 & 0 \\
-1.27e-3 & -4.9 \\
0 & 0
\end{bmatrix}
\begin{bmatrix} \delta_T \\ \delta_s \end{bmatrix}
\qquad (5.194)
$$

and

$$
\begin{bmatrix} \dot{v} \\ \dot{p} \\ \dot{r} \\ \dot{\phi} \\ \dot{\psi} \\ \dot{r}^{s/p} \\ \dot{\psi}^{s/p} \end{bmatrix}
=
\begin{bmatrix}
-0.193 & 2.24 & -3.28 & 1.20 & 0 & -5.9e-4 & -0.0913 \\
-0.0197 & -0.0624 & 0.773 & -1.20 & 0 & 9.8e-4 & 0.0784 \\
20.3 & -152.6 & -123 & 2.6e-4 & 0 & 7.9e-3 & 0.643 \\
0 & 1 & 0.0604 & 0 & 0 & 0 & 0 \\
0 & 0 & 1 & 0 & 0 & 0 & 0 \\
-20.3 & 153 & 123 & -0.481 & 0 & -0.313 & -26.1 \\
0 & 0 & 0 & 0 & 0 & 1 & 0
\end{bmatrix}
$$

$$
\times
\begin{bmatrix} v \\ p \\ r \\ \phi \\ \psi \\ r^{s/p} \\ \psi^{s/p} \end{bmatrix}
+
\begin{bmatrix} 0.0247 \\ -0.006 \\ 529 \\ 0 \\ 0 \\ -529 \\ 0 \end{bmatrix}
\delta_a
\qquad (5.195)
$$

[TE deflection inputs are in radians, and engine throttle input is in Newtons with no TE deflection and about 190 N (43 lb) input required for a straight and level flight.]

To compute the parameters of the store, the following equations for the longitudinal and lateral-directional channels were employed, respectively [Ochi et al.

2009; Ochi and Watanabe 2011]:

$$\mathbf{x}_s^{\text{lon}} = [u_s, w_s, q_s, \theta_s, q^{s/P}, \theta^{s/P}]^T = \mathbf{T}^{\text{lon}}\mathbf{x}^{\text{lon}} \tag{5.196}$$

with

$$\mathbf{T}^{\text{lon}} = \begin{bmatrix} 0.997 & 0.0842 & 6.83 & 0 & 0.332 & -1.70 \\ -0.0842 & 0.997 & -0.549 & 0 & 0 & 9.39 \\ 0 & 0 & 1 & 0 & 1 & 0 \\ 0 & 0 & 0 & 1 & 0 & 1 \\ 0 & 0 & 0 & 0 & 1 & 0 \\ 0 & 0 & 0 & 0 & 0 & 1 \end{bmatrix} \tag{5.197}$$

and

$$\mathbf{x}_s^{\text{lat}} = [v_s, p_s, r_s, \phi_s, \psi_s, r^{s/P}, \psi^{s/P}]^T = \mathbf{T}^{\text{lat}}\mathbf{x}^{\text{lat}} \tag{5.198}$$

with

$$\mathbf{T}^{\text{lat}} = \begin{bmatrix} 1 & -6.85 & -0.028 & 0 & 0 & -0.028 & -9.22 \\ 0 & 0.997 & 0.0842 & 0 & 0 & 0.0842 & 0 \\ 0 & -0.0842 & 0.997 & 0 & 0 & 0.997 & 0 \\ 0 & 0 & 0 & 0.999 & 0 & 0 & 0.0603 \\ 0 & 0 & 0 & -0.0842 & 1 & 0 & 0.997 \\ 0 & 0 & 0 & 0 & 0 & 1 & 0 \\ 0 & 0 & 0 & 0 & 0 & 0 & 1 \end{bmatrix} \tag{5.199}$$

In these equations matrices $\mathbf{T}^{\text{lon}}$ and $\mathbf{T}^{\text{lat}}$ were obtained by linearizing equations of motion transferring the components of the airspeed and angular velocity vectors from the canopy to store coordinate frame. These nonlinear motion transferring equations were derived in a manner similar to that of Sec. 5.2.4, but for the eight-DoF model. Ochi et al. [2009] used the linearized equations to develop a PID controller based on the measurements provided by AGU sitting atop the store (Fig. 5.43). The outputs to be controlled were chosen to be the descent rate $h_s = w_s - u_0\theta_s$ forward speed u_s and yaw angle ψ_s, so that the output matrix $\mathbf{C}$ in Eq. (5.175) was

$$\mathbf{C} = \begin{bmatrix} 0 & 1 & 0 & -u_0 & 0 & 0 \\ 1 & 0 & 0 & 0 & 0 & 0 \end{bmatrix} \tag{5.200}$$

for the longitudinal motion, and

$$\mathbf{C} = [0 \quad 0 \quad 0 \quad 0 \quad 1 \quad 0 \quad 0] \tag{5.201}$$

for the lateral-directional motion.

Linearized equations (5.182) and (5.183), (5.185) and (5.186), (5.189) and (5.190), and (5.194) and (5.195) included not only PADS dynamics, developed in Secs. 5.2.2–5.2.5, but linearized rotational kinematics given by Eq. (5.15) as well. Let us develop an analytical solution linearizing Eq. (5.15) with respect to some nominal flight. Adding small variations to all parameters of

$$
\begin{bmatrix} \dot{\phi} + \delta\dot{\phi} \\ \dot{\theta} + \delta\dot{\theta} \\ \dot{\psi} + \delta\dot{\psi} \end{bmatrix} = \begin{bmatrix} 1 & \sin(\phi + \delta\phi)\tan(\theta + \delta\theta) & \cos(\phi + \delta\phi)\tan(\theta + \delta\theta) \\ 0 & \cos(\phi + \delta\phi) & -\sin(\phi + \delta\phi) \\ 0 & \sin(\phi + \delta\phi)/\cos(\theta + \delta\theta) & \cos(\phi + \delta\phi)/\cos(\theta + \delta\theta) \end{bmatrix}
$$
$$
\times \begin{bmatrix} p + \delta p \\ q + \delta q \\ r + \delta r \end{bmatrix} = \Phi^\delta \begin{bmatrix} p + \delta p \\ q + \delta q \\ r + \delta r \end{bmatrix} \tag{5.202}
$$

and then subtracting Eq. (5.15) (with a matrix on the right-hand side denoted as Φ) results in

$$
\begin{bmatrix} \delta\dot{\phi} \\ \delta\dot{\theta} \\ \delta\dot{\psi} \end{bmatrix} = (\Phi^\delta - \Phi) \begin{bmatrix} p \\ q \\ r \end{bmatrix} + \Phi^\delta \begin{bmatrix} \delta p \\ \delta q \\ \delta r \end{bmatrix} \tag{5.203}
$$

Now we linearize Eq. (5.203) about $\phi_0 = 0$, $\theta_0 \neq 0$, $\psi_0 = 0$, $p_0 = 0$, $q_0 = 0$, and $r_0 = 0$. Neglecting the second-order terms in the matrix Φ^δ leads to

$$
\begin{bmatrix} \delta\dot{\phi} \\ \delta\dot{\theta} \\ \delta\dot{\psi} \end{bmatrix} = \begin{bmatrix} 1 & \delta\phi\tan(\theta_0) & \delta\theta + \tan(\theta_0) \\ 0 & 1 & -\delta\phi \\ 0 & \delta\phi/\cos(\theta_0) & 1/\cos(\theta_0) \end{bmatrix} \begin{bmatrix} \delta p \\ \delta q \\ \delta r \end{bmatrix} \tag{5.204}
$$

and finally to

$$
\begin{bmatrix} \delta\dot{\phi} \\ \delta\dot{\theta} \\ \delta\dot{\psi} \end{bmatrix} = \begin{bmatrix} 0 & 0 & 0 & 1 & 0 & \tan(\theta_0) \\ 0 & 0 & 0 & 0 & 1 & 0 \\ 0 & 0 & 0 & 0 & 0 & 1/\cos(\theta_0) \end{bmatrix} \begin{bmatrix} \delta\phi \\ \delta\theta \\ \delta\psi \\ \delta p \\ \delta q \\ \delta r \end{bmatrix} \tag{5.205}
$$

To conclude this section, let us develop a relatively simple linearized model that will further be used to develop a controller (Sec. 7.8.5) and demonstrate one of the parameter identification techniques (Sec. 11.5.6). As seen from Eq. (5.205), the pitch motion is decoupled from roll and yaw; hence, we can write

$$
\begin{bmatrix} \delta\dot{\phi} \\ \delta\dot{\psi} \end{bmatrix} = \begin{bmatrix} 0 & 0 & 1 & \tan(\theta_0) \\ 0 & 0 & 0 & 1/\cos(\theta_0) \end{bmatrix} \begin{bmatrix} \delta\phi \\ \delta\psi \\ \delta p \\ \delta r \end{bmatrix} \tag{5.206}
$$

The vector $\mathbf{x} = [\delta\phi, \delta\psi, \delta p, \delta r]^T$ on the right-hand side constitutes the state vector for the reduced-order linearized model we want to develop.

To develop equations for δp and δr, we turn to the fourth and sixth rows of Eq. (5.88). Assuming that the components of airspeed vector $\mathbf{V}_a$ are constant, we can write

$$\left[\mathbf{I} + \mathbf{I}'_{a.i.} - \mathbf{S}(\mathbf{r}_{BM})\mathbf{I}'_{a.m.}\mathbf{S}(\mathbf{r}_{BM})\right]\begin{bmatrix} \dot{p} \\ \dot{q} \\ \dot{q} \end{bmatrix} = \mathbf{B} \tag{5.207}$$

where

$$\mathbf{B} = \mathbf{M}_a - \left[\mathbf{S}(\boldsymbol{\omega} + \delta\boldsymbol{\omega})(\mathbf{I} + \mathbf{I}'_{a.i.}) - \mathbf{S}(\mathbf{r}_{BM})\mathbf{S}(\boldsymbol{\omega})\mathbf{I}'_{a.m.}\mathbf{S}(\mathbf{r}_{BM})\right]\begin{bmatrix} p \\ q \\ r \end{bmatrix}$$

$$- \mathbf{S}(\mathbf{r}_{BM})\mathbf{S}(\boldsymbol{\omega} + \delta\boldsymbol{\omega})\mathbf{I}'_{a.m.}\left(\begin{bmatrix} u \\ v \\ w \end{bmatrix} - {}^b_n\mathbf{RW}\right) \tag{5.208}$$

Writing Eqs. (5.207) and (5.208) in variations, subtracting the original Eq. (5.206), and neglecting the higher-order terms yield

$$\left[\mathbf{I} + \mathbf{I}'_{a.i.} - \mathbf{S}(\mathbf{r}_{BM})\mathbf{I}'_{a.m.}\mathbf{S}(\mathbf{r}_{BM})\right]\begin{bmatrix} \delta\dot{p} \\ \delta\dot{q} \\ \delta\dot{q} \end{bmatrix}$$

$$= \mathbf{M}_a^\delta - \mathbf{S}(\mathbf{r}_{BM})\mathbf{S}(\delta\boldsymbol{\omega})\mathbf{I}'_{a.m.}\begin{bmatrix} V_a\cos(\alpha) \\ 0 \\ V_a\sin(\alpha) \end{bmatrix} \tag{5.209}$$

Because of symmetry, the matrix on the left-hand side of Eq. (5.205) has the form

$$\hat{\mathbf{I}} = \begin{bmatrix} \hat{I}_{xx} & 0 & \hat{I}_{xz} \\ 0 & I_{yy} & 0 \\ \hat{I}_{xz} & 0 & \hat{I}_{zz} \end{bmatrix} \tag{5.210}$$

The inverse of this matrix will have this form as well:

$$\mathbf{J} = \frac{1}{\hat{I}_{xx}\hat{I}_{zz} - \hat{I}_{xx}^2}\begin{bmatrix} \hat{I}_{zz} & 0 & -\hat{I}_{xz} \\ 0 & (\hat{I}_{xx}\hat{I}_{zz} - \hat{I}_{xz}^2)I_{yy}^{-1} & 0 \\ -\hat{I}_{xz} & 0 & \hat{I}_{xx} \end{bmatrix} \tag{5.211}$$

Having in mind that the first and third rows of vector $\mathbf{M}_a^\delta$ expressed from Eq. (5.79) are

$$\mathbf{M}_a^\delta = QSb \begin{bmatrix} \dfrac{b}{2V_a}C_{lp}\delta p + \dfrac{b}{2V_a}C_{lr}\delta r + C_{l\delta_a}\delta\bar{\delta}_a \\ \cdots \\ \dfrac{b}{2V_a}C_{np}\delta p + \dfrac{b}{2V_a}C_{nr}\delta r + C_{n\delta_a}\delta\bar{\delta}_a \end{bmatrix} \qquad (5.212)$$

we can now combine Eq. (5.206) with the first and third rows of Eq. (5.209) to get

$$\begin{bmatrix} \delta\dot{\phi} \\ \delta\dot{\psi} \\ \delta\dot{p} \\ \delta\dot{r} \end{bmatrix} = \begin{bmatrix} 0 & 0 & 1 & \tan(\theta_0) \\ 0 & 0 & 0 & 1/\cos(\theta_0) \\ 0 & 0 & 0.25\rho V_a Sb^2 J_{xx}C_{lp} & 0.25\rho V_a Sb^2 J_{xz}C_{nr} \\ 0 & 0 & 0.25\rho V_a Sb^2 J_{xz}C_{lp} & 0.25\rho V_a Sb^2 J_{zz}C_{nr} \end{bmatrix} \begin{bmatrix} \delta\phi \\ \delta\psi \\ \delta p \\ \delta r \end{bmatrix}$$

$$+ \begin{bmatrix} 0 \\ 0 \\ QSb(J_{xx}C_{l\delta_a} + J_{xz}C_{n\delta_a}) \\ QSb(J_{xz}C_{l\delta_a} + J_{zz}C_{n\delta_a}) \end{bmatrix} \delta\bar{\delta}_a \qquad (5.213)$$

To eliminate the center-of-pressure location from the dynamic equations while maintaining the PADS tendency to glide with no roll during neutral control, Slegers and Costello [2005] introduced one additional term to the first row of the aerodynamic moment vector of Eq. (5.212), $C_{l\phi}\delta\phi$. As a result, Eq. (5.213) takes the final form:

$$\begin{bmatrix} \dot{\phi} \\ \dot{\psi} \\ \dot{p} \\ \dot{r} \end{bmatrix} = \begin{bmatrix} 0 & 0 & 1 & \tan(\theta_0) \\ 0 & 0 & 0 & 1/\cos(\theta_0) \\ QSbJ_{xx}C_{l\phi} & 0 & 0.25\rho V_a Sb^2 J_{xx}C_{lp} & 0.25\rho V_a Sb^2 J_{xz}C_{nr} \\ QSbJ_{xz}C_{l\phi} & 0 & 0.25\rho V_a Sb^2 J_{xz}C_{lp} & 0.25\rho V_a Sb^2 J_{zz}C_{nr} \end{bmatrix} \begin{bmatrix} \phi \\ \psi \\ p \\ r \end{bmatrix}$$

$$+ \begin{bmatrix} 0 \\ 0 \\ QSb(J_{xx}C_{l\delta_a} + J_{xz}C_{n\delta_a}) \\ QSb(J_{xz}C_{l\delta_a} + J_{zz}C_{n\delta_a}) \end{bmatrix} \bar{\delta}_a \qquad (5.214)$$

(As before, we dropped the deltas for simplicity.)

TABLE 5.9   SUMMARY OF EQUATIONS FOR A THREE-DoF MODEL

	Kinematics	Dynamics
Translational motion of $\{b\}$ with respect to $\{I\}$	x, y [Eq. (5.29)]	—
Rotational motion of $\{b\}$ with respect to $\{I\}$ expressed in $\{b\}$	—	ψ [Eq. (5.38)]

5.5   SUMMARY AND EXTENSION TO POWERED SYSTEMS

This section summarizes the models developed in this chapter and shows how to extend them for other systems involving powered and unpowered payload of a complex shape.

5.5.1   SUMMARY OF THE DEVELOPED MODELS

Chapter 5 started from presenting two simple models that describe two translational/one rotational DoF (Table 5.9) and two translational/two rotational DoF (Table 5.10).

It was shown that for this simple analysis PADS motion in the vertical channel can be either completely decoupled and defined by a simple Eq. (5.35) or coupled using the experimental so that the descent rate depends on the yaw rate Eqs. (5.42) and (5.171).

The higher-fidelity models included the six-DoF model (Table 5.11), the nine-DoF model (Table 5.12), the seven-DoF model (Table 5.13), and the eight-DoF model (Table 5.14).

Linearized models of different complexity are presented in Sec. 5.4.

5.5.2   INCORPORATING COMPLEX-SHAPE (POWERED) PAYLOAD

To complete this chapter, the high-fidelity models developed in Sec. 5.2 for unpowered PADS can easily be extended to the powered systems. Examples of such systems include Snowgoose PADS mentioned in Chapter 1 (see Fig. 1.10) and

TABLE 5.10   SUMMARY OF EQUATIONS FOR A FOUR-DoF MODEL

	Kinematics	Dynamics
Translational motion of $\{b\}$ with respect to $\{I\}$	x, y [Eq. (5.28)]	u, w [Eq. (5.46)]
Rotational motion of $\{b\}$ with respect to $\{I\}$ expressed in $\{b\}$	ψ [Eq. (5.49)]	ϕ [Eq. (5.45)]

TABLE 5.11 SUMMARY OF EQUATIONS FOR THE SIX-DoF MODEL

	Kinematics	Dynamics
Translational motion of $\{b\}$ with respect to $\{I\}$	x, y, z [Eq. (5.28)]	u, v, w [Eq. (5.88)]
Rotational motion of $\{b\}$ with respect to $\{I\}$ expressed in $\{b\}$	ϕ, θ, ψ [Eq. (5.15)]	p, q, r [Eq. (5.88)]

PPG mentioned in Sec. 5.4 (Fig. 5.43). Figure 5.44 shows a zoom-in view of powered payload of Snowgoose PADS (Fig. 5.44a) and two PPG, or Paramotoring, where a pilot wears a motor on his/her back (Fig. 5.44b) or sits in a trike (Fig. 5.44c). In all cases a motor provides enough thrust to take off and stay in the air longer.

Another type of system, utilizing ram-air parachutes, is a parafoil-aircraft system (Fig. 5.45). Parafoil can be used for just a recovery, that is, without power (Fig. 5.45a), or for an extended-duration slow flight with a running engine (Fig. 5.45b). In the latter case aircraft propeller adds power allowing to climb, cruise, and descend. It can be controlled in a usual manner using aircraft controls. This scheme can also eliminate the need for a runway and add to system's effectiveness. It shortens landing distance and speed to allow for example unmanned aerial vehicle (UAV) shipboard recovery (Fig. 5.46).

There are several efforts dedicated to derive equations of motions for the aforementioned systems from scratch (e.g., [Hur 2005; Watanabe and Ochi 2008] for PPG, [Redelinghuys 2007; Wise 2006] for an aircraft-parafoil system), but, in fact, the six-/seven-/eight-/nine-DoF models developed in Sec. 5.2 can be easily adapted to include a (powered) payload of a complex shape. Besides the fact that the store inertia matrix I_s will not be descried by Eq. (5.110) but

TABLE 5.12 SUMMARY OF EQUATIONS FOR THE NINE-DoF MODEL

	Kinematics	Dynamics
Translational motion of the connection point C with respect to $\{I\}$	x_c, y_c, z_c [Eq. (5.109)]	
Rotational motion of $\{b\}$ with respect to $\{I\}$ expressed in $\{b\}$	ϕ, θ, ψ [Eq. (5.15)]	$u_c, v_c, w_c, p, q, r, p_s, q_s, r_s$ [Eq. (5.121)]
Rotational motion of $\{s\}$ with respect to $\{I\}$ expressed in $\{s\}$	ϕ_s, θ_s, ψ_s [Eq. (5.130)]	

TABLE 5.13   SUMMARY OF EQUATIONS FOR THE SEVEN-DoF MODEL

	Kinematics	Dynamics
Translational motion of the connection point C with respect to $\{I\}$	x_c, y_c, z_c [Eq. (5.109)]	
Rotational motion of $\{b\}$ with respect to $\{I\}$ expressed in $\{b\}$	ϕ, θ, ψ [Eq. (5.15)]	$u_c, v_c, w_c, p, q, r, r_s$ [Eq. (5.154)]
Rotational motion of $\{s\}$ with respect to $\{I\}$ expressed in $\{s\}$	p_s, q_s, ψ_s [Eq. (5.137)]	

TABLE 5.14   SUMMARY OF EQUATIONS FOR THE EIGHT-DoF MODEL

	Kinematics	Dynamics
Translational motion of the connection point C with respect to $\{I\}$	x_c, y_c, z_c [Eq. (5.109)]	
Rotational motion of $\{b\}$ with respect to $\{I\}$ expressed in $\{b\}$	ϕ, θ, ψ [Eq. (5.15)]	$u_c, v_c, w_c, p, q, r, q_s, r_s$ [Eq. (5.159)]
Rotational motion of $\{s\}$ with respect to $\{I\}$ expressed in $\{s\}$	p_s, θ_s, ψ_s [Eq. (5.144)]	

a)   b)   c)

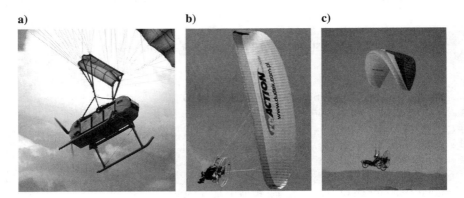

Fig. 5.44   Examples of powered systems: a) SnowGoose PADS, b) PPG with a wearable motor, and c) PPG with a trike [SnowGoose UAV 2014; Aerolight 2014; Power Pack 2014].

a)   b)

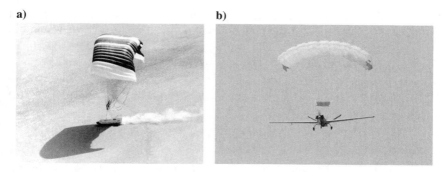

Fig. 5.45   Parafoil systems: a) X-38 recovery system and b) Australian Army tactical UAV I-View system [X-38 2014, I-View 2004].

Fig. 5.46   Raytheon's Air & Sea Cargo Transport concept (TEMP) [TEMP 2014].

rather have a more sophisticated form involving one more off-main-diagonal element I_{xz}, the only changes in the model will involve more sophisticated expressions for the aerodynamic force and moment acting on the store compared to that of Eq. (5.148) and $\mathbf{M}_a^s = 0$ assumed for a payload represented by a solid rectangular cuboid or cylinder (Fig. 5.25).

Depending on a specific payload configuration (powered/unpowered, simple/complex shape), the aerodynamic force $\mathbf{F}_a^s$ will be described by

$$\mathbf{F}_a^s = Q^s S_{s_w}^s \mathbf{R} \begin{bmatrix} C_{D0}^s + C_{D\alpha^2}^s \alpha_s^2 + C_{D\delta_e}^s \overline{\delta}_e^s + C_{D\delta_T} \overline{\delta}_T \\ C_{Y\beta}^s \beta_s + C_{Y\delta_a}^s \overline{\delta}_a^s + C_{D\delta_r}^s \overline{\delta}_r^s \\ C_{L0}^s + C_{L\alpha}^s \alpha_s + C_{L\delta_e}^s \overline{\delta}_e^s + C_{Y\delta_T} \overline{\delta}_T \end{bmatrix} \qquad (5.215)$$

where $\overline{\delta}_e^s$ is the store elevator deflection, $\overline{\delta}_a^s$ is the store aileron deflection, $\overline{\delta}_r^s$ is the store rudder deflection (in the case of any aircraft controls are used), $\overline{\delta}_T$ is the engine thrust (for powered systems), and all coefficients pertain to the payload aerodynamics. Similarly, for the now nonzero moment vector $\mathbf{M}_a^s$, we would use

$$\mathbf{M}_a^s = Q^s S_{s_w}^s \mathbf{R} \begin{bmatrix} b_s \left(C_{l\beta}^s \beta_s + \dfrac{b_s}{2V_a^s} C_{lp}^s p_s + \dfrac{b_s}{2V_a^s} C_{lr}^s r_s + C_{l\delta_a}^s \overline{\delta}_a^s + C_{l\delta_r}^s \overline{\delta}_r^s \right) \\ c_s \left(C_{m0}^s + C_{m\alpha}^s \alpha_s + \dfrac{c_s}{2V_a^s} C_{mq}^s q_s + C_{m\delta_e}^s \overline{\delta}_e^s + C_{m\delta_T} \overline{\delta}_T \right) \\ b_s \left(C_{n\beta}^s \beta_s + \dfrac{b_s}{2V_a^s} C_{np}^s p_s + \dfrac{b_s}{2V_a^s} C_{nr}^s r_s + C_{n\delta_a}^s \overline{\delta}_a^s + C_{n\delta_r}^s \overline{\delta}_r^s \right) \end{bmatrix} \qquad (5.216)$$

For the six-DoF model, similar terms should be added to Eqs. (5.74) and (5.79).

CHAPTER 6

Stability and Steady-State Performance

Oleg Yakimenko[*]
Naval Postgraduate School, Monterey, California

Thomas Jann[†]
Institute of Flight Systems of DLR, Braunschweig, Germany

Gil Iosilevskii[‡]
Technion—Israel Institute of Technology, Haifa, Israel

Peter Crimi[§]
Andover Applied Sciences, Inc., West Boxford, Massachusetts

Glen Brown[¶]
Vorticity US, Santa Cruz, California

Following the derivation of equations of motions in Chapter 5, the standard procedure is to study system's stability and controllability. Obviously, this subject is of a major importance while designing a new canopy and considering the overall mass-geometry properties of PADS, but once the cargo system is designed it seems like there is less interest in studying its stability properties in a traditional manner. (The number of papers published on this subject is much less compared to ones developing ADS models or GNC algorithms for PADS.) The main reason is that inherently stable PADS usually assumes no stability augmentation system, must account for the winds rather than consider them as disturbance, and has a very slow dynamics anyway. Nevertheless, for completeness this chapter aims at briefly overviewing parafoil-payload system stability issues with the emphasis on the differences as compared to a common aircraft. Based on earlier publications, it is organized as follows. Section 6.1 discusses an eigenvalue obtained for a linearized six-DoF model of typical PADS numerically. Section 6.2 proceeds with establishing the c.g. limits ensuring a longitudinal static stability of a cargo PADS. Section 6.3 employs the simpler models to be able to have a detailed

---

[*]Professor.
[†]Research Scientist.
[‡]Associate Professor.
[§]President (retired).
[¶]Director of Engineering.

close look at lateral-directional stability and steady turn performance emphasizing the scaling effects. This latter subject has an immediate impact when developing GNC algorithms for different-eight PADS, and so Sec. 6.4 offers yet another look at this issue.

6.1 STABILITY ANALYSIS OF THE LINEARIZED MODEL

Stability (and controllability) analysis of a linearized PADS model involves reducing the nonlinear mathematical models of Chapter 5 to the state-space form of Eqs. (5.174) and (5.175). Then, eigenvectors of state matrix **A** define dynamic properties of the linearized system. For example, linearizing the six-DoF model of Eq. (5.91) complemented with rotational kinematics of Eq. (5.15) results in the linearized differential equation with the nine-state vector $\mathbf{x} = [u, v, w, p, q, r, \phi, \theta, \psi]^T$ and two-state input vector $\mathbf{u} = [\delta_s, \delta_a]^T$ [Eqs. (5.176) and (5.177)]. The eigenvalues of matrix **A** estimated for the steady glide of ALEX PADS (see Fig. 1.31) flying at 500 m (1,640 ft) AGL and brake setting to 0 with a forward speed of about 13 m/s (25.3 kt) are shown in Fig. 6.1a. (Most examples in this section are taken from [Jann 2004]). For qualitative comparison, Fig. 6.2b shows eigenvalues of a similar model (with the same nine states)

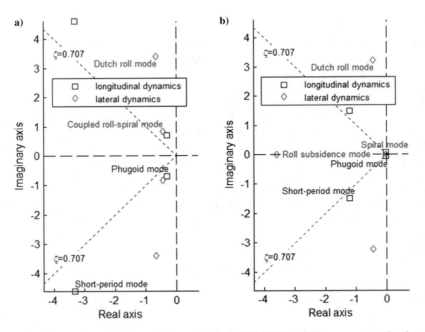

Fig. 6.1   Eigenvalues of a linearized model of a) ALEX PADS and b) a representative jet aircraft.

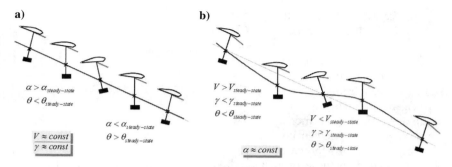

Fig. 6.2   Representation of the a) short-period and b) phugoid modes.

estimated for a jet aircraft flying straight and level at about 150 m/s (290 kt). [For simplicity, Fig. 6.1 does not show one eigenvalue at [0;0] corresponding to a pure integrator resulting from linearizing the last equation of Eq. (5.15) as seen from the last equation in Eq. (5.206).] Both systems shown in Fig. 6.1 are stable (the real parts of all eigenvalues are negative) and feature similar modes in both longitudinal and lateral-directional channels. ALEX PADS eigenvalues along with period and damping ratio of oscillations are shown in Table 6.1.

In the longitudinal motion, a parafoil-based ADS consists of two distinct oscillations, the well-known short-period and phugoid (long-period) modes, characterized by two pairs of complex-conjugate poles. The schematic representations of these two modes is given in Figs. 6.2a and 6.2b, respectively.

As shown in Fig. 6.2a, the short-period motion involves rapid changes in the angle of attack α and pitch attitude θ at roughly constant airspeed V_a and flight-path angle γ. Similar to an aircraft, this mode is highly damped for PADS as well. The short-period frequency is strongly related to the PADS' static margin, which will be further investigated in Sec. 6.2.1. It is proportional to $(C_{m\alpha}/C_L)^{0.5}$.

The long-period of phugoid mode (Fig. 6.2b) involves a trade between kinetic and potential energy. In this mode PADS glides at nearly constant angle of attack, pitches up and slows, then dives, losing altitude at a higher rate while picking up speed. The period of this motion is independent of PADS character-istics and altitude and depends only on the trimmed airspeed and as such is of

TABLE 6.1   EIGENVALUES OF THE ALEX PADS (LINEARIZED ABOUT A STEADY GLIDE)

Motion Mode	Eigenvalues	Period T, s	Damping Ratio ζ
Short-period mode	$-3.33 \pm i\,4.63$	1.1	0.58
Phugoid mode	$-0.30 \pm i\,0.69$	8.4	0.40
Dutch roll mode	$-0.67 \pm i\,3.40$	1.8	0.19
Coupled roll-spiral mode	$-0.44 \pm i\,0.83$	6.7	0.47

Fig. 6.3   Illustration of Dutch-roll mode.

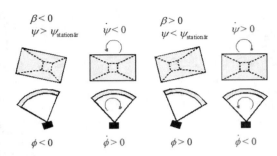

the order of magnitude lower for PADS compared to an aircraft. (As seen from Fig. 6.1, two eigenvalues corresponding to a PADS phugoid mode are farther to the left from the origin of the complex plane compared to those of an aircraft.) The damping ratio is inversely proportional to the trimmed airspeed squared.

As opposed to an aircraft with the lateral-directional motion usually described by three dynamic modes which include Dutch roll mode, roll subsidence mode, and spiral mode (Fig. 6.1b), the dynamics of ALEX PADS features two distinct oscillations, one of which is the same Dutch roll mode and another is a coupled roll-spiral mode (Fig. 6.1a and Table 6.1) [Kempel 1971]. The coupled oscillatory roll-spiral mode is mainly caused by the arched shape of canopy and combines the fast asymptotic roll subsidence and slow spiral modes of aircraft. (As will be shown later in this section and also in Sec. 6.3.1, a parafoil-based PADS can exhibit the three lateral-directional modes of airplane as well.)

The Dutch roll mode (Fig. 6.3) is a coupled roll and yaw motion that is not sufficiently damped. Figure 6.1 features almost exactly the same parameters of this mode for the PADS and an aircraft. High directional stability $C_{n\beta}$ tends to stabilize the Dutch roll mode while reducing the stability of the spiral mode. Conversely, large effective dihedral (rolling moment due to sideslip $C_{l\beta}$) stabilizes the spiral mode while destabilizing the Dutch roll motion (to be discussed later in this section).

The coupled roll-spiral mode (Fig. 6.4) involves roll oscillations accompanied by the yaw-angle oscillations of about 60% amplitude of that of roll oscillations. In contrast to the Dutch roll mode, the yaw-angle motion exhibits about 90 deg lead compared to the roll motion. The sideslip angle remains practically unchanged. The frequency of the coupled roll-spiral mode depends on the distance between canopy c.p. and c.g. $(g/l)^{0.5}$.

As a typical example, Tables 6.2 and 6.3 show amplitudes and phase angles for all modes of motion as applied to ALEX PADS simulation (with α, θ, β, and ϕ being a primary modes with

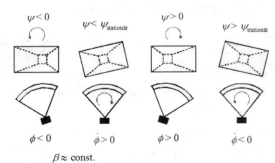

Fig. 6.4   Illustration of a coupled roll-spiral mode.

TABLE 6.2 AMPLITUDE AND PHASE ANGLE OF PARAMETERS OF LONGITUDINAL MOTION

	Short-Period Motion			Phugoid (Long-Period) Motion	
	Amplitude	Phase Angle		Amplitude	Phase Angle
u	0.5 m/s	−175 deg	u	2.1 m/s	100 deg
w	2.6 m/s	−0.3 deg	w	0.5 m/s	97 deg
q	50 deg/s (0.87 rad/s)	88 deg	q	7.5 deg/s (0.13 rad/s)	113 deg
θ	8 deg (0.15 rad)	−38 deg	θ	10 deg (0.18 rad)	0 deg
V	0.5 m/s	−5 deg	V	2.1 m/s	100 deg
α	10 deg (0.18 rad)	0 deg	α	0.9 deg (0.016 rad)	−74 deg

an amplitude of 10 deg and a phase angle of 0 deg, respectively), and Fig. 6.5 features a graphical representation of six-DoF model parameters involved into these different modes of motion. Note that in the model, angular values are expressed in radians, and eigenvectors are normalized to the length of 1.

Using a standard root-locus technique applied to a linearized PADS model, one can explore the influence of different parameters onto location of state

TABLE 6.3 AMPLITUDE AND PHASE ANGLE OF PARAMETERS OF LATERAL-DIRECTIONAL MOTION

	Dutch Roll Motion			Coupled Roll-Spiral Motion	
	Amplitude	Phase Angle		Amplitude	Phase Angle
v	2.6 m/s	0 deg	v	0.1 m/s	3 deg
p	11 deg/s (0.2 rad/s)	95 deg	p	9.4 deg/s (0.16 rad/s)	115 deg
r	33 deg/s (0.57 rad/s)	278 deg	r	5.8 deg/s (0.1 rad/s)	26 deg
ϕ	4.1 deg (0.07 rad)	−6 deg	ϕ	10 deg (0.18 rad)	0 deg
ψ	9.5 deg (0.17 rad)	177 deg	ψ	6.2 deg (0.1 rad)	268 deg
V	0 m/s	0 deg	V	0 m/s	3 deg
β	10 deg (0.18 rad)	0 deg	β	0.2 deg (0.004 rad)	3 deg

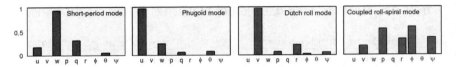

Fig. 6.5   Magnitudes of eigenvectors for the different modes of motion.

matrix **A** eigenvalues (the dynamic system poles) numerically. For example, Fig. 6.6 shows root locus while varying the anhedral angle.

Figure 6.6 was obtained for a fixed canopy (with a fixed wing span b), and because the anhedral angle depends on the ratio of span b to line length R [Eqs. (2.15) and (4.1)], the change of anhedral angle was achieved by varying R between 16.6 and 2.8 m (54.5 and 9.2 ft). Varying line length R (anhedral angle ε_b) affected PADS aerodynamic parameters (as described in Sec. 4.1.2). As shown in Fig. 6.6, increasing canopy curvature (shortening suspension line length) results in increase of the frequency of oscillations of the angle of attack (short-period mode) and sideslip angle (Dutch roll mode). It also leads to decrease of damping of these two modes as well as a coupled roll-spiral mode (roll angle). Hence, increasing canopy's curvature has a destabilizing effect. A very small canopy curvature (anhedral angle) results in the coupled roll-spiral mode split in to two traditional aperiodic modes, roll subsidence mode and spiral mode.

Figure 6.7 shows root locus while changing the rigging angle μ. In this case, the aerodynamic coefficients stay the same, but the rigging angle has an effect through the rotation matrix ${}^p_b\mathbf{R}$ of Eq. (5.83). As seen, increasing a rigging angle has a stabilizing effect because it increases damping of all modes. The frequencies

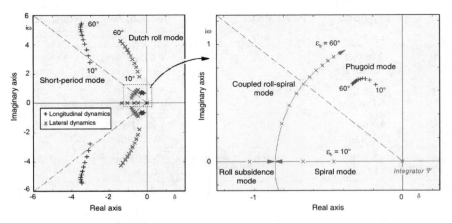

Fig. 6.6   ALEX PADS model root locus while varying the anhedral angle $\varepsilon_b = 10 \ldots 60$ deg.

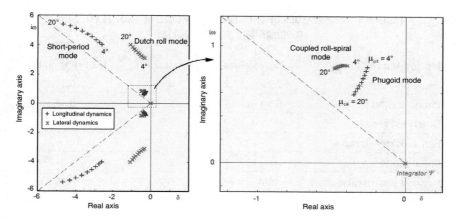

Fig. 6.7   ALEX PADS model root locus while varying the rigging angle $\mu = 4 \ldots 20$ deg.

of angle of attack (short-period mode) and sideslip angle (Dutch-roll mode) oscillations increase as well, and the frequency of the phugoid decreases (while having
almost no effect on the coupled roll-spiral mode).

Finally, Fig. 6.8 shows ALEX PADS model root locus when accounting for
the apparent mass and inertia tensor terms as described in Sec. 5.1.4 (although
the previous simulations did not take them into account). In this particular
example, increasing the effect of these terms from 0 (disregarding these terms)
to 100% (fully accounting for these terms) leads to a slightly increased damping
in the short-period mode (accompanied by slight decrease of the natural

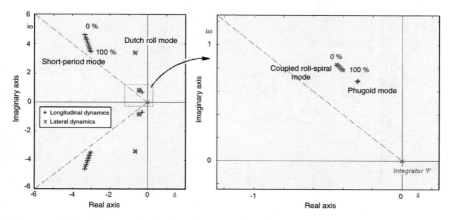

**Fig. 6.8   ALEX PADS model root locus while varying the effect of apparent mass and
moments of inertia.**

**TABLE 6.4   EIGENVALUES OF THE SEVEN-DOF MODEL OF SNOWFLAKE PADS
[GORMAN AND SLEGERS 2011B]**

Motion Mode	Eigenvalues	Period T, s	Damping Ratio ζ
Short-period mode	$-2.39 \pm i\,6.72$	0.88	0.34
Phugoid mode	$-0.59 \pm i\,1.12$	5	0.47
Dutch roll mode	$-0.55 \pm i\,5.22$	1.2	0.1
Roll subsidence mode	-4.32	0.23*	—
Spiral mode	-0.37	2.7*	—
Payload yaw oscillations	$-0.44 \pm i\,6.9$	0.9	0.062

*Time constant.

frequency), but seems to have almost no effect on other modes. Of course, for larger systems the effect might be different.

For the sake of comparison, Table 6.4 shows the eigenvalues of the state matrix **A** computed for the longitudinal and lateral-directional motion based off of the seven-DoF model of Snowflake PADS (see Fig.1.30a) described by Eqs. (5.185)

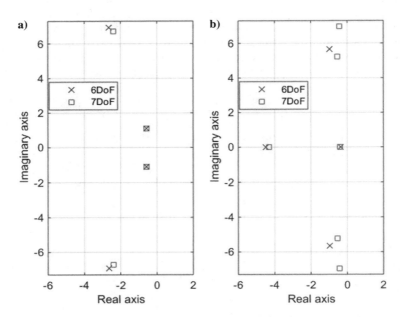

Fig. 6.9   Eigenvalues of a linearized model of Snowflake PADS in a) longitudinal and b) lateral-directional motion.

and (5.186). (For simplicity, the pole at zero is not shown.) As seen, the parameters of the longitudinal motion and Dutch-roll mode are of the same order of magnitude as those of a much bigger ALEX PADS (Table 6.1). Snowflake, however, does not exhibit a coupled oscillatory roll spiral, but has the traditional (for aircraft) roll subsidence and spiral modes. Table 6.4 also shows a payload yawing mode.

Eigenvalues of Table 6.4 are also shown in Fig. 6.9 along with the eigenvalues based off of the six-DoF model of Snowflake PADS described by Eqs. (5.182) and (5.183).

6.2 LONGITUDINAL STATIC STABILITY ANALYSIS

In the context of longitudinal control, two flying qualities of a gliding parachute seem peculiar. One is that for a longitudinally stable parachute, the lift coefficient increases with TE deflected downward symmetrically. The other is that, judging from the top speed data of currently flying designs, there seems to exist a minimal lift coefficient (roughly, 0.5) at which a conventional gliding parachute can be possibly trimmed, regardless of the particularities of its design. In this section, a standard longitudinal static stability analysis is used to show that the minimal lift coefficient limit does exist; moreover, it is a consequence of requiring the lift coefficient to be an increasing function of symmetric TE deflection. Specifically, it is shown that, in apparent contrast with a conventional airplane, a loss of longitudinal static stability and a loss of control authority each impose a limit on the most forward c.g. position, and, concurrently, on the minimal lift coefficient possibly attainable at trim. It is also shown that, with a proper design, requiring the lift coefficient to be an increasing function of symmetric TE deflection is sufficient to ensure longitudinal static stability; in this case, both the forward c.g. limit and the minimal lift coefficient limit are associated with a loss of control authority. Under several simplifying assumptions—linear lift-curve slope, parabolic drag polar, fixed-shape canopy, fixed-lengths lines, and plain flaps for elevens—the following analysis suggests a very simple expression for the minimal lift coefficient limit. The value of this limit depends, mainly, on the relative flap chord and on the lines-to-chord lengths ratio; specifically, it decreases with an increase in either parameter. A sample calculation of the minimal lift coefficient limit for a typical gliding parachute yields, indeed, about 0.5.

6.2.1 TRIM AND STABILITY

Consider a gliding parachute in a symmetric, unaccelerated flight. Following conventional definitions of aerodynamic coefficients, the projected wing area and the mean aerodynamic chord will serve as the respective references. Drag coefficients C_D^c, C_D^l, and C_D^s represent contributions of canopy, suspension lines, and payload (store), respectively; C_L and C_{m0}^c are the lift and pitching-moment coefficients of

Fig. 6.10   Notation and the reference frame.

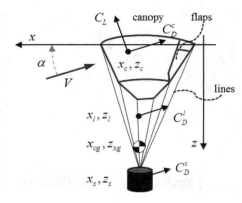

canopy about its aerodynamic center. The right-handed orthogonal reference frame will have its x, y, and z axes pointing forward, right, and downward, respectively. In particular, the x axis will connect the TE and leading edge of canopy's midsection, and the y axis will be parallel to the line connecting its left and right tips. Using the mean aerodynamic chord as a unit of length, $[x_c, y_c, z_c]^T$, $[x_{cg}, y_{cg}, z_{cg}]^T$, $[x_l, y_l, z_l]^T$, and $[x_s, y_s, z_s]^T$ will be the respective dimensionless coordinates of the wing's aerodynamic center, PADS' c.g., suspension lines' c.p., and payload's c.p., respectively (see Fig. 6.10), and α will be the angle of attack, measured between the direction of the flow and the x_b axis.

The pitching-moment coefficient of the parachute about its c.g. is

$$C_m = C_{m0}^c + \left[C_L \cos(\alpha) + C_D^c \sin(\alpha)\right](x_c - x_{cg})$$
$$+ \left[C_L \sin(\alpha) - C_D^c \cos(\alpha)\right](z_c - z_{cg}) + C_D^l \sin(\alpha)(x_l - x_{cg})$$
$$- C_D^l \cos(\alpha)(z_l - z_{cg}) + C_D^s \sin(\alpha)(x_s - x_{cg}) - C_D^s \cos(\alpha)(z_s - z_{cg}) \qquad (6.1)$$

[compare Eq. (2.31)]. Assuming that all drag coefficients are of the order αC_L and that $\alpha \ll 1$, Eq. (6.1) reduces to

$$C_m \approx C_{m0}^c + C_L(x_c - x_{cg}) + (C_L\alpha - C_D^c)(z_c - z_{cg}) - C_D^l(z_l - z_{cg}) - C_D^s(z_s - z_{cg}) \qquad (6.2)$$

It will be assumed that the lift coefficient is linear with the angle of attack

$$\alpha = (C_L - C_{L0})/C_{L\alpha} \qquad (6.3)$$

where $C_{L0} = -C_{L\alpha}\alpha_0$ is the lift coefficient at zero angle of attack, α_0 is the angle of attack at zero lift, and $C_{L\alpha}$ is the lift-slope coefficient; for an arched wing it can be roughly approximated by Eq. (4.38). It also will be assumed that the drag coefficients of the lines and payload are independent of the lift coefficient whereas the drag coefficient of the wing is given by the parabolic polar [Perkins and Hage 1957]

$$C_D^c = C_{D0}^c + KC_L^2 \qquad (6.4)$$

[compare Eq. (2.8)].

Using Eqs. (6.3) and (6.4), Eq. (6.2) becomes

$$C_m = C_{m0} + C_L \frac{1}{C_{L\alpha}} \left[C_{L\alpha}(x_c - x_{cg}) - C_{L0}(z_c - z_{cg}) \right] + C_L^2 \frac{1}{C_{L\alpha}} (1 - C_{L\alpha}K)(z_c - z_{cg})$$

(6.5)

where

$$C_{m0} = C_{m0}^c - C_{D0}^c(z_c - z_{cg}) - C_D^l(z_l - z_{cg}) - C_D^s(z_s - z_{cg})$$ (6.6)

A necessary condition for equilibrium (trim) is

$$C_m = 0$$ (6.7)

With Eq. (6.5), Eq. (6.7) furnishes a quadratic equation

$$C_{L\alpha}C_{m0} + C_L \left[C_{L\alpha}(x_c - x_{cg}) - C_{L0}(z_c - z_{cg}) \right] + C_L^2(1 - C_{L\alpha}K)(z_c - z_{cg}) = 0$$ (6.8)

for the lift coefficient C_L. The two obvious solutions of Eq. (6.8) are

$$C_{L\pm} = \frac{-C_{L\alpha}(x_c - x_{cg}) + C_{L0}(z_c - z_{cg})}{2(1 - C_{L\alpha}K)(z_c - z_{cg})}$$

$$\pm \frac{\sqrt{\left[C_{L\alpha}(x_c - x_{cg}) - C_{L0}(z_c - z_{cg}) \right]^2 - 4(1 - C_{L\alpha}K)(z_c - z_{cg})C_{L\alpha}C_{m0}}}{2(1 - C_{L\alpha}K)(z_c - z_{cg})}$$ (6.9)

Among these two solutions, if they exist, we will choose the one at which the parachute is stable, that is, we choose the solution for which

$$\partial C_m / \partial C_L < 0$$ (6.10)

If the canopy and the lines do not deform with the change in the lift coefficient, then from Eq. (6.5) we have

$$\frac{\partial C_m}{\partial C_L} = x_c - x_{cg} - C_{L0} \frac{1}{C_{L\alpha}} (z_c - z_{cg}) + C_L \frac{2}{C_{L\alpha}} (1 - C_{L\alpha}K)(z_c - z_{cg})$$ (6.11)

At $C_L = C_{L\pm}$, it yields

$$\frac{\partial C_m}{\partial C_L} = \pm \sqrt{\left[x_w - x_{cg} - C_{L0} \frac{1}{C_{L\alpha}} (z_c - z_{cg}) \right]^2 - \frac{4}{C_{L\alpha}} (1 - C_{L\alpha}K)(z_c - z_{cg})C_{m0}}$$

(6.12)

To satisfy Eq. (6.10), the sign in Eq. (6.12), and hence in Eq. (6.9), has to be negative; consequently, the trim is stable at

$$C_L = C_{L-}$$ (6.13)

Because C_{m0} and C_{L0} vary with (generalized) symmetric TE deflection, δ_s, C_{L-} varies with both x_{cg} and δ_s. When needed, this notion will be emphasized by writing $C_{m0}(\delta_s)$, $C_{L0}(\delta_s)$, and $C_{L-}(x_{cg}, \delta_s)$; in most cases, however, the arguments x_{cg} and δ_s will be tacitly omitted.

6.2.2 FORWARD CENTER-OF-GRAVITY LIMIT

The solution C_{L-}, as defined in Eq. (6.9), exists only if

$$\left[(x_w - x_{cg})C_{L\alpha} - (z_c - z_{cg})C_{L0}\right]^2 \geq 4(1 - C_{L\alpha}K)(z_c - z_{cg})C_{L\alpha}C_{m0} \qquad (6.14)$$

The left-hand side of Eq. (6.14) is nonnegative whereas the right-hand side is either positive or negative, depending on signs of the respective multipliers. Specifically, the term $z_w - z_{cg}$ is negative by the choice of the frame of reference (Fig. 6.10), and $C_{L\alpha}$ is positive as long as the canopy does not stall whereas $1 - C_{L\alpha}K$ is invariably positive [Iosilevskii 1995] (see Appendix F). Hence, if $C_{m0} \geq 0$, then Eq. (6.14) holds unconditionally; that is, no direct limitations are imposed on the longitudinal c.g. position. If, on the other hand, $C_{m0} < 0$, then there exists

$$x_{\pm} = x_c - \frac{1}{C_{L\alpha}}C_{L0}(z_c - z_{cg}) \pm \frac{1}{C_{L\alpha}}\sqrt{4(1 - C_{L\alpha}K)(z_c - z_{cg})C_{L\alpha}C_{m0}} \qquad (6.15)$$

such that Eq. (6.14) holds if either

$$x_{cg} \leq x_- \qquad (6.16)$$

or

$$x_{cg} \geq x_+ \qquad (6.17)$$

According to Eq. (6.9), when $x_{cg} = x_+$, the absolute value of C_{L-} is

$$C_{L1} = \sqrt{\frac{C_{L\alpha}C_{m0}}{(1 - C_{L\alpha}K)(z_c - z_{cg})}} \qquad (6.18)$$

Hence,

$$C_{L-} \geq C_{L1} \qquad \text{for } x_{cg} \leq x_- \qquad (6.19)$$

$$C_{L-} \leq -C_{L1} \qquad \text{for } x_{cg} \geq x_+ \qquad (6.20)$$

Because negative lift coefficients are irrelevant, Eq. (6.17) is ruled out by Eq. (6.20); whence, according to Eqs. (6.16) and (6.19), c.g. should be located behind x_-, and, there exists a lower bound on the lift coefficient possibly attainable at trim. Because both x_- and C_{L1} are implicit functions of symmetric TE deflection δ_s, this notion will further be emphasized by writing $x_-(\delta_s)$ and $C_{L1}(\delta_s)$.

According to Eqs. (6.11) and (6.15), a gliding parachute with $C_{m0} < 0$ is neutrally stable at $x_{cg} = x_-$. Because it is supposed to be stable at $x_{cg} < x_-$, moving c.g. backwards seems to have a stabilizing effect. This can be formally shown by substituting Eq. (6.15) in Eq. (6.12), which yields

$$\frac{\partial C_m}{\partial C_L} = -\sqrt{(x_- - x_{cg})\left[x_- - x_{cg} + \frac{2}{C_{L\alpha}}\sqrt{4(1 - C_{L\alpha}K)(z_c - z_{cg})C_{L\alpha}C_{m0}}\right]} \qquad (6.21)$$

An increase in $x_- - x_{cg}$ indeed makes the derivative $\partial C_m/\partial C_L$ more negative.

Using the typical values for all pertinent parameters appearing on the right-hand side of Eq. (6.6) [Iosilevskii 1995], one can see that the sum of drag contributions to C_{m0} is positive. Hence, at least in principle, by flattening the profile, it is possible to design a parachute with $C_{m0} > 0$. As just mentioned, it will have no apparent limitations on its longitudinal c.g. position, and therefore it can be designed to fly at any lift coefficient below stall [see Eq. (6.9)]. At the same time, it seems improbable that one can design a conventional TE-controlled gliding parachute in such a way that it will have an adequate range of accessible lift coefficients on the one hand and positive C_{m0} for all possible TE deflections on the other. But in order to trim the parachute with negative C_{m0}, the most forward c.g. position should be limited by Eq. (6.16). Thus, given the range $D = (\delta_{min}, \delta_{max})$ of usable TE deflections, and the range $D_1 \subset D$ of TE deflections for which $C_{m0} \leq 0$ (see the following), the requirement that a trim condition should exist for each $\delta_s \in D$, limits the c.g. position with

$$x_{cg} < x_1 \qquad (6.22)$$

where

$$x_1 = \min_{\delta_s \in D_1} x_-(\delta_s) \qquad (6.23)$$

Toward the following analysis, let us make two simplifying assumptions:

$$\partial C_{m0}/\partial \delta_s < 0 \quad \text{for each } \delta_s \in D \qquad (6.24)$$

$$f = \frac{\partial C_{m0}/\partial \delta_s}{\partial C_{L0}/\partial \delta_s} < 0 \text{ and } \partial f/\partial \delta_s \geq 0 \quad \text{for each } \delta_s \in D \qquad (6.25)$$

In the context of Eq. (6.24), from Eq. (6.6) we note that $\partial C_{m0}/\partial \delta_s = \partial C^c_{m0}/\partial \delta_s - (\partial C^c_{D0}/\partial \delta_s)(z_c - z_{cg})$. Given that the canopy is not stalled, the first term here is always negative. At the same time, the second term, although making a positive contribution, vanishes for small TE deflections [McCormick 1979]. Hence, at least in some cases, the assumption presented by Eq. (6.24) holds. In the context of Eq. (6.25), if the canopy can be represented by an equivalent rigid wing equipped with a plain flap, then for small TE deflections the ratio $(\partial C_{m0}/\partial \delta_s)/(\partial C_{L0}/\partial \delta_s)$ is independent of δ_s and $\partial C_{L0}/\partial \delta_s$ is positive. Hence, at least in some cases, Eq. (6.25) holds as well.

Under Eq. (6.24), there exists δ_0 (not necessarily admissible), for which $C_{m0}(\delta_0) = 0$. Because no forward limit bounds the c.g. position when $C_{m0} \geq 0$, it is tacitly assumed that $\delta_0 \leq \delta_{max}$. Let $\delta_1 = \max\{\delta_0, \delta_{min}\}$ be the minimal admissible deflection for which $C_{m0} \leq 0$. From Eq. (6.15), it follows that x_- is continuous on $D_1 = (\delta_1, \delta_{max})$. Hence, the minimum of x_- on D_1, x_1, corresponds either to an extremum, or to one of the endpoints, δ_1 and δ_{max}.

At the extremum of x_-,

$$\partial x_-/\partial \delta_s = 0 \qquad (6.26)$$

Substituting Eq. (6.15) in Eq. (6.26), one will find that the respective TE deflection ∂_{ex} is a solution of the equation

$$C_{m0} = \frac{(1 - C_{L\alpha}K)C_{L\alpha}}{z_c - z_{cg}}\left(\frac{\partial C_{m0}/\partial \delta_s}{\partial C_{L0}/\partial \delta_s}\right)^2 \qquad (6.27)$$

With $z_c - z_{cg} < 0$, the expression on the right-hand side is a negative-valued non-decreasing function of δ_s by Eq. (6.25). The expression on the left is a decreasing function of δ_s by Eq. (6.24); it is negative when $\delta_s > \delta_0$. Hence, under both assumptions given in Eqs. (6.24) and (6.25), Eq. (6.27) has a single solution. Thus, $x_1 = x_-(\delta_{ex})$ if $\delta_{ex} \in D_1$ whereas $x_1 = x_-(\delta_1)$, or $x_1 = x_-(\delta_{max})$ if $\delta_{ex} < \delta_1$ or $\delta_{ex} > \delta_{max}$, respectively.

6.2.3 CONTROL DERIVATIVES

Consider the (control) derivatives $\partial C_{L-}/\partial \delta_s$ and $\partial C_{L-}/\partial x_{cg}$. Differentiating Eq. (6.8) with respect to δ_s and x_{cg}, and using Eq. (6.8) in the resulting expressions again, one will find

$$\frac{\partial C_{L-}}{\partial \delta_s} = \frac{(z_c - z_{cg})C_{L-}^2 \frac{\partial C_{L0}}{\partial \delta_s} - C_{L\alpha}C_{L-}\frac{\partial C_{m0}}{\partial \delta_s}}{-C_{L\alpha}C_{m0} + (1 - C_{L\alpha}K)(z_c - z_{cg})C_{L-}^2} \qquad (6.28)$$

$$\frac{\partial C_{L-}}{\partial x_{cg}} = \frac{C_{L\alpha}C_{L-}^2}{-C_{L\alpha}C_{m0} + (1 - C_{L\alpha}K)(z_w - z_{cg})C_{L-}^2} \qquad (6.29)$$

The expression appearing in the denominator on the right-hand side of either Eqs. (6.28) or (6.29) is negative for each $x_{cg} < x_1$ by Eqs. (6.18), (6.19), and (6.23). Hence,

$$\partial C_{L-}/\partial x_{cg} < 0 \quad \text{for each } x_{cg} < x_1 \qquad (6.30)$$

That is, moving the c.g. forward reduces the lift coefficient at trim as in a conventional airplane.

Predictable longitudinal control requires that the lift coefficient at trim should be a monotonic function of controls deflection (actually, this is a definition of speed stability). For a gliding parachute, this requirement can be specified as

$$\partial C_{L-}/\partial \delta_s > 0 \quad \text{for each } \delta_s \in D \tag{6.31}$$

To satisfy (6.31) for some $x_{cg} < x_1$, one needs

$$(z_c - z_{cg})\partial C_{L0}/\partial \delta_s C_{L-} - C_{L\alpha}\partial C_{m0}/\partial \delta_s < 0 \quad \text{for each } \delta_s \in D \tag{6.32}$$

by Eq. (6.28). Here, $\partial C_{L0}/\partial \delta_e$ is normally positive whereas $z_c - z_{cg}$ is negative. Hence, with

$$C_{L2}(\delta_s) = \frac{C_{L\alpha}}{z_c - z_{cg}} \frac{\partial C_{m0}/\partial \delta_s}{\partial C_{L0}/\partial \delta_s} \tag{6.33}$$

Eq. (6.31) imposes a restriction on the lift coefficient,

$$C_{L-}(x_{cg}, \delta_s) > C_{L2}(\delta_s) \quad \text{for each } \delta_s \in D \tag{6.34}$$

With equality sign instead of ">", $\partial C_{L-}/\partial \delta_s$ vanishes.

If Eq. (6.31) holds, then $C_{L-}(x_{cg}, \delta_s)$ is an increasing function of δ_s, and hence, $C_{L-}(x_{cg}, \delta_{min}) \leq C_{L-}(x_{cg}, \delta_s)$, for each $\delta_s \in D$. If Eq. (6.25) holds, then $C_{L2}(\delta_s)$ is a nonincreasing function of δ_s, and hence $C_{L2}(\delta_{min}) \geq C_{L2}(\delta_s)$ for each $\delta_s \in D$. Consequently, Eq. (6.34) can be reduced to

$$C_{L-}(x_{cg}, \delta_{min}) > C_{L2}(\delta_{min}) \tag{6.35}$$

In turn, Eq. (6.35) poses a restriction on the forward c.g. position. In fact, with

$$x_{trim}(C_L, \delta_s) = x_c + \frac{C_{m0}(\delta_s)}{C_L} + \frac{1}{C_{L\alpha}}(z_c - z_{cg})[(1 - C_{L\alpha}K)C_L - C_{L0}(\delta_s)] \tag{6.36}$$

the longitudinal c.g. position needed to trim the parachute for given C_L and δ_s, and

$$x_2 = x_{trim}(C_{L2}(\delta_{min}), \delta_{min}) \tag{6.37}$$

Equation (6.35) is equivalent to

$$x_{cg} < x_2 \tag{6.38}$$

note that, by definition, $x_{trim}(C_{L-}(x_{cg}, \delta_s), \delta_s) = x_{cg}$; moreover, according to Eq. (6.30) $x_{trim}(C_L', \delta_s) < x_{trim}(C_L'', \delta_s)$ for any $C_L'' < C_L'$. Consistent with the preceding, $\partial C_{L-}/\partial \delta_s$ vanishes at $\delta_s = \delta_{min}$ and $x_{cg} = x_2$.

Substituting Eq. (6.27) in Eq. (6.18), and using Eq. (6.33), one will find

$$C_{L1}(\delta_{ex}) = C_{L-}(x_-(\delta_{ex}), \delta_{ex}) = C_{L2}(\delta_{ex}) \tag{6.39}$$

Moreover, it follows from Eq. (6.18) that under Eq. (6.24)

$$C_{L1}(\delta_s) > C_{L1}(\delta_{ex}) \quad \text{for each } \delta_s > \delta_{ex} \qquad (6.40)$$

it follows from Eqs. (6.18), (6.19), and (6.27) that

$$C_{L-}(x_{cg}, \delta_{ex}) > C_{L1}(\delta_{ex}) \quad \text{for each } x_{cg} < x_-(\delta_{ex}) \qquad (6.41)$$

If $\delta_{ex} = \delta_{\min}$, then $x_1 = x_-(\delta_{\min}) = x_{\text{trim}}(C_{L2}(\delta_{\min}), \delta_{\min}) = x_2$ by Eq. (6.37). In other words, Eqs. (6.38) and (6.22) are equivalent. If $\delta_{ex} > \delta_{\min}$, then $x_1 = x_-(\delta_{ex})$. Using Eqs. (6.35), (6.31), and (6.39), in that order, we form

$$C_{L-}(x_{cg}, \delta_{\min}) > C_{L2}(\delta_{\min}) \geq C_{L2}(\delta_{ex}) = C_{L-}(x_1, \delta_{ex}) > C_{L-}(x_1, \delta_{\min}) \qquad (6.42)$$

Because $C_{L2}(\delta_{\min}) > C_{L-}(x_1, \delta_{\min})$, as manifested in the second and fifth terms in Eq. (6.42), Eq. (6.38) implies Eq. (6.22); that is, $x_1 > x_2$. If $\delta_{ex} < \delta_{\min}$, then $x_1 = x_-(\delta_{\min})$. Using Eqs. (6.19), (6.41), (6.39), and (6.40), in that order, we form

$$C_{L-}(x_{cg}, \delta_{\min}) > C_{L1}(\delta_{\min}) > C_{L1}(\delta_{ex}) = C_{L2}(\delta_{ex}) \geq C_{L2}(\delta_{\min}) \qquad (6.43)$$

The relation between the first and the last term recovers Eq. (6.35). Consequently, Eq. (6.22) implies Eq. (6.38); that is, $x_1 < x_2$.

6.2.4 MINIMAL LIFT COEFFICIENT AT TRIM AND AFT CENTER-OF-GRAVITY LIMIT

Concluding the results of Secs. 6.2.2 and 6.2.3, the forward c.g. limit, and, concurrently, the minimal lift coefficient are consequences of requirements imposed by Eqs. (6.22) and (6.31). From the design point of view, the requirement imposed by Eq. (6.22) appears weaker than the one imposed by Eq. (6.31). In fact, as cited in the discussion of Sec. 6.2.2, by flattening the wing's camber, one can design a parachute with positive $C_{m0}(\delta_{\min})$, in which case $\delta_{ex} > \delta_0 > \delta_{\min}$. Following the results of Sec. 6.2.2, the forward c.g. limit in this case is x_2 whereas the respective minimal lift coefficient is $C_{L2}(\delta_{\min})$; both are consequences of Eq. (6.31).

In view of Eq. (6.33), $C_{L2}(\delta_{\min})$ depends on several design parameters, of which the most noticeable are the length of the lines and the part of the chord occupied by the flaps. An increase in either of these two parameters typically reduces $C_{L2}(\delta_{\min})$. Increasing the line length relative to the mean aerodynamic chord increases $z_{cg} - z_c$ (see Sec. 6.2.2) whereas increasing the flaps chord relative to the mean aerodynamic chord reduces the ratio of the derivatives $\partial C_{m0}/\partial \delta_s$ and $\partial C_{L0}/\partial \delta_s$. The last point is best elucidated by considering the limit where the flaps extend over the entire chord. In this case, TE deflections do not affect C_{m0} at all, and $C_{L2} = 0$ by Eq. (6.32). In other words, if $C_{m0} > 0$, no limit is imposed on the minimal lift coefficient at trim. Because C_{L2} is clearly positive in the case where the flaps occupy the TE area only, the initial assertion follows. In this context, also note that application of the all-wing flaps is analogous to the change in the c.g. position (relative to the canopy-fixed coordinate system depicted in Fig. 6.10). Hence, the present conclusion that minimal lift coefficient at trim is unbounded for a parachute having $C_{m0} > 0$ and equipped with full-chord flaps agrees with the comparable conclusion of Sec. 6.2.2.

Line length is an inflexible parameter, as it affects most stability derivatives of the parachute. In fact, the ratio between their length and the wingspan is almost the same in all present designs. With this said, parachutes with high-aspect wings (having short aerodynamic chord and hence large $z_{cg} - z_c$) should have higher top speed than those with low-aspect wings. This conclusion is in apparent agreement with present design trends. Large flaps require large control forces. They can complicate actuators' design as all of the lines connected to them will need to be shortened, in turn, from the TE forward. Yet, they hold the potential of a gliding parachute with aerodynamically unlimited minimal lift coefficient.

From the preceding discussion, it follows that no stability [Eq. (6.22)] or control [Eq. (6.31)] requirements, bound the most rear c.g. position. Nonetheless, for a given symmetric TE deflection, the lift coefficient at trim increases when the c.g. moves backwards; see Eq. (6.30). Consequently, there exists a certain aft c.g. position where the canopy will stall at the maximal (mechanical) symmetric TE deflection. Hence, the requirement that the maximal TE deflection $\delta_{\max}$ should be attained without stalling the canopy imposes a restriction on the aft c.g. position. Specifically, with $C_{L\max}(\delta_s)$, the lift coefficient at stall with TE deflected at δ_s, this requirement implies

$$x_{cg} < \max_{\delta_s \in D}[x_{\text{trim}}(C_{L\max}(\delta_s), \delta_s)] \qquad (6.44)$$

by Eqs. (6.30) and (6.36).

6.3 LATERAL-DIRECTIONAL MOTION

This section considers several aspects of lateral-directional motion. Specifically, Sec. 6.3.1, based on a slightly edited version of [Crimi 1990], considers lateral-directional stability in a steady glide, and Sec. 6.3.2 based on [Brown 1993] deals with a steady turn response featuring two different-size systems. More discussion about scale effects will be given in Sec. 6.4.

6.3.1 LATERAL-DIRECTIONAL STABILITY IN A STEADY GLIDE

Let us establish the analytical relationships of aerodynamic and inertial parameters to parafoil stability that in general can be characterized by the yaw static-stability coefficient $C_{n\beta}$ [Zimmerman 1937]. Consider Fig. 6.11 that reintroduces the states involved into the lateral-directional motion. In this figure, X and Z denote the wind-fixed axes, and Y represents a body-fixed axis, all originating at the body mass center. The variables governing lateral-directional response are the

Fig. 6.11   Body-fixed coordinate system.

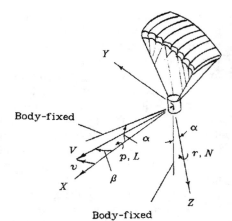

roll angle ϕ, roll rate $p = \dot{\phi}$, yaw angle ψ, yaw rate $r = \dot{\psi}$, and sideslip angle β. The relevant aerodynamic loads are side force F_Y, rolling moment L, and yawing moment N. With the composite mass center a distance s_l below and Δ forward of the wing mass center, the sideslip angle at the wing is $\beta + s_l p / V - \Delta r / V$. As a result, the corresponding aerodynamic coefficients for F_Y, L, and N are

$$C_Y = \frac{2F_Y}{\rho V^2 S} = C_{Y\beta}\left(\beta + \frac{s_l p}{V} - \frac{r\Delta}{V}\right) \tag{6.45}$$

$$C_l^* = \frac{2L}{\rho V^2 Sb} = C_{lp}\frac{pb}{2V} + \left(C_{l\beta} + \frac{s_l}{b}C_{Y\beta}\right)\left(\beta + \frac{s_l p}{V} - \frac{r\Delta}{V}\right) + C_{lr}\frac{rb}{2V} \tag{6.46}$$

$$C_n^* = \frac{2N}{\rho V^2 Sb} = C_{np}\frac{pb}{2V} + \left(C_{n\beta} - \frac{\Delta}{b}C_{Y\beta}\right)\left(\beta + \frac{s_l p}{V} - \frac{r\Delta}{V}\right) + C_{nr}\frac{rb}{2V} \tag{6.47}$$

In Eqs. (6.46) and (6.47), C_l^* and C_n^* are coefficients of the roll and yaw moments, respectively, about the composite mass center while C_l and C_n are coefficients of the roll and yaw moments, respectively, about the wing mass center. Let us neglect canopy flexibility and consider the yaw static-stability coefficient $C_{n\beta}$ as a dependent variable and approximate other coefficients in Eqs. (6.45–6.47) in terms of wing and flight parameters as follows [Zimmerman 1937; Perkins and Hage 1957]:

$$C_{Y\beta} = -\varepsilon^2 C_{L\alpha}, \quad C_{l\beta} = -\varepsilon C_{L\alpha}/4, \quad C_{lp} = -(C_{L\alpha} + C_{Dw})/8, \quad C_{lr} = C_L/3,$$
$$C_{np} = -(C_L - C_{Dw\alpha})/8, \quad C_{nr} = -C_{Dw}/4 \tag{6.48}$$

The dihedral angle ε can be computed according to Eq. (2.15). $\varepsilon = \varepsilon_b / 2$

The distance of the wing mass center aft of the composite mass center is determined by the longitudinal trim of the parachute. Summing moments about the wing mass center, it can be found that

$$\frac{\Delta}{b} = \frac{C_m - C_{DL}\dfrac{s_l}{b}}{C_L} - \frac{s_l}{b}\tan(\gamma) \tag{6.49}$$

where C_m is the wing pitching-moment coefficient. Note that the static stability in yaw, characterized by the coefficient $C_{n\beta}$, derives from the term $C_{n\beta} - (\Delta/b)C_{Y\beta}$

in Eq. (6.47); in turn, according to Eq. (6.49), the ratio Δ/b is a strong function of both s_l/b and the glide slope $-\tan(\gamma)$.

Using the coordinates defined in Fig. 6.11, the equations governing the lateral-directional response are

$$v\frac{\partial F_Y}{\partial v} + p\frac{\partial F_Y}{\partial p} + r\frac{\partial F_Y}{\partial r} + W[\phi\cos(\gamma) + \psi\sin(\gamma)] = m(\dot{v} + ru) \tag{6.50}$$

$$v\frac{\partial L}{\partial v} + p\frac{\partial L}{\partial p} + r\frac{\partial L}{\partial r} = I_{xx}\dot{p} + I_{xz}\dot{r} \tag{6.51}$$

$$v\frac{\partial N}{\partial v} + p\frac{\partial N}{\partial p} + r\frac{\partial N}{\partial r} = I_{xz}\dot{p} + I_{zz}\dot{r} \tag{6.52}$$

The total mass m in Eq. (6.50) includes the apparent mass of the canopy, which for a ram-air parachute is about half the mass of the displaced air [Goodrick 1979b]. In a steady glide, $W\cos(\gamma) = 1/2\rho V^2 SC_L$. Also, $\beta \approx v/V$, $p = \dot{\phi}$, $r = \dot{\psi}$, and $u \approx V$. To simplify derivations, let us introduce the following quantities:

$$\tau = \frac{\rho SV}{m}t, \quad \xi = \frac{m}{\rho Sb}, \quad h_{xx} = \frac{4I_{xx}}{mb^2}, \quad h_{zz} = \frac{4I_{zz}}{mb^2}$$

$$h_{xz} = \frac{4I_{xz}}{mb^2}, \quad y_s = \frac{C_{Y\beta}}{\xi}\frac{s_l}{b}, \quad y_\delta = \frac{C_{Y\beta}}{\xi}\frac{\Delta}{b} + 2 \tag{6.53}$$

$$l_\beta = C_{l\beta} + \frac{s_l}{b}C_{Y\beta}, \quad l_p = C_{lp} + 2\frac{s_l}{b}l_\beta, \quad l_r = C_{lr} - 2\frac{\Delta}{b}l_\beta$$

$$n_\beta = C_{n\beta} - \frac{\Delta}{b}C_{Y\beta}, \quad n_p = C_{np} + 2\frac{s_l}{b}n_\beta, \quad n_r = C_{nr} - 2\frac{\Delta}{b}n_\beta \tag{6.54}$$

Using these relations and substituting expressions for F_Y, L, and N from Eqs. (6.45–6.47), the equations of motion (6.50–6.53) can be written as follows:

$$C_{Y\beta}\beta - 2\beta'_\tau + C_L\phi + y_\delta\phi'_\tau + C_L\psi\tan(\gamma) - y_s\psi'_\tau = 0 \tag{6.55}$$

$$2\mu l_\beta\beta + l_p\phi'_\tau - h_{xz}\phi''_{\tau\tau} + l_r\psi'_\tau - h_{zz}\psi''_{\tau\tau} = 0 \tag{6.56}$$

$$2\mu n_\beta\beta + n_p\phi'_\tau - h_{xz}\phi''_{\tau\tau} + n_r\psi'_\tau - h_{zz}\psi''_{\tau\tau} = 0 \tag{6.57}$$

If $\beta = \bar{\beta}\exp(\lambda\tau)$, $\phi = \bar{\phi}\exp(\lambda\tau)$, and $\psi = \bar{\psi}\exp(\lambda\tau)$ are substituted in these equations, the characteristic equation is found to be

$$A\lambda^4 + B\lambda^3 + C\lambda^2 + D\lambda + E = 0 \tag{6.58}$$

where

$$A = h_{xx}h_{zz} - h_{xz}^2$$

$$B = -0.5C_{Y\beta}A - h_{xx}n_r - h_{zz}l_p + h_{xz}(l_r + n_p)$$

$$C = 0.5C_{Y\beta}\left[h_{xx}n_r + h_{zz}l_p - h_{xz}(l_r + n_p)\right]$$

$$\quad - n_pl_r + l_pn_r - \xi l_\beta(y_sh_{zz} + y_\delta h_{xz}) + \xi n_\beta(y_sh_{xz} + y_\delta h_{xx}) \qquad (6.59)$$

$$D = 0.5C_{Y\beta}(n_pl_r - l_pn_r) + \xi l_\beta\left[y_sn_r + y_\delta n_p - h_{zz}C_L + h_{xz}C_L\tan(\gamma)\right]$$

$$\quad - \xi n_\beta\left[y_sl_r + y_\delta l_p - h_{xz}C_L + h_{xx}C_L\tan(\gamma)\right]$$

$$E = \xi C_L\left\{l_\beta\left[n_r - n_p\tan(\gamma)\right] - n_\beta\left[l_r - l_p\tan(\gamma)\right]\right\}$$

The characteristic equation (6.58) generally yields two real roots and a pair of complex conjugate roots. As is the case with conventional airplanes, one of the real roots has a large negative value and is not of concern with regard to stability. The other three roots might or might not cause divergent lateral-directional response. For a given set of geometric, inertial, and aerodynamic parameters, the response is typically found to be characterized by a range of values for $C_{n\beta}$ within which the response is positively damped. If $C_{n\beta}$ exceeds an upper bound (positive $C_{n\beta}$ implies static stability in yaw), the characteristic equation (6.58) has a positive real root that results in what is termed spiral divergence [Perkins and Hage 1957]. If $C_{n\beta}$ is reduced below a lower bound, a complex pair of roots with a positive real part is found, resulting in an oscillatory instability.

The location of the spiral divergence boundary can be examined analytically because neutral stability results when the constant E in the characteristic equation (6.58) vanishes. By setting the expression for E [the last expression in Eq. (6.59)] to zero and accounting for Eqs. (6.54)

$$l_\beta\left[\left(C_{nr} - 2\frac{\Delta}{b}n_\beta\right) - \left(C_{np} + 2\frac{s_l}{b}n_\beta\right)\tan(\gamma)\right] - n_\beta\left[l_r - l_p\tan(\gamma)\right] = 0 \quad (6.60)$$

we find

$$n_\beta = \frac{l_\beta C_{nr} - C_{np}\tan(\gamma)}{2l_\beta\dfrac{\Delta}{b} + 2l_\beta\dfrac{s_l}{b}\tan(\gamma) + \left[l_r - l_p\tan(\gamma)\right]}$$

$$\quad = \frac{\left[C_{l\beta} + (s_l/b)C_{Y\beta}\right]C_{nr} - C_{np}\tan(\gamma)}{C_{lr} - C_{lp}\tan(\gamma)} \qquad (6.61)$$

Hence, the relation for the value of $C_{n\beta}$ to cause spiral divergence becomes

$$C_{n\beta} = n_\beta + \frac{\Delta}{b}C_{Y\beta} = \frac{\left[C_{l\beta} + (s_l/b)C_{Y\beta}\right]C_{nr} - C_{np}\tan(\gamma)}{C_{lr} - C_{lp}\tan(\gamma)} + \frac{\Delta}{b}C_{Y\beta} \qquad (6.62)$$

From this expression, it is evident that the stability boundary is a strong function of s_l, the glide slope $-\tan(\gamma)$, and the dihedral angle ε [because coefficients $C_{Y\beta}$ and $C_{l\beta}$ depend on it as shown in Eq. (6.48)]. Similarly, the lift coefficient C_L has considerable effect too because C_{lr} is proportional to C_L [see Eq. (6.48)]. Note that the mass ratio ξ, which is proportional to the wing loading [see Eq. (6.53)], has no effect on the spiral divergence boundary, nor do the moments of inertia.

Let us now analyze lateral-directional stability for some nominal PADS with

$$\xi = 20, \quad h_{xx} = 0.27, \quad h_{zz} = 0.038, \quad h_{xz} = -0.1\frac{s_l}{b}\frac{\Delta}{b} \qquad (6.63)$$

$$C_L = 0.6726, \quad C_{L\alpha} = 2.86, \quad C_{Dw\alpha} = 0.618, \quad C_{Dw} = 0.0315, \quad C_m = -0.0164 \qquad (6.64)$$

by varying only three parameters: s_l/b, the dihedral angle, and glide slope. All other parameters entering the coefficients (6.13) of characteristic equation (6.58) are computed using these 12 basic quantities. Having numerical values for coefficients of Eq. (6.58), the roots of the characteristic equation define lateral-directional stability.

The effect of dihedral angle on stability is shown in Fig. 6.12 for $s_l/b = 1.2$ and a glide slope of 1/3. For ε between -12 deg and 0, the static stability derived by suspending the payload is not sufficient to overcome the destabilizing effect of anhedral, and so the vehicle is unstable, regardless of the value of $C_{n\beta}$. As ε decreases below -12 deg or increases above zero, stable flight results with $C_{n\beta}$ values below the spiral divergence boundary and above the oscillatory stability boundary. Note that, for typical parafoil configurations with large effective anhedral,

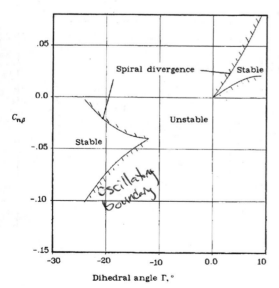

Fig. 6.12   Effect of suspension line length on oscillatory response for $-\tan(\gamma) = 1/3$, $s_l/b = 1.2$.

Fig. 6.13   Effect of suspension line length on oscillatory response for $-\tan(\gamma) = 1/3$, $\varepsilon = -18$ deg.

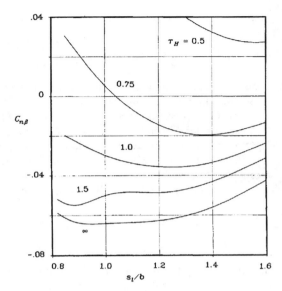

the canopy itself must have negative $C_{n\beta}$, that is, it must be statistically unstable in yaw, to preclude spiral divergence, because of the static yaw stability afforded by the mass center offset Δ.

The effect of suspension line length on lateral-directional stability is shown in Figs. 6.13 and 6.14, for a glide slope of $\frac{1}{3}$ and a dihedral angle of -18 deg. The oscillatory stability boundary and lines of constant dimension-less time τ_H to damp to half-amplitude are plotted in Fig. 6.13, and the spiral divergence boundary and lines of constant time to double amplitude τ_D are plotted in Fig. 6.14, against s_l/b. [For the second-order dynamic system, time to halve τ_H or double τ_D the amplitude is given by $t_{1/2;2} = \ln(2)/(\omega_n \zeta)$]. Note that

s_l is the distance of the wing mass center above the composite mass center and so includes the distance from the composite mass center to the suspension line attachment to the payload. The stability of the oscillatory mode is seen to be relatively insensitive to s_l/b, but increasing s_l/b for a given value of $C_{n\beta}$ markedly improves the stability of the spiral divergence mode. At s_l/b of 0.81, the boundaries

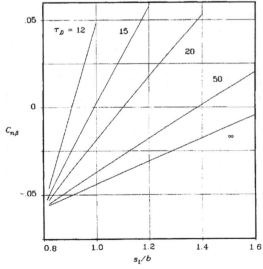

Fig. 6.14   Effect of suspension line length on spiral divergence for $-\tan(\gamma) = 1/(3)$, $\varepsilon = -18$ deg.

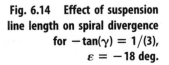

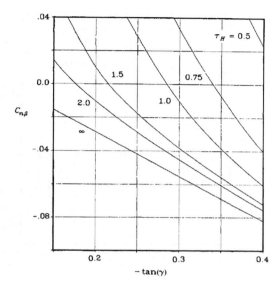

Fig. 6.15 Effect of glide slope on oscillatory response for $s_l/b = 1.2$, $\varepsilon = -18$ **deg.**

of spiral divergence and oscillatory instability cross. As a result, the character of the solution changes for s_l/b less than 0.81. The system is then unstable, regardless of the value of $C_{n\beta}$, the reduced pendular stability in roll being insufficient to overcome the effects of anhedral.

The effect of glide slope is shown in Figs. 6.15 and 6.16 for $s_l/b = 1$ and a dihedral angle of -18 deg. Increasing the glide slope (i.e., decreasing the ratio of lift to drag) increases stability for both oscillatory response, Fig. 6.14, and spiral divergence, Fig. 6.16. Note, from Fig. 6.17, where the stability boundaries are replotted on the same graph, that as the glide slope is reduced, the range of values of $C_{n\beta}$ within which the system is stable in both modes becomes smaller. If the glide slope is less than about 0.15, spiral divergence is unavoidable if the oscillatory mode is positively damped.

As shown, the lateral-directional stability characteristics of gliding parachutes are, in some respects, similar to those of conventional fixed-wing aircraft. Both systems exhibit an oscillatory mode of response and a tendency toward spiral divergence. Also, inertial characteristics, including apparent mass effects and wing loading, have little or no

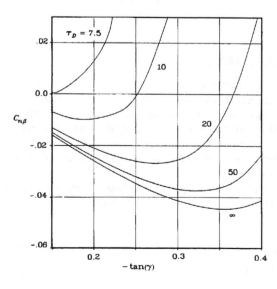

Fig. 6.16 Effect of glide slope on spiral divergence for $s_l/b = 1.2$, $\varepsilon = -18$ **deg.**

Fig. 6.17 Effect of glide slope on stability boundaries for $s_l/b = 1.2$, $\varepsilon = -18$ deg.

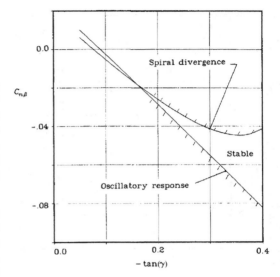

effect on stability boundaries in both cases. The static stability in roll afforded by suspending the payload distinguishes gliding parachutes, however, with the following relations observed from the calculation results:

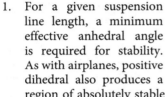

1. For a given suspension line length, a minimum effective anhedral angle is required for stability. As with airplanes, positive dihedral also produces a region of absolutely stable response.

2. Increasing suspension line length is stabilizing for spiral divergence but has little effect on oscillatory response. There is a minimum suspension line length below which the system is unstable, regardless of the static stability in yaw.

3. Decreasing the glide slope is destabilizing for both oscillatory response and spiral divergence. For a very small glide slope, spiral divergence is unavoidable if the oscillatory response is positively damped.

6.3.2 STEADY TURN RESPONSE TO CONTROL INPUT

Compared to conventional aircraft, parafoil maneuvering differs in that the control input (asymmetric flap deflection) produces turn rate, rather than roll rate, and also in that turning is associated with side-slip or "skid." This results in extremely high sensitivity to steering inputs with small, highly loaded parafoils and very sluggish response with large parafoils. This section presents an analytical steady-turn model to account for parafoil specifics and illustrate effects of scale and wing loading on turn rate and turn radius.

To derive a simplified yet meaningful model, the following assumptions that are most nearly correct at low roll angles and turn rates are made:

*   Parafoil aerodynamics is linear.

*   The turn equations are independent of the steepness of the spiraling descent.

*   Rate derivatives are negligible (especially $C_{nr} = 0$).

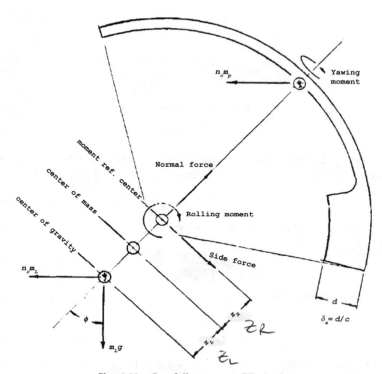

Fig. 6.18   Parafoil geometry (TE view).

Referring to Fig. 6.18, the equations for horizontal force equilibrium in a steady turn can be written as

$$n_c g \frac{m_L + m_P}{QS} = C_L \sin(\phi) + C_{Y\beta}\beta \cos(\phi) \tag{6.65}$$

Here, $n_c = V^2 Rg$ is the lateral-directional projection of a load factor with R being a turn radius, m_L denote a total real mass of payload and parafoil, and m_P represents the apparent mass together with included mass of the air inflating the parafoil (even though they are of entirely different nature).

For the vertical equilibrium,

$$m_L g = \left[C_L \cos(\phi) - C_{Y\beta}\beta \sin(\phi) \right] QS \tag{6.66}$$

For convenience, we define a mass ratio:

$$R_m = \frac{m_L + m_P}{m_L} \tag{6.67}$$

Then, combining Eqs. (6.65) and (6.66) yields an equation for lateral-directional acceleration in terms of roll and sideslip:

$$n_c = R_m \frac{C_L \sin(\phi) + C_{Y\beta}\beta \cos(\phi)}{C_L \cos(\phi) - C_{Y\beta}\beta \sin(\phi)} \qquad (6.68)$$

For $\beta = 0$ and $R_m = 1$, this reduces to the known relation

$$n_c = \tan(\phi) \qquad (6.69)$$

Assuming that only sideslip β and asymmetric flaps deflection δ_a produce yawing moments (i.e., neglecting yaw damping), yaw equilibrium gives

$$-C_{n\beta}\beta = C_{n\delta_a}\delta_a \qquad (6.70)$$

from which sideslip derives from control deflection via

$$\beta = -\frac{C_{n\delta_a}}{C_{n\beta}}\delta_a \qquad (6.71)$$

We derive the equilibrium condition in roll by summing all rolling-moment contributions:

$$-z_L m_L g \sin(\phi) + z_R C_{Y\beta}\beta QS + C_{l\beta}\beta QSb + C_{l\delta_a}\delta_a QSb = 0 \qquad (6.72)$$

Here, we exclude the $n_c m_L$ and $n_c m_P$ contributions because they balance about the c.m., by definition, for lateral motion. The $m_L g$ term is not balanced by a corresponding m_P term, however, because gravity does not act on apparent mass. Dividing by Eq. (6.72) by QSb puts the equation in a more familiar dimensionless form:

$$-\frac{z_L}{b}\frac{m_L g}{QS}\sin(\phi) + \frac{z_R}{b}C_{Y\beta}\beta + C_{l\beta}\beta + C_{l\delta_a}\delta_a = 0 \qquad (6.73)$$

Using Eq. (6.66) to replace $m_L g$ removes all explicit reference to mass:

$$-\frac{z_L}{b}C_L \cos(\phi)\sin(\phi) + \frac{z_R}{b}C_{Y\beta}\beta \sin^2(\phi) + \frac{z_R}{b}C_{Y\beta}\beta + C_{l\beta}\beta + C_{l\delta_a}\delta_a = 0 \quad (6.74)$$

We apply Eq. (6.71) to replace all β, arriving at an equation of the form

$$A \cos(\phi)\sin(\phi) + B\delta_a \sin^2(\phi) + C\delta_a = 0 \qquad (6.75)$$

where

$$A = -\frac{z_L}{b}C_L, \quad B = \frac{z_R}{b}\frac{C_{n\delta_a}}{C_{n\beta}}C_{Y\beta}, \quad C = C_{l\delta_a} - \frac{C_{n\delta_a}}{C_{n\beta}}C_{l\beta} - \frac{z_R}{b}\frac{C_{n\delta_a}}{C_{n\beta}}C_{Y\beta} \qquad (6.76)$$

This can be solved as a quadratic in $\sin(2\phi)$ yielding

$$\phi = \frac{1}{2}\sin^{-1}\left[\delta_a \frac{-A(B+2C) - \sqrt{A^2(B+2C)^2 - (A^2 + B^2\delta_a^2)(4BC + 4C^2)}}{A^2 + B^2\delta_a^2}\right]$$

(6.77)

This allows computation of the roll angle ϕ as a function of control deflection δ_a. Equation (6.68) then completes the picture of the skidding turn by computing lateral-directional acceleration for the calculated roll angle.

Referring to the vertical equilibrium in turning flight [Eq. (6.66)], the dynamic pressure Q can be expressed as

$$Q = \frac{m_L g}{S} \frac{1}{C_L \cos(\phi) - C_{Y\beta}\beta \sin(\phi)}$$

(6.78)

Compared to a similar expression for an aircraft in a coordinated turn

$$Q = \frac{W}{S} \frac{1}{C_L \cos(\phi)}$$

(6.79)

where C_L would be increased as necessary through the application of elevator control to maintain constant Q. In the case of a parafoil though, C_L is increased by an amount that is not under independent control, but that we have assumed to be (weakly) proportional to turn control deflection. An additional term in denominator of Eq. (6.35) compared to that of Eq. (6.79) caused by a side force due to yaw gaining a vertical component with roll that subtracts from the diminishing vertical component of lift results in an increase of an airspeed with roll angle, given by

$$\frac{V_{\text{turn}}}{V_{\text{level}}} = \sqrt{\frac{C_L}{C_L \cos(\phi) - C_{Y\beta}\beta \sin(\phi)}}$$

(6.80)

Consequently, the glide ratio and path angle in the spiraling turn can be expressed as

$$GR = \frac{1}{\tan(\phi)} = \frac{C_L \cos(\phi) - C_{Y\beta}\beta \sin(\phi)}{C_D}$$

(6.81)

In Eqs. (6.80) and (6.81), C_L, C_D, β, and ϕ are all functions of δ. Although the dependencies for β and ϕ were derived earlier in Eqs. (6.71) and (6.77), respectively, the expressions for C_L and C_D can be represented as

$$C_L = C_{L0} + C_{L\delta_a}\delta_a$$

(6.82)

and

$$C_D = C_{D0} + C_{D\delta_a}\delta_a + \frac{C_L^2}{e\pi AR} \tag{6.83}$$

(In these relations the effect of control deflection is introduced explicitly.)

To analyze the turn performance of PADS described by Eqs. (6.77) and (6.81), a model PADS with the 55.7-m^2 (600-ft^2) reference wing area, 12.3-m (40.3-ft) span, and 4.54-m (14.9-ft) chord (corresponding to AR = 2.7) was considered. The aerodynamic derivatives for this 460-kg (1,014-lb) PADS shown in Table 6.5 were obtained using LinAir software [Kroo 1987] in a manner similar to that shown in Figs. 5.13 and 5.14. Control derivatives were calculated for both a steering input δ_a and a "brake" input δ_s. The high value of a parasitic drag required as an input into LinAir was determined to cause the predicted rate of turn that matched that observed in flight. This drag value corresponds to approximately $C_p = -0.4$ acting over the streamwise projected area of the control surface element. Given flight-test results that include (as a minimum) trimmed airspeed, glide ratio, and turn rate at a known control deflection,

TABLE 6.5   AERODYNAMIC AND CONTROL DERIVATIVES OBTAINED USING LINAIR SIMULATIONS

Longitudinal Derivatives		Lateral-Directional Derivatives	
Derivative	Value	Derivative	Value
C_{D0}	0.091	$C_{Y\beta}$	−0.55
$C_{L\alpha}$	2.88	C_{Yr}	−0.006 $[rb/(2V_a)]^{-1}$
$C_{L\,max}$	1.0	$C_{l\beta}$	−0.08
$C_{L\,min}$	0.4	C_{lp}	−0.133 $[pb/(2V_a)]^{-1}$
α_0	−5.11 deg	C_{lr}	0.01 $[rb/(2V_a)]^{-1}$
C_{m0}	0.25	$C_{n\beta}$	0.0287
$C_{m\alpha}$	−0.9	C_{np}	−0.013 $[pb/(2V_a)]^{-1}$
C_{mq}	−1.86 $[qc/(2V_a)]^{-1}$	C_{nr}	−0.035 $[rb/(2V_a)]^{-1}$
$C_{L\delta_a}$	0.235	$C_{Y\delta_a}$	0.137
$C_{D\delta_a}$	0.0957	$C_{l\delta_a}$	−0.0063
$C_{m\delta_a}$	0.294	$C_{n\delta_a}$	0.0155
$C_{L\delta_s}$	0.467	$C_{Y\delta_s}$	0
$C_{D\delta_s}$	0.190	$C_{l\delta_s}$	0
$C_{m\delta_s}$	0.0298	$C_{n\delta_s}$	0

LinAir calculates a self-consistent set of aerodynamic coefficients and derivatives that match the observed results.

The components of the apparent mass tensor $\mathbf{I}_{a.m.} = \text{diag}([A, B, C])$ [see Eq. (5.62)] are calculated using the equations suggested by [Lissaman and Brown 1993] and presented in Sec. 5.1.4. Specifically, the mass components in the chordwise, spanwise, and wing-normal directions (principle axes of the body frame) are given by Eqs. (5.63) and (5.68) with coefficients defined in Eq. (5.69) [or Eq. (5.71)]. Included mass is calculated using the approximation

$$m_I = \rho b \frac{ct}{2} \tag{6.84}$$

The apparent and included mass in the direction of flight m_P entering Eq. (6.65) can then be calculated using the rotation matrix ${}^w_b\mathbf{R} = {}^b_w\mathbf{R}^T$. Using Eq. (5.75), we arrive at

$$m_P = {}^w_b\mathbf{R}\,\text{diag}([A, B, C]) + m_I$$
$$= A\cos(\alpha)\cos(\beta) + B\cos(\beta) + C\cos(\alpha) + m_I \tag{6.85}$$

where angle of attack is calculated as

$$\alpha = \alpha_0 + \frac{C_L}{C_{L\alpha}} \tag{6.86}$$

Figure 6.19 shows the calculated parafoil turn response for parafoils of 2, 20, 200, and 2,000 m^2 (21.5; 215; 2,153; and 21,528 ft^2) at a wing loading of 10 kg/m^2 (2 lb/ft^2). The top curve in each graph is $\tan^{-1}(n_c)$, which is the bank angle of an aircraft in a coordinated turn at the same airspeed and turn rate as the parafoil. The curve just below this is the roll angle of the parafoil ϕ from Eq. (6.77). The vertical distance between these curves indicates the degree of "skid" in the turn.

In physical terms, the apparent mass of the parafoil increases faster than the payload mass (wing loading held constant), causing an increasing displacement between the c.g. and the center of mass z_L (Fig. 6.18). This is the source of an anti-roll moment that is the primary effect causing decreasing turn response and increased "skidding" with increased scale, as shown in the graphs.

Although we have included the variation of apparent mass with angle of attack and sideslip as it is affected by control input, we find that the total variation is approximately 1%, meaning that m_P could have been treated as a constant for each case without loss of accuracy. However, the large difference in A and C (two orders of magnitude) means that angle of attack must be estimated with care and that changes in camber or trim will have strong effects on the apparent mass in the direction of flight and therefore on turn response.

The very bottom curve in the figures is the path angle calculated using Eq. (6.81). The degradation of the glide angle is underestimated here because the added drag due to sideslip $C_{D\beta}\beta$ is not included in Eq. (6.83).

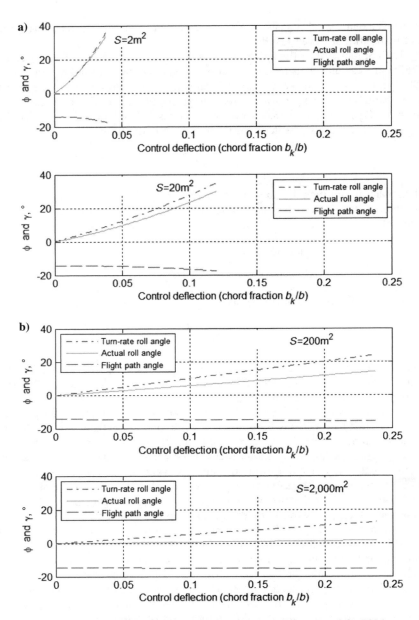

Fig. 6.19   Turn response and scale effect for four different weight PADS.

These results show a striking change in turn response with scale. The small 2-m^2 (21.5-ft^2) parafoil shows extreme sensitivity to control input and a turn that is very nearly coordinated. The very large $2{,}000\text{-m}^2$ ($21{,}528\text{-ft}^2$) parafoil, in contrast, shows low sensitivity to control response and turns with almost no bank angle.

At some control deflection producing a bank angle of less than 40 deg in each case, Eq. (6.56) gives complex results. We will not attempt to give a physical interpretation of this. We do observe that skydivers often perform steep spiraling maneuvers with extreme bank angles. The analytical model used here does not apply well to such cases, and, as we find in these examples, fails to offer any solution at all for flight at high bank angles. Hence, meaningful results are limited to bank angles of less than 30 deg. Results are also sensitive to some parameters that are difficult to determine with accuracy.

6.4 SCALING EFFECTS

Following the preceding section, this one continues exploring the scaling effects on performance of ram-air airfoils based on a work by [Goodrick 1984]. Using nine different canopy areas with $AR = 2$, lateral-directional and longitudinal channel performance were explored in computer simulations and real drops [Goodrick 1979b; Goodrick 1981]. The values of lift and drag coefficients were taken as $C_L = 0.7$ and $C_D = 0.2...0.27$ (depending on a payload drag), respectively. The mass-geometry data on these nine canopies utilized to carry different-weight payload so that the wing loading varies from about 3 to 20 kg/m^2 (0.6 to 4 lb/ft^2) are given in Appendix G.

Figures 6.20–6.26 show parameters of different models used in [Goodrick 1984] graphically. Figure 6.20 gives an idea about payload mass used for 20 different PADS setups that used the aforementioned nine different-size canopies. It also features three lines corresponding to the three values of wing loading: $\xi = [3; 10; 20]$ kg/m^2 (0.6, 2, and 4 lb/ft^2).

The length (total height) of the entire PADS, l, is proportional to the length of suspension lines R, that is, the distance from the canopy to the confluence point (see Fig. 2.13). In turn, the length R is primarily determined (limited) by the canopy's span b. Assuming the same anhedral angle for all nine canopies, we can write

$$l \sim R \sim 0.75b = 0.75\sqrt{AR \cdot S} \qquad (6.87)$$

[Usually, the suspension lines' length is within [70; 90]% range (see data in Tables C.1–C.3 of Appendix C)]. The height of PADS is usually extended using risers. (See typical examples in Figs. 1.6, 1.16a, 1.25, and 1.26.) Figure 6.21 shows the total PADS height for all 20 setups of a Goodrick's model corresponding to 70, 90, and 110% of a wingspan so that the maximum height of PADS in these

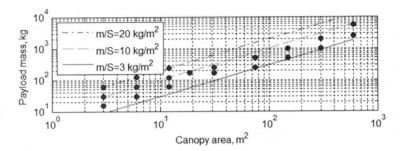

Fig. 6.20   Modeled ADS payload mass vs canopy area.

models was brought up to

$$l \approx 1.1b \qquad (6.88)$$

[as opposed to $l \sim 0.75b$, as in Eq. (6.87)]. As discussed in Sec. 6.1, this length (length between the centers of mass of the canopy and payload) affects dynamic stability about the roll and pitch axes.

By design, the mass of payload (store) in these models is proportional to the canopy size (area) while the mass of the canopy with an air mass trapped inside is proportional to the canopy's volume (span) plus about a triple canopy's area (mass of the canopy itself). Hence,

$$m_s = \xi S, \quad m_c = \rho AR^{-0.5} h S^{1.5} + 3\kappa S \qquad (6.89)$$

where $\xi = [3; 10; 20]$ kg/m^2 ([0.6; 2; 4.1] lb/ft^2), $\rho \approx 1.225$ kg/m^3 (0.0765 lb/ft^3) (at the sea level), $h \approx 0.15$ is the inlet height, and $\kappa = 0.037$ kg/m^2 (1.1 oz/yd^2) is a nylon mass per area. Using Eqs. (6.88) and (6.89), the location of PADS c.g.

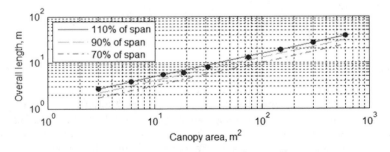

Fig. 6.21   Modeled PADS height vs canopy area.

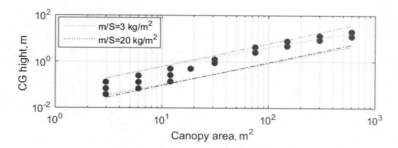

Fig. 6.22   PADS c.g. height above payload c.g. vs canopy area.

above the payload's c.g. can be estimated as

$$h_{cg} = \frac{m_c l}{m_s + m_c} \approx \frac{(\rho AR^{-0.5} h S^{1.5} + 3\kappa S)(AR \cdot S)^{0.5}}{\xi S + \rho AR^{-0.5} h S^{1.5} + 3\kappa S}$$

$$= \frac{AR^{0.5}(\rho AR^{-0.5} h S^2 + 3\kappa S^{1.5})}{(\xi + 3\kappa)S + \rho AR^{-0.5} h S^{1.5}} \approx \frac{\rho h S^2}{\xi S + \rho AR^{-0.5} h S^{1.5}} \sim \frac{\rho h}{\xi} S \quad (6.90)$$

The ratio of h_{cg} to the overall ADS length is then

$$\bar{h}_{cg} = \frac{h_{cg}}{l} = \frac{\rho AR^{-0.5} h S^{1.5} + 3\kappa S}{(\xi + 3\kappa)S + \rho AR^{-0.5} h S^{1.5}} \approx \frac{\rho AR^{-0.5} h S^{1.5}}{\xi S + \rho AR^{-0.5} h S^{1.5}} \sim \frac{\rho h}{\xi \sqrt{AR}} \sqrt{S} \quad (6.91)$$

Figures 6.22 and 6.23 show both dependencies [for more accurate and simplified relations described by Eqs. (6.90) and (6.91)] vs PADS canopy area for two wing loading quantities to demonstarte that all 20 setups of a Goodrick's model fall in between these two dependencies. As seen from Fig. 6.23, with increase of the

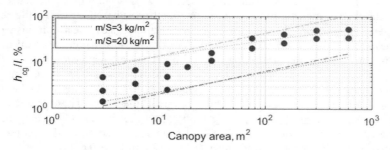

Fig. 6.23   Ratio of ADS c.g. height above payload c.g. to overall length of PADS vs canopy area.

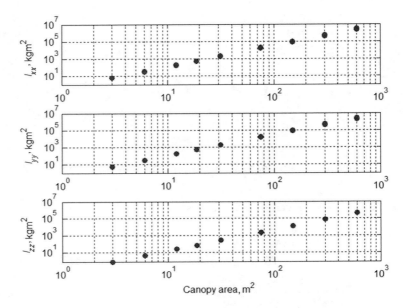

Fig. 6.24   PADS moments of inertia vs canopy area.

canopy size (volume of entrapped air) the PADS c.g. moves up toward a canopy, which justifies a necessity of using longer risers.

For these setups, the principal moments of inertia in the coordinate frame $\{b\}$ are presented in Fig. 6.24. Figure 6.25 shows the same moments of inertia normalized by $m_s h_{cg}^2$.

Having the aforementioned 20 setups for his model, Goodrick considered two methods of turn control. The universally accepted method today that involves downward deflection of a portion of TE results in a yaw and sideslip toward the side of deflection. As discussed in Sec. 6.3.2, this sideslip generates a side force that banks the canopy. An alternative method is based on tilting the resultant aerodynamic force vector into the turn directly. This can be accomplished by deflecting the inboard tip of a canopy (More details on possible mechanization, that is, the appropriate shortening of certain groups of suspension lines, will be given in Sec. 8.2.4.) To achieve the same effectiveness as TE deflection, the desired tilt happens to be very small [Goodrick 1979]. To this end, Fig. 6.26 shows matching simulations for both control methods. These simulations involve wing loading of 4.9 kg/m^2 (1 lb/ft^2), and the deflected TE being 12.5% of the span located from 50 to 75% half-span measured from the center. (In terms of Fig. 2.46, it corresponds to $b_k/b = 0.125$ and $\bar{l} \approx 0.625$.)

As seen from Fig. 6.26, as wing area decreases, the turn rate increases (for the same control input). Depending on wing loading, there is a limiting value of turn

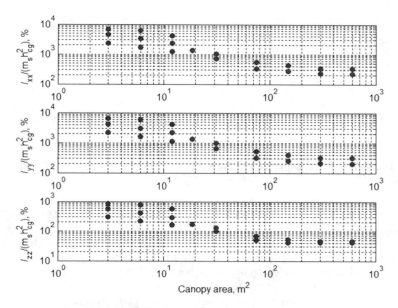

Fig. 6.25   Normalized moments of inertia vs canopy area.

rate above which the turn becomes divergent and PADS enters a spiral, which tightens to become a spin. That's why the small values of b_k/b and $\bar{l}$ were chosen for these simulations. For a large canopy (the last but one data point in Fig. 6.26), deflecting the outer 25% of the span leads to a quadruple increase of the turn rate from 2.3 deg/s up to 9.5 deg/s. Compatible performance would be obtained for 9-deg tilt (4.5-fold increase compared to 2 deg). Goodrick [1984] shows that the turn rate increases with tilt almost linearly for both small and

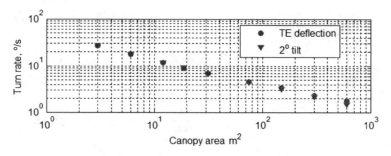

Fig. 6.26   Control authority vs canopy area.

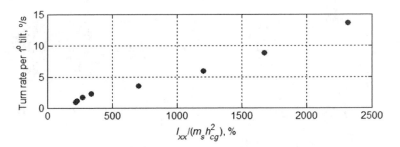

Fig. 6.27 Control authority per 1-deg tilt vs normalized roll moment of inertia.

large systems. A linear relation is also observed between the turn rate per 1-deg tilt and $I_{xx}/(m_s h_{cg}^2)$ (Fig. 6.27).

In his study, Goodrick observed both short-period and long-period (phugoid) motion and also noticed that although for small systems the amplitude of the short-period oscillations is negligible, it becomes larger for large systems. The phugoid motion is significant for all-size systems. As shown in Fig. 6.1, it appears in both roll and pitch affecting turn and glide dynamics, respectively. As a result, the turn response to the control inputs as well as all sensor

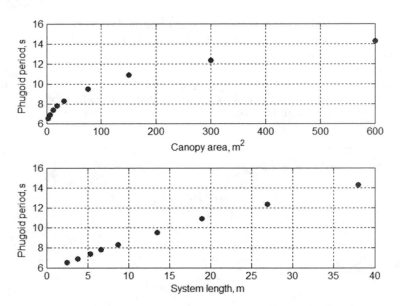

Fig. 6.28 Phugoid period vs PDS canopy area and length.

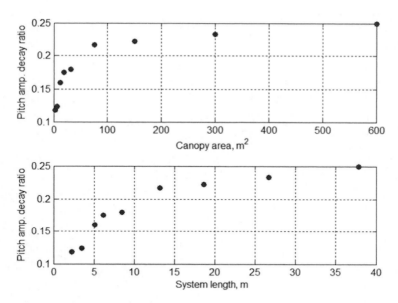

Fig. 6.29   Pitch amplitude decay ratio vs PADS canopy area and length.

measurements by AGU will be affected (i.e., it needs to be taken into account in GNC algorithms). The two characteristics of the phugoid motion are period and amplitude decay ratio. Whereas the period varies linearly with the system length (proportional to the square root of canopy area) as shown in Fig. 6.28, the decay ratio varies with the system size (length) nonlinearly (Fig. 6.29).

CHAPTER 7

Guidance, Navigation, and Control

Oleg Yakimenko*

Naval Postgraduate School, Monterey, California

Thomas Jann[†]

Institute of Flight Systems of DLR, Braunschweig, Germany

As indicated in Chapter 1, all self-guided PADS carry aerial guidance unit (AGU), which is responsible for understanding where the parafoil system is with respect to a desired landing point, what environment (winds aloft) it flies through, generating a desired path that leads to the desired landing point and generating and applying control actions to drive a parafoil along the desired trajectory. Although these tasks are common tasks for any autopilot, because of specific properties of PADS there are some features that distinguish them from those of other autonomous systems (underwater, surface, ground, aerial).

Although general guidance strategy realized in autopilots of PADS described in Chapter 1 is well understood, the specifics are usually not known. In this chapter an attempt is made to classify different guidance, navigation, and control (GNC) approaches and provide the details on those that have been published. As such, this chapter is based on [Calise and Preston 2008; Calise and Preston 2009; Carter et al. 2007; Culpepper et al. 2013; Hattis et al. 2006; Jann 2001; Jann 2004; Jann 2005; Kaminer et al. 2005; Prakash and Ananthkrishnan 2006; Slegers and Costello 2005; Slegers and Yakimenko 2009; Slegers and Yakimenko 2011; Yakimenko and Slegers 2011] and is organized as follows. It starts from a general discussion of GNC system structure and algorithms as applied to the self-guided PADS. It then proceeds with the navigation tasks that need to be addressed in flight. Next, several guidance concepts are introduced. Finally, several control approaches to follow the desired path are presented.

7.1 BACKGROUND AND GENERAL TERMS

The typical trajectory of a self-guided PADS is shown in Fig. 7.1. Upon deployment and acquisition of the GPS signal, it glides towards a landing area, the stage commonly called homing. Upon arrival to a vicinity of the landing area,

---

*Professor.
[†]Research Scientist.

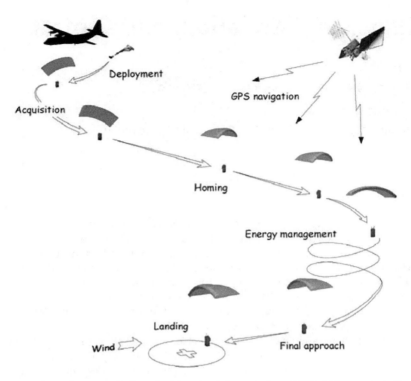

Fig. 7.1   General representation of a sequence of events from PADS deployment until soft landing [Wegereef and Jentink 2003].

also referred to as intended point of impact (IPI) or simply a target, PADS, dissipates an excess of potential energy, the stage called energy management (EM), and then performs a final maneuver attempting to land into the wind as close to IPI as possible.

To assist PADS in performing these tasks, AGU runs a variety of GNC algorithms. A functional representation of AGU is shown in Fig. 7.2. The two major objectives of navigation are understanding the current PADS position and attitude with respect to IPI and also estimating atmospheric conditions, including winds. The guidance block accepts this information to provide with the best possible solution allowing getting from the current PADS position to IPI. Control block ensures stable flight behavior, satisfactory tracking of a trajectory generated by the guidance block, and timely implementation of the flare maneuver for the soft landing. Some GNC concepts consider the guidance and control blocks as a single block that generates some predetermined maneuvers depending on the outputs of the navigation block (i.e., there is no path generation and path following stages involved). The inner loop of the control block generates its commands

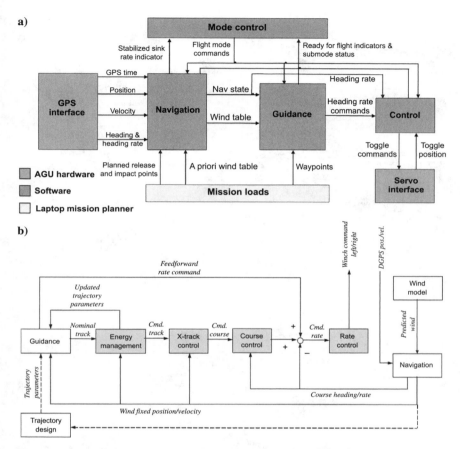

Fig. 7.2 Examples of a functional representation of GNC concept for a) MegaFly and b) X-38 [Carter et al. 2009; Hattis et al. 2006; Soppa et al. 2005].

in terms of a desired yaw rate $\dot\psi_c$, and the outer loop then converts these commands into the desired inputs to two motors to deflect left and right flaps of a canopy (outer portions of the trailing edge). A priori known data may include winds aloft, loaded into the board computer as a look-up table, other atmospheric parameters, including a barometric pressure at IPI, desired waypoints (WPs), and/or topographic maps.

For determining a horizontal PADS position, relative to IPI, the earlier studies of the pre-GPS era could only rely on using multiple radio beacons [Goodrick 1969; Goodrick 1970; Goodrick et al. 1973; Goodrick 1979a; Murphy 1971]. Several well-known navigation aids were considered. For example, Fig. 7.3a features a scheme relying on measuring bearings to two

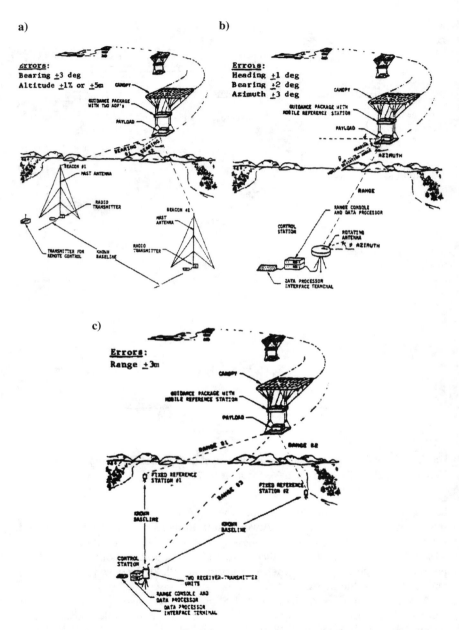

Fig. 7.3   Navigation schemes: a) theta-theta, b) rho-theta, and c) rho-rho-rho [Goodrick 1979a].

ground beacons, Fig. 7.3b incorporates measuring the bearing and range to a single beacon, and Fig. 7.3c shows measuring a range to three stations. The first two schemes resemble multiple short-range VOR radio beacons and a combined VOR/DME radio navigation stations used in modern aviation; the latter one can be considered as a GPS predecessor. Obviously, not only the accuracy of these navigation aids as applied to PADS was sufficient to build robust AGU algorithms, but also these schemes could not support PADS autonomy. (Nevertheless, Sec. 7.3.2 addresses terminal guidance suggested by these early studies in more details.)

Thanks to GPS, a new push to develop GNC for autonomous PADS began in the mid-1990s. These days AGUs of the vast majority of commercial PADS rely on GPS data and GPS data only (Fig. 7.2a). These low-rate data include synchronized PADS position in $\{I\}$, components of the groundspeed vector $\mathbf{V}$ (also in $\{I\}$), and a ground track angle χ (refer to Fig. 5.1c). Some PADS are also equipped with barometric pressure sensor to allow more accurate estimation of the altitude above IPI. For the very same reason, some attempts were made to possibly incorporate a SODAR (sonic detection and ranging) or LIDAR (light detecting and ranging) as remote-sensing technologies that enable more accurate measurements of an attitude above the ground (IPI). To improve PADS autonomy and robustness, an inexpensive inertial measurement unit (IMU) can also be incorporated, as in ALEX and Snowflake PADS do. In this case, additional high-rate measurement data also include linear accelerations and angular rates (measured in $\{b\}$ or $\{s\}$), and possibly Euler angles, roll, pitch, and yaw. Larger experimental platforms (like ALEX or X-38 CRV) can also be equipped to measure components of the airspeed vector usually measured as the airspeed V_a, angle of attack α, and angle of sideslip β.

To conclude this section, let us emphasize that low-speed unpowered PADS strongly rely on the knowledge of the winds. First, knowing the winds aloft ensures that PADS released at a certain altitude can possibly reach the IPI area. Second, for autonomous landing the autopilot must know the direction of the wind. This is because landing against the wind reduces the groundspeed preventing payload roll-over. It also enables a flare maneuver reducing the vertical and horizontal speed at impact. Finally, some assumption on the surface-layer winds model is necessary to plan and execute a terminal landing maneuver. That is why among other navigational tasks addressed in the next section estimation of winds and accommodation of these estimates in the guidance block plays a key role.

7.2 NAVIGATION SOLUTIONS

As presented in the preceding section, the navigation block provides the guidance block with PADS current position. These data are directly available from an onboard GPS receiver along with three components of the groundspeed vector V_x, V_y, $V_z = \dot{h}$ (we assume NED coordinate frame to be our $\{I\}$). To support the control block, the navigation block must also provide the current heading rate $\dot{\chi}_a$. However with GPS only, this information might not be readily available. The only assumption that can be made to simplify derivation of estimate of $\dot{\chi}_a$ is that in a steady flight the sideslip angle is small, $\beta \approx 0$, leading to PADS heading being approximately equal to the yaw angle (Fig. 5.1c)

$$\chi_a \approx \psi \tag{7.1}$$

so that

$$\dot{\chi}_a \approx \dot{\psi} \tag{7.2}$$

Yet, in order to define angle ψ and consequently its derivative $\dot{\psi}$, we need to know several other parameters, V_h, w_x, and w_y. Having a pitot gauge or simply knowing the horizontal component of the airspeed vector V_h from the previous tests does help, but we still need the estimates of horizontal components of the wind, w_x, w_y. This section addresses the question of estimation of ψ, $\dot{\psi}$, V_h, w_x, and w_y having the different sets of data available.

7.2.1   ESTIMATION OF YAW/HEADING

We start from the first two equations of Eq. (5.30) rewritten with account of Eq. (7.1)

$$\begin{aligned} V_x &= V_h \cos(\psi) + w_x \\ V_y &= V_h \sin(\psi) + w_y \end{aligned} \tag{7.3}$$

that can be resolved for ψ as

$$\hat{\psi} = \tan^{-1}\left(\frac{V_y - \hat{w}_y}{V_x - \hat{w}_x}\right) \tag{7.4}$$

(hereinafter the symbol "$\hat{\ }$" is used to represent an estimated quantity). Hence, once the estimates of w_x and w_y are available, knowing the north and east components of the groundspeed vector $\mathbf{V}$, V_x and V_y, from GPS, we obtain an estimate of the current yaw angle. Its derivative can be estimated by differentiating Eq. (7.4) [Jann 2005]

$$\hat{\dot{\psi}} = \frac{\dot{V}_y(V_x - \hat{w}_x) - \dot{V}_x(V_y - \hat{w}_y)}{(V_x - \hat{w}_x)^2 + (V_y - \hat{w}_y)^2} \tag{7.5}$$

Because differentiation of V_x and V_y might not be a good idea, the estimate of a yaw rate $\dot\psi$ can also be obtained by applying any backward difference approximation formula, for example,

$$\hat{\dot\psi}_i = \frac{3\hat\psi_i - 4\hat\psi_{i-1} + \hat\psi_{i-2}}{2\Delta t} \tag{7.6}$$

(where Δt is a sampling time) for the three-point backward difference approximation, or

$$\hat{\dot\psi}_i = \frac{11\hat\psi_i - 18\hat\psi_{i-1} + 9\hat\psi_{i-2} - 2\hat\psi_{i-3}}{6\Delta t} \tag{7.7}$$

for the four-point backward difference approximation.

If compass reading ψ and IMU data r are available, then the yaw angle ψ and yaw rate $\dot\psi$ can be estimated directly. Compass heading can be improved to obtain the yaw angle by using a complementary filter, which eliminates the errors during a turn maneuver with a large bank angle. These errors are caused by an inclinometer trying to correct the magnetometer raw data by using the apparent instead of the real perpendicular. The classical linear complementary filter (LCF) with a constant feedback gain is shown in Fig. 7.4a. Even a better robustness can be achieved with a nonlinear complementary filter (NCF) with a feedback gain adaptive to the yaw rate r (Fig. 7.4b).

Figure 7.5 shows the results of implementing both filters of Fig. 7.4 to process raw data collected on the ALEX PADS. Using LCF with a feedback gain of 0.1, the maximum error during maneuvering was reduced to less than 30 deg instead of more than 100 deg, and the root mean squared error (RMSE) was reduced to 13.4 deg as opposed to 24.5 deg. On the other hand, NCF allowed reducing the maximum errors during dynamic turns to 20 deg and the RMSE—down to 7.6 deg.

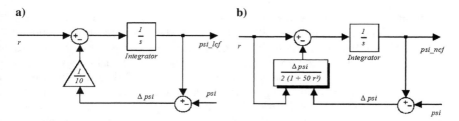

Fig. 7.4   Block diagrams for a) linear and b) nonlinear complementary filters [Jann 2005].

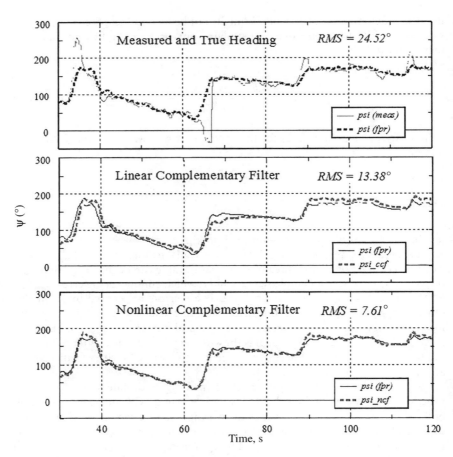

Fig. 7.5 Heading (yaw angle) reconstruction with LCF and NCF [Jann 2005].

7.2.2 ESTIMATION OF CURRENT WINDS AND AIRSPEED

Equation (7.4) relies on wind estimates, so let us consider how these estimates could be obtained using the different sets of data coming from onboard sensors. Under the assumption that wind profile and airspeed can be treated as constant, the wind can easily be estimated using GPS data only. Consider PADS flying with the wind (which in aviation is called downwind direction) having a horizontal component of an airspeed vector V_h. Then the ground speed will be measured by GPS as $V_{\text{downwind}}^{\text{GPS}} = V_h + W$ where $W = |\mathbf{W}| \approx (w_x^2 + w_y^2)^{0.5}$ is a magnitude of a wind's horizontal component. If PADS flies against the wind (upwind), GPS senses $V_{\text{upwind}}^{\text{GPS}} = V_h - W$. From here both horizontal component of an airspeed vector and horizontal component of wind can be estimated as

$$V_h = 0.5(V_{\text{downwind}}^{\text{GPS}} + V_{\text{upwind}}^{\text{GPS}})$$
$$W = 0.5(V_{\text{downwind}}^{\text{GPS}} - V_{\text{upwind}}^{\text{GPS}})$$

(7.8)

In a similar manner, while executing a constant-control-input turn under no wind conditions produces a right circle; in a windy environment it results in a stretched cycloid (Fig. 7.6). Hence, when the vehicle has completed its 360-deg

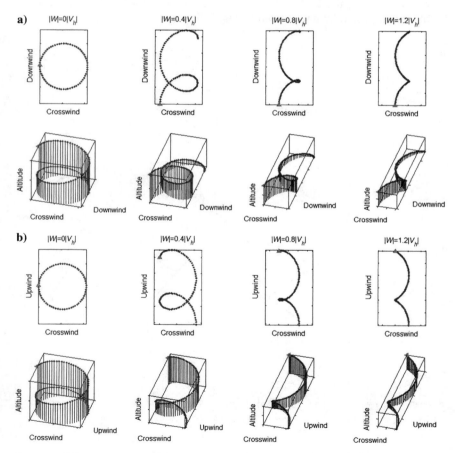

Fig. 7.6   PADS trajectories when a turn maneuver starts in the a) downwind and b) upwind direction with a different magnitude of wind.

turn heading in the same direction a maneuver started from, the wind drift can be read off directly from GPS position data

$$W = \frac{\sqrt{(x_f - x_0)^2 + (y_f - y_0)^2}}{t_f - t_0}$$

$$\chi_W = \tan^{-1}\left(\frac{y_f - y_0}{x_f - x_0}\right)$$

(7.9)

Equation (7.9) can also be used for an "eight" maneuver (Fig. 7.7) although generally speaking the PADS airspeed is a function of the (symmetric) control inputs (as suggested by data shown in Fig. 5.39).

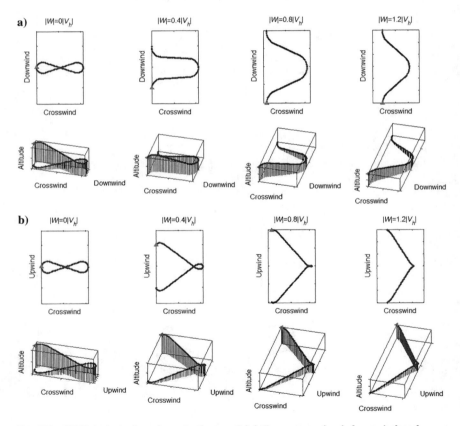

Fig. 7.7 PADS trajectories when starting an "eight" maneuver in a) downwind and b) upwind direction with different magnitudes of wind.

An alternative to Eq. (7.9) would be measuring the minimum $V_{min}^{GPS} = V_h - W$ and maximum $V_{max}^{GPS} = V_h + W$ speed while executing maneuvers involving a complete 360-deg turn. Then,

$$V_h = 0.5(V_{max}^{GPS} + V_{min}^{GPS})$$

$$W = 0.5(V_{max}^{GPS} - V_{min}^{GPS}) \qquad (7.10)$$

$$\chi_W = 0.5(\chi_{V_{max}^{GPS}} + \chi_{V_{min}^{GPS}} + \pi)$$

Without executing a complete 360-deg turn (circles or eights), say while PADS glides at a constant and small turn rate (to eliminate dependence on the varied control inputs), we can employ a general parameter estimation procedure as follows. Starting from an obvious relation

$$V_h^2 = (V_x - w_x)^2 + (V_y - w_y)^2 \qquad (7.11)$$

we can rewrite it as a function of three unknown (constant) variables

$$f(w_x, w_y, V_h) = V_h^2 - (V_x - w_x)^2 - (V_y - w_y)^2 \qquad (7.12)$$

Linearization of Eq. (7.12) using the Taylor-series expansion yields

$$\delta f = f(w_x, w_y, V_h) = \left[2(V_x - w_x), 2(V_y - w_y), 2V_h\delta V_h\right]$$

$$= \begin{bmatrix} \delta w_x \\ \delta w_y \\ \delta V_h \end{bmatrix} = \mathbf{h}\delta\mathbf{x} \qquad (7.13)$$

Having k consecutive measurements of V_x and V_y, the problem reduces to solving a set of k nonlinear algebraic equations

$$f_i(w_x, w_y, V_h) = V_h - \sqrt{(V_{x;i} - w_x)^2 + (V_{y;i} - w_y)^2} = 0, \quad i = 1, \ldots, k \qquad (7.14)$$

Equation (7.13) takes the form

$$\delta\mathbf{f}_{n\times1} = \mathbf{H}_{n\times3}\delta\mathbf{x}_{3\times1} \qquad (7.15)$$

where

$$\mathbf{H} = \begin{bmatrix} h_{1;1} & h_{2;1} & h_{3;1} \\ h_{1;2} & h_{2;2} & h_{3;2} \\ \vdots & \vdots & \vdots \\ h_{1;k} & h_{2;k} & h_{3;k} \end{bmatrix} \qquad (7.16)$$

The iterative procedure for solving the set of Eq. (7.14) then looks as follows:

1. Pick initial guess for w_x, w_y, and V_h ($j = 1$) $\hat{\mathbf{x}}_j$

2. Form a discrepancies vector $\delta\mathbf{z}_j = -\mathbf{f}(\hat{\mathbf{x}}_j)$

3. Compute the least-squares solution for Eq. (7.15) $\delta\hat{\mathbf{x}}_j = (\mathbf{H}^T\mathbf{H})^{-1}\mathbf{H}^T \delta\mathbf{f}_j$

4. Update the last guess $\hat{\mathbf{x}}_{j+1} = \hat{\mathbf{x}}_j + \delta\hat{\mathbf{x}}_j$

5. Go to step 2 and repeat until convergence   (until $\|\hat{\mathbf{x}}_{j+1} - \hat{\mathbf{x}}_j\| \leq \varepsilon$)

If the airspeed V_h is measured by a pitot gauge or known somehow else (and still assumed constant), then the nonlinear function in Eq. (7.12) becomes a function of only two variables

$$f(\mathbf{w}) = f(w_x, w_y) = V_h^2 - (V_x - w_x)^2 - (V_y - w_y)^2 \qquad (7.17)$$

and the measurement matrix $\mathbf{H}$ of Eq. (7.16) "loses" the third column. The remainder of the aforementioned procedure is the same.

A similar procedure, excluding the iterative process and utilizing a set of linear algebraic equations in w_x and w_y, was also suggested in [Ward et al. 2012]. Having k consecutive measurements of V_x and V_y, we can rewrite Eq. (7.11) for the ith measurement as

$$\begin{aligned}
V_h^2 &= (V_{x;i} - w_x)^2 + (V_{y;i} - w_y)^2 \\
&= (V_{x;i}^2 + V_{y;i}^2) + (w_x^2 + w_y^2) - 2(V_{x;i}w_x + V_{y;i}w_y)
\end{aligned} \qquad (7.18)$$

Considering that the control input is not changing, and assuming constant airspeed and wind components, Eq. (7.18) can be simplified by subtracting the expected values of the measured quantities

$$V_h^2 - E(V_h^2) = 0 = V_i^2 - E(V_i^2) - 2\{[V_{x;i} - E(V_{x;i})]w_x + [V_{y;i} - E(V_{y;i})]w_y\} \qquad (7.19)$$

In Eq. (7.19), $V_i^2 = V_{x;i}^2 + V_{y;i}^2$ and expected values are estimated by sample means

$$V_i^2 = \mu_{V^2}, \quad E(V_{x;i}) = \mu_{V_x}, \quad E(V_{y;i}) = \mu_{V_y} \qquad (7.20)$$

Having multiple measurements results in a overdetermined system of linear algebraic equations

$$\begin{bmatrix} V_{x;1} - \mu_{V_x} & V_{y;1} - \mu_{V_y} \\ \vdots & \vdots \\ V_{x;k} - \mu_{V_x} & V_{y;k} - \mu_{V_y} \end{bmatrix} \begin{bmatrix} w_x \\ w_y \end{bmatrix} = \frac{1}{2} \begin{bmatrix} V_1^2 - \mu_{V^2} \\ \vdots \\ V_k^2 - \mu_{V^2} \end{bmatrix} \qquad (7.21)$$

that can be easily solved in a least-squares sense. The steady-state value of the horizontal component of airspeed is obtained as the average of the airspeed estimates over a particular constant-control flight segment

$$\hat{V}_{h;i} = \sqrt{(V_{x;i} - w_x)^2 + (V_{y;i} - w_y)^2}, \quad \hat{V}_h = \mu_{V_h} \tag{7.22}$$

Having more GPS readings of the ground-speed vector components during a constant-rate turn results in a better-conditioned system (7.21). Ward et al. [2012] determined the upper bound on the quality of a solution by a range of heading change during a maneuver $\Delta\psi$

$$\sigma(\hat{V}_h) < \frac{\sigma(V^{\text{GPS}})}{\sin[0.25 \min(|\Delta\psi|, 2\pi)]^2} \tag{7.23}$$

Obviously, a complete 360-deg turn results in a best dilution of precision of 1 as shown in Fig. 7.8. In this case the airspeed estimate error is bounded by the accuracy of GPS sensor $\sigma(V^{\text{GPS}})$. For a 180-deg turn, dilution of precision doubles, and decreasing $\Delta\psi$ even further leads to more uncertainty.

Obviously, if the onboard sensor suite includes not only GPS sensor, but also IMU, compass, and a pitot gauge, then the wind components can be estimated directly. [Again, we are operating under the assumption that Eq. (7.1) holds.]

$$\begin{aligned} w_x &= V_x - V_h \cos(\psi) \\ w_y &= V_y - V_h \sin(\psi) \end{aligned} \tag{7.24}$$

For estimating a constant wind profile from a turbulent environment with noisy measurements, the recursive mean value of the wind components can be computed as follows [Jann 2005]:

$$\begin{aligned} \bar{w}_{x;k+1} &= (k\,\bar{w}_{x;k} + w_x)/(k+1) \\ \bar{w}_{y;k+1} &= (k\,\bar{w}_{y;k} + w_y)/(k+1) \end{aligned} \tag{7.25}$$

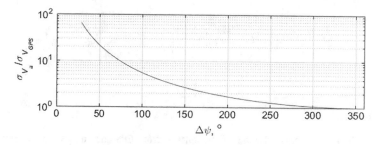

Fig. 7.8   Dilution of precision $\sigma(\hat{V}_h)/\sigma(V_{\text{GPS}})$ vs $\Delta\psi$.

In Eq. (7.25), k is the continuously increasing number of samples since the beginning of the measurements. As an example, Fig. 7.9 shows the true and estimated wind components for one simulated flight of the ALEX PADS obtained using Eqs. (7.24) and (7.25) (further referred to as the classical method).

For the sake of comparison, Fig. 7.9b shows an example of estimating wind components using a nonlinear estimation filter (NEF) shown in Fig. 7.10 that utilizes GPS data only.

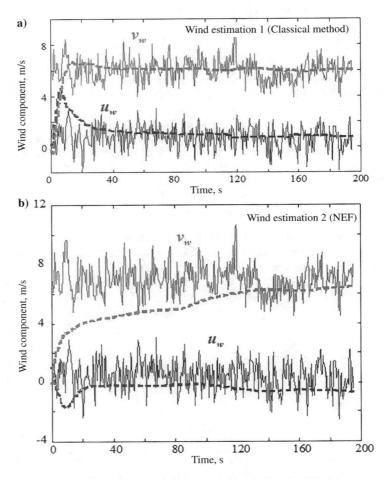

Fig. 7.9   True and estimated wind components a) from GPS and compass measurements and b) using NEF [Jann 2005].

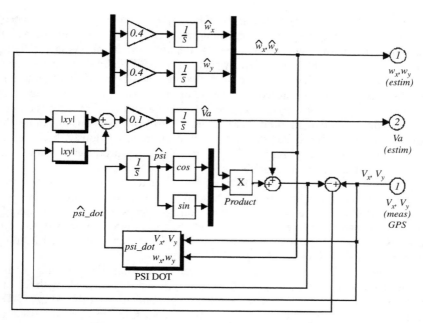

Fig. 7.10 NEF for wind and airspeed [Jann 2005].

Instead of using Eq. (7.24), this filter utilizes the following equations:

$$\hat{w}_x = 0.4 \int \left[V_x^{GPS}(t) - \hat{V}_h(t) \cos \hat{\psi}(t) \right] dt$$

$$\hat{w}_y = 0.4 \int \left[V_y^{GPS}(t) - \hat{V}_h(t) \sin \hat{\psi}(t) \right] dt \qquad (7.26)$$

Here an estimate of the heading angle $\hat{\psi}$ is obtained by integrating Eq. (7.5) in the PSI DOT block of the Simulink block diagram shown in Fig. 7.10, and the estimate of the horizontal component of airspeed is obtained as

$$\hat{V}_h = 0.1 \int \left(\left\| \left[V_x^{GPS}(t), V_x^{GPS}(t) \right] \right\| - \left\| \left[\hat{V}_x(t), \hat{V}_y(t) \right] \right\| \right) dt \qquad (7.27)$$

where

$$\hat{V}_x(t) = \hat{V}_h(t) \cos \hat{\psi}(t) + \hat{w}_x$$

$$\hat{V}_y(t) = \hat{V}_h(t) \sin \hat{\psi}(t) + \hat{w}_x \qquad (7.28)$$

As seen from Fig. 7.9b, using the GPS-only filter of Fig. 7.10 takes more time to converge to the correct values of w_x and w_y as compared to Fig. 7.9a [based on Eqs. (7.24) and (7.25) utilizing the GPS/IMU/compass/pitot gauge sensor suite]. Speed and quality of the estimation depend very much on the actual flight pattern of the vehicle. Best results are obtained if the system is flying circles or S-turns (which is proved by Fig. 7.8). If PADS glides along a straight path for a long time, then GPS data do not contain sufficient information to distinguish between motion caused by the wind or the own airspeed of the vehicle. In this case, the quantities are correlated and cannot be estimated reliably. If a horizontal projection of an airspeed vector in Eq. (7.26) is set to a constant value, the correlations are reduced but still not resolved entirely. Obviously, because the wind estimation is rather sensitive to V_h, the filter only produces accurate estimates if this parameter is accurate.

Figure 7.11 shows the time history of reconstructed heading $\hat{\psi}$ for the same data set of Fig. 7.9. Time delay, noise, and sampling are results of the GPS sensor model used. The bias, which temporarily reaches even 30 deg, is mainly due to the errors in the estimated wind. Because these errors are reduced towards the end of the flight as shown in Fig. 7.9b, the reconstructed heading becomes more accurate.

To demonstrate the importance of accurate estimation of wind components, airspeed and yaw rate, the preceding two approaches (classical method and NEF) were incorporated into a self-guided flight simulation of the ALEX PADS (Fig. 7.12) employing the identical guidance algorithm (further discussed in Sec. 7.4.1).

To evaluate the effects of two wind estimation paradigms (of Fig. 7.9) being a navigational data source, Monte Carlo simulations using a block diagram of Fig. 7.12a were executed. Figure 7.13a demonstrates a general (benchmark) performance of GNC algorithm while varying initial position within ±500 m (1,640 ft), altitude within 1,000 ± 200 m (3,280 ± 656 ft), and winds within of 0 to 8 m/s (0 to 15.6 kt) coming from a different known direction. As seen, all landings occur within the 50-m (164-ft) radius while the landing direction deviates less than ±10 deg from the exact direction against the wind. Figure 7.13b shows how PADS performance degrades when sensor, wind, and system parameter errors are introduced. Sensor- and wind-parameter stochastic errors used in these simulations are presented in Table 7.1. Even though the landing precision degrades, because the optimal trajectory is continuously updated, 60% of the landings still occur within 50 m (164 ft), and less than 10% leave the 100-m (328-ft) radius. On several occasions, unfavorable combinations of errors and initial conditions prevent PADS from reaching IPI.

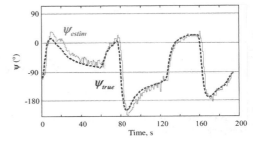

Fig. 7.11   True and reconstructed heading [Jann 2005].

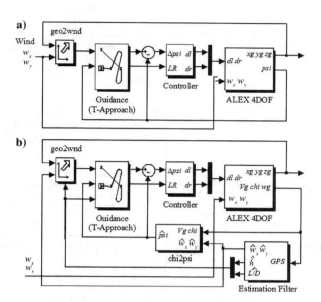

Fig. 7.12   Simulation block diagram a) without and b) with onboard wind estimation using NEF [Jann 2005].

For the very same simulations with errors described in Table 7.1, Figs. 7.13c and 7.13d show implementation of the two different wind estimation schemes. (Both simulations utilized a block diagram of Fig. 7.12b.) Figure 7.13c shows the results of the scheme utilizing GPS/IMU/compass/pitot gauge suit. As seen, using the enhanced sensor package allows mitigation of the errors' effects and bringing the overall performance to a circular error probable (CEP) of less than 50 m (164 ft). Figure 7.13d shows the results of utilizing NEF of Fig. 7.10. Initial conditions were set to zero for both wind components and to a $\pm 20\%$ erroneous value for the airspeed. The results shown in Fig. 7.13d demonstrate that about 50% of landings occur within the 50-m (164-ft) and 90% within the 100-m (328-ft) radius around IPI. Hence, landing accuracy is almost the same as for case 2 (Fig. 7.13b) although, in contrast to case 2, no initial wind information was required at all.

Concerning the landing direction, about 80% of landings in these simulations were within the ± 30-deg sector. However, most of the 20% off-direction landings happened at very low wind conditions, where small errors in the estimated wind components caused strong deviations in the estimated wind direction. On the other hand, the importance of the landing direction is much less in such low wind conditions than with a stronger wind.

To conclude this section, let us show one more algorithm used by the Draper Laboratory [Carter et al. 2009]. Along with the measurements of V_x and V_y,

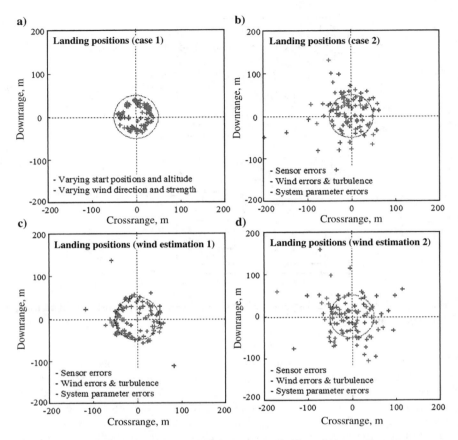

Fig. 7.13   Miss distance a) without errors in sensors, wind knowledge, and system parameters; b) with errors given in Table 7.1; c) with errors but good estimates of winds using classical method; and d) with errors and less precise estimates of winds using NEF [Jann 2005].

provided by the onboard GPS receiver, this algorithm also relies on knowing the descent rate $-\dot{h}$, also measured by GPS, and lift-to-drag ratio, assumed to be known from the previous flight tests.

The estimation of a vector $\mathbf{w} = [w_x, w_y]^T$ (assumed to be constant) is based on minimizing a nonlinear function of Eq. (7.17) not using an iterative procedure outlined after Eq. (7.16) or closed solution of Eq. (7.21) but rather using the iterative process based off the Newton–Raphson formula

$$\mathbf{w}_{k+1} = \mathbf{w}_k - \alpha \left[\frac{\partial f(\mathbf{w})}{\partial \mathbf{w}}\right]^+_{\mathbf{w}_k} f(\mathbf{w}_k) \qquad (7.29)$$

TABLE 7.1   ERRORS IN MEASURING THE PADS STATES

Parameter	Error
GPS error (position/altitude)	$\pm 5/10$ m (16.4/32.8 ft)
Compass error (bias/noise)	$\pm 3/5$ deg
Wind-speed error	± 2 m/s (3.9 kt)
Wind-direction error	± 20 deg
Turbulence	$\sigma = 0 \ldots 1.6$ m/s (5.2 ft/s)
Glide ratio	$\pm 20\%$
Sink rate	± 1 m/s (3.3 ft/s)

with the gain α less than unity. The Moore–Penrose pseudo inverse of the gradient of function has the form

$$\left[\frac{\partial f(\mathbf{w})}{\partial \mathbf{w}}\right]^{+}_{\mathbf{w}_k} = \frac{-1}{2[(V_{x;k} - w_x)^2 + (V_{y;k} - w_y)^2]} \begin{bmatrix} V_{x;k} - w_x \\ V_{y;k} - w_y \end{bmatrix} \tag{7.30}$$

The values of $V_{x;k}$ and $V_{y;k}$ are supplied by GPS (at a rate of 1 Hz), and the value of $V_{h;k}$ [entering $f(\mathbf{w}_k)$] is computed as

$$\hat{V}_{h;k} = \frac{L}{D}\hat{V}_{v;k} \tag{7.31}$$

where the estimate of the vertical component of airspeed $\hat{V}_v = -\dot{h}$ is obtained from a filtered GPS solution as well.

Obviously, a capability to estimate key PADS parameters onboard during the flight [e.g., airspeed, descent rate, glide ratio (GR)] not only simplifies the preparation of airdrop missions (because a lesser amount of data needs to be entered into AGU prior to flight), but also makes the GNC more robust because system changes or failures (e.g., in the case of not completely disreefed or damaged canopy) affecting these parameters can be taken into account.

For the high-altitude drops, the airspeed estimates obtained at the higher altitudes $\hat{V}_{h_1}(h_1)$ should be corrected to the lower-altitude air density using either a standard atmosphere approximation $\rho_h = \rho(h)$ or actual (measured) air density $\rho_h = \rho_k$ of a nonstandard atmosphere (if available). Similar to Eq. (3.17) or (3.36), we can write

$$\hat{V}_h = \sqrt{\frac{\rho_{h_1}}{\rho_h}}\hat{V}_{h_1} \tag{7.32}$$

Calise and Preston [2009] suggested correcting the airspeed V and its horizontal component V_h for the bank angle (if the airspeed estimates were obtained in a constant turn), but this correction seems to be applicable to the smaller PADS only while the larger PADS do not exhibit large bank angle doing a skid turn instead, so that the horizontal component V_h does not change much (Fig. 5.39b).

Onboard wind estimation algorithms allow obtaining reasonable estimates that can be used to extrapolate a wind profile until touchdown. Remaining uncertainties in the estimated profile should then be compensated by the adaptive path planning.

7.2.3   DETERMINATION OF ALTITUDE ABOVE SURFACE

Obviously, having an error in estimating the height above the ground (above IPI) has a negative effect on the miss distance

$$Miss = GR\Delta h \qquad\qquad (7.33)$$

Data presented in Fig. 7.14 allow the performing of a rough estimate of this effect for PADS having different GR. While data shown assume no ground wind condition, landing into the strong wind mitigates this effect, and landing with a wind worsens it.

Also shown in Fig. 7.14 with the horizontal lines are typical errors that are expected when using just a GPS receiver or just a barometric altimeter. Before it was turned off on 2 May 2000, typical selective availability (SA) errors were about 50 m (164 ft) horizontally and about 100 m (328 ft) vertically. (Because of the dilution of precision caused by GPS constellation geometry, the vertical accuracy is on the average of twice as worse as that of the horizontal one.) However, the advancement of technology means that today civilian GPS fixes under a clear view of the sky are on average accurate to about 5 m (16 ft) horizontally and 10 m (33 ft) vertically (and that is what is shown in Fig. 7.14) [GPS 2014; GPS Error Analysis 2014]. Real-world data from the Federal Aviation Administration (FAA) show that their high-quality civilian GPS service (SPS) receivers provide even better horizontal accuracy of the order of 3 m (9.8 ft). Surely, because the military GPS service (PPS) broadcasts on two frequencies

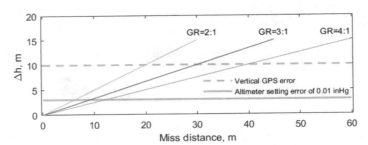

Fig. 7.14   Miss distance as a function of the high above IPI error.

(as opposed to SPS) to perform ionospheric correction, a technique that reduces radio degradation caused by the Earth's atmosphere, the PPS accuracy is reduced even further. On the contrary, adverse weather conditions, mountainous topography (or urban environment), intentional blocking, jamming, or interfering with the GPS signals may dramatically degrade accuracy.

Higher accuracy is attainable by using GPS in combination with augmentation systems, nationwide and global differential GPS system (NDGPS and GDGPS), wide-area augmentation system (WAAS), continuously operating reference stations (CORS), international GNSS service (IGS), and others. It is questionable, however, that much better accuracy could be achieved in the areas not covered by these services. That is why there is an interest to employ other high-accuracy sensors (radar, sodar, and lidar) onboard PADS [Dietz et al. 2007; Ulich et al. 2003]. Of course, the cheapest option would be to use DGPS or simply a barometric pressure. The bad thing is that they both require some additional setups in the vicinity of IPI to provide either a differential GPS signal or a barometric pressure on the ground.

The pressure altimeter is an aneroid barometer that measures the pressure P of the atmosphere at the level where the altimeter is located and returns an altitude that corresponds to this pressure for standard atmosphere. Air is denser at sea level than aloft, so as altitude increases, atmospheric pressure decreases. This difference in pressure at various levels causes the altimeter to indicate changes in altitude. To be more specific, the equation of state for an ideal gas (the ideal gas law) can be written as

$$\frac{dP}{P} = -\frac{dh}{RT} \tag{7.34}$$

where $R = (C_P - C_V)g^{-1} = 29.27 \text{ m/K}$ is the ratio of the specific gas constant for air to acceleration due to gravity (C_P is the specific heat for a constant pressure and C_V is the specific heat for a constant volume), and T is temperature in degrees Kelvin. For troposphere [when a geometric altitude above MSL $h < 11.019 \text{ km}$ (36,150 ft)], a nonisothermal layer where most of the cargo drops occurs, the standard atmosphere model assumes

$$T(h) = T_0 + \beta_T h \tag{7.35}$$

(For troposphere, the temperature lapse rate $\beta_T = -0.0065°\text{C/m}$ ($-0.002°\text{C/ft}$).] Substituting Eq. (7.35) into Eq. (7.34) results in

$$\frac{P(h)}{P_0} = \left(1 + \frac{h}{T_0}\beta_T\right)^{-\frac{1}{R\beta_T}} \tag{7.36}$$

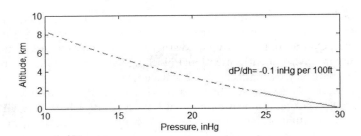

Fig. 7.15   The pressure model the barometric altimeter is based upon.

Integrating both sides yields

$$h = \frac{T_0}{\beta_T} \left[\left(\frac{P(h)}{P_0} \right)^{-R\beta_T} - 1 \right]$$ (7.37)

In Eqs. (7.35–7.37), $T_0 = 288.15\,\text{K}$, and $P_0 = 29.92\,\text{inHg}$. Barometric altimeters utilize Eq. (7.37). According to Eq. (7.36) for low altitudes [below 2 km (6,562 ft) MSL], air pressure drops about 0.1 inHg per every 30.5 m (100 ft) above sea level (see a fit in Fig. 7.15). This gradient decreases with altitude. As a practical example, Fig. 7.16a shows an altimeter setting (in Las Vegas, NV)

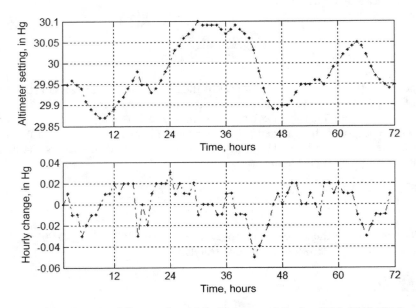

Fig. 7.16   Altimeter drift over three days [Jackson and Crocker 2014; NOAA 2014].

collected over the course of three days. Even though an hourly change does not usually exceed just 0.01–0.02 inHg (Fig. 7.16b), which corresponds to 3...6 m (10...20 ft), it is clear that the usage of several-hour-old data may cause real problems.

Obviously, the altitude readings are correct only when the sea-level barometric pressure, free air temperature, and temperature lapse rate are standard. Yet, this is never the case. Let us take a variation from both sides of Eq. (7.37)

$$\delta h = \frac{\delta T_0}{\beta_T}\left(\frac{P(h)}{P_0}^{-R\beta_T} - 1\right) - \frac{T_0}{\beta_T^2}\left(\frac{P(h)}{P_0}^{-R\beta_T} - 1\right)\delta\beta_T$$
$$+ \frac{T_0}{\beta_T}\left[-\frac{P(h)^{-R\beta_T}}{P_0^2}\delta P_0 - R\delta\beta_T\frac{P(h)^{-R\beta_T}}{P_0}\log\left(\frac{P(h)}{P_0}\right)\right] \quad (7.38)$$

Substituting Eq. (7.37) into Eq. (7.38) yields

$$\delta h = \frac{\delta T_0}{T_0}h - \frac{\delta\beta_T}{\beta_T}h$$
$$+ \frac{T_0}{\beta_T}\left[-\left(\frac{\beta_T}{T_0}h + 1\right)\frac{\delta P_0}{P_0} + \frac{\delta\beta_T}{\beta_T}\left(\frac{\beta_T}{T_0}h + 1\right)\log\left(\frac{\beta_T}{T_0}h + 1\right)\right] \quad (7.39)$$

Because $\beta_T T_0^{-1}h \ll 1$, we can neglect all but one term in the parentheses, and after dividing both sides of Eq. (7.39) by h obtain

$$\frac{\delta h}{h} \approx \frac{\delta T_0}{T_0} - \frac{\delta\beta_T}{\beta_T} - \frac{T_0}{h\beta_T}\frac{\delta P_0}{P_0} \quad (7.40)$$

This equation shows what happens if one (or all) of the aforementioned parameters varies. For example, temperatures warmer than standard will make the altimeter read lower than the actual altitude (this and other effects of a nonstandard atmosphere as applied to PADS are discussed in Watkins [2011]). However, there are no means to correct barometric altimeter readings for a nonstandard temperature!

The only adjustment to nonstandard conditions can be realized by correcting Eq. (7.37) to the current sea-level pressure [as seen from Eq. (7.40) the pressure variance may have a major effect, especially at low altitudes]. If not available at IPI, this corrected barometric pressure can be obtained from a flight services station or weather broadcast. Obviously, the further from IPI where the altimeter setting was determined (in altitude, distance, and time), the less accurate the altimeter readings become.

Combining both GPS and barometric altimeters allows obtaining the most accurate altitude readings. Absolute location is originally provided by GPS to help autocalibrate the barometric altimeter, then the barometric altimeter is used

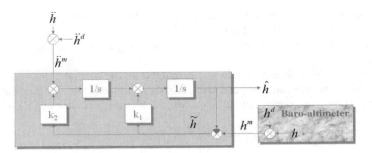

Fig. 7.17   Complementary filter blending $\ddot{h}$ signal with barometric altimeter sensor data.

to provide a more stable altitude change. In this case the barometric altimeter provides altitude readings even if a GPS signal is not available.

If along with the GPS or barometric pressure sensor AGU includes a three-axis accelerometer sensor, then the data quality may be improved by utilizing a complimentary filter similar to that of Fig. 7.4a. Figure 7.17 features a theoretical situation just to show that

$$\hat{h} = h + \frac{k_1 s + k_2}{s^2 + k_1 s + k_2} h^d + \frac{1}{s^2 + k_1 s + k_2} \ddot{h}^d \qquad (7.41)$$

so that the steady-state estimate carries all disturbances

$$\left. \hat{h} \right|_{s \to 0} = h + h^d + \frac{1}{k_2} \ddot{h}^d \qquad (7.42)$$

[in Fig. 7.17 subindex "d" stands for "disturbance" (bias) and "m" for "measured"]. A practical implementation shown in Fig. 7.18 involves additional

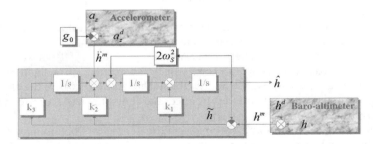

Fig. 7.18   Practical implementation of a complementary filter for altitude estimation.

integrator to kill accelerometer bias, accounts for acceleration due to gravity (g_0) and utilizes Schuler tuning [$\omega_S = (g_0/R_0)^{0.5}$ with R_0 being the Earth radius]. The estimate of the altitude becomes

$$\hat{h} = h + \frac{k_1 s^2 + k_2 s + k_3}{s^3 + k_1 s^2 + (k_2 - 2\omega_S^2)s + k_3} h^d + \frac{s}{s^3 + k_1 s^2 + (k_2 - 2\omega_S^2)s + k_3} a_z^d$$

(7.43)

with

$$\hat{h}\big|_{s \to 0} = h + h^d$$

(7.44)

7.2.4 WIND ACCOMMODATION

Obviously, for the high-altitude airdrops having some information about the winds aloft is crucial, especially if the winds are strong. As shown in Chapter 3, these winds define the area of possible PADS deployment so that it could reach the IPI area. These winds may be available for the broader area from the very detailed meteorological models for weather observation and forecast. Of course, these prognoses are never 100% accurate, but having a general idea about winds aloft allows planning PADS trajectory in the wind-fixed coordinate system $\{w\}$ so that the actual PADS position is shifted to a virtual position by the predicted winds drift (Fig. 7.19) [Soppa 1997; Jann 2005]

$$\Delta \mathbf{s}^W = \begin{bmatrix} \Delta x^W \\ \Delta y^W \end{bmatrix} = \int\limits_0^h \begin{bmatrix} w_x(z) \\ w_z(z) \end{bmatrix} \frac{dz}{V_v + w_z(z)} \approx \frac{h W^{\text{bal}}(h)}{V_v} \begin{bmatrix} \cos(\chi_W(h)) \\ \sin(\chi_W(h)) \end{bmatrix} \quad (7.45)$$

[See Sec. 3.1.3 introducing the ballistic winds and specifically Eq. (3.6)].

In $\{w\}$ the trajectory can be planned as if there would be no wind. Also, the coordinate system is rotated in order to align the x axis with the wind direction, so landing against the wind always means landing in x direction (downrange).

The sink rate in Eq. (7.45) varies during the descent as well, and even its mean value cannot be computed precisely. As shown in Eq. (5.171), it increases during the turn maneuvers. The amount depends on the asymmetric TE deflection and the time the turn is maintained. In a spiral dive, the descent rate can easily exceed twice of the usual value. Because the number and type of turns is never known, there may be a significant error in the mean descent rate and therefore in Eq. (7.45). In fact, all parameters on the right-hand side of Eq. (7.45) carry certain errors; hence, the wind drift cannot be compensated precisely. Therefore, PADS GNC should be robust and adaptive in order to manage remaining uncertainties at the final stages of descent.

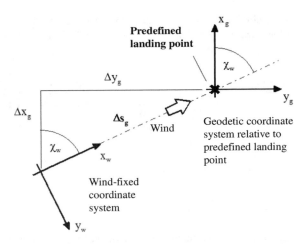

Fig. 7.19   Correcting the IPI position with account of effect of expected winds [Jann 2005].

Provided that it is adequately instrumented, PADS can also do an onboard estimation of the current wind by itself. However, for trajectory planning not only the actual wind but the wind profile ahead (below) for the remaining flight is required. This can be a problem mainly in meteorologically complicated regions like mountains and also higher altitudes.

Because this new estimate of winds (at a current altitude) is different from what may have been stored in AGU's memory as a look-up table for winds heading and direction vs altitude, the wind profile from the current altitude $\tilde{h}$ to the very next flight level h^*, where the mission loaded forecast wind $\hat{\mathbf{W}}^*(h^*)$ is available, may be assumed to be linear

$$\hat{\mathbf{W}}(h) = \lambda\hat{\mathbf{W}}(\tilde{h}) + (1 - \lambda)\hat{\mathbf{W}}^*(h^*), \quad \tilde{h} > h > h^* \qquad (7.46)$$

As suggested by Carter et al. [2009], the weighting coefficient

$$\lambda = \exp[-(\tilde{h} - h)/h_s] \qquad (7.47)$$

includes some reprogrammable correlation height h_s. Figure 7.20 shows an example of dependence $\lambda = \lambda(h - \tilde{h})$ for several h_s. [Carter et al. [2009] reported $h_s = 600$ m (1,969 ft) to work the best for large PADS.]

To estimate the vector $\mathbf{x} = [a, b]^T$ describing a logarithmic winds model near the ground

$$W(h) = [\ln(h), 1]\mathbf{x} = \mathbf{Hx} \qquad (7.48)$$

[see Eq. (3.40) and Fig. 3.41d].

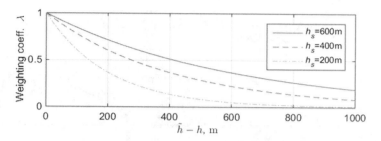

Fig. 7.20   Weighting coefficient for wind corrections.

The onboard recursive least-squares (RLS) adaptive filtering algorithm can be applied [Hewgley 2014; Hewgley and Yakimenko 2011]. With the measurements of h and $W(h)$ coming through while descending, this algorithm minimizes the total error given by the following sum of weighted error squares:

$$J(\mathbf{x}) = \sum_{i=1}^{n} \lambda^{n-i}(W_i - \mathbf{H}_i \hat{\mathbf{x}}_n)^2 \tag{7.49}$$

Here λ is the forgetting factor usually equal to or greater than 0.9 [Manolakis et al. 2005]. The RLS adaptive filtering algorithm is summarized as follows:

1. Pick initial guess for a and b $\hat{\mathbf{x}}_0 = [a, b]^T$

2. Initialize the inverse correlation matrix $\mathbf{P}_0 = \varepsilon \mathbf{I}_{2 \times 2}$ (ε is a small positive constant)

3. Compute the adaptation gain vector ($n = 1$) $\mathbf{k}_n = \dfrac{\lambda^{-1}\mathbf{P}_{n-1}\mathbf{H}_n^T}{1 + \lambda \mathbf{H}_n \mathbf{P}_{n-1}\mathbf{H}_n^T}$

4. Compute the a priori error value $\delta \mathbf{z}_n = W_n - \mathbf{H}_n \hat{\mathbf{x}}_{n-1}$

5. Update the last guess $\hat{\mathbf{x}}_{n+1} = \hat{\mathbf{x}}_n + \mathbf{k}_n \delta \mathbf{z}_n$

6. Update the inverse correlation matrix $\mathbf{P}_n = \lambda^{-1}\mathbf{P}_{n-1} - \lambda^{-1}\mathbf{k}_n \mathbf{H}_n \mathbf{P}_{n-1}$

7. Increment a discrete counter $n = n + 1$, obtain new measurements of altitude h and $W(h)$, h_{n+1} and W_{n+1} (which is used in forming the measurement matrix $\mathbf{H}_{n+1}$), and go to step 3.

Figure 7.21 shows an example of a logarithmic wind profile obtained using the outlined procedure compared to the wind estimates obtained by the Snowflake PADS all of the way to touchdown. In this example, as Snowflake begins a final maneuver, the wind profile estimate produced by the final calculated values of a and b is used for the GNC purposes. For this particular text the logarithmic model agrees with the measured data reasonably well.

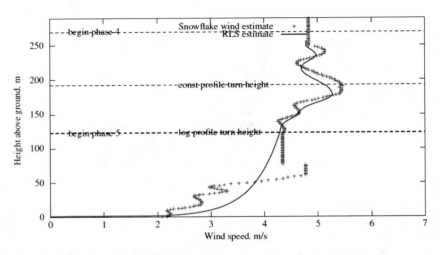

Fig. 7.21   Comparison of logarithmic wind model with the actual wind estimates [Hewgley 2014].

7.3   OPTIMAL GUIDANCE AND PRACTICAL APPROACHES

This and the following three sections discuss the algorithms implemented in the guidance block of AGU (see Fig. 7.2). As observed by Rademacher [2009a], the practical algorithms to generate PADS trajectories typically utilize one of the following approaches. Most of them use WP-based algorithms to generate and utilize a sequence of WPs to manage an excess altitude. They differ however by the exit criteria and final approach (FA) maneuver. Maneuver-based algorithms developed for the final stages of descent generate a reference glide slope (GS) to IPI (a plane tilted by the GS angle), usually shifted upwind to allow for wind uncertainty, and assume performing a sequence of maneuvers to stay close to (above) this reference GS. Path-based algorithms generate and constantly update a continuous feasible reference trajectory connecting the PADS current position to IPI. The trajectory is these methods usually parameterized by time or altitude, but in the most advanced methods a virtual argument is used to add more flexibility in path variations. Most algorithms conduct planning in the horizontal plane and remove the influence of the wind by working in a wind-fixed coordinate frame {w} as shown in Eq. (7.45) and Fig. 7.19. Some algorithms can handle additional constraints placed on the trajectory, including the addition of specific geographic WPs, terrain or geographic area avoidance, and final heading constraints. This section reviews practical algorithms as implemented on different PADS but starts from the basics of the optimal guidance theory to establish the common grounds.

7.3.1   OPTIMAL GUIDANCE

As explained in Sec. 7.3.4, most of guidance algorithms accommodate the best-known wind profile by planning a trajectory in the {w} with a shifted relative initial position [Eq. (7.45)]. Having this in mind, let us consider an optimal path planning based on the Pontrjagin Maximum Principle. PADS kinematics in the horizontal plane with a yaw rate being a control input can be described with as little as three differential equations

$$\begin{bmatrix} \dot{x} \\ \dot{y} \\ \dot{\psi} \end{bmatrix} = \begin{bmatrix} V_h \cos(\psi) \\ V_h \sin(\psi) \\ u \end{bmatrix} \tag{7.50}$$

The first two equations are similar to those of Eq. (5.30), but with no wind and $\beta = 0$, so that $\chi = \chi_a = \psi$. We need to bring PADS from some initial conditions $\mathbf{x}(0) = [x_0, y_0, \psi_0]^T$ to the final conditions

$$\mathbf{x}(t_f) = [x_f, y_f, \psi_f]^T \tag{7.51}$$

by varying a bounded control input $|u| \leq u_{\max}$.

The Hamiltonian for system (7.50) can be written as

$$H = p_x V_h \cos(\psi) + p_y V_h \sin(\psi) + p_\psi u \tag{7.52}$$

where p_x, p_y, and p_ψ are the so-called costate variables with dynamics described by

$$\begin{bmatrix} \dot{p}_x \\ \dot{p}_y \\ \dot{p}_\psi \end{bmatrix} = - \begin{bmatrix} \partial H/\partial x \\ \partial H/\partial y \\ \partial H/\partial \psi \end{bmatrix} = \begin{bmatrix} 0 \\ 0 \\ V_h(p_x \sin(\psi) - p_y \cos(\psi)) \end{bmatrix} \tag{7.53}$$

This problem is very well described in the literature and for the minimum-time problem, that is, minimizing the performance index

$$J = \int_0^{t_f} dt \tag{7.54}$$

results in solving

$$H = p_x V_h \cos(\psi) + p_y V_h \sin(\psi) + p_\psi u - 1 = 0 \tag{7.55}$$

[Note that compared to the Minimum Principle, the last term in Eq. (7.55) has a negative sign.]

The optimality condition of maximizing the derivative $\partial H / \partial u$ allows reducing the problem to

$$u_{\text{opt}} = \max_{|u| \le u_{\text{max}}} (p_\psi u) \qquad (7.56)$$

(in the Minimum Principle we are searching for u_{opt} as $\min_{|u| \le u_{\text{max}}} (p_\psi u)$), which allows synthesizing a simple control scheme

$$u_{\text{opt}} = \begin{cases} u_{\text{max}}, & \text{when} \quad p_\psi > 0 \\ 0, & \text{when} \quad p_\psi = 0 \\ -u_{\text{max}}, & \text{when} \quad p_\psi < 0 \end{cases} \qquad (7.57)$$

Further analysis reveals that an optimal trajectory consists of at most three motion primitives. Depending on the initial and final conditions, two of these three primitives may be the left (L) or right (R) turns, with $u = -u_{\text{max}}$ and $u = u_{\text{max}}$, respectively, and the third primitive represents a so-called singular arc, which in our case is a straight line (S) with $u = 0$. This type of a solution is also known as a bang–bang control [provided that the control u appears in Hamiltonian (7.55) linearly]. This solution is identical to the shortest path problem with the performance index

$$J = \int_0^{t_f} \sqrt{\dot{x}^2 + \dot{y}^2} \, dt \qquad (7.58)$$

The latter performance index was used in a known Dubins car model, which is described by the same models of Eq. (7.50) [Dubins 1957], so that the possible optimal combinations of motion primitives just introduced are also known as *Dubins curves*.

Figures 7.22 and 7.23 show a complete set of possible combinations of motion primitives. PADS being deployed far away from IPI (Fig. 7.22) would need to turn towards it, fly a straight path (singular arc), and then make a turn to land with a given heading ψ_f (into the wind). Depending on the given boundary conditions (BC), it means flying LSL (Fig. 7.22a), SL (Fig. 7.22b), RSL (Fig. 7.22c), or S (Fig. 7.22d) segments. (Figure 7.22 omits two more segments, RSR, SR, and LSR, which are symmetric to LSL, SL, and RSL, respectively.)

If starting from a close vicinity of IPI, a few more possibilities become vital. First, RSL or LSR segments may lose the S primitive (Fig. 7.23a). Second, BC may only require a single turn, L (Fig. 7.23b) or R. Third, the optimal path might require a composition of three turn segments, RLR (Fig. 7.23c) or LRL.

Even though the trajectories shown in Figs. 7.22 and 7.23 seem to be very simple to generate; in practice, deciding which specific combination of these three primitives to follow turns out to be quite a difficult task [Bui 1994; Bui et al. 1994]. To illustrate this, Fig. 7.24 shows three situations when the maneuver

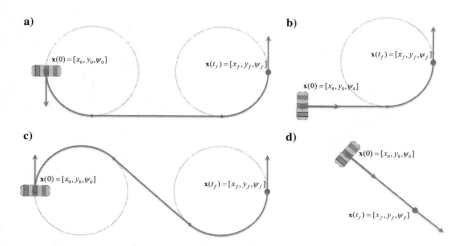

Fig. 7.22 Time-minimum optimal trajectories composed of a) LSL, b) SL, c) RSL, and d) S segments.

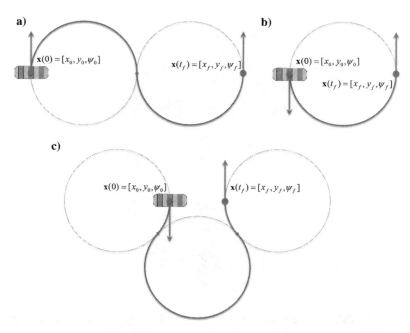

Fig. 7.23 Time-minimum optimal trajectories composed of a) RL, b) L, and c) RLR segments.

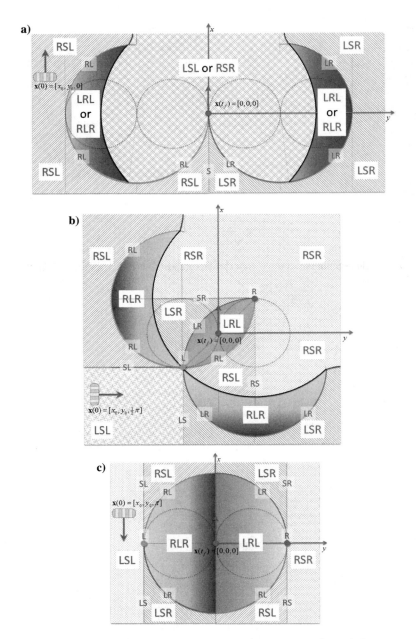

Fig. 7.24   Guidance triplet partitions for a) $\psi_0 = 0$, **b)** $\psi_0 = \pi/2$, **and c)** $\psi_0 = \pi$.

starts with $\psi_0 = 0$ (Fig. 7.24a), $\psi_0 = 0.5\pi$ (Fig. 7.24b), and $\psi_0 = \pi$ (Fig. 7.24c) somewhere in the vicinity of IPI and terminates with $\psi_f = 0$. The areas with a different shading show partitions calling for a different sequence of motion primitives if the initial point (with the headings as just discussed) happens to fall within. Clearly, choosing an optimal sequence when starting far away from IPI poses no problem. (Generally, it is one of RSR, RSL, LSR, LSL, or even simpler combinations, which are very easy to compute.) However, when a trajectory needs to be generated (regenerated) from a point closer to IPI, it is not so obvious what guidance strategy to choose. What makes it so difficult is that the boundaries between partitions are described by transcendental equations that are not easy to solve. Hence, implementing Dubins curves for PADS terminal guidance is problematic.

Another problem with formulation of Eqs. (7.54) or (7.58) is that neither of them accounts for a vertical motion. The solutions depicted in Figs. 7.22 and 7.23 result in a specific arrival time t_f, which does not necessarily coincides with the time of descent T_f, which can be roughly estimated as

$$\hat{T}_f = (\hat{h}_f - h_0)\hat{V}_v^{-1} \tag{7.59}$$

(In practice, it is not known precisely.)

To ensure arrival to the final state precisely at $t_f = \hat{T}_f$, we would need to modify the formulation of Eq. (7.54) or Eq. (7.58) and consider another performance index. For example, considering a minimum control-effort problem with

$$J = \int_0^{\hat{T}_f} u^2 dt \tag{7.60}$$

leads to

$$H = p_x V_h \cos(\psi) + p_y V_h \sin(\psi) + p_\psi u - u^2 = 0 \tag{7.61}$$

The optimality condition $\partial H/\partial u = 0$ implies

$$u_{opt} = \begin{cases} u_{max}, & \text{when } p_\psi > 2u_{max} \\ 0.5\,p_\psi, & \text{when } |p_\psi| \leq 2u_{max} \\ -u_{max}, & \text{when } p_\psi < -2u_{max} \end{cases} \tag{7.62}$$

Now, the optimal control structure excludes singular arcs, that is, straight lines, and allows turning with a variable turn rate (not necessarily the maximum turn rate) leading to a capability of satisfying final conditions at any given (fixed) time $\hat{T}_f$. For the specific set of BC, the two-point boundary-value (TPBV) problem with three varied parameters (the initial values of costate variables p_x, p_y, and p_ψ) has to be solved. [In practice, keeping in mind Eq. (7.61) allows reducing the number of varied parameters to just two.]

This later formulation with a simple model represented by Eq. (7.50) and a performance index similar to Eq. (7.60) is what started the optimal control of a gliding PADS research in the early 1970s. The only difference in the optimal problem formulation was that instead of trying to satisfy the terminal conditions of Eq. (7.51) they were added to the performance index [Goodrick et al. 1973; Pearson 1972]

$$J = 0.5\left[(x_f - x_f^{des})^2 + (y_f - y_f^{des})^2\right] + w_1 0.5(\psi_f - \psi_f^{des})^2 + w_2 0.5 \int_0^{t_f} u^2 dt \quad (7.63)$$

Back then, it seemed to be more efficient from the computational standpoint because in this case, instead of attempting to satisfy the terminal conditions (7.51), the zero terminal conditions on three costate variables (which followed directly from the transversality conditions) needed to be enforced.

Once formulated and studied in [Pearson 1972], it was immediately recognized that the problem has no analytical solution and is very difficult to solve numerically. The first attempt to address this TPBV problem numerically (on a time-sharing system CP/CMS of IBM/360 with 128 Kb memory) utilized a second-order differential dynamic programming algorithm [Mårtensson 1973; Wei and Pearson 1974]. The scenario assumed landing a generic PADS with a horizontal component of an airspeed of 9 m/s (18 kt) into an easterly wind of magnitude of 6 m/s (12 kt) ($x_f^{des} = y_f^{des} = 0$, $\psi_f^{des} = 0.5\pi$) in 100 s. The problem was normalized, and a solution was obtained in 100 nodes. With the weighting coefficients of $w_1 = 1$ and $w_2 = 0.1$, the relative tolerance on minimizing the performance index (7.62) allowed a touchdown error of no more than 4.6 m (45 ft) and deviation from the upwind direction of 8.5 deg. The computations were carried from the 60 initial points (located on 12 radial rays, extending from IPI, with five points on each) varying the initial heading. Figure 7.25 shows a region (relative to IPI) where the solution was obtained regardless of the initial heading. In the nonshaded area the solutions were obtained only for some initial headings.

While this original research ([Pearson 1972; Wei and Pearson 1974]) used the aforementioned TPBV problem just for a feasibility study, to find a reachability region only, another attempt was made to use it while developing a closed-loop solution. This attempt utilized the performance index of Eq. (7.60) and assumed a feedback optimal control in the form of [Koopersmith and Pearson 1975]

$$u_{opt} = a_1 x + a_2 y + a_3 \quad (7.64)$$

To determine three varied parameters in Eq. (7.64), a_1, a_2, and a_3, two different approaches were explored. One utilized nonlinear algebraic equations involving elliptical integrals of the first and second kinds, and another included optimization via minimization of the terminal error. Both approaches proved to be computationally ineffective. For the first one, the solution could only be

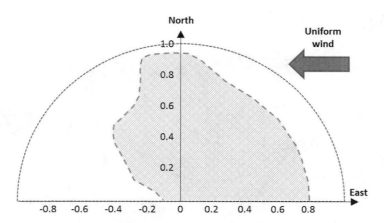

Fig. 7.25   Summary chart of feasible region in the upper half of unit circle [Wei and Pearson 1974].

obtained for several particular cases and was tedious. The second one required integrating the total of 12 differential equations at each iteration of optimization procedure, which slowed down the convergence of the method. Eventually, the performance index (7.60) was abandoned, and a geometric approach (Dubins curves) was explored instead [Koopersmith and Pearson 1975]. The intuitive perception was that the Dubins-curves approach could allow the open-loop solution becoming in effect a feedback control with a stream of changing initial conditions serving as the current state (i.e., open-loop solution would need to be recomputed at each control cycle). As just explained (Fig. 7.24), this paradigm cannot be utilized in practice.

Another attempt to deal with the performance index (7.63) was undertaken two decades later, in the late 1990s, when more powerful computers became available. Optimal solutions were explored in a more systematic manner; however, utilizing this approach in practice was again proved not to be feasible [Gimadieva 2001; Gimadieva 2006]. The TPBV problem in this attempt was attacked by using the method of Krylov and Chernous'ko [1963]. The latest attempt was undertaken in late 2000s [Rademacher 2009; Rademacher et al. 2009].

Figure 7.26 shows examples of implementation of a control architecture of Eq. (7.62). Specifically, Fig. 7.26a shows what happens to a LSL time-optimum trajectory of Fig. 7.22a when the arrival time needs to be greater than the minimum time (PADS happen to be higher than needed). Rademacher [2009a] utilized a so-called altitude margin to characterize an excess of time needed to be spent to arrive to IPI at $\hat{T}_f$. This unitless quantity is basically the difference $\hat{T}_f - t_{min}$ expressed in the units of time required to complete one full 360-deg turn at a maximum turn rate of u_{max}. It is called the altitude margin rather than time margin, because for the constant-descent-rate flight the ratio involving altitudes instead of times would

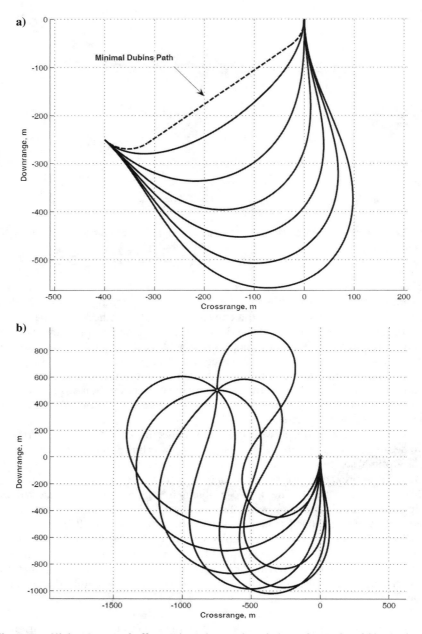

Fig. 7.26 Minimum control-effort trajectories varying a) time of arrival and b) initial heading [Rademacher et al. 2009b].

produce the same result. Back to Fig. 7.26a, it shows several trajectories with the margin varied from 0.2 (the closest to the Dubins solution) to 2.2 in increments of 0.4.

Figure 7.26b demonstrates a family of RSR-like to LSL-like trajectories of Fig. 7.22c. Each trajectory has the same staring position but different initial heading. All trajectories have the altitude margin of three. For the smaller margin, they would look more like LSL and RSL trajectories. Figure 7.27a shows yet another extreme example when the initial point is located right above IPI with the initial heading opposite to that of the required final heading and an altitude margin of 10. Interestingly, the numerical solution of the TPBV problem in this case produces one of the known excess EM pattern, figure "eight." As will be shown in Sec. 7.4, in practical algorithms this figure "eight" is rotated by 90 deg around IPI and shifted downwind, which ensures more robustness in the case of descent rate miscalculation.

This latter approach could have been implemented in AGU. Neglecting some disadvantages of unconstrained minimum-control-effort problem formulation, it allows solving a three-dimensional (3D) problem compared to just a two-dimensional (2D) problem in a horizontal plane. The problem with utilizing the performance index of Eq. (7.60), however, is in generally poor convergence properties. The procedure for solving the TPBV problem for a specific set of BC involves 1) choosing the initial values for p_x, p_y, and p_ψ; 2) integrating Eqs. (7.50) and (7.53) until T_f; 3) checking whether the final conditions $\mathbf{x}(t_f) = [x_f, y_f, \psi_f]^T$ are satisfied; and 4) repeating all of the steps all over again until the final conditions are satisfied (to a certain tolerance). The problem is that 1) there is no guarantee that the iterative process converges, and 2) even if it does the CPU time required for convergence to some reference trajectory may be too large. Specifically, as reported in [Rademacher et al. 2009b], with the large initial offsets and large altitude margin the convergence is better, but when it comes to reoptimizing a trajectory in the close vicinity of IPI with small-altitude margins the algorithm may not converge with a rate required for the real-time implementation. Figure 7.27b shows a possible solution when PADS initially steers to the closer vicinity of IPI area using the minimum-time solution (RSR in this particular case) to ensure the largest possible altitude margin and then switches to the minimum-control-effort solution, which could be obtained while executing the first part. However, in practice due to unmodeled dynamics and wind disturbances blowing PADS away from this reference trajectory, there will be a need to recompute it very quickly with small-altitude margins that will cause problems because the solution for the performance index of Eq. (7.60) cannot possibly converge to the solution for the performance index of Eq. (7.54).

As a result, instead of using the indirect method (Pontrjagin's Maximum Principle) some practical applications rather rely on direct methods. In direct methods, a candidate reference trajectory is given upfront, analytically, and the search of a quasi-optimal trajectory is conducted by varying a few parameters

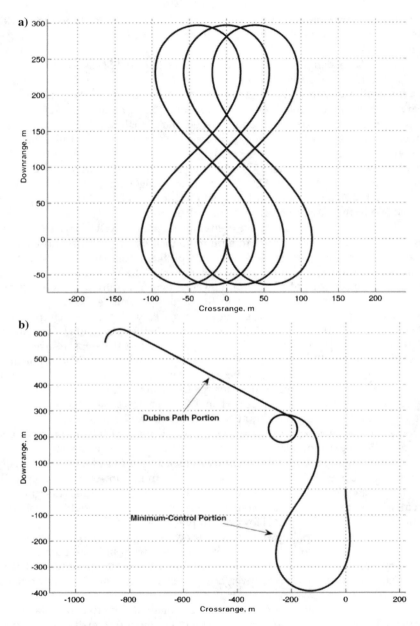

Fig. 7.27   Example of a) an EM maneuver and b) combined minimum-time and minimum control-effort trajectory [Rademacher et al. 2009b].

defining this trajectory so that PADS could follow this trajectory not violating constrains imposed on its dynamics. Two such approaches will be presented in Secs. 7.5–7.7.

The overall strategy for all known practical guidance algorithms is to bring PADS closer to the target and then mitigate uncertainties close to the IPI area while executing EM maneuvers. That includes using some heuristic rules mimicking the behavior of a human jumper to safely land into the wind. This approach could also involve executing some preset maneuvers (with preset controls actions) or establishing a set of varied condition-based WPs allowing to intercept GS in a more systematic way. While the terminal guidance strategies of most known self-guided systems seem to be maneuver-based, an example of a WP approach will be considered in Sec. 7.4.

To conclude the discussion on optimal guidance, let us mention the results presented by [Cleminson 2013], who used deterministic dynamic programming to generate and analyze feasible trajectories accounting for PADS kinematics in a wind field. The optimization was based on PADS model represented by Eqs. (5.30) and (5.31) with time replaced by measure of the distance travelled relative to the air. The continuous state space was discretized to a finite x-y-ψ lattice of discrete points (nodes) so that any path could be represented by a sequence of nodes, each node being selected from the lattice, accounting for PADS kinematics. The cost function accrued while traveling through the path nodes was represented by

$$C_i = \begin{cases} (x_i - x_{\text{IPI}})^2 + (y_i - y_{\text{IPI}})^2 & h_i < 300 \text{ lattice units} \\ (y_i - y_{\text{IPI}})^2 & \text{otherwise} \end{cases} \quad (7.65)$$

where i is the node index.

Figure 7.28 shows an example of two optimal trajectories obtained using the cost (7.65) in no wind conditions (Fig. 7.28a) and 50% tailwind (Fig. 7.28b). The same two trajectories are shown in 3D in Figs. 7.29 and 7.30. Even though a computationally heavy numerical algorithm was never intended to be implemented onboard but be rather used as a tool for studying feasible trajectories, it produced two types of EM maneuvers, circles, and eights, which are actually used by most of known PADS in practice (presented in Sec. 7.3.3)!

7.3.2   EARLIER MANEUVER-BASED STRATEGIES

Let us begin an overview of practical PADS guidance strategies by noting that in aviation the power-off emergency maneuvers are well known. Figure 7.31a depicts this maneuver graphically. No wonder that this maneuver resembles the final portion of trajectory shown in Fig. 7.29. An aircraft speed is typically much higher than that of PADS, and so the winds can be considered negligible for an aircraft in Fig. 7.31a, which coincides with no-wind condition for PADS in Fig. 7.29. For comparison, Fig. 7.31b shows the standard landing pattern. Every pilot is proficient to execute a power-off landing starting from a point when it

Fig. 7.28   Plan view of the a) zero-wind case and the b) 50%-wind case (i.e., wind speed is 50% of airspeed) [Cleminson 2013].

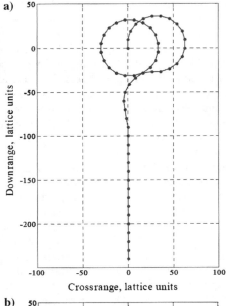

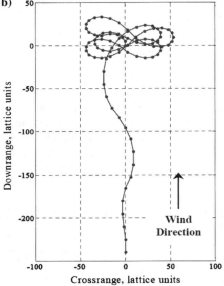

is downwind on a pattern abeam a control tower (mid-downwind). Of course, in the case of engine failure, an aircraft still has a control over its glide angle (GA) to correct for an excess of altitude (during the final approach leg), which PADS usually does not have.

Figure 7.32a shows a typical recommended landing pattern for an unmanned aerial vehicle (UAV) that also features a helical EM maneuver downwind from a runway. (To avoid confusion in terminology, Fig. 7.32b shows definitions of downwind and upwind as related to a position relative to IPI, as opposed to an aircraft position on a standard landing pattern of Fig. 7.31b.)

The overall idea of landing a nonpowered PADS right at IPI is that it captures the unique GS defined by PADS GR and near-surface winds at the correct altitude. If PADS that has no GA control is above it, it will overshoot IPI; if it arrives on final below that unique GS, it will land short, even if it has a GA control. [As explained in Chapter 8, usually GS angle (GSA) control is only capable of increasing GA.] This is different from human jumpers (see discussion in Sec. 1.1.1 and Fig. 1.4), who are using GSA control very actively, so that their goal is to arrive to the IPI area high and then gradually "slide" to the surface defined by the nominal GS (Fig. 7.33a).

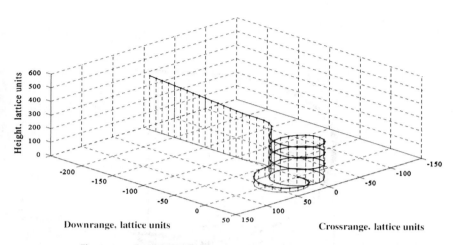

Fig. 7.29   Trajectory of the zero-wind case [Cleminson 2013].

Within the PADS GNC concept, the guidance stage is divided into three phases: *homing*, *EM*, and *landing*. Homing, which is simply speaking flying towards the landing site, that is, towards the middle of the inverted cone (Fig. 7.33b), maximizes the altitude reserve. This altitude reserve is then reduced by the following EM. During this second phase, the trajectory must be planned in a way that the vehicle finally reaches the correct position for the landing approach. To land straight against the wind, an additional position downwind offset from IPI must

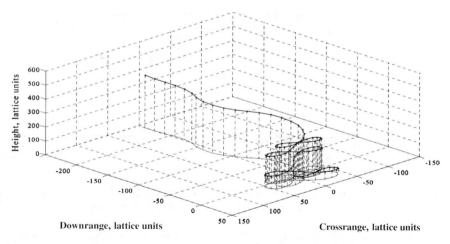

Fig. 7.30   Trajectory of the 50%-wind case [Cleminson 2013].

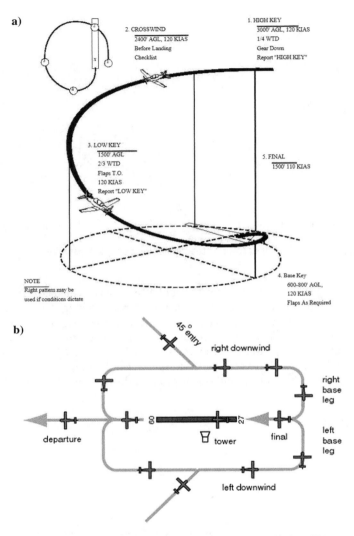

Fig. 7.31   a) Emergency landing pattern and b) standard visual traffic pattern.

be added [Jann 2005]. The various procedures presented in this and the following sections mainly differ in how the EM and the transition between the phases occur.

The earlier work on designing GNC algorithms for autonomous parafoils relied on a possible use of pre-GPS navigational aids (Fig. 7.3) and was primarily

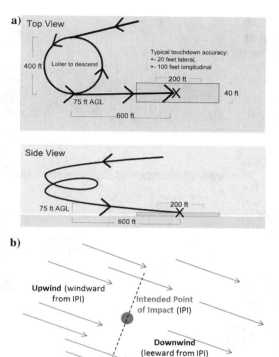

a) Top View

400 ft   Loiter to descend

Typical touchdown accuracy:
+- 20 feet lateral,
+- 100 feet longitudinal

200 ft

40 ft

75 ft AGL

600 ft

Side View

75 ft AGL

200 ft

600 ft

b)

Upwind (windward from IPI)

Intended Point of Impact (IPI)

Downwind (leeward from IPI)

Direction of wind

Fig. 7.32   a) Recommended pattern for UAV and b) illustration of windward (upwind) and leeward (downwind) position with respect to IPI.

concerned with the homing phase [Goodrick 1979a]. Given guidance geometry shown in Fig. 7.34 (with IPI at the origin and axes aligned with the wind), four guidance strategies were studied and compared in computer simulations.

The first strategy (method A) employed a nonproportional radial homing guidance, where PADS turns with the fixed angular increments $\Delta\psi\,\mathrm{sign}(-\eta)$ (when the homing error $|\eta|$ exceeds a certain predefined value η_{max}) and endeavors to the point towards IPI at all times. (Figure 7.35a shows three of these increments.) Method B extends the previous approach to include a "cone of silence," which is essentially the cone shown in Fig. 7.33b with the slope equal to the effective (wind-dependent) GS. PADS is programmed to turn in a direction opposite to that in which it was when crossing the cone of silence and to maintain the turn until PADS exits it. Figure 7.35a features three of such turnaway maneuvers resulting in guaranteed downwind arrival to the IPI area.

While the first two approaches rely on an estimate of a relative IPI bearing, method C employs a conical homing strategy also based on the estimates of the range to IPI $p = \|\mathbf{p}\|$ and its derivative $\dot{p}$. The goal of GNC in this case was to maintain an equality between range to altitude ratio and range derivative to rate of descent ratio, so that PADS could glide along the surface of the cone of silence. It was accomplished by varying angle η so that [Goodrick et al. 1973]

$$-\frac{\dot{p}}{\dot{h}} = \frac{p}{h} + \left(k_C \left| \cos\left(\eta - \frac{\dot{p}}{V_h} \right) \right| - 1 \right) \qquad (7.66)$$

Fig. 7.33   a) Human jumper trajectory utilizing turn and brake control [Bergeron et al. 2011] and b) three PADS guidance phases [Jann 2005].

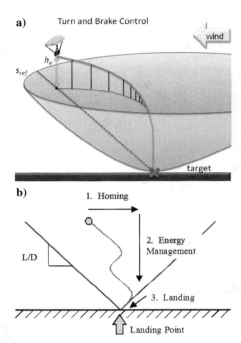

Fig. 7.33   a) Human jumper trajectory utilizing turn and brake control [Bergeron et al. 2011] and b) three PADS guidance phases [Jann 2005].

Finally, method D enforced angle η to be

$$\eta = k_D(\Phi - 0.5\pi) \qquad (7.67)$$

ensuring a right base arrival to the IPI area. For a uniform constant wind, the exact value of k_D bringing PADS to IPI (as shown in Fig. 7.35a) depends on the initial conditions (Φ_0, p_0) and the required flight time T [Eq. (7.59)], but in a practical situation with any fixed-value empirical k_D PADS may end up far away from IPI. (The effects of the winds were also studied in [Goodrick 1970; Murphy 1971].)

Figure 7.35a also features method E, which is a manual control by operator having all information to make the control inputs. Multiple computer simulations conducted using the aforementioned pre-GPS-era approaches showed large dispersion of landing accuracy with varying winds and basically proved these approaches impractical requiring further studies and modifications [Goodrick et al. 1973].

If arrived to the IPI area with an excess of altitude, the EM phase suggested "circling" down pretty much following the pattern shown in Figs. 7.31a. This terminal guidance strategy was studied quite intensively ([Goodrick 1969; Goodrick 1970; Goodrick et al. 1973]) and produced simple estimates of touchdown performance. The motion of PADS after reaching the IPI area (with an excess of an altitude) was assumed to be governed by the following equations:

$$x(t) = R[\sin(V_h t/R) - Wt]$$
$$y(t) = -R[\cos(V_h t/R) - 1] \qquad (7.68)$$

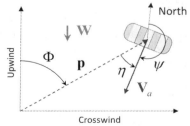

Fig. 7.34   Geometric relations for radial homing guidance.

for $0 \le t \le T_0$ and

$$x(t) = WT_0 + (t - T_0)(V_h - W)$$
$$y(t) = 0 \tag{7.69}$$

$T_0 < t \le T$. [In Eq. (7.68) R is the turn radius.] This motion consists of a 360-deg rotation from a position over IPI (heading upwind) to a position downwind of IPI executed in $T_0 = 2\pi R/V_h$ seconds followed by a straight homing flight back to IPI (Fig. 7.35b) so that the full cycle takes $T = 2\pi R/(V_h - W)$ seconds. Depending on the wind penetration parameter $\lambda = V_h/W$, the EM pattern is either a perfect circle ($\lambda = \infty$) or one cycle of a stretched cycloid of Fig. 7.6b. [Figure 7.35b is produced for the airspeed $V_h = 10\,\text{m/s}$ (19.4 kt) and turn radius of $R = 50\,\text{m}$ (164 ft).]

Because the initial altitude above IPI is not known, PADS may land at any point along the trajectories shown in Fig. 7.35b. With the time instances uniformly

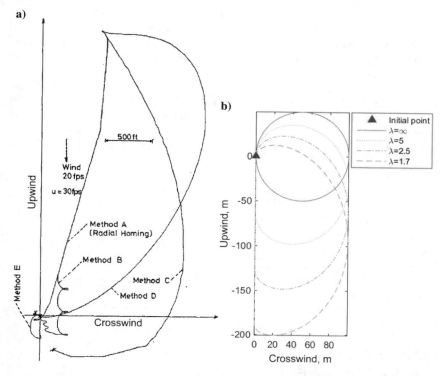

Fig. 7.35   Examples of a) simulated trajectories and b) terminal guidance strategy [Goodrick et al. 1973].

distributed on $[0; T]$, the landing points produce the CEP radius estimate [Good-rick 1969; Goodrick 1970; Goodrick et al. 1973]

$$\frac{CEP}{R} = \frac{1.7 - 0.84\lambda}{1.27 - \lambda} \tag{7.70}$$

(this formula is approximate) and probability of landing with a ground speed less than V_h (i.e., $\dot{x} \le V_h$)

$$P^{\text{upwind}} = \frac{1 + \lambda}{2} - (1 - \lambda)\frac{\sin^{-1}(\lambda)}{\pi} \tag{7.71}$$

Equations (7.70) and (7.71) are presented graphically in Fig. 7.36. As seen from Fig. (7.36a), in no-wind conditions ($\lambda = \infty$ or $\lambda^{-1} = 0$) the CEP is equal to $1.34R$ and increases with decreasing the wind penetration parameter (becomes infinity when $\lambda \le 1$). In fact, for the $\lambda = \infty$ case the CEP should be equal to $\sqrt{R}$, but Eq. (7.70), which is a regression for a wide range of parameter λ, produces a slightly smaller number. The probability P^{upwind} starts at 0.5 at $\lambda = \infty$ and increased to 1 with $\lambda \to 1$. Hence, the suggested approach can guarantee good landing performance neither in no-wind conditions (does not guarantees landing into the wind) nor in a windy conditions (increasing CEP well beyond that of the turn radius).

These earlier studies showed an importance of estimating winds aloft and surface winds and also a necessity to include a FA maneuver to break the EM pattern in a very precise moment to ensure landing into the wind in the close vicinity of IPI. The lessons learned during the earlier GNC studies were then accounted for when developing practical algorithms for the GPS-era PADS presented next [Jann 2004].

The GNC approach developed for the NASA Spacewedge in 1992–1994 was the first published GPS-based guidance strategy for gliding parachute load systems [Murray et al. 1994; Sim et al. 1994]. In this procedure the transitions between phases were

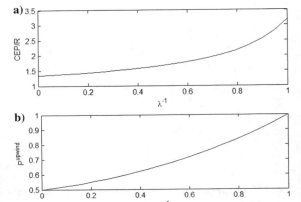

Fig. 7.36   a) The CEP/R ratio and b) probability to land upwind vs the inverse of wind penetration parameter.

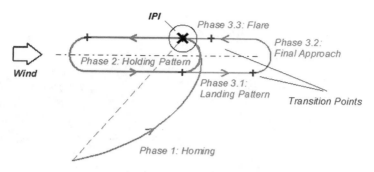

Fig. 7.37 Guidance strategy for the NASA Spacewedge.

controlled by the barometric height. The goal was to land at a distance of less than 450 m ($\frac{1}{4}$ n mile) from IPI. Although wind direction and speed were given at initialization, wind drift was taken into account only for landing. The landing phase was divided into three substages, resulting in a total of five flight phases, for which the guidance strategy foresees the following procedure:

Phase 1, *Homing*: flight to the nominal landing point (Fig. 7.37 features an example of such a curve, radiodrome, or so-called "dog curve," which is a result of winds not being considered here).

Phase 2, *Holding pattern* (=EM): When reaching the IPI area (i.e., distance falls below a certain value), algorithm first verifies whether excess altitude must be reduced. If this is the case, PADS turns against the wind and flies one or more oval trajectories that are aligned along the wind. The turning point for this "holding pattern" is computed in order to produce an altitude loss of 152 m (500 ft) for one complete circulation.

Phase 3.1, *Enter the landing pattern*: The landing phase is initiated when the altitude falls below 91 m (300 ft) above ground. If the system is already flying with the wind, the direction is retained, and the system passes the landing point to some point downwind; otherwise, it performs a 180-deg turn. The decision height for the transition to the next phase of 46 to 61 m (150 to 200 ft) is determined by the wind speed and the distance from IPI. The accuracy depends on the correct timing of the transition to the next phase because at this point no future correction can be made.

Phase 3.2, *FA*: PADS makes a 180-deg turn and flies against the wind towards IPI. An ultrasonic height sensor, which is used to trigger the subsequent flare maneuver, is activated.

Phase 3.3, *Flare*: The flare maneuver is initiated at an altitude of 8 m (26 ft) above ground. Both steering lines are pulled down simultaneously at a maximum rate. The landing occurs 4 s later.

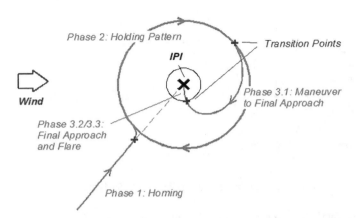

Fig. 7.38   Guidance strategy for the Draper Laboratory PGAS.

As first published guidance for PADS, the aforementioned guidance approach inspired some other developments, in which this technique was adopted with some modifications. For example, for the Pegasus PADS (Fig. 1.11) phases 2 and 3 were adopted virtually unchanged. Phase 1 was expanded, however, with a possibility to fly to certain WPs [Aguilar and Figueiredo 1995]. The guidance strategy developed by Shaw [1994] at the University of Cranfield was also based on the same principle, but performed an onboard wind estimation used during phases 1 and 2.

In 1994–1996, the Draper Laboratory conducted an experimental study called PGAS (precision guided airdrop system) for the U.S. Army [Hattis and Benney 1996; Hattis et al. 1997]. This study was based on the work of NASA and also used their Spacewedge PADS for the test flights. The main objective was the improvement of navigation algorithms, in order to ensure safe and precise landing for military applications even in the case of erroneous GPS reception. This was achieved through the use of IMU coupled with GPS. The guidance strategy was also revised to meet the requirement for a better landing accuracy of 50-m (164-ft) CEP. The wind drift was addressed using a priori known wind model, which was supplemented by an online wind estimate during the flight. Surprisingly, the landing itself did not necessarily occur against the wind (Fig. 7.38). Another feature of this approach was using an automatic trim by identifying the actuator asymmetry before entering the first phase of flight.

As it was the case for the NASA Spacewedge, PGAS utilized five phases as well (Fig. 7.38):

Phase 1, *Maneuver to WP* (=homing): flight towards WP (normally the nominal landing point or IPI). The point was shifted with the expected wind drift, that is, defined in {*w*} (see Fig. 7.19).

Phase 2, *Holding pattern* (=EM): As soon as a certain distance to WP is reached, the system makes a 90-deg turn to the left and initiates a circular flight around IPI with a radius of 610 m (2,000 ft) in clockwise direction. This circular flight continues until the decision altitude of 305 m (1,000 ft) above ground is reached.

Phase 3.1, *Maneuver to FA*: In this phase, the radial velocity component, that is, the direction of flight relative to the landing point, is controlled depending on the altitude and distance to IPI. This results in a spiral curve, which eventually turns towards IPI.

Phase 3.2, *FA*: To prevent PADS from landing in the turn, FA starts approximately 100 m (328 ft) from IPI. PADS guides straight towards IPI. Because of the short distance remaining, the wind drift cannot be fixed.

Phase 3.3, *Flare*: When the height above ground falls below a preset value [e.g., 4.6 m (15 ft)], both steering lines are fully retracted at a maximum speed initiating the flare maneuver. The height above ground is measured using an ultrasonic sensor.

In 1995–1997, the European Space Agency (ESA) carried out the parafoil technology demonstrator (PTD) project aimed at developing an autonomous precision landing technology for large gliding parachutes [Petry et al. 1997; Petry et al. 1999]. The prie contractor at the time was the DASA (Daimler-Benz Aerospace) group, which was also responsible for development of an autonomous GNC system. Within this effort, a complete trajectory leading to IPI was calculated and updated continuously during the flight. This nominal trajectory, which was composed of the circular arcs and straight sections (Dubins' primitives), was tracked by the control block of AGU [Soppa et al. 1997]. The trajectory was defined in {w} that accounted for the wind drift. The wind profile was measured experimentally before each flight test.

Compared to the preceding approach, the PTD guidance strategy broke phases 2 and 3 into two subphases. In this approach though, the nominal trajectory was computed backwards from the IPI and intended landing direction using known parameters of PADS like the GR, sink rate, and expected turn radius. Hence, the following five phases refer to the sections of the nominal trajectory rather than to the phases of GNC (Fig. 7.39):

Phase 1, *Acquisition phase* (=homing): PADS flies along the shortest distance to the entry point into the EM pattern. The entry point is determined by maximizing the duration of EM for the expected total flight distance until landing.

Phase 2.1, *Pattern entry phase*: Entry into the EM pattern is done through a turn with a defined radius. The turn direction arises from the preceding calculation of the entry point.

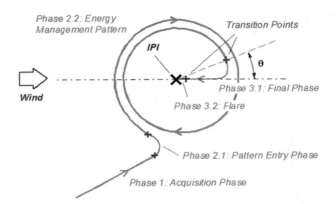

Fig. 7.39   PTD guidance strategy (schematic).

Phase 2.2, *EM pattern*: Excess altitude is removed by circling around IPI (in the wind-fixed coordinate system). While the remaining flight distance is calculated constantly, the turn radius of the EM pattern (500 to 1,000 m [1,640 to 3,280 ft)] is adjusted in such a way that PADS is able to enter the next phase at the correct height and position relative to IPI.

Phase 3.1, *Final phase*: The beginning of the landing phase is defined by an angular range $\pm\theta$ relative to the intended landing direction (see Fig. 7.39). When PADS enters this range during its last circle, it turns towards IPI for the last phase.

Phase 3.2, *Flare*: The flare maneuver is initiated at a given altitude above the ground by pulling both control lines simultaneously with the maximum speed. The altitude above ground is measured by a laser altimeter.

Although during the PTD project the aforementioned approach was actually tested only once, it paved the road to various enhancements. The PTD guidance strategy was also considered in the Capree study, which was conducted by Astrium and others on behalf of ESA [Klotz et al. 1999; Markus 1999]. In the American–European cooperation within the X-38 project (Fig. 1.8b), this method was perfected and has eventually replaced the former guidance strategy, which was originally adopted from the Orion PADS (Fig. 1.6). The improved guidance algorithm was successfully proven during several X-38 flight tests [Gründer 2001; Klädtke et al. 1999].

The reference trajectory for the NASA's X-38 PADS is shown in Fig. 7.40a [Stein et al. 2005; Soppa et al. 2005]. The first guidance phase is the target acquisition turn. The guidance then exits the turn and proceeds to a homing leg, which is a straight flight portion headed towards the EM circle (EMC) entry point. While on the EMC, the guidance modulates the diameter of EMC (Fig. 7.40b) to regulate range error caused by wind or parafoil performance dispersions. At the

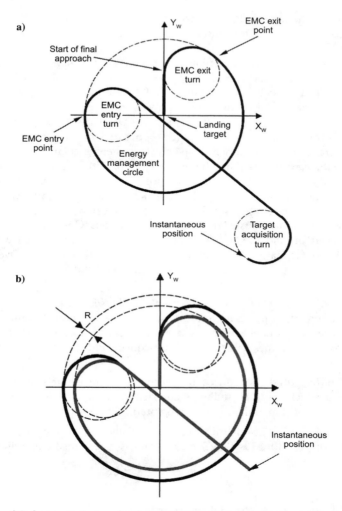

Fig. 7.40   a) Reference trajectory for the X-38 PADS and b) example of reconstructed ground track for one of the tests [Soppa et al. 2005].

appropriate location, guidance initiates an EMC exit turn onto the predetermined final heading (to land into the surface winds). While on FA, PADS continues to minimize a cross-track error. Ten seconds prior to flare, a preflare setting of 40% flaps is commanded to minimize an effect of nonlinear control dynamics (Fig. 5.40a) before going into flare. When a laser altimeter detects the predetermined altitude, AGU initiates the flare, and the parafoil flaps are reeled in to a maximum at an average rate of 0.46 m of control line per second (1.5 ft/s).

Performance depends on PADS aerodynamic characteristics, such as the GR, turn performance, and flare timing, derived from the model and incorporated into AGU.

In 1996–1997, the Institute of Flight Mechanics of German Aerospace Center (DLR) worked on a guidance strategy that was initially planned for the ALEX PADS (Fig. 1.31) [Gockel 1997b; Gockel 1998]. The CARESI (capsule recovery simulation) strategy was to adopt proven principles in aircraft control (model following control). The control scheme was based on two cascades, the inner and outer loops. While the outer loop cared for returning PADS to a nominal trajectory, the inner loop created the necessary control inputs to achieve the yaw rates required by the outer loop. Both control loops contained feed forward elements, whose parameters were derived from the PADS model.

The nominal flight path, defining the outer-loop commands, was defined in $\{w\}$. The guidance strategy for the nominal trajectory was to use a circular EM pattern with its radius adjusted depending on deviations. Unlike PTD, the circling did not occur around the nominal landing point, but was shifted downwind (Fig. 7.41) to facilitate transitions between the different phases. Overall, the CARESI guidance strategy was composed of the following four phases:

Phase 1, *Acquisition phase*: The entry point for the second phase is determined by variation of turn radius for EM in order to tangentially enter the EM circle after the straight flight. The initial vehicle position relative to the wind axis determines whether the subsequent EM is done in the left or right turns.

Phase 2, *EM*: Altitude is reduced by executing clockwise or counterclockwise spiral, in which the turn radius is adjusted continuously in order to obtain the correct position and altitude when entering the FA phase.

Phase 3.1, *FA*: Exiting EM phase occurs tangentially to the last circle, with PADS heading upwind towards IPI. The exit point is determined depending on the

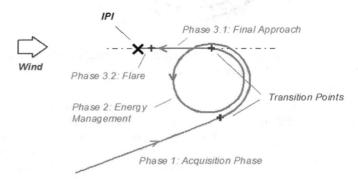

Fig. 7.41   CARESI guidance strategy (schematic).

altitude for the landing approach of 100 m (328 ft), PADS GR, and the wind conditions.

Phase 3.2, *Flare*: The flare maneuver is initiated at a certain altitude above ground and carried out pulling both brakes simultaneously.

Because of the high complexity, the two-cascade CARESI strategy has never been tested in real flight.

7.3.3 PRECISION PLACEMENT GUIDANCE ALGORITHMS

In the 2000s, GNC algorithms presented in the preceding section were advanced even further to achieve a better landing accuracy. As discussed in Sec. 1.2, several different weight PADS were developed, tested, and demonstrated at PATCAD events in 2001–2009. Thousands of drops were executed in between PATCADs to tune GNC algorithms and make them robust. The GNC algorithms for a few PADS were developed and documented by either universities or government laboratories and therefore are available. These algorithms will be discussed in Secs. 7.4–7.7. Most of PADS demonstrated at PATCAD events however were developed by private companies, and as a result the specific details on their GNC approaches are not known. However, general ideas behind the guidance approaches implemented by these PADS can be understood by analyzing published test results. To this end, Figs. 7.42 and 7.43 present a bird's-eye view of typical trajectories flown by different PADS.

Some PADS employ either WP navigation or combination of heuristic rules that guide them to the IPI area, usually downwind from IPI to execute the EM phase and then land into the wind. Guidance may rely on some preset maneuvers (circles or figure-eight) and/or mimic a behavior of a human jumper. For example, the trajectory shown in Fig. 7.42a leads to an area 650 m (1,233 ft) downwind of IPI and then utilizes a figure-eight EM maneuver to estimate current winds and manage an excess of altitude before breaking this maneuver and proceeding with FA trying to stay right above GS. While steering to IPI, the wind estimation continues, and because the actual winds in this particular drop are sensed to be weaker than assumed, the GS angle gets corrected (decreased), so that PADS needs to lose some altitude to avoid an overshoot. It does it by making a side-wave maneuver to slide onto a new GS. For this particular case, the surface winds could be stronger than anticipated, or the side-wave maneuver could result in losing too much of an altitude, or there could be a wind downdraft closer to the surface, so that because of any of these reasons PADS lands a little bit short of IPI. A trajectory shown in Fig. 7.42b seems to utilize either figure-eight or helix as an EM maneuver. In windy conditions these maneuvers look like trajectories presented earlier in Figs. 7.6 and 7.7. It first flies directly towards IPI and then executes an EM maneuver until it captures GS. The actual near-surface winds happen to be stronger than predicted resulting in landing short of IPI. A trajectory of Fig. 7.42c might utilize a similar strategy performing a constant

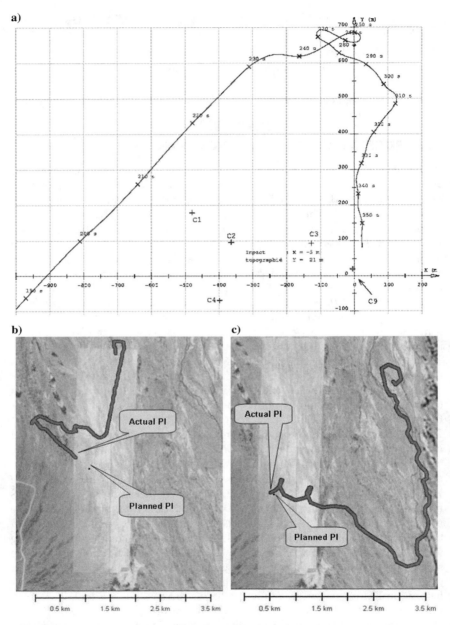

Fig. 7.42   Examples of a flight path of the a) SPADeS, b) Panther, and c) Sherpa PADS [de Lassat de Pressigny 2009; PATCAD 2007].

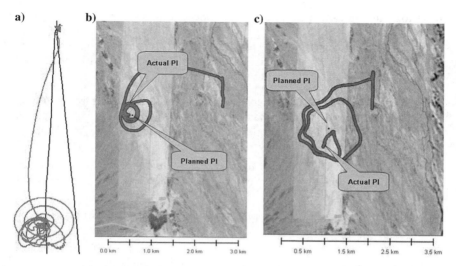

a)   b)   c)

Actual PI

Planned PI

Planned PI

Actual PI

0.0 km   1.0 km   2.0 km   3.0 km   0.5 km   1.5 km   2.5 km   3.5 km

Fig. 7.43   Examples of a flight path of the a) Mosquito, b) Onyx, and c) Screamer PADS [Benney et al. 2009a; PATCAD 2007].

turning maneuver to manage an excess of altitude. However, either because of miscalculation of the height above IPI or weaker than expected winds, it flies too far upwind IPI and fails to make FA into the wind. These are sample trajectories, and other trajectories by the same PADS may look differently, but the point to make here is that these systems employ maneuver–based guidance approach and incorporate certain heuristic rules to accommodate uncertainties, primarily unknown winds below the altitude PADS is currently at.

Figure 7.43 presents another approach pursued by three other companies utilizing high-wing-loading parafoils featuring much larger airspeed compared to other PADS. The guidance logic here is to bring PADS to a vertical line extending upwards from IPI and further spiral it down around this vertical line. At a certain altitude (based on a nominal GS computed for the current conditions), guidance algorithm commands to exit a spiral and head towards IPI. The two-stage PADS (like Screamer shown in Figs. 1.18, 1.19, or Onyx 500 shown in Fig. 1.20) do not glide all of the way down under the high wing loading parafoil, but use a secondary (main parachute) that is deployed at as low altitude as possible to bring payload down to the ground uncontrollably. Theoretically, in this case a miss distance is roughly limited by the radius of a spiral R. However, in practice, the strong winds take their toll and either prevent PADS from executing a balanced spiral (like in the example shown in Fig. 7.43a) or steer it away from IPI when a main canopy deploys (Fig. 7.43c). Of course, using this guidance strategy does not guarantee landing into the wind. Moving the helix downwind of IPI (by say $2R$) could allow having a FA stage, but would involve adding some exiting logics to assure the same landing accuracy [Gockel 1997a; Gockel 1997b; Kaminer and Yakimenko 2003].

Formally, the guidance strategy for the high-wing-loading system may consist of the following flight phases [Calise and Preston 2008; Calise and Preston 2006; Calise et al. 2007; Calise and Preston 2008]:

Phase 1, *Initial homing*: Upon deployment, turn towards IPI, and fly straight for a specified time period to estimate a glide speed V_a and flight-path angle γ_a. (For Onyx 500 and Onyx 2200, it is done by averaging GPS sensor data, that is, assuming that $V_a \approx V$ and $\gamma_a \approx \gamma$.)

Phase 2, *Short steep descent*: Turn for a specified time period with a maximum asymmetric trailing-edge deflection $\delta_{a_{max}}$ to estimate the rate of descent and the turn rate.

Phase 3, *Homing*: Exit the previous phase with a heading towards IPI biased by the turn radius.

Phase 4, *Transition to the IPI site*: This phase is divided into four subphases, recomputed at every guidance update for the reminder of the flight.

Phase 4.1, *Nominal altitude computation*: Use the estimates from the first two phases [propagated all the way down to the surface according to Eq. (7.32)] to compute a desired altitude for the current range to IPI. This computation is accomplished iteratively based on a preset wind profile (ballistic winds).

Phase 4.2, *Altitude adjustment*: Execute a number of 360-deg turns until PADS arrives near, but not below, the desired altitude. The desired altitude is biased upward to ensure against ending below it. The bias linearly decreases with the range to IPI but never goes to zero to allow compensating for unknown winds.

Phase 4.3, *Resetting altitude error*: Command weaves as needed to maintain the altitude error (compared to that computed at phase 4.1) below a prescribed limit. These weaves are executed in lieu of the full 360-deg turns of phase 4.2. The weave command forces PADS to turn and fly perpendicularly to the desired heading maintaining it until the altitude error is reduced to 50% of its original value. Then, PADS is commanded to return to the original heading.

Phase 4.4, *Heading adjustment*: Command a turn rate proportional to the heading error.

Phase 5, *Final spiral descent*: Brief spiral around IPI with $\delta_{a_{max}}$ to dissipate any excess altitude that might remain.

Another way of planning a trajectory was realized within the so-called T-Approach that was developed by DLR [Jann 2001; Jann 2004; Jann 2005]. Based on the actual position and available information, the nominal trajectory is continuously updated until landing. By this means, deviations due to unknown winds and uncertainties in the system parameters can be compensated. The continuous recalculation of trajectory implies that steady changes in the actual position must result in steady changes in the planned trajectory and the generated commands. Otherwise, decision ambiguities cannot be excluded. Instead of using circles, WPs are distributed along a 'T'-formed pattern so that flying along it allows reducing the existing surplus in altitude (Fig. 7.44). The three guidance phases for T-approach work as follows:

Phase 1, *Homing*: From the actual position EMC is approached if possible. If the remaining distance is not long enough to reach IPI afterwards, WP 1 is moved closer to the final turn point (FTP). If altitude and subsequently remaining distance are still too small to reach the landing point, backup mode is entered.

Phase 2, *EM*: After reaching EMC, a surplus altitude is reduced by flying S-patterns to the EM turn points (EMTP). If the remaining distance becomes too small to reach IPI, the EMTP is shifted along the EM axis towards the EMC.

Phase 3, *Landing*: The landing phase is divided into three subphases, specifically.

Phase 3.1, *Approach FTP*: Fly towards FTP after passing over the last WP of EM pattern.

Phase 3.2, *Turn into wind*: After reaching FTP, prepare to land against the wind.

Phase 3.3, *Flare*: Triggering of the flare maneuver in a predetermined altitude above ground.

Backup mode: PADS keeps heading towards FTP until a specified decision altitude is reached. Then, PADS turns against the wind independently of its position and prepares for landing.

More details on the T-approach guidance are provided in Sec. 7.4.

Because it is well understood that terminal guidance determines touchdown accuracy, it is crucial to adapt the path generation algorithm to the changing

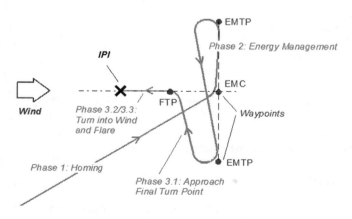

Fig. 7.44 T-approach guidance concept.

winds, that is, to allow constant real-time recomputation of the final turn maneuver to accommodate wind disturbances as much as possible. One such attempt realized on large PADS was a GNC unit developed by Draper Laboratory [Hattis 2007; PATCAD 2007].

The guidance strategy in this effort included several modes corresponding to the different flight phases and was organized as shown in Fig. 7.45. The primary GNC modes were *Preflight, Trimflight, Autoflight,* and *Flare* [Carter et al. 2007; Hattis et al. 2006]. In *Preflight* mode, PADS attempts to navigate while waiting for exit from a carrier aircraft. This mode with no control commands ends after successful deployment of the main canopy (sensed by the stabilized descent rate). The *Trim* mode is dedicated to tuning the control block parameters to correct for any initial bias in the turning rate. This mode is followed by *Autoflight* mode, which makes up most of the flight. The *Autoflight* mode consists of three flight phases. Far from IPI, the guidance block (shown in Fig. 7.2a) generates the *homing* commands, that is, heading rate commands that point the PADS velocity vector towards IPI (which is at the origin of Fig. 7.45). Closer to IPI, PADS steers to fly a figure-eight EM pattern oriented transverse to the desired landing direction to lose altitude while not straying too far from IPI. When PADS descents low enough, the FA maneuver is executed.

During the final portion of descent, AGU steers PADS to intersect IPI at a heading, which is specified in the mission file (preferably landing into the

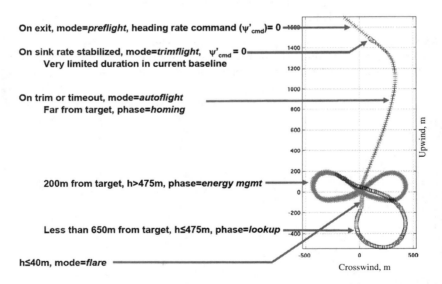

Fig. 7.45   Guidance modes for MegaFly and GigaFly in the wind-fixed frame [Carter et al. 2007; Hattis et al. 2006].

predicted wind to minimize a speed at impact). Near the ground, the control lines are fully retracted to further reduce speed at impact (which constitutes a flare).

The terminal guidance problem then can be stated as follows: specify commanded heading rate as a function of navigated position, heading, and heading rate to reach a prescribed final position, heading, and heading rate. The terminal guidance was first implemented as an optimized look-up table to choose a specific terminal flight path while entering a terminal area based on the current conditions. It utilized a large family of pre-mission-computed heading rate commands stored in flash memory. The look-up table turning commands were indexed by altitude, along-track and cross-track position with respect to IPI and heading. Each of the look-up table trajectories either hits the target or, if that is not possible from the given initial state, minimizes a function of position and heading error at impact. However, being unable to optimize a maneuver in real time, it was later replaced with a more robust approach, band-limited guidance (BLG), involving a direct method of calculus of variations to be able to produce a quasi-optimal trajectory in real time [Carter et al. 2007; Hattis et al. 2006]. These efforts are discussed in more detail in Sec. 7.5.

Another direct-method approach, based on the inverse dynamics in the virtual domain (IDVD) method to deal with uncertainties during the terminal guidance phase, was implemented on the Snowflake PADS (Fig. 1.30). As opposed to all other approaches, the terminal guidance in this case was accomplished in the inertial (local) system of coordinates and was determined by five parameters: IPI coordinates, loiter pattern offset distance (away distance), cycle distance, turn diameter, and wind heading angle. These five parameters define five tracking points (fixed in the inertial coordinate frame), which are the target, A, B, C, and D shown in Fig. 7.46 [Slegers and Yakimenko 2009].

Using these five points, the PADS precision placement objectives for six phases are defined as follows (Fig. 7.46):

Phase 0, *Deployment and stabilization*: After canopy opening, guidance is delayed to allow oscillations to cease.

Phase 1, *Steering to IPI area*: This phase is used to reach the loitering area (point A).

Phase 2, *Loitering*: While loitering AGU estimates, winds, descent rate, and the altitude at which to turn towards IPI, the loiter area is bounded by four points A, B, C, and D. As just stated, these four points are determined by the target location, wind direction, away distance, cycle distance, and turn diameter. During loiter, winds are estimated when traveling from A to B and from C to D (long parallel legs). Ideally, at least one complete loop around A, B, C, and D is completed. Winds, distance to the target, and vertical velocity are used to determine a switching altitude to start the downwind leg towards IPI. When the measured altitude reaches this altitude, phase 2 is terminated, and the system turns towards IPI.

Phase 3, *Exiting the loitering*: Wind estimation is halted as the loiter area is exited.

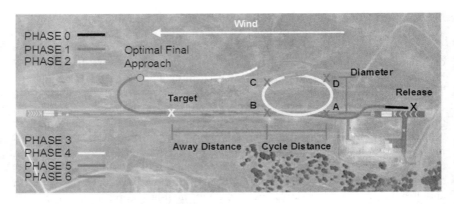

Fig. 7.46   Guidance strategy for the Snowflake PADS [Slegers and Yakimenko 2009].

Phase 4, *Downwind leg*: Wind estimate is reinitiated as the system travels towards the target. All estimated data are used to determine a distance past the target to make a final turn and approach. When this distance is reached, phase 5 is initiated.

Phase 5, *Base turn*: A 180-deg turn to capture the FA GS.

Phase 6, *FA and landing*: FA leg and landing into the wind. (For larger PADS, the FA would end up with a flare phase, but for the very small Snowflake PADS the flare option was not available.

The last three phases just introduced (phases 4–6) are the most critical stages of PADS guidance. As shown in Fig. 7.46, PADS arrives to a loiter pattern upwind of IPI to ensure IPI can be reached even with high winds exceeding V_a. The precision placement algorithm is considered in more detail in Secs. 7.6 and 7.7.

In each phase, the current position and desired tracking points are used to determine a desired PADS heading. Prior to terminal guidance (phases 0 through 4), the desired heading is found simply by heading directly to the next tracking point. However, in the critical final phases the desired heading is found by rapidly and continuously recalculating an optimal trajectory as detailed in the next section. A desired heading is tracked using a model predictive control described in Sec. 7.8.3.

7.4   T-APPROACH GUIDANCE

This section provides more details on the WP-based T-approach guidance concept introduced in Fig. 7.44 of Sec. 7.3.3 and also mentions some other concepts employing S-shape maneuvers to manage an excess of altitude close to the IPI area [Jann 2004; Jann 2005].

7.4.1 OVERALL T-APPROACH STRATEGY

As mentioned in Sec. 7.3.3, within the T-approach the nominal trajectory is recalculated repeatedly at specific time intervals until landing. The process leads to an ongoing adjustment of the trajectory with the aim to direct the system at the correct altitude at the position for the beginning the landing approach.

For convenience, Fig. 7.47 basically repeats Fig. 7.44. As seen, because the landing should occur straight upwind, the last WP is placed in some distance downwind from IPI. The distance between FTP and nominal IPI is determined by the altitude where FA shall begin. To ensure a stable flight state after the last turn, the altitude for FTP should not be too low, for example, for the ALEX PADS 50 m (164 ft) was selected. The distance is then determined using the average GR of the system.

If the altitude is sufficiently large, an EM maneuver is performed. One important issue of the T-approach is that a new WP always originates from FTP or the current WP and is distributed from there. In this way, the flight distance and direction changes steadily if WPs added, moved, or deleted, thereby ensuring the adaptability of the trajectory. The T-distribution is not necessarily required but has advantages with regard to the reliable reach of FTP before landing. The principle of cross approaches is considered as safe and well-controlled, which is the reason it is taught to beginners in skydiving for their first landings [Burke 1997; Hogue et al. 1993]. If required, other free WP formations and distributions rather than letter T are possible (see Sec. 7.4.2).

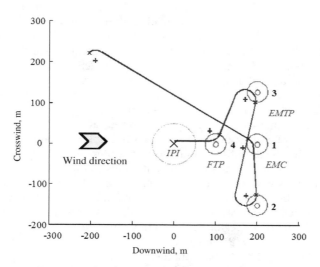

Fig. 7.47 T-approach guidance strategy (schematic).

From higher altitudes, the EM within the T-approach is done by repeatedly flying figure-eight (aka S-patterns) between EMTP (Fig. 7.47). The distance between EMTPs with EMC in the middle should be selected to assure a good compromise between the desired control activity, proximity to the landing site, and the dynamic limits of the system. For the ALEX PADS with approximately 120 kg (265 lb), a value of 150 m (492 ft) was used; for larger and more sluggish systems, a larger value would probably be more appropriate. Because EMC and EMTPs are located downwind relative to FTP at a certain distance, turning in the direction of FTP is possible from all positions during the EM phase. The transverse distribution of WPs relative to the wind direction has phase advantage; at the most only a 90-deg turn is required to turn into the wind. For the same reason the U-turn maneuvers should always be made in the direction of IPI, that is, against the wind.

The lower the altitude is at the beginning, the more the EM phase is reduced. This is done by shifting the EMTPs in reverse order of their creation, first along the EM axis towards EMC and then to FTP. When WP coincides with FTP, it can be deleted. In this manner, the nominal trajectory is constantly readjusted during the flight depending on the current position and the remaining estimated flight distance or path length. Because the position of FTP is fixed, the path length can be adjusted only by shifting the EMTPs and EMC. The last WP before the FTP determines the landing precision.

During the homing phase, the system heads towards EMC and starts the EM after reaching it. If there are no EMTPs, EMC is shifted towards FTP, which further reduces the path length (Fig. 7.48a). When EMC and FTP coincide, the result is the backup mode.

The starting point for computation of the WP configuration and the corresponding nominal trajectory is the remaining altitude above IPI [Eq. (7.59)]. In a normal glide, the relationship between flight distance and altitude is given by PADS GR. For a given PADS, its GR is approximately constant and only changes by about 20% with symmetrical flap deflection (Fig. 5.38a). On the other hand, because the sink rate increases in turns by up to about 40% (Fig. 5.40b), the average GR for the whole trajectory might be significantly smaller (Fig. 5.38b). This depends on the number and type of turns, that is, the turn radius and, due to dynamics of the system, also on the duration of the turn. Thus, the total flight distance and the wind influence on the flight are not determined solely by the altitude, but also on the number and the type of maneuvers. However, because of its ability of continuously adjusting the trajectory, the T-approach works well with the assumption of a constant GR. For this case, the altitude can be directly transformed into a remaining flight distance.

The estimated flight distance or path length must approximately equal to the sum of straight or curved segments the nominal trajectory is composed of. During trajectory planning, the WP configuration is varied iteratively until this condition is fulfilled (Fig. 7.48). The WPs are numbered in the order of their approach. Their positions determine the direction and distance the vehicle has to fly. Reaching a

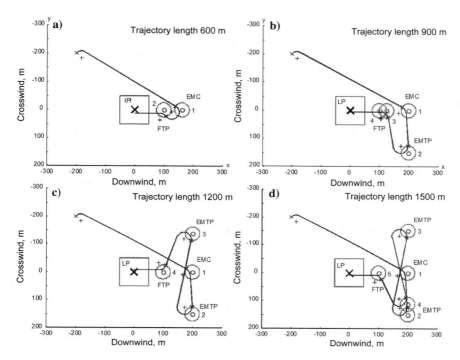

Fig. 7.48   T-approach for different routes (in {w}).

specified area around the aimed WP, the heading is changed in order to approach the next one. The guidance output is the commanded heading computed from the direct connection between the actual position and the next WP.

7.4.2   COMPUTER SIMULATIONS

To evaluate the T-approach guidance concept from the standpoint of the effects of parameter changes, unknown winds and sensor errors, a computer simulation study has been conducted. These simulations used an identified four-DoF model of PADS forming the closed GNC control loop presented in Fig. 7.12a and featuring the update interval of 1 s, mean GR of 2, mean sink rate of 5 m/s (16.4 ft/s), mean turn radius of 25 m (82 ft), radius of WP area of 25 m (82 ft), height for landing approach of 50 m (164 ft), FTP-EMC distance of 100 m (328 ft), and maximum EMC-EMTP distance of 150 m (492 ft).

Figure 7.49a shows an example of a nominal trajectory with a heading control (see Sec. 7.8.1) and with no errors in wind assumptions, system parameters, sensor outputs, or actuator responses. The simulation starts with PADS flying north at 600 m (1,959 ft) altitude at a position 200 m (656 ft) North and 200 m (656 ft)

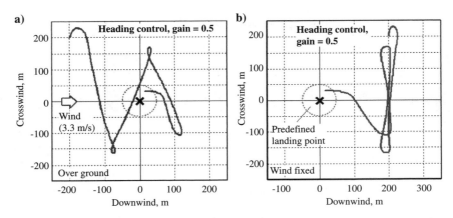

Fig. 7.49   Simulated trajectory A in the a) local tangent and b) wind-fixed coordinates.

West of IPI. The wind is coming directly from the west with 3.3 m/s (6.4 kt). Figure 7.49b shows the same trajectory, which is projected into {w}. The shape of the T-approach trajector can be immediately recognized.

The trajectory in Fig. 7.50 presents a simulation with another control scheme, course control, using GPS measurements only (Sec. 7.8.1). Applying the course control paradigm for the ALEX PADS model, the scope of suitable controller gains is very narrow. If the gain is too high, the vehicle tends to overshoot; if the gain is too small, the reaction becomes too slow to follow the nominal trajectory. Hence, in general, the course control is better suited for the larger sluggish PADS compared to a relatively small and agile PADS like ALEX.

The T-approach guidance can be configured in such a way that instead of figure-eight turns, PADS will implement the spiraling down EM. As discussed in Sec. 7.3.1, the advantages of the steep descent mode are that 1) the actuator activity is usually much less for circles than for figure-eight pattern, which allows saving battery power and weight, and 2) the higher descent rate, which shortens the flight time for PADS to be exposed to the winds. On the other hand, because of the higher speed, the landing is more challenging. Figure 7.51a

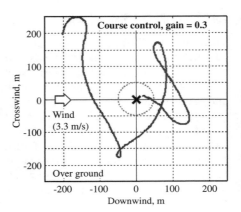

Fig. 7.50   Simulated trajectory B.

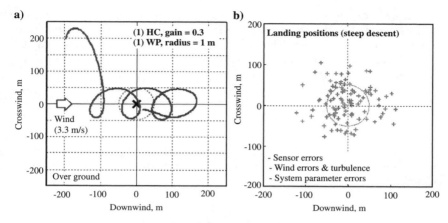

Fig. 7.51 **a) Steep-descent trajectory and b) the results of steep-descent Monte Carlo analysis.**

shows an example of the steep-descent mode achieved as result of two parameter changes:

1.   The controller gain is set to a smaller value (e.g., 0.3 for the heading control), which increases the turn radius of the vehicle.

2.   The radius of the WP area is set to a very small value that can hardly be utilized by the real PADS [e.g., less than 1 m (3.3 ft)].

As a result of these changes, PADS tries to approach EMC, but never enters the WP area and consequently circles around WP. However, because WPs are recalculated continuously, the method is still capable to adapt to the changing wind drift and finally exit EM to proceed with landing approach phase.

Monte Carlo simulations have shown that the steep-descent mode works well even under adverse conditions (Fig. 7.51b). Compared to the results of Fig. 7.13b, the landing accuracy is within the same range. However, because the four-DoF model only simulates the coordinated turns rather than higher-order dynamics (see Sec. 5.2.6), the significance of the spiral-dive simulation is somewhat limited [Jann 2004; Jann and Strickert 2005].

Original T-approach guidance worked with WPs that were distributed along the letter T pattern. Nevertheless, the method is compatible to other WP configurations, provided that there is enough reserve (surplus) altitude for the maneuvers. This feature enhances the flexibility of the method and makes it possible to include geographical WPs to the flight path. (Consideration of additional geographical WPs can be advantageous, if geographical areas have to be included or excluded from the flight path.)

The only caveat is that to plan the trajectory and to compute the adequate heading command, the geographical WP coordinates have to be transformed into the wind-fixed frame. This, however, cannot be done by applying Eq. (7.45) directly because the altitude loss depends on the effective distance in the wind-fixed frame, which again depends on the wind drift

$$d^{\text{eff}} = \sqrt{(\Delta x + \Delta x^W)^2 + (\Delta y + \Delta y^W)^2} \qquad (7.72)$$

[In Eq. (7.72), $\Delta \mathbf{s} = [\Delta x, \Delta y]^T$ is the vector from the actual position to the aimed WP position in $\{I\}$.] Assuming a constant GR (L/D), the altitude loss can be calculated by

$$\Delta h = \frac{d^{\text{eff}}}{L/D} \qquad (7.73)$$

Substituting Eq. (7.72) into Eq. (7.73) and replacing Δx^W and Δy^W with the corresponding expressions from Eq. (7.45) leads to a quadratic equation in Δh:

$$(L/D\,\Delta h)^2 = \left(\Delta x + \frac{W_x^{\text{bal}}}{V_v}\Delta h\right)^2 + \left(\Delta y + \frac{W_y^{\text{bal}}}{V_v}\Delta h\right)^2 \qquad (7.74)$$

If the roots of this quadratic equation are complex, the WP cannot be reached. The sum of both vectors $\Delta \mathbf{s}$ and $\Delta \mathbf{s}^W$ yields a virtual WP position in $\{w\}$ that is used to compute the corresponding heading command

$$\psi_{\text{cmd}} = \tan^{-1}\left(\frac{\Delta y + \Delta y^W}{\Delta x + \Delta x^W}\right) \qquad (7.75)$$

Within GNC algorithm, this feature can be integrated by splitting up the guidance in the two modules: 1) geographical WP approach and 2) T-approach, which are executed either one or another. Provided that the mission can be fulfilled completely, PADS guidance switches from 1) to 2) after reaching the last WP from the geographical WP list that must be defined prior to the flight. Figure 7.52 shows an example of simulated trajectory with two geographical WPs added to the standard T-approach WP set.

7.4.3   GUIDANCE STRATEGIES UTILIZING AN S-TURN PATTERN

As mentioned already, the T-approach guidance strategy happens to be a well-known approach commonly used by human skydivers and paragliders loitering

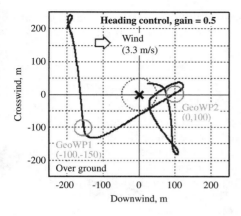

Fig. 7.52 Simulated trajectory with two additional geographical WPs.

just downwind of IPI and performing a series of figure-eight (*S*) turns. The only difference is that while in the T-approach guidance the figure-eight turns are the result of the WP guidance, so that PADS steers from one WP to another; paragliders execute a figure-eight maneuver intuitively as a predetermined maneuver requiring no WPs. The turns are always made into the wind to allow entering FA quickly, with minimum control efforts, in case conditions change rapidly. Figure 7.53 shows an example of a simulated flight trajectory starting near the end of the loitering phase and featuring a perfect exit leading right to IPI.

In practice, exiting a figure-eight loiter pattern is accomplished by constantly comparing a current altitude above IPI to that required to break a loiter pattern and make FA. As suggested by Culpepper et al. [2013], the altitude required to break a loiter pattern and complete FA should include two terms

$$h_{\text{req}} = \frac{\hat{V}_v}{V_h} d^{\text{eff}} + T_{\text{turn}} \hat{V}_v \qquad (7.76)$$

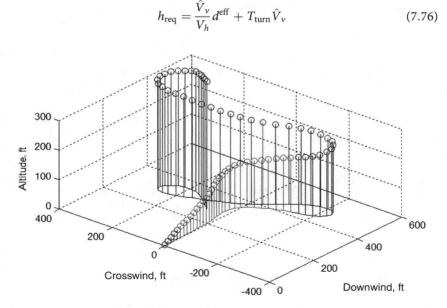

Fig. 7.53 Ideal landing trajectory [Culpepper et al. 2013].

[compare Eq. (7.73)] because some additional altitude is required to make the turn to IPI, which is assumed to occur at a constant nominal turn rate

$$T_{\text{turn}} = \left| \frac{\psi_0 - \tan^{-1}(\Delta y / \Delta x)}{\dot{\psi}} \right|$$

(7.77)

[Equation (7.76) utilizes Eq. (7.31) to replace the L/D ratio with the estimates of horizontal and vertical components of the airspeed.] Hence, the altitude margin is defined as the difference between the current altitude and the altitude required to reach IPI h_{req}. When the altitude margin falls below a specified value, guidance algorithm switches from the loiter phase to the final approach phase.

In a two-stage FA suggested in [Culpepper et al. 2013], on its way to IPI PADS first tracks an offset target (Fig. 7.54a). This offset target is placed downwind of IPI, and the altitude of this offset target is set to lie just above the nominal

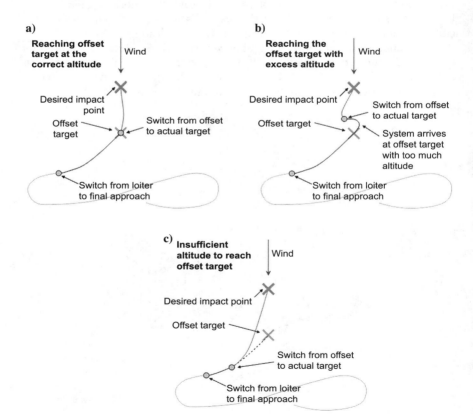

Fig. 7.54   Two-stage FA examples [Culpepper et al. 2013].

glide path to the actual IPI. While homing to the offset target, AGU computes the altitude margin for reaching IPI. When the altitude margin for reaching IPI reaches zero, PADS begins homing to the actual target. If PADS reaches the offset target with the correct amount of altitude, it then flies to the actual IPI and lands into the wind (Fig. 7.54a). If the system reaches the offset target with excess altitude, it loiters over the offset target until the excess altitude is burned off (Fig. 7.54b). Finally, if PADS runs out of altitude margin on the way to the offset target, the offset target is abandoned, and PADS flies straight to IPI (Fig. 7.54c). This allows the final approach trajectory to adapt to possible changes in the wind during FA.

Guidance strategy for the ParaLander PADS reported in [Altmann 2011] also includes figure-eight (S) turns at the terminal phase (Fig. 7.55). Upon canopy deployment and stabilization, PADS relies on the best known winds prior to deployment and therefore steers towards the wind drift trajectory (which would lead to IPI with no control efforts) (Fig. 7.55a). Given that the winds-aloft data are accurate enough, this maximizes the chances of reaching the IPI area. Once PADS intercepts the wind drift trajectory, it steers towards an area located slightly downwind of IPI, followed by the S-turns pattern for surface-layer wind identification and altitude control. Knowing the surface winds (depicted with the "Inflight identification wind" arrow in Fig. 7.55a), ParaLander AGU plans the terminal maneuver. As shown in Fig. 7.55a, rather than exiting straight onto FA when appropriate as it was the case in Figs. 7.53 and 7.54, ParaLander PADS starts a standard landing pattern (downwind—base—FA). Figure 7.55 features a left landing pattern, but depending on the altitude where the decision to brake the S-turns pattern is made, it might be a right landing pattern as well. Figure 7.55b shows an example of practical realization of this approach.

7.5   BANK OF TRAJECTORIES AND BAND-LIMITED GUIDANCE

The guidance strategy shown in Fig. 7.45 assumed online computation of the final turn and approach maneuver following exiting of the EM pattern. From the standpoint of the TPBV problem, this terminal guidance means selecting a function $\dot{\psi}_{cmd}(h)$ to bring PADS from any initial condition defined by the state vector $[x_0, y_0, h_0, \psi_0, \dot{\psi}_0]^T$ to the final conditions defined by $[0, 0, 0, 0, 0]^T$. Assuming Eq. (7.1) still holds, the five-state equations of motion given by Eqs. (5.55) and (5.57) can be rewritten as [Carter et al. 2007]

$$dx = -V[\cos(\gamma_a)\cos(\psi) + w_x][V\sin(\gamma_a) + w_z]^{-1}dh$$

$$dy = -V[\cos(\gamma_a)\sin(\psi) + w_y][V\sin(\gamma_a) + w_z]^{-1}dh$$

$$d\psi = -\dot{\psi}[V\sin(\gamma_a) + w_z]^{-1}dh \tag{7.78}$$

$$d\dot{\psi} = -(\dot{\psi}_{cmd} - \dot{\psi})T_{\psi}^{-1}[V\sin(\gamma_a) + w_z]^{-1}dh$$

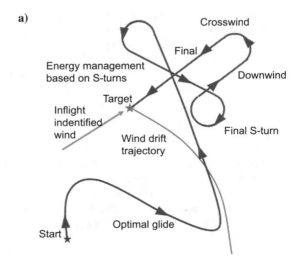

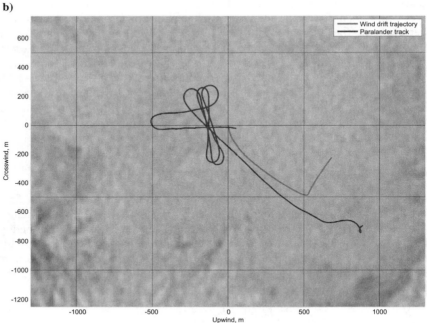

Fig. 7.55 a) ParaLander PADS guidance strategy and b) bird's-eye view of one of the flight tests [Altmann 2011].

Airspeed is considered to be a known function of heading rate $\dot{\psi}$ and altitude h, $V = V(\dot{\psi}, h)$, flight-path angle—a known function of $\dot{\psi}$, $\gamma_a = \gamma_a(\dot{\psi})$, the wind components $w_x = w_x(x, y, h)$, $w_y = w_y(x, y, h)$, and $w_z = w_z(x, y, h)$—known functions of x, y, and h.

One approach to address this problem relied on off-line computation of a large number of candidate trajectories, starting at different points on a 5D grid covering a sufficiently large neighborhood of IPI (see Table 7.2). In each node, the value of $\dot{\psi}_{cmd}(x_i, y_j, h_k, \psi_l, \dot{\psi}_m)$ is chosen by addressing the following optimization problem:

$$\dot{\psi}_{cmd}(x_i, y_j, h_k, \psi_l, \dot{\psi}_m) = \min_{\dot{\psi}_{cmd}}\left(C(x_i + dx, y_j + dy, \psi_k + d\psi, \dot{\psi}_l + d\dot{\psi})\right) \quad (7.79)$$

Here dx, dy, $d\psi$, $d\dot{\psi}$ are computed according to Eq. (7.78) for the altitude resolution dh of 25 m (82 ft) (which corresponds to an altitude loss in 5 s, representing a typical time between control actuations by a MegaFly PADS controller), and the cost function $C(x, y, h, \psi, \dot{\psi})$ determines the closeness to IPI and desired final heading with the appropriate weighting coefficients w_1 and w_2

$$C(x, y, \psi, \dot{\psi}) = x^2 + y^2 + w_1\tan^2(\psi) + w_2\dot{\psi}^2 \quad (7.80)$$

The optimal value of $\dot{\psi}_{cmd}$ for each node is stored in 1 byte (utilizing the int8 format), thus requiring the total of 124 Mb. Figure 7.56 shows an example of computer simulation utilizing the bank of commanded heading rates (featuring the same guidance modes as in Fig. 7.45).

Ideally, the bank for $\dot{\psi}_{cmd}$ should be created enroute making use of the best information about the winds, so that the most current wind velocity components $w_x = w_x(x, y, h)$, $w_y = w_y(x, y, h)$, and $w_z = w_z(x, y, h)$ could be used in Eq. (7.78) no matter how they were obtained (mesascale wind modeling,

TABLE 7.2   SIZE OF A MESH GRID [CARTER ET AL. 2007]

Coordinate	Range	Resolution	Number of Gridpoints
Along-track position	−800 to +400 m (−2,625 to +1,312 ft)	8 m (26 ft)	151
Cross-track position	0 to 400 m (0 to 1,312 ft)	8 m (26 ft)	51
Altitude	25 to 500 m (82 to 1,640 ft)	25 m (26 ft)	20
Heading	−180 to +180 deg	5 deg	73
Heading rate	−15 to +15 deg/s	3 deg/s	11
Total number of nodes			123,678,060

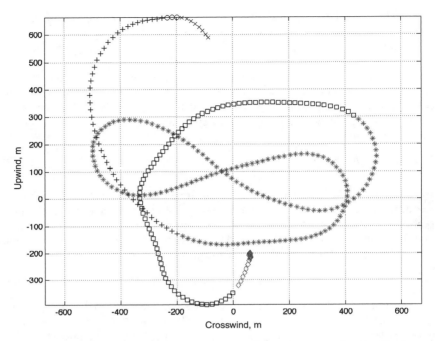

Fig. 7.56   Computer simulation utilizing the bank of $\dot{\psi}_{cmd}$ [Carter et al. 2007; Hattis et al. 2006].

dropsonde or balloon data). However, in practice to avoid computational burden the dynamic programming calculation of the guidance table is done only once, with an assumption that the speed is zero. That is why (due to symmetry) Table 7.2 is only computed for a positive cross-track position. Whatever wind data are available before the drop all mesh grid is simply shifted against the wind, as explained by Eq. (7.45), that is, again, the wind-fixed coordinates are used for guidance. As shown in [Yakimenko 2000], using a trajectories bank for making an educated initial guess on the values of varied parameters allows conducting real-time optimization as opposed to off-line optimization with the arbitrary values of varied parameters. In this particular case though, simple interpolation of the value of $\dot{\psi}_{cmd}$ within the hypercube determined by Table 7.2 was conducted every 5 s to produce a step-like control input $\dot{\psi}_{cmd}(h)$.

Another approach, BLG, relies on the direct method of calculus of variations. In this approach, the search of a desired (commanded) heading rate profile is limited by the following parameterization:

$$\psi'(h) = \sum_{k=0}^{M} \psi'_k \frac{\sin(\xi_k)}{\xi_k} \qquad (7.81)$$

with

$$\xi_k = \frac{\pi(h - k\Delta h)}{\Delta h} \qquad (7.82)$$

[To obtain the heading rate command $\dot{\psi}_{cmd}$, Eq. (7.81) needs to be multiplied by $\hat{V}_y$.] The idea is to ensure that control system is able to track the heading rate command profile accurately. The form of parameterization (7.81) allows explicitly restricting the heading rate profiles by the frequency that is significantly less than the bandwidth of the closed-loop system [Carter et al. 2009].

For example, for the MegaFly PADS [with $V_v^* = 5\,\text{m/s}$ (16 ft/s)] a final turn and approach maneuver begins at 800 m (2,625 ft) AGL, followed by the application of full brakes (flare) at 60 m (197 ft) AGL [Carter et al. 2009]. Hence, if we choose $\Delta h = 200\,\text{m}$ (656 ft), that makes $M = 800\,\text{m}/200\,\text{m} = 4$. This amounts to sampling the heading rate command once every $200/5 = 40$ s, so that the band limit (Nyquist rate) is approximately $1/80$ Hz (0.08 rad/s), well below the system closed-loop bandwidth (of about 0.53 rad/s). We may choose $\Delta h = 100\,\text{m}$ (328 ft), which makes $M = 800\,\text{m}/100\,\text{m} = 8$ and results in a double-band limit of ~ 0.16 rad/s. For $\Delta h = 50\,\text{m}$ (164 ft) $(M = 16)$ the band limit of ~ 0.32 rad/s would become dangerously close to PADS closed-loop bandwidth.

Varied parameters ψ_k', which represent turning rates at consecutive multiples of Δh, are defined by solving the following unconstrained optimal terminal control problem:

$$\min_{\psi_k', k=0,\dots,4} (J) \qquad (7.83)$$

where the performance index J represents a weighted sum of a squared miss distance and squared heading error [similar to that of Eq. (7.80)]

$$J = (1 - w_1)\left[(x_f - x_T)^2 + (y_f - y_T)^2\right] + w_1 \tan^2(\psi_f - \psi^{\text{des}}) \qquad (7.84)$$

(ψ^{des} is the desired final heading). The final values x_f, y_f, and ψ_f are to be computed via integration of four kinematic equations

$$\begin{aligned}
x' &= (V_x + w_x)V_z^{-1} \\
y' &= (V_y + w_y)V_z^{-1} \\
V_x' &= -\psi'(h)V_y \\
V_y' &= \psi'(h)V_x
\end{aligned} \qquad (7.85)$$

The major goal of this approach was to make the trajectory optimization routine fast enough to be able to recompute trajectories faster than in real time. That is why the form (7.85) was chosen for integration—it eliminates the necessity to deal with trigonometric functions. Also, the number M in Eq. (7.81) needs to be small, so for practical application it was chosen to be less than 4. Finally, a precomputed look-up table was used to store the

Fig. 7.57 Example of a flight path of MegaFly PADS with a) BLG and b) BLG statistics [Carter et al. 2009].

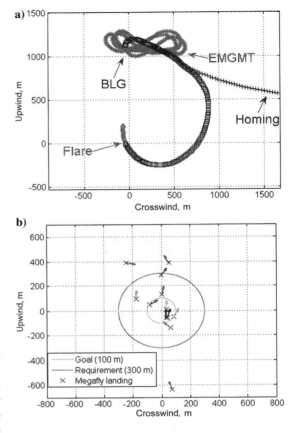

values of the normalized cardinal sine function $\sin c(x) = \sin(\pi x)/(\pi x)$ used in parameterization (7.81). As stated by Carter et al. [2009], integration of kinematic equations was done using fixed-point arithmetic for efficiency. Using the simplex search Nelder–Mead algorithm (MATLAB's *fminsearch* function) for optimization allowed re-solving MegaFly FA guidance at 1-Hz rate (using each preceding solution as a new initial guess). With the imposed 100 limit on the number of iterations per 1-Hz cycle, the convergence to an optimal solution was obtained only gradually (not necessarily within one control cycle). An example of the last portion of a guided descent is presented in Fig. 7.57a. Figure 7.57b shows the spread of the touchdown locations for 12 consecutive drops, executed back in 2009.

7.6 QUASI-OPTIMAL TERMINAL GUIDANCE

This and the next section provides more details on the terminal guidance concept developed for the Snowflake PADS and introduced in Fig. 7.46 of Sec. 7.3.3. This section closely follows [Slegers and Yakimenko 2009; Slegers and Yakimenko 2011].

7.6.1 CLOSED-FORM SOLUTION

To begin with, let us assume that there is no crosswind component, that is, the ground winds uploaded to AGU before deployment have not changed. An ideal

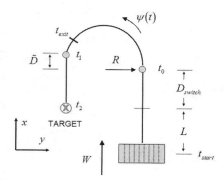

Fig. 7.58   Terminal guidance maneuver.

terminal guidance trajectory can then be defined as outlined in Fig. 7.58. Note, this figure represents the left landing pattern, but everything (except the heading sign) will be the same for the right pattern as well.

The terminal guidance problem can be formulated as follows (Fig. 7.58). For PADS subject to a constant wind

$$\mathbf{W} = [W, 0, 0]^T \tag{7.86}$$

at altitude h and a distance L from the IPI line, find the distance D_{switch} to the final 180-deg turn initiation point (TIP), required to travel past the target for an ideal impact at IPI t_2. In Fig. 7.58, R is the turn radius, $\psi(t)$ is the heading time history to track during the final turn, t_{start} is the time corresponding to the beginning of the downwind leg (corresponding to phase 4 in Fig. 7.46), t_0 is the time corresponding to the beginning of the final turn (phase 5), t_{exit} is the time when AGU switches from phase 5 to phase 6, t_1 is the time corresponding to intercepting glide path (phase 6), and

$$\tilde{D} = WT_{turn} \tag{7.87}$$

is the distance defined by asymmetry of the final turn due to the wind.

Let us assume that 1) the turn rate is slow, so that the roll and sideslip angles can be ignored [which again means that Eq. (7.1) holds], and 2) the descent rate V_v^* and airspeed V_h^* are viewed as nearly constant (* denotes the steady-state values). Then, the problem reduces to handling a simple kinematic model represented by three components of the ground-speed vector derived from Eq. (5.30) with account of Eq. (7.86) as

$$\begin{bmatrix} \dot{x} \\ \dot{y} \\ \dot{z} \end{bmatrix} = \begin{bmatrix} V_h^* \cos(\psi) + W \\ V_h^* \sin(\psi) \\ V_v^* \end{bmatrix} \tag{7.88}$$

From the last equation it immediately follows that

$$t_2 = t_{start} - \frac{z_{turn}}{V_v^*} \tag{7.89}$$

Now, the reference turn function $\psi(t)$ can be chosen as any function that satisfies BC of $\psi(t_0) = 0$ and $\psi(t_1) = \mp\pi$ (minus corresponds to the left approach pattern shown in Fig. 7.58 and plus—to the right pattern). For example, for a constant turn rate, we can write

$$\dot{\psi}_{\text{cmd}} = \mp \frac{V_h^*}{R} \tag{7.90}$$

and

$$T_{\text{turn}} = t_1 - t_0 = \frac{\pi R}{V_h^*} \tag{7.91}$$

After the final turn PADS travels directly to IPI, the FA leg lasts

$$T_{FA} = t_2 - t_1 \tag{7.92}$$

Assuming a constant turn rate of Eq. (7.90), integration of inertial velocities along axes x and y from t_{start} to t_2 (phases 4–6) yields two simple equalities:

$$D_{\text{switch}} - \int_{t_0}^{t_1} \dot{x}\,dt - \int_{t_1}^{t_2} \dot{x}\,dt = D_{\text{switch}} + WT_{\text{turn}} - (V_h^* - W)T_{FA} = 0 \tag{7.93}$$

$$z_{\text{start}} + V_v^* \left(\frac{L + D_{\text{switch}}}{V_h^* + W} \right) + V_v^* T_{\text{turn}} + V_v^* T_{FA} = 0 \tag{7.94}$$

Resolving them with respect to D_{switch} and a T_{app} results in

$$D_{\text{switch}} = -WT_{\text{turn}} + \left(\frac{V_h^{*2} - W^2}{2V_h^*} \right) \left(\frac{-z_{\text{start}}}{V_v^*} - T_{\text{turn}} - \frac{L - WT_{\text{turn}}}{V_h^* + W} \right) \tag{7.95}$$

$$T_{FA} = \left(\frac{V_h^* + W}{2V_h^*} \right) \left(\frac{-z_{\text{start}}}{V_v^*} - T_{\text{turn}} \right) - \frac{L - WT_{\text{turn}}}{2V_h^*} \tag{7.96}$$

From Eqs. (7.95) and (7.96), it follows that the higher the altitude z_{start} is, the larger D_{switch} and T_{FA} become. As the parafoil loiters upwind of IPI, the loiter exit altitude z_{start} can be found by using a desired FA time T_{FA}^{des}. The switching altitude to achieve T_{FA}^{des} is then given by solving Eq. (7.96) for z_{start}, which yields

$$z_{\text{start}} = -V_v^* \left(\frac{2V_h^*}{V_h^* + W} T_{FA}^{\text{des}} + \frac{L - WT_{\text{turn}}}{V_h^* + W} + T_{\text{turn}} \right)$$

$$= -V_v^* \frac{L + V_h^*(T_{\text{turn}} + 2T_{FA}^{\text{des}})}{V_h^* + W} \tag{7.97}$$

Once the system is traveling towards IPI, the goal is to bring it to the point defined by $x_T = D_{switch}$ and $y_T = \pm 2R$ (for the left and right turn, respectively). The distance $\hat{D}_{switch}$ is estimated during the downwind leg continuously using the analog of Eq. (7.95)

$$
\hat{D}_{switch} = -\hat{W}T_{turn} + \left(\frac{\hat{V}_h^{*2} - \hat{W}^2}{2\hat{V}_h^*}\right)\left(\frac{-\hat{z}}{\hat{V}_v^*} - T_{turn} - \frac{\hat{x} - \hat{W}T_{turn}}{\hat{V}_h^* + \hat{W}}\right)
$$
$$
= -\frac{\hat{z}(\hat{V}_h^{*2} - \hat{W}^2) + \hat{V}_h^*\hat{V}_v^*T_{turn}(\hat{V}_h^* + \hat{W}) + \hat{x}\hat{V}_v^*(\hat{V}_h^* - \hat{W})}{2\hat{V}_h^*\hat{V}_v^*} \qquad (7.98)
$$

where $\hat{V}_h^*$, $\hat{V}_v^*$, and $\hat{W}$ are the estimates of corresponding parameters at the current position $(\hat{x}, \hat{z})$. Note, Eq. (7.98) produces the value of $\hat{D}_{switch}$ in the assumption that V_h^*, V_v^*, and W remain constant from the current altitude all of the way down.

Figure 7.59 demonstrates the simulation results for the ideal case when all parameters (horizontal airspeed, descent rate, wind) are assumed to be estimated with no errors, there are no disturbances, and the commanded yaw rate of Eq. (7.90) is tracked precisely. Specifically, for the horizontal airspeed of 6.8 m/s (13.3 kt), descent rate of 3 m/s (10 ft/s), wind of 3 m/s (−6 kt), turn radius of 38 m (125 ft), and the desired FA time T_{FA}^{des} of 7.5 s, the maneuver starts at the altitude of 171 m (561.4 ft) [$L = 229$ m (750 ft)]. The distance D_{switch} is estimated to be −25 m (−82.3 ft), it takes 17.5 s to make a full turn, and PADS touches down precisely on target in 45.65 s.

In practice, sensor errors, variable winds (magnitude and direction), and imperfect control will certainly disturb the ideal touchdown depicted in Fig. 7.59. To this end, Table 7.3 provides measurement errors typical for the GNC unit used by Snowflake AGU. On top of that, the surface layer wind is not constant either. Figure 7.60 illustrates how inaccuracies in measuring PADS position and variable winds affect the touchdown accuracy.

Specifically, Fig. 7.60 shows two Monte Carlo simulations. For the simulation shown in Fig. 7.60a, it was assumed that the downwind leg ends with some error in the PADS horizontal position and heading. The standard deviation of horizontal position error in each coordinate was assumed to be 6 m (20 ft), and the standard deviation in heading 10 deg. As seen, these errors result in spreading the arrival to the glide-path intercept point. The situation is even worse when unaccounted winds are acting on PADS during the final turn. For simulations shown in Fig. 7.60b, the standard deviation for wind disturbances applied at each integration step of 0.05 s was 0.9 m/s (1.8 kt) (with a zero mean value) in both x and y direction, but as shown it led to a drastic degradation of overall performance. Obviously, if we add a constant (unmeasured wind) that acts on the system all of the way down in any direction, it will simply blow the PADS off

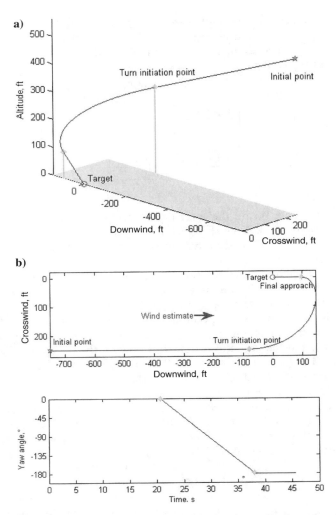

Fig. 7.59   Simulated guidance in ideal conditions: a) 3D trajectory horizontal projection and b) the time history of the yaw angle.

the desired trajectory. This would be the worst-case scenario, but that is exactly what ADS deal with in practice.

As a reminder, the navigation block of AGU only provides indirect estimates of the wind based on the difference between the measured ground-speed vector and PADS steady-state speed components V_v^* and V_h^* (Sec. 7.2.2). Moreover, these estimates, good or bad, are only valid for the current altitude and provide

TABLE 7.3   ERRORS IN MEASURING THE PADS STATES

Parameter	Bias	Standard Deviation
Altitude z	3 m (10 ft)	0.3 m (1 ft)
Descent rate $\dot{z}$	0	0.6 m/s (2 ft/s)
Roll angle ϕ	1.5 deg	1.5 deg
Yaw angle ψ	3 deg	1.5 deg
Roll rate p	1 deg/s	1.5 deg/s
Yaw rate r	1 deg/s	1.5 deg/s

no prediction of future winds at lower altitudes (Sec. 3.1.1). To mitigate the effect of unknown and varied surface-layer winds, a direct-method-based optimization procedure described in the next section was employed.

7.6.2   TWO-DoF TERMINAL GUIDANCE

To be able to quickly generate an inertial reference trajectory that brings PADS from its current state (influenced by unaccounted winds) to a certain point on the glide path as shown in Fig. 7.58, the TPBV problem can be formulated as follows. Staring at some initial point at $t = t_c$ (using notations of Fig. 7.58 $t_0 \le t_c \le t_1$) with the state vector defined as $\mathbf{x}_c = [x_c, y_c, \psi_c]^T$, PADS subject to Eqs. (5.36) and (5.37) with the wind vector (7.86) needs to be brought to another point

$$\mathbf{x}_f = \left[(V_h^* - W)T_{FA}^{\text{des}}, \ 0, \ -\pi \right]^T \tag{7.99}$$

(for consistency with the preceding section we will consider a left landing pattern) in exactly T_m seconds. Hence, we need to find the optimal control $\psi_{\text{opt}}(t)$ subject to

$$|\dot{\psi}| \le \dot{\psi}_{\max} \tag{7.100}$$

that minimizes the performance index

$$J = (t_f - t_c - T_m)^2 + w_p\left\{[x(t_f) - x_f]^2 + [y(t_f) - y_f]^2\right\} + w_\psi[\psi(t_f) - \psi_f]^2 \tag{7.101}$$

where w_p and w_ψ are weighting coefficients. [If needed, the performance index can also include the additional terms related to the compactness of the maneuver and/or control expenditure like in Eq. (7.63).] The duration of the maneuver is defined as

$$T_m = \frac{h_c}{V_v^*} - T_{FA}^{\text{des}} \tag{7.102}$$

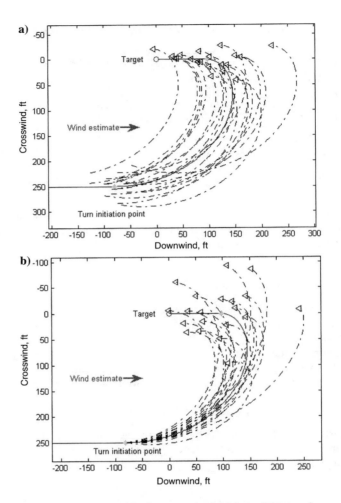

Fig. 7.60   Final turn trajectories with a) not precise initial conditions and b) unaccounted winds.

where h_c corresponds to $t = t_c$ and PADS altitude at $t = t_f$ is defined as

$$h_f = V_v^* T_{FA}^{des} = h_c - T_m V_v^* \qquad (7.103)$$

Obviously, while trying to execute the reference maneuver, the key assumption of Eq. (7.86) will be violated. Therefore, the winds that were unaccounted for, $\mathbf{w}^{dist} = \begin{bmatrix} w_x, & w_y, & 0 \end{bmatrix}^T$ (the vertical component can be accounted for implicitly via estimating the current $\hat{V}_v$ or explicitly as explained in Sec. 7.7.3), will not allow exact tracking of the calculated optimal trajectory. The idea is to constantly recompute the remaining portion of the final turn, each time starting from the current (off the original trajectory) initial conditions. The remaining time until arrival at the top of FA will be constantly updated as well

$$T_m = -\frac{\hat{z}}{\hat{V}_v} - T_{FA}^{des} \tag{7.104}$$

The estimate of the switching altitude z_{start}, as defined by Eq. (7.97), is based on the certain yaw rate profile of Eq. (7.90). While choosing the turn radius R, we should make sure that

$$R \gg V_h^* \dot{\psi}_{max}^{-1} \tag{7.105}$$

to allow satisfying Eq. (7.100) while recomputing the reminder of the turn with the actual winds greater than assumed.

The development of the IDVD-based optimal 2D trajectory generation algorithm relies on the first two equations of Eq. (7.88). From these equations, it follows that

$$\psi = \tan^{-1}\left(\frac{\dot{y}}{\dot{x} - W}\right) \tag{7.106}$$

Differentiating (7.106) yields the yaw rate required to follow the reference final turn trajectory in presence of the constant wind W

$$\dot{\psi} = \frac{\ddot{y}(\dot{x} - W) - \ddot{x}\dot{y}}{(\dot{x} - W)^2 + \dot{y}^2} \tag{7.107}$$

The x–y projection of the ground-speed vector [Eq. (5.4)] depends on the curzrent yaw angle as well

$$V_G = \sqrt{\dot{x}^2 + \dot{y}^2} = \sqrt{V_h^{*2} + W^2 + 2V_h^* W \cos(\psi)} \tag{7.108}$$

Now, following the general idea of direct methods of calculus of variations, we will assume the solution of the TPBV problem to be represented analytically as the functions of some scaled abstract argument $\bar{\tau} = \tau/\tau_f \in [0; 1]$

$$x(\bar{\tau}) = P_1(\bar{\tau}) = a_0^1 + a_1^1\bar{\tau} + a_2^1\bar{\tau}^2 + a_3^1\bar{\tau}^3 + b_1^1 \sin(\pi\bar{\tau}) + b_2^1 \sin(2\pi\bar{\tau})$$
$$y(\bar{\tau}) = P_2(\bar{\tau}) = a_0^2 + a_1^2\bar{\tau} + a_2^2\bar{\tau}^2 + a_3^2\bar{\tau}^3 + b_1^2 \sin(\pi\bar{\tau}) + b_2^2 \sin(2\pi\bar{\tau}) \tag{7.109}$$

The coefficients a_i^η and b_i^η ($\eta = 1, 2$) in these formulas are defined by BC up to the second-order derivative at $\tau = 0$ and $\tau = \tau_f$ ($\bar{\tau} = 1$). According to the problem formulation and the first two equations of Eq. (7.88), these BC are as follows:

$$\begin{bmatrix} x \\ y \end{bmatrix}_{\tau=0} = \begin{bmatrix} x_c \\ y_c \end{bmatrix}, \quad \begin{bmatrix} \dot{x} \\ \dot{y} \end{bmatrix}_{\tau=0} = \begin{bmatrix} V_h^* \cos(\psi_c) + W \\ V_h^* \sin(\psi_c) \end{bmatrix}, \quad \begin{bmatrix} \ddot{x} \\ \ddot{y} \end{bmatrix}_{\tau=0} = \begin{bmatrix} -\dot{\psi}_c V_h^* \sin(\psi_c) \\ \dot{\psi}_c V_h^* \cos(\psi_c) \end{bmatrix}$$

$$(7.110)$$

$$\begin{bmatrix} x \\ y \end{bmatrix}_{\tau=\tau_f} = \begin{bmatrix} (V_h^* - W) T_{FA}^{des} \\ 0 \end{bmatrix}, \quad \begin{bmatrix} \dot{x} \\ \dot{y} \end{bmatrix}_{\tau=\tau_f} = \begin{bmatrix} -V_h^* + W \\ 0 \end{bmatrix}, \quad \begin{bmatrix} \ddot{x} \\ \ddot{y} \end{bmatrix}_{\tau=\tau_f} = \begin{bmatrix} 0 \\ 0 \end{bmatrix}$$

$$(7.111)$$

Similar to the cost function of Eq. (7.80), the second-order derivatives at the final point are zeroed for a smooth arrival. Compared to formulation of Eq. (7.80) though where it is one of the objectives during an optimization process, in the current approach this is guaranteed upfront.

While the final conditions of Eq. (7.111) are constant, the initial conditions will reflect the current state of PADS at each cycle of optimization (which generates a new reference trajectory starting from the current state). Let us now differentiate Eqs. (7.109) twice with respect to τ to get

$$\tau_f P_\eta'(\bar{\tau}) = a_1^\eta + 2a_2^\eta \bar{\tau} + 3a_3^\eta \bar{\tau}^2 + \pi b_1^\eta \cos(\pi\bar{\tau}) + 2\pi b_2^\eta \cos(2\pi\bar{\tau})$$
$$\tau_f^2 P_\eta''(\bar{\tau}) = 2a_2^\eta + 6a_3^\eta \bar{\tau} - \pi^2 b_1^\eta \sin(\pi\bar{\tau}) - (2\pi)^2 b_2^\eta \sin(2\pi\bar{\tau})$$

$$(7.112)$$

Equating these derivatives at the terminal points to the known BC in Eqs. (7.110) and (7.111) yields a system of linear algebraic equations to solve for coefficients a_i^η and b_i^η ($\eta = 1, 2$). For instance, for the x coordinate we obtain

$$\begin{bmatrix} 1 & 0 & 0 & 0 & 0 & 0 \\ 1 & 1 & 1 & 1 & 0 & 0 \\ 0 & 1 & 0 & 0 & \pi & 2\pi \\ 0 & 1 & 2 & 3 & -\pi & 2\pi \\ 0 & 0 & 2 & 0 & 0 & 0 \\ 0 & 0 & 2 & 6 & 0 & 0 \end{bmatrix} \begin{bmatrix} a_0^1 \\ a_1^1 \\ a_2^1 \\ a_3^1 \\ b_1^1 \\ b_2^1 \end{bmatrix} = \begin{bmatrix} x_0 \\ x_f \\ x_0' \tau_f \\ x_f' \tau_f \\ x_0'' \tau_f^2 \\ x_f'' \tau_f^2 \end{bmatrix}$$

$$(7.113)$$

Being resolved, the system (7.113) yields

$$a_0^1 = x_0, \quad a_1^1 = -(x_0 - x_f) + \frac{(2x_0'' + x_f'')\tau_f^2}{6}$$

$$a_2^1 = \frac{x_0'' \tau_f^2}{2}, \quad a_3^1 = -\frac{(x_0'' - x_f'')\tau_f^2}{6}$$

$$b_1^1 = \frac{2(x_0' - x_f')\tau_f + (x_0'' + x_f'')\tau_f^2}{4\pi}$$

$$b_2^1 = \frac{12(x_0 - x_f) + 6(x_0' + x_f')\tau_f + (x_0'' - x_f'')\tau_f^2}{24\pi}$$

(7.114)

The only problem is that the derivatives in Eq. (7.114) are taken in the virtual domain while actual BC are given in the physical domain. Mapping between the virtual domain $[0; \tau_f]$ and physical domain $[t_c; t_f]$ is addressed by introducing a speed factor λ [Taranenko 1968; Yakimenko 2000]

$$\lambda = \frac{d\tau}{dt}$$

(7.115)

Using this speed factor, we may now compute corresponding derivatives in the virtual domain using the obvious differentiation rules valid for any time-variant parameter ξ

$$\dot{\xi} = \lambda \xi', \quad \ddot{\xi} = \lambda(\lambda' \xi' + \lambda \xi'')$$

(7.116)

Inverting Eq. (7.116) yields

$$\xi' = \lambda^{-1}\dot{\xi}, \quad \xi'' = \lambda^{-2}\ddot{\xi} - \lambda'\lambda^{-1}\dot{\xi}$$

(7.117)

Note that we only need to use Eqs. (7.117) once to transfer BC. Because the speed factor λ simply scales the entire problem—the higher speed factor λ is, the larger τ_f it results in. Hence, we may let

$$\lambda_{0;f} = 1 \quad \text{and} \quad \lambda'_{0;f} = 0$$

(7.118)

which means that instead of Eq. (7.117) we can safely assume [Taranenko 1968]

$$\xi' = \dot{\xi}, \quad \xi'' = \ddot{\xi}$$

(7.119)

Now let us describe the numerical procedure for finding the optimal solution among all candidate trajectories described by Eqs. (7.109). First, we guess on the value of the only varied parameter τ_f and compute the coefficients of the candidate trajectory using Eqs. (7.114) with BC of Eqs. (7.110) and (7.111) converted to the virtual domain via Eqs. (7.119) [accounting for Eqs. (7.118)]. For the initial value of τ_f when starting optimization for $t = t_0$, we can take the length of the circumference connecting the two terminal points

$$\tau_f = \frac{\pi}{2}\sqrt{(x_f - x_0)^2 + (y_f - y_0)^2} \tag{7.120}$$

[When recomputing a trajectory for $t = t_0$, we may take $\tau_f(t_c - t_0)(t_1 - t_0)^{-1}$.]

Having an analytical representation of the candidate trajectory, Eqs. (7.90) and (7.93), define the values of x_j, y_j, x_j', and y_j', $j = 1, \ldots, N$ over a fixed set of N points spaced evenly along the virtual arc $[0; \tau_f]$ with the interval

$$\Delta\tau = \tau_f(N - 1)^{-1} \tag{7.121}$$

so that

$$\tau_j = \tau_{j-1} + \Delta\tau, \quad j = 2, \ldots, N, (\tau_1 = 0) \tag{7.122}$$

Then, for each node $j = 2, \ldots, N$, we compute

$$\Delta t_{j-1} = \sqrt{\frac{(x_j - x_{j-1})^2 + (y_j - y_{j-1})^2}{V_h^{*2} + W^2 + 2V_h^* W \cos(\psi_{j-1})}} \tag{7.123}$$

$(\psi_1 \equiv \psi_0)$, and

$$\lambda_j = \Delta\tau \Delta t_{j-1}^{-1} \tag{7.124}$$

The yaw angle ψ can now be computed using the virtual domain version of Eq. (7.106)

$$\psi_j = \tan^{-1}\left(\frac{\lambda_j y_j'}{\lambda_j x_j' - W}\right) \tag{7.125}$$

Finally, the yaw rate $\dot{\psi}$ is evaluated using Eq. (7.107) [with time derivatives evaluated using inverse of Eq. (7.119)] or simply as a two-point backward difference approximation of the first-order derivative

$$\dot{\psi}_j = (\psi_j - \psi_{j-1})\Delta t_{j-1}^{-1} \tag{7.126}$$

When all parameters (states and controls) are computed in each of the N points, we can compute the performance index

$$J = \left(\sum_{j=1}^{N-1} \Delta t_j - T_{\text{turn}}\right)^2 + k_{\psi}\Delta \tag{7.127}$$

where

$$\Delta = \max_j \left(0; |\dot{\psi}_j| - \dot{\psi}_{\text{max}}\right)^2 \tag{7.128}$$

with k_{ij} being the weighting coefficient for a penalty term that was added to be able to address the problem using an unconstraint optimization method. Note that compared to Eq. (7.101) all but the first term have disappeared because they were satisfied by default when constructing the candidate reference trajectory with the terminal conditions defined by Eq. (7.111). Now, the problem can be solved with as simple optimization function as *fminbnd* of MATLAB.

As a demonstration of the IDVD-based approach just presented, Fig. 7.61 shows an example of the optimal final turn trajectory, which is essentially the same as in Fig. 7.59b. However, as opposed to Fig. 7.59b, the yaw angle changes smoothly at departure from TIP and arrival to the top of FA. As a result, the turn trajectory is slightly different from that of Fig. 7.59b, but what is of the most importance is that PADS captures the glide slope in exactly T_{turn} seconds as Fig. 7.59b trajectory does. So now have a tool allowing us to construct the optimal trajectory from any initial point to the predetermined final point to be achieved in a certain time. To this end, even if we get to the TIP with errors as shown in Fig. 7.59, we can easily accommodate these errors and compute the optimal control profile for each of these cases, so that we will still be at the top of FA in the predetermined time T_{turn} (see Fig. 7.62). Similarly, in case of wind disturbances (Fig. 7.60b) the ability to recompute the trajectory from the current (off the original trajectory) conditions to the same final conditions as often as practical seems to alleviate the problem.

The numerical algorithm just described allows computing the optimal turn trajectory very fast, even with the nongradient optimization routine *fminbnd* based on the straightforward golden section search and parabolic interpolation algorithm. To be more specific, a 16-bit 80-MHz onboard processor (used by

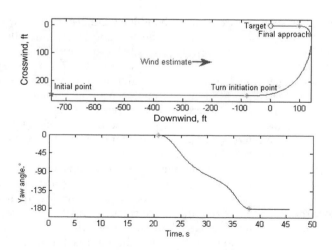

Fig. 7.61   Optimal guidance in the perfect conditions as in Fig. 7.59b.

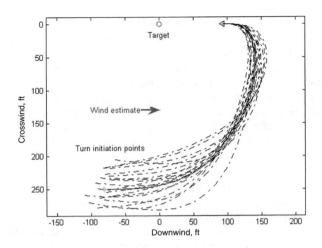

Fig. 7.62   Optimal guidance with an error at TIP.

Snowflake PADS) allows computing a 17.5-s turn maneuver with 20 nodes in just 10 iterations, which takes only 0.07 s altogether. This means that with the controls update rate of 0.25 s the reference trajectory can be updated as often as every control cycle. However, in practice there is no need to update trajectory as often; recomputing the trajectory every 5 s or so (to allow the tracking error to build up) seems to be enough. Alternatively trajectory updates can be scheduled so that the specified number of updates occurs at equal intervals over the period from t_0 to t_{exit} (see Fig. 7.58).

Figure 7.63 presents two examples of handling wind disturbances with the trajectory update rate set as a function of the tracking error. Once the tracking error (while following the previously generated trajectory) exceeds a certain threshold [3 m (10 ft) and 4.6 m (15 ft) in Figs. 7.63a and 7.63b, respectively], a new trajectory is generated. In Fig. 7.63, multiple reference trajectories are shown with the solid lines while a continuous parafoil trajectory is depicted with a dashed line. The points where the old reference trajectory was abandoned are depicted with the circles, and the corresponding locations of the parafoil blown away from the reference trajectory by strong (up to 15% of the forward speed) unaccounted winds is marked with triangles.

Although the optimal terminal guidance algorithm works well, several further adjustments could to be made to make it more robust. The kinematic model of Eq. (7.88) does not account for parafoil turning dynamics and assumes that the sideslip and roll angles are small. When the turn rate is sufficiently small or the radius R is large, the model described by Eq. (7.88) provides sufficient accuracy. However, to track the desired trajectory for a wide range of R, the error from sideslip and turning dynamics can be compensated

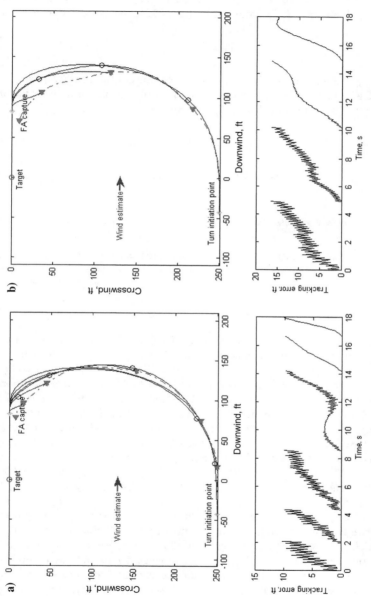

Fig. 7.63 Optimal guidance with wind disturbance with a a) 10-ft and b) 15-ft cap on a tracking error.

by adding an additional commanded yaw rate input for the first t_{pre} seconds after the TIP has been reached. Specifically, for the first t_{pre} seconds of the maneuver the commanded yaw rate ψ_{cmd} produced by Eq. (7.126) is augmented as follows:

$$\tilde{\dot{\psi}}_{cmd} = \dot{\psi}_{cmd} \mp K_{turn} \frac{V_h^*}{R} \qquad (7.129)$$

Here, K_{turn} is the correction gain, and $\mp V_h^*/R$ is the constant turn-rate command input of Eq. (7.90). Alternatively, turning dynamics could be accounted for by using Eq. (5.38) instead of Eq. (5.37) to begin with. In this case,

$$\tilde{\dot{\psi}}_{cmd} = \dot{\psi}_{cmd} + T_\psi \ddot{\psi}_{cmd} \qquad (7.130)$$

where the second-order derivative can be estimated using the three-point backward difference approximation

$$\ddot{\psi}_j = [\psi_j \Delta t_{j-2} - \psi_{j-1}(\Delta t_{j-1} + \Delta t_{j-2}) + \psi_{j-2}\Delta t_{j-1}]\Delta t_{j-1}^{-1} \Delta t_{j-2}^{-1} \qquad (7.131)$$

Another practical adjustment corrects the glide-path capture location defined by T_{FA}^{des}. The values of h_f given by Eq. (7.103) and x_f given by Eq. (7.99) are corrected to

$$h_f = V_v^* T_{FA}^{des} \varepsilon_{FA} \quad \text{and} \quad x_f = (V_h^* - W)T_{FA}^{des} \varepsilon_{FA} \qquad (7.132)$$

where $\varepsilon_{FA} < 1$ is the glide-path tracking efficiency. The rationale is that at FA wind disturbances result in deviation from the shortest path, and some extra time $T_{FA}^{des}(1 - \varepsilon_{FA})$ should be allowed to correct them.

Figures 7.64 and 7.65 show two examples of how the practical additions to the guidance law [described by Eqs. (7.129) and (7.132)] work for desired turn radii of 38 m (125 ft) and 99 m (325 ft), respectively. In these figures, the full non-linear Snowflake PADS model is simulated with the model predictive control (MPC) based controller (to be described in Sec. 7.8.5) to track the desired yaw angle. In both cases, $T_{FA}^{des} = 7.5$ s, $\varepsilon_{FA} = 0.95$, $t_1 - t_{exit} = 3$ s, $t_{pre} = 6$ s, $K_{turn} = 1$, $V_h^* = 6.8$ m/s (13.3 kt), $W = 3$ m/s (5.9 kt), $\dot{\psi}_{max} = 20$ deg/s, and simulation starts at TIP with $[x_0, y_0, h_0] = [-27, 76, 95]$m ([−88.3, 250, 311]ft) for Fig. 7.64 and $[x_0, y_0, h_0] = [-111, 198, 200]$m ([−363, 650, 658]ft) for Fig. 7.65. In both cases PADS initially tracks the optimal trajectory; however, errors slowly build up due to mismatch in the actual dynamics and the model. (In Fig. 7.65, the longer turn allows more error to accumulate.) At each of one (Figs. 7.64a and 7.65a) or three (Figs. 7.64b and 7.65b) updates, a new trajectory is generated starting from the current state and still leading to the FA capture point. Figures 7.64 and 7.65 show PADS trajectory passing the FA capture point to demonstrate that in all cases PADS model hits IPI within 0.4-m (1.3-ft) accuracy.

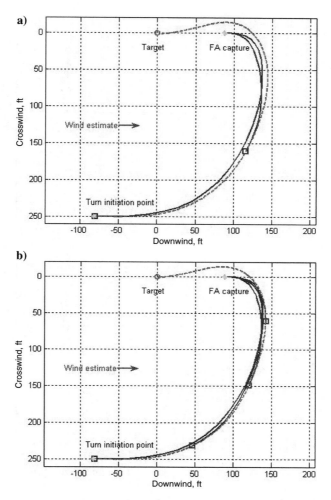

Fig. 7.64 Optimal real-dynamics-augmented guidance with a) one and b) three trajectory updates with a 125-ft turn radius.

Simulation involving all phases of flight (as shown in Fig. 7.46) is presented in Fig. 7.66. Fully deployed at $[x, y, h] = [-366; -61; 670]$m ($[-1,200; -200; 2,200]$ft), the PADS model reaches the loiter area and loiters in phase 2 continuously estimating wind $\hat{W}$, horizontal projection of airspeed $\hat{V}_h^*$, and descent rate $\hat{V}_v^*$. The altitude at which to exit loitering and start towards the target z_{start}, found by using Eq. (7.97), assumes that PADS is already traveling towards the target. While loitering in phase 2, there will be a delay between reaching

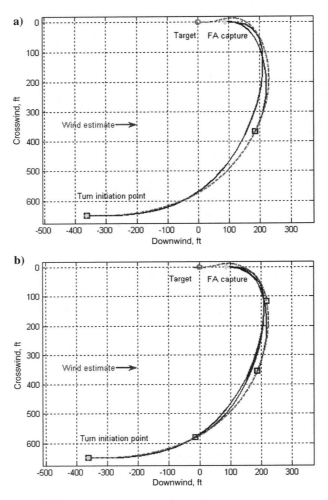

Fig. 7.65 Optimal real-dynamics-augmented guidance with a) one and b) three trajectory updates with a 325-ft turn radius.

the altitude z_{start} and when PADS can turn to face IPI. To achieve the desired FA time T_{FA}^{des}, PADS must exit loitering T_{delay} seconds early. The altitude to exit loitering can then be estimated by using the same Eq. (7.97) but using the combined time $T_{delay} + T_{FA}^{des}$ rather than the ideal T_{app}^{des}. Upon exiting the loiter area, phase 4 is entered, and the parafoil travels towards IPI updating estimates of $\hat{D}_{switch}$ [Eq. (7.98)]. The location where the parafoil reaches $\hat{D}_{switch}$ is defined as the TIP (the beginning of phase 5). After exiting the loiter pattern, the approach time T_{FA} is determined by Eq. (7.96). Because of disturbances, measurement error, and tracking error, T_{FA} may not match T_{FA}^{des}. The optimal final turn trajectory is determined using the last estimates of $\hat{W}$, $\hat{V}_h^*$, $\hat{V}_v^*$, and T_{FA} as outlined earlier in this section. While in phase 5, the optimal final turn trajectory is continually updated until time t_{exit}, after which phase 6 begins commences.

This simulation used the same precision placement parameters as in Figs. 7.64 and 7.65, $R = 38$ m (125 ft) and $T_{delay} = 6.5$ s. The away distance for the loiter area was set to 228.6 m (750 ft) and the cycle distance to 121.9 m (400 ft) (refer to Fig. 7.46). During the final turn in phase 5, two updates of the optimal trajectory were used. In all phases of guidance, the MPC algorithm based on the linear Snowflake PADS approximation was used to track the desired yaw angle. Progression of algorithm phases is shown by triangles representing the transition to the next guidance phase, with the first transition marking the beginning of phase 2. The trajectory is shown in Fig. 7.66 with the four loiter region points marked by x's, optimal trajectory updates marked by squares, and the final target by a circle. Upon entering phase 2, the parafoil circles the loiter region in a clockwise direction.

Figure 7.67 shows the estimation of the exit altitude z_{start} from Eq. (7.97) based on current wind estimates, the approach time T_{FA}^{des}, and transition delay T_{delay}. Transition from phase 2 to 3 occurs when the altitude reaches z_{start} at 120 s at an altitude of 223 m (732 ft). Figure 7.68 shows that T_{FA} ($T_{FA}^{des} + T_{delay}$) is 14 s when phase 3 is entered. Transition to phase 4 occurs at 133 s at which time T_{FA} has decreased to 6.9 s due to the time required to make the turn exiting the loitering region. During phase 4, the required T_{FA} continues to vary as tracking error and wind estimates change. In phase 4, the transition to the final turn occurs when $\hat{D}_{switch}$, the estimated distance to pass the target, is reached. Figure 7.67 shows $-\hat{D}_{switch}$, where the value is positive when the parafoil must turn before reaching IPI line and negative if the turn is to occur after passing

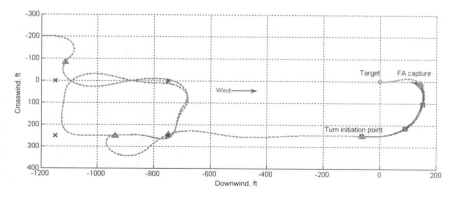

Fig. 7.66 Precision placement trajectory.

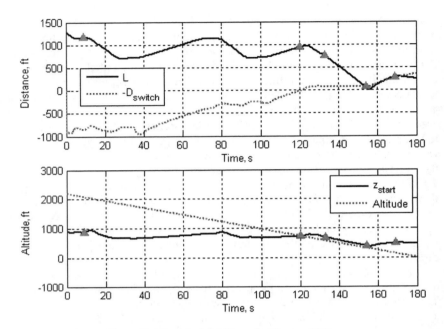

Fig. 7.67   Terminal guidance decision variables.

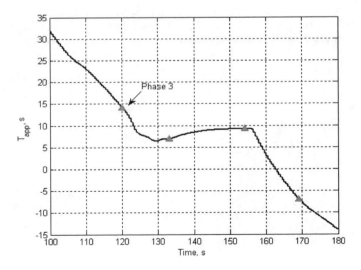

Fig. 7.68   Final approach time monitoring.

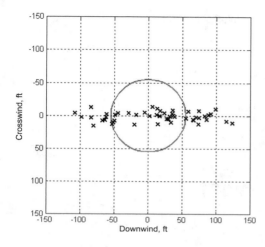

**Fig. 7.69   Monte Carlo
simulation dispersion.**

the target. Early in the trajectory, $-\hat{D}_{\text{switch}}$ is much less than zero demonstrating if a turn to IPI was performed at that time PADS would need to pass the PIP line by a large amount before turning because of its current altitude. Transition from phase 4 to 5 occurs when the distance to target L reaches $-\hat{D}_{\text{switch}}$ at 154 s with PADS 18.9 m (62 ft) ahead of IPI line. Two updates of the optimal final turn trajectory occur before transition to phase 6 at 169 s. Final impact occurs at 0.6 m (2 ft) cross range and 2.4 m (8 ft) downrange at 180 s.

Figure 7.69 is composed of the touchdown error dispersion results for 50 Monte Carlo simulations similar to that of Fig. 7.66 (utilizing the complete precision placement algorithm). Gaussian noise was injected into measured position, altitude, and inertial sensors. In addition to sensor errors, two sources of wind variation were added to the simulation: varying direction and varying ground wind magnitude. Wind was assumed to have a constant magnitude prior to terminal guidance. Wind magnitude varied linearly from W at the TIP altitude to W_F at the ground. Prevailing wind was assumed by the system to travel downrange at a heading of 0 deg while the true wind varied in its direction. Sensor noise and wind variation statistics are listed in Table 7.4. For all simulations, IPI was set as the origin. The resulting CEP depicted by the circle is 16.8 m (55.1 ft).

The actual flight tests prove the correctness of the preceding CEP estimation. The actual CEP for over 200 drops conducted over the period of May 2008–January 2013 is of the order of 9 m (30 ft). Figure 7.70 shows two examples of typical performance of the Snowflake PADS.

Let us compare two direct-method-based approaches implemented in real PADS and thoroughly tested: BLG approach of Sec. 7.5 and the IDVD-method-based approach of the current section. Table 7.5 presents the results of a side-by-side comparison.

As seen, the BLG approach can limit the control bandwidth explicitly while IDVD can only do it indirectly by penalizing $\dot{\psi}$. On the other hand, IDVD allows satisfying all BC automatically and guarantees fast convergence. Moreover, dealing with a trajectory parameterization [Eq. (7.109)] rather than with control time-history parameterization [Eq. (7.81)], IDVD-based approach has more

TABLE 7.4   ERROR STATISTICS

Parameter	Mean	Standard Deviation
Initial position x	-457.2 m(-1500 ft)	76.2 m (250 ft)
Initial position y	0	76.2 m (250 ft)
Initial position z	-914.4 m (-3000 ft)	76.2 m (250 ft)
GPS x bias	0	0.6 m (2 ft)
GPS y bias	0	0.6 m (2 ft)
GPS x deviation	0.46 m (1.5 ft)	0
GPS y deviation	0.46 m (1.5 ft)	0
Altitude bias	0	2.3 m (7.5 ft)
Altitude variation	0.46 m (1.5 ft)	0
Roll, pitch, and yaw bias	0	2 deg
Roll, pitch, and yaw deviation	1 deg	0
u, v, and w bias	0	0.076 m/s (0.25 ft/s)
u, v, and w deviation	0.15 m/s (0.5 ft/s)	0
p, q, and r bias	0 deg	1 deg
p, q, and r deviation	1 deg	0
W	-3 m/s (-5.9 kt)	
W_F	-3 m/s (-5.9 kt)	1 m/s (1.9 kt)
Wind heading error	0 deg	15 deg

potential by adjusting the terminal conditions according to the current accuracy of reference trajectory tracking.

To conclude this section, it should be noted that typically, having one varied parameter for the parameterization (7.109), the length of the virtual length τ_f is more than enough. However, if more flexibility is required the IDVD-method-based algorithm presented in this section can be extended to incorporate more varied parameters, for example, the final heading. In this case, Eqs. (7.99), (7.111), and (7.127) will take the form

$$\mathbf{x}_f = \left[(V_h^* \cos(\psi_f) - W)T_{FA}^{\mathrm{des}}, \; -V_h^* \sin(\psi_f)T_{FA}^{\mathrm{des}}, \; \psi_f \right]^T \qquad (7.133)$$

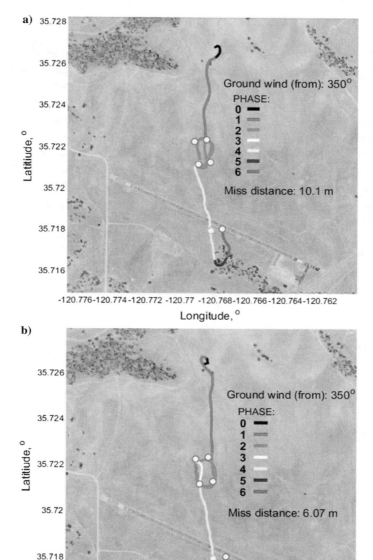

Fig. 7.70 A bird's-eye view of two typical Snowflake PADS trajectories.

TABLE 7.5 COMPARISON OF IDVD AND BLG

Characteristic	IDVD	BLG
Computing the coordinates	Parameterized analytical function	Via integrating equations of motion
Computing the heading rate	Via inverse dynamics	Parameterized analytical function
Satisfying terminal conditions on coordinates	Automatically	Via optimization [Eq. (7.84)]
Satisfying terminal conditions on heading	Automatically	Via optimization [Eq. (7.84)]
Satisfying terminal conditions on heading rate	Automatically	Not capable
Satisfying initial conditions on heading rate	Automatically	Not capable
Satisfying the control bandwidth constraint	Via optimization [Eq. [7.128]]	Automatically
Number of varied parameters	One	Five
Minimization engine	*fminbnd*	*fminsearch*
Convergence	Guaranteed	Not guaranteed
Number of iterations to converge	<10	<100*
Onboard CPU	16-bit 80 MHz	8-bit 44.2 MHz [George et al. 2005]
CPU time to converge	0.07 s	<1 s*

*Means if converges to the solution

$$
\begin{bmatrix} x \\ y \end{bmatrix}_{\tau=\tau_f} = \begin{bmatrix} (V_h^* \cos(\psi_f) - W) T_{FA}^{\text{des}} \\ -V_h^* \sin(\psi_f) T_{FA}^{\text{des}} \end{bmatrix}
$$
$$
\begin{bmatrix} \dot{x} \\ \dot{y} \end{bmatrix}_{\tau=\tau_f} = \begin{bmatrix} V_h^* \cos(\psi_f) - W \\ V_h^* \sin(\psi_f) \end{bmatrix}, \quad \begin{bmatrix} \ddot{x} \\ \ddot{y} \end{bmatrix}_{\tau=\tau_f} = \begin{bmatrix} 0 \\ 0 \end{bmatrix}
$$
(7.134)

$$
J = \left(\sum_{j=1}^{N-1} \Delta t_j - T_{\text{turn}} \right)^2 + k_{\dot{\psi}} \Delta + k_{\psi}(\psi_N + \pi)^2
$$
(7.135)

To make a candidate trajectory even more flexible, the order of parameterization (7.109) for the y coordinate can be increased by two (by adding two more monomials), so that the third-order derivatives of the y coordinate (jerks) at the terminal points could be used as two additional varied parameters [Yakimenko 2000]. The remainder of the algorithm would still be the same. Of course, in this case, instead of *fminbnd*, another optimization routine allowing two (or four) varied parameters would be used. For this purpose the Hook–Jeeves method [Hook and Jeeves 1961] proved to be more effective compared to the *fminsearch* function of MATLAB [Yakimenko 2011].

7.7   ACCOUNTING FOR THE COMPLEX SURFACE-LAYER WIND MODELS

The goal of this section is to show how easy it is to incorporate wind data gathered as discussed in Sec. 3.1.2 into the IDVD-method-based terminal guidance presented in Sec. 7.6 at the stage of generating the reference control (neither a maneuver-based nor BLG approach can really take advantage of knowing these winds upfront). Following [Yakimenko and Slegers 2011], this section starts with providing with some test data showing how unmodeled varied winds affect the PADS touchdown accuracy. This is somewhat a continuation of a discussion in Sec. 3.1.1. Then, it shows how to incorporate different surface-layer wind models presented in Secs. 3.1.3 and 3.1.5, which includes crosswinds and vertical wind disturbances (if known).

7.7.1   EFFECT OF UNSETTLED SURFACE-LAYER WINDS

Consider two samples of data measured/estimated and recorded onboard the Snowflake PADS presented in Fig. 7.71. The plots in the left column (Fig. 7.71a) belong to one drop terminated with a miss distance of 10 m (33 ft) (Fig. 7.70a) and in the right column (Fig. 7.71b)—to a second drop that resulted in a miss distance of 6 m (20 ft) (Fig. 7.70b). The first row of the plots presents the altitude vs time profiles. The second row shows an estimate of downwind component of the wind (which changes all of the time based on the latest observations of the ground speed). The third row shows a vertical speed of PADS also measured by GPS and smoothed (filtered) by the onboard IMU. The last row of the plots presents different stages of flight when these data were collected.

It is clearly seen that in practice, neither V_v^* nor W used in Eq. (7.88) are constant. While the first set of data (Fig. 7.71a) exhibits a kind of gradual decrease of the downwind component W with time (altitude), the second set of data (Fig. 7.71b) features more or less constant winds up to about 200 m (656 ft) altitude AGL with a sudden halved decrease below this altitude. For convenience of analysis, Fig. 7.72 presents the values of V_v^* and W for the same sets of data vs altitude, rather than time. Looking at both sets of data together, one can notice that the descent rate (negative of the vertical speed) varies from 0 m/s (as a result of some updraft during the downwind leg in

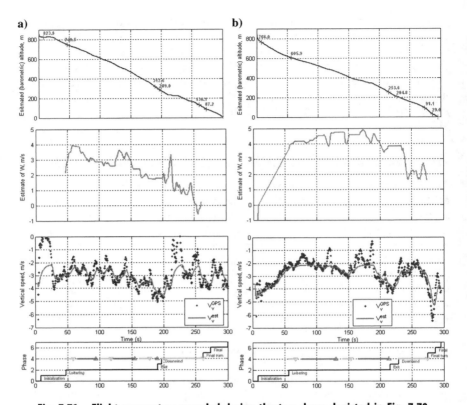

Fig. 7.71   Flight parameters recorded during the two drops depicted in Fig. 7.70.

Fig. 7.72a) all of the way up to 6 m/s (19.7 ft/s) (affected by downdraft during the final turn for the data set of Fig. 7.72b). It is indicative that both sets of data belong to the same PADS dropped at the same location less than an hour apart. No wonder that any PADS being influenced by varying winds exhibit varying performance.

Figures 7.73 and 7.74 show the same data presented in Figs. 7.71 and 7.72, but zoomed in to the final 300 m (986 ft) of altitude, which includes a downwind leg, to show surface-layer winds and variations in parameter estimates while making a decision to start executing the final turn and approach.

Obviously, in a general case the unaccounted winds may have components in all three directions

$$\mathbf{w}^{\text{dist}}(h) = \begin{bmatrix} w_x, & w_y, & w_z \end{bmatrix}^T \tag{7.136}$$

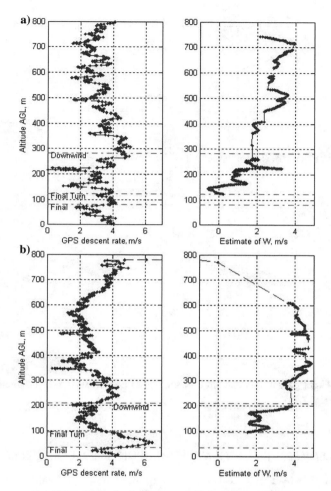

Fig. 7.72   Altitude dependences of the PADS estimates.

[Here, w_x denotes a downwind component, not included in Eq. (7.86), while w_z is considered positive for downdrafts to be consistent with the descent rate sign convention.] With disturbances of Eq. (7.136), the kinematic equations (7.88) become truly three-dimensional:

$$\begin{bmatrix} \dot{x} \\ \dot{y} \\ \dot{h} \end{bmatrix} = \begin{bmatrix} V_h^* \cos(\psi) + W + w_x \\ V_h^* \sin(\psi) + w_y \\ -V_v^* - w_z \end{bmatrix} \qquad (7.137)$$

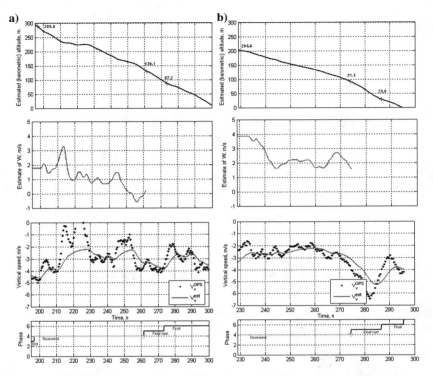

Fig. 7.73   Flight parameters exhibited at the 300-m (986-ft) surface layer (zoomed-in versions of Fig. 7.71).

7.7.2 ACCOMMODATING LINEAR AND LOGARITHMIC SURFACE-LAYER WIND MODELS

Assume that instead of a constant x component of the prevailing wind W vs altitude h (Fig. 3.41a), we have a linear profile (Fig. 3.41c)

$$W(h) = W_G + k_W h \qquad (7.138)$$

[compare Eq. (3.39)] where W_G is a known ground wind and coefficient $k_W = (W_0 - W_G)h_0^{-1}$ is defined by the ground wind W_G and wind W_0 measured at an altitude h_0 (corresponding to TIP). In terms of Eq. (7.136), it means that we are trying to model the downwind disturbance as $w_x(h) = W_G - W_0 + k_W h$. Such a profile might be based on the known ground winds (available from the nearby airport, measured by the target ground station, etc.) uplinked to the descending PADS [Bourakov et al. 2009; Yakimenko et al. 2009].

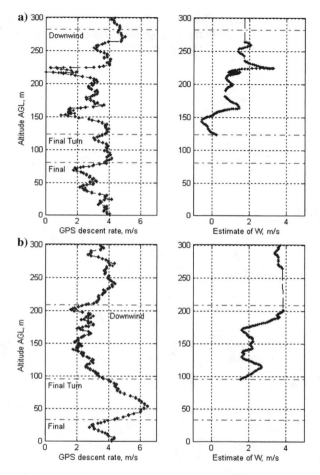

Fig. 7.74   Altitude dependences of the PADS' estimates between the surface and 300-m (986-ft) AGL altitude (zoomed-in versions of Fig. 7.72).

Then, the horizontal kinematics of Eq. (7.88) should be reformulated as

$$\begin{bmatrix} \dot{x} \\ \dot{y} \end{bmatrix} = \begin{bmatrix} V_h^* \cos(\psi) + W(h) \\ V_h^* \sin(\psi) \end{bmatrix} \tag{7.139}$$

and a new terminal point defined as

$$\mathbf{x}_f = \left[(V_h^* - W_G)T_{FA}^{\text{des}} - \frac{1}{2}k_W V_v^*(T_{FA}^{\text{des}})^2, 0, -\pi \right]^T \tag{7.140}$$

with D_{switch} computed slightly differently compared to Eq. (7.95) (to be addressed at the end of this section). To compute the offset in Eq. (7.140), we used the fact that FA starts at the altitude $V_v^* T_{FA}^{des}$, so that using an obvious relation

$$dt = -\frac{dh}{V_v^*} \tag{7.141}$$

we may write

$$x_f = \int_0^{T_{FA}^{des}} [V_h^* - W(h)]dt = \int_{V_v^* T_{FA}^{des}}^0 [V_h^* - W(h)]\frac{-dh}{V_v^*}$$

$$= \int_0^{V_v^* T_{FA}^{des}} [V_h^* - W(h)]\frac{dh}{V_v^*} = (V_h^* - W_G)T_{FA}^{des} - \frac{1}{2}k_W V_v^*(T_{FA}^{des})^2 \tag{7.142}$$

In this case, inverse kinematics

$$\psi = \tan^{-1}\left(\frac{\dot{y}}{\dot{x} - W(h)}\right) \tag{7.143}$$

features an altitude-dependent wind profile $W(h)$ [compare Eq. (7.106)], so that its differentiation with account of Eq. (7.138) results in a slightly different equation for the turn rate

$$\dot{\psi} = \frac{\ddot{y}[\dot{x} - W(h)] - [\ddot{x} - \dot{W}(h)]\dot{y}}{[\dot{x} - W(h)]^2 + \dot{y}^2} = \frac{\ddot{y}[\dot{x} - W(h)] - (\ddot{x} - k_W V_v^*)\dot{y}}{[\dot{x} - W(h)]^2 + \dot{y}^2} \tag{7.144}$$

[compare Eq. (7.107)].

The only modifications the numerical algorithm described in Sec. 7.6.2 requires in this case are that it should involve the new BC

$$\begin{bmatrix} \dot{x} \\ \dot{y} \end{bmatrix}_{\tau=0} = \begin{bmatrix} V_h^* \cos(\psi_0) + W_0 \\ V_h^* \sin(\psi_0) \end{bmatrix}, \quad \begin{bmatrix} \ddot{x} \\ \ddot{y} \end{bmatrix}_{\tau=0} = \begin{bmatrix} k_W V_v^* + \dot{\psi}_0 V_h^* \sin(\psi_0) \\ \dot{\psi}_0 V_h^* \cos(\psi_0) \end{bmatrix} \tag{7.145}$$

$$\begin{bmatrix} x \\ y \end{bmatrix}_{\tau=\tau_f} = \begin{bmatrix} (V_h^* - W_G)T_{FA}^{des} - \frac{1}{2}k_W V_v^*(T_{FA}^{des})^2 \\ 0 \end{bmatrix}$$

$$\begin{bmatrix} \dot{x} \\ \dot{y} \end{bmatrix}_{\tau=\tau_f} = \begin{bmatrix} -V_h^* + W_G + k_W V_v^* T_{FA}^{des} \\ 0 \end{bmatrix} \tag{7.146}$$

as well as computation of an altitude

$$h_j = h_{j-1} - V_v^* \Delta t_{j-1}, \quad j = 2, \ldots, N, \ (h_1 \equiv h_0) \tag{7.147}$$

and the corresponding wind magnitude at each computational node

$$W_j = W_G + k_W h_j \tag{7.148}$$

The latter two values are then used to compute the time intervals between two computational nodes

$$\Delta t_{j-1} = \sqrt{\frac{(x_j - x_{j-1})^2 + (y_j - y_{j-1})^2}{V_h^{*2} + 0.25(W_j + W_{j-1})^2 + V_h^*(W_j + W_{j-1})\cos(\psi_{j-1})}} \tag{7.149}$$

and heading

$$\psi_j = \tan^{-1}\left(\frac{\lambda_j y_j'}{\lambda_j x_j' - W_j}\right) \tag{7.150}$$

In the case when the ground winds are not available, a general logarithmic wind profile of Eq. (3.40) (Fig. 3.41d) might be used. In this case some of the preceding equations will be replaced with the new ones:

$$x_f = \int_0^{V_v^* T_{FA}^{des}} (V_h^* - W)\frac{dh}{V_v^*} = (V_h^* + a - b)T_{FA}^{des} - aT_{FA}^{des}\ln(V_v^* T_{FA}^{des}) \tag{7.151}$$

$$\begin{bmatrix} \ddot{x} \\ \ddot{y} \end{bmatrix}_{\tau=0} = \begin{bmatrix} -\dot{\psi}_0 V_h^* \sin(\psi_0) - a h_0^{-1} V_v^* \\ \dot{\psi}_0 V_h^* \cos(\psi_0) \end{bmatrix} \tag{7.152}$$

$$\begin{bmatrix} x \\ y \end{bmatrix}_{\tau=\tau_f} = \begin{bmatrix} (V_h^* + a - b)T_{FA}^{des} - aT_{FA}^{des}\ln(V_v^* T_{FA}^{des}) \\ 0 \end{bmatrix}$$

$$\begin{bmatrix} \dot{x} \\ \dot{y} \end{bmatrix}_{\tau=\tau_f} = \begin{bmatrix} -V_h^* + b + a\ln(V_v^* T_{FA}^{des}) \\ 0 \end{bmatrix} \tag{7.153}$$

and

$$W_j = b + a\ln(h_j) \tag{7.154}$$

The remaining equations of Sec. 7.6.2 will still be the same.

Obviously, different wind models require changing the initial conditions to initiate the final turn as well. Similar to Eq. (7.93), the x- (downwind direction) budget equation for the last two phases, the base turn ($t \in [t_0; t_1]$) and FA ($t \in [t_1; t_2]$), can be represented as

$$D_{\text{switch}} = -\int_{t_0}^{t_1} W(h)dt + \int_{t_1}^{t_2} [V_h^* - W(h)]dt$$

$$= V_h^* T_{FA} - \int_{t_0}^{t_2} W(h)dt = V_h^* T_{FA} - (T_{\text{turn}} + T_{FA})\bar{W}|_{h_0}^0 \qquad (7.155)$$

Here $\bar{W}|_{h_0}^0$ denotes a downwind component of the wind averaged within the altitude range $h \in [0; h_0]$.

The altitude budget equation for these two plus the portion of the downwind leg phase starting at some altitude h at a distance x from the target's traverse is

$$h = V_v^* \left(\frac{-x + D_{\text{switch}}}{V_h^* + \bar{W}|_h^{h_0}}\right) + V_v^* T_{\text{turn}} + V_v^* T_{FA} \qquad (7.156)$$

[compare Eq. (7.94)]. Ideally, when $x = D_{\text{switch}}$ the altitude h should be equal to $V_v^*(T_{\text{turn}} + T_{FA})$. In practice however, it might not be the case. To eliminate the altitude error, we might want to adjust the actual FA time T_{FA}.

Resolving Eqs. (7.155) and (7.156) with respect to D_{switch} and T_{FA} yields

$$D_{\text{switch}} = \frac{hV_h^*\left(V_h^* - \bar{W}|_{h_0}^0 + \bar{W}|_h^0\right) + h\bar{W}|_{h_0}^0\left(\bar{W}|_{h_0}^0 - \bar{W}|_h^0\right)}{V_v^*\left(2V_h^* - 2\bar{W}|_{h_0}^0 + \bar{W}|_h^0\right)}$$

$$- \frac{-x\left(V_h^* + \bar{W}|_{h_0}^0\right) + T_{\text{turn}}V_h^*\left(V_h^* - \bar{W}|_{h_0}^0 + \bar{W}|_h^0\right)}{2V_h^* - 2\bar{W}|_{h_0}^0 + \bar{W}|_h^0} \qquad (7.157)$$

$$T_{FA} = \frac{h\left(V_h^* - \bar{W}|_{h_0}^0 + \bar{W}|_h^0\right)}{V_v^*\left(2V_h^* - 2\bar{W}|_{h_0}^0 + \bar{W}|_h^0\right)} - \frac{-x + T_{\text{turn}}\left(V_h^* - 2\bar{W}|_{h_0}^0 + \bar{W}|_h^0\right)}{2V_h^* - 2\bar{W}|_{h_0}^0 + \bar{W}|_h^0} \qquad (7.158)$$

(here we used an obvious relation $\bar{W}|_h^{h_0} = \bar{W}|_h^0 - \bar{W}|_{h_0}^0$).

In the case of $W = \bar{W}|_h^0 = \bar{W}|_{h_0}^0 = const(h)$, Eqs. (7.157) and (7.158) are simplified to Eqs. (7.95) and (7.96). In all other cases, Eqs. (7.157) and (7.158) have to be used. Specifically, for the linear and logarithmic surface-layer wind models of Eqs. (7.138) and (7.154), the averaged winds from some current altitude h down to the ground can be computed as

$$\bar{W}|_h^0 = \frac{1}{h}\int_0^h (W_G + k_W h)\mathrm{d}h = W_G + \frac{1}{2}k_W h \qquad (7.159)$$

$$\bar{W}|_h^0 = \frac{1}{h}\int_0^h [b + a\ln(h)]\mathrm{d}h = b + a\ln(h) - a \qquad (7.160)$$

respectively. Substituting $\bar{W}|_h^0$ and $\bar{W}|_{h_0}^0$, computed using Eq. (7.159) or Eq. (7.160), into Eqs. (7.157) and (7.159) results in a wind-model-specific values for D_{switch} and T_{app}. (They are quite bulky and are not given here.)

Alternatively, the values for $\bar{W}|_h^0$ and $\bar{W}|_{h_0}^0$ can be substituted with the corresponding downrange component of the ballistic winds [computed in accordance with Eq. (3.9) for h and h_0, respectively]

$$\bar{W}|_h^0 = W^{\mathrm{bal}}(h)\cos\left[\chi_W^{\mathrm{bal}}(h)\right], \quad \bar{W}|_{h_0}^0 = W^{\mathrm{bal}}(h_0)\cos\left[\chi_W^{\mathrm{bal}}(h_0)\right] \qquad (7.161)$$

7.7.3 ACCOMMODATING CROSSWINDS AND VERTICAL WIND DISTURBANCES

The optimization routine of Sec. 7.6.2 can also accommodate crosswinds if they are known in one form or another. This capability may be useful for landing onto a ship's deck, not necessarily aligned with the wind [Hewgley and Yakimenko 2011]. Consider

$$w_x(h) = f_x(h), \quad w_y(h) = f_y(h) \qquad (7.162)$$

to be x (downwind) and y (crosswind) components of a horizontal wind profile approximated with some analytical dependence [e.g., of the form of Eqs. (7.138) and (7.154)]. Then, we can write

$$\begin{bmatrix} \dot{x} \\ \dot{y} \end{bmatrix} = \begin{bmatrix} V_h^* \cos(\psi) + w_x(h) \\ V_h^* \sin(\psi) + w_y(h) \end{bmatrix} \qquad (7.163)$$

Note that instead of the first two equations of Eq. (7.88) we are now using the first two equations of Eq. (7.137), emphasizing that the entire trajectory is not intentionally aligned with a major wind component, so that $w_y(h)$ may actually be even larger than $w_x(h)$ (i.e., we do not have the prevailing winds anymore letting $W \equiv 0$).

Accounting for the new kinematics described by Eq. (7.163), the final point can now be defined as

$$\mathbf{x}_f = \begin{bmatrix} x_f, & y_f, & -\pi \end{bmatrix}^T \qquad (7.164)$$

where the offsets in x and y direction are computed as

$$x_f = V_h^* T_{FA}^{des} - \int_0^{V_v^* T_{FA}^{des}} w_x(h)\frac{dh}{V_v^*}, \quad y_f = -\int_0^{V_v^* T_{FA}^{des}} w_y(h)\frac{dh}{V_v^*} \tag{7.165}$$

The inverse kinematics (heading angle) equation becomes

$$\psi = \tan^{-1}\left(\frac{\dot{y} - w_y(h)}{\dot{x} - w_x(h)}\right) \tag{7.166}$$

while its derivative is now presented by

$$\dot{\psi} = \frac{[\ddot{y} - w_y'(h)V_v^*][\dot{x} - w_x(h)] - [\ddot{x} - w_x'(h)V_v^*][\dot{y} - w_y(h)]}{[\dot{x} - w_x(h)]^2 + [\dot{y} - w_y(h)]^2} \tag{7.167}$$

(where $w_x' = dw_x/dh$ and $w_y' = dw_y/dh$). The horizontal component of the ground speed takes the following form:

$$|V_G| = \sqrt{\dot{x}^2 + \dot{y}^2}$$

$$= \sqrt{V_h^{*2} + [w_x(h) + w_y(h)]^2 + 2V_h^*[w_x(h)\cos(\psi) + w_y(h)\sin(\psi)]} \tag{7.168}$$

The new BC are

$$\begin{bmatrix} x \\ y \end{bmatrix}_{\tau=0} = \begin{bmatrix} x_0 \\ y_0 \end{bmatrix}, \quad \begin{bmatrix} \dot{x} \\ \dot{y} \end{bmatrix}_{\tau=0} = \begin{bmatrix} V_h^*\cos(\psi_0) + w_x(h_0) \\ V_h^*\sin(\psi_0) + w_y(h_0) \end{bmatrix}$$

$$\begin{bmatrix} \ddot{x} \\ \ddot{y} \end{bmatrix}_{\tau=0} = \begin{bmatrix} -\dot{\psi}_0 V_h^*\sin(\psi_0) + w_x'(h)V_v^* \\ \dot{\psi}_0 V_h^*\cos(\psi_0) + w_y'(h)V_v^* \end{bmatrix} \tag{7.169}$$

$$\begin{bmatrix} x \\ y \end{bmatrix}_{\tau=\tau_f} = \begin{bmatrix} x_f \\ y_f \end{bmatrix}, \quad \begin{bmatrix} \dot{x} \\ \dot{y} \end{bmatrix}_{\tau=\tau_f} = \begin{bmatrix} -V_h^* + w_x(h) \\ w_y(h) \end{bmatrix}, \quad \begin{bmatrix} \ddot{x} \\ \ddot{y} \end{bmatrix}_{\tau=\tau_f} = \begin{bmatrix} 0 \\ 0 \end{bmatrix} \tag{7.170}$$

and wind components computed at each step

$$w_{xj} = f_x(h_j) \quad \text{and} \quad w_{yj} = f_y(h_j) \tag{7.171}$$

should be used in the numerical procedure of Sec. 7.6.2. The remaining equations will still be valid. That includes Eqs. (7.157) and (7.158) because computation of D_{switch} relies on the downrange winds only.

If the original algorithm involves varying the arrival heading ψ_f, which allows cutting the final turn maneuver and steering directly to IPI with $|\psi_f| \neq \pi$ [as described by Eqs. (7.133–7.135)], then Eq. (7.146) should be replaced with

$$
\begin{bmatrix} x \\ y \end{bmatrix}_{\tau=\tau_f} = \begin{bmatrix} (V_h^* \cos(\psi_f) - W_G)T_{FA}^{\text{des}} - \dfrac{1}{2}k_W V_v^*(T_{FA}^{\text{des}})^2 \\ -V_h^* \sin(\psi_f)T_{FA}^{\text{des}} \end{bmatrix}
$$

$$
\begin{bmatrix} \dot{x} \\ \dot{y} \end{bmatrix}_{\tau=\tau_f} = \begin{bmatrix} V_h^* \cos(\psi_f) + W_G + k_W V_v^* T_{FA}^{\text{des}} \\ V_h^* \sin(\psi_f) \end{bmatrix}
\tag{7.172}
$$

Eq. (7.153) with

$$
\begin{bmatrix} x \\ y \end{bmatrix}_{\tau=\tau_f} = \begin{bmatrix} (V_h^* \cos(\psi_f) + a - b)T_{FA}^{\text{des}} - aT_{FA}^{\text{des}} \ln(V_v^* T_{FA}^{\text{des}}) \\ -V_h^* \sin(\psi_f)T_{FA}^{\text{des}} \end{bmatrix}
$$

$$
\begin{bmatrix} \dot{x} \\ \dot{y} \end{bmatrix}_{\tau=\tau_f} = \begin{bmatrix} V_h^* \cos(\psi_f) + b + a\ln(V_v^* T_{\text{app}}^{\text{des}}) \\ V_h^* \sin(\psi_f) \end{bmatrix}
\tag{7.173}
$$

and Eq. (7.170) with

$$
\begin{bmatrix} x \\ y \end{bmatrix}_{\tau=\tau_f} = \begin{bmatrix} \sqrt{x_f^2 + y_f^2}\,\cos(\psi_f) \\ -\sqrt{x_f^2 + y_f^2}\,\sin(\psi_f) \end{bmatrix}, \quad \begin{bmatrix} \dot{x} \\ \dot{y} \end{bmatrix}_{\tau=\tau_f} = \begin{bmatrix} V_h^* \cos(\psi_f) + w_x(h) \\ V_h^* \sin(\psi_f) + w_y(h) \end{bmatrix}
\tag{7.174}
$$

Finally, let us address the vertical component of the winds. According to data in Fig. 7.71, the vertical winds do affect the overall performance. Suppose we managed to have two PADS, so that while descending the first one produces and passes the winds estimates on up to the second one. Such estimates will be represented by the GPS time-stamped quadruples: h_k, $w_{x;k}$, $w_{y;k}$, and $w_{z;k}$, $k = 2, 3, \ldots, M$ ($k = 1$ corresponds to $h = 0$). As shown in Sec. 7.7.1, even if triplets h_k, $w_{x;k}$, $w_{y;k}$ are available, accounting for these data while generating a reference trajectory may still pose a computational problem. Accounting for the vertical component of the wind $w_z(h)$, that is, updrafts and downdrafts, is even more complicated. In a nonstable atmosphere this component of the winds may cause the same type of problem as horizontal winds unaccounted for simply because it changes the descent time forcing PADS to land sooner (shorter of IPI) or later (resulting in the overshoot).

For example, consider a sudden updraft on the downwind leg (Figs. 7.73a and 7.74a) or downdraft while PADS executes the final turn (Figs. 7.73b and 7.74b). Obviously, such events may cause a serious problem. At the downwind leg, a vertical motion of the air mass causes a violation of the altitude budget equation (7.156). The final turn maneuver is therefore the last chance to mitigate this violation. However, updrafts and downdrafts at this phase of descent mess the optimal solution obtained assuming a certain time of maneuver T_{turn} [Eq. (7.102)]. Even though the IDVD-method-based algorithm is capable of recomputing the turn maneuver at each control cycle (if needed), downdrafts may cause a major problem because they can decrease the time of the final-turn maneuver at once, leaving no time to recover. Hence, a capability to account for the vertical winds is quite beneficial.

Suppose that the vertical component of the winds is known. Again, it most likely comes as a look-up table, w_{zk} vs h_k, but theoretically we could use a low-order polynomial regression to approximate it with some analytical dependence $w_z(h)$. In this case this dependence might be used to modify the vertical motion equation of Eq. (7.141) to

$$dt = -\frac{dh}{V_v^* + w_z(h)} \tag{7.175}$$

This equation is then to be used in Eqs. (7.142), (7.151), (7.155), (7.159), (7.160), and (7.165). Obviously, depending on the specific analytical representation $w_z(h)$, the resulting equations may be very bulky, so the alternative approach would be based on the analogous of the ballistic winds concept (introduced in Sec. 3.1.3), which does not require analytical regression but can rather utilize (h_k, w_{zk}) pairs explicitly.

Following Eqs. (3.7) and (3.8), let us introduce

$$\bar{w}_z|_h^0 = \frac{1}{h}\int_0^h w_z(h)dh \approx \frac{1}{2h}\sum_{k=2}^M (h_k - h_{k-1})(w_{zk} + w_{z,k-1}) \tag{7.176}$$

which denotes a downdraft component of the wind averaged within the altitude range $h \in [0; h]$. Using this average downdraft, we can rewrite Eq. (7.175) as

$$h = T\left(V_v^* + \bar{w}_z|_h^0\right) \tag{7.177}$$

This equation (implicit in h) allows estimating time T needed to descent from the altitude h. Now let us use Eq. (7.177) to correct Eq. (7.102). To this end, let us use Eq. (7.177) to replace Eq. (7.103) with

$$(V_v^* + \bar{w}_z|_{h_{FA}}^0)T_{FA}^{\text{des}} = h_0 - T_{\text{turn}}(V_v^* + \bar{w}_z|_{h_{\text{TIP}}}^{h_{FA}}) \tag{7.178}$$

Noting that $\bar{w}_z|_{h_{\text{TIP}}}^{h_{FA}} = \bar{w}_z|_{h_{\text{TIP}}}^0 - \bar{w}_z|_{h_{FA}}^0$ we arrive to

$$T_{\text{turn}} = \frac{h_0 - (V_v^* + \bar{w}_z|_{h_{FA}}^0)T_{FA}^{\text{des}}}{V_v^* + \bar{w}_z|_{h_{\text{TIP}}}^0 - \bar{w}_z|_{h_{FA}}^0} \tag{7.179}$$

Using this corrected value of the final turn maneuver allows AGU to produce a more balanced control input $\dot{\psi}_{opt}(t)$.

Figure 7.75 shows an example of a reference trajectory that accommodates wind data available from a dropsonde (see sample of dropsonde data in Fig. 3.13). As seen from Fig. 7.75a, the trajectory satisfies the boundary conditions and descends from the TIP altitude to the FA altitude in exactly 18.4 s as determined by the estimated descent rate of the Snowflake PADS. The wind arrow shown in Fig. 7.75a indicates a direction of a ballistic wind computed in accordance with the procedure laid down in Sec. 3.1.3 for the TIP altitude. The

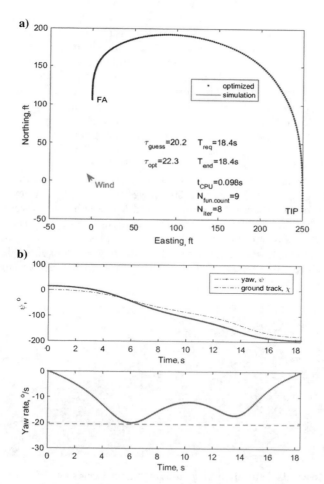

Fig. 7.75   Example of a reference trajectory optimized using the actual winds.

magnitude of the ballistic wind at this altitude was half of that of the forward speed of Snowflake. Figure 7.75b shows that the constraint imposed on the yaw rate is satisfied. It also shows the ground track angle for the consequent simulation that utilizes the same winds. As seen, the ground track angle is different from that of the heading angle [which in lieu of Eq. (7.1) equals yaw angle]. The yaw rate shown in Fig. 7.5b is to be executed by the model predictive controller discussed in Sec. 7.8.4.

Also shown in Fig. 7.75a is a CPU time required to complete trajectory optimization in the interpretative environment of MATLAB using the fminbnd function. Clearly, it is very small, and the number of iterations (and function evaluations) is very low as well. As mentioned in Sec. 7.6.2 already, it allows recomputing the reference trajectory to accommodate changing initial conditions caused by the unmodeled winds as often as feasible (see Table 7.5). Figure 7.76a shows another example of touchdown performance, while Fig. 7.76b demonstrates that the IDVD-based guidance strategy can accommodate changing the final condition for the terminal maneuver as well. In this particular experiment, a radio beacon was installed on a minivan driving along an airstrip. Snowflake PADS received its signal and continuously estimated its relative position with respect to the moving target, regenerating the reference trajectory every few seconds. The reference trajectory generation accounted for the surface-layer winds as discussed in in this section. The landing was planned along the runway and therefore was accomplished with some crosswind. One of potential applications for such a capability is vertical replenishment of naval vessels [Hewgley and Yakimenko 2009; Hewgey et al. 2011; Hewgey 2014].

As mentioned in Sec. 1.2, several UL- and ML-weight PADS were developed within JPADS to fill a niche for delivering the small items like sensors, ammunition, ground robots, etc. Figure 7.77 shows examples of two more novel applications. Small-weight PADS (redesigned for the lower stratosphere deployment) can be used as a final stage of the multistage system to retrieve small articles

Fig. 7.76   Snowflake PADS a) utilizing precision terminal guidance and control and b) landing at the moving IPI.

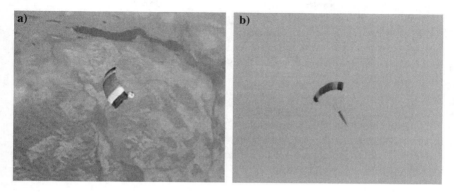

Fig. 7.77 **Prototype system a) for final stage of retrieving small articles from the international state station and b) for return of the sensor suite of the sounding rocket.**

from the international space station (Fig. 7.77a) and as a means to retrieve sensor suite of the weather sounding rocket (Fig. 7.77b) [Benton and Yakimenko 2013; Yingling et al. 2011]. Autonomous aerial payload delivery system "Blizzard" that combines a Snowflake PADS derivative with a long endurance (over 16 h) and up to 45-kg (100-lb) payload capacity Arcturus T20 UAV has even more potentials and possible applications [Yakimenko et al. 2011]. In this system cargo resigns in a detachable pod (Fig. 7.78a) and UAV then carries two of them to the DZ (Fig. 7.78b).

7.8 PADS CONTROL SCHEMES

As mentioned in Sec. 7.1, GNC algorithms for most of PADS rely on a series of predefined maneuvers to be executed depending on the estimates of 1) the

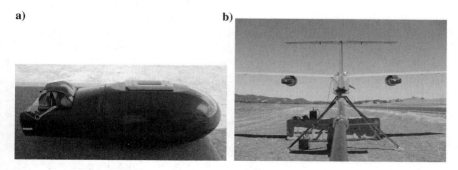

Fig. 7.78 **a) Prototype of a Blizzard system cargo container and b) Arcturus T20 UAV ready for launch with two cargo pods under the wings.**

current position and attitude with respect to IPI and 2) the winds. For these PADS, the control problem reduces to eliminating the difference between the desired (commanded) and actual (estimated) heading. In view of Eq. (7.1), the AGU controllers usually deal with the yaw-angle inputs so that the error signal takes the form

$$\Delta \psi = \psi_{cmd} - \hat{\psi} \qquad (7.180)$$

This error signal is usually handled by traditional proportional, derivative, and integral (PID) controllers. For those PADS involving generation of a reference trajectory, the AGU controller takes care of accurately tracking this trajectory by zeroing either Eq. (7.180) or

$$\Delta \dot{\psi} = \dot{\psi}_{cmd} - \hat{\dot{\psi}} \qquad (7.181)$$

These latter PADS may have a need to utilize more sophisticated controllers. In what follows, several different controllers including those implemented and thoroughly tested on the ALEX, DragonFly, MegaFly, Onyx, and Snowflake PADS are described.

7.8.1   PID CONTROLLERS

Following earlier work described in Sec. 7.3.2, the generalized concept of a controller for autonomous PADS was laid down in [Gockel 1997a; Gockel 1997b]. The two-cascade control architecture included the outer loop for the slowly changing heading angle to produce the ψ_{cmd} command based on error given by Eq. (7.180) and the inner loop for a faster changing yaw rate to produce the δ_{cmd} command based on error of Eq. (7.181). The essential part of Gockel's control concept was to utilize the PADS model to include feedforward control in both independent loops, to ensure less activity of feedback loops and easier refinement of the controller after flight-test results analysis.

Gockel [1997] performed several sensitivity studies (towards unknown winds, PADS model uncertainty, and measurement uncertainty). He used the developed general concept to perform simulations of the entire mission for the six-DoF model of the ALEX PADS. For the practical application however, the ALEX PADS used a simple proportional controller [Jann 2005]. The commanded heading [Eq. (7.75)] was compared to the actual one giving the amount of error the controller has to reduce. Depending on the sign of the heading error, either the left or the right actuator was activated, leading to a left or a right turn. As an output, the desired actuator position was computed by a saturation-limited proportional controller (Fig. 7.79).

While small heading offsets can be clearly assigned to a left or right turn, for the larger offsets close to 180 deg additional information from the guidance block is required to resolve for ambiguity. In the block diagram of Fig. 7.79, this is done inside the PSI CON module. With the gain of 0.6, a commanded turn of 180 deg

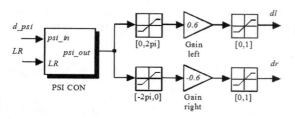

Fig. 7.79 Heading controller for the ALEX PADS.

(π) first activates the corresponding actuator to full stroke until the azimuth difference falls below 95 deg and the proportional part of the controller takes over. To accommodate a nonlinear control characteristic (as shown in Fig. 5.41) for small differences between commanded and true heading, this block changes the trim of the control lines and sets the zero point at 30% deflection. Hence, a commanded stroke from 0 to 100% results in a deflection from 30 to 100%.

Having the appropriate wind measurement or estimate, the heading can also be computed from the speed and course over ground. In contrast to the heading control of Fig. 7.79, the course (or track) derived control enables the control task to work with GPS data only

$$\chi_a \approx \tan^{-1}\left(\frac{V_G \sin(\chi) - \hat{w}_y}{V_G \cos(\chi) - \hat{w}_x}\right) \qquad (7.182)$$

Because of the slow GPS sampling rate of usually 1 Hz and the GPS time delay of about 1 s, the control gain for such a controller has to be halved (reduced to 0.3) slowing down the reaction of the complete system. Figure 7.80 shows the step response for a commanded 90-deg turn. The yaw motion is simulated using a 3-DoF model presented in Sec. 5.1.3. In real systems, where yaw motion is much more complex, the control gain should be reduced even more.

As shown in Sec. 5.2.6, if measurements are acquired at payload (AGU sits atop the payload), they are overlaid by the relative canopy-payload motion. To avoid unwanted excitation of the relative yawing, natural frequencies of PADS and controller must not come close to each other. This can be achieved by using the even smaller control gains.

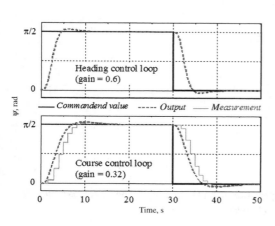

Fig. 7.80 Simulated yaw command step responses for the heading and course-derived control loops.

To develop the commanded heading rate controller for the DragonFly 10K PADS, the Draper Laboratory performed system identification, resulting in a standard second-order model

$$G_{\dot\psi}(s) = \frac{K_{\dot\psi}\omega_n^2}{s^2 + 2\zeta\omega_n + \omega_n^2} \qquad (7.183)$$

where $\omega_n = 0.44\,\text{rad/s}$, $\zeta = 0.6$, and $K_{\dot\psi} = 9.5\,\text{deg/s}$ (the data collected showed a near linear relation between differential toggle and turn rate) [Carter et al. 2005; Hattis et al. 2006]. Figure 7.81 shows the measured responses and how they were compared to the matching model of Eq. (7.183).

Based on this model, a PID controller $C(s)$ was designed via loop-shaping procedure to do the following:

• Provide at least 60 deg of a phase margin (to accommodate uncertainty in plant dynamics) and maximize the bandwidth of the closed-loop transfer function

$$T(s) = \frac{C(s)G_{\dot\psi}(s)}{1 + C(s)G_{\dot\psi}(s)} \qquad (7.184)$$

• Minimize the gain of the input transfer function $D(s) = G_{\dot\psi}(s)[1 + C(s)G_{\dot\psi}(s)]^{-1}$ to limit the impact of the payload oscillation frequency

• Maintain low sensitivity

$$S(s) = \frac{1}{1 + C(s)G_{\dot\psi}(s)} \qquad (7.185)$$

at all operational frequencies.

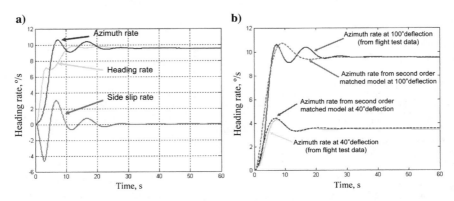

Fig. 7.81   Measured DragonFly response to a) a unit-step input and b) comparison to a matched model responses [Carter et al. 2005].

As reported in [Carter et al. 2005; Hattis et al. 2006], the optimization procedure resulted in the following gains:

$$C(s) = K_p + \frac{K_i}{s} + K_d s = 1.65 + \frac{1.07}{s} + 3.0 \text{ s} \qquad (7.186)$$

Figure 7.82 shows the unit-step response of the original [Eq. (7.183)] and compensated model. This figure also shows the rise and settling times.

The basic PID design of Eq. (7.186) was further augmented with a configurable low-pass filter for the heading rate error, and a heading acceleration estimator (from a steady state Kalman filter) was used to estimate the heading rate derivative needed by the PID controller (Fig. 7.83). The controller output was clipped so as not to exceed the control toggle position limits. Finally, a configurable dead band with hysteresis was used to reduce the servo motor duty cycles.

As seen from Figs. 1.3a and 1.26c and d, the AGU for the DragonFly PADS resides at the confluence point, that is, not sitting atop the payload as for some other PADS (see Table 1.7). However, even though the canopy itself did not exhibit significant oscillations, the payload experienced a pendular motion beneath the AGU, and these oscillations were picked up by navigation sensors of AGU and appeared in the feedback signal. Hence, the loop shaping procedure for designing a controller was chosen to synthesize the control gains so that the control line deflection command sent to the motors was desensitized to this disturbance signal and did not attempt to correct for this spurious measurement [Carter et al. 2005].

As reported in [Carter et al. 2007] the latter design for a lager system, the MegaFly 30 K PADS (Fig. 1.28b), involved a possibility to control either heading rate or heading depending on the phase of flight. Figure 7.84 shows the controller architecture in this case.

The outer-loop controller in Fig. 7.84 is used during the homing phase, when the PADS deployed away from IPI steers towards it (or intermediate WP). In this case, the desired yaw rate command

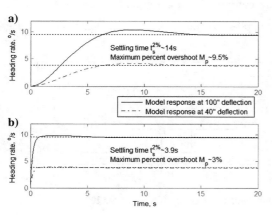

a)

Settling time $t_s^{2\%} \sim 14$s
Maximum percent overshoot $M_p \sim 9.5\%$

—— Model response at 100" deflection
– – – Model response at 40" deflection

b)

Settling time $t_s^{2\%} \sim 3.9$s
Maximum percent overshoot $M_p \sim 3\%$

Time, s

Fig. 7.82 Step response of the a) original and b) compensated matching model.

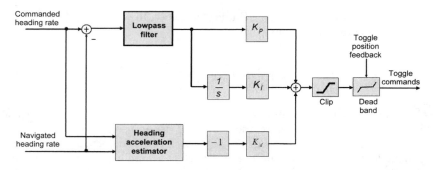

Fig. 7.83   DragonFly heading rate controller architecture [Hattis et al. 2006].

$\dot{\psi}_{\text{des}}$ is computed from heading error in the outer loop

$$\dot{\psi}_{\text{des};k} = K_p^H (\psi_{\text{cmd}} - \hat{\psi}_k) + K_d^H \left(\frac{d\psi_{\text{cmd}}}{dt} - \hat{\dot{\psi}}_k \right) \qquad (7.187)$$

(here k denotes a discrete instance of time).

The inner-loop PID controller has the following form:

$$\delta_{\text{toggle};k} = K_p (\dot{\psi}_{\text{des};k} - \hat{\dot{\psi}}_k) + K_i \sum_{j=1}^{k} (\dot{\psi}_{\text{des};j} - \hat{\dot{\psi}}_j) - K_d \frac{d\hat{\dot{\psi}}_k}{dt} \qquad (7.188)$$

As shown in Fig. 7.84, in the homing phase, $\dot{\psi}_{\text{des}}$ is computed from heading error in the outer loop, whereas when in any other phase of flight it is not in the homing mode, $\dot{\psi}_{\text{des}} = \dot{\psi}_{\text{cmd}}$. The practical realization includes antiwindup protection for the integral term like in a standard PID controller in Simulink (Fig. 7.85). The derivatives $\hat{\dot{\psi}}_j$ are computed using a lateral dynamics estimator, and the derivative term is only applied to the feedback state signal $\hat{\dot{\psi}}_k$ because

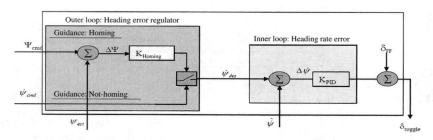

Fig. 7.84   MegaFly controller architecture [Carter et al. 2007].

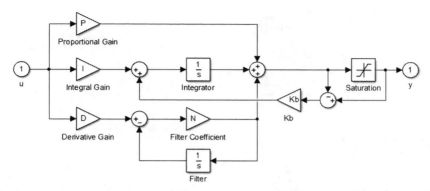

Fig. 7.85   Standard PID controller in Simulink with antiwindup protection.

the guidance commands $\ddot{\psi}_{des}$ are mostly the step commands, so that the second-order derivative $\ddot{\psi}_{des}$ is described by the delta function (mostly zero except nontrackable short-duration spikes) [Carter et al. 2007].

As in the preceding realization, the gains of controller described by Eq. (7.188) were obtained via a frequency-domain loop shaping. Specifically, it was done by minimizing the weighted performance index

$$ J = \min_{K_p, K_i, K_d} \left\{ w_1 \sum_{i=1}^{M} [T(j\omega_i) - T^{des}(j\omega_i)]^2 + w_2 \sum_{i=N_1}^{N_2} [S(j\omega_i) - 1]^2 \right\} \quad (7.189) $$

to best match a combination of a desired closed-loop transfer function for bandwidth and stability at the lower frequencies [the first sum in Eq. (7.189)] and a desired sensitivity function [described by Eq. (7.185)] for disturbance rejection at the higher frequencies [the second sum in Eq. (7.189)]. Figure 7.86 shows the closed-loop transfer function of a tuned controller based on the MegaFly model similar to that of Eq. (7.183).

The inner-loop controller (7.188) was further augmented with the feedforward term. The feedforward signal was computed as follows [Carter et al. 2007]:

- Given a yaw rate command from a guidance block $\dot{\psi}_{cmd}$, invert an idealized plant model of Eq. (7.183) to compute the expected steady-state control deflection, that is, $\delta_{FF} = \dot{\psi}_{cmd} K_{\chi}^{-1}$.

- Drive a plant model $G_{\dot{\chi}}(s)$ with δ_{FF} to compute the expected transient yaw rate at the next time step (due to the feedforward term) $\dot{\psi}_{FF}$.

- Drive the feedback controller with this computed $\dot{\psi}_{FF}$.

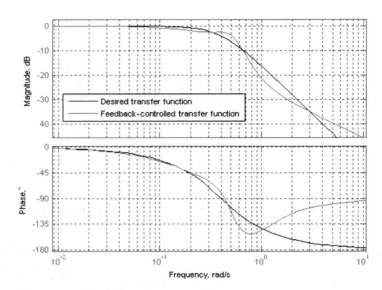

Fig. 7.86   Bode plot of the tuned MegaFly closed-loop transfer function $T(s)$ matching the desired one $T^{des}(s)$ [Carter et al. 2007].

Thus, the complete heading-rate controller took the form

$$\delta_{\text{toggle};k} = K_{FF}\,\delta_{FF;k} + (1 - K_{FF})\left[K_p(\dot{\psi}_{FF;k} - \hat{\dot{\psi}}_k) + K_i\sum_{j=1}^{k}(\dot{\psi}_{FF;j} - \hat{\dot{\psi}}_j) - K_d\frac{d\hat{\dot{\psi}}_k}{dt}\right]$$

$$(7.190)$$

where the varied gain $K_{FF} \in [0; 1]$ reflects how much feedforward input is needed.

Further investigation revealed that in practice the feedback signal also carries a torsional motion of a payload relative to canopy (the AGU on the MegaFly PADS also resides at the confluence point as shown in Fig. 1.28b), that is, the torsion mode frequency of 0.57 rad/s was very close to the model natural frequency of 0.53 rad/s [Carter et al. 2009]. Carter et al. [2009] reported redesigning PID controller to match a desired closed-loop reference model for the heading rate dynamics

$$G_{\text{ref}}(s) = \frac{\omega_n^2}{s^2 + 2\zeta\omega_n + \omega_n^2}$$

$$(7.191)$$

with $\omega_n = 0.2\,\text{rad/s}$ (well below the torsion mode frequency) and $\zeta = 0.8$. Optimization routine involved just the first term in the performance index (7.190) and resulted in a close match of the closed-loop responses as shown in

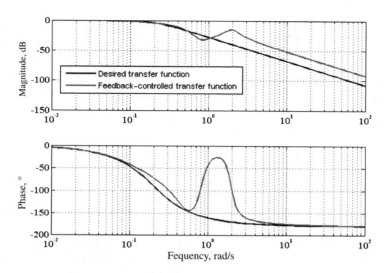

Fig. 7.87   Bode plot of the tuned MegaFly closed-loop transfer function $T(s)$ matching the desired one $T^{des}(s)$ [Carter et al. 2009]. (See the discussion on the PADS modes in Sec. 6.1.)

Fig. 7.87. This design ensured the closed-loop gain being very close to 1 (0 dB) for frequencies less than 0.1 rad/s, and less than 0.18 (−15 dB) for frequencies above 0.5 rad/s covering the payload torsional and canopy Dutch roll modes.

7.8.2   ADVANCED PID CONTROLLERS

Working on developing GNC algorithm for the Pegasus PADS, Kaminer and Yakimenko [2003] also utilized PID controllers (Fig. 7.88).

The integrated guidance and control algorithm was based on generating and tracking an inertial reference trajectory. In [Kaminer and Yakimenko 2003], the authors explored the guidance strategy similar to that of CARESI shown in

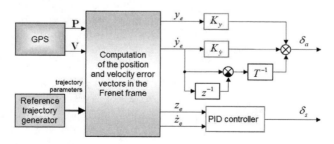

Fig. 7.88   PID controller implementation for Pegasus PADS [Kaminer and Yakimenko 2003].

**Fig. 7.89   Problem
geometry [Kaminer
et al. 2005].**

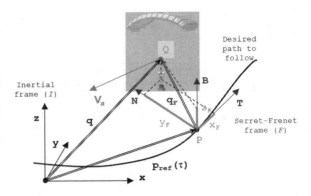

Fig. 7.44, so that the
reference trajectory con-
sisted of straight lines
and helix. From the con-
trols standpoint the key
ideas of the design meth-
odology included the fol-
lowing steps:

- Reparameterize reference trajectory using the arclength s, thus eliminating time as an independent variable.

- Form error dynamics for PADS consisting of the trajectory and parafoil model, where the position and velocity error states are resolved in Frenet frame (Fig. 7.89).

- Design a linear tracking controller for the linearization of the system along the reference trajectory.

Later on having in mind coordinated payload delivery by multiple PADS, the reference trajectory was replaced with a spatial differentiable trajectory, and a virtual arc τ was used as an independent variable [Kaminer et al. 2005]. The trajectory generation routine involved the IDVD method and was somewhat similar to that described in Sec. 7.6.2, but the overall control architecture remained the same. Hence, converting errors in the Frenet frame remained being a key feature of implementing PID controller in this case as well.

Using the notations of Fig. 7.89, let $\mathbf{p}_{ref}(\tau)$, where τ denotes the arc length introduced in Sec. 7.6.2, be the PADS path to be followed, and let Q denote the current PDS position. Further, let P be an arbitrary point on the path that plays the role of a virtual (ideal) PADS to be followed. Let a Serret–Frenet frame $\{F\}$ be attached to the desired PADS path and the wind frame $\{w\}$ be attached to the parafoil-payload system. We denote by $\boldsymbol{\omega}_{ref}^F$ denote the angular velocity of $\{F\}$ with respect to $\{I\}$ resolved in $\{F\}$. Point Q can be resolved in $\{I\}$ as $\mathbf{q} = [x, y, z]^T$ or in $\{F\}$ as $\mathbf{q}^F = [x_1, y_1, z_1]^T$. The simplified PADS kinematic equations can be written as

$$\begin{bmatrix} \dot{x} \\ \dot{y} \\ \dot{z} \end{bmatrix} = V_a \begin{bmatrix} \cos(\gamma_a)\cos(\psi) \\ \cos(\gamma_a)\cos(\psi) \\ \sin(\gamma_a) \end{bmatrix} + \mathbf{W}$$

$$\dot{\gamma}_a = u_1$$

$$\dot{\psi} = u_2$$

(7.192)

where kinematic equations are the same as in Eq. (5.29) except that they are written in the north-west-up coordinate frame and utilize assumption of Eq. (7.1), while u_1, u_2 denote the virtual control inputs, pitch rate and yaw rate, respectively. Following standard nomenclature, the orthonormal unit tangent and normal and binormal basis vectors of the Frenet frame $\{F\}$ are computed as (see Fig. 7.89)

$$\mathbf{T}(\tau) = \frac{\dfrac{d\mathbf{p}_{\text{ref}}(\tau)}{ds}}{\left\|\dfrac{d\mathbf{p}^{\text{ref}}(\tau)}{d\tau}\right\|}, \quad \mathbf{N}(\tau) = \frac{\dfrac{d\mathbf{T}(\tau)}{d\tau}}{\left\|\dfrac{d\mathbf{T}(\tau)}{d\tau}\right\|}, \quad \mathbf{B}(\tau) = \frac{\mathbf{T}(\tau) \times \mathbf{N}(\tau)}{\|\mathbf{T}(\tau) \times \mathbf{N}(\tau)\|} \quad (7.193)$$

The rotation matrix from $\{F\}$ to $\{I\}$ is defined as

$$_F^I\mathbf{R} = [\mathbf{T}\,\mathbf{N}\,\mathbf{B}] \quad (7.194)$$

It is well-known that

$$\boldsymbol{\omega}_{\text{ref}}^F = [\varsigma\dot{\tau},\ 0,\ \kappa\dot{\tau}]^T \quad (7.195)$$

where $\kappa(\tau) = \|\frac{d\mathbf{T}(\tau)}{d\tau}\|$ is the curvature of $\mathbf{p}_{\text{ref}}(\tau)$ and $\varsigma(\tau) = \pm\|\frac{d\mathbf{B}(\tau)}{d\tau}\|$ is its torsion. With these notations we can write

$$\mathbf{q} = \mathbf{p}_{\text{ref}}(\tau) + _F^I\mathbf{R}\mathbf{q}_F \quad (7.196)$$

Next, let us write an obvious relationship

$$_I^F\mathbf{R}\begin{bmatrix}\dot{x}\\\dot{y}\\\dot{z}\end{bmatrix} = _w^F\mathbf{R}_I^w\mathbf{R}\begin{bmatrix}\dot{x}\\\dot{y}\\\dot{z}\end{bmatrix} = _F^w\mathbf{R}^T\begin{bmatrix}V_a\\0\\0\end{bmatrix} + _I^F\mathbf{R}\mathbf{W} \quad (7.197)$$

that uses the rotation matrix $_F^w\mathbf{R}$ relating $\{F\}$ to $\{w\}$. This matrix is similar to that of Eq. (5.10) and utilizes the corresponding Euler angles, ϕ_e, θ_e, and ψ_e (here e stands for the error). Now, applying the differentiation rule in the rotating coordinate frame [see Eq. (5.20)], we can differentiate Eq. (7.196) to obtain

$$\dot{\mathbf{q}} = _F^I\mathbf{R}\begin{bmatrix}\dot{\tau}\\0\\0\end{bmatrix} + _F^I\mathbf{R}\begin{bmatrix}\dot{x}_F\\\dot{y}_F\\\dot{z}_F\end{bmatrix} + _F^I\mathbf{R}\left(\boldsymbol{\omega}_{\text{ref}}^F \times \begin{bmatrix}x_F\\y_F\\z_F\end{bmatrix}\right) \quad (7.198)$$

It then follows that

$$_I^F\mathbf{R}\dot{\mathbf{q}} = _I^F\mathbf{R}\begin{bmatrix}\dot{x}\\\dot{y}\\\dot{z}\end{bmatrix} = \begin{bmatrix}\dot{x}_F + \dot{\tau}(1 - \kappa y_F)\\\dot{y}_F + \dot{\tau}(\kappa x_F - \varsigma z_F)\\\dot{z}_F + \varsigma\dot{\tau}y_F\end{bmatrix} \quad (7.199)$$

which using Eq. (7.197) can be rewritten as

$$
\begin{bmatrix} \dot{x}_F \\ \dot{y}_F \\ \dot{z}_F \end{bmatrix} = \begin{bmatrix} -\dot{\tau}(1 - \kappa y_F) \\ -\dot{\tau}(\kappa x_F - \varsigma z_F) \\ -\varsigma \dot{\tau} y_F \end{bmatrix} + {}^w_F \mathbf{R}^T \begin{bmatrix} V_a \\ 0 \\ 0 \end{bmatrix} + {}^F_I \mathbf{R} \mathbf{W}
$$

$$
= \begin{bmatrix} -\dot{\tau}(1 - \kappa y_F) + V_a \cos(\theta_e) \cos(\psi_e) \\ -\dot{\tau}(\kappa x_F - \varsigma z_F) + V_a \cos(\theta_e) \sin(\psi_e) \\ -\varsigma \dot{\tau} y_F - V_a \sin(\theta_e) \end{bmatrix} + {}^F_I \mathbf{R} \mathbf{W} \qquad (7.200)
$$

Following Eq. (5.15), kinematics for the Euler angles ϕ_e, θ_e, and ψ_e can be represented as

$$
\begin{bmatrix} \dot{\phi}_e \\ \dot{\theta}_e \\ \dot{\psi}_e \end{bmatrix} = \begin{bmatrix} 1 & \sin(\phi_e)\tan(\theta_e) & \cos(\phi_e)\tan(\theta_e) \\ 0 & \cos(\phi_e) & -\sin(\phi_e) \\ 0 & \dfrac{\sin(\phi_e)}{\cos(\theta_e)} & \dfrac{\cos(\phi_e)}{\cos(\theta_e)} \end{bmatrix} \boldsymbol{\omega}^w_{wF} = \mathbf{Q}\boldsymbol{\omega}^w_{wF} \qquad (7.201)
$$

The expression for $\boldsymbol{\omega}^w_{wF}$ can be obtained as

$$
\boldsymbol{\omega}^w_{wF} = \boldsymbol{\omega}^w_{wI} - \boldsymbol{\omega}^w_{\text{ref}} = \boldsymbol{\omega}^w_{wI} - {}^w_F \mathbf{R} \boldsymbol{\omega}^F_{\text{ref}} \qquad (7.202)
$$

In Eq. (7.202), $\boldsymbol{\omega}^w_{wI}$ can be derived in a manner similar to that of Eq. (5.13)

$$
\boldsymbol{\omega}^w_{wI} = \begin{bmatrix} 0 \\ \dot{\gamma}_a \\ 0 \end{bmatrix} + \begin{bmatrix} \cos(\gamma_a) & 0 & -\sin(\gamma_a) \\ 0 & 1 & 0 \\ \sin(\gamma_a) & 0 & \cos(\gamma_a) \end{bmatrix} \begin{bmatrix} 0 \\ 0 \\ \dot{\psi} \end{bmatrix} = \begin{bmatrix} -\dot{\psi}\sin(\gamma_a) \\ \dot{\gamma}_a \\ \dot{\psi}\cos(\gamma_a) \end{bmatrix} \qquad (7.203)
$$

Substituting Eqs. (7.202) and (7.203) to Eq. (7.201) and keeping in mind that

$$
\mathbf{Q}^w_F \mathbf{R} = \begin{bmatrix} \dfrac{\cos(\psi_e)}{\cos(\theta_e)} & \dfrac{\sin(\psi_e)}{\cos(\theta_e)} & 0 \\ -\sin(\psi_e) & \cos(\psi_e) & 0 \\ \cos(\psi_e)\tan(\theta_e) & \sin(\psi_e)\tan(\theta_e) & 1 \end{bmatrix} \qquad (7.204)
$$

yields

$$
\begin{bmatrix} \dot{\phi}_e \\ \dot{\theta}_e \\ \dot{\psi}_e \end{bmatrix} = \begin{bmatrix} \dot{\gamma}_a \sin(\phi_e)\tan(\theta_e) + \dot{\psi}(\cos(\gamma_a)\cos(\phi_e)\tan(\theta_e) - \sin(\gamma_a)) - \dot{\tau}\varsigma \dfrac{\cos(\psi_e)}{\cos(\theta_e)} \\ \dot{\gamma}_a \cos(\phi_e) - \dot{\psi}\sin(\phi_e)\cos(\gamma_a) - \dot{\tau}\varsigma\sin(\psi_e) \\ \dot{\gamma}\dfrac{\sin(\phi_e)}{\cos(\theta_e)} + \dot{\psi}\cos(\gamma_a)\dfrac{\cos(\phi_e)}{\cos(\theta_e)} - \dot{\tau}(\varsigma\cos(\psi_e)\tan(\theta_e) + \kappa) \end{bmatrix}
$$

$$(7.205)$$

The last two equations in Eq. (7.205) can be further rewritten as

$$
\begin{bmatrix} \dot{\theta}_e \\ \dot{\psi}_e \end{bmatrix} = \begin{bmatrix} \dot{\tau}\varsigma\sin(\psi_e) \\ -\dot{\tau}(\varsigma\cos(\psi_e)\tan(\theta_e) + \kappa) \end{bmatrix} + \begin{bmatrix} \cos(\phi_e) & -\sin(\phi_e)\cos(\gamma_a) \\ \dfrac{\sin(\phi_e)}{\cos(\theta_e)} & \cos(\gamma_a)\dfrac{\cos(\phi_e)}{\cos(\theta_e)} \end{bmatrix} \begin{bmatrix} \dot{\gamma}_a \\ \dot{\psi} \end{bmatrix}
$$

$$
= \mathbf{d} + \mathbf{G}\begin{bmatrix} \dot{\gamma}_a \\ \dot{\psi} \end{bmatrix}
$$

$$(7.206)$$

Because $\mathbf{G}$ is a nonsingular matrix for all for all $\gamma \neq \pm 0.5\pi$, we can invert Eq. (7.206) to get controls for the model of Eq. (7.193):

$$
\begin{bmatrix} u_1 \\ u_2 \end{bmatrix} = \begin{bmatrix} \dot{\gamma}_a \\ \dot{\psi} \end{bmatrix} = \mathbf{G}^{-1}\left(\begin{bmatrix} u_\theta \\ u_\psi \end{bmatrix} - \mathbf{d} \right)
$$

$$(7.207)$$

Combining Eqs. (7.200) and (7.206) yields equations for the (path following) error dynamics:

$$
\begin{bmatrix} \dot{x}_F \\ \dot{y}_F \\ \dot{z}_F \end{bmatrix} = -\dot{\tau}\begin{bmatrix} 1 \\ 0 \\ 0 \end{bmatrix} + \begin{bmatrix} 0 & -\kappa & 0 \\ \kappa & 0 & -\varsigma \\ 0 & \varsigma & 0 \end{bmatrix}\begin{bmatrix} x_F \\ y_F \\ z_F \end{bmatrix} + V_a \begin{bmatrix} \cos(\theta_e)\cos(\psi_e) \\ \cos(\theta_e)\sin(\psi_e) \\ -\sin(\theta_e) \end{bmatrix} + {}^F_I\mathbf{RW}
$$

$$\dot{\theta}_e = u_\theta$$

$$(7.208)$$

$$\dot{\psi}_e = u_\psi$$

Notice that the rate of progression of a point $\mathbf{p}_{\text{ref}}(\tau)$ along the path, speed factor $\dot{\tau}$, is a variable that can be manipulated at will [see Eq. (7.115)]. If the winds $\mathbf{W}$ are not known, $\mathbf{W}$ in Eq. (7.208) is set to zero. (They are considered as disturbance.) However, as we know from Sec. 7.2.2, some estimates of the winds aloft are available so that we could incorporate them in the control algorithm. The IDVD based approach for generating inertial reference trajectories allows incorporating winds data in $\mathbf{p}_{\text{ref}}(\tau)$ as well (Sec. 7.7). The error dynamics of Eq. (7.208) was developed for the general, 3D case, with two controls as in Eq. (7.192), but if a glide angle control is not applicable, then it can easily be reduced to the 2D projection with only one control δ_a.

A globally asymptotically stable control law of Eq. (7.208) drives all error variables in Eq. (7.208) to zero using u_θ and u_ψ [Eq. (7.206)] as control inputs. Figure 7.90 shows an example of a batch of simulations with realistic winds (considered as disturbances) and errors in estimating GPS position and airspeed proving a feasibility of the developed approach. Unfortunately, these algorithms have not been tested in the real flight yet.

Another advanced PID controller was developed by Calise and Preston [2006, 2008] and implemented on the Onyx PADS. The outer (guidance) loop to track trajectories similar to those shown in Fig. 7.90 generated the commanded yaw rate as

$$\dot{\psi}_{cmd} = K_p^\psi (\psi_{cmd} - \hat{\psi}) \tag{7.209}$$

(with $K_p^\psi = 0.28$ and 0.5-rad/s limit) [Calise and Preston 2008]. Then, two implementations of inner-loop PID controllers were tested using the same PADS dynamics model, but forming the feedback signal slightly differently. Both models were based on Eq. (5.19). From the last two equations of Eq. (5.19), it follows that in a steady-state turn ($\dot{V}_a = 0$, $\dot{\gamma}_a = 0$)

$$\dot{\chi}_a = \frac{g \tan(\phi_a)}{V_a} \tag{7.210}$$

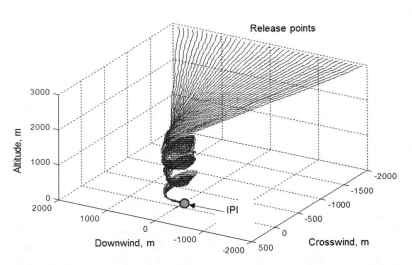

Fig. 7.90   Simulation results of the reference trajectory tracking in presence of winds [Kaminer and Yakimenko 2003].

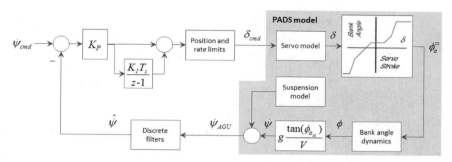

Fig. 7.91 Basic PI controller with a yaw rate feedback [Calise and Preston 2008].

Hence, in lieu of Eq. (7.1)

$$\dot{\psi} = \frac{g \tan(\phi_a)}{V_a} \quad (7.211)$$

The first model used to tune the PI controller for Onyx 500 and Onyx 2200 is shown in Fig. 7.91 (the PADS model is depicted by grey background). In this model the bank angle dynamics is modeled as a standard second-order system response [compare Eq. (5.45)]

$$T_\phi \ddot{\phi}_a + = -2\zeta\omega_n \dot{\phi}_a - \omega_n^2[\phi_a - \phi_a^{ss}(\delta_a)] \quad (7.212)$$

(with $\zeta = 0.95$ and $\omega_n = 2$ rad/s obtained from previous flight tests), which is then passed on in Eq. (7.21). To model a two-point suspension scheme for Onyx PADS AGU (Figs. 1.20–1.22), the suspension model block in Fig. 7.91 added a 2-Hz sinusoid with the 0.3-rad/s amplitude. The output of the model is the yaw rate as measured by AGU. The discrete filters block includes two identical first-order filters with the time constant of 0.45 s to remove noise and AGU oscillations. The filtered yaw rate signal was then used as feedback.

The servo model incorporated the first-order response [with a position limit of 23 cm (9 in.) and rate limit of 3.8 cm/s (1.5 in./s)] with a time constant of 0.2 s. The $\phi_a^{ss}(\delta_a)$ model included a dead zone of 2.5 cm (1 in.) (to account for slack in the suspension lines), a low-gain region up to 7.6-cm (3-in.) stroke [where the turn rate is 0.14 rad/cm (20 deg/in.)], a high-gain region up to 10-cm (4-in.) stroke [where the turn rate is 0.55 rad/cm (80 deg/in.)], 5-deg canopy bias at zero stroke and a saturation limit [Calise and Preston 2008; Calise and Preston 2006; Calise and Preston 2008].

As a result of simulations to track the desired path, the PI controller gains for the model in Fig. 7.91 were tuned to $K_p = 4$ and $K_i = 0.4$ (to eliminate canopy bias) [Calise and Preston 2008]. However, this design (yaw rate feedback) seems to suffer from the presence of the digital filters and was not able to follow the guidance commands. Hence, another design was developed.

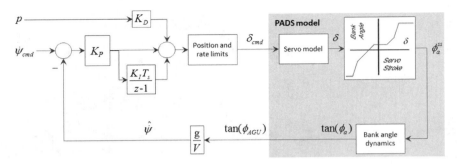

Fig. 7.92   Basic PID controller with a bank angle feedback [Calise and Preston 2006].

The PID controller design with $K_p = 5$, $K_i = 1$, and $K_d = 5$ shown in Fig. 7.92 is based on roll-angle output from AGU. Even though this design feeds back the same dynamic representation of PADS dynamics described by Eq. (7.211), it has several advantages [Calise and Preston 2006]. The major advantage of this design compared to that of Fig. 7.91 is that it is not hampered by digital filters, which increases the bandwidth. Second, the indicated roll angle from IMU is less sensitive to yaw oscillations of AGU. (It is integrated, i.e., naturally filtered, from the roll rate, which is perpendicular to AGU oscillations axis.) Third, the derivative term can now be implemented using the relatively clean roll rate. Last, tangent of roll angle is a natural output of a strapdown INS utilizing quaternions. For this implementation, the outer-loop command of Eq. (7.209) was substituted by

$$\dot{\psi}_{\mathrm{cmd}} = K_p^{\psi}(\psi_{\mathrm{cmd}} - \hat{\psi}) + K_i^{\psi} \int (\psi_{\mathrm{cmd}} - \hat{\psi}) \, \mathrm{d}t \qquad (7.213)$$

with the same $K_p^{\psi} = 0.28$ and $K_i^{\psi} = 0.004$ to offset canopy bias in the guidance command (to eliminate situation when PADS has a zero bank but nonzero turn rate).

To mitigate AGU oscillations in the design of Fig. 7.91, Calise and Preston [2006] also tried an adaptive architecture shown in Fig. 7.93. In this figure the nominal controller block includes all but the shaded background blocks and discrete filters block shown in Fig. 7.91. The reference model and neural-network blocks were added to augment the original design. As described in [Calise and Preston 2006; Calise and Preston 2008], the reference model represents the first-order model of PADS dynamics compared to the actual yaw rate filtered with the same discrete filters as in the original design. The error is used to adapt the single hidden-layer weights in the neural-network block. Vector **x** consists of the heading rate $\dot{\psi}$, model output ψ_m, and one step-delayed value of δ_{cmd}. The adaptive correction δ_{ad} is added to δ_{cmd} to keep the error signal e

close to zero. In the process of reducing this error, the effect of disturbances is mitigated as well. An example of the original design output and adaptive design is presented in Fig. 7.94.

To conclude the PID control section, let us address the issue of effect the winds have on PADS guidance and control. For those guidance schemes that rely on path planning in the wind-fixed coordinate system $\{w\}$ (Fig. 7.19), this may not be an issue, but for those relying on path generation in the inertial frame the winds might have a destabilizing effect [Calise and Preston 2009].

In no-wind condition, a control input δ_a results in a certain heading rate $\dot{\chi}_a$, which in the light of Fig. 5.1c and Eq. (7.2) can be expressed as

$$\dot{\chi} = \dot{\chi}_a \approx \dot{\psi}_{ss} = K(\delta_a)\delta_a \tag{7.214}$$

(see Fig. 5.41). The gain $K(\delta_a)$ here can be referred to as the PADS gain. Knowing the required ground track rate $\dot{\chi}_{cmd}$ (the guidance loop output), the servo command is obtained as

$$\delta_a = \dot{\chi}_{cmd}K^{-1}(\delta_a) \tag{7.215}$$

However, in a windy condition the ground track rate is not equal the heading rate, and as a result while the servo command of Eq. (7.215) still produces an expected result in terms of $\dot{\chi}_a$, it does not produce a desired result in terms of $\dot{\chi}$ (see examples in Figs. 7.6 and 7.7). With a tailwind, $\dot{\chi} < \dot{\chi}_a$, and the system's response will feel sluggish, and with a headwind $\dot{\chi} > \dot{\chi}_a$ the response will be perceived as excessive, which may lead to instability. This is a well-known fact for all pilots practicing ground reference maneuvers, like S-turns, eights on pylon or turns around the point (Fig. 7.95). To compensate for the winds, pilots vary a bank angle increasing it to a maximum on a downwind leg and decreasing to the shallowest bank angle on an upwind portion of a maneuver. In a similar

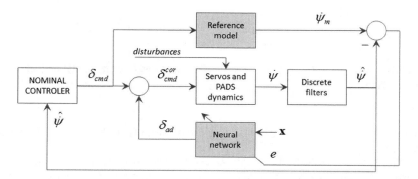

Fig. 7.93   Adaptive heading rate controller [Calise and Preston 2008].

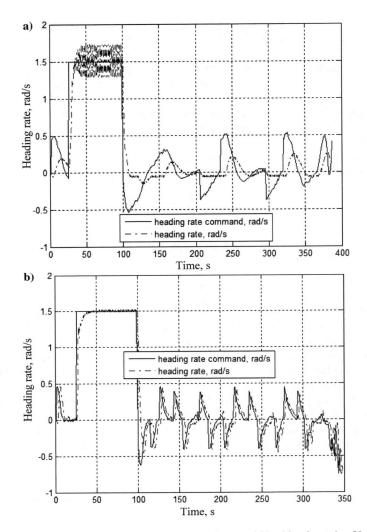

Fig. 7.94 Example of tracking performance a) without and b) with adaptation [Calise and Preston 2008].

manner, Calise and Preston [2009] suggested multiplying PADS gain $K(\delta_a)$ by the $\dot{\chi}/\dot{\chi}_a$ ratio.

This ratio depends on the relative direction of the wind with respect to the airspeed vector and can be estimated as follows. Referring to Fig. 7.96 (which is a cleaned version of Fig. 5.1c with two auxiliary angle notations) and following

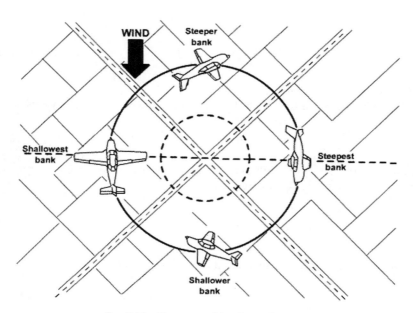

Fig. 7.95   Turn-around-a-point techniques.

[Calise and Preston 2009], we may write

$$\chi = \chi_a + \eta \qquad (7.216)$$

In aviation angle, η is referred to as a wind correction angle (WCA) and angle $\pi - \eta - \sigma$ as a wind-to-track angle (WTA). We can also write that

$$\sigma = \pi - \chi_W + \chi_a = \pi - \delta\psi \qquad (7.217)$$

where

$$\delta\psi = \chi_W - \chi_a \qquad (7.218)$$

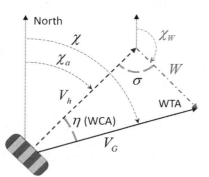

is the angle between the airspeed vector and the wind vector. With a headwind $\delta\psi = \pm\pi$; with a tailwind $\delta\psi = 0$. From kinematics it follows that

$$\dot{\sigma} = \dot{\chi}_a \qquad (7.219)$$

Differentiating Eq. (7.216) yields

$$\dot{\chi} = \dot{\chi}_a + \dot{\eta} \qquad (7.220)$$

Fig. 7.96   Airspeed–ground-speed diagram.

or

$$\frac{\dot{\chi}}{\dot{\chi}_a} = 1 + \frac{\dot{\eta}}{\dot{\chi}_a} \tag{7.221}$$

Hence, we need to find $\dot{\eta} = \dot{\eta}(\dot{\chi}_a)$.

Resolving the wind triangle in Fig. 7.96 using the law of cosines yields

$$2V_h V_G \cos(\eta) = V_h^2 + V_G^2 - W^2 \tag{7.222}$$

Taking a derivative from both sides of Eq. (7.222) and remembering that we assume $W = const$, $V_h = const$ results in

$$\dot{\eta} = -\frac{V_G - V_h \cos(\eta)}{V_h V_G \sin(\eta)} \dot{V}_G \tag{7.223}$$

On the other hand, we can apply the law of cosines to write

$$V_G^2 = V_h^2 + W^2 - 2V_h W \cos(\sigma) \tag{7.224}$$

Differentiating it and accounting for Eqs. (7.217) and (7.219) yield

$$2V_G \dot{V}_G = 2V_h W \sin(\sigma)\dot{\sigma} = 2V_h W \sin(\pi - \delta\psi)\dot{\chi}_a = 2V_h W \sin(\delta\psi)\dot{\chi}_a \tag{7.225}$$

Expressing $\dot{V}_G$ from Eq. (7.225) and substituting it in Eq. (7.223) results in a ratio we were looking for:

$$\frac{\dot{\eta}}{\dot{\chi}_a} = -\frac{W}{V_G} \left|\frac{\sin(\delta\psi)}{\sin(\eta)}\right| \left(1 - \frac{V_h}{V_G}\cos(\eta)\right) \tag{7.226}$$

(An absolute value was used to avoid a sign confusion in a situation when $\chi_a > \chi$.) Substituting Eq. (7.226) in Eq. (7.221) results in the PADS adjustment gain [Calise and Preston 2009]

$$\frac{\dot{\chi}}{\dot{\chi}_a} = 1 - \frac{W}{V_G} \left|\frac{\sin(\delta\psi)}{\sin(\eta)}\right| \left(1 - \frac{V_h}{V_G}\cos(\eta)\right) \tag{7.227}$$

Figure 7.97 shows this dependence for two ratios of the wind-to-horizontal component of airspeed. Specifically, as seen in Fig. (7.97b), the change of the required adjustment gain from headwind to tailwind may be quite big (raising the stability concerns when the wind magnitude approaches horizontal component of airspeed).

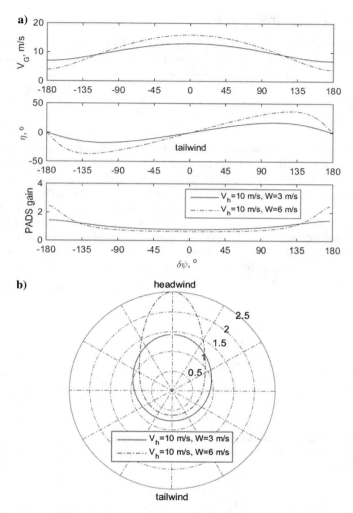

Fig. 7.97   Genesis of dependence of the a) PADS adjustment gain on $\delta\psi$ and b) PADS gain in the polar coordinates.

Note that implementation of Eq. (7.227) in practice would lead to an indeterminate value of the gain in a headwind situation. Calise and Preston [2009] evaluated this gain at $\delta\psi \to \pm\pi$ as $V_h V_G^{-1}$ and used a conservative approximation of Eq. (7.227) in Onyx PADS AGU to mitigate the distortion of a spiral descent over IPI site in a high-wind condition (to avoid situation shown in example of Fig. 7.43a).

7.8.3 MODEL PREDICTIVE CONTROLLERS

Another approach was developed in [Slegers 2004; Slegers and Costello 2005] and used on the Snowflake PADS AGU. Consider a simple single-input single-output (SISO) discrete system described in state-space form as

$$\mathbf{x}_{k+1} = \mathbf{G}\mathbf{x}_k + \mathbf{H}u_k$$
$$y_k = \mathbf{C}\mathbf{x}_k \tag{7.228}$$

Here $\mathbf{G}$, $\mathbf{H}$, and $\mathbf{C}$ are discretized system matrices, based on the linearized continuous models of Sec. 5.4, $\mathbf{x}_k$ is the state vector, u_k is the control input, and y_k is the output at time k. The model just described can be used to estimate the future state of the system. Assuming a desired trajectory w_k is known, an estimated error signal $\hat{e}_k = z_k - \hat{y}_k$ is computed over a finite set of future time instants called the prediction horizon h_p. In MPC, the control computation problem is cast as a finite-time discrete optimal control problem. To compute the control input at a given time instant, a quadratic performance index is minimized through the selection of the control history over the control horizon. The performance index can be written as

$$J = (\mathbf{Z} - \hat{\mathbf{Y}})^T \mathbf{Q}(\mathbf{Z} - \hat{\mathbf{Y}}) + \mathbf{U}^T \mathbf{R}\mathbf{U} \tag{7.229}$$

where

$$\mathbf{Z} = [z_{k+1}, z_{k+2}, \ldots, z_{k+h_p}]^T \tag{7.230}$$

$$\mathbf{U} = [u_k, u_{k+1}, \ldots, u_{k+h_p-1}]^T \tag{7.231}$$

$$\hat{\mathbf{Y}} = \mathbf{K}_{CG}\mathbf{x}_k + \mathbf{K}_{CGH}\mathbf{U} \tag{7.232}$$

and both $\mathbf{R}$ and $\mathbf{Q}$ are symmetric positive semidefinite matrices of size $h_p \times h_p$. Equation (7.232) is used to express the predicted output vector $\hat{\mathbf{Y}}$ in terms of system matrices

$$\mathbf{K}_{CG} = \begin{bmatrix} \mathbf{CG} \\ \mathbf{CG}^2 \\ \vdots \\ \mathbf{CG}^{h_p} \end{bmatrix}, \quad \mathbf{K}_{CGH} = \begin{bmatrix} \mathbf{CH} & 0 & 0 & 0 & 0 \\ \mathbf{CGH} & \mathbf{CH} & 0 & 0 & 0 \\ \mathbf{CG}^2\mathbf{H} & \mathbf{CGH} & \mathbf{CH} & 0 & 0 \\ \vdots & \vdots & \vdots & \ddots & 0 \\ \mathbf{CG}^{h_p-1}\mathbf{H} & \cdots & \mathbf{CG}^2\mathbf{H} & \mathbf{CGH} & \mathbf{CH} \end{bmatrix} \tag{7.233}$$

The first term in Eq. (7.229) penalizes tracking error while the second term penalizes control action. Equations (7.229) and (7.232) can be combined resulting in the performance index

$$J = (\mathbf{Z} - \mathbf{K}_{CG}\mathbf{x}_k - \mathbf{K}_{CGH}\mathbf{U})^T \mathbf{Q}(\mathbf{Z} - \mathbf{K}_{CG}\mathbf{x}_k - \mathbf{K}_{CGH}\mathbf{U}) + \mathbf{U}^T \mathbf{R}\mathbf{U} \tag{7.234}$$

that is now expressed in terms of the system state $\mathbf{x}_k$, desired trajectory $\mathbf{W}$, control vector $\mathbf{U}$, and system matrices $\mathbf{G}$, $\mathbf{H}$, $\mathbf{C}$, $\mathbf{Q}$, and $\mathbf{R}$.

The control $\mathbf{U}$, which minimizes Eq. (7.234), can be found analytically as

$$\mathbf{U} = \mathbf{K}(\mathbf{Z} - \mathbf{K}_{CG}\mathbf{x}_k) \tag{7.235}$$

where

$$\mathbf{K} = \left(\mathbf{K}_{CGH}^T \mathbf{Q} \mathbf{K}_{CGH} + \mathbf{R}\right)^{-1} \mathbf{K}_{CGH}^T \mathbf{Q} \tag{7.236}$$

Equation (7.236) contains the optimal control inputs over the entire control horizon; however, at time k only the first element u_k is needed. The first element u_k can be extracted from Eq. (7.236) by defining $\mathbf{K}_1$ as the first row of $\mathbf{K}$. The optimal control over the next time sample becomes

$$u_k = \mathbf{K}_1(\mathbf{Z} - \mathbf{K}_{CG}\mathbf{x}_k) \tag{7.237}$$

Here, calculation of the first element of the optimal control sequence requires the desired trajectory $\mathbf{Z}$ over the prediction horizon and the current state $\mathbf{x}_k$.

Let us consider two practical examples. To track a commanded heading angle [see Eq. (7.1)], Culpepper et al. [2013] assumed the turn rate response being related to the differential brake deflection by the first-order linear model represented by Eq. (5.39), which takes the following state-space form:

$$\dot{\mathbf{x}} = \mathbf{A}\mathbf{x} + \mathbf{B}u \tag{7.238}$$

with

$$\mathbf{x} = \begin{bmatrix} \psi \\ \dot{\psi} \end{bmatrix}, \quad \mathbf{A} = \begin{bmatrix} 0 & 1 \\ 0 & -T_\psi^{-1} \end{bmatrix}, \quad \mathbf{B} = T_\psi^{-1}\begin{bmatrix} 0 \\ K_\psi \end{bmatrix}, \quad u = \delta_a \tag{7.239}$$

The discretized versions of matrices $\mathbf{A}$ and $\mathbf{B}$, which are necessary for the MPC controller [matrices $\mathbf{G}$ and $\mathbf{H}$ in Eq. (7.228)], are computed as

$$\mathbf{G}(\Delta t) = e^{\mathbf{A}\Delta t} = \mathbf{I} + \mathbf{A}\Delta t + \frac{1}{2!}\mathbf{A}^2\Delta t^2 + \cdots = \sum_{i=0}^{\infty} \frac{1}{i!}\mathbf{A}^i\Delta t^i \tag{7.240}$$

$$\mathbf{H}(\Delta t) = \left(\int_0^{\Delta t} e^{\mathbf{A}\lambda}\mathrm{d}\lambda\right)\mathbf{B} \tag{7.241}$$

where Δt is the sampling time of the discretized system in seconds (the update interval for the controller). Hence, the discretized version of takes the form of Eq. (7.228) with

$$\mathbf{x}_k = \begin{bmatrix} \psi_k \\ \dot{\psi}_k \end{bmatrix}, \quad \mathbf{G}(\Delta t) = \begin{bmatrix} 1 & \Delta t - \varepsilon \\ 0 & 1 - \Delta t T_\psi^{-1} + \varepsilon T_\psi^{-1} \end{bmatrix}$$

$$(7.242)$$

$$\mathbf{H}(\Delta t) = \begin{bmatrix} K_\psi \varepsilon \\ K_\psi \Delta t T_\psi^{-1} - K_\psi \varepsilon T_\psi^{-1} \end{bmatrix}, \quad u_k = \delta_{a;k}, \quad \mathbf{C} = [1 \quad 0]$$

In this equation, the term ε is of the order of $0.5\Delta t^2 T_\psi^{-1}$, and for $\Delta t < 1$ s it can be neglected, thus leaving

$$\mathbf{G}(\Delta t) = \begin{bmatrix} 1 & \Delta t \\ 0 & 1 - \Delta t T_\psi^{-1} \end{bmatrix}, \quad \mathbf{H}(\Delta t) = \begin{bmatrix} 0 \\ K_\psi \Delta t T_\psi^{-1} \end{bmatrix}, \quad \mathbf{C} = [1 \quad 0] \quad (7.243)$$

Matrices $\mathbf{G}$, $\mathbf{H}$, and $\mathbf{C}$ of Eq. (7.243) describing the internal model are then used to generate a commanded brake differential δ_a given a current state estimate from navigation $\mathbf{x}_k$ and a vector of the desired heading commands $\boldsymbol{\psi}_{\text{des}}$. Equation (7.237) becomes

$$\delta_{a;k} = \mathbf{K}_1(\boldsymbol{\psi}_{\text{cmd}} - \mathbf{K}_{CG}\mathbf{x}_k)$$

$$(7.244)$$

For a practical application, this command was complemented as

$$\delta_a = \delta_{a;k} + \delta_{a;k}^{\text{bias}}$$

$$(7.245)$$

where

$$\delta_{a;k}^{\text{bias}} = \delta_{a;k-1}^{\text{bias}} + K_I(\dot{\psi}_{\text{cmd}} - \dot{\psi})K_\psi^{-1}$$

$$(7.246)$$

This additional term takes care of a typical random turn bias (nonzero turn rate at zero differential brake input as shown in Fig. 5.41) by estimating it with a heavily damped filter for the actual turn rate [Culpepper et al. 2013].

In a more general case, because for parafoils pitch and speed are not typically controllable, a simple two-DoF model involving the roll and yaw dynamics should be used. Starting from Eq. (5.214) describing the Snowflake PADS reduced-order dynamics and converting it to the discretized form with $\Delta t = 0.5$ s (using the [G,H] = c2d(A,B,0.5) command in MATLAB) yield

[Slegers and Yakimenko 2011]

$$
\begin{bmatrix} \phi \\ \psi \\ p \\ r \end{bmatrix}_{k+1} = \begin{bmatrix} 0.962 & 0 & 0.153 & 0.012 \\ 0.0078 & 1 & -0.011 & 0.043 \\ -0.103 & 0 & 0.033 & 0.004 \\ 0.0191 & 0 & -0.0023 & -0.003 \end{bmatrix} \begin{bmatrix} \phi \\ \psi \\ p \\ r \end{bmatrix}_k + \begin{bmatrix} -0.006 \\ 0.0501 \\ -0.0131 \\ 0.1098 \end{bmatrix} \delta_{a;k}
$$

(7.247)

where the state vector **x** is composed of the linearized roll, yaw, roll rate, and yaw rate. The output matrix is

$$
\mathbf{C} = \begin{bmatrix} 0 & 1 & 0 & 0 \end{bmatrix}
$$

(7.248)

The optimal differential brake deflection δ_a to follow the desired yaw profile is then defined as stated in Eqs. (7.244–7.246).

7.8.4 NONLINEAR DYNAMIC INVERSION CONTROLLER

The idea of Gockel [1997a; 1997b] to use as much information about PADS dynamics as possible was realized in theoretically studied by Prakash and Ananthkrishnan [2006]. Having a nonlinear nine-DoF model of PADS expressed in the general form of Eq. (5.172), they implemented a nonlinear dynamic inversion (NDI) controller. Such a controller intends to cancel the original dynamics of the system and replace it with the desired dynamics. An advantage is that developing this controller does not necessarily require linearizing PADS dynamics; hence, the controller ensures the desired performance at all regimes. The disadvantage though is that dynamic inversion requires equal number of inputs and states, which is not the case for underactuated PADS with only one control, asymmetric TE deflection.

Hence, Prakash and Ananthkrishnan [2006] utilized a simple cascade control system involving two control loops (inner and outer) as shown in Fig. 7.98. The outer loop calculates the commanded yaw rate r_{cnd} based on PADS yaw angle command ψ_{cmd}, and the inner loop calculates the desired TE deflection δ_a^{des} based on r_{cnd} generated by the outer loop, pretty much as it was presented in Fig. 7.80. Each loop assumes NDI.

The inner-loop inversion is based on a single equation for $\dot{r}$ extracted from the nine-DoF model of Eq. (5.121) and represented as

$$
\dot{r} = \mathbf{a}_{6,:}(\mathbf{B}_0 + \mathbf{B}_{\delta_a}\delta_a)
$$

(7.249)

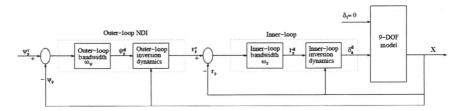

Fig. 7.98   Block diagram of a NDI controller [Prakash and Ananthkrishnan 2006].

Here, $\mathbf{a}_{6,:}$ is the sixth row of the inverted state matrix of Eq. (5.121), and the vector on the right-hand side of Eq. (5.121) is split between the part not depending on the control input $\mathbf{B}_0$ and the part depending on the control input via aerodynamic forces and moments [see Eqs. (5.74) and (5.79)]$\mathbf{B}_{\delta_a}\delta_a$. The inversion of Eq. (7.249) yields

$$\delta_a = \frac{\dot{r} - \mathbf{a}_{6,:}\mathbf{B}_0}{\mathbf{a}_{6,:}\mathbf{B}_{\delta_a}} \tag{7.250}$$

If $\dot{r}$ on the right-hand side of Eq. (7.250) is the desired yaw acceleration (error) modeled as a first-order response

$$\dot{r}_{\text{des}} = \omega_r(r_{\text{cmd}} - r) \tag{7.251}$$

where ω_r is the bandwidth; then on the left-hand side we have the desired TE deflection to achieve $\dot{r}_{\text{des}}$, δ_a^{des}. Substituting the expression for δ_a^{des} back to Eq. (7.249) results in dynamics we wanted:

$$\dot{r} = \omega_r(r_{\text{cmd}} - r) \tag{7.252}$$

The equation for the yaw angle rate is extracted from Eq. (5.15)

$$\dot{\psi} = \frac{\sin(\phi)}{\cos(\theta)}q + \frac{\cos(\phi)}{\cos(\theta)}r \tag{7.253}$$

Inverting it yields

$$r = \left(\dot{\psi} - \frac{\sin(\phi)}{\cos(\theta)}q\right) \bigg/ \left(\frac{\cos(\phi)}{\cos(\theta)}\right) \tag{7.254}$$

Following the same procedure and substituting $\dot{\psi}$ in Eq. (7.254) with the desired dynamics

$$\dot{\psi}_{\text{des}} = \omega_\psi(\psi_{\text{cmd}} - \psi) \tag{7.255}$$

where ω_ψ is the bandwidth, we obtain the commanded yaw rate to achieve $\dot{\psi}_{des}$, r_{cmd}. Substituting r_{cmd} back to Eq. (7.256) results in

$$\dot{\psi} = \omega_\psi(\psi_{cmd} - \psi) \qquad (7.256)$$

Obviously, because the NDI inner loop works by inverting $\dot{r}$ alone; in other states the nine-DoF model responds according to their open-loop dynamics. To ensure that the inner loop has the faster response compared to the outer loop, Prakash [2006] attributed a higher bandwidth to it. Specifically, he set $\omega_\psi = 2$ (which corresponds to the time constant of 0.5 s) and $\omega_r = 5$ (which corresponds to the time constant of 0.2 s).

CHAPTER 8

Glide-Angle Control

James Moore[*]
JM Technologies LLC, Tucson, Arizona

Oleg Yakimenko[†]
Naval Postgraduate School, Monterey, California

Deflection of the trailing edge (TE) is typically the major and only method used for control. Although asymmetric deflection of TE provides effective lateral control, symmetric deflection provides effective airspeed control with little to no effect on the glide angle (GA). (By definition, GA is simply the negative of flight-path angle.) Without an effective GA control mechanism, these systems are forced to rely on carefully planned lateral control maneuvers to place the system on final glide at the correct altitude and distance from the target. Human jumpers are able to achieve a pinpoint accuracy using these techniques, but performing these lateral maneuvers precisely with an autonomous system is a very challenging problem. The problem is further complicated by deviations from the assumed wind during landing approach, necessitating frequent recalculation of the approach trajectory. Hence, the limited control authority and sensing capability of current airdrop systems make it extremely difficult for an autonomous system to recover from unanticipated changes in the wind profile near the ground. The need for better landing precision has been the primary motivation for the development of methods to increase the range of GA control. However, there are also changes in airspeed associated with glide angle control that can improve PADS maximum range especially when considering the effects of wind uncertainty. If the PADS encounters wind conditions approaching or exceeding the horizontal speed of the vehicle, severe crosswind constraints (i.e., the inability to achieve certain ground track directions) occur. Depending upon the relative speed ratio and the direction of travel with respect to the wind, the ability to increase airspeed as part of glide angle control can play an important part in the ability to achieve a landing in the target area. As will be described in in this chapter, using a speed system for GA control is a well-established technology in the commercial paraglider industry. Several methods for controlling the glide angle have been studied. This chapter is based on these studies and organized as follows. Section 8.1 starts from introducing a concept of so-called drag-assisted GA control featuring deployment of a drogue chute on a final approach. While it

---

[*]President.
[†]Professor.

is possible implement continuous GA control using a mechanical actuator on a drogue kill line, this initial study focuses on a simpler, single-stage deployment, which implies an accurate computation of a drogue chute deployment time. After drogue deployment, the PADS would continue with conventional TE control, albeit with much increased GA. Next, two concepts of a continuous GA control are introduced. The first one, presented in Sec. 8.2, relies on mechanically varying a rigging angle of PADS. The second approach, introduced in Sec. 8.3, suggests varying GA aerodynamically using upper surface spoilers. In both cases traditional flaps might be eliminated completely because enabling asymmetric rigging or applying spoilers asymmetrically seem to provide sufficient authority of a turn rate control as well.

8.1   DROGUE-ASSISTED GLIDE-ANGLE CONTROL

This section is based on [Moore 2013] and describes a method for GA control intended for the "on final" terminal guidance phase of PADS while gliding and into wind. The GA control is accomplished by drogue attachment and deployment, which reduces the lift-to-drag or glide ratio (GR) and airspeed of a parafoil while not fully arresting the gliding flight. Photographic results of a prototype drogue deployment for a 1.05-m^2 (11.3-ft^2) subscale parafoil are used to demonstrate the validity of the approach. It is also validated by computer simulations based on a parametric model of a 998-kg (2,200-lb) capable PADS.

8.1.1   MAJOR LIMITATION OF PADS PRECISION LANDING STRATEGIES

As discussed in Chapter 1, PADS are designed with a high GR to support requirements for greater than 3:1 (distance over altitude) deployment offset. The typical parafoil system uses TE brakes to decrease the airspeed and GR on terminal approach. However, for a heavily loaded canopy in low wind conditions, TE braking provides little reduction in GA. A high GR is detrimental to terminal accuracy as the along-track miss distance is proportional to the product of vertical altitude uncertainties and the terminal inertial GSA. (Inertial GA means negative of the Earth-referenced flight-path angle γ, as opposed to γ_a.) The lack of effective GA control contributes directly to disproportionately larger along-track miss errors as compared to cross-track miss errors. Ineffective GA control has contributed to a shift in the PADS industry away from precision into-wind landings and continues to be a challenge for a parafoil-based system.

A survey of guidance methods presented in Chapter 7 revealed a variety of guidance approaches used by different developers of PADSs. Some systems do not even pursue soft into-wind landing. These systems feature a spiral descent at the target, which only guide directly to the target after being commanded at low altitude to exit the spiral. Some high-glide PADS utilizing this guidance strategy deploy a ballistic round parachute to fully arrest the gliding parafoil flight near the ground. As a result of the uncontrolled descent of the ballistic trajectory, uncertainty in the low-level winds can contribute significantly to miss distance. One of such systems, the JPADS-XL Screamer, is shown in Fig. 8.1 transitioning a) from guided flight b) to recovery chute landing. The Screamer development has laid the groundwork for a class of PADS alternatives using drogue glide augmentation for a precision into-wind landing.

The corridor of feasible trajectories for a parafoil on final approach to an IPI is constrained by the glide ability of the parafoil and the prevailing wind conditions. To reduce the inertial velocity at touchdown, an into-wind approach is most desirable. A downwind approach will not only increase the horizontal airspeed at impact, it makes the timing of turns much more critical. Small timing variations become magnified by the increased ground speed of a downwind trajectory, resulting in larger position errors.

Let us define the glide slope (GS) as an inclined surface that contains a desired glide path (GP) and hence intersects IPI. Then, the GS angle (GSA) defines the inclination of this plane (GS) with respect to the local tangent plane. For a set of prevailing wind conditions, the glide slope intercept (GSI) in a point in space from which the parafoil while operating into wind at 50% brake will intercept the desired target point with a straight-line trajectory. Generally, the GSI altitude is limited to a few hundred feet to as much as 328 m (1,000 ft) above the target. The start of the final approach occurs when the parafoil reaches a GSI.

The trajectory before the parafoil reaching the GSI is less constrained. Any set of maneuvers that will reliably ensure that the parafoil reaches the GSI in the face of unknown and potentially varying wind conditions should be equivalent with no residual error remaining at the landing. The lower the altitude, the more constrained the set of feasible solutions to an accurate landing are. The typical precision approach to a target area evolves from a more or less unconstrained

a)   b)

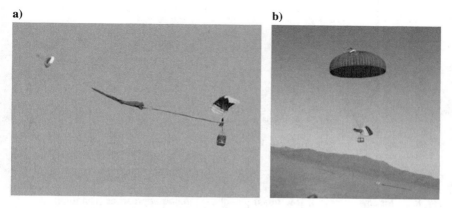

Fig. 8.1   JPAD-XL Screamer's transition from a) guided flight and b) recovery chute landing.

energy management phase where parafoil altitude is bled off until the parafoil is sufficiently low to initiate the landing approach.

Over the last 40 years, the hang-gliding and paragliding sports as well as the parachuting industry have developed best practices for training pilots in the technical elements of landing that best deal with target area constraints and wind uncertainties. As discussed in Chapter 7, there are basically two approved landing approaches. The first is the aircraft approach, and the second is the figure-eight approach. These two approaches are shown schematically in Fig. 8.2 with the inclusion of both a GSI and energy management elements.

The aircraft approach consists of three legs: downwind, base, and final. It is generally the first landing technique taught to paraglider pilots. The technique is relatively simple to learn, but is generally only applicable to low-wind conditions. As wind speeds increase, the ability to achieve the GSI is greatly affected by the timing of the into-wind turn. Late timing on the into-wind turn will result in the parafoil being below GS, and the impact will be downwind of the intended target. Conversely, an early turn will result in a landing too far in front of the target. For these reasons, more experienced pilots use the figure-8 approach.

The figure-eight approach is similar to the aircraft approach, except that the into-wind turn of the base leg is replaced with two steps: a turn into-wind and a series of figure-eight altitude control maneuvers. Because this approach provides for a succession of figure-eights, the timing of the into-wind turn is not as critical.

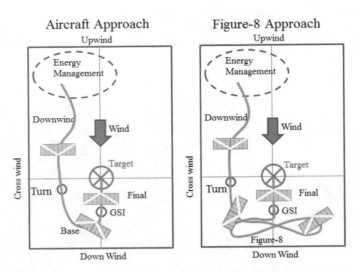

Fig. 8.2   Schematic view of traditional paraglider landing approaches (variations of precision into-wind landings).

In this approach, a position error from the into-wind turn occurs at a higher altitude where there is sufficient time to compensate for the inaccuracy of a mistimed turn. For moderate wind conditions that are generally less than the parafoil airspeed, the figure-eight maneuvers are completed so that the parafoil remains downwind of GSI. In higher winds, the figure-eight pattern migrates in front of the target and requires a backing-in approach to the target.

These concepts are codified in the hang-glider proficiency requirements for landing are listed in Table 8.1 [SOP 2013]. To achieve an H-3 (intermediate) or H-4 (advanced) rating, the pilot must demonstrate the ability to land within 15.2 m (50 ft) and 7.6 m (25 ft) of a designated target, respectively. Hang gliders and paragliders generally have much higher GRs than the typical PADS. As will be shown, a high GR does exacerbate the problem of landing precision. Despite this, glider pilots routinely achieve landing accuracy far better than auto land guidance for the typical PADS. One primary advantage of the human pilot is better knowledge of ground winds. There is usually a windsock located in the landing area, and the direction of other pilots approaches can provide a distinct advantage to determining how to best land under prevailing wind conditions.

The paraglider landing approaches can be viewed as a set of maneuvers that ensure a final turn into a GSI. If properly positioned at the GSI, the parafoil will be on GS to the target at the midrange of the GA control range (e.g., 50% TE braking). The main challenges with achieving the GSI are changing wind conditions and a lack of knowledge of the winds below the GSI. The figure-eight approach allows the GSI to be at a lower altitude, and therefore there is less uncertainty in where the GSI should be. The aircraft approach will usually require a higher GSI, which will subject the parafoil to a longer duration exposed to wind variations. On the GSI, heading control is generally limited to maintain a ground track straight to the target. Braking control is used to maintain a position

TABLE 8.1 HANG-GLIDING LANDING PROFICIENCY RATINGS SYSTEM

SOP 12-02 Paragraph	Hang-Gliding Rating	Pilot Proficiency System
12-02.07	H-2 Novice	Must have logged a minimum of 25 flights with a required ability to demonstrate successful T-landing or aircraft landing approach
12-02.08	H-3 Intermediate	Demonstrate three consecutive landings that average less than 50 ft from a target after a flight that requires turns on approach
12-02.09	H-4 Advanced	Demonstrate three consecutive landings that average less than 25 ft from a target after a flight that requires turns on approach

on GS. If the PADS is above GS, brakes are applied. If the parafoil is below GS, then the brakes are let off to the flatten GA.

8.1.2 MOTIVATION AND PERFORMANCE OF PARAFOIL GLIDE AUGMENTATION

As just mentioned, a parafoil generally has very good lateral control (i.e., heading control), but will suffer from an inability to generate a wide range of longitudinal control. The conventional parafoil uses TE steering line brakes to decrease airspeed and GR. A typical 3:1 full glide parafoil will drop to a 2:1 GR at maximum braking. Overcontrolling the brakes to further decrease the GR invites the potential for stall. Excessive TE brakes can result in sudden and unpredictable stall conditions. Depending upon the type of parafoil, the onset of stall can start slowly with the parafoil rocking back in positive pitch motion and falling backwards in stall. The recovery can be dramatic as the system has to regain forward speed by rocking back to a large negative pitch angle. Stall can also result in a sudden loss of altitude by dropping nearly vertical. Alternatively, when coupled with uneven braking, what is known as a stall turn is more likely. The stall turn occurs when one side of the canopy experiences a large stall-induced drag. The stall turn results in a sudden rotation of the parafoil about the yaw axis usually resulting in a downwind turn. Significant steering line deflection margins are required to preclude potential stall conditions. A common occurrence in low-altitude turbulence is for a braking canopy at high angle of attack (AoA) to stall due to sudden changes in the wind gradient.

Even though the PADS air mass lift-to-drag (L/D) does not change significantly with symmetrical brakes, braking is effective in modulating inertial GA when operating against a headwind. The reduced airspeed under brakes provides substantially more inertial GA control into-wind than in no-wind conditions. Compensation for vertical errors is constrained by limited GA control. This is made worse in no-wind conditions because the inertial GR is highest.

Whether using the aircraft approach or the figure-eight approach, the PADS should be able to effectively use lateral control to come in close proximity to the GSI (x, y) location. From the GSI, any horizontal errors typically only require minor heading adjustments to generate a ground track direct to the target (x, y) position. However, the altitude errors of the parafoil when at the GSI (x, y) generally contribute to miss errors that the braking control cannot overcome.

Figure 8.3 shows how the vertical GSI errors scale by the inertial GA resulting in projected along-track (ATRK) errors. Relatively small elevation errors are multiplied by a shallow GA to generate much larger horizontal miss errors. It follows that the same vertical error at the GSI does not contribute to miss if the inertial GA is vertical. This is the prime motivation for generating an increasingly vertical descent to the target as it minimizes the effect of vertical guidance and navigation errors.

In addition to the error contributions due to inexact guidance and control, uncertainty about the winds on the final leg to the target contribute to GSA

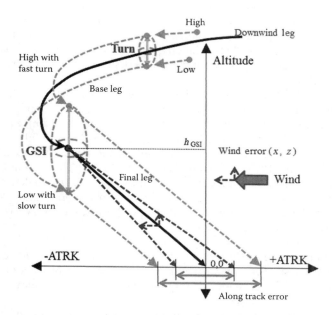

Fig. 8.3   Final approach error contributors.

variation and therefore variation in where the true GSI is. To the extent that the limited braking control can compensate for these errors, there may be no effect. However, as has usually been the case, the root-mean-square (RMS) of all of the error contributions can easily exceed the limited PADS braking control authority.

The application of TE brakes tends to increase lift. An increase in lift also has an associated increase in induced drag. Therefore, to a point, TE brakes tend to both decrease the GR and decrease the airspeed. Beyond traditional braking, there are two primary methods to augment parafoil glide: lift augmentation and drag augmentation. Lift augmentation can be embodied in various forms. Generally, when lift is decreased, there is also a drop in induced drag, and airspeed tends to increase at a lower GR. For example, a decrease in AoA will reduce the lift. However, lowering this AoA too much may have a destabilizing effect. To avoid canopy collapse, the section leading-edge (LE) stagnation point must be kept within the ram-air inlet. Significant changes in parafoil lift and AoA must be made with care to avoid LE canopy collapse.

As parafoil drag is increased, there is also a decrease in GR. For moderate increases in drag, the airspeed will remain relatively constant as the parafoil is trading horizontal airspeed for vertical airspeed. Drogue GA augmentation is a reliable and pilot-tested hang-glider method for achieving improved landing accuracy. Drogue augmentation does not tend to change the main gliding decelerator

AoA or lift characteristics. The hang-glider drogue chute is said to "transform a high-performance glider into a low-performance glider." The benefits are to assist the pilot in the difficulties of landing a glider with a high GR. This is essentially the same issue being faced by the PADS designers as was described in the preceding section.

The Orion PADS (Fig. 1.6), qualified by the U.S. Army in 1996, featured the first into-wind landing algorithm [Patel et al. 1997]. The pinnacle of performance for Orion occurred in approximately 1999–2000 when a fully adaptive figure-eight approach was used to demonstrate a contractual 52-m (171-ft) CEP for a 998-kg (2,200-lb) capable system. (There is little published evidence of this early system other than in [Allen 1995; Wailes and Hairington 1995].) This system included a finely perfected onboard stall detection algorithm that was able to detect stall and release brakes so as to limit stall turns on final to a maximum heading change of 45 deg. Vertical impact velocity of 8.5 m/s (28 ft/s) was sometimes exceeded if the stalls occurred at an inopportune time.

Improving the performance of this weight class of system is pivotal as the higher airspeeds and slower turn times make terminal accuracy even more difficult than for smaller systems. Figure 8.4 shows a concept for a drogue augmentation of a 998-kg (2,200-lb) class system with two-point drogue bridle attachment at AGU and payload swivel.

Figure 8.5 shows the simulated steady-state trim conditions for a 2,200-lb capable 111.5-m^2 (1,200-ft^2) PADS both with and without a drogue. For this example, we have made the drogue reference area S^d the same as the parafoil S^c. The coefficient C_D^d of the drogue is assumed to be 0.6. A PADS is typically characterized by a relatively large C_{D0}^c drag increment due to the combination of payload,

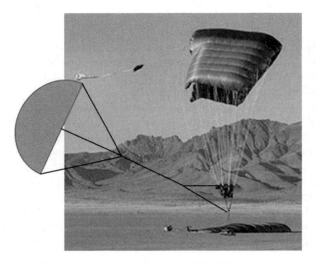

Fig. 8.4   Proposed concept for 2K PADS add-on.

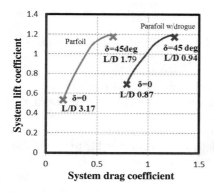

Fig. 8.5   The 2K PADS drag polar with simulated steady-state trim conditions with variable symmetrical brake deflection.

suspension line, and ram-air inlet drag. As is true for any gliding decelerator, there is a quadratic relationship between induced drag and lift. Deep brakes at high AoA (greater than 20 deg) are required to achieve the lowest GR. Inclusion of the drogue shifts the entire polar to the right resulting in a GR less than 1:1. The TE brakes further complement the drogue by creating even higher aerodynamic coefficients that further reduce airspeed.

Now, consider an arbitrary combination of a lifting parafoil combined with a drogue drag element. We can determine the GA $(-\gamma_a)$ for any combination of parafoil and drogue from a condition that in steady-state and straight-line flight the sum of the total lift and drag vectors must equal the total system weight

$$L = mg \cos(\gamma_a) = QS^c C_L \tag{8.1}$$

$$D = -mg \sin(\gamma_a) = Q(S^c C_D^c + S^d C_D^d) \tag{8.2}$$

$$\gamma_a = -\tan^{-1}\left(\frac{D}{L}\right) = -\tan^{-1}\left(\frac{S^c C_D^c + S^d C_D^d}{S^c C_L}\right) \tag{8.3}$$

In these equations

$$C_D^c = C_{D0}^c + k_{C_L^2} C_L^{c2} \tag{8.4}$$

and $Q = 0.5\rho V_a^2$ is the dynamic pressure. Equation (8.1) can be resolved for V_a to yield

$$V_a = \sqrt{\frac{2mg \cos(\gamma_a)}{\rho S^c C_L}} \tag{8.5}$$

Combining these results with the PADS drag polar, we can compare the inertial GR of lift and drag augmentation with and without winds. Figure 8.6 shows the range of inertial GR control available for the 2K PADS. With standard braking and no wind, the GR varies from approximately 3:1 to 2:1 for full glide and maximum braking respectively. In 5.1 m/s (9.9 kt) of ground wind, the GR improves to just below 2:1 and 1:1 for full glide and maximum braking, respectively.

For purposes of comparison, we show what occurs under glide augmentation where the C_{Lc} is reduced to $\frac{3}{4}$ and $\frac{1}{2}$ of C_L trim. At $\frac{3}{4}$ C_L trim, the GR does not quite achieve a 2:1 GR in still air. With $\frac{1}{2}$ C_L trim, GR drops considerably to approximately 1.5:1. However, in both cases the airspeed also increases appreciably. Drogue augmentation is shown as being able to attain a 1:1 GR in still air. In 5.1 m/s (10 kt), the GR is well below 0.5:1. Drag augmentation provides lower parafoil GRs with lower airspeeds as compared to any form of lift reduction.

Drag augmentation with a drogue of reference area comparable to that of the parafoil can achieve a GA control range corresponding to a 3:1 GR in full glide down to less than 1:1 GR in still air. As a rule of thumb, along-track errors should be halved in comparison to PADS with just conventional drag brakes. The improved GR not only improves along track, but it also will allow better control of the cross-track (XTRK) landing errors and angle to the wind (ATW) as there is less lateral "S"ing control required to combat the high glide of the parafoil alone.

Subscale testing was performed with a 1.05-m² (11.3-ft²) Flexifoil traction kite. The parafoil was rerigged for a 6-deg rigging angle. A cross-parachute drogue design was selected for the prototype for the ease of construction and the well-understood drag performance [Ludtke 1971]. The primary parafoil and drogue parameters are listed in Table 8.2.

A drogue bridle was constructed with a four-point attachment. The lower attachment points were at the simulated payload. The upper attachment points were at 30% of the rigging line length (canopy lower edge to confluence). A pulley system was used to apply 19 cm (7.5 in.) of TE deflection with drogue line tension. Two different cross drogues were tested.

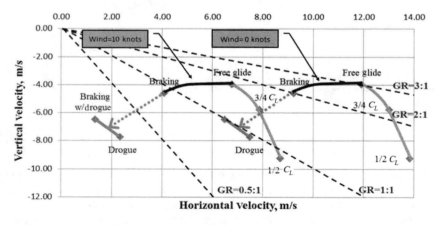

Fig. 8.6   A 2K parafoil glide augmentation.

The parafoil was ground launched with a two-line kite bar (Fig. 8.7). As a result of having only moderate ground winds, no more than a 1.36-kg (3-lb) payload could be lofted. The drogue was deployed by the third line attached to a drogue stowage sock. The video frames shown are at the times indicated relative to drogue sock removal. After the initial success with the small drogue, testing immediately progressed to the larger drogue. Figure 8.7 shows a video sequence of the larger 1.1-m^2 (11.8-ft^2) drogue. The drogue deployed surprisingly well considering the lightly loaded canopy and the heavy braking input. The drogue appeared to have a stabilizing effect despite achieving what looked like a 1:1 gliding descent. The tell-tail attached to the TE of the canopy shows evidence of unstable separated flow; however, there was no obvious effect on the parafoil attitude.

8.1.3 DROGUE DEPLOYMENT STRATEGY AND COMPUTER SIMULATIONS

The development of the Orion system was facilitated by a detailed six-DoF PADS simulation that served as the foundation for all guidance, navigation, and control (GNC) activities. The simulation includes detailed models of the parafoil aerodynamics, navigation sensors, winds, actuators, and interfaces between the GNC software and the embedded software environment of the flight management system (FMS). Drogue forces resolved into bridle tensions, which are applied in the canopy dynamics model (CDM) moment equations. The CDM is not only used as part of a what-if and Monte Carlo test-bed, it is also integrated into the navigation solution to improve dead-reckoning and aiding of the PADS navigation state where sensor data may have become invalid.

During the Guided Parafoil Air Delivery System (GPADS) programs for light, medium, and heavy as well as the NASA X-38 program [Harrington and

TABLE 8.2   SUBSCALE TEST PARAMETERS

Parafoil Parameter	Value	Drogue Parameter	Value
Aspect ratio	3.2	Width/length	0.3
Span	1.8 m (6.0 ft)	Reference area (large drogue)	1.1 m^2 (11.8 ft^2)
Chord (midspan)	0.76 m (2.5 ft)	Reference area (small drogue)	0.45 m (4.8 ft^2)
Area	1.05 m^2 (11.3 ft^2)	Reefing line length-to-canopy diameter ratio	0.5
Rigging angle	6.0 deg	Suspension lines	16
Rigging line ratio	0.8		

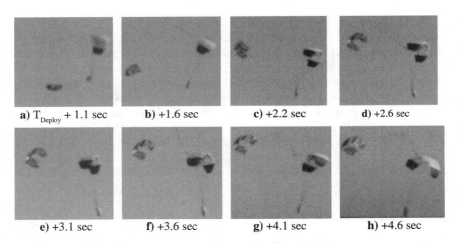

a) T_{Deploy} + 1.1 sec   **b)** +1.6 sec   **c)** +2.2 sec   **d)** +2.6 sec

e) +3.1 sec   **f)** +3.6 sec   **g)** +4.1 sec   **h)** +4.6 sec

Fig. 8.7   Subscale drogue deployment.

Doucette 1999], there was an opportunity to test a succession of cargo parafoils from all of the primary U.S. manufacturers. Canopies have ranged from the 34.4-m² (370-ft²) Paraflite MT1-X up to the largest Pioneer 929-m² (10,000-ft²) GPADS-H. A free-body diagram of the CDM including the recently added drogue model is shown in Fig. 8.8.

To make effective use of the drogue, a strategy for drogue deployment had to be developed. Depending upon how much increase in drag is provided by the drogue as well as the ground wind conditions, the GR will be severely restricted following the deployment of the drogue. Careful algorithm development is required to avoid turning a high GR problem into a more severe low GR problem. The scope of this initial analysis was to consider the feasibility of an effective accuracy enhancement by timing single drogue deployment as compared to a continuously variable drag approach using a reefable drogue.

The strategy employed in the PADS GNC is to allocate the solution into two GNC functions, a GSA monitor and a drogue deployment detector. In the final phase of the terminal approach and at each iteration of the guidance update, the PADS trajectory is projected forward to the target under two hypotheses. Both cases use the same projected winds to the ground. The first hypothesis is that the standard precision approach with 50% brakes is employed. The second case is that the drogue is deployed with 50% brakes. The concept is shown diagrammatically in Fig. 8.9 with trajectory projections occurring at times T_1, T_2, T_3, and T_{deploy}. The PADS continuously updates the three-dimensional projections to the target assuming both brakes and drogue. At T_3 drogue deployment is armed because braking error exceeds drogue error. At T_{deploy} drogue error begins to increase and the drogue is deployed.

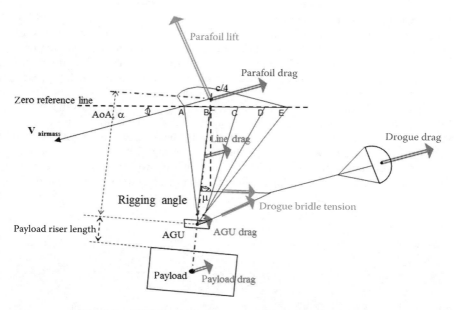

Fig. 8.8   PADS free-body diagram with parafoil and drogue.

The drogue deployment detector arms the logic to deploy the drogue when the projected miss using brakes exceeds the projected error using the drogue. After being armed, the detector monitors the projected miss under drogue and deploys the drogue when the error begins to increase from a minimum. This algorithm is much more robust than a simple deployment scheme based on altitude as it accounts for many more variable that would affect the

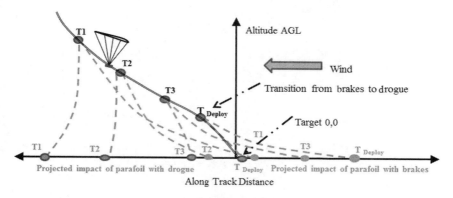

Fig. 8.9   Drogue deployment strategy.

deployment decision. An example of drogue deployment, which is executed just prior to landing, is shown in Fig. 8.10. When deploying so close to the ground, the projections to the target under drogue must incorporate the time required for deployment, line stretch, and drogue inflation to accurately impact the target. Final stages of a figure-eight approach show turn in on final approach, with drogue deployment timed for best intercept of target.

The PADS Monte Carlo tool was used to exercise the CDM with the drogue deployment algorithm. The baseline system is the 2001 Orion^{TM} production standard that had been optimized for the 2:1 braking of the Strong C1200 canopy. No changes were made to the guidance algorithms. The comparison system is as shown in Fig. 8.4. The results of the two 10-run test cases are shown in Figs. 8.11 and 8.12 for baseline and drogue augmentation, respectively.

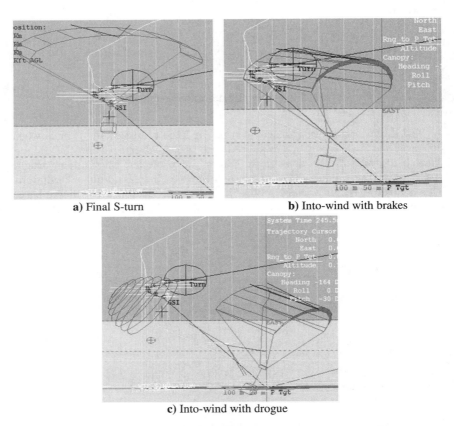

a) Final S-turn   b) Into-wind with brakes

c) Into-wind with drogue

Fig. 8.10   Three-dimensional visualization of drogue-assisted precision approach.

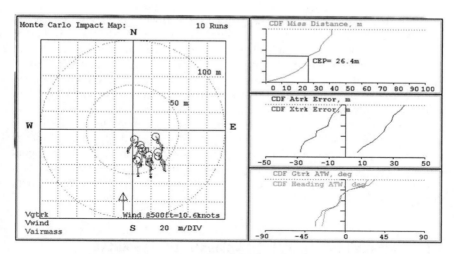

Fig. 8.11   Baseline 2K PADS Monte Carlo simulation results.

The nominal winds are moderate with approximately 5.1 m/s (10 kt) of southerly wind at 152.4 m (500 ft).

Both systems have good CEP measures for the benign wind conditions. The baseline system has a CEP of 26.4 m. With limited braking, landings are persistently long. The drogue add-on system CEP is slightly less at 22.6 m (74 ft) with impacts falling nominally short and north of the IPI. Both systems exhibit

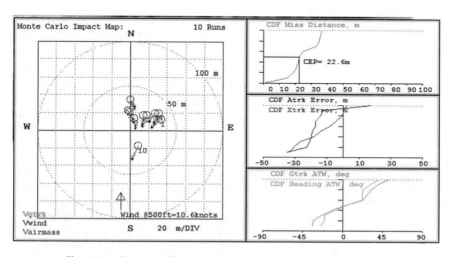

Fig. 8.12   Drogue add-on 2K PADS Monte Carlo simulation results.

similar distributions of heading and ground track angles to the wind (ATW) as well as similar XTRK error distributions. The primary difference is in the change in the along-track miss errors. While the baseline system tended to overrun the target each time, the add-on system now tends to fall short. Run 10 of the drogue add-on set did land long, but in reviewing the data it was determined that the drogue deployment algorithm did not release the drogue and the system landed as it would have in the baseline set.

The Monte Carlo test runs were produced to demonstrate the use of the simulation tool as well as the functionality of the drogue deployment. A better comparative assessment would involve optimization of the drogue guidance algorithms to take full advantage of the improved GA control and then the completion of a much broader set of wind-speed test cases to validate the algorithms. Then, this optimized system could be compared to the earlier baseline results.

8.2   GLIDE-ANGLE CONTROL WITH VARIABLE RIGGING ANGLE

This section explores another approach for GA control by changing the rigging angle of a canopy μ or in other words by rotating the canopy around the quarter-chord point in the body coordinate frame $\{b\}$. The effect of varying angle μ is very similar to the effect of the elevator on a fixed-wing aircraft in glide.

8.2.1   RIGGING ANGLE VARYING EFFECTS

Consider Fig. 8.13 reintroducing some of the angular quantities of the longitudinal channel and also defining the dimensionless glide coefficient (GC) to be used as a control input to control GA [Slegers et al. 2008]

$$\mathrm{GC}_a = -1/\tan(\gamma_a), \quad \mathrm{GC} = -1/\tan(\gamma) \tag{8.6}$$

By definition, GC_a and GC vary between 0 (corresponding to $\gamma_a = -90\,\mathrm{deg}$ and $\gamma = -90\,\mathrm{deg}$, respectively) and ∞ (corresponding to $\gamma_a = 0\,\mathrm{deg}$ and $\gamma = 0\,\mathrm{deg}$, respectively). Essentially, Fig. 8.13 repeats Fig. 2.23, but following Chapter 5 takes into account a negative sign of the rigging and flight path angles (Fig. 2.23 utilizes the glide angle, which is negative to the flight path angle, and unsigned rigging angle). Figure 8.14 provides a visual representation of Eq. (8.6).

Just as the elevator on a fixed-wing aircraft alters the trim AoA of the wing, altering the rigging angle produces a change in the trim AoA of the parafoil canopy [Ward 2012]. Assuming that the TE deflections are small and therefore pitching moments are low, the suspended weight tends to hang directly below the quarter-chord, which means the pitching angle is negligibly small. Then, the sum of the rigging angle and the AoA magnitudes are approximately the flight-path angle magnitude. This means that there is a unique curve of flight-path angle vs AoA for a given setting of rigging angle (Fig. 8.15). Similarly, there is a unique lift-to-drag ratio (GR) vs AoA curve determined by the aerodynamic

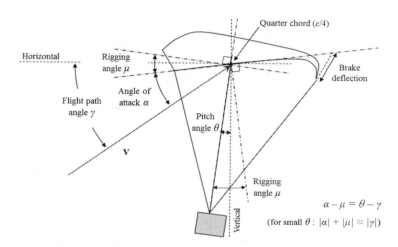

Fig. 8.13   Relation between the rigging angle of attack, pitch, and flight-path angles.

characteristics of a given canopy. (Figure 8.15 features this concept for two notional canopies.) The intersection of the flight-path angle curve for a given rigging angle setting and the GC curve for a given canopy represents the trimmed flight condition for that combination of rigging angle and canopy.

As seen from Fig. 8.15a, the first canopy has a peak GC of 2.5, while the second canopy has a peak GC of 4.5. This plot provides some important insight into the use of rigging angle as a GA control mechanism. In particular, the GS range is increased for a more efficient (higher GR) canopy, the sensitivity of GS to rigging angle is highest just below the peak GR trim point, and the sensitivity of GC to rigging angle is dramatically reduced at low GR. Rigging angle has a more direct influence on GA than GR, so it is important to keep in mind the nonlinear relationship between GC and GA. As shown in Fig. 8.14, a reduction in GC from 3 to 2 represents an 8-deg change in flight-path angle whereas a reduction in

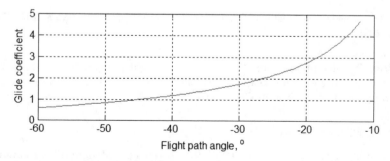

Fig. 8.14   Relationship of GC to flight-path angle [defined by Eq. (8.6)].

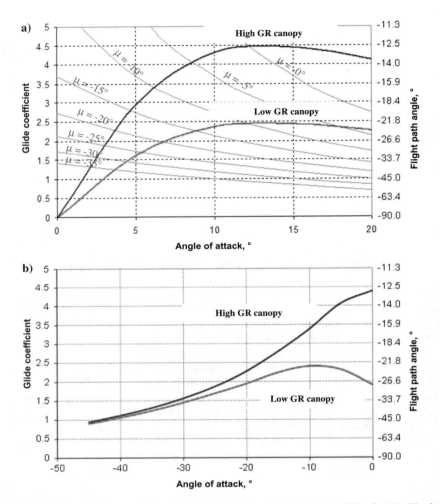

Fig. 8.15   Canopy trim conditions defined for different rigging angles [Ward 2012; Ward et al. 2013].

GC from 2 to 1 represents an 18-deg change in flight-path angle. For this simple reason, it is more efficient to apply variable rigging angle as a GA control mechanism on canopies with high GR [Ward 2012].

8.2.2   EXPERIMENTS WITH ULW ADS

By trimming the A lines in concert with the brakes, a pure longitudinal rotation of the canopy to different rigging angles can be achieved as shown in Fig. 8.16. As

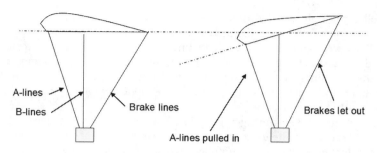

Fig. 8.16   Rigging angle control [Ward 2012].

shown in the preceding section, this provides direct control over the trim AoA, allowing the full range of the canopy's lift-to-drag ratio to be utilized in flight. This section, based on a slightly edited Flight Results section of [Slegers et al. 2008], describes the results of testing UL PADS built to investigate this issue.

This experimental PADS is shown in Figs. 8.17a and 8.17b with the canopy deployed and undeployed, respectively. The payload of the system consisted of a 15 × 15 × 46 cm (6 × 6 × 18 in.) cardboard box with two avionic boxes on either end. The upper avionics box consisted of a data logger with battery, canopy pack, and a Hitec HS-311 servo used to release the packed parafoil. Sensors included in the upper avionics are three accelerometers, gyroscopes and magnetometers, a global positioning system, and barometric altimeter. The upper box was designed to allow the top flaps of the cardboard box to flare out at a 45-deg angle to allow the undeployed system to be cone stabilized (see Fig. 8.17). The lower avionics box contained three Hitec HS-785 HB sail winches, a Hitec Electron 6 FM receiver, and a battery pack. Sail winch 1 and 2

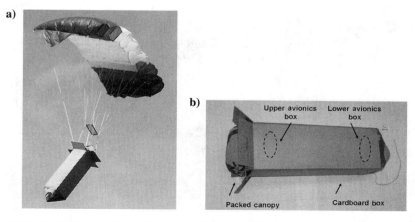

Fig. 8.17   Test system with a) deployed and b) undeployed parafoil [Slegers et al. 2008].

controlled the right and left brake lines, while sail winch 3 controlled the front lines of the parafoil. The sail winch signals were mixed together so that as the front lines were pulled in, the brake lines were let out and vice versa. This allowed the geometry of the canopy to accommodate different rigging angles. Note the rear lines of the canopy remained fixed to the upper box while the brake lines and front lines run through the upper avionics box, to the sail winches in the lower avionics box.

As just mentioned; the rigging angle of the parafoil was changed using the three sail winches in the lower avionics box. An example of this rigging change is shown in Fig. 8.18a. Two parafoil systems, differing mainly in their canopy thickness, LE geometry and payload weight, were used in this study. The parameters of both systems are shown in Fig. 8.18b and outlined in Table 8.3.

Four flight tests were conducted in low winds; each system was tested under two conditions (see Table 8.4). System 1 was configured with a nominal rigging μ_1 of -6 deg, and once equilibrium was achieved, the rigging angle was changed. In the first flight of System 1, the canopy was rotated down to a rigging μ_2 of -24 deg whereas during the second flight the canopy was rotated up to a rigging μ_3 of 10 deg. System 2 was configured with a nominal rigging μ_4 of -24 deg and was subsequently rotated down to a rigging μ_5 of -44 deg for both the third and fourth flight tests. Figure 8.19 shows results for the flight path where both altitude and distance have been nondimensionalized with respect to the initial altitude.

As seen from Table 8.4, System 1 responds to a decrease in rigging from -6 to -24 deg with a 70% increase in GR, from 1.45 to 2.46. Increasing the rigging from -6 to 10 deg results in a stalled condition where the GC is decreased 89%. System 2 responds in an opposite manner with a decrease in rigging from -24 to -44 deg resulting in a 48% decrease in GR from 3.70 to 1.94. Differences

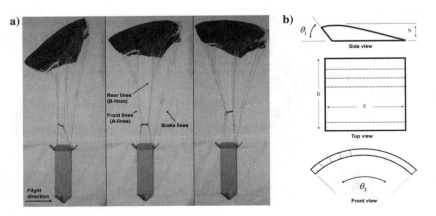

Fig. 8.18   a) Parafoil rigging angle change and b) canopy geometry [Slegers et al. 2008].

TABLE 8.3   SYSTEM CHARACTERISTICS [SLEGERS ET AL. 2008]

Parameter	System 1	System 2
θ_1	80 deg	50 deg
θ_2	45 deg	45 deg
h	0.11 m (0.35 ft)	0.05 m (0.17 ft)
b	1.37 m (4.5 ft)	1.52 m (5.0 ft)
c	0.64 m (2.1 ft)	0.4 m (1.3 ft)
Weight	2.37 kg (5.23 lb)	0.72 kg (1.59 lb)

are also observed in the vertical and forward velocity trends where for System 1 changing the rigging results in large forward speed changes with vertical speed remaining nearly unchanged whereas for System 2 the opposite is true. Results from System 1 were used to estimate C_L and C_D curves for the combined system including payload. The C_L and C_D curves are approximated by a cubic and quadratic curves defined by C_{L0}, $C_{L\alpha}$, $C_{L\alpha^3}$, C_{D0}, and $C_{D\alpha^2}$. Using results for System 1 in Table 8.4, the coefficients are estimated as 0.28, 0.68, −0.35, 0.135, and 0.95, respectively. Figure 8.20 shows the estimated curves compared with the measured results for System 1 including canopy and payload. The estimated values are consistent with results from Ware and Hassell [1969] who observed maximum lift-to-drag ratios near 2.5, high profile drag, and low maximum lift coefficients when compared to a standard rigid wing. As demonstrated by results from Ware and Hassell [1969], the C_L curve is typically flat near stall with the exact AoA at stall difficult to define. The C_L curve is approximated well by a cubic function prior to and poststall; however, a higher-order function is required to approximate the stall region. The estimated C_L curve in Fig. 8.20 is valid at angles of attack lower than 30 deg and higher than 70 deg the location of the maximum C_L can only be identified as occurring within that. Estimation of the stall region is unnecessary because all GA control and simulations occur prior to this region. Simulations of the estimated system GC are shown in Fig. 8.21 and are consistent with test data.

TABLE 8.4   FLIGHT-TEST SUMMARY [SLEGERS ET AL. 2008]

	System 1 $\mu_1 = -6$ deg	System 1 $\mu_2 = -24$ deg	System 1 $\mu_3 = 10$ deg	System 2 $\mu_4 = -24$ deg	System 2 $\mu_5 = -44$ deg
α	28 deg	6 deg	70 deg	10 deg	5 deg
GR	1.45	2.46	0.28	3.70	1.94
Speed	7.9 m/s (15.4 kt)	10.7 m/s (20.8 kt)	4.9 m/s (9.5 kt)	6.4 m/s (12.4 kt)	7.9 m/s (15.4 kt)
C_L	0.56	0.35	0.49	0.45	0.27

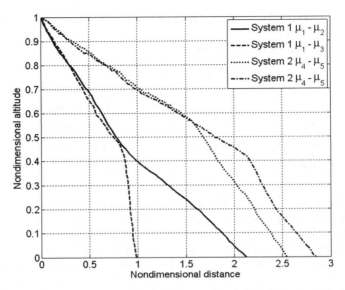

Fig. 8.19   Altitude vs distance traveled for four flight tests [Slegers et al. 2008].

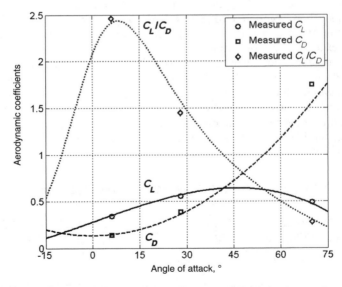

Fig. 8.20   Estimated lift coefficient, drag coefficient, and C_L/C_D for System 1 including payload [Slegers et al. 2008].

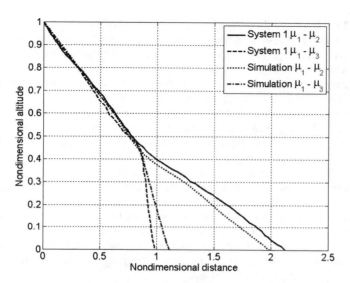

Fig. 8.21   Comparison of simulated and measured GA for System 1 [Slegers et al. 2008].

As seen from Fig. 8.21, System 1 operates to the right of its maximum lift-to-drag ratio for all rigging angles resulting in increased GC as the rigging is decreased. Decreasing the rigging of System 2 resulted in a decreased GC demonstrating System 2 operates to the left of its maximum lift-to-drag ratio. Noting that the maximum L/D occurs well before stall, its AoA can be estimated as

$$\alpha_{(L/D)_{\max}} = \sqrt{\left(\frac{C_{L0}}{C_{L\alpha}}\right)^2 + \frac{C_{D0}}{C_{D\alpha^2}}} - \frac{C_{L0}}{C_{L\alpha}} \tag{8.7}$$

This equation follows from maximizing (taking a derivative of) an expression for the GR

$$\frac{L}{D} = \frac{C_{L0} + C_{L\alpha}\alpha}{C_{D0} + C_{D\alpha^2}\alpha^2} \tag{8.8}$$

resulting in a quadratic equality

$$\alpha^2 + 2\frac{C_{L0}}{C_{L\alpha}} - \frac{C_{D0}}{C_{D\alpha^2}} = 0 \tag{8.9}$$

Equation (8.8) shows the maximum L/D AoA increases as $C_{L\alpha}$ increases and $C_{D\alpha2}$ decreases, both occurring as the lifting surface efficiency factor increases. This is consistent with System 2 having a small thickness-to-chord ratio, larger AR, and more rounded nose. Parafoils of higher efficiency will be able to operate to the left of their maximum L/D whereas an inefficient canopy may operate to the right. It is demonstrated by Systems 1 and 2 that a parafoil system can effectively operate on either side of the maximum L/D AoA and have effective GA control. A system however can be arranged such that it operates near its maximum L/D; in such a case minimal GA control will result from the small slope of the L/D curve in this vicinity.

Simulations were completed for System 1 where the rigging was decreased from −6 to −12 deg, −18 and −24 deg at 15-s intervals with each change occurring linearly over a second. Figure 8.22 shows the glide slope dynamics persist for 5 s after each change in rigging. The GC initially decreases in response to the decreased lift from decreasing AoA before it increases as the speed increases. Changes in AoA and velocities are shown in Fig. 8.23. As rigging is decreased, the AoA decreases approaching the maximum L/D where GA control authority diminishes. A nearly linear GC mechanism can be implemented for System 1 by designing the nominal rigging to be −12 deg so that ±25% changes in GC can be achieved over a −6- to −18-deg rigging angle range. If maximum L/D flight is desired, System 1 can be flown at a rigging of −18 deg; however, the GC can only be effectively decreased.

The GA control can be treated separately from heading tracking and may be implemented similar to proportional navigation of guided missiles. A diagram of the GA guidance is shown in Fig. 8.24, where $\mathbf{r}_{LOS}$ is the line of sight vector from the parafoil to IPI.

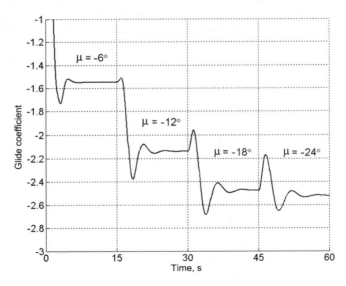

Fig. 8.22   Simulated GC varying rigging μ of System 1 [Slegers et al. 2008].

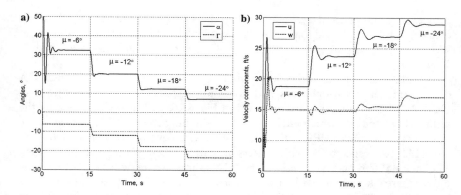

Fig. 8.23   a) Simulated AoA and b) velocities varying rigging μ of System 1 [Slegers et al. 2008].

As the parafoil approaches the target, any misalignment of the velocity vector and $\mathbf{r}_{LOS}$ will result in $\mathbf{r}_{LOS}$ rotating with the angular velocity ω_{LOS}

$$\omega_{LOS} = \frac{V}{|\mathbf{r}_{LOS}|} \sin(\theta_{LOS} + \lambda) \qquad (8.10)$$

For the parafoil to impact the target, the angular velocity of the line-of-sight vector ω_{LOS} must be zero; if the parafoil is falling too fast or too slow, ω_{LOS} will be positive or negative, respectively. A discrete PI controller

$$\delta_{\mu;k} = \delta_{\mu;k-1} + \left(K_p + K_i \Delta T/2\right) \omega_{LOS,k} - \left(K_p - K_i \Delta T/2\right) \omega_{LOS,k-1} \qquad (8.11)$$

uses rigging angle ($\delta_\mu \equiv \mu$) to track zero ω_{LOS}, thus placing the system on the required GS to impact the target. In Eq. (8.11), the angular velocity ω_{LOS} is sampled at intervals of ΔT.

Monte Carlo simulations of 100 drops were completed using the precision placement algorithm with MPC. Noise was injected into GPS, altitude, and IMU sensors. In addition to sensor errors, three sources of wind variation were added to the simulation: wind shear, varying magnitude, and direction. The wind was divided into two segments varied independently, namely, wind above 30.5 m (100 ft) and wind below 305 m (1,000 ft) in order to simulate inconsistent wind

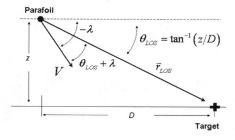

Fig. 8.24   Glide-slope guidance geometry [Slegers et al. 2008].

profiles. Prevailing wind was assumed by the system to come from a heading of 0 deg while true wind varied in its direction. For all simulations, the target was set as the origin. Sensor noise and wind variation statistics are listed in Table 8.5.

Monte Carlo simulations were first completed with and without GA control including sensor errors and no wind. Dispersion results are shown in Fig. 8.25 while histograms are provided in Fig. 8.26. CEP is defined here by the radius about the IPI, which encloses 50% of the impacts. The CEP for each case is shown by a circle. CEP with and without GA control are 3 m (9.8 ft) and 4 m (13.2 ft), respectively, with dispersion patterns being similar in both cases. The main difference is found in the histograms where without GA control impacts are skewed toward larger errors, where 5% of impacts have more than 9.14 m (30 ft) of error. With GA control no impact has more than 9.14 m (30 ft) of error.

TABLE 8.5   ERROR STATISTICS [SLEGERS ET AL. 2008]

Parameter	Mean	Standard Deviation
Initial condition position x	1,067 m (3,500 ft)	228.6 m (750 ft)
Initial condition position y	0	228.6 m (750 ft)
Initial condition position z	1,372 m (4,500 ft)	228.6 m (750 ft)
GPS x bias	0	0.91 m (3 ft)
GPS y bias	0	0.91 m (3 ft)
GPS x deviation	0.31 m (1 ft)	0
GPS y deviation	0.31 m (1 ft)	0
Altitude bias	0	1.52 m (5 ft)
Altitude variation	0.31 m (1 ft)	0
Roll, pitch, and yaw bias	0	1.7 deg
Roll, pitch, and yaw deviation	1.7 deg	0
u, v, and w bias	0	3 cm/s (0.1 ft/s)
u, v, and w deviation	21 cm/s (0.7 ft/s)	0
p, q, and r bias	0	1.7 deg
p, q, and r deviation	1 deg	0
Wind 1	3 m/s (5.9 kt)	0.9 m/s (1.8 kt)
Wind 2	3 m/s (5.9 kt)	0.9 m/s (1.8 kt)
Wind heading error	0	11 deg

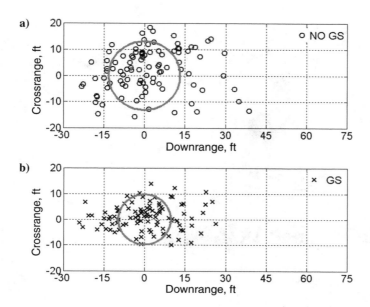

Fig. 8.25 Dispersion for all sensor errors and no wind a) without and b) with GA control [Slegers et al. 2008].

Results including both sensor errors and wind variations are shown in Figs. 8.27 and 8.28. The 50% CEP with and without a GA control are 5.1 m (16.7 ft) and 22.1 m (72.4 ft), respectively. Including winds, the GA control CEP increased by only 70% while the CEP without GA control increased by 450%. Including GA control, a reduction by more than a factor of three is achieved in CEP, and sensitivity to winds is reduced. Dispersion patterns also differ significantly. With GA control, the dispersion is mainly in range with 97% of the cases having less than 6.1 m (20 ft) of cross-range error. Swerving required without GA control increased dispersion in cross range. GA control also reduced the number of errors greater than 61 m (200 ft) from 11 to 1.

8.2.3 EXPERIMENTS WITH A LARGER PADS

This section, based on a slightly edited work by [Ward et al. 2011b; Ward 2012], presents the results of more rigging-angle control experiments conducted on larger PADS with two different aspect-ratio (AR) canopies (Fig. 8.29). The canopy geometry, rigging geometry, and mass properties of the low and medium AR test systems are given in Table 8.6. The medium AR canopy is slightly larger than the low AR canopy, so to maintain a similar wing loading, when flying the medium AR canopy ballast was added. The ballast consisted of lead

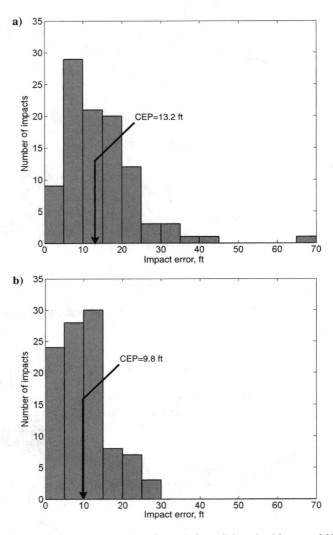

Fig. 8.26   Histogram for sensor errors and no-wind condition a) without and b) with a GA control [Slegers et al. 2008].

plates mounted to the estimated location of the center of gravity of the payload.

The flight tests were focused on obtaining steady-state values of airspeed and GC as a function of rigging angle and brake deflection. The flight-test procedure began by climbing under power up to the test altitude [normally

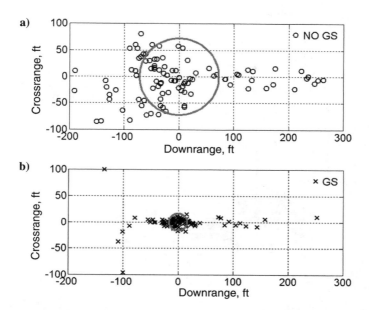

Fig. 8.27   Dispersion for all sensor errors and varying wind a) without and b) with GA control [Slegers et al. 2008].

457.2 m (1,500 ft) AGL] where the power to the motor was cut off. The rigging angle and symmetric brake level were set to preprogrammed settings, the data logger was switched on, and GPS and barometric altimeter data were recorded for approximately 20 s of gliding flight. The system was then sent back up to the test altitude to repeat the procedure for the next control setting. For each setting of rigging angle and symmetric brake, a small amount of asymmetric brake was applied to produce a noticeable turn rate (normally 5–15 deg/s) to expose the wind effects. If a noticeable turn rate could not be achieved with less than 2 cm of brake differential, then the constant control segment was interrupted after approximately 10 s, the pilot took control and turned the system manually through approximately 180 deg, and the constant control segment was continued for roughly another 10 s on the new heading angle.

Estimates of the atmospheric wind and forward airspeed were generated based on the vector diagram of Fig. 5.1c. The airspeed and wind vector were assumed constant for each segment of the flight where a constant control deflection was held. The airspeed and wind vector are estimated simultaneously for each constant control segment using an optimizer to minimize the difference between the measured ground track velocity V_G and the estimated ground-track velocity (computed as the sum of the estimated airspeed and wind vectors). This process worked well when each constant control segment covers a large change

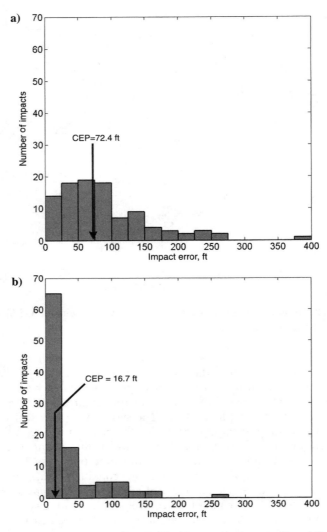

Fig. 8.28   Histogram for sensor errors and varying winds a) without and b) with GA control [Slegers et al. 2008].

in azimuth to expose the wind (e.g., if a control input was held long enough to fly a complete circle, the airspeed is just the average speed measured over the circle, and the wind vector is determined from the drift of the circle). The estimation process failed if a constant control segment did not contain enough azimuthal variation (e.g., if the vehicle flew in a straight line during the constant control segment, it

a)   b)

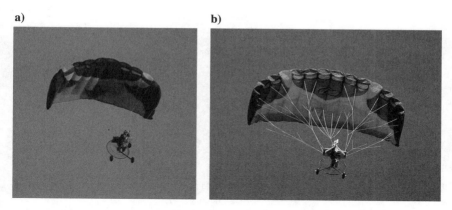

Fig. 8.29   Comparison of low and medium AR canopies in flight [Ward 2012].

was impossible to extract separate estimates of the airspeed and wind vector). This was handled by appending a penalty to the optimization cost function proportional to the difference in the estimated wind vector between concurrent flight segments. In other words, if there was no unique airspeed and wind vector combination that could be extracted from a given flight segment, then the optimizer set the wind vector to match adjacent flight segments. Figure 8.30 shows a sample GPS ground track for a constant control flight segment. Notice

TABLE 8.6   CANOPY, RIGGING, AND PAYLOAD PARAMETERS FOR FLIGHT-TEST
VEHICLE [WARD 2012]

	Low AR	Mid AR
Aspect ratio	2.79	3.35
Area	2.1 m^2 (22.6 ft^2)	2.72 m^2 (29.3 ft^2)
Span	2.4 m (7.9 ft)	3 m (9.8 ft)
Mean chord	0.88 m (2.9 ft)	0.91 m (3 ft)
Canopy arc radius	1.68 m (5.5 ft)	2.1 m (6.9 ft)
Projected aspect ratio	2.01	2.39
Projected area	1.7 m^2 (18.3 ft^2)	2.23 m^2 (24 ft^2)
Total rigging line length	26 m (84 ft)	57 m (187 ft)
Mass (weight)	3.7 kg (8.1 lb)	4.72 kg (10.4 lb)
Wing loading	1.76 kg/m^2 (0.36 lb/ft^2)	1.74 m^2 (0.35 lb/ft^2)
Mass ratio	1.01	0.88

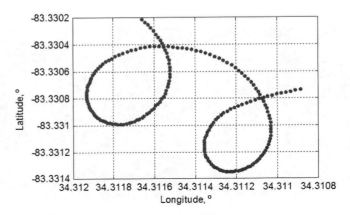

Fig. 8.30   GPS track for constant control segment [Ward 2012].

the gentle turn rate and the drift of the system over ground due to wind. Figure 8.31 shows the measured ground speed, the estimated airspeed, and the ground speed reconstructed from the airspeed and wind estimates. Figure 8.32 shows the descent rate derived from the barometric altimeter reading during the flight segment. The descent rate estimate is obtained as the median of the measured descent rate.

Each segment of constant control gliding flight results in a single data point of forward airspeed $\hat{V}_h$, descent rate $\hat{\dot{z}}$, and turn rate $\hat{\omega}$ for a particular combination of rigging angle and symmetric brake. Brake deflection is normalized to the mean canopy chord c.

$$\delta_s = 0.5(\delta_r + \delta_l) \tag{8.12}$$

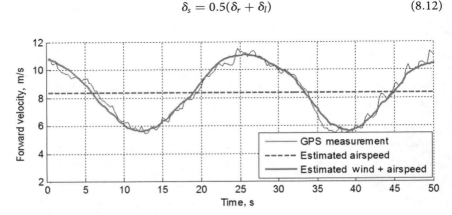

Fig. 8.31   Extracting forward airspeed from GPS ground speed [Ward 2012].

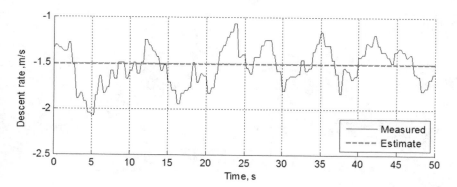

Fig. 8.32   Descent rate estimate from constant control segment [Ward 2012].

Using geometry of Fig. 8.33, we can write

$$L = \sqrt{L'^2 + (m\hat{V}_h\hat{\omega})^2} = \sqrt{(mg\cos(\gamma_a))^2 + (m\hat{V}_h\hat{\omega})^2} \qquad (8.13)$$

$$D = L\tan(\gamma_a) = L(\hat{\dot{z}}/\hat{V}_h) \qquad (8.14)$$

The change in canopy rigging angle produces a variation in AoA α. The AoA can be approximated as the difference between the rigging angle and flight-path angle γ_a and the rigging angle (Fig. 8.13). Using the estimates for lift [Eq. (8.13)], drag [Eq. (8.14)], and AoA extracted from fight-test data, the aerodynamic lift and drag coefficients are estimated as

$$C_L = C_{L0} + C_{L\alpha}\alpha + C_{L\alpha^3}\alpha^3, \quad C_D = C_{D0} + C_{D\alpha^2}\alpha^2 \qquad (8.15)$$

In the flight-test experiments, the low and medium AR canopies were flown at varying rigging angles with zero symmetric brake. In other words, the canopies were rotated through a variety of rigging angles with the brakes trimmed to keep a flat TE. The extracted lift and drag coefficient vs AoA behaviors for the low and medium AR canopies are shown in Fig. 8.34, and the identified aerodynamic parameters are shown in

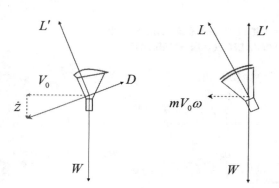

Fig. 8.33   Estimating lift and drag from forward speed, descent rate, and turn rate [Ward 2012].

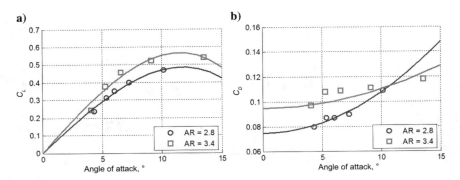

Fig. 8.34   a) Lift and b) drag coefficients vs AoA for low and medium AR canopies [Ward 2012].

Table 8.7. As expected, the lift-curve slope for the medium AR canopy is higher than the low AR canopy. However, the medium AR canopy appears to have a higher profile drag coefficient than the low AR canopy. This may be because of the increased complexity of the rigging for the medium AR canopy. Referring to Table 8.6, the reference area of the medium AR canopy is only 30% larger than the low AR canopy, but there is 120% more rigging line (and line drag) for the medium AR canopy.

The GA control achieved by varying rigging angle for these two canopies is shown in Fig. 8.35. This plot shows that dramatic and effective GA control can be achieved by varying the canopy rigging angle. The low AR canopy has a peak GC of 4.4, and the medium AR canopy has a peak GC of 4.9. The lower limit of GC for the canopies is not well established. There is a minimum AoA required to keep the canopies inflated, so testing near the lower limit of GC risks a severe frontal collapse of the canopy.

The flight tests were conducted for the low AR canopy at three levels of symmetric brake. The extracted lift and drag vs AoA behavior is shown in Fig. 8.36,

TABLE 8.7   IDENTIFIED LIFT AND DRAG PARAMETERS WITH ZERO BRAKE DEFLECTION [WARD 2012]

Parameter	AR = 2.8	AR = 3.4
C_{L0}	0	0
$C_{L\alpha}$	3.56	4.23
$C_{L\alpha^3}$	−28	−35
C_{D0}	0.074	0.095
$C_{D\alpha^2}$	1.12	0.496

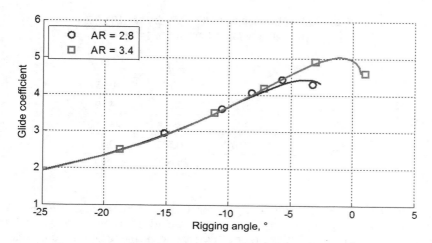

Fig. 8.35   GC vs rigging angle for low and medium AR canopies [Ward 2012].

and the identified aerodynamic parameters are given in Table 8.8. The variable rigging angle provides insight into the effect of symmetric braking that is not normally available from parafoil flight tests. The effect of symmetric brake is typically modeled as producing an increment in both lift and drag. It appears that in addition to this incremental effect, the symmetric brakes also increase the slopes of the lift and drag curves.

The effect of rigging angle on GC at the three symmetric brake levels is shown in Fig. 8.37. Symmetric braking produces only a modest effect on GC. This is consistent with typical airdrop systems in that little change in GC is normally achieved with symmetric braking until the system nears stall.

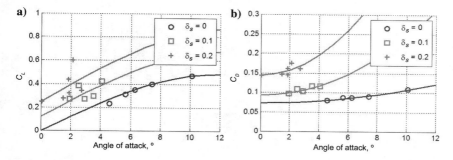

Fig. 8.36   a) Lift and b) drag behavior vs AoA and symmetric brake for the low AR canopy [Ward 2012].

TABLE 8.8   IDENTIFIED LIFT AND DRAG CHARACTERISTICS
FOR THE LOW AR CANOPY [WARD 2012]

Parameter	$\delta_s = 0$	$\delta_s = 0.1$	$\delta_s = 0.2$
C_{L0}	0	0.125	0.251
$C_{L\alpha}$	3.56	3.87	4.19
$C_{L\alpha^3}$	-28	-28	-28
C_{D0}	0.074	0.103	0.155
$C_{D\alpha^2}$	1.12	2.09	3.52

Figure 8.38 shows the effect of rigging angle and symmetric brake on airspeed. This plot shows that rigging angle produces a dramatic effect on airspeed as well as GC. Though symmetric braking is not effective in controlling GC, it is quite effective in controlling airspeed. The relationship between rigging angle and symmetric brake produces the envelope of possible combinations of airspeed and GC shown in Fig. 8.39. This is very interesting from a guidance and control perspective because it means that glide slope and airspeed can be controlled independently (within the constraints of the envelope) by modulating rigging angle and symmetric brake together.

Variation of the canopy rigging angle can create substantial changes in the GC of a parafoil with respect to the air mass. However, it is the PADS GC with respect to the ground $[1/\tan(\gamma)]$ that must be controlled to improve landing accuracy. As explained in Sec. 5.1.1, the Earth-referenced flight-path angle γ is defined by the forward speed over ground V_G and the descent rate V_d, where the forward speed over ground is determined by adding the horizontal wind speed with the horizontal component of airspeed (Fig. 8.40).

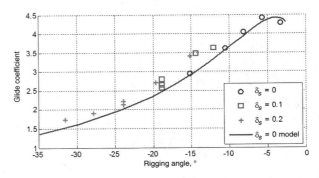

Fig. 8.37   GC vs rigging for a low AR canopy with varying symmetric brake [Ward et al. 2011b].

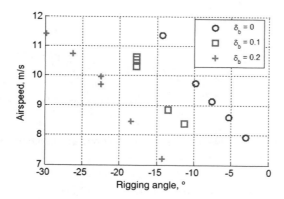

Fig. 8.38   Airspeed vs rigging angle for a low AR canopy [Ward et al. 2011b].

This is an important point because in any amount of wind, the GC over ground behaves in a very different manner than the GC with respect to the air. The variation in canopy rigging angle is used to vary the AoA of the parafoil canopy. The minimum rigging angle results in the minimum AoA, which also corresponds to the minimum aerodynamic GR but also the maximum airspeed. As rigging angle and, hence, AoA are increased, the GC is increased while the airspeed is decreased. The consequence of this inverse relationship between aerodynamic glide angle and airspeed in terms of GC over ground is shown conceptually

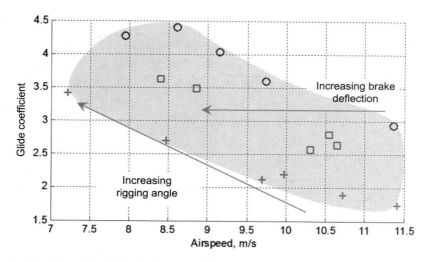

Fig. 8.39   GC and airspeed envelope for a low AR canopy [Ward et al. 2011b].

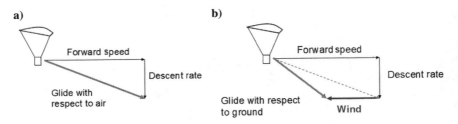

Fig. 8.40   Glide slope angle with respect to a) air and b) over ground [Ward 2012].

in Fig. 8.41. In a zero-wind environment, increasing rigging angle results in an increasing GC over ground. As the wind is increased, the effect of variable rigging angle on GC over ground is diminished. In fact, there is a particular wind speed for which the variation in rigging angle will produce no change in the GC over ground. Beyond this wind speed, the effect of rigging angle on glide slope over ground is reversed, so that the maximum GC over ground is now achieved at the minimum rigging angle setting.

The use of symmetric TE brake deflection to provide airspeed control in conjunction with variable rigging angle can dramatically improve the range of control of GC over ground. Figures 8.42–8.45 show the range of GC over ground that can be achieved with rigging-angle variation alone and with rigging-angle variation in conjunction with symmetric brake deflection. These results are based on the flight characteristics of the medium AR canopy used

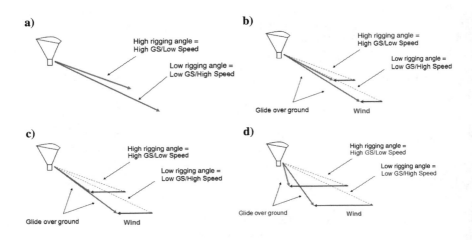

Fig. 8.41   Behavior of GC over ground vs rigging angle [Ward 2012].

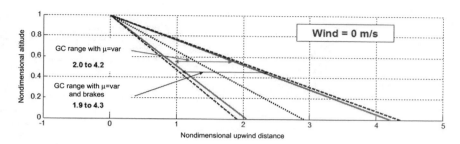

Fig. 8.42 Range of GC over ground in no wind [Ward 2012].

for the variable rigging angle flight tests just discussed. As shown in Figs. 8.42 and 8.40, the variation of the canopy rigging angle provides a significant range of control of GC over ground in zero and light wind conditions, while the deflection of TE brakes provides almost no effect on GC control over ground. This is because TE deflection provides a change in airspeed with little change in the aerodynamic glide angle of the parafoil canopy.

However, as shown in Fig. 8.44, when the wind increases to the point where the variation of rigging angle produces no change in the GC over ground, the use of TE brake deflection to alter speed can produce a significant range of control in the GC over ground. Furthermore, as shown in Fig. 8.45, the range of GC control in stronger wind conditions can be dramatically increased by the use of trailing brakes in conjunction with rigging-angle variation.

To summarize, the variation of canopy rigging angle produces a significant change in the aerodynamic GC of a parafoil, but the actual change in GC over ground can be dramatically reduced in certain wind conditions. The use of TE brake deflection in conjunction with rigging-angle variation can compensate for the reduced control authority in these wind conditions can dramatically improve the control authority in windy conditions.

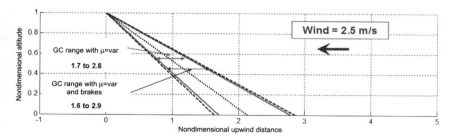

Fig. 8.43 Range of GC over ground in light wind [Ward 2012].

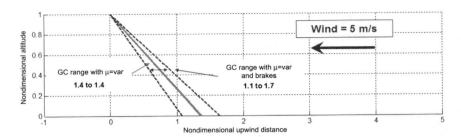

Fig. 8.44   Range of glide slope over ground in moderate wind [Ward 2012].

8.2.4   LARGER PADS MECHANIZATION

The majority of this chapter has focused on methods of controlling GA for the purpose of achieving improved terminal accuracy. However, there is a required precondition for terminal accuracy, which is that the PADS achieve the target area in the first place. Most all of the traditional PADS have been configured with a fixed parafoil rigging angle, deployment brakes, TE control, and are designed as a compromise of deployment reliability, glide capability, and landing accuracy. With rigging-angle control, the PADS can be adapted to the optimal rigging angle depending on the phase of the mission or the conditions encountered.

The GA control has the ability to provide a much wider performance envelope for the parafoil allowing overall system performance to be increased in terms of deployment reliability, standoff range, adverse wind penetration, and overall landing accuracy. For example, the ability to control airspeed through glide trim control has the primary benefit of increasing the probability of achieving the target area in the face of uncertain and unexpected wind conditions. This is a primary factor in overall achievable offset range, which is limited by wind uncertainty. If wind uncertainty is large with respect to PADS airspeed, then offset ranges have to be limited to ensure sufficient altitude to achieve the target. The

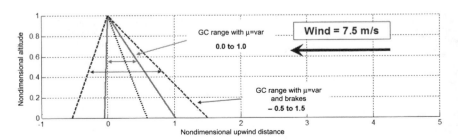

Fig. 8.45   Range of DC over ground in strong wind [Ward 2012].

ability to increase airspeed increases reliable standoff range and is a fundamental part of PADS mission planning. Within limits, rigging-angle control for deployment and glide angle trim control for range both play an important role in increasing mission success rates.

The TE control does have an effect on glide as well, and often canopy designers build in a certain amount of initial deflection to improve full glide L/D. However, the combined design objectives for safe deployment and high glide can only be met to a limited extent through the use of a fixed rigging-angle design with TE preset. As has been demonstrated in Fig. 8.6, standard TE control has limited effect on glide control, and terminal accuracy suffers. For these reasons, we are lead to the other possibilities of lift and drag augmentation. In fact, there are other fundamental limitations of TE control. These limitations have the largest impact on autonomous PADS systems and have profound effects on overall mission reliability of the PADS. We will describe these issues in order to provide more motivation for making changes to the standard PADS TE control.

While TE control is almost universally employed in the skydiving and paragliding sport markets, there are many limitations to this type of control when considering parafoil deployment and flight conditions and the payload configurations of the wide range of weight classes of PADS. It is well understood that TE braking has various control deficiencies when the PADS system does not exhibit good symmetry and balance. A primary influence on parafoil asymmetry that can manifest itself in several forms is canopy damage caused during deployment. Parafoil deployment is a high-energy event owing in large part to the high forces leading to high deceleration that occurs as the parafoil is inflating. There are a range of things that can occur including broken steering lines that cause a total or partial loss in steering control, asymmetric canopy deployment damage that causes differential drag on the parafoil resulting in a directional steering bias, and additional elements of rigging errors that can either cause deployment damage themselves or impede the directional control (e.g., disconnected or reversed control lines).

Even if deployment damage has not hindered the ability of the PADS to glide, depending upon the effectiveness of drag based steering of TE controlled systems, even relatively minor damage, can render the PADS incapable of achieving the target. For example, one actuator might be completely ineffective in overcoming a differential turn due to damage induced differential drag. The same damage will typically induce a dramatically increased effectiveness in the other turn direction. Guidance and control systems can be designed to adapt to these situations to optimize what residual controllability exists, but the ability to overcome a loss in control in a system that is already subject to a minimum of controllability usually has a strong diminishing return and can be easily catastrophic to the mission.

The second class of influence is through the weight and CG balance of the payload being carried under the parafoil. If the CG of the payload is not directly under the center line of the parafoil, the canopy will be under the influence of a roll moment that must be countered with a yawing moment from the TE control. Lateral acceleration due to roll angle is proportional to lift whereas lateral acceleration due to yaw is proportional to differential drag. Because L/D is typically in the 3:1 range, it is easy to see how CG imbalance can play havoc with TE brake. During early testing of the Orion PADS, it was discovered that it was possible for the payload to start spinning during flight resulting in a yaw moment coming through the twisted payload risers applying a yaw moment to the parafoil that could not be overcome from TE steering commands. This is the reason that most of all but the smallest PADS systems typically have a payload swivel to avoid this payload wrap-up torque.

A worse situation exists when the payload CG may be centered, but tends to rock under the payload due to compliance in the primary riser system that connects the payload to the parafoil. This is generally termed *weight shift* and can happen if the payload can roll with respect to the parafoil frame of reference (i.e., rigging compliance). With larger payloads and large control line forces, the side of the payload that has the active actuator can lift due to the control line force required to deflect the TE. The lifting of one side of the payload causes a weight shift of the entire payload weight to the opposite turn direction. With sufficient compliance, the payload tilt may completely counteract the intended TE turn input. The worst situation in this regard is in a typical configuration where a vehicle is being carried on a platform where the rigging attachment points are below the CG of the vehicle. This is an unstable configuration that is only controlled by cross bracing of the supporting rigging lines. When excessive compliance exists with these low CG attachment points, the payload can literally lurch from one maximum tilt angle to a maximum tilt angle on the other side. The results can be quite stunning especially when observing a large 6,800-kg (15,000-lb) PADS system from near distance.

One particular GA control approach, which embodies the high lateral control authority of differential lift control as well as symmetrical control for longitudinal trim, is the Variable AoA Control System (VACS) proposed by Moore [2012]. At the core of the technology is a pulley riser system that under automatic control can effect both symmetrical and lateral AoA changes to the PADS. A typical configuration with four parafoil line groups is shown in Fig. 8.46.

VACS is generally applicable to any size and weight range of PADS and easily extends to the largest systems that have been tested. VACS does not require any decelerator changes although arrangement of line groups with the appropriate pulley sets is required for minimizing airfoil distortion as rigging angle is adjusted. In addition to rigging-angle control, by appropriate design of the pulley ratios, parafoil camber can also be dynamically changed as part of the in-flight GA control. With dual actuator controls and appropriate pulley arrangements, the technology can be applied to any flexible suspended weight gliding decelerator. For example, by the selection of 2:1, 3:1, and 6:1 pulley ratios corresponding to the A, B, and C riser respectively, the equally spaced chord-wise line groups can be controlled in length to accurately alter the effective rigging angle. In the configuration shown, the C line 6:1 is implemented by cascading the 2:1 C line

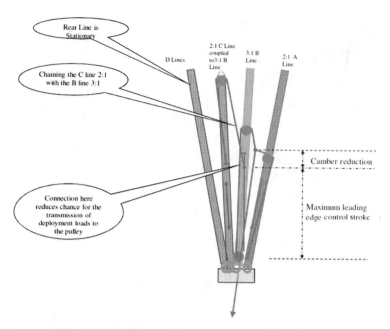

Fig. 8.46   Typical parafoil VACS with A-B-C-D line groups.

off of the 3:1 B line. Additionally, proper placement of the A group pulley with respect to the B line top pulley affords the opportunity to alter the parafoil camber as the single control line is retracted. Flare is provided by simply releasing the previously stored control load. The construction of such a system seems to be simple, reliable, and cost effective. During deployment, a slackened control line removes the entire pulley riser system from the deployment load path.

The VACS technology is a derivative of foot-controlled "speed systems" that are widely used on sport paragliders, an example of which is shown in Fig. 8.47. Speed systems have evolved from the more basic linear rigging angle control mechanisms to rigging systems that are integral components of refined high L/D gliders. The sport speed systems have fractional line ratios designed to affect AoA and camber working in concert with the canopy line groups. The speed system and canopy are designed together for optimum high glide performance. Paraglider operators typically employ the speed system in combination with TE controls. Using arms, feet, and weight shift, the pilot is able to better effect the glider and optimize the maneuvering and glide performance demanded during sport contests.

A pilot-operated speed system is primarily used for rigging angle control. In contrast, VACS for PADS applications typically benefit from using the speed system for both longitudinal and lateral control. A typical configuration would

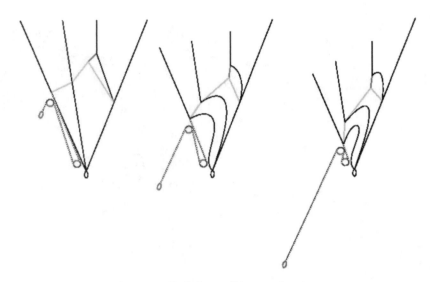

Fig. 8.47   Typical paraglider speed system.

have only two actuators, one actuator to control the left and one for the right side
AoAs of the parafoil. The VACS controls is typically applied to a parafoil with
spanwise line groups and other considerations depending upon the desired
AoA and camber changes. With a dual-actuator VACS, the combination of
both symmetrical and differential AoA control reduces the need for TE controls,
and this traditional control can be eliminated entirely.

Figure 8.48 shows other potential configurations, each having some advantages and disadvantages. Figure 8.47a represents pivoting about a center line,
which requires opposite controls (front and rear), Fig. 8.47b shows pivoting
about a front line that requires power to increase AoA for flare, and Fig. 8.47c
depicts pivoting about a rear line allowing unpowered release to increase AoA
for flare. Figure 8.47c also represents the system just described in Fig. 8.46.

Because VACS lateral control creates a differential AoA to induce a differential
lift, a strong roll moment is induced in the gliding decelerator. This strong control
authority of VACS is due to the AoA change of the entire side of the wing under
control. This type of control was investigated by Goodrick [1984], and comparison with the TE control is shown in Fig. 6.27. VACS is capable of overcoming
virtually any canopy asymmetry or payload balance effect described earlier. For
example, to initiate a positive turn (to the right), the right actuator pulls on the
right pulley system reducing the AoA of the right ride (Fig. 8.49). This reduced
AoA reduces the lift on that side resulting in a differential lift with positive
(clockwise moment). A small reduction in the AoA creates a large lift differential
and corresponding roll moment.

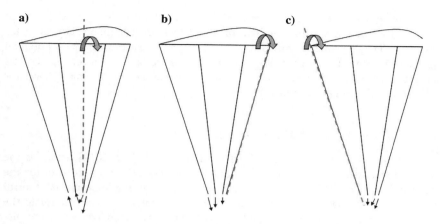

Fig. 8.48   Rotation axis tradeoffs.

While control forces for the VACS actuators are expected to be higher than those of TE braking (e.g., at least twice as much), the control stroke of VACS should be much less (e.g., one-quarter), and so the average power requirement for just lateral control is expected to be half as much as lateral TE control. Lateral control is also expected to be enhanced as direct roll-angle control is faster than skidding yaw turns, which induce coordinated turn roll angles. The typical actuator power requirement for a TE system is dominated by symmetrical control and especially if a dynamic flare is required. Symmetrical

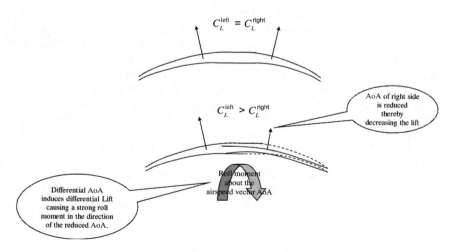

Fig. 8.49   Differential AoA turn control.

trim changes to VACS should be similar to lateral turn changes as the speed of control is reduced to one-half of the lateral requirement. However for dynamic flare, a properly configured VACS PADS relies completely on potential energy with no actuator control requirements. The release of the actuator creates an immediate AoA change with no actuation power. This is the main driver for selecting the configuration in Fig. 8.48c.

8.2.5   TERMINAL RIGGING-ANGLE CONTROL

This section, based on slightly edited Chapter 6 of [Ward 2012], explores the ways to implement variable rigging angle and brakes deflections control for the terminal stage of PADS descent. The simplest strategy for longitudinal control in terminal approach is to implement a PID controller that will modulate the canopy rigging angle to maintain an intercept path with IPI. Given a current distance estimate from IPI ($\Delta x = |x - x_{IPI}|$) and altitude estimate ($\Delta z = |z - z_{IPI}|$), the error between the desired glide path to IPI and estimated GC is determined for a particular time instant k

$$e_{GC,k} = \Delta x_k/\Delta z_k - \dot{x}_k/\dot{z}_k \qquad (8.16)$$

Then, the PID controller in the discrete form can be implemented as

$$\delta_{\mu,k} = K_p e_{GC,k} + K_d(e_{GC,k} - e_{GC,k-1})/\Delta t + K_i \sum_{j=0}^{k} e_{GC,j}\Delta t \qquad (8.17)$$

[compare a discrete PI controller of Eq. (8.11)].

An improved strategy, however, may utilize the coupling of rigging angle and symmetric brake to attempt to control airspeed and GC simultaneously. This later strategy can be separated into two distinct problems: 1) developing an optimal trajectory of GC and airspeed and 2) tracking this trajectory in time using a coupled controller on the rigging angle and symmetric brakes. To maximize the ability of the system to respond to sudden changes in the wind, it is desirable to maintain the vehicle on a nominal flight path with the controls in the center of their ranges. The optimal trajectory can then by defined as the trajectory that will take the system from its current state to a nominal glide path to the target with the minimum control input.

This problem is formulated by assuming a linear relationship between the control inputs (rigging angle δ_μ and symmetric brake δ_s) and the desired outputs (horizontal component of an airspeed V_h and the air-relative GC)

$$V_h = \partial V_{h,s}\delta_s + \partial V_{h,\mu}\delta_\mu + V_0 \qquad (8.18)$$

$$GC_a = \partial GC_{a,s}\delta_s + \partial GC_{a,\mu}\delta_\mu + GC_0 \qquad (8.19)$$

The nominal forward speed and GC with the controls centered are denoted V_0 and GC_0, respectively. The speed and GC over ground on final approach (when PADS glides precisely against the wind) are related to the air-relative quantities by the wind speed W so that using simple planar geometry yields

$$V = V_h - W \tag{8.20}$$

$$GC = GC_a(1 - W/V_h) \tag{8.21}$$

The nominal forward speed and GC over ground corresponding to zero control input are

$$V_{nom} = V_0 - W \tag{8.22}$$

$$GC_{nom} = GC_0(1 - W/V_0) \tag{8.23}$$

Similar to Eq. (8.19), the GC over ground is related to the control inputs as

$$GC = \partial GC_s \delta_s + \partial GC_\mu \delta_\mu + GC_{nom} \tag{8.24}$$

where the corresponding sensitivities can be defined using Eq. (8.21) as follows:

$$\partial GC_\mu = \partial GC_{a,\mu}(1 - W/V_h) + \partial V_{a,\mu} GC_a W/V_h^2 \tag{8.25}$$

$$\partial GC_s = \partial GC_{a,s}(1 - W/V_h) + \partial V_{a,s} GC_a W/V_h^2 \tag{8.26}$$

For a commanded GC over ground GC_{cmd}, the optimal control inputs result from addressing the following minimization problem:

$$\min(\delta_\mu^2 + \delta_s^2), \quad \text{such that } GC_{cmd} - GC_{nom} = \partial GC_s \delta_s + \partial GC_\mu \delta_\mu \tag{8.27}$$

The solution to this problem is given by

$$\delta_s = \frac{(GC_{cmd} - GC_{nom})\partial GC_s}{\partial GC_s^2 + \partial GC_\mu^2}, \quad \delta_\mu = \frac{(GC_{cmd} - GC_{nom})\partial GC_\mu}{\partial GC_s^2 + \partial GC_\mu^2} \tag{8.28}$$

The corresponding optimal forward speed and GC commands are given by substitution of the optimal controls (8.28) into the assumed linear models (8.18) and (8.19)

$$V_{h,cmd} = (GC_{cmd} - GC_{nom})\frac{\partial V_{h,s}\partial GC_s + \partial V_{h,\mu}\partial GC_\mu}{\partial GC_s^2 + \partial GC_\mu^2} + V_0 \tag{8.29}$$

$$GC_{a,cmd} = (GC_{cmd} - GC_{nom})\frac{\partial GC_{a,s}\partial GC_s + \partial GC_{a,\mu}\partial GC_\mu}{\partial GC_s^2 + \partial GC_\mu^2} + GC_0 \tag{8.30}$$

Equations (8.29) and (8.30) are used to determine optimal forward airspeed and GC commands based on a commanded GC over ground, which aims to bring the PADS onto a nominal flight path to the target

$$
GC_{com} = \begin{cases} GC_{nom}(1 - e_{GC}^2) + e_{GC}^2 GC_{max}, & e_{GC} < 0 \\ GC_{nom}(1 - e_{GC}^2) + e_{GC}^2 GC_{min}, & e_{GC} \geq 0 \end{cases} \tag{8.31}
$$

Equation (8.31) utilizes a normalized error between nominal and required GC

$$
e_{GC} = 2 \frac{GC_{nom} - \Delta x / \Delta h}{GC_{max} - GC_{min}} \tag{8.32}
$$

When the glide path error e_{GC} is 0, PADS is on an intercept course with the target on the nominal glide path. When the glide path error is 1, PADS will hit the target with the controls set for minimum GC over ground, and when the glide path error is -1, the system will hit the target with the controls set for maximum GC over ground. To minimize control inputs near the nominal glide path, the GC commands are made proportional to the square of the glide path error. In practice, an additional parameter e_{sat} is used define the magnitude of glide path error at which the controls saturate. This parameter is set to a value less than one so that the controls will saturate before the system reaches the minimum and maximum GC boundaries. Hence, the corrected error becomes

$$
\tilde{e}_{GC} = \begin{cases} \min(e_{GC}/e_{sat}, -1), & e_{GC} < 0 \\ \max(e_{GC}/e_{sat}, 1), & e_{GC} \geq 0 \end{cases} \tag{8.33}
$$

and Eq. (8.31) used $\tilde{e}_{GC}$ instead of e_{GC}.

An example of the GC commands generated with this method is shown in Fig. 8.50. For this scenario, the minimum GC over ground is set at 1, the maximum is set at 3, and the normalized error at which the controls saturate e_{sat} is set at 0.5. The plot shows how the commanded GC is generated to bring the system smoothly onto an intercept course with the target on the nominal glide path. If the system is outside the boundaries set by the e_{sat} parameter, the maximum or minimum glide over ground is commanded to bring the system back towards the nominal glide path, and if the system is outside the minimum and maximum GC boundaries, it will not be able to reach the target.

Sample mappings were created to fit the flight-test data vehicle used for autonomous landings. Contours of constant GC over ground for these example mappings are shown in Fig. 8.51 vs normalized rigging angle and brake deflection. There is one plot corresponding to each of 0-, 2-, 4-, and 6-m/s (0-, 3.9-, 7.8-, 11.7-kt) wind speeds. These normalized deflections

$$
\tilde{\delta}_s = 2(\delta_s - \delta_{s,min})/(\delta_{s,max} - \delta_{s,min}) - 1
$$
$$
\tilde{\delta}_\mu = 2(\delta_\mu - \delta_{\mu,min})/(\delta_{\mu,max} - \delta_{\mu,min}) - 1 \tag{8.34}
$$

span the range -1 to 1. In practice, the limits of rigging angle are actually a function of the level of brake deflection. The limit on rigging angle is assumed to be

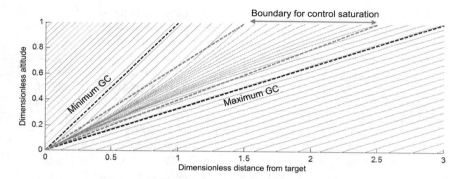

Fig. 8.50   Visualization of commanded GC logic [Ward 2012].

linear function of brake input

$$\delta_{\mu,\min} = \delta_{\mu 0} + \delta_{\mu\delta s}\delta_s \qquad (8.35)$$

These plots demonstrate how very different the effect of the control inputs on GC over ground can be in different wind conditions. As just stated, the effect of

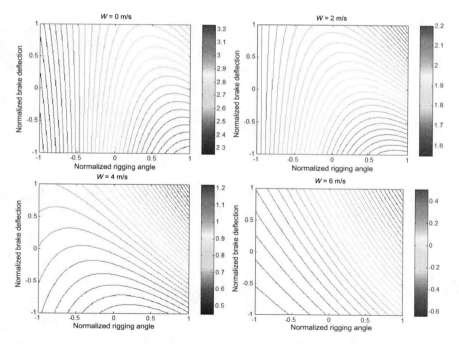

Fig. 8.51   Glide slope over ground vs rigging angle and brake at increasing wind levels [Ward 2012].

rigging angle on GC over ground can be completely opposite in different wind conditions. There is a complex interaction between rigging angle and TE brake in determining the GC over ground in different wind conditions.

It is also clear from the plots in Fig. 8.51 that a wide range of control inputs can produce the same GC over ground. The goal is to choose the "optimal" set of control inputs, which achieves the commanded GC over ground. It is an interesting area for further research to explore different definitions of the "optimal" control logic for inverting the GC mapping. For instance, the controls that either minimize or maximize the airspeed for a specified GC over ground could be chosen. Alternatively, the controls could be chosen such that the system is able to move from the maximum to minimum GC setting in the minimum time. The controls are restricted to lie on a line drawn on the GC mapping from the maximum glide point to the minimum glide point. This ensures that the full range of glide over ground is achieved, and the problem of inverting the nonlinear mapping to obtain the controls to achieve a given GC command is reduced to a line-search problem. The attraction of this approach is the simplicity of implementation and minimal computation time required.

These lines of optimal control inputs are plotted on top of the GC contours in Fig. 8.52. The line-search problem is solved with successive three-point

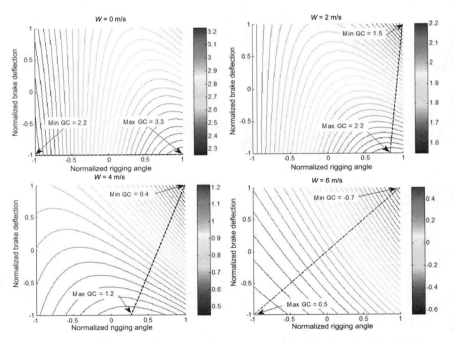

Fig. 8.52   Range of "optimal" control inputs at increasing wind levels [Ward 2012].

quadratic approximations. In fact, because the minimum and maximum GC configurations always lie on the boundaries of the mapping, these quantities can be determined using the same line-search algorithm.

Simulation results of an example final approach from an altitude of 100 m (328 ft) are shown in Fig. 8.53 Two cases are shown: one case uses the coupled GC controller just described, and the second case has the controls fixed. The minimum, maximum, and nominal GC lines shown in Fig. 8.53 were determined based on the average winds during the final approach. The wind profiles used for the example approach are shown in Fig. 8.54. The average wind is 5 m/s (9.7 kt), and the standard deviation of the vertical wind component used for the Dryden turbulence model is 0.8 m/s (1.6 kt). Initially, the system is nearly on the nominal glide path with the controls centered. Once the altitude drops below approximately 40 m (131 ft), the head wind weakens, and, to make matters worse, a positive vertical wind component picks up. As shown by the controls-fixed flight path, this combination of changes in the wind would normally cause the system to overshoot by nearly 40 m (131 ft). As shown in Fig. 8.55, the GC controller reacts to this change in the wind by quickly applying a large amount of TE brake and increasing the rigging angle to the maximum setting. This causes a large reduction in forward flight speed and a significant

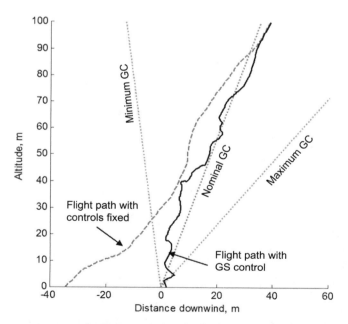

Fig. 8.53   Final approach trajectories with and without (coupled) GC control [Ward 2012].

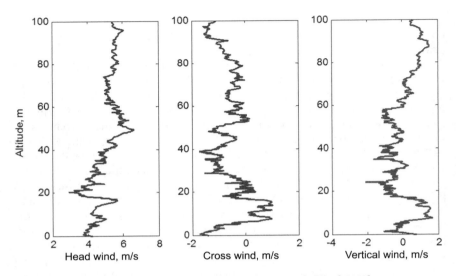

Fig. 8.54   Wind profiles for example approach [Ward 2012].

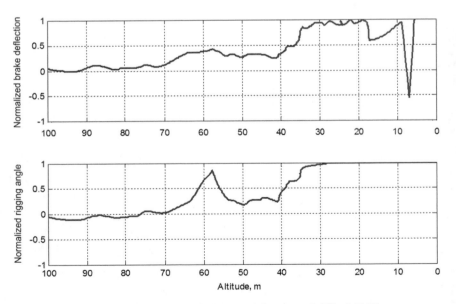

Fig. 8.55   Control inputs during example approach [Ward 2012].

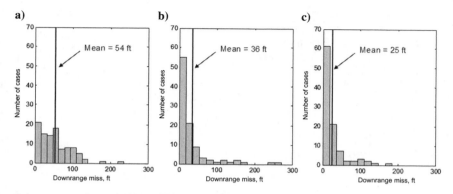

Fig. 8.56 Downrange miss distance histograms for a) no, b) δ_μ, and c) δ_μ/δ_s control [Ward et al. 2011b].

reduction in the GC over ground, allowing the system to stay on the nominal glide path to the target.

To compare the landing performance with and without GC control, a series of Monte Carlo simulations were carried out. To generate a large quantity of simulation results quickly, only the final approach portion of the flight is simulated. The vehicle is placed at an initial altitude of 100 m (328 ft) downwind of the target. The downwind distance is calculated so that the system will glide exactly to the target at the nominal control setting. This distance is shifted based on the mean wind speed, and, to generate a realistic starting condition, this wind shift is calculated with a random error in the assumed wind speed of ± 0.5 m/s (± 1 kt). Three cases, with no GC control, PID control of Eq. (8.17), and coupled rigging angle/brakes control of Eq. (8.28), were considered.

TABLE 8.9   CEP/AVERAGE MISS DISTANCE AS A FUNCTION OF A WIND SPEED [WARD ET AL. 2011b]

	Wind = 0	Wind 3.5 m/s (6.8 kt)	Wind 5.25 m/s (10.2 kt)	Wind = 7 m/s (13.6 kt)
No GC control	12.5/14.6 m (41/48 ft)	17.4/18.3 m (57/60 ft)	18/21 m (59/69 ft)	19.2/22.9 m (63/75 ft)
δ_μ PID GC control	5.2/6.4 m (17/21 ft)	4.3/7.3 m (14/24 ft)	7.3/12.8 m (24/42 ft)	14/17.1 m (46/56 ft)
Coupled δ_μ/δ_s GC control	4.9/6.1 m (16/20 ft)	4.6/7.3 m (15/24 ft)	7.3/17.4 m (24/57 ft)	21.6/37.5 m (71/123 ft)

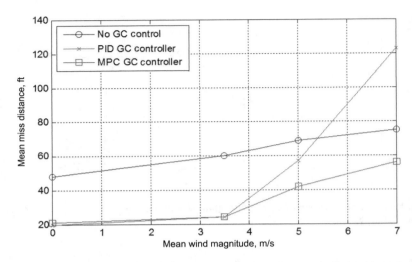

Fig. 8.57 Simulated landing accuracy vs mean wind speed using PID (δ_μ) and coupled (δ_μ/δ_s) controllers [Ward et al. 2011b].

For the first set of simulations, simulated landing accuracies were generated assuming a uniformly varying mean wind profile from zero up to the nominal flight speed of the vehicle. One-hundred cases were run for each control strategy. The simulated impact points are compared in Fig. 8.56. A second series of cases was run at particular mean wind speeds to demonstrate the benefit of using the symmetric brakes in conjunction with rigging angle to control GC. The results are shown in Table 8.9 and Fig. 8.57.

To summarize, a nonlinear proportional GC controller was developed to utilize TE deflection in conjunction with rigging-angle variation to control GC over ground during a straight-line final approach. The controller consists of two parts, a GC over ground command logic designed to keep the system on a nominal flight path to the target and an inversion of the control to GC mapping to obtain the optimal control input based on the commanded GC and wind conditions.

8.3 GLIDE-ANGLE CONTROL USING UPPER-SURFACE SPOILERS

This section, based on slightly edited work by Gavrilovski et al. [2012] and Ward and Costello [2013], explores yet another way of controlling PADS' GC using the upper-surface aerodynamic spoilers. To enable it, a spanwise slit is introduced across a number of cells in the center section of the upper surface of the canopy [Higgins 1979]. The slit location is determined using computational-fluid-

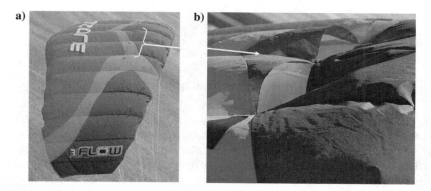

Fig. 8.58   a) Unactuated and b) actuated upper-surface slit spoiler on test canopy [Ward et al. 2011a].

dynamics simulations as a guide and corresponds to the minimum pressure point on the upper surface. All of the cells that contain a slit have a control line attached to the LE side of the slit. These lines pass through the bottom surface and run down to the payload where they are controlled by winches. When the control line is actuated, the material ahead of the slit is deflected downward. Because of the internal pressure of the canopy, the remainder of the cell on the TE side of the slit remains unperturbed. This causes an airflow bubble on the upper surface, which distorts the airflow, much like conventional aircraft spoilers except that this slit spoiler configuration uses vented ram air as a spoiler rather than a mechanical flap. When the slit is not actuated, the spanwise tension in the canopy is sufficient to keep the slit closed. Figure 8.58 shows the slit control mechanism as implemented on the 2.7-m^2 (29-ft^2) canopy of a Buckeye PADS used for testing. This section introduces the aforementioned concept and investigates a capability using only upper-surface spoiler control (i.e., excluding traditional TE flaps). This PADS control method is evaluated in its ability to follow spatial reference trajectories and ensure GC control on the final approach stage along with sufficient control to produce a required turn rate.

8.3.1   SYMMETRIC SPOILER CONFIGURATIONS

A parametric study of symmetric spoiler configurations was performed by varying the construction, span, and chordwise position of the upper-surface slits. The construction of the slits was varied by moving the attachment points from the LE side of the slits to the TE side and by testing the effect of adding sealing flaps to the slits. All of these configurations have slits across the center eight cells at the 30% chord location. The span of the spoilers was varied by testing configurations with slits across the center two, four, six, and eight cells, resulting in spoiler widths of 14, 28, 42, and 56% of the constructed span. All of these slits were made at the

a)   b)

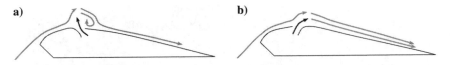

Fig. 8.59   a) LE vs b) TE deflection of slit [Gavrilovski et al. 2012].

30% chord line. The effect of chordwise location was tested using configurations with slits across the center eight cells at 15, 30, and 50% chord locations.

A limit on control travel is imposed by the local cell height at the spoiler location. When the upper surface is maximally actuated, the upper and lower surfaces of the canopy come in contact. It was observed in the flight test that actuation up to approximately 90% of this limit began to cause large canopy deformations (of the lower surface, as well as the upper surface) and possibly stall, so the actual limit on spoiler deflection is configuration dependent and is discussed next. All spoiler deflections are presented as fractions of the cell height at the chordwise location of the spoiler slit. The cell heights at the 15, 30, and 50% chord locations are 17.2 cm (6.8 in.), 17.7 cm (7 in.), and 13.5 cm (5.3 in.), respectively.

The first basic construction variation tested was the location of the spoiler actuation line attachment. The actuation of the LE side of the slit causes the fabric in front of the slit to deflect downward while leaving the fabric aft of the slit in its original shape, as shown in Fig. 8.59a. This causes the air to be vented from the canopy in an opposing direction to the freestream airflow. The actuation of the TE side of the slit causes the fabric aft of the slit to be deformed while leaving the fabric in front of the slit in its original shape, venting air along the direction of the freestream flow. This situation is depicted in Fig. 8.59b.

The second construction variation was the addition of internal flaps to help seal the slits when the spoilers are not actuated. The sealing flaps effectively extend the fabric on the LE side of the slit as shown in Fig. 8.60. When the spoilers are not actuated, the internal pressure presses the flap against the slit and creates a seal, preventing any air from venting into the freestream and disturbing the airflow over the upper surface. When actuated, the shape of the opening is very similar to that of the simple slit, except that slightly more actuation is required to make up for the extra length of the flap.

Flight-test results comparing these variations in spoiler construction are shown in Fig. 8.61. Overall, the deflection of the spoiler produces a reduction in

a)   b)

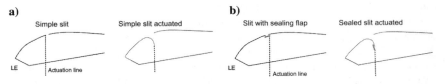

Fig. 8.60   Addition of b) sealing flaps to a) simple slits [Gavrilovski et al. 2012].

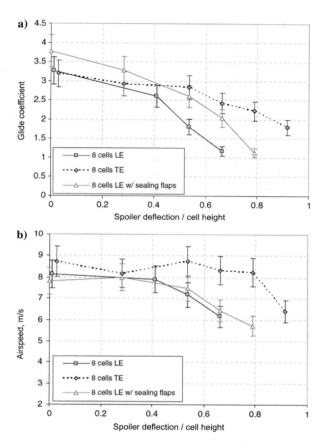

Fig. 8.61 Effects of varying spoiler construction [Gavrilovski et al. 2012].

lift, resulting in a lower value for GC. Glide-ratio reduction occurs in a smooth fashion up to the maximum deflection. The descent rate is increased, and the forward velocity is decreased, resulting in a net reduction in airspeed.

Comparing the results for LE vs TE actuation, it is clear that actuation of the LE of the slit is more effective in reducing GC. The actuation of the LE allows a 64% reduction in GC compared to a 43% reduction when actuated the TE of the slit. Also, the canopy maintains a significantly higher airspeed under the actuation of the TE until the actuation reaches 90% of the cell height, where there is a sudden drop in airspeed. This is most likely due to large-scale canopy deformation caused by the extreme actuation of the spoiler.

Comparing the simple slit and sealed slit construction with the actuation of the LE side, it is apparent that the nominal GC corresponding to zero spoiler actuation is slightly reduced due to air leaking through the simple slits. When sealing flaps are incorporated into each slit, the nominal GC loss is recovered, and the zero spoiler deflection GC increases from 3.2 to 3.8. As expected, the minimum GC achieved with the sealed slit design is the same as that achieved with the simple slits, though a slightly higher spoiler deflection is required to take up the extra fabric introduced by the sealing flaps. By improving the seal of the closed slits, the sealing flaps increase the effective control authority of the spoilers from a 64% reduction in GC to a 70% reduction.

The effect of spoiler span on the GC and airspeed control authority is shown in Fig. 8.62. All of these configurations use simple, unsealed slits with LE actuation. As expected, the effectiveness of the spoiler increases as the span is increased. With two cells, the nominal GC of 3.2 can be reduced to 2.5, and with eight cells

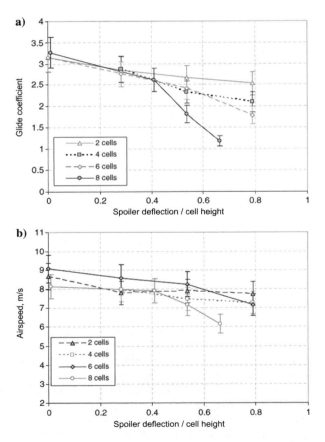

Fig. 8.62   Effect of varying spoiler span [Gavrilovski et al. 2012].

the GC can be reduced to 1.1. The general trend in airspeed is that there is a reduction in airspeed proportional to the reduction in GC. With eight cells actuated to produce a 64% reduction in GC, there is a corresponding reduction in airspeed of 25%. For comparison, with the heavy deflection of the TE brakes, the same canopy experiences a 30% reduction in airspeed with only a 0.5% reduction in GC.

Figure 8.63 depicts the growth in spoiler effectiveness by increasing the number of cells containing actuated slits. The maximum spoiler deflection for the two- to six-cell configurations was determined by the observation of significant canopy deformation. For the eight-cell configuration, the maximum spoiler deflection was determined by the flight condition. At lower GCs, the system became difficult to control and prone to stall, implying that a minimum practical GC limit for this particular canopy was reached. This means that increasing the span of the spoiler beyond eight cells will not increase the range of GC control authority. In addition, the canopy becomes increasingly sensitive to small variations in the slit and actuation line construction as the spoiler span is increased, which suggests that, practically speaking, the spoiler span should not be increased beyond the minimum span necessary to reach the minimum GC limit of a particular canopy's flight envelope.

The final variation in symmetric spoiler configuration examined was the chordwise location of the upper-surface slits. The effect of moving the slit locations aft from 30 to 50% of the canopy chord on GC and airspeed is shown in Fig. 8.64. Both configurations had simple, unsealed slits across eight cells. The slits at 50% chord are less effective than the slits at 30% chord. A similar reduction in GC and corresponding reduction in airspeed are obtained with the slits at 50% chord, but a much higher spoiler deflection is required. The simple explanation is that the flow over a smaller portion of the wing is affected when the slits are moved aft, so the spoiler becomes less effective.

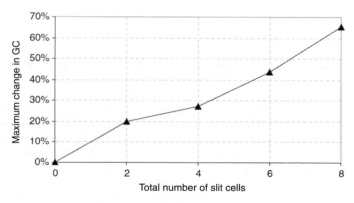

Fig. 8.63   Glide rate control authority vs spoiler span [Gavrilovski et al. 2012].

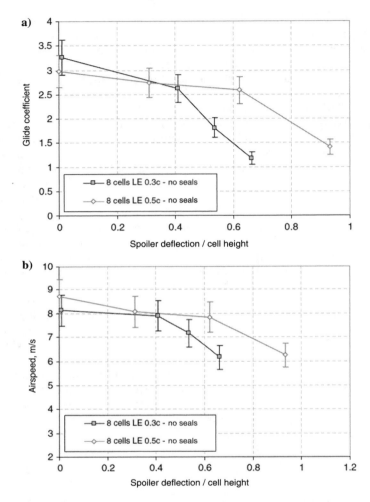

Fig. 8.64 **Effect of moving spoilers aft of nominal location [Gavrilovski et al. 2012].**

The slit locations were also moved forward to 15% of the canopy chord. This configuration was originally flown with simple, unsealed slits, but the small amount of air leakage and canopy deformation caused by the slits was so detrimental to the canopy performance that the system was unable to climb under power. With sealing slits added, the performance of the canopy improved enough to allow the system climb and obtain data. The effectiveness of the spoiler with slits at 15% of the canopy chord is compared to the spoiler with slits at 30% chord in Fig. 8.65. Both of these configurations use sealing flaps, but a maximum GC of only 3.0 is achieved with the slits at the 15% chord location,

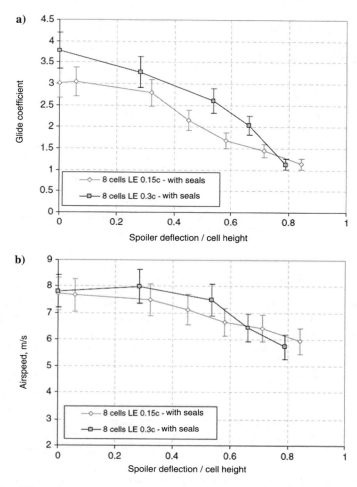

Fig. 8.65   Effect of moving spoilers forward of nominal location [Gavrilovski et al. 2012].

while the full nominal GC of 3.8 was achieved with the sealed slits at the 30% chord location. This indicates that the canopy is extremely sensitive to any modifications near the LE. Furthermore, under large deflections of the slits at the 15% chord location, the entire section of the canopy in front of the slit actually collapses, and the spoiler opening becomes the new ram-air inlet of the cell. This situation produces a similar minimum GC of 1.1 to that achieved with the slits at 30% chord, but the LE collapse results in a severely deformed canopy shape, which makes the system extremely difficult to control. The canopy does not recover from this collapse when the spoiler actuation line is released.

Fig. 8.66   Glide ratio control authority vs chordwise location [Gavrilovski et al. 2012].

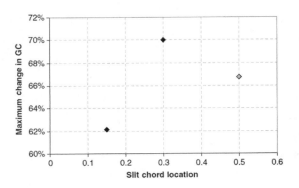

The overall effectiveness of the upper-surface spoiler in reducing GC as a function of chordwise location is summarized in Fig. 8.66. To make a fair comparison, the effectiveness of the spoiler with the slits at the 50% chord location was adjusted by assuming that the nominal GC of 3.8 would be reached with the addition of sealing flaps.

8.3.2   ASYMMETRIC SPOILER CONFIGURATIONS

Asymmetric upper-surface spoiler configurations were evaluated for their utility as a lateral control mechanism. All asymmetric spoiler configurations have a simple slit with LE control line attachments and no sealing mechanism. These configurations varied in size from one to four cells on one side of the canopy. Configurations with less than four cells were tested as "inboard" spoilers, where the modified cells begin at the centerline and count outward, and also as "outboard" spoilers, where the modified cells begin with the fourth cell from the centerline and count inward. Figure 8.67 shows the steady-state turn rate achieved vs spoiler deflection for all of the tested asymmetric spoiler configurations. Error bars are not shown in this plot because the estimated error for each data point is on the order of 1 deg/s, which is approximately the size of the markers used for the data points. Increasing the spoiler size produces a direct increase in turn rate authority and spoilers that are at a greater distance from the centerline ("outboard" vs "inboard" spoilers) generate higher turn rates at lower slit deflections. The control authority of the majority of the asymmetric spoiler configurations tested is sufficient to achieve a spiral dive. This condition produces a turn rate of 50 deg/s for this particular parafoil and payload system, which is considered as the upper limit of the useful turn rate.

The maximum turn rate achieved for each asymmetric spoiler configuration is plotted as a function of the number of actuated cells in Fig. 8.68. The ability to achieve a spiral dive indicates that a particular asymmetric spoiler configuration is able to generate the maximum practical turn rate, which also means that a particular asymmetric spoiler configuration is able to generate the same lateral control authority as the TE brakes. For this particular canopy, the actuation of the slits on only two canopy cells is required to achieve the full turn rate

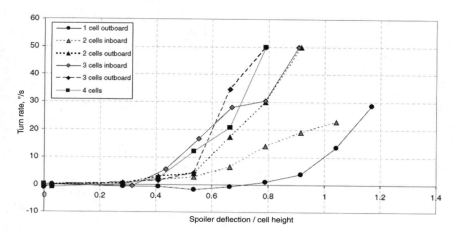

Fig. 8.67 Turn rate vs asymmetric spoiler deflection for different configurations [Gavrilovski et al. 2012].

capability of this particular parafoil and payload aircraft. Overall, the asymmetric actuation of the upper-surface spoilers is clearly an effective lateral control mechanism.

Asymmetric upper-surface spoiler configurations were evaluated for their utility as a lateral control mechanism. These configurations varied in size from one to four cells on one side of the canopy. Configurations with less than four cells were tested as "inboard" spoilers, where the modified cells begin at the centerline and count outward, and also as "outboard" spoilers, where the modified

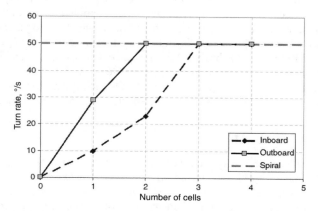

Fig. 8.68 Lateral control authority vs asymmetric spoiler configuration [Gavrilovski et al. 2012].

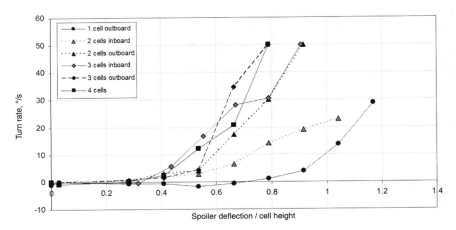

Fig. 8.69   Turn rate vs asymmetric spoiler deflection for different configurations [Gavrilovski et al. 2012].

cells begin with the fourth cell from the centerline and count inward. Figure 8.69 shows the steady-state turn rate achieved vs spoiler deflection for all of the tested asymmetric spoiler configurations. Increasing the spoiler size produces a direct increase in turn rate authority, and spoilers that are at a greater distance from the centerline ("outboard" vs "inboard" spoilers) generate higher turn rates at lower slit deflections. The control authority of the majority of the asymmetric spoiler configurations tested is sufficient to achieve a spiral dive. This condition produces a turn rate of 50 deg/s for this particular parafoil and payload system, which is considered as the upper limit of useful turn rate. The ability to achieve a spiral dive indicates that a particular asymmetric spoiler configuration is able to generate the maximum practical turn rate, which also means that a particular asymmetric spoiler configuration is able to generate the same lateral control authority as the TE brakes. For this particular canopy, actuation of the slits on only two canopy cells is required to achieve the full turn rate capability of this particular parafoil and payload aircraft. Overall, the asymmetric actuation of the upper-surfaces spoilers is clearly an effective lateral control mechanism.

8.3.3   SIMULATION RESULTS

Based on the studies presented in the two preceding sections, the longitudinal and lateral control effectiveness was tested on a model of Buckeye PADS (see Figs. 1.9 and 8.29) with a spoiler configuration shown in Fig. 8.70. As seen, the center eight cells of the canopy are actuated. The center four cells are used for longitudinal control only, while the outer four cells are used for both lateral and longitudinal control. In straight flight, all eight cells are actuated together to achieve

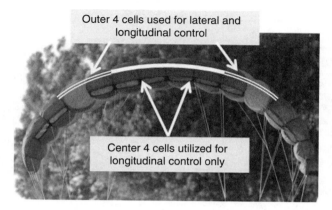

Fig. 8.70   Spoiler control configuration [Ward and Costello 2013].

GC control. A lateral control command is applied as a differential actuation of the outer cell groups.

Computer simulations were based on the six-DoF model of Sec. 5.1.5. Spoiler actuation in the simulation model produces changes to the lift-and-drag characteristics of the canopy. These control characteristics were set to match the flight-test data presented in the two preceding sections. Longitudinal actuation of the spoilers produces a nonlinear reduction in forward speed and an increase in descent rate as shown in Fig. 8.71.

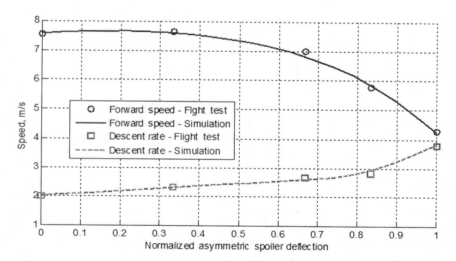

Fig. 8.71   Response to symmetric spoiler deflection [Ward and Costello 2013].

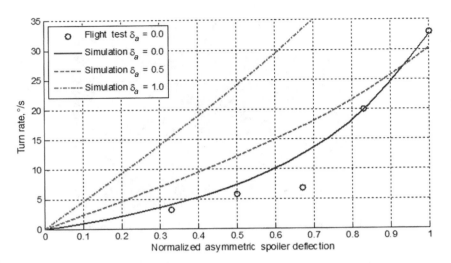

Fig. 8.72   Response to asymmetric spoiler deflection [Ward and Costello 2013].

Asymmetric deflection of the spoilers produces the steady-state turn rate behavior shown in Fig. 8.72. The turn rate response to asymmetric spoiler deflection
is nonlinear and heavily dependent on the level of symmetric spoiler deflection.

The guidance approach used for these simulations utilized the IDVD-based
final turn as described in Sec. 7.6 with MPC of Sec. 7.8.3. Lateral control was
provided by a nonlinear proportional controller tracking a desired heading
angle χ_{des} and heading angle rate $\dot{\chi}_{des}$. The controller uses an internal model of
the steady-state turn rate behavior as a function of asymmetric and symmetric
spoiler deflection to determine the appropriate control input

$$\dot{\chi}_{cmd} = K_P(\chi_{des} - \chi) + \dot{\chi}_{des} \tag{8.36}$$

$$\delta_a = f(\dot{\chi}_{cmd}, \delta_s) \tag{8.37}$$

Longitudinal control was based on the GC error of Eq. (8.33). For a particular
wind condition, there is a one-to-one mapping of spoiler deflection to GC over
ground, so the appropriate spoiler deflection can be determined uniquely by
inversion of the nonlinear, polynomial-based mapping for a given spoiler deflection. The GC over ground vs spoiler deflection behavior in a variety of wind conditions is shown in Fig. 8.73.

Two example simulation flights were generated with the wind profile shown in
Fig. 8.74. The wind aloft is blowing 4 m/s (7.8 kt) to the North on average, and the
ground wind is blowing 1.5 m/s (3 kt) to the North. Moderate turbulence is
included in all three axes. The transition from the lower to the upper wind
layers occurs over a region from 50 m (162 ft) to 75 m (246 ft) in altitude over
the ground.

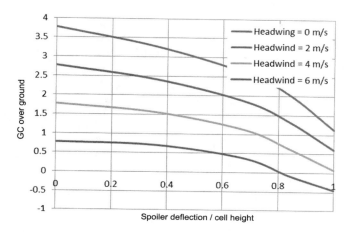

Fig. 8.73   Effect of spoiler on GC over ground [Ward and Costello 2013].

The first example flight utilized the spoilers for lateral control only, while the second flight utilized the spoilers for both lateral and longitudinal control. The ground tracks for the final approach of each trajectory are shown in Fig. 8.75, and a side view of the final approach trajectories is shown in Fig. 8.76.

The initially planned approach trajectory is shown as a dotted line in Fig. 8.76. Both systems set up on final approach so that they glide into the target facing

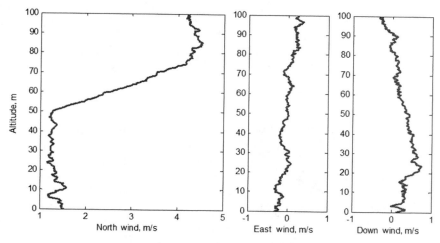

Fig. 8.74   Wind profile for example cases [Ward and Costello 2013].

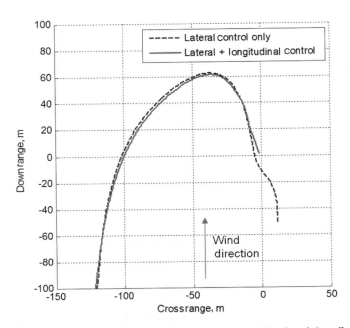

Fig. 8.75   Ground tracks for example approach trajectories [Ward and Costello 2013].

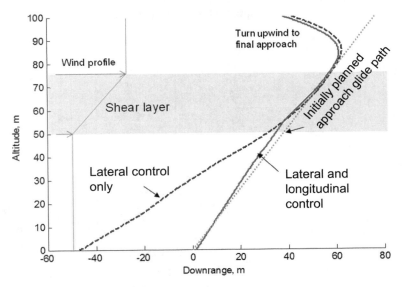

Fig. 8.76   Approach path profiles for example trajectories [Ward and Costello 2013].

into the wind on this planned approach path. As the wind speed decreases, the system begins to overshoot the target. Utilizing only lateral control, the system makes a slight s-turn to mitigate the overshoot, but the system is too low and too close to the target to compensate for the change in wind. For the case utilizing lateral and longitudinal control, the spoilers are simply opened to maintain the system on the desired glide path all the way down to the target. The control inputs during each of these example cases are shown in Fig. 8.77.

A set of Monte Carlo simulations was conducted over a variety of wind conditions to evaluate PADS landing precision utilizing the spoilers for lateral control only as well as for both lateral and longitudinal control. The mean wind speed was allowed to vary uniformly from 0–5 m/s (0–9.7 kt), and the turbulence level (defined as the standard deviation of the vertical gust component in the Dryden turbulence model) was varied from 0–0.75 m/s (0–1.5 kt). A total of 250 landings were simulated for each control case.

Landing dispersions from the Monte Carlo simulations are shown in Fig. 8.78. Circles are drawn over the dispersions, which encompass 50 and 90% of the landings (50 and 90% CEP). For the case of lateral control only, 90% of the landings are within 68 m of the target, which is typical for a small guided airdrop system

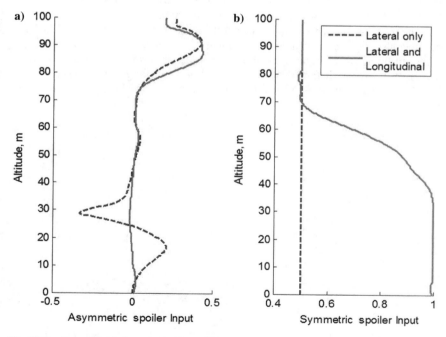

Fig. 8.77   Normalized a) asymmetric and b) symmetric control inputs for example trajectories [Ward and Costello 2013].

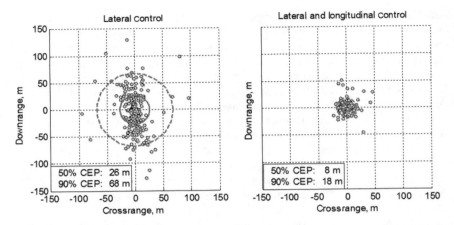

Fig. 8.78   Landing dispersions from 250 landing simulations [Ward and Costello 2013].

using the conventional mechanism of TE brake control. When using the spoilers for both lateral and longitudinal control, 90% of the landings are within 18 m (60 ft) of the target, nearly a factor of 4 improvement.

The miss distances are plotted against the level of turbulence in Fig. 8.79. Each point is also shaded according to the mean wind during that particular flight. When utilizing only lateral control, higher levels of turbulence cause very large increases in miss distance. With the addition of longitudinal control, the landing error is much less sensitive to the turbulence level. Also, the few cases

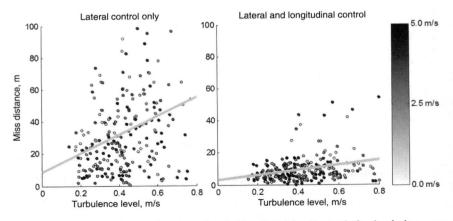

Fig. 8.79   Miss distance as a function of turbulence level for Monte Carlo simulations [Ward and Costello 2013].

of miss distances greater than 40 m when using the spoilers for longitudinal control occurred with higher levels of turbulence in stronger wind conditions. These represent cases where an unexpected change in wind is simply too large to be compensated with the spoilers.

Flight-test results on using the upper-surface spoilers on a small 2-kg (4.4-lb) PADS under an 1-m^2 (10.8-ft^2) canopy (a miniature version of PPG shown in Fig. 8.29) are presented in [Scheuermann et al. 2015].

CHAPTER 9

Control of Non-Gliding Parachute Systems

Oleg Yakimenko[*]
Naval Postgraduate School, Monterey, California

Travis Fields[†]
University of Missouri-Kansas City, Kansas City, Missouri

As shown in the preceding chapters, quite a few efforts were devoted to the development of high-landing accuracy self-guided aerial delivery systems based on relatively high-glide-ratio ram-air parafoils. For completeness, this chapter presents another approach that is based on the usage of standard non-gliding parachutes upgraded with the control units enabling a limited control authority either by disturbing the shape of their canopy to make the glide or varying its reefing. In the first case an asymmetric canopy shape creates a side force resulting in about 0.5:1 to 0.8:1 glide ratio permitting a limited steering control authority. In the second case, changing the effective drag area results in a different descent rate thus permitting PADS to take advantage of the varying winds aloft. Obviously, because of a very limited control authority, both approaches rely heavily on wind predictions. On the positive side however, compared to the ram-air parafoil-based PADS, both aforementioned approaches are more cost effective because they allow the use of much cheaper canopies. This chapter is primarily based on a previous work on the now fielded Affordable Guided Airdrop System (AGAS) and more recent mostly theoretical work on the controlled reefing-disreefing with some preliminary test results. Accordingly, this chapter is subdivided into two sections: Sec. 9.1 is devoted to guidance and control of a flat circular parachute using canopy distortion and Sec. 9.2—using a reefing control. Section 9.1 also mentions some potential developments of achieving better touchdown accuracy with cross-type canopies and parachute clusters.

9.1 NON-GLIDING PARACHUTE GUIDANCE USING CANOPY DISTORTION

This section represents a slightly edited previous work [Dobrokhodov et al. 2003; Yakimenko et al. 2004] and addresses the development and testing of GNC

---

[*]Professor.
[†]Assistant Professor.

algorithms for AGAS that integrates a low-cost guidance and control system into fielded cargo ADS (Fig. 1.12). First, this section presents the underlying AGAS concept, system's architecture, and components. It then proceeds with a synthesis of a classical optimal control based on Pontrjagin's maximum principle followed by the development and testing of a practical control algorithm. Finally, it presents a six-DoF model of circular controlled ADS used in computer simulations and GNC algorithms development.

9.1.1 CONCEPT OF OPERATIONS

The challenges of high-altitude deployment of conventional cargo ADS are well known. Figure 9.1 shows a schematics of cargo deployment for a container delivery system (CDS) taken from [CARP 2005]. This document, Air Force Policy Directive 11-231, prescribes standard methods and terminology for employment of the Computed Air Release Point (CARP) system. This system governs aircrew involved in computing CARP data during employment phases of aerial delivery operations.

As shown in Fig. 9.1, the CARP solution (usually computed for the first load to exit the aircraft) is based on many factors but mainly on average parachute ballistics and fundamental dead-reckoning principles. Each parachute has its own peculiarities. The ballistics (vertical distance to fully deploy and stabilize canopy, deceleration quotient, descent rate, etc.) is based on testing conducted by select government organizations. CARP [2005] contains different parachute ballistics given as acceptable averages. It also contains C-130, C-5, C-17, and C-141 aerial delivery airspeeds, altitudes, and heavy equipment exit times. Other factors influencing CARP determination are actual aircraft altitude above the DZ, which needs to be corrected for nonstandard atmosphere (see [Watkins 2011]) and winds. The actual ground pattern of sequentially airdropped loads also depends on aircraft track from "green light" to "red light" and time lapse between "green light" and time of exit of last item.

The primary solution for computing CARP these days is the computer solution where all requirements are solved using the MB-4 or other approved software. (In this case, chute ballistic data are contained in a database within the computer.) Alternatively, AF Form 4018 (CARP Computation) or AF Form 4013 (Modified CARP Solution), AF Form 4015 [High Altitude Release Point (HARP) Computations], or AF Form 4017 (Modified HARP Computations) could be used for manual CARP computation. Software is constantly updated based on new data, standards and high-fidelity simulations. For example, Boggs [2015] suggests a way to better collect, process, store, and distribute parachute ballistic data, and Jann [2015] presents a tool for 2D and 3D cargo airdrop and parachute simulations.

The drift effect that is probably a major uncertainty contributor is computed using a composite of altitude, winds aloft, and surface winds (if known). The most accurate drop wind is the one obtained at drop altitude and airspeed

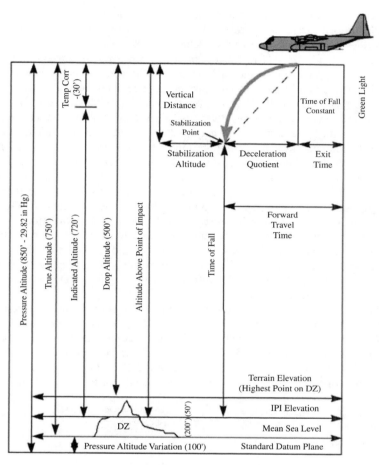

Fig. 9.1   Sample CARP diagram.

shortly before the drop, but rather than that wind data (even when available) are never 100% accurate. No wonder that wind modeling was one of the major drivers behind developing JPADS-MP, as presented in Sec. 3.3.1 [Hattis et al. 2001].

The two points to make from the preceding discussion are that 1) the knowledge of the winds plays a crucial role in CARP determination and 2) no matter how accurate CARP solution is claimed to be, no unguided ADS, even released precisely at CARP, will ever hit IPI. However, if CARP is produced by using relatively recent and accurate wind data, a little control authority allowing steering ADS in the horizontal plane might make a big difference in terms of precise payload delivery. Adding such a limited (and therefore inexpensive) control authority to existing cargo ADS was the key idea behind the AGAS concept developed in the late 1990s [Dellicker 1999; Dellicker et al. 2001].

The first step in the AGAS concept of operations is for the cargo requester to broadcast a supply request that includes information on where and when it is needed on the ground (Fig. 9.2a; compare to Fig. 1.5). Upon arrival in the vicinity of the assigned drop zone (DZ), a support aircraft (possibly an unmanned air vehicle) drops a wind dropsonde (alternatively a weather balloon could be released by the requestor). The wind profile acquired during either dropsonde descent or weather balloon ascent allows computation of the reference trajectory (RT) and CARP (Fig. 9.2b). The delivery aircraft will then be navigated to that point for air delivery of the materiel (payload) (Fig. 9.2c). Should the wind estimate and calculation of the CARP be perfect, and should the aircrew get the aircraft to this point precisely, then the parachute would fly along the RT towards the DZ with no control inputs required. However, as mentioned, wind estimation is not a precise science. Furthermore, calculation of the CARP relies on average, sometimes even not perfect, estimates of ADS aerodynamics, and the flight crews cannot precisely hit the CARP for each airdrop mission [especially in the case of massive (multiple) deliveries]. Therefore, the AGAS GNC system needs to overcome these potential errors.

The ultimate goal of the AGAS system is to allow delivery aircraft to accurately drop payloads at or above 5,500 m (18,000 ft), keeping the aircraft out of the range of shoulder-fired ground-to-air missiles. Another benefit of the system is the ability to pre-address each bundle in a load and to guide the individual bundles to their own preprogrammed DZs. Obviously, in order to accomplish these goals, the AGAS system needs to be simple, affordable, durable, and reusable. (It should survive multiple drops without any repairs.) It should not require major modifications to the standard delivery system's harness or bundle, major modifications to the cargo parachute, or a significant amount of rigger training. No changes to the parachute or cargo system were allowed. As a result, the AGAS design concept employed a commercial GPS

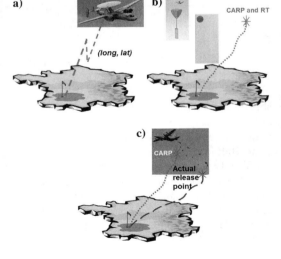

**Fig. 9.2   AGAS concept
of operations.**

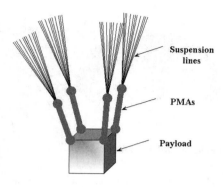

Suspension
lines

PMAs

Payload

Fig. 9.3   Payload suspension scheme.

receiver and a heading reference from navigation sensors, an inexpensive guidance computer to determine and activate the desired control inputs, and application of four pneumatic muscle actuators (PMAs) to generate control inputs. (For the fielded implementation, the PMAs were replaced with an electrical-motor-based system allowing step lengthening of the risers.) The navigation system and guidance computer are secured to an existing container delivery system while PMAs are attached to each of four parachute risers and to the container. Control is affected by lengthening one or two adjacent risers (Fig. 9.3). Hence, upon deployment of the system from the aircraft, the guidance computer steers the system along the preplanned RT. The AGAS concept relies on sufficient control authority to be produced to overcome errors in wind estimation and in the point of release of the system from the aircraft.

In general, AGAS may be implemented on any circular parachute (a flat circular parachute is the one that when laid out on the ground forms a circle). That includes a commonly used G-12 cargo parachute. This 150-m^2 (1,615-ft^2) nylon cargo parachute, with 64 suspension lines (SLs), weighs 60 kg (132 lb) and is capable of carrying a payload of up to one ton in weight [with a descent rate of around 9 m/s (30 ft/s)] [Knacke 1992]. A cargo box employed for the AGAS is a prototype adopted from the standard A-22 1.82 m^3 (64 ft^3) almost cubic delivery container with PMA instrumentation and AGU atop of it. All SLs are assembled into eight link assembles. Each pair of assemblies is attached to one of four risers. At the other end, the risers are coupled to the payload at four dispersed points (see Fig. 9.3).

The PMAs, developed by Vertigo, Inc., were braided fiber tubes with neoprene inner sleeves that could be pressurized by nitrogen [Benney et al. 1999]. Upon pressurization, the PMA contracted in length from 7.6 to 5.8 m (25 to 19 ft) and expanded in diameter. Upon venting, it did the opposite (lengthened by 30%). When three of four PMAs are pressurized (filled) and one is activated (vented), this action "deforms" the parachute creating an asymmetrical shape, essentially shifting the center of pressure, and providing a drive or slip condition. This forces the parachute to glide in the direction opposite the control action (vented PMA). Two adjacent PMAs can be activated simultaneously. Figure 9.4 shows both possibilities (one and two PMAs activated) realized in a CFD-based simulation [Mosseev 2001a; Mossev 2001b] and observed in the flight tests. Thus, AGAS can be forced to fly in any one of eight directions (relative to the body frame $\{b\}$).

The AGU consisted of two accumulator tanks valves and pressure circuits and resided in a specially designed container that occupied the space atop a

a)   b)

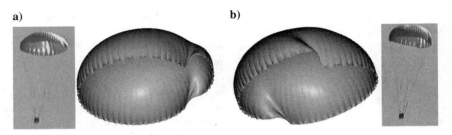

Fig. 9.4   a) One and b) two PMAs actuated (vented).

cargo container and also included the GNC electronics package. The volume of the onboard nitrogen tanks limited the number of possible fills for all four PMAs to 32 per drop. PMA fill and vent times remained a constant 5 s throughout each drop (regardless of the volume of gas remaining in the tanks). The full weight of the AGAS package to be added to the standard parachute (including AGU with PMAs, batteries, sensor suite, and GNC computer) was about 80 kg (176 lb). When fully charged with gas, the system weighed about 11 kg (22 lb) more.

Obviously, circular parachutes have a very limited ability to overcome winds. This underscores the importance of having a reference trajectory that closely matches the flight path of a parachute during an uncontrolled drop. Therefore, there is a strong need to use the latest available wind profile in the DZ to precompute the CARP and RT. As seen from Fig. 3.58, a delay in parachute deployment from the time a CARP was computed can cause significant degrada-

tion in accuracy, even for the case of a controlled drop. Compared to Fig. 3.58, Fig. 9.5 represents an inverted data set to show a footprint of impact points of ADS released from the same point hourly with CARP computed at hour 0. Obviously, impact points for later releases do not coincide with IPI as it was the case at hour 0 release (the CARP was computed for).

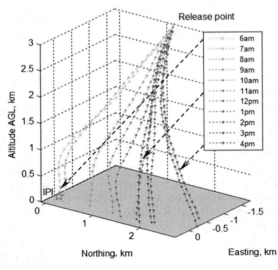

Fig. 9.5   Uncontrolled trajectories with the different wind profiles.

For this particular set of wind profiles, a touchdown accuracy gradually degraded to more than 3,000 m (9,840 ft).

9.1.2 SYNTHESIS OF THE OPTIMAL CONTROL

Based on the AGAS concept just introduced, the optimal control problem for determination of the parachute trajectories from an actual release point (RP) to IPI can be formulated as follows: *among all admissible trajectories that satisfy the system of differential equations, given initial and final conditions and constraints on control inputs, determine the optimal trajectory that minimizes a cost function of state variables* **z** *and control inputs* **u**

$$J = \int_0^T f_0(t, \mathbf{z}, \mathbf{u})\, dt \tag{9.1}$$

and compute the corresponding optimal control. In Eq. (9.1), T is a descent time that is not known a priori not only due to atmospheric turbulence but also due to the fact that a descent rate also depends on control state (activation of each PMA decreases it by approximately 3%). For the AGAS, the most suitable cost function J is the number of PMA activations. However, this cost function cannot be formulated analytically in the form given by Eq. (9.1). Therefore, some other well-known cost functions were investigated, and the results obtained were used to determine the most suitable cost function for the problem at hand.

To determine the optimal control strategy, Pontrjagin's maximum principle [Pontrjagin et al. 1969] was applied to a simplified planar kinematic (three-DoF) model of the parachute. The control objective is to steer the parachute to a single stationary point on a horizontal plane. Obviously, this should be done for the final time t_f less than T.

The simplest model describing parachute kinematics in the horizontal plane with four equal on–off controllers (Fig. 9.6) can be written as

$$\dot{\mathbf{P}} = \mathbf{R}\mathbf{U}, \quad \dot{\psi} = C + \zeta(t) \tag{9.2}$$

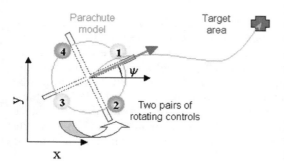

In this equation, $\mathbf{P} = [x, y]^T$ is the vector of position errors in the horizontal plane that has to be driven to zero using the control vector $\mathbf{U} = [u, v]^T$, ψ is the parachute yaw angle

Fig. 9.6 Projection of the optimization problem onto the horizontal plane.

(angle between the body x axis and the x axis of the LTP), C is a constant, function $\zeta(t)$ represents yaw rate disturbance, and $\mathbf{R} = {}^n_b\mathbf{R}$ is a rotation matrix from the body $\{b\}$ to the navigational plane $\{n\}$

$$\mathbf{R} = \begin{bmatrix} \cos(\psi) & -\sin(\psi) \\ \sin(\psi) & \cos(\psi) \end{bmatrix} \tag{9.3}$$

Compared to the $\mathbf{R}_\psi$ matrix of Eq. (5.9), a positive yaw angle corresponds to the counterclockwise rotation (because of the flipped x and y axes constituting a left-handed coordinate frame). Activation of PMAs lead to discrete control inputs $u, v \in [-V; 0; V]$, where for the G-12 based AGAS V was on the order of 4 m/s.

According to Pontrjagin's maximum principle (see Sec. 7.3.1), the Hamiltonian for the system (9.2) can be written in the following form:

$$H = \begin{bmatrix} p_x, & p_y \end{bmatrix}\mathbf{RU} + p_\psi[C + \zeta(t)] - f_0 \tag{9.4}$$

where differential equations for the costate variables p_x, p_y, and p_ψ are given by

$$\dot{p}_x = \dot{p}_y = 0, \quad \dot{p}_\psi = \begin{bmatrix} p_x, & p_y \end{bmatrix}\begin{bmatrix} u\sin(\psi) + v\cos(\psi) \\ -u\cos(\psi) + v\sin(\psi) \end{bmatrix} \tag{9.5}$$

We consider two cost functionals,

$$f_0 \equiv 1 \quad \text{and} \quad f_0 \equiv |u| + |v| \tag{9.6}$$

typical for the minimum-time and minimum fuel problems. For the model given by Eq. (9.2), the minimum-time problem implies that the parachute must be driven to the origin in minimum time given the constraints on the control vector $\mathbf{U}$. Also, in this application, the second cost function defines the system's momentum or energy rather than fuel because AGAS uses gas only to activate PMAs. (There is no gas expenditure needed to maintain PMA filled/vented.)

Importantly, with four on–off controllers available, a circular parachute can move in only one of eight possible directions with respect to the body frame. This makes performance of a control algorithm very sensitive to the rotation of the parachute or lack thereof. Specifically, if the parachute is not rotating, there only exists a single initial condition for which the TPBV problem can be solved. This is the reason for introducing a nonzero yaw rate in Eq. (9.2). Furthermore, because the yaw rate can never be precisely known, the disturbance term $\zeta(t)$ was also added. The two main sources of uncertainty in yaw rate include wind disturbance and yaw moment due to adjacent PMA activations (see Sec. 9.1.5).

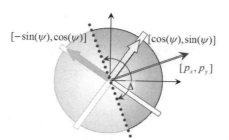

Fig. 9.7 Time-optimal control.

According to the Pontrjagin's maximum principle, the optimal control is determined as $\mathbf{u}_{\text{opt}} = \text{argmax}\,(H(\mathbf{p}, \mathbf{z}, \mathbf{u}))$. Therefore, for the time-minimum problem, the optimal control is given by

$$u = V\text{sign}\left([p_x,\ p_y] \begin{bmatrix} \cos(\psi) \\ \sin(\psi) \end{bmatrix} \right) \quad \text{and} \quad v = V\text{sign}\left([p_x,\ p_y] \begin{bmatrix} -\sin(\psi) \\ \cos(\psi) \end{bmatrix} \right) \quad (9.7)$$

Figure 9.7 shows the graphical interpretation of these expressions. In general, the vector $[p_x,\ p_y]$ defines a direction towards the DZ and establishes a semiplane perpendicular to itself that defines the nature of control actions. Specifically, if the PMA happens to be located within a certain operating angle (OA) Δ with respect to the vector $[p_x,\ p_y]$, it should be activated. For a time-optimum problem, $\Delta = \pi$. Therefore, two PMAs will always be active as determined by the parachute rotation attitude. (We do not address the case of singular control, which in general is possible if the parachute is required to satisfy a final condition for yaw angle.)

Figure 9.8 shows an example of a time-optimal trajectory. It consists of several arcs and a sequence of activations. For the sake of simplicity, $\dot\psi = 2\ \text{deg}/\text{s}$ was taken for this simulation as observed in one of the earliest flight tests. (In practice, control activations themselves are changing canopy azimuth all the time.) Figure 9.9 demonstrates two Hamiltonian isoline patterns corresponding to two instances of time. Because isolines for the time-optimal problem are straight lines rotating counterclockwise with 2-deg/s angular velocity, the optimal solution on this graph (Hamiltonian maximum) can be located in one of four corners of the controls envelope.

Similarly, for the "fuel"-minimum problem, we synthesize the following optimal control structure:

$$\begin{aligned}
p_x\cos(\psi) + p_y\sin(\psi) \geq 1 &\quad \Rightarrow \quad u = V \\
-1 < p_x\cos(\psi) + p_y\sin(\psi) < 1 &\quad \Rightarrow \quad u = 0 \\
p_x\cos(\psi) + p_y\sin(\psi) \leq -1 &\quad \Rightarrow \quad u = -V \\
-p_x\sin(\psi) + p_y\cos(\psi) \geq 1 &\quad \Rightarrow \quad v = V \\
-1 < -p_x\sin(\psi) + p_y\cos(\psi) < 1 &\quad \Rightarrow \quad v = 0 \\
-p_x\sin(\psi) + p_y\cos(\psi) \leq -1 &\quad \Rightarrow \quad v = -V
\end{aligned} \qquad (9.8)$$

In this case, each PMA will be employed only when aligned with the certain direction, meaning $\Delta \to 0$. Figure 9.10 shows that in this case Hamiltonian isolines are represented by a rhomboid figure, which maintains an orientation so that one of its vertexes touches the control envelope at $u = 0$ (as shown in Fig. 9.10 for the particular instance of time) or $v = 0$.

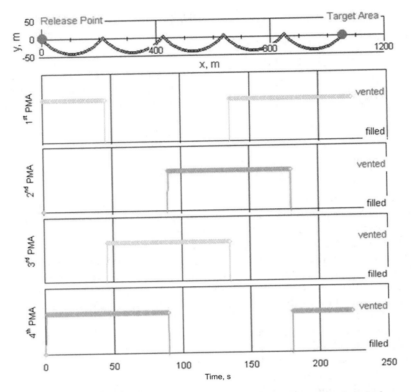

Fig. 9.8   Example of the time-optimal trajectory and time-optimal controls.

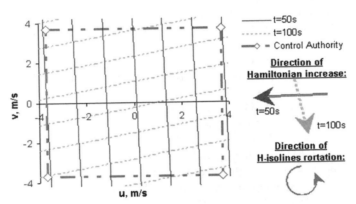

Fig. 9.9   Hamiltonian isolines for the time-optimal control problem.

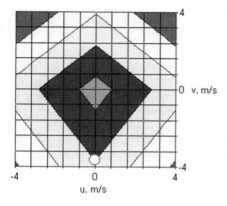

Fig. 9.10   Hamiltonian isolines for the "fuel"-minimum problem.

In general, any cost function other than minimum time will require an operating angle $\Delta \leq \pi$ (Fig. 9.11). (Note that any control with $\Delta < 0.5\pi$ may not work at all if the parachute is not rotating.)

Figure 9.12 shows the effect of OA's magnitude on the flight time, "fuel," and number of PMA activations (from "vented" to "filled" state). It is clearly seen that the nature of the dependence of the number of activations on the OA is the same as that of the time of flight. This implies that by solving the time-minimum problem, we automatically ensure a minimum number of activations. Moreover, it is also seen that the slope of these two curves in the interval $\Delta \in [0.5\pi, \pi]$ is flat. This implies that small changes of OA from its optimal value will result in negligible impact on the number of activations. Therefore, changing the OA to account for the realistic PMA model (see Sec. 9.1.5) will not change the number of activations significantly.

Figure 9.13 demonstrates the influence of constant yaw rate on different OAs. The results were obtained for the time-optimal control problem illustrated in Fig. 9.8. Obviously, the smaller the yaw rate is, the smaller the number of activations. Decreasing the OA for the same yaw rate leads to an increase in the number of PMA activations.

Figure 9.14 includes simulation results for the case where yaw angle from a flight test was used to drive the first equation in Eq. (9.2) while optimal control was computed using Eq. (9.7). As can be seen, the flight-test yaw angle is not smooth. Neither is it monotonic. Although a synthesized optimal control drives the model of the parachute towards a target area (TA), because of the erratic yaw the number of PMA activations increases to 35 (vs 12 with the monotonic 2-deg/s yaw rate as seen from Fig. 9.12). For this particular simulation, the OA was equal to 2.5 radians. This example illustrates the sensitivity of the optimal control algorithm to uncertainties in yaw angle. Therefore, the flight control algorithm must be more

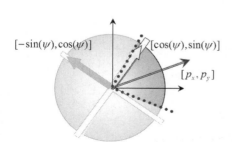

Fig. 9.11   Generalized case of optimal control.

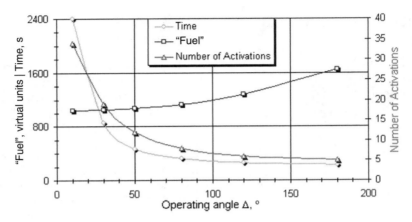

Fig. 9.12   Influence of the OA's magnitude.

robust to these uncertainties to prevent a significant increase in the number of PMA activations.

9.1.3   PRACTICAL CONTROL ARCHITECTURE

As discussed in Sec. 9.1.1, the activation box for PMAs is only capable of producing a bang-bang control. Optimal control analysis of a simplified parachute model discussed in Sec. 9.1.2 suggests that for the cost functions of Eq. (9.6), bang-bang is also the optimal control strategy. Furthermore, this analysis led to an important concept of an operating angle, which was used to define the basic control concept for AGAS. Because for a given OA the bang-bang control strategy was shown to minimize the number of activations for a planar model,

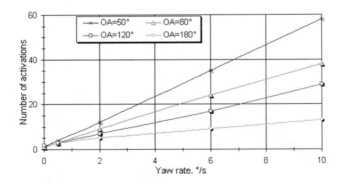

Fig. 9.13   Influence of a constant yaw rate.

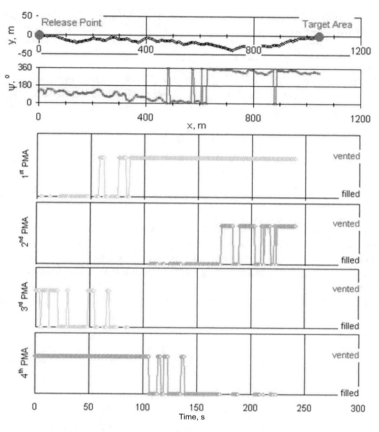

Fig. 9.14   Flight path computed with usage of a real yaw-angle profile.

this strategy was employed to get the parachute to within a predefined altitude-dependent TA (defined by inner and outer cones discussed next) and then for the remainder of descent to stay within this area. In addition, this basic strategy must be made robust to uncertainties in yaw motion. (As just mentioned, a sensor suite was installed on top of a payload, so that some discrepancy exists between the measured heading reference and the real attitude of the canopy.) These considerations were used to develop the practical flight control algorithm for AGAS as follows.

Considering the relatively low glide ratio (GR) of circular parachutes, the AGAS can only overcome less than 4-m/s wind. It is therefore imperative that the control system steers the parachute along a prespecified RT obtained from most recent wind prediction. This can be done by comparing the current

GPS position of the parachute with the desired one on the RT at a given altitude h to obtain the position error

$$\mathbf{P}(h) = \mathbf{p}_{AGAS}(h) - \mathbf{p}_{RT}(h) \qquad (9.9)$$

This position error $\mathbf{P}(h)$ is computed in LTP frame with an origin in the TA and is then converted to the body axis using an Euler angle rotation $\mathbf{R}^T$. The resulting body-axis error vector

$$\mathbf{P}^b = \mathbf{R}^T \mathbf{P} \qquad (9.10)$$

is then used to identify the error angle (EA) λ

$$\lambda = \arg\left(\mathbf{P}^b\right) \qquad (9.11)$$

As opposed to parafoils, circular parachute angular dynamics is characterized by coning motion. Coning motion is represented by coupled oscillations in pitch and roll. Therefore, when averaged over one period, pitch and roll angles are zero, and it is sufficient to use heading angle only when computing rotation matrix $\mathbf{R}$ in Eq. (9.10).

In turn, the EA is then used to define what PMA ($i = 1, \ldots, 4$) must be activated:

$$i = \begin{cases} 1, & \text{if } \lambda \leq \dfrac{\Delta}{2} \vee \lambda \geq 2\pi - \dfrac{\Delta}{2} \\[2mm] 2, & \text{if } \lambda \in \left[\dfrac{3\pi}{2} - \dfrac{\Delta}{2}; \dfrac{3\pi}{2} + \dfrac{\Delta}{2}\right] \\[2mm] 3, & \text{if } \lambda \in \left[\pi - \dfrac{\Delta}{2}; \pi + \dfrac{\Delta}{2}\right] \\[2mm] 4, & \text{if } \lambda \in \left[\dfrac{\pi}{2} - \dfrac{\Delta}{2}; \dfrac{\pi}{2} + \dfrac{\Delta}{2}\right] \end{cases} \qquad (9.12)$$

[By definition, λ is counted from PMA #3 counterclockwise, that is, in the situation shown as an example in Fig. 9.15, PMAs #2 and #3 would be activated (vented).]

As suggested in Sec. 9.1.2, to account for the refill time and sensors errors, the operating angle was set to $\Delta \approx 2.5$ radians instead of $\Delta = \pi$ (whereas on the earliest AGAS versions, refill time was not even constant and was equal to about 20 s towards the end of the descent so that the yaw rate of 2 deg/s resulted in a 40-deg yaw angle offset.) This still allows the activation of a single control input or two

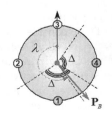

Fig. 9.15   Control-activation rule.

simultaneous control inputs without significant degradation of AGAS performance (see Fig. 9.12). Furthermore, it is greater than $\pi/2$ and, therefore, within the stability range for the OA.

For better robustness, the aforementioned basic control architecture was complemented with a few additional terms, which include outer and inner tolerance cones and a prediction term. To begin, the initial error after deployment should not exceed a certain value because of AGAS's limited control authority. This area of attraction has the radius R_A around the RT that can be roughly estimated by a simple formula

$$R_A(h) = 0.8 k_\Delta \mathrm{GR}_{\max} h, \quad \text{where } k_\Delta \approx \Delta \pi^{-1} \tag{9.13}$$

The coefficient k_Δ is approximated by using the data given in Fig. 9.12, maximum glide ratio $\mathrm{GR}_{\max}$ for the G-12-based AGAS with two adjacent PMAs activated is equal to 0.5 [Dellicker et al. 2000], and the coefficient 0.8 accounts for real-world yaw profile. (Note, the maximum glide ratio with only one PMA activated is slightly higher and of the order of 0.8 [Dellicker et al. 2003].)

To eliminate unnecessary activations of PMAs, a tolerance (outer) cone was established (Fig. 9.16) [Dellicker 1999]. Its radius at the CARP [at an altitude of 3,000 m (9,840 ft)] is $\Re_{\mathrm{outer}}(3,000) = 200\,\mathrm{m}$ (656 ft), and it decreases linearly to $\Re_{\mathrm{outer}}(0) = 100\,\mathrm{m}$ (328 ft) radius circle at the TA (at ground level). Should the magnitude of the position error in the lateral plane $\left|\mathbf{P}^b(h)\right|$ be outside of this tolerance cone

$$\left|\mathbf{P}^b(h)\right| > \Re_{\mathrm{outer}}(h) \tag{9.14}$$

a control is activated to steer the system back to the planned RT.

When the system is within the inner cone $\Re_{\mathrm{inner}}$

$$\left|\mathbf{P}^b(h)\right| < \Re_{\mathrm{inner}} \tag{9.15}$$

(which is set to 60-m radius regardless of altitude) the control is disabled, and the parachute drifts with the wind ($\Re_{\mathrm{inner}}$ was selected to account for the refill time) until the outer cone is reached and control is activated again. These inner and outer cones can be interpreted as altitude-dependent hysteresis surfaces.

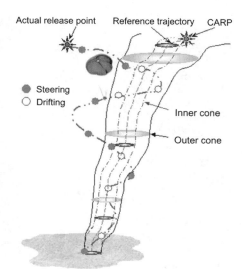

Fig. 9.16 Outer and inner cones.

Actual release point   Reference trajectory   CARP

Steering
Drifting

Inner cone

Outer cone

Fig. 9.17 a) "Positive" and b) negative effect of PMA transition moment.

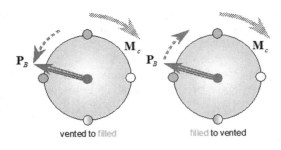

vented to filled   filled to vented

The basic control strategy uses the following activation rule: both the tolerance cone and the operating angle constraints must be active for a given PMA to be activated.

The control algorithm just outlined was flight tested at YPG. As expected, the number of commands to activate PMAs was unacceptably high. This resulted in a premature emptying of onboard nitrogen tanks, thus leaving the AGAS with no control authority at the bottom of descent. Analysis of flight-test data indicated that this was caused by frequent yaw angle changes and that these changes occurred when one of the adjacent PMAs was actuated while the other one was in transition from vent to full or vice versa.

Figure 9.17 explains this phenomenon. If one PMA is activated (vented), and an adjacent PMA is performing a transition from one state to another, this causes a yaw moment M_c. This moment can be "useful" (when the direction of rotation of the vector P^b is opposite to the direction of M_c), or "harmful" (vice versa). In the latter case, the rotation of the parachute under the action of M_c causes a deactivation command to the PMA that was just activated. Moreover, during this deactivation the useful moment in turn makes the situation even worse. This case is shown in Fig. 9.18a, where shaded circles denote activated state of any PMA.

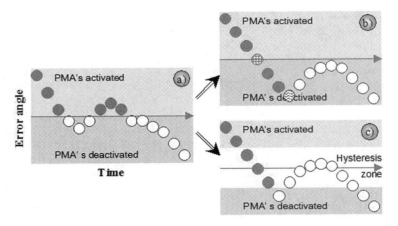

Fig. 9.18   Two ways of decreasing the influence of yaw oscillations.

To eliminate unnecessary activations, delay logic was introduced in each PMA channel (Fig. 9.18b). Any new command that requires change in the PMA state triggers the delay timer (circle with a grid inside in Fig. 9.18b). While the delay timer is active, no command is executed including the triggering command. At the end of the delay, the timer is reset, and the first available command (circle with waves inside in Fig. 9.18b) is executed until the next command that requires change in the PMA state triggers the delay timer again. (This simple logic was used in lieu of a filter due to programming constraints.)

Another approach that helps reduce the number of unnecessary activations and that meets programming constraints is to use hysteresis as shown in Fig. 9.18c. The size of a hysteretic zone is adjusted to exclude recurring activations.

Both delay and hysteresis angle values can be adjusted as a function of system dynamics, and, in principle, they achieve the same result.

Another approach that drastically improves robustness in the presence of yaw oscillations is to introduce a derivative term into the control logic of Eq. (9.10) as follows. First, let us change inequalities (9.14) and (9.15) to

$$\left| \mathbf{R}^T \left(\mathbf{P} - k_c \dot{\mathbf{P}} \right) \right| > \Re_{\text{outer}}, \quad \left| \mathbf{R}^T \left(\mathbf{P} - k_c \dot{\mathbf{P}} \right) \right| < \Re_{\text{inner}} \qquad (9.16)$$

where coefficient k_c should be adjusted to provide better performance (smallest overshoot). This softens the outer and inner cone edges. For example, if AGAS is approaching the inner cone with high planar velocity rather than slowly drifting into it is better to deactivate (fill) all PMAs earlier than would be done by the control strategy based on Eq. (9.15). On the other hand, if AGAS is leaving the outer cone with a high planar velocity, it is worth activating (venting) appropriate PMA(s) earlier than it would be done by Eq. (9.14) to prevent further rapid increase of the radial error.

Second, redefine the EA to be

$$\lambda = \arg\left\{ \mathbf{P}^T \left(\mathbf{P} + k_r \dot{\mathbf{P}} \right) \right\} \qquad (9.17)$$

where k_r determines the delay in the execution of next command defined by Eq. (9.14), similar to the one discussed earlier in this section [compare the definition of the EA in Eq. (9.17) with that in Eq. (9.11)]. Note that Eqs. (9.16) and (9.17) have opposite signs for the derivative term. In Eq. (9.16), the negative sign accelerates control action whereas in Eq. (9.17) it does the opposite, therefore reducing sensitivity to the oscillations in yaw.

For the case when available wind prediction is either too old or nonexistent, an alternative to tracking an RT is proposed. For this purpose, Eq. (9.9) should be replaced by the following one

$$\mathbf{P}(h) = \mathbf{p}_{\text{AGAS}}(h) - k_w \mathbf{p}_{\text{RT}}(h) \qquad (9.18)$$

a)   b)   c)   d)

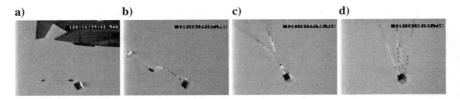

Fig. 9.19   AGAS deployment and risers untwisting sequence.

When wind prediction is available $k_w = 1$. When wind prediction is either too old or nonexistent $k_w = 0$, Eq. (9.18) becomes

$$\mathbf{P}(h) = \mathbf{p}_{AGAS}(h) \tag{9.19}$$

An appropriate value of k_w can be determined by comparing real-time motion of AGAS during a drop with its predicted response generated by the onboard model. In fact, assuming the model is sufficiently accurate, it can be used to determine errors in the predicted wind profile.

As a safety precaution, the GNC system starts implementing control commands 25 s after the initial deployment. This time is needed in order for the AGAS to be released, the main canopy to be fully deployed, and risers to be untwisted as shown in Fig. 9.19.

By design, the initial shock during deployment is absorbed by Kevlar load lines. So all PMAs are initially vented (when vented, they are longer than Kevlar load lines). The first command sent and executed after the 25-s deployment delay is to fill all PMAs (as shown in Fig. 9.20). After an additional 5 s, any other command can be executed.

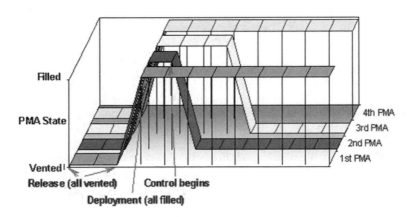

Fig. 9.20   Control actions history right after release.

9.1.4   COMPUTER SIMULATION AND FLIGHT-TEST RESULTS

An extensive simulation analysis was done to test the flight control algorithm, to determine the accuracy requirements for the sensor suite and the control authority requirements for AGU, and to estimate AGAS overall performance. These simulations used a complete nonlinear six-DoF model of a controlled G-12 parachute presented in detail in Sec. 9.1.5. It also included a model of the PMA dynamics. Figure 9.21 shows an example where the 3D position of AGAS from a flight test is compared to that generated by the model. The model output matches flight-test data fairly well, with only a 15-m (49-ft) difference between impact points (IPs). The number of PMA activations is the same and in this case is equal to 14.

Figure 9.22 illustrates the influence of the OA's magnitude on control performance. In this case, only the basic control algorithm was tested, that is, no cones, delay, hysteresis, or any other additional features designed to minimize the number of activations discussed in the preceding section were included. Each graph represents radial error vs current altitude during the simulated drops. The target is at (0,0) on the graph.

While simulation with the OA = 180 deg ensures the best accuracy of the predefined RT tracking, it also requires 53 activations (compare with only 35 activations obtained in simulation with the same yaw profile in Fig. 9.16 when using three-DoF model.) With the decrease of OA, the number of activations

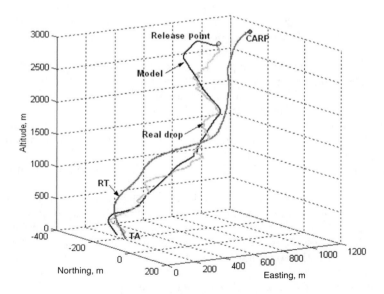

Fig. 9.21   AGAS model vs real AGAS drop.

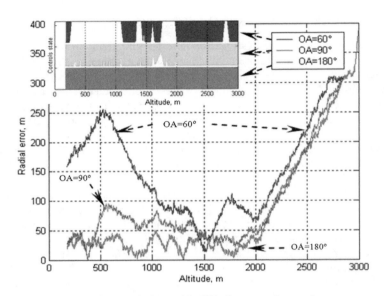

Altitude, m

Fig. 9.22   Simulations with different operating angles.

also decreases (17 for $\Delta = 90$ deg and 14 for $\Delta = 60$ deg). However, touchdown accuracy degrades as well.

The grayscale bar in the top-left portion of Fig. 9.22 indicates whether any PMA was activated during the simulated drop for each OA. Obviously with $\Delta < 90$ deg (top strip), blind sectors with no PMA activation become possible (see Sec. 9.1.2).

Figure 9.23 illustrates the impact of introducing outer and inner cones on the control performance. The operating angle in this simulation and hereafter was $\Delta = 143$ deg, and the number of activations was nine (as opposed to 24 without cones). Obviously, this was achieved by not activating PMAs while between cones (see the bar similar to one in Fig. 9.22 in the top-left corner of Fig. 9.23). This also results in a slight degradation of the touchdown accuracy. [Starting from about 700 m (2,300 ft) above DZ, the wind blew the parachute away from TA; however, neither PMA was activated because the parachute was inside the outer cone.]

Although nine activations seem to be an excellent result (for a given CARP generated using the most recent wind profile), further analysis revealed that almost half of the activations were due to yaw oscillations. As discussed in Sec. 9.1.3, unnecessary activations can be reduced by introducing delay or hysteresis in PMA activation. Figure 9.24 gives an idea of how these features affect the performance of the system. In addition, it was discovered that while the delay does not affect the number of PMA activations (for this simulation it remained approximately the same 9, 9 and 10 for 0-, 5-, and 10-s delays,

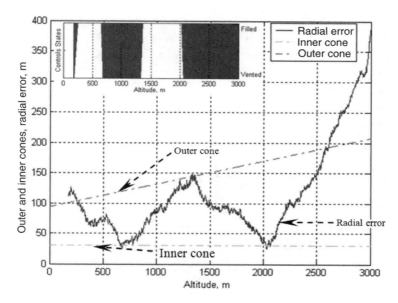

Fig. 9.23   Introducing the cones.

respectively), the EA hysteresis not only improves performance but decreases the number of activations as well: nine for 0 deg, and five for ±5 and ±10 deg. Therefore, all of the following simulation results include the EA hysteresis of ±5 deg.

Figure 9.25 shows the effect of introducing a prediction term on Eq. (9.39). As expected, the overall performance improves (trajectory stays strictly between cones). However, this term leads to a certain increase in the number of PMA

a)   b)

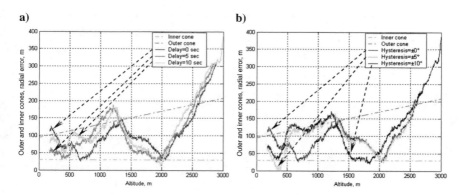

Fig. 9.24   Introducing a) delay and b) hysteresis to fight yaw oscillations.

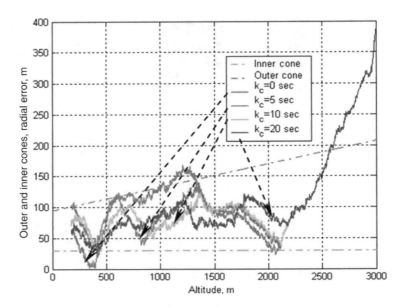

Fig. 9.25   Introducing a prediction term.

activations (the dead zone between two "soft-edge" cones is smaller than between "solid-edge" ones). For this particular simulation, the number of activations was equal to five with no prediction $k_c = 0$ s, 10 with 5- to 10-s prediction, and 14 with $k_c = 20$ s.

Figure 9.26 summarizes the preceding discussion and shows the decrease in total number of PMA activations when more sophisticated control logic is

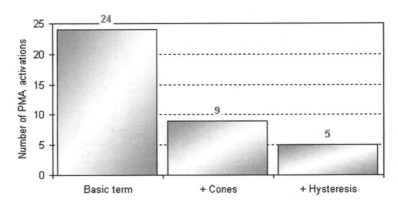

Fig. 9.26   Number of PMA activations decrease.

employed. As seen, the more complex control scheme produced a significant reduction in the number of PMA activations in comparison with the basic "classical" control logic. This ensured an almost seven-fold increase in the reserve with respect to available control authority (32 activations).

Figure 9.27 shows an example of Monte Carlo simulation with the same wind profile and two control strategies. The first one employs basic term, cones, and 10-s delay. The second one uses basic term, 5-deg hysteresis instead of delay, and prediction term in Eq. (9.39) with $k_c = 25$ s. The thick central curves on both graphs represent the RT. By observing these two plots and analyzing the results of simulations, it can be stated that the second algorithm performs much better. Not only is the mathematical expectation of the final position error for the second algorithm twice as small as for the first one, but the standard deviation decreases by a factor of three. While replacing the 10-s delay with a 5-deg hysteresis leads to the four-fold decrease of the number of PMA activations without a deterioration of the overall performance, introducing a prediction term costs a two-fold increase of PMA activations so that on average the first algorithm required twice as many activations as the second.

Finally, Fig. 9.28 includes the results of a simulation where $\mathbf{P}_I$ is defined using Eq. (9.19), also known as a target-seek control strategy. These results are compared with the standard RT-tracking strategy given by Eq. (9.17). All inputs are the same for both simulations including $\Delta = 143$ deg, solid-edge cones (around the RT in the first case and around vertical line stretching upward from the target), and ± 5-deg hysteresis. Both algorithms perform well requiring only five PMA activations each.

Figure 9.28 also includes the simulation run where no control was used to steer the parachute. Even when being released exactly at the CARP (with no initial error), the uncontrolled parachute flies away from the RT ending up with a touch-down error of nearly 400 m (1,312 ft). Clearly, the reason for this is that the predicted wind profile was not sufficiently accurate. However, the controlled parachute with a reasonable algorithm suggested by Eq. (9.41) handles this situation fairly well.

A total of 11 controlled drops were executed to test the AGAS concept and control algorithm. The final demonstration took place at PATCAD 2001 (see Table 1.2). During preliminary tests, a ground station was used to control AGAS via a wireless modem. The AGAS sent its current position and yaw angle to the ground station. The ground station processed the data using the flight control algorithm and then issued appropriate commands to the AGAS GNC. For the final drops, all GNC algorithms were executed aboard AGAS. The downlink message was used for real-time monitoring during the drop. Figure 9.29a shows AGAS system rigged for flight featuring an A-22 container, pneumatically driven AGAS AGU with PMAs and a canopy on top.

Table 9.1 summarizes the results AGAS drops performed during the PATCAD demonstration. The rigged weight of all AGAS was 1,600 lb. On

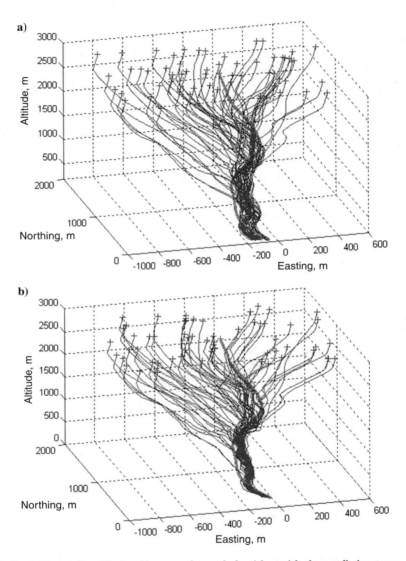

Fig. 9.27   a) Simplified vs b) more advanced algorithm with the prediction term.

September 14, AGAS-3 quit working halfway down because of a valve system malfunction, so only AGAS-1, AGAS-2, and AGAS-4 should be considered as fully operational systems. As seen, the miss distance for three operable AGAS systems was less than 78 m (256 ft) as opposed to 142–1,372 m (466–4,500 ft) for uncontrolled G-12 cargo parachute systems.

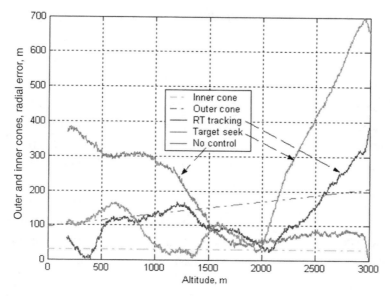

Fig. 9.28 The RT-tracking vs target-seek trajectories.

The same control algorithm being employed on AGAS-1 and AGAS-2 led to impact of the two systems during the first drop as is seen in Fig. 9.29b. For the second drop, different target areas were input into the GNC systems of the two parachutes to avoid possible collision.

Figure 9.30a demonstrates the integral data for two first-day drops from the altitude of 3,000 m (9,840 ft). The 30-min old wind data were used to compute the RT. Despite a large initial error, both AGAS steered to the TA fairly well:

a)   b)

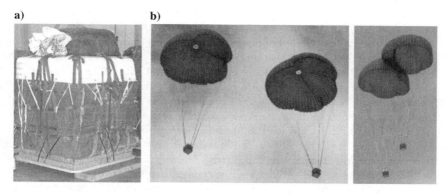

Fig. 9.29   a) Rigged AGAS and b) two AGAS steering towards the same RT.

TABLE 9.1   PATCAD 2001 RESULTS [PATCAD 2001A; PATCAD 2001B]

Test Item	Mass	IPI Miss
13 September 2001		
WindPack	21 kg (46 lb)	515 m (1690 ft)
Standard G-12	724 kg (1600 lb)	512 m (1680 ft)
Standard G-12	773 kg (1700 lb)	142 m (466 ft)
AGAS-1	**726** kg (1600 lb)	**76** m (250 ft)
AGAS-2	**726** kg (1600 lb)	**78** m (256 ft)
14 September 2001		
WindPack	21 kg (46 lb)	1049 m (3440 ft)
Standard G-12	726 kg (1600 lb)	1372 m (4500 ft)
AGAS-3	**726** kg (1600 lb)	**347** m (1140 ft)
AGAS-4	**726** kg (1600 lb)	**56** m (182 ft)

17- and 18-PMA activations were needed to hit the target, with approximately the same miss distance. Figure 9.30b presents the same set of data for the second drop of the AGAS on September 14 [released at 4,500 m (14,760 ft)]. The wind profile used for this drop was 2 h old. Observe that AGAS-4 drifts away from the CARP due to a bad wind estimate for the first 1,000 m (3,280 ft). However, upon leaving the outer cone, it is steered back inside. As soon as the PMAs inflate upon entering the inner cone, the AGAS proceeds to drift out again. Figure 9.31 shows the control-related data for the AGAS-4. The total of 28 PMA fills were needed to hit the target with a 55-m (180-ft) miss.

After the PATCAD 2001 event, more tests of AGAS have been done. Twelve more successful drops with a rigging weight of 590 to 860 kg (1,300 and 1,900 lb) occurred in early 2003 in preparation for PATCAD 2003 [Dellicker et al. 2003]. Altogether the 15 tests (12 plus three PATCAD 2001 drops) exhibited a 69-m (226-ft) CEP (Fig. 9.32a). The AGAS performance at PATCAD 2003, also shown in Fig. 9.32a, demonstrated an even better result with a CEP of only 35 m.

Later on, the AGAS underwent a major upgrade resulting in replacement of PMAs requiring bulky and heavy gas containers with four Kevlar risers routed through corner guide slots in the AGU top cover and wound onto a hoist spool (Fig. 9.33). Some drop data for the modified electromechanically driven AGAS are shown in Fig. 9.32b (by 2005 over 150 AGAS drops were executed) [Jorgensen and Hickey 2005]. These drops exhibited a 38-m (125-ft) CEP. Also shown in Fig. 9.32b are the results of seemingly successful drops of the modified AGAS with a rigged weight ranging from 1,500 to 2,000 lb as demonstrated at PATCAD 2005 [PATCAD 2005]. The demonstrated CEP of 49.9 m (163.7 ft) is

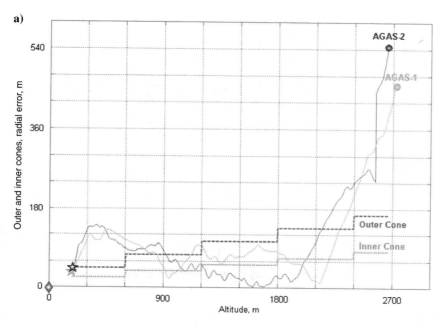

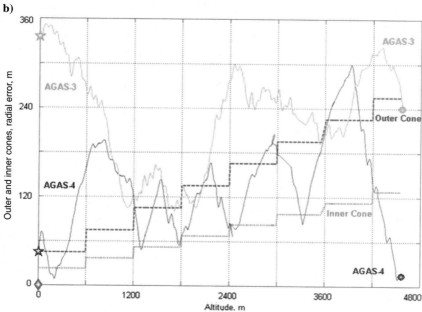

Fig. 9.30   a) September 13th 3,000-m drops and b) September 14th 4,500-m drops.

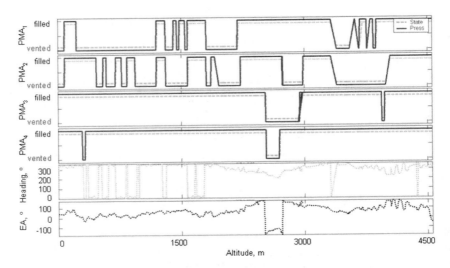

Fig. 9.31   AGAS-2 control profile vs altitude.

slightly worse than that demonstrated in the preliminary drops; however, it is still within a design goal of 50 m (164 ft). This superb performance can only be achieved with a good knowledge of the current winds, that is, when recent wind-sonde or weather balloon data are available. As shown in Jorgensen [2005], if CARP is computed based on the aged wind data or forecast winds, this performance degrades dramatically.

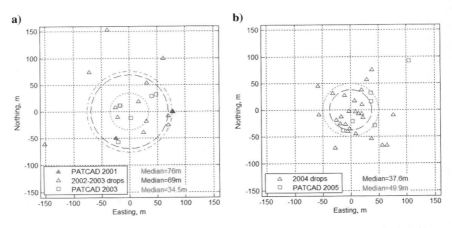

Fig. 9.32   Touchdown errors for the a) 2001–2003 and b) 2004–2005 series of AGAS drops.

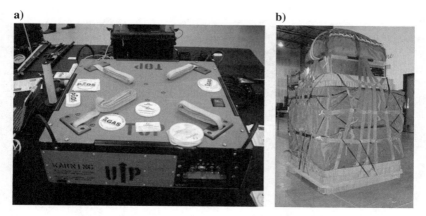

Fig. 9.33   Electromechanically driven AGU of a) AGAS-2000 and b) AGAS-2000 rigged for flight [Jorgensen 2005].

Capewell Components Company also downscaled the original AGAS system, AGAS 2K, capable of carrying up to 2,200-lb weight payload under a 64-ft-diam G-12 canopy, to AGAS 500 to be used with smaller, flat, circular canopies, 7.3-m (24-ft) T-10R and 8.5-m (28-ft) C-9 canopies with a payload up to 136 kg (300 lb) and 10.7-m (35-ft) T-10D canopy with a payload up to 227 kg (500 lb). Moving up, it could be modified to AGAS 5K to carry up to 2.3 tons (5,000 lb) under a 30.5-m (100-ft) G-11 canopy or even AGAS 10K to carry up to 4.5 tons (10,000 lb) under a 42-m (137-ft) triconical parachute [Jorgensen and Hickey 2005].

9.1.5   SIX-DoF MODEL OF CONTROLLABLE CIRCULAR PARACHUTE

This and the next section present a six-DoF model of a controllable parachute used to develop and tune the control algorithms discussed in Sec. 9.1.4. Linear position of the PADS is computed with respect to an ENU local tangent plane $\{u\}$ (Fig. 9.34a). Its positive y direction is aligned with local North, the positive x direction points East, and the positive z direction points up. All other computations are performed in the body-fixed coordinate frame $\{b\}$. Body origin is attached to the center of the open-end plane of the canopy. The x and y body axes lie in the plane parallel to the canopy's base, and z is aligned with the imaginary axis extending towards the centroid of the payload.

In different aerohydrodynamic studies, the origin of $\{b\}$ is sometimes placed at the canopy's centroid. The undisturbed canopy is always assumed to be a planetary ellipsoid, but it may have different ratio of minor and major axes. That makes the location of the origin of $\{b\}$ to be conditional from the concrete parachute design. Moreover, the z coordinate of canopy centroid can be determined

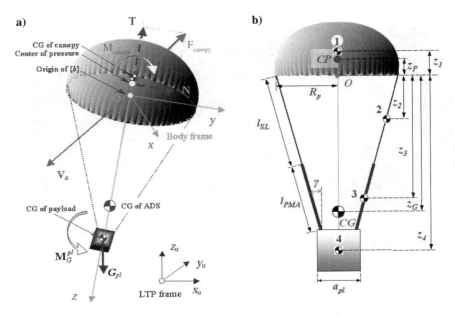

Fig. 9.34   a) Coordinate frames and b) ADS geometry .

only approximately because in the general case there are no analytical formulas for ellipsoidal shells. Originally, the six-DoF model presented here was also developed with this setup of the $\{b\}$ frame, but later on to make equations of motion more universal all derivations were repeated with the origin at the center of the open-end plane of the canopy.

For simplicity we will omit subscript "b" in the further analysis, assuming that all variables and aerodynamic coefficients when applicable are defined in $\{b\}$.

To develop the model, the following assumptions mostly adopted from [Tory and Ayres 1977] were used:

• Because of the predetermined architecture, the parachute and payload are considered to be rigidly connected to each other (Fig. 9.3).

• During the airdrop, these two rigidly connected parts are assumed to experience only gravity and aerodynamic forces.

• The canopy experiences all aerodynamic forces and moments about its center of pressure.

• The aerodynamic forces generated by the payload are negligible.

• The undistorted canopy has an axial symmetry about z_b axis.

Distortion of the canopy's shape (caused by the lengthening of one or two adjacent risers) introduces asymmetric forces and moments allowing steering of the PADS in a certain direction in a horizontal plane. Moreover, it obviously makes PADS not symmetrical. (One plane of symmetry still remains, but its location is not constant with respect to the body frame.) However, we will assume that the effect of risers lengthening and canopy distortion from the standpoint of changing tensors of inertia and apparent masses is negligibly small (which will be proven later in this section). Therefore, even for a controlled circular parachute, these two tensors will be assumed to have the same form as for an uncontrolled (symmetrical) parachute.

Figure 9.34b introduces a set of parameters defining the parachute's geometry. Numbers 1–4 denote the PADS's components: canopy, rigging (suspension) lines, actuators (PMAs), and payload. The position of their centroids Z_i is determined with respect to the body frame origin (point O in Fig. 9.34b).

On this figure, $R_p = \frac{2}{3}R_0$ denotes the radius of inflated parachute (R_0 is a nominal or reference radius of uninflated canopy), l_{SL} the length of SLs, l_{PMA} the nominal length of PMAs, and a_{pl} the dimension of a container. (Without loss of generality, all three dimensions of a cargo container were considered to be the same.)

As just mentioned, the shape of undisturbed canopy is a half of a planetary ellipsoid (hemispheroid) meaning that it is circular in plan when viewed along z_b direction and elliptical when viewed from the side. The ratio of the minor to major axes (canopy shape ratio) is denoted as ε. This ratio for a flat circular canopy is typically equal to 0.5 [Tory and Ayres 1977]. Other researchers have assumed the canopy to be a hemisphere ($\varepsilon = 1$) [Yavuz 1981; Yavus 1989; Cockrell and Doherr 1981]. For the PADS at hand, ε is equal to 0.82.

Following basic principles of analytical mechanics, the total energy T of a whole descending system relative to body frame can be given by

$$T = T^{ADS} + T^{air} \qquad (9.20)$$

where T^{ADS} is the kinetic energy of the parachute and T^{air} is the kinetic energy of surrounding air [see Eq. (5.59)]. For an ideal fluid, the kinetic energy can be determined in terms of velocity potential. However, it is common practice to assume that the kinetic energy of real fluid can be defined similarly [Cockrell and Doherr 1981]. Using the Lagrange approach, the basic equations of motion for parachute-air system in $\{b\}$ can be obtained as follows:

$$\frac{d}{dt}\frac{\partial T}{\partial \mathbf{v}} + \boldsymbol{\omega} \times \frac{\partial T}{\partial \mathbf{v}} = \mathbf{F} \qquad (9.21)$$

$$\frac{d}{dt}\frac{\partial T}{\partial \boldsymbol{\omega}} + \boldsymbol{\omega} \times \frac{\partial T}{\partial \boldsymbol{\omega}} + \mathbf{v} \times \frac{\partial T}{\partial \mathbf{v}} = \mathbf{M} \qquad (9.22)$$

Here $\mathbf{F}$ and $\mathbf{M}$ are the vectors of external force and moment that act on ADS, while $\mathbf{v} = [u, v, w]^T$ and $\boldsymbol{\omega} = [p, q, r]^T$ are the vectors of linear and angular velocity [compare Eqs. (5.20) and (5.24)]. After substituting the expression for kinetic energy of the body and the air into Eqs. (9.21) and (9.22), applying appropriate apparent mass tensor for a symmetric body, and some algebra, the final form of equations of motion will appear to be as follows [Cockrell and Doherr 1981]:

$$
\mathbf{F} = \begin{bmatrix}
(m + \alpha_{11})(\dot{u} - vr) + (m + \alpha_{33})wq + (K + \alpha_{15})(\dot{q} + rp) \\
(m + \alpha_{11})(\dot{v} + ur) - (m + \alpha_{33})wp - (K + \alpha_{15})(\dot{p} - qr) \\
(m + \alpha_{33})\dot{w} - (m + \alpha_{11})(uq - vp) - (K + \alpha_{15})(p^2 + q^2)
\end{bmatrix}
\tag{9.23}
$$

$$
\mathbf{M} = \begin{bmatrix}
(I_{xx} + \alpha_{44})\dot{p} - (K + \alpha_{15})(\dot{v} - wp + ur) - (I_{yy} + \alpha_{44} - I_{zz} - \alpha_{66})qr + (\alpha_{33} - \alpha_{11})vw \\
(I_{yy} + \alpha_{44})\dot{q} + (K + \alpha_{15})(\dot{u} + wq - vr) + (I_{yy} + \alpha_{44} - I_{zz} - \alpha_{66})pr - (\alpha_{33} - \alpha_{11})uw \\
(I_{zz} + \alpha_{66})\dot{r} + (I_{yy} - I_{xx})pq
\end{bmatrix}
$$

$$\tag{9.24}$$

In Eqs. (9.23) and (9.24), $m = \sum_{i=1}^{4} m_i$ denotes the mass of the system that includes the mass of canopy m_1, rigging lines m_2, actuators m_3 and payload m_4, $K = \sum_{i=1}^{4} m_i z_i = m z_G$, where Z_G is the static center of gravity (c.g.) of the overall PADS with respect to the point O (see Fig. 9.34b). The other notations denote the diagonal components of PADS's inertia tensor I_{xx}, I_{yy}, I_{zz} and apparent mass tensor's components α_{mn}. The only asymmetry left in these equations is due to possible asymmetry of payload (I_{xx} in general may not be equal to I_{yy}).

Equations (9.23) and (9.24) are very similar to those used to model a rigid-body motion [$\mathbf{F} = m\mathbf{v} + m(\boldsymbol{\omega} \times \mathbf{v})$ and $\mathbf{M} = \mathbf{I}\dot{\boldsymbol{\omega}} + \boldsymbol{\omega} \times \mathbf{I}\boldsymbol{\omega}$], and in the vector form they can be rewritten as

$$
\mathbf{F} = \mathbf{M}_m \dot{\mathbf{v}} + \boldsymbol{\Lambda}_c \hat{\mathbf{M}}_m \quad \text{and} \quad \mathbf{M} = \mathbf{I}\dot{\boldsymbol{\omega}} + \mathbf{H}_c \hat{\mathbf{I}} + \mathbf{M}_{cr}
\tag{9.25}
$$

In the latter expression

$$
\mathbf{M}_m = \text{diag}([m + \alpha_{11}, m + \alpha_{11}, m + \alpha_{33}])
$$

$$
\hat{\mathbf{M}}_m = [m + \alpha_{11}, m + \alpha_{33}, K + \alpha_{15}]^T, \quad \mathbf{M}_{cr} = \begin{bmatrix} 0, 0, (I_{yy} - I_{xx})pq \end{bmatrix}^T
$$

$$
\boldsymbol{\Lambda}_c = \begin{bmatrix}
-vr & wq & \dot{q} + rp \\
ur & -wp & qr - \dot{p} \\
vp - uq & 0 & -(p^2 + q^2)
\end{bmatrix}
$$

$$
\mathbf{H}_c = \begin{bmatrix}
wp - ur - \dot{v} & -qr & vw \\
wq - vr + \dot{u} & pr & -uw \\
0 & 0 & 0
\end{bmatrix}
$$

$$
\mathbf{I} = \text{diag}\big([I_{xx} + \alpha_{44}, I_{yy} + \alpha_{44}, I_{zz} + \alpha_{66}]\big)
$$

$$
\hat{\mathbf{I}} = \begin{bmatrix} K + \alpha_{15}, I_{yy} + \alpha_{44} - I_{zz} - \alpha_{66}, \alpha_{33} - \alpha_{11} \end{bmatrix}^T
$$

Being resolved with respect to $\dot{\mathbf{v}}$ and $\dot{\boldsymbol{\omega}}$, Eq. (9.25) yields

$$
\dot{\mathbf{V}} = \mathbf{M}_m^{-1}(\mathbf{F} - \boldsymbol{\Lambda}_c \hat{\mathbf{M}}_m) \quad \text{and} \quad \dot{\boldsymbol{\omega}} = \mathbf{I}^{-1}(\mathbf{M} - \mathbf{H}_c \hat{\mathbf{I}} - \mathbf{M}_{cr})
\tag{9.26}
$$

[compare Eq. (5.91)]. The attitude of the PADS is determined by the Euler angles ϕ, θ, and ψ, and their dynamics is defined by Eq. (5.15). The local tangent plane coordinates $\mathbf{p} = [x, y, z]^T$ of the origin of $\{b\}$ are obtained by integrating Eq. (5.28).

The static mass center and moments of inertia are determined based on the weight and dimensions of each component. First, the z coordinate of each component centroid with respect to one of its surfaces $\tilde{z}_i$ and individual central moments of inertia for each component in the axes of correspondent centroid $\tilde{I}^i_{jj}$ were derived. (Magnitudes of Z_i can then be computed with account of $\tilde{z}_i$.) Table 9.2 contains formulas for each ADS component. The approximate formula for a hemispheroidal shell was specifically derived to match the real values in the range of $\varepsilon \in [0.5; 1]$. In Table 9.2, R_* is a radius of the shell measured at z coordinate of its centroid $\{R_* = R_p - 0.5 l_{SL} \sin(\gamma)$ for SLs and $R_* = 0.5 [\sqrt{2} a_{pl} + l_{PMA} \sin(\gamma)]$ for PMAs}. The cone half-angle γ for the considered ADS configuration can be computed from the geometric relation $\sin(\gamma) = (R_p - 0.5 \sqrt{2} a_{pl}) / (l_{SL} + l_{PMA})$.

Next, the individual inertia components computed according to relations of Table 9.2 are transferred to the origin of $\{b\}$ using the parallel axis theorem

$$I^i_{jj} = \tilde{I}^i_{jj} + m_i z^2_i, \quad i = 1, \ldots, 4 \tag{9.27}$$

Finally, the moments of inertia for the whole ADS are computed as a sum of inertias of corresponding PADS components $I_{jj} = \sum^4_{i=1} I^i_{jj}$.

The geometry-mass properties of all AGAS components along with the values for the moments of inertia of each component are given in Appendix H. Although the major contributor to the magnitudes of I_{xx} and I_{yy} is obviously the payload (more than 98%), the moment I_{zz} is primarily affected by the canopy, SLs and PMAs, with less than 15% contribution from the payload. That proves the assumption about neglecting the effect of ADS asymmetry while lengthening risers to be quite reasonable. Another feature is that the symmetry of the cargo box with respect to the axis z simplifies Eq. (9.26) zeroing vector $\mathbf{M}_{cr}$.

Similar to the rigid-body mass tensor, the apparent (virtual) mass tensor $\mathbf{A}$ has $6 \times 6 = 36$ elements. For an ideal fluid, however, $\mathbf{A}$ is a symmetrical tensor, leaving a maximum of 21 distinct terms. In the case of a body with two planes of symmetry and the coordinate frame origin located somewhere on the axis of symmetry, tensor $\mathbf{A}$ can be further reduced to

$$\mathbf{A} = \begin{bmatrix} \alpha_{11} & 0 & 0 & 0 & \alpha_{15} & 0 \\ 0 & \alpha_{22} & 0 & \alpha_{24} & 0 & 0 \\ 0 & 0 & \alpha_{33} & 0 & 0 & 0 \\ 0 & \alpha_{24} & 0 & \alpha_{44} & 0 & 0 \\ \alpha_{15} & 0 & 0 & 0 & \alpha_{55} & 0 \\ 0 & 0 & 0 & 0 & 0 & \alpha_{66} \end{bmatrix} \tag{9.28}$$

TABLE 9.2   RELEVANT FORMULAS OF MOMENTS OF INERTIA FOR PARACHUTE COMPONENTS

Component	Canopy (1) [Favorin 1977]	Suspension lines (2) and PMAs (3)	Payload (4) [Tuma 1987]
Geometry			
$\tilde{z}_i = \left\| \overrightarrow{BC} \right\|$	$-\dfrac{R_p}{2} \varepsilon^{0.83}$	$\dfrac{L \cos \gamma}{2}$	$\dfrac{a_{pl}}{2}$
$\tilde{I}_{aa}^i = \tilde{I}_{bb}^i$	$0.248 m_1 R_p^2 e^{0.52\varepsilon} 0.246 m_1 R_p^2 e^{\varepsilon} \ (I_{xx}^1 = I_{yy}^1)$	$\dfrac{m_i}{2} \left[\dfrac{L^2(1 + \cos^2 \gamma)}{12} + R_*^2 \right]$	$\dfrac{m_4 a_{pl}^2}{6}$
$\tilde{I}_{cc}^i$	$\dfrac{2}{3} m_1 R_p^2 (1 + 0.143 \ln \varepsilon)$	$m_i \left(\dfrac{L^2 \sin^2 \gamma}{12} + R_*^2 \right)$	$\dfrac{m_4 a_{pl}^2}{6}$

Here the first three diagonal elements represent apparent masses of the air virtually stagnant within and around (below) canopy, the next three are corresponding apparent moments of inertia, and off-diagonal air mass/inertia elements contribute to the coupling motion. Because of axial symmetry of the circular canopy $\alpha_{22} = \alpha_{11}$, $\alpha_{55} = \alpha_{44}$, and $\alpha_{24} = -\alpha_{15}$. That leaves only three distinct elements, which are α_{11}, α_{33}, α_{44}, α_{15}, and α_{66} [that is why dynamic equations have their appearance as in Eqs. (9.23) and (9.24)].

In the earlier studies to represent a flow around a fully deployed canopy, the latter was represented as a spheroid. In this case the reference air mass and moments of inertia correspond to those of the air displaced by the body

$$m_a = m_a^s = \frac{4}{3} \pi \rho R_p^3 \varepsilon \tag{9.29}$$

$$I_{xx}^{air} = \frac{1}{5} m_a R_p^2 (1 + \varepsilon^2), \quad I_{zz}^{air} = \frac{2}{5} m_a R_p^2 \tag{9.30}$$

(see conventions of Table 9.2). Today, it is usual practice to refer to the air trapped within a hemispheroid. In this case, air mass makes the half of spheroid

$$m_a = 0.5 m_a^s \tag{9.31}$$

and formulas for the moments of inertia are the same as in Eq. (9.30). If computed with respect to centroid axes, the z coordinate for hemispheroid's centroid equal to $z_P = -3 \varepsilon R_p / 8$ needs to be taken into consideration.

Generally speaking, apparent mass terms depend on canopy's configuration, porosity, acceleration, and spatial angle of attack. To be more specific, apparent masses may drop their values more than 20 times with a porosity increase from 0 to 40% [Ibrahim 1965a; Ibrahim 1965b]. Apparent moments of inertia also decrease their values by the factor of 2.75. The coefficient α_{33} decreases 4.5 times with increase of the angle of attack from 0 to 40 deg [Yavus and Cockrell 1981; Yavus 1989]. While experiencing steady acceleration, the apparent mass terms might change as much as by a factor of five. Nevertheless, the most commonly used form assumes an explicit dependence of the apparent mass terms on the air density only while all other possible effects are represented by constant multipliers

$$\alpha_{11} = k_{11} m_a, \quad \alpha_{33} = k_{33} m_a, \quad \alpha_{44} = k_{44} \tilde{I}_{xxa}, \quad \alpha_{66} = k_{66} \tilde{I}_{zza}, \quad \alpha_{15} = k_{15} m_a z_P \tag{9.32}$$

For the sake of comparison, Table 9.3 presents the values of these multipliers used in different studies over the course of 60 years. The current model uses $k_{11} = 0.5$, $k_{33} = 1.0$, $k_{44} = 0.24$, and $k_{15} = 0.75$, the values that are very close to those of Doherr and Saliaris [1981]. The slight difference is caused by a necessity to match notations of Eqs. (9.30) and (9.31). The coefficient k_{66} responsible for damping yaw oscillations was set to zero. Also, the distance from the frame $\{b\}$ origin to the point of application of the translational apparent mass component

TABLE 9.3   VALUES OF APPARENT MASS COEFFICIENTS

##	Researcher	Model	Origin of $\{b\}$	Ref. Body	ε	k_{11}	k_{33}	k_{44}	k_{15}
1	Henn [1944]	3-DoF	O	Spheroid	0.5	0.5 [0–0.6]	0.5 [0–1.0]		
2	Ludwig and Heins [1963]	3-DoF	O	Spheroid	0.5	1.0 [0.6–1.4]	1.0 [0.6–1.4]		
3	White and Wolf [1968]	5-DoF	Sys. cg	Sphere	1	0.7	0.7	0.23	
4	Tory and Ayres [1977]	6-DoF	Can. cg	Spheroid	0.5	1.31	2.12	1.34	
5	Eaton [1982]	6-DoF	Can. cg	Spheroid	0.5	0.2 [0–0.5]	0.4 [0–1.0]	Unclear	
6	Yavuz and Cockrell [1981]; Yavuz [1989]	6-DoF	Can. cg	Hemisphere	1	[1.31–6]	[2.12–10]	1.5–3.0	
7	Cockrell and Doherr [1981]; Doherr and Saliaris [1981]	6-DoF	O	Hemisphere	1	0.5	1.0	0.24	0.75
8	Present	6-DoF	O	Hemispheroid	0.82	0.5	1.0	0.24	0.75

(canopy's center of pressure) z_P was utilized as opposed to the expression $l_B = \sqrt{(l_{SL} + l_{PMA})^2 - R_p^2}$ (see Fig. 9.34b) used in other studies.

For the G-12 based AGAS with $m_a = 472$ kg (1,040 lb) $[\rho = 1 \text{ kgm}^{-3}$ (0.062 lb/ft³)], $\alpha_{11} = 236$ kg (520 lb) ($\sim$22% of m), $\alpha_{33} = 472$ kg (1,040 lb) ($\sim$44% of m), $\alpha_{44} = 1,600$ kgm² (37,970 lb/ft²) (less than 0.5% of I_{xx}), and $\alpha_{15} = 707$ kgm (5,114 lb/ft) ($\sim$3% of K). By looking at these numbers, it becomes clear that α_{11} and α_{33} are evidently of most importance. The difference between them affects the dynamic behavior of the system [see Eq. (9.24)].

9.1.6 IDENTIFICATION OF AERODYNAMIC FORCES AND MOMENTS

The total external force and moment acting on the system (Fig. 9.34a) are caused by the aerodynamic effects and the weight of each system component. Thus, we can write

$$\mathbf{F} = \mathbf{F}^{a/d} + \mathbf{G}, \quad \mathbf{M} = \mathbf{M}^{a/d} + \sum_i \mathbf{M}_G^i \tag{9.33}$$

where $\mathbf{F}^{a/d}$ and $\mathbf{M}^{a/d}$ are the total aerodynamic force and total aerodynamic moment, $\mathbf{G} = {}_b^u \mathbf{R}^T [0, 0, mg]^T$ [the apparent (virtual) masses do not contribute to the weight of the system], and $\mathbf{M}_G^i$ are the moments due to the weight of each component when being translated to the origin of the body frame. In turn, for a controlled ADS

$$\mathbf{F}^{a/d} = \mathbf{F}_{\text{canopy}} + \mathbf{F}_{\text{risers}}, \quad \mathbf{M}^{a/d} = \mathbf{M}_{\text{canopy}} + \mathbf{M}_{\text{risers}} \tag{9.34}$$

where $\mathbf{F}_{\text{canopy}}$ and $\mathbf{M}_{\text{canopy}}$ denote aerodynamic force and moment acting on an undistorted canopy, while $\mathbf{F}_{\text{risers}}$ and $\mathbf{M}_{\text{risers}}$ denote the aerodynamic force and moment caused by the change in a riser length.

The aerodynamic force vector $\mathbf{F}_{\text{canopy}}$ in Eq. (9.34) depends on the spatial angle of attack α_{sp} and dynamic pressure $[Q = 0.5\rho\|\mathbf{V}_a\|^2$ and can be presented as follows:

$$\mathbf{F}_{\text{canopy}} = C_D(\alpha_{sp}) Q S_0 \frac{\mathbf{V}_a}{\|\mathbf{V}_a\|} \tag{9.35}$$

Here $\|\cdot\|$ denotes the Euclidian norm of a vector, and $S_0 = \pi R_0^2$ is the canopy's reference area.

The spatial angle of attack α_{sp} and its components, angle of attack α and sideslip angle β, can be computed using $\{b\}$-frame components of an airspeed vector $\mathbf{V}_a$ as follows (see Fig. 9.35):

$$\alpha_{sp} = \tan^{-1}\left(\frac{\sqrt{v_x^2 + v_y^2}}{v_z}\right), \quad \alpha = \tan^{-1}\left(\frac{v_x}{v_z}\right), \quad \beta = \tan^{-1}\left(\frac{v_y}{\sqrt{v_x^2 + v_z^2}}\right) \tag{9.36}$$

Fig. 9.35   Flight angles' determination.

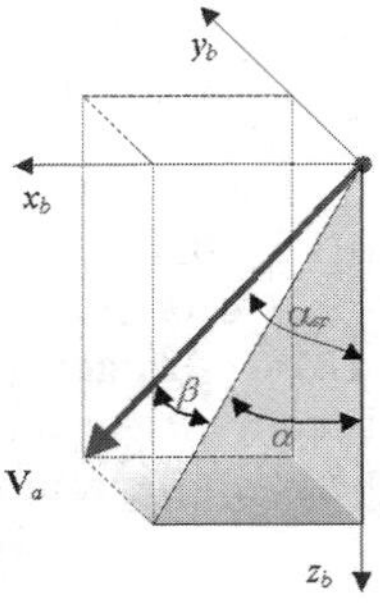

Obviously, these definitions are slightly different from those for the parafoil [Eq. (5.77)] because they are defined off the z_b axis as opposed to the x_b axis. The components of the airspeed vector in $\{b\}$ are defined exactly the same as for a parafoil [Eq. (5.78)].

The general shape of dimensionless aerodynamic coefficients of the undistorted circular parachute vs angle of attack, for example, $C_D(\alpha_{sp})$, was adopted from [Knacke 1992], with initial aerodynamic coefficient estimates specific to G-12 AGAS based on the CFD simulations. For example, Fig. 9.36 shows the functional dependence of two aerodynamic coefficients on the angle of attack [Mosseev 2001a; Mosseev 2001b]. In this figure, $C_D^*(\alpha_{sp})$ denotes the aerodynamic drag coefficient, and $C_m^*(\alpha_{sp})$ denotes the total aerodynamic moment coefficient both depending on a spatial angle of attack α_{sp}. [Asterisk in $C_D^*(\alpha_{sp})$ and $C_m^*(\alpha_{sp})$ means that these are computed values that need to be validated using the results of the real drops.]

The longitudinal and the lateral motion of a symmetric parachute in glide plane are sufficiently uncoupled [White and Wolf 1968]. Therefore, the longitudinal and lateral motion can be studied separately, similar to studying linearized dynamics of an aircraft. Because the roll and pitch motion of the ADS have the same moment characteristics, that is, $C_{\text{roll}} = C_m(\beta)$, $C_{\text{pitch}} = C_m(\alpha)$, and the vector of aerodynamic moment becomes

$$\mathbf{M}_{\text{canopy}} = 2QS_0 R_0 \begin{bmatrix} C_{\text{roll}} \\ C_{\text{pitch}} \\ C_n \end{bmatrix} \tag{9.37}$$

[The moments are dimensionalized by a parachute diameter—that is why Eq. (9.37) features a factor of 2.]

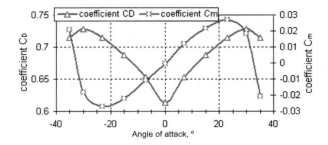

Fig. 9.36   G-12 parachute aerodynamics.

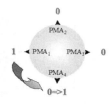

Fig. 9.37 Yaw rotation scheme.

In Eq. (9.37), C_n denotes the yaw moment due to yaw acceleration, and for a symmetric body it is equal to zero. However, in the case of transition of one of the risers from one state to another while one of the adjacent risers has been already lengthened, this moment is not equal to zero.

Figure 9.37 clarifies this situation. In this figure, 0 stands for a shortened riser ($\bar{l}_k = 0$), and 1 denotes a lengthened riser ($\bar{l}_k = 1$). Typically, while riser transition (4...5 s), a yaw angle of 15...20 deg is accrued (Fig. 9.38).

Therefore, if the kth riser undergoes transition, the following relation is valid:

$$C_n = \text{sign}(\bar{l}_{k-1} - \bar{l}_{k+1})C_n^{\bar{l}}(\bar{l})\bar{l}_k + C_n^r r \qquad (9.38)$$

In this equation $\bar{l}_k = (l_k - l_{min})/(l_{max} - l_{min})$, the difference $(\bar{l}_{k-1} - \bar{l}_{k+1})$ defines the sign of the moment, and C_n^r is the damping moment coefficient. By analyzing data from 20 flight tests, the functional dependence $C_n^{\bar{l}}(\bar{l})$ was found to be as presented in Fig. 9.39, and the coefficient C_n^r to be constant and equal to 2 s.

When $\bar{l}$ or $1 - \bar{l}$ is small, as happens at the very beginning and the very end of the actuation process, the magnitude of the rotation is low because the canopy symmetry distortion is minimal. The rotation is at its highest at the middle of the actuation process when the shape distortion is maximal (see Fig. 9.38).

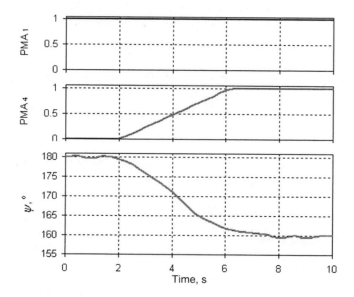

Fig. 9.38 Effect of changing riser length.

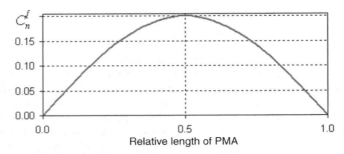

Fig. 9.39   Dependence $\bar{C}_n^{\bar{l}}(\bar{l})$.

Next, let us proceed with the second terms in Eq. (9.34). Upon venting, PMA increases its length by approximately one-third from $l_{PMA} = l_{min}$ to $l_{PMA}^* = l_{max} = l_{PMA} + \Delta l_{PMA}$ and compresses in diameter [the dependence $\bar{l}_{PMA} = f(P_{PMA})$ where P_{PMA} stands for the PMA pressure is shown in Fig. 9.40]. This action "deforms" the parachute canopy creating an unsymmetrical shape, essentially shifting the center of pressure, and providing a drive or slip condition in the opposite direction of the control action. With four independently controlled actuators, two of which can be activated simultaneously, eight distinct control actions can be generated.

The change in the aerodynamic force due to PMA activation $\mathbf{F}_{risers}$ was modeled as a function of PMA's relative length $\bar{l}$, number of PMA actuated n, and involved actuation system dynamics with a transition time τ [Benney et al. 1999]

$$||\mathbf{F}_{risers}(t)|| = QS_0 f(\bar{l}, n, \tau) \tag{9.39}$$

Figure 9.41 shows the steady-state values of dependence $f(\bar{l}, n, \infty)$.

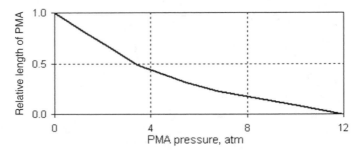

Fig. 9.40   Dependence $\bar{l}_{PMA} = f(P_{PMA})$.

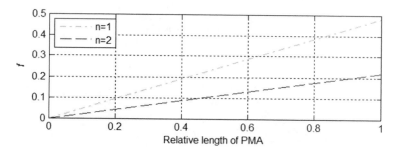

Fig. 9.41   Dependence $f(\bar{l}, n, \infty)$.

In turn, the actuator moment may be computed as $\mathbf{M}_{risers} = \mathbf{P}_{CP} \times \mathbf{F}_{risers}$, where $\mathbf{P}_{CP} = [0, 0, z_P]^T$.

A standard nonlinear system identification technique was applied to tune the CFD dependences $C_D^*(\alpha_{sp})$ and $C_m^*(\alpha_{sp})$ so that the output of the developed six-DoF model for a controlled circular parachute matched flight-test data. (Chapter 11 addresses different parameter identification techniques in more detail.) To this end, expressions for $C_d(\alpha_{sp})$ and $C_m(\alpha_{sp})$ were parameterized as

$$C_D(\alpha_{sp}) = k_{C_{D0}} \tilde{C}_{D_{\alpha=0}} + k_{C_{D\alpha}} \left[C_D^*(\alpha_{sp}) - C_D^*(0) \right] \tag{9.40}$$
$$C_m(\alpha_{sp}) = k_{C_m} C_m^*(\alpha_{sp}) \tag{9.41}$$

yielding three optimization parameters $k_{C_{D0}}$, $k_{C_{D\alpha}}$ and k_{C_m}. The initial value of $\tilde{C}_{D_{\alpha=0}} = mg(QS_0)^{-1}$ was obtained from the obvious equation for a steady descent rate.

The objective of parameter identification was to vary these three parameters to minimize a cost function

$$J = \int_0^{t_f} ||\mathbf{p}_u(t) - \hat{\mathbf{p}}_u(t)|| \, dt \tag{9.42}$$

where $\mathbf{p}_u(t)$ is the inertial position of PADS obtained in flight test, $\hat{\mathbf{p}}_u(t)$ denotes the estimated position of ADS obtained in simulation, and t_f stands for the flight time.

From the analysis of the flight-test data, it became obvious that there was a significant difference in the AGAS dynamics during controlled and uncontrolled drops. Hence, the identification algorithm was first applied to the data obtained from the uncontrolled drop. The resulting values of $k_{C_{D0}}$, $k_{C_{D\alpha}}$, and k_{C_m} were then used to initialize the second step, where the same technique was applied to a controlled drop. While values of optimization parameters obtained in the first step provided estimates of ADS aerodynamics around a zero angle of attack, their adjustment at the second step characterized ADS dynamics at higher

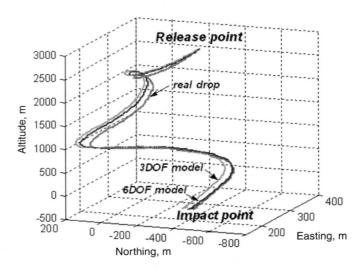

Fig. 9.42   Three-dimensional projections.

angles of attack with the nonzero control inputs. Averaging these values allowed matching both controlled and uncontrolled drop sets.

Figures 9.42 and 9.43 show the 3D and 2D plots of the trajectories obtained in one of the real flights along with the matching three-DoF and six-DoF simulations. This particular data set was selected due to the richness of the frequency spectrum of the parachute motion involved, including three instances where parachute's direction of motion changes by more than 90 deg. Because the ADS model assumes a fully deployed canopy, all simulations were

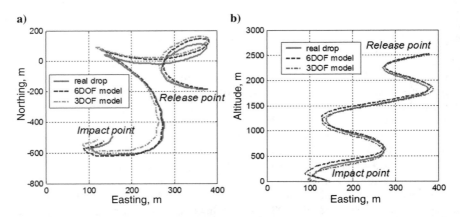

Fig. 9.43   a) Horizontal and b) vertical plane projections of a trajectory shown in Fig. 9.41.

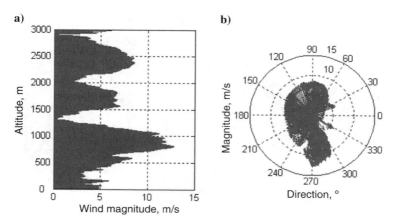

Fig. 9.44 a) Measured wind magnitude vs altitude profile and b) scattergram of wind vectors.

initialized at the point on a real drop trajectory where the canopy was fully deployed. All trajectories are plotted in the East-North-Up coordinate frame. Figure 9.44 shows the wind profile obtained prior to the AGAS drop. It clearly indicates that both wind direction and its magnitude experience significant changes with time.

Because during the uncontrolled drop the angle of attack and sideslip angle are close to zero, the value of k_{C_m} in Eq. (9.41) was set to zero. This choice implies that the amplitude of coning motion may differ from the flight-test data; however, for uncontrolled drops its natural frequency does not depend on k_{C_m}. The optimal values $k_{C_{d0}}$ and $k_{C_{D\alpha}}$ were obtained by applying a simple search technique to minimize a cost function of Eq. (9.42).

Figures 9.45 and 9.46 show the plots of x, y, and z components of the inertial velocity obtained from the matching three-DoF and six-DoF simulations as compared to the ones recorded during the drop (for the three-DoF simulation the descent rate remains set as a constant). Clearly, minimization of the cost function over a set of two varied parameters gave a fairly good result, so that the matching six-DoF model follows flight-test data reasonably well.

Figure 9.47 presents the power spectrum density (PSD) plots for the Euler angles taken from the flight test and the matching six-DoF model run, and Table 9.4 shows the numerical values of principal eigenfrequencies for each channel obtained via analysis of the PSD plots. Similar results for the principal eigenfrequencies were obtained independently for another set of AGAS flight, which indicates the consistency of ADS flight-test data taken in the presence of the different wind profiles. Thus, Table 9.4 data reflect the inherent properties of ADS rather than the wind spectrum. To summarize, the developed six-DoF model catches the dynamics of a real uncontrolled AGAS fairly well.

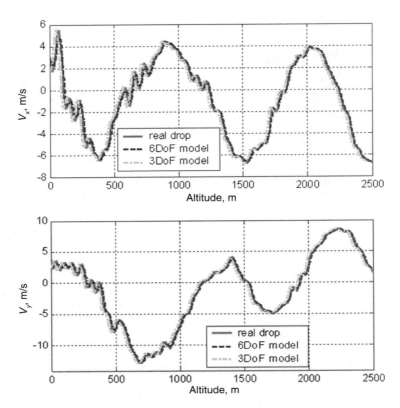

Fig. 9.45   Comparison of the a) *x* and b) *y* components of the ground-speed vector.

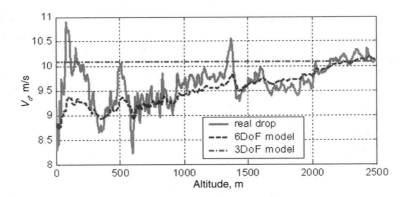

Fig. 9.46   Comparison of the descent rate.

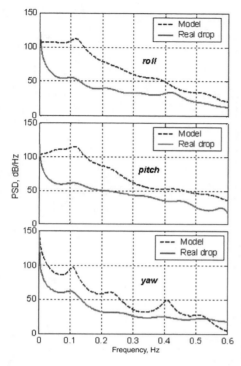

Fig. 9.47   Euler angles PSD analysis.

To investigate the influence of the control inputs on the AGAS dynamics, the following approach was used. First, the wind profile measured during the controlled drop was used to simulate the uncontrolled AGAS model verified at the previous step as just described. Then, this simulated data were subtracted from the flight-test data, and this difference was used as an input for the system identification algorithm. (This implies that the control's influence on the descent rate is minor, which is pretty much the case.) An example of trajectories at the presence of strong wind is shown in Fig. 9.48a.

The result of the identification procedure for the data set shown in Fig. 9.48a is presented in Fig. 9.48b. Here, the *"Control + NoWind"* plot represents a position obtained by driving six-DoF model with the control inputs recorded in the real drop in the no-wind conditions (because they were accounted for already). Figure 9.48c compares controlled drop trajectories obtained in a simulation with double tuned aerodynamics and the flight test. The final values of three varied parameters are shown in Table 9.5.

TABLE 9.4   EIGENFREQUENCIES AND THE CORRESPONDING PERIODS IDENTIFIED FROM PSD ANALYSIS

Channel	Real Drop		Six-DoF Model	
	Frequency, Hz	Period, s	Frequency, Hz	Period, s
Roll	0.109	9.2	0.114	8.8
Pitch	0.112	8.9	0.111	9.0
Yaw	0.12, 0.22, and 0.406	8.3, 4.5, 2.5	0.13, 0.22, and 0.41	7.7, 4.5, 2.4

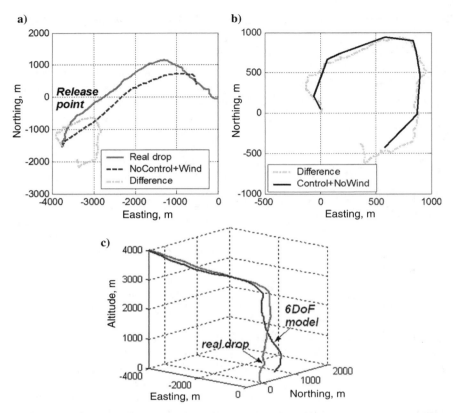

Fig. 9.48   a) Extraction of control actions contribution and b, c) real drop data vs matched simulation comparison.

9.1.7   CONTROL OF CROSS-TYPE PARACHUTES AND PARACHUTE CLUSTERS

Because of their known advantages, cruciform or cross-type canopies might be another alternative decelerator system within the AGAS-like framework. These

TABLE 9.5   VALUES OF OPTIMIZATION PARAMETERS

	Initial Value	Uncontrolled Drop	Controlled Drop
$k_{C_{D0}}$	1	1.2	1.2
$k_{C_{D\alpha}}$	1	1.005	1.25
k_{C_m}	1	0	0.6

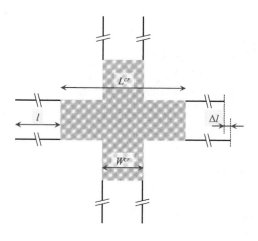

Fig. 9.49   Specifications of a cruciform parachute [Jorgensen and Cockrell 1981].

parachutes originally developed by Forichon in 1961 consist of two identical cloth rectangles, crossed and joined to each other at the square intersection to form a flat surface having four equal arms [Forichon 1961]. Suspension lines are attached to the outer edges of the four arms (Fig. 9.49). Forichon suggested using tie cords between corners of the adjacent arms, but designs described here omit them.

Because of their simple geometric configuration, parachutes having a cruciform or cross design are notable for their simplicity in construction and, therefore, low cost. Cruciform canopies feature good drag to canopy area ratio and good static and dynamic stability characteristics and exhibit small effect of AoA on the axial force. These parachutes have been used for aircraft braking and decelerating other vehicles at ground level, for the stabilization and deceleration of weapons and for recovery of high-altitude probe equipment, for which a low rate of descent is required [Jorgensen and Cockrell 1981].

Aerodynamic characteristics of cross-type canopies are well studied [Nice et al. 1965; Ludtke 1972; Jorgensen 1982; Shen and Cockrell 1988; Levin and Shpund 1997; Han et al. 2013]. That includes the effect of changing the ratio of the arm length L^{cr} and arm width W^{cr} (within 2.4 to 4 range), changing the ratio of the suspension line length l and arm length L^{cr} (within 0.67 to 2 range), varying cloth porosity, and arm shape, that is, the ratio of trapezoidal-shape arms a to b in Fig. 9.50a (within the 0.5 to 2 range).

Symmetrically designed cross-type parachutes, however, have a tendency to rotate during descent, which also affects longitudinal stability. The spin rate depends on the amount of asymmetry, that is, manufacturing inaccuracy, payload-induced asymmetry, initial deployment

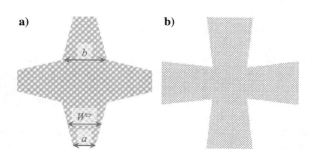

Fig. 9.50   Modifications of geometry of cross-type parachutes [Levin and Shpund 1997].

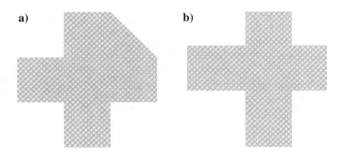

Fig. 9.51   Examples of asymmetric cross-type parachutes.

conditions, and the spin damping of the canopy. The modified design shown in Fig. 9.50a seems to be less susceptible to manufacturing inaccuracies while the design of Fig. 9.50b offers a better spin rate control [Levin and Shpund 1997].

Spin can also be caused by intentionally introduced differences in suspension line lengths Δl, as shown in Fig. 9.49. If the spin rate can be controlled, the cross-type parachutes could be used similarly to AGAS. The (limited) gliding capability could be achieved by modifying their geometry and/or retrimming selected suspension lines (shortening or even removing certain suspension lines). Figure 9.51a features a canopy built out of a cruciform parachute and incorporating one triangular panel (the Hycross parachute tested by Potvin et al. [2003] employed two panels). Figure 9.51b shows asymmetric shape studied by Shpund and Levin [1997]. Potvin et al. [2003] conducted the first field study assessing the effect of different retrimming schemes on gliding capabilities and spin rate by performing drop testing with a variety of static suspension line lengths in a myriad of configurations.

As mentioned in Sec. 9.1.4, the original AGAS employed a single G-12 flat circular parachute to carry up to 1-ton (2,200-lb) payload. To carry heavier payloads, AGAS-5K utilized a larger G-11 canopy. To carry even heavier payloads, clusters of parachutes might be required. To this end, Table 9.6 shows a maximum payload that can be carried by major cargo parachutes used singularly or in clusters configurations [Cargo Parachutes 2012; CARP 2005]. These cargo parachute assemblies were developed in the late 1940s and early 1950s and until today remain the mainstay of airdrop and recovery operations (in the mid-1960s G-14 replaced a smaller G-13 canopy). Surely, up until now these cargo parachute assemblies were not controllable and were modeled as such [Ke et al. 2009; Ray and Morris 2011]. However, now that AGAS proved that physical modifications of a single canopy (canopy deformation or openings on the canopy) can provide some glide capability, it is worth investigating the potential of using the same control concept on cluster assemblies. For completeness, this section briefly introduces one of such initial efforts.

TABLE 9.6   SUSPENDED WEIGHT IN POUNDS OF CARGO PARACHUTES USED
SINGULARLY OR IN CLUSTERS

Type	Diameter	Parachutes	Minimum	Maximum
G-14	34 ft (10.3 m)	1	100 lb (45.5 kg)	500 lb (226 kg)
		2	500 lb (226 kg)	1,000 lb (453 kg)
		3	1,000 lb (453 kg)	1,500 lb (6,003 kg)
G-12E	64 ft (19.5 m)	1	501 lb (227 kg)	2,200 lb (997.9 kg)
		2	2,270 lb (1,030 kg)	3,500 lb (1,587.5 kg)
G-11B	100 ft (30.3 m)	1	2,270 lb (1,030 kg)	5,000 lb (2,267 kg)
		2	5,001 lb (2,268 kg)	10,000 lb (4,535 kg)
		3	10,001 lb (4,356 kg)	15,000 lb (6,003 kg)
		4	15,001 lb (6,004 kg)	20,000 lb (9,071 kg)
G-11C	100 ft (30.3 m)	5	20,001 lb (9,072 kg)	25,000 lb (11,339 kg)
		6	25,001 lb (11,340 kg)	30,000 lb (13,607 kg)
		7	30,001 lb (13,608 kg)	35,000 lb (15,875 kg)
		8	35,001 lb (12,867 kg)	40,000 lb (18,144 kg)

Obviously, individual glide and steering methods applied to each canopy will not necessarily work on a cluster assembly. This is illustrated in Fig. 9.52 showing typical geometry of the clusters of two to eight circular canopies. Canopies will obviously move around their nominal locations and interact between themselves making it impossible to achieve glide and control of the cluster assembly as a whole even for the simplest two-canopy assembly. The pendulum motion of parachute clusters has its consequences as well [Machin and Ray 2015].

To circumvent this difficulty, Lee and Buckley [2004] suggested partially connecting the canopies of a cluster, specifically two- and three-canopy assemblies as shown in Fig. 9.53. [Inspired by the work of Boca, Render and Coulter [1993] studied drag and stability characteristics of a cluster of three small cruciform parachutes (Fig. 9.54).]

As shown in Fig. 9.53a, a two-parachute cluster assumes two canopies to be connected at the skirt between the points A and D. In this particular example, one-quarter of a skirt circumference is connected. For a three-canopy cluster assembly (Fig. 9.53b) to maintain a more streamlined configuration as compared to that of Fig. 9.52, canopies 1 and 3 are suggested to be connected to canopy 2 at an angle of 30 deg with respect to the center diameter EH of canopy 2. As a result, connections A–E and L–H utilize one-sixth of a skirt circumference of each canopy.

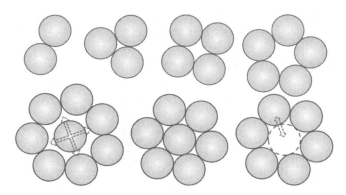

Fig. 9.52   Nominal locations of parachutes in a cluster composed of two to eight canopies [Baca 1984].

To control these two assemblies, the skirt of the leading gores between the points C–D and D–E in Fig. 9.53a and between the points D–E, F–G, and H–I in Fig. 9.53b should be pulled down toward the confluence point. A suggested amount of pull-down is about 5% of the canopy circumferential diameter. When the leading gores are pulled down, the inflated canopy is deformed to create an air jet emerging from the opposite side. A circumferential rectangular opening cut or radial slots (with a total vent area of about 2.5% of the canopy surface) near the skirt directly opposite to the pulldown gores in both configurations are supposed to enhance this jet effect. These jets produce a side force to glide the entire cluster assembly forward. If a turn is desired, the pull-down of outwards cluster canopies will be executed asymmetrically. As a result, the cluster assembly of circular canopies will glide in a manner similar to that of parafoil as opposed to AGAS (which has no turn control).

Figure 9.55a shows a perspective view of a steerable cluster of two parachutes featuring two pull-down/release mechanisms M1 and M2. In fact,

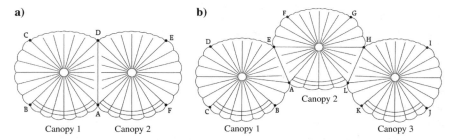

Fig. 9.53   Example of steerable cluster designs composed of a) two and b) three parachutes [Lee and Buckley 2004].

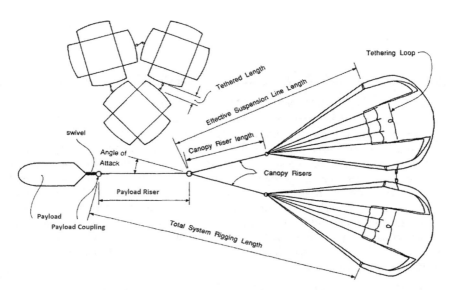

Fig. 9.54 Example of cruciform parachute cluster design [Render and Coulter 1993].

to control a two-canopy assembly AGU of AGAS (Fig. 9.33a) can probably be employed with only minor mechanical modifications. (The GNC paradigm should obviously be different.) Figures 9.53a and 9.53b show two snapshots of a flight test of the prototype two-canopy assembly composed of one-quarter-scale G-12 canopies featuring a steady-state glide (Fig. 9.55a) and turn (Fig. 9.55b), respectively. In these initial tests the encouraging values of up to 0.9:1 (fluctuating between 0.3:1 and 0.9:1) glide ratio (pretty much the same

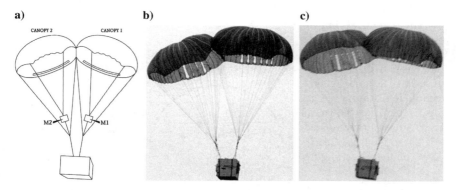

Fig. 9.55 A steerable cluster of two parachutes: a) perspective view, b) steady glide, and c) turn [Lee and Buckley 2004].

as for a single G-12 AGAS) and 2.5-deg/s turn rate were achieved [Lee and Buckley 2004].

9.2   NON-GLIDING PARACHUTE GUIDANCE USING REEFING CONTROL

This section is based on [Fields et al. 2011; Fields et al. 2012; Fields et al. 2013] and describes yet another approach to control a circular parachute via changing its descent rate by varying the effective drag area (controlled reefing-disreefing). The following presents a general approach for this method starting with a simple 1-DoF guidance, which attempts to intercept some finite line segment on the ground surface, like a road, by computing the single optimal reefing of a canopy based on the best knowledge of the current winds aloft. Then, a more sophisticated approach is described in which the canopy reefing occurs continuously during descent, and therefore is capable of achieving two-DoF guidance capabilities. Although these approaches have not been fully tested, computer simulations and some preliminary flight tests suggest that this approach is worth further exploration.

9.2.1   ONE-DoF GUIDANCE

The suggested approach is based on varying the descent rate of circular parachutes via reefing/disreefing of the skirt resulting in changes to the effective drag area. The continuous control line-based skirt reefing methodology was noted by Knacke [1992] and was implemented by Sadeck and Lee [2009]. However, these systems were developed for passive continuous disreef capabilities, not for reversible reefing of a parachute canopy. Figure 9.56 presents the general approach of using the suspension lines as the reefing mechanism. As the suspension lines are reeled in, the effective canopy drag area is reduced. This technique may be more appropriate for small canopies due to the power demands of the actuation system.

The technical implementation of a reversible reefing system for circular canopies can vary. To this end, several reefing techniques alternative to the simple one shown in Fig. 9.56 were investigated [Fields and Basore 2015]. Some of these techniques

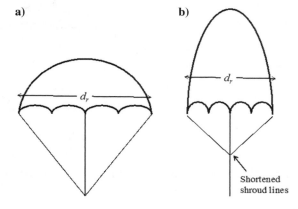

Fig. 9.56   Parachute a) without and b) with reefing via shroud line choking.

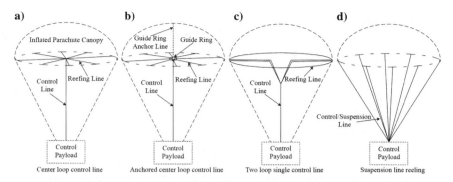

Fig. 9.57 Four reversible reefing techniques with solid lines indicating control and/or reefing lines.

are presented in Fig. 9.57. The four reefing strategies shown in Fig. 9.57 stemmed from a single control-line approach in which a single control line can manipulate the parachute size/shape directly while carrying only a small portion of the suspension line load. Center loop control-line reefing (Fig. 9.57a) requires small lines to be affixed to the skirt at the suspension-line skirt attachment point. These small lines are then attached to a single control line, which, when actuated, forces the skirt closed. In an attempt to keep the small reefing lines as close to perpendicular to the canopy as possible, a small guide ring was attached to the canopy via another short line (Fig. 9.57b). The two loop control-line technique (Fig. 9.57c) incorporates two loops woven through the suspension-line attachment points and connected to a single control line in the center of the canopy. The suspension-line reefing technique (Fig. 9.57d) uses a winch actuator to simply reel the suspension lines in or out to achieve the desired descent speed similar to the technique shown in Fig. 9.56.

To get a sense of a rate of descent control authority, Fig. 9.58 shows the results from drop testing of the two-loop system (which provides the largest descent rate control authority while also requiring the lowest actuation force) on the 1.2-m flat-circular canopy with two different payload configurations. The 100% reefing corresponds to the maximum control line deflection of 0.63 m (2.1 ft). The results adhere to an approximately linear trend with a relatively large controllable descent range. The linearity of the double-loop reefing technique is an ideal candidate for in-flight navigation to be discussed next.

Figure 9.59 depicts the projection of the descent path onto the ground p in relation to the targeted "finish line," which takes the mathematical form of a line between two points, (x_1, y_1) and (x_2, y_2). The idea of the one-DoF guidance is to control the descent rate so that it lands somewhere along this finish line. The descent path projection p originates at the projected release point (x_0, y_0) and

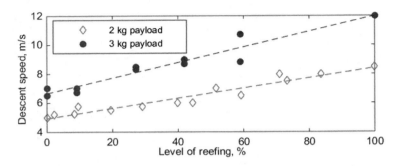

Fig. 9.58 Two loop skirt reefing drop-test data for two payload configurations.

terminates at (x_f, y_f), where the projected path's start and end are connected by the range vector r (of length $r = \|\mathbf{r}\|$), at an azimuth angle θ from the release point.

When only actuating the parachute once (which means keeping drag area S constant for remainder of descent), θ remains constant. Thus, the landing location (or finish line) can be reached if the range vector r intersects the finish line assuming the landing location is within the performance characteristics of the parachute. Then, by choosing the appropriate parachute area (thereby controlling the descent rate), the range can be scaled so that the descent vehicle will land

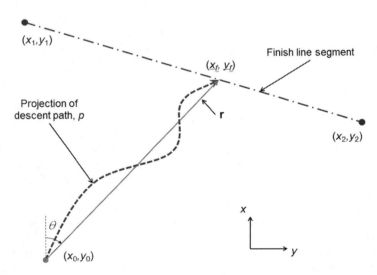

Fig. 9.59 Projection of descent path p onto ground surface with target finish line segment.

on the finish line. Hence, the one-DoF GNC algorithm includes the following components:

1. Obtaining wind field data (direction and speed data as a function of altitude) $W(h)$ (This can come from forecast data or can be directly measured immediately prior to descent as discussed in Sec. 3.1.2.)

2. Identifying the current location and IPI coordinates

3. Using an initial estimate of the desired descent profile and the wind field data $W(h)$ and calculating the *reference predicted descent path* $\hat{P}$ projected onto the ground (The *shape* of this path will be invariant with parachute reefing and can thus be directly scaled for different values of the parachute area S.)

4. Determining the necessary parachute area S that scales the reference descent path so that its end falls along the targeted finish line

5. Adjusting the reefing of parachute to achieve the necessary parachute area S

In practice, this calculation can be periodically updated (looping steps 2–5), and adjustments to the parachute trim can be made to adapt to changing conditions during the descent.

The procedure to find the appropriate reference area S to guide ADS to the target line may rely on using the ballistic winds introduced in Sec. 3.1.3. Having some sort of a winds-aloft profile (measured or predicted), the ballistic winds can be computed according to Eqs. (3.9). For ADS dropped at altitude H, these winds provide with $W^{\text{bal}}(H)$ and $\chi_W^{\text{bal}}(H)$ defining the azimuth angle θ in Fig. 9.59. This azimuth angle defines the length of range vector $\mathbf{r}$ to intersect with the target line. The ratio $r/W^{\text{bal}}(H)$ defines the time T needed to reach the target line intercept point. This time is uniquely related to the descent rate. For example, if using a standard atmosphere, by inverting Eq. (3.20) for the tropospheric drops (below 11 km) and Eq. (3.35) for the high-altitude drops, we obtain

$$V_{v0} = -14{,}171 \frac{1}{T} \left[(1 - 0.00002256 H_d)^{3.128} - (1 - 0.00002256 H_{DZ})^{3.128} \right]$$

$$\text{(9.43)}$$

$$V_{v0} = -13{,}700 \frac{1}{T} \left(e^{-0.000073 H_d} - e^{-0.000073 H_{DZ}} \right) \tag{9.44}$$

respectively. Figure 9.60 shows both dependences graphically for several canopy deployment altitudes and $H_{DZ} = 0$ (we also assumed $V_d \approx V_v$). Knowing speed V_{v0} and utilizing Eq. (3.15) yield

$$S = \frac{2mg}{\rho_0 V_{v0}^2 C_D} \tag{9.45}$$

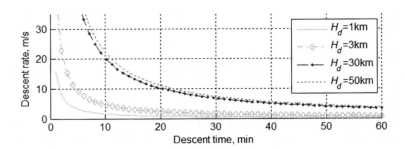

Fig. 9.60   Descent rate required to stay in the air for a certain amount of time.

As mentioned in Sec. 3.1.4 [Eqs. (3.36–3.38)], if air density is provided along with the winds aloft, then a numerical integration procedure may give more accurate results. In this case, a descent rate at each altitude level $k = 1, \ldots, M$ where the air density measurements (estimates) are available ($H_d \leq h_M$) can be computed using Eq. (3.36) or

$$V_{v;k} = \sqrt{\frac{2mg}{\rho(h_k)SC_D}} \tag{9.46}$$

Using these data, the density vs altitude profile is then represented by some approximation $P_{V_v}(h)$ (which may be as simple as a piece-wise linear interpolation). With this approximation, the change in altitude can be determined by integrating the last of three equations of Eq. (3.1)

$$h_n = h_{n-1} - P_{V_v}(h_{n-1})\Delta t \tag{9.47}$$

where $n = 1, 2, 3, \ldots$, $h_0 = H_d$, and integration stops at $n = K$ when $h_K \leq H_{DZ}$.

Utilizing approximations $P_{w_x}(h)$ and $P_{w_y}(h)$ [created similarly to that of $P_{V_v}(h)$], integration of the first two equations of Eq. (3.1) yields

$$x_n = x_{n-1} + P_{w_x}(h_{n-1})\Delta t, \quad y_n = y_{n-1} + P_{w_y}(h_{n-1})\Delta t \tag{9.48}$$

The triad $\{x_n, y_n, h_n, n = 1, \ldots, K\}$ constitutes a descent path prediction $\hat{P}$.

By varying the parachute surface area S, the various possible landing locations can be determined. This process is depicted visually in Fig. 9.61, where five different values of S for a simulated parachute were considered to produce the desired range value of $r = 10.3$ km (6.4 miles).

To test the predictive portion of the algorithm just outlined, a set of simulations was conducted using experimental data obtained by dropping GPS logging payloads brought up to altitudes ranging from 10 to 30 km (32,800 to

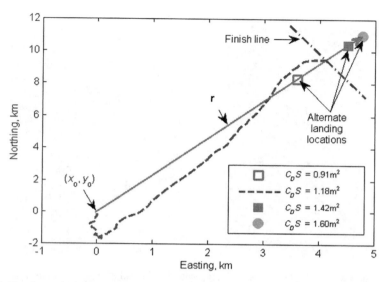

Fig. 9.61 Ground projection of a typical simulated descent profile for various parachute throttling values (reference areas).

98,400 ft) using conventional latex weather balloons. The GPS location and altitude data during an ascent were also recorded and telemetered to the ground tracking station. Because atmospheric parameters were not measured, a standard atmosphere was assumed. The ADS mass ranged from 1.67 to 6.14 kg (3.7 to 13.5 lb). A typical experimental setup, similar to that shown in Fig. 3.36a, is presented in Fig. 9.62. A total of 19 experiments were conducted.

Because there were no other electronics aboard the ADS for these tests, horizontal speed vs altitude data gathered by GPS loggers was assumed to be equal to that of the horizontal wind speed with the corrections for the inertia-induced wind measurement [Hock and Franklin 1999]

$$w_{x;n} \approx \dot{x}_n - \ddot{x}_n \dot{h}_n g^{-1}, \quad w_{y;n} \approx \dot{y}_n - \ddot{y}_n \dot{h}_n g^{-1} \qquad (9.49)$$

The ascent phase (wind characterization) of a typical test lasts 60–90 min, after which ADS deployment occurs, usually when the balloon bursts, or the payload is remotely released from the balloon from the ground via radio link. Descent typically takes 45–60 min depending on parachute configuration and release altitude. After landing, the payload was recovered, and the data describing the position as a function of time were recovered from the data logger (radio downlink of these data served as a backup). The ascent phase data were used to construct a $\mathbf{W}(h)$ wind data set, and the descent phase data served as a comparison test for the parachute used in the experiment (with unchanging reference area).

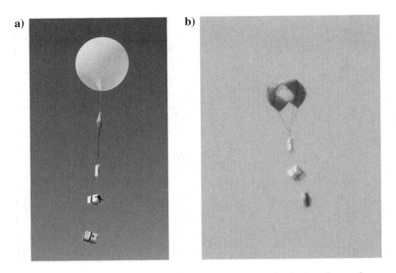

Fig. 9.62   a) High-altitude balloon payload train and b) deployed parachute.

This provided the actual descent path data P for comparison with the simulated path prediction $\hat{P}$ generated by the algorithm (conducted postmission). Unlike the full-fledged algorithm in these experiments, the parachute size was fixed at the start of the descent, and therefore the simulation algorithm was run with a similar constraint. The simulated landing location was compared with the experimentally measured location.

Obviously, the GNC algorithm just outlined relies on a good knowledge of all parameters entering the right-hand side of Eq. (9.46), specifically the aerodynamic drag coefficient C_D. Because variations of air density and vertical component of the winds aloft were unknown (the standard atmosphere and $w_z \approx 0$ were assumed), any variation in the descent rate was attributed to variations in C_D. In other words, for convenience all accrued errors were simply placed in the coefficient of drag. Figure 9.63 shows the computed values of coefficient C_D vs parachute loading (mg/S) and deployment altitude. As seen, the average values vary within the range of 0.7–1.5, and most of the time exhibit $\sigma_{C_D} \sim 0.2$ standard deviation. Because no clear trend was established between the tests, the typical value of C_D for a $S = 1.59$ m^2 (17 ft^2) canopy that was used in experiments was set up to 1.0.

To compensate for unknown w_z and variations in air density and $w_z \approx 0$, an adaptive algorithm to adjust the value of C_D was developed. To begin, the value of $C_D = 1$, known mass, parachute area, and a standard atmosphere were used to generate a descent rate profile

$$\hat{V}_v(h) = \sqrt{\frac{2mg}{\rho^{SA}(h)SC_D}} \tag{9.50}$$

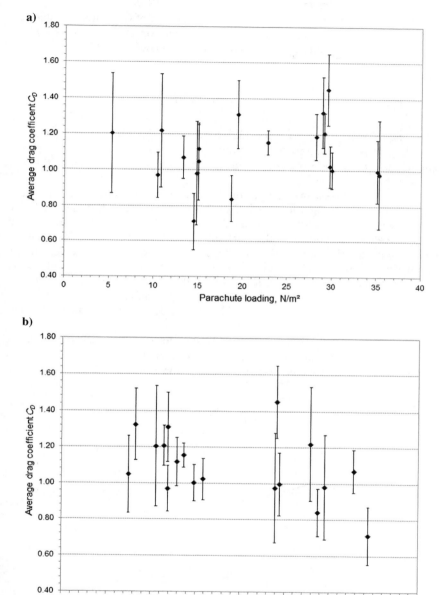

Fig. 9.63   Average coefficient of drag vs a) parachute loading and b) deployment altitude.

Equation (9.47) was then used to produce a prediction of the altitude profile $\hat{h}(t)$ with a time horizon of 5 min. The actual descent rate collected during these 5 min $V_v(h)$ was then used retroactively to adjust the drag coefficient C_D

$$\hat{C}_D = \sqrt{\frac{2mg}{\rho^{SA}(h)SV_v^2(h)}} \qquad (9.51)$$

The root mean squared error (RMSE) between the actual $h(t)$ and predicted altitude profile $\hat{h}(t)$ was used to monitor a goodness of adaptation algorithm. This C_D update routine continued during the remaining descent all of the way to the ground.

Figure 9.64 shows an example of the improved predictive capability that results from estimating the actual C_D middescent using actual descent data. [Figure 9.64 gives four snapshots of adaptation at 28.7, 13, 7.29, and 2.55 km (94,000; 42,600; 24,000; and 8,400 ft) altitude MSL.] The rapid decrease in RMSE indicates the predicted descent profile $\hat{h}(t)$ quickly converges to the actual profile $h(t)$. Surely, because the coefficient of drag accommodates all errors, not necessarily in C_D itself, it varies throughout a descent. Using this adaptation algorithm allows predicting the descent profile $\hat{h}(t)$ more accurately, which also contributes to more accurate knowledge of the predicted descent path $\hat{P}$.

Obviously, the error in determining a landing location (intersect with a target line) is predominantly due to errors in either the descent profile $\hat{h}(t)$ or differences between predicted (available) $W(h)$ wind data and actual wind data. Using the

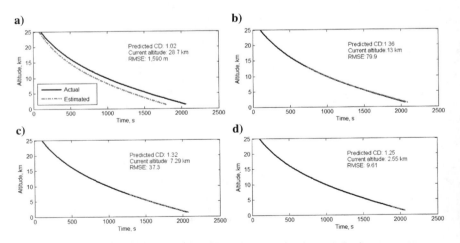

Fig. 9.64 Sequence of four descent profile simulations at a) 0, b) 10, c) 20, and d) 30 min, respectively.

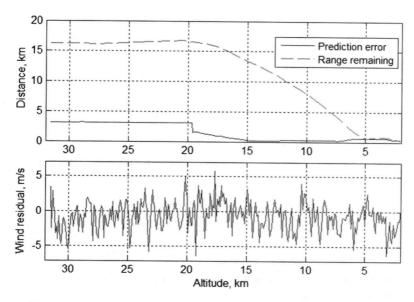

Fig. 9.65   Predictive improvements during a descent with residual winds.

adaptive algorithm to improve the descent profile reduces only one contribution to the error in landing location. Figure 9.65 shows the current landing
location prediction error compared with the distance from the projection of the
current ADS location onto the surface to the landing location. As seen, the
initial error remains constant until the adaptive algorithm begins using previous
data to improve the descent profile [which occurs at approximately 20 km
(65,617 ft) attitude]. Immediately after the adaptive algorithm initializes, the
error quickly decreases down to the minimum [at about 15 km (45,213 ft) alti-
tude]. However, farther down the error begins to increase again. This increase
in error is due to the other main cause in error: variable winds. By the time
ADS descends to the lower altitude, the winds collected at this altitude by the
ascending balloon are about 1 h old. This error is quantified in the lower plot
in Fig. 9.65, in which the wind residual (difference between predicted and
actual horizontal wind speed available because it was recorded at both ascent
and descent) maintains some negative value. This causes the prediction to
deteriorate as the simulation progresses to lower altitudes as only poor wind
data remain. The larger the percentage of poor wind data, the larger the error
will be. It is important to note the residual errors are an artifact of the wind predicting technique (using a high-altitude balloon); however, similar errors can
occur whenever the predicted wind data are reported from either a different
time and/or location (i.e., reported data from NOAA).

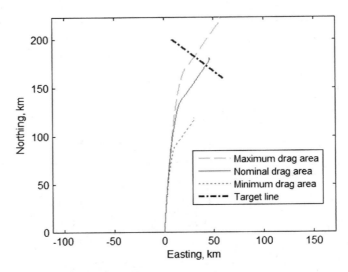

Fig. 9.66   Projection of descent path prediction for maximum, nominal, and minimum drag area configurations.

That is where varying the area of ADS canopy in-flight can aid in reducing wind-based landing location errors. To determine the desired drag area setting, three descent path predictions are calculated for three distinct settings: $f_{max} = (C_D S)_{max}$, $f_{nom} = (C_D S)_{nom}$, and $f_{min} = (C_D S)_{min}$ (which is done to account for possible nonlinear effects). These settings result in a horizontal range the ADS can cover during its descent r_{max}, r_{nom}, and r_{min}, respectively. An example of these three descent path predictions is shown in Fig. 9.66.

By determining the intersection between the downrange azimuth vector and the target line segment, the desired range can be determined. This range represents the distance traveled downwind (in the ballistic winds sense) resulting in successful landing on the target line

$$r_d = \sqrt{x_d^2 + y_d^2} \qquad (9.52)$$

where

$$x_d = \lambda(x_{max} - x_{min}) + x_{min}, \quad y_d = \lambda(y_{max} - y_{min}) + y_{min} \qquad (9.53)$$

and $\lambda \in [0; 1]$ is the length multiplier denoting the scaling between the maximum and minimum drag areas.

The parameter λ is solved for by calculating the intersection point between the maximum and minimum landing locations and the target line meaning that the point (x_d, y_d) should belong to the target line defined by two points (x_1^{tl}, y_1^{tl}) and (x_2^{tl}, y_2^{tl})

$$\frac{y_2^{tl} - y_1^{tl}}{x_2^{tl} - x_1^{tl}} = \frac{y_d - y_1^{tl}}{x_d - x_1^{tl}} \tag{9.54}$$

Substituting Eq. (9.53) into Eq. (9.54) and resolving it for λ yield

$$\lambda = \frac{(y_{\min} - y_1^{tl})(x_2^{tl} - x_1^{tl}) - (x_{\min} - x_1^{tl})(y_2^{tl} - y_1^{tl})}{(x_{\max} - x_{\min})(y_2^{tl} - y_1^{tl}) - (y_{\max} - y_{\min})(x_2^{tl} - x_1^{tl})} \tag{9.55}$$

Developing a quadratic interpolation $f = g(r)$ based on the maximum, nominal, and minimum ADS drag area settings and substituting the desired range of Eq. (9.52) result in the desired ADS setting

$$
\begin{aligned}
f_d =\, & f_{\min} + \left(\frac{f_{\max}}{r_{\max} - r_{\min}} + \frac{f_{\text{nom}}}{r_{\text{nom}} - r_{\min}} - \frac{f_{\max} - f_{\text{nom}}}{r_{\max} - r_{\text{nom}}} \right)(r_d - r_{\min}) \\
& + \frac{f_{\max}(r_{\text{nom}} - r_{\min}) - f_{\text{nom}}(r_{\max} - r_{\min})}{(r_{\max} - r_{\min})(r_{\text{nom}} - r_{\min})(r_{\max} - r_{\text{nom}})}(r_d - r_{\min})^2 \tag{9.56}
\end{aligned}
$$

9.2.2 ONE-DoF GNC SIMULATIONS AND FLIGHT TEST

The ADS drag setting control approach outlined in Sec. 9.2.1 was verified in Monte Carlo simulations and real flight drops. For the computer simulations, a Gaussian noise with a 2-m/s standard deviation was added to a (GPS-derived) experimentally measured true wind data. The noise-free data were used in the calculations of the actual ADS motion. The noise amplitude was estimated using short-term variations in experimentally measured wind data, consistent with other reported approaches [Quadrelli et al. 2001]. Two-hundred-and-fifty simulated descents were performed with the target at the origin and the release point at $[x_0; y_0; z_0] = [-5.5; 12; 10]$km ([$-3.4$ miles; 7.5 miles; 32,800 ft]). The target line passing through origin of the LTP coordinate frame was chosen to be approximately perpendicular to the dominant wind direction to simulate a worst-case scenario in changing wind magnitudes, as the finite control authority of ADS can potentially render the vehicle incapable of reaching the landing target, depending on the magnitude of the noise. Additionally, noise in the wind direction produces scattering laterally along the target line. These simulations represent the case where a dropsonde or radiosonde is released nearby, immediately before the deployment of ADS, and the applied noise represents typical measurement precision and scatter levels in the wind data.

Simulation results shown in Fig. 9.67 clearly indicate that the controlled parachutes performed much better than the uncontrolled parachutes. For

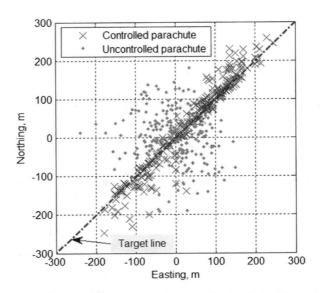

Fig. 9.67   Distribution of touchdown errors in Monte Carlo simulation of one-DoF guidance.

controlled ADS, the average touchdown error with respect to the target line happened to be $\bar{e} = 14$ m (46 ft) with $\sigma_e = 17$ m (56 ft), which is only one-third of that of uncontrolled ADS featured $\bar{e} = 54$ m (177 ft) and $\sigma_e = 41$ m (135 ft), respectively. A paired two-sided t test also confirms that at a 95% confidence level there is a difference in both accuracy and precision of the controlled and uncontrolled systems (because of the Gaussian nature of the superimposed noise the landing locations are distributed normally).

To test the feasibility of a reversible reefing parachute control platform during a real drop, a prototype ADS was developed. Its mechanical system is shown in Fig. 9.68a. Conceptually, the reeling system resembles a large electric "fishing reel" capable of winding up the parachute shroud lines on a spool. The spool diameter can be adjusted to limit the required motor torque associated with various payload masses. The motor used is a high-torque 12-V right angle gearbox motor assembly by Robot Marketplace. With a no-load rotation rate of 116 rpm, an electronic speed controller was added to accurately control the speed of the motor. The rotation of the spool is measured using a 10-bit continuous-rotation absolute encoder. The housing of the reeling system is a fiberglass pod courtesy of Arcturus UAV. The fully assembled system with a quarter-spherical cross-based canopy is shown in Fig. 9.68b.

A Monkey Cortex Navigation Platform (Ryan Mechatronics) provides the computing platform for the descent vehicle. This platform includes an ARM Cortex processor, GPS, six servo outputs, barometric pressure sensor, micro SD

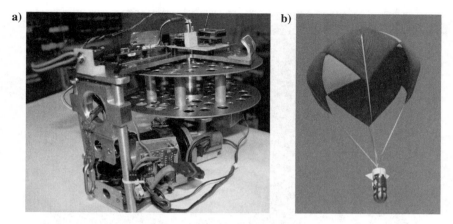

Fig. 9.68   a) Assembled hardware with mounted electronics and b) side view of fully operational ADS in flight.

card logging, two user serial ports, and an optional IMU upgrade. Programming is done in C++; however, communications with all sensors are preprogrammed. This microcomputer platform provides a very fast, affordable means of controlling small robotics payloads.

In addition to the central processing platform, several accessory electronics were incorporated (Fig. 9.69). An electronic speed controller (ESC) is used to power the spool motor. A continuous-turn absolute encoder is used to track the motion of the spool. Air-to-ground communications is performed with a XBEE-PRO XSC RF modem (Digi). These modems provide communication ranges of up to 45 km (28 miles). The ground station uses a modified version of a DIY Drones Ardupilot open-source ground station software running in National Instruments LabVIEW (Fig. 9.70). This software shows real-time location information using Google Earth and displays/records ADS system information for use by the ground crew.

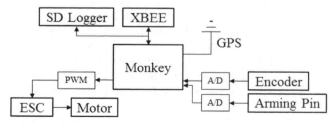

Fig. 9.69   Flowchart showing communication between all electrical components of the ADS.

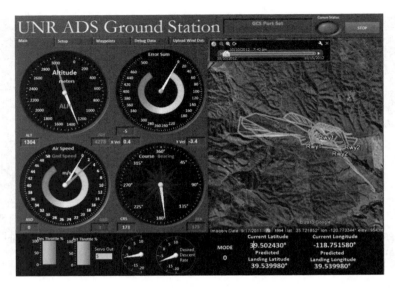

Fig. 9.70   Ground control station GUI.

A flowchart of the computational routine is shown in Fig. 9.71, outlining the software architecture. Corrections to the desired descent speed were calculated every 5 s, with sensors read at a frequency of 4–10 Hz, depending on the specific sensor. Motor control was performed at a frequency of 10 Hz to minimize errors associated with parachute-induced reeling effects.

The controller dynamics assumed to be described by a first-order-plus-dead-time model

$$G(s) = \frac{K}{\tau s + 1} e^{-Ls} \tag{9.57}$$

The input to the control system is the motor angle, with feedback provided by the GPS-derived descent speed. The model parameters were estimated empirically using the tangent method [Fadali and Visioli 2009] based on a step input response. Specifically, once deployed, ADS maintained a fully open canopy for 10 s to reach a steady state. After 10 s, the desired motor angle was set to the maximum (corresponding to about 1,400 deg of reel rotation), thereby completely reefing the canopy. The ADS response from which the model presented in Eq. (9.57) was derived is shown in Fig. 9.72. Prior to landing, the canopy was opened back completely to slow the descent.

Simple PI and PID controllers were used to control the descent rate within achievable limits. Their gains were tuned using the Ziegler–Nichols

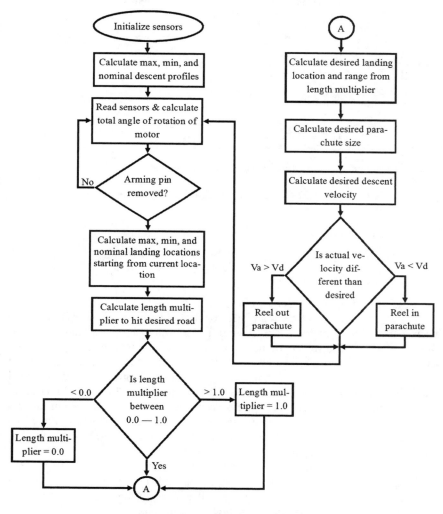

Fig. 9.71 Software control routine.

tuning rules providing the desired disturbance rejection characteristics, albeit at the cost of some overshoot [Visioli and Zhong 2011], Specifically, for the PI controller

$$K_p = 0.9 \frac{\tau}{KL}, \quad K_i = 0.9 \frac{\tau}{3KL^2} \qquad (9.58)$$

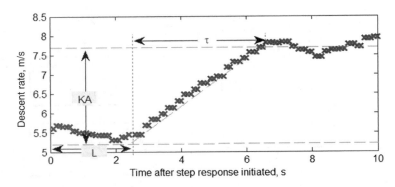

Fig. 9.72   Step response of ADS at 700 m AGL.

and for a PID controller

$$K_p = 1.2\frac{\tau}{KL}, \quad K_i = 0.6\frac{\tau}{KL^2}, \quad K_d = 0.6\frac{\tau}{K} \tag{9.59}$$

The resulting numerical values are shown in in Table 9.7.

The culminating flight-testing phase consisted of deploying the ADS from an aircraft (not from a balloon) and having it autonomously manipulate the parachute size so that the vehicle maneuvers to a desired landing line. The ADS parameters used in testing are shown in Table 9.8. Prior to release, the calculated aerial release point (CARP) for the average parachute drag area was determined using the predicted wind data. Release altitude was about 3 km (9,840 ft) AGL. To determine the effectiveness of the descent vehicle, the landing accuracy was compared to a simulated uncontrolled parachute "released" from the same location at the same time. Wind data collected during the ADS descent were used to simulate the descent path of the uncontrolled parachute.

At this point, there were just a few low-altitude drops conducted, so that the GNC algorithm was not tuned to the optimal level. However, these initial test drops allowed checking the overall concept and revealing some pitfalls to be fixed in the following studies. Also, flight testing consisted of first collecting wind data from a GPS dropsonde at the vicinity of where the descent vehicle is

**TABLE 9.7   EMPIRICALLY TUNED PI AND PID
CONTROLLER GAINS**

Controller	K_p	K_i	K_d
PI	787	105	
PID	1050	210	1310

TABLE 9.8   ADS AND CONTROL PARAMETERS USED IN FLIGHT TESTING

Parameter	Value
ADS mass	5.45 kg (12 lb)
Maximum drag area S_{max}	4.56 m^2 (49 ft^2)
Nominal drag area S_{nom}	2.50 m^2 (27 ft^2)
Minimum drag area S_{min}	1.30 m^2 (14 ft^2)
Numerical integration step Δt	7.0 s

to be released. The onboard ADS software was then updated with the predicted wind profile. Finally, the ADS was sealed up and carried aboard a small piloted aircraft to be released as close to CARP as possible (which of course is quite difficult to achieve with poorly instrumented general aviation aircraft). The total time to collect the dropsonde wind data, enter it into the ADS software, and reach the desired release location was approximately 2 h (making the predicted wind profile data 2 h old). Obviously, it took its toll on touchdown performance.

Figure 9.73 shows the results of one of unsuccessful drops. In this particular drop, not only controlled ADS did not reach the target line, but also it performed worse than a simulated uncontrolled ADS. Detailed analysis revealed that for this drop the landing location error was predominantly due to the large difference between the predicted and actual winds as illustrated in Fig. 9.74. It was observed that the wind conditions during the ADS tests were stronger than those measured by the dropsonde (in the direction of the target road). These stronger than predicted winds caused the vehicle to be carried farther downwind than predicted. As the range of descent speeds achievable at the low altitude for the given parachute sizes is limited to 5...10 m/s (16...33 ft/s), the vehicle was simply unable to overcome the errors in the predicted wind profile. The average wind speed perpendicular to the road for the actual and predicted wind profiles was −1.0 and −2.6 m/s (−1.9 and −5 kt), respectively (i.e., off by more than a factor of 2). During low-wind scenarios like this, average errors between actual and predicted of 1.6 m/s (3 kt) cause significant errors when path planning. In a higher wind testing scenario, errors of 1.6 m/s (3 kt) would have caused correspondingly smaller landing location errors. Wind prediction errors propagate beyond simple path planning errors. CARP itself would have also needed to shift significantly to account for the stronger wind conditions. Overall, this drop test demonstrates the necessity for obtaining accurate wind data prior to payload release and a general lack of control authority in low-altitude releases under low-wind speed conditions. This result agrees with the effects of large spatial and/or temporal differences between predicted and actual wind conditions on the landing location accuracy of a reversibly reefed circular canopy.

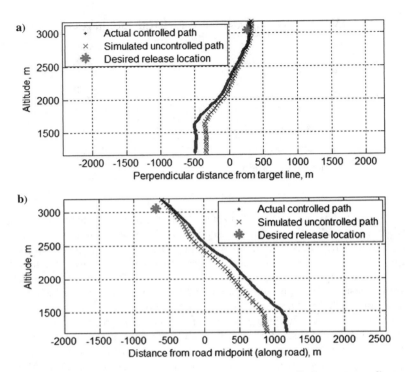

Fig. 9.73   Controlled and uncontrolled descent paths a) perpendicular to target line and b) along the target line for drop 1.

The results of another, more successful, drop are shown in Fig. 9.75. This time the controlled ADS was able to land closer to the target line than the simulated uncontrolled ADS. The major problem for this drop was that ADS was released 250 m (820 ft) away from CARP perpendicular to the target line and 400 m (1,312 ft) away from it parallel to the target line (although downrange release errors do not affect accuracy). However, the controllable range perpendicular to the target road for this drop test was only 130 m (427 ft), that is, way outside of the controllable range so ADS could not potentially make it to the desired landing location anyway.

The differences between the actual and predicted wind data for the second drop test were smaller than those shown in Fig. 9.74. Average wind speeds perpendicular to the target line for the actual and predicted wind profiles were −0.89 and −0.85 m/s (−1.73 and −1.65 kt), respectively. As the vehicle was significantly farther upwind of the target line than expected, it maximized the parachute size for most of the flight. Below 1,700 m (5,580) MSL, the wind conditions shifted significantly, and, as a result, the vehicle minimized the parachute size to reduce

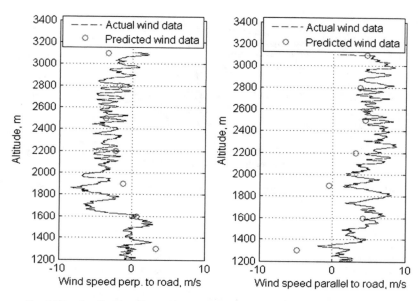

Fig. 9.74   Predicted and actual winds experienced during the first drop test.

translations away from the target road. This minimized the distance traveled away from the target line for the remainder of the descent. The small controllable range of landing locations of 130 m (427 ft) is attributed to the extremely small average wind speeds perpendicular to the target road, and the relatively low release altitude.

The initial flight testing of a one-DoF guidance concept revealed several key problems when attempting to deploy and steer a circular parachute. They included reliable parachute deployment, shroud line entanglement during dereefing, and controllability of descent speed via reefing. With regard to the safe parachute deployment, the problem was fixed by utilizing a static line attached to the parachute bag. To facilitate full parachute/shroud line extension prior to canopy inflation, the shroud lines were bound in several locations with rubber bands, and a 10-lb break line was attached between the top of the parachute and the parachute bag. The parachute bag was inserted into the pod and held inside the pod with the masking tape. This provided a secure way of holding the parachute in the pod (to protect against premature inflation) while still providing a means of pulling the parachute bag out of the pod during desired parachute deployment.

Another major issue was parachute shroud line entanglement. If the parachute is completely reeled in, it no longer has enough moment of inertia about the z axis (as the parachute diameter is reduced significantly). Any instability in the payload

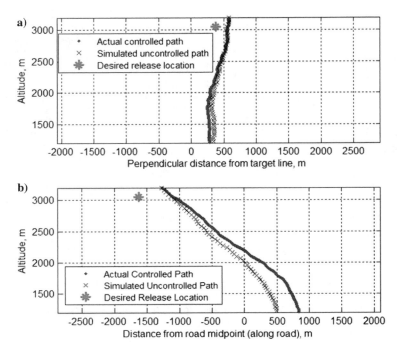

Fig. 9.75 Controlled and uncontrolled descent paths a) perpendicular to target line and b) along target line for drop 2.

that induce rotations causes the parachute shroud lines to tangle. Then, when dereefing the parachute, the canopy will be unable to inflate. To overcome this issue, the payload was statically balanced, and stabilizing fins were added to the rear end of the payload (as seen in Fig. 9.68b). The four shroud line parachute design also helped minimize line tangling (several other parachute designs were abandoned in part because of this particular reason).

Finally, the shape and type of the parachute were found to have a significant impact on the controllability of reversibly reefing the canopy. A standard circular parachute with 12 shroud lines was initially tested. Drop testing revealed that the canopy has a large nonlinearity in the descent rate as a function of reefing applied (when reefed with the reefing system as described earlier). As the shroud line constriction approaches the canopy, sudden canopy collapse can occur. Similarly, a ring-slot parachute was tested, with comparable nonlinearities during reefing. By using a quarter-spherical cross-based canopy as shown in Fig. 9.68b, parachute inflation was fully maintained regardless of the reefing applied. This provided a predictable relationship between the motor angle (reefing) and the vehicle descent speed.

9.2.3   PERFORMANCE ENVELOPE FOR TWO-DoF GUIDANCE

Being encouraged by a relative success of testing one-DoF guidance with a goal of intercepting a target line (anywhere along the line), an effort was undertaken to extend this idea to two-DoF, so that a specific IPI could be reached.

The first step in developing the two-DoF control scheme is to determine the performance envelope of the descent rate controlled ADS. The envelope, whose shape and size evolves throughout the descent, encompasses the set of all points from which IPI can be reached as a result of control action. This envelope can be envisioned as a horn–shaped volume extending upwards and upwind from the IPI location on the ground. This performance envelope is determined using given wind conditions and is calculated for a single azimuth direction at a time (the probe direction, $\mathbf{B}_i$, $i = 1, \ldots, L$). Within each control layer, the descent rate is configured to maximize the horizontal drift vector's component along the probe direction.

If the wind is counter to the probe direction, then the control objective is to minimize its influence. The net drift over all M layers is then totaled as the maximum possible drift in that azimuth direction. By repeating the process for all azimuth directions, the bounding envelope can be calculated. The bounding envelope calculation starts at the IPI location and works upward (backward) through the M wind layers, as collected by the weather balloon or dropsonde, each of thickness $\Delta h_k = h_k - h_{k-1}, k = 2, \ldots, M$ ($\Delta h_1 = h_1 - h_{\mathrm{IPI}}$) to the release altitude. The elapsed time to fly through each layers

$$\Delta t_{ik} = \Delta h_k \sqrt{\frac{\rho_k f_{ik}}{2mg}} \qquad (9.60)$$

were the drag area setting $f_{ik} = (C_D S)_{ik}$ dependents on both the current probe direction i and wind layer k

$$f_{ik} = \begin{cases} (C_D S)_{\max} & \text{if } \mathbf{B}_i \mathbf{W}_k \geq 0 \\ (C_D S)_{\min} & \text{if } \mathbf{B}_i \mathbf{W}_k < 0 \end{cases} \qquad (9.61)$$

Similar to Eq. (9.48), we can write

$$x_k = x_{k-1} - w_{x;k}\Delta t_{ik}, \quad y_k = y_{k-1} - w_{y;k}\Delta t_{ik} \qquad (9.62)$$

(a negative sign is caused by the fact that the construction of the performance envelope is carried from the ground up). Note that within a given layer k, Δt_{ik} will vary and take one of two distinct values, depending on the angle between the wind and probe directions.

With Eq. (9.61), a favorable wind condition exists in an individual layer k when $\mathbf{W}_k$ is within ± 90 deg of the current probe direction $\mathbf{B}_i$. In such a case, the maximum drag area is used to slow the ADS's descent within this layer and take maximum advantage of the favorable wind. Additionally, between layers,

Δt_{ik} will also vary due to air density changes. For the special case when the wind in a layer is zero ($W_k = 0$), the previous (below) drag area is maintained.

An example of the boundary envelope construction for the winds of Fig. 9.76 and eight probe directions $L = 8$ is depicted in Fig. 9.77. The procedure starts at the ground layer ($k = 0$) by initializing the envelope to the origin (IPI). In this layer, the envelope consists of a single point at (0, 0), with the eight probe directions shown by their respective arrows. The next layer ($k = 1$) incorporates the first decision-making process in which each probe direction is compared with the layer's wind direction (depicted by the vector plotted in the center). Positions that resulted from control decisions to select the maximum drag area are plotted with an X while decisions to use the minimum drag area are plotted with a circle.

The envelope for the first layer (above the ground) takes the form of a line connecting the two distinct boundary points resulting from consideration of the eight probe directions. Note that in this first layer, there can be only two possible new positions (resulting from either maximum or minimum drag area); thus, the eight different probe directions are grouped into two coincident clusters at each end of the line segment. The second layer ($k = 2$) further unfolds the performance envelope into a two-dimensional shape. As the wind has changed direction significantly from 142 deg in layer 1 to 19.9 deg in layer 2, different drag areas are selected in comparison to the previous layer. Probe directions 8, 1, and 2 all favored the maximum drag area configuration for layer 1, but chose the minimum drag area for layer 2. Extending the illustration with one more layer ($k = 3$), it is seen that it is possible to have fewer unique boundary points than the maximum (2^k).

Note that the farthest distances achievable when probing both the $\mathbf{B}_1$ and $\mathbf{B}_2$ directions result at the same point. This illustrates that for a finite number of wind layers (and control actuations), it is not possible to generate a continuous set of points defining the boundary envelope. Instead, this construction method makes a "best attempt" to evaluate lateral motion in a given probe direction and will guide the parachute as close to that direction as is possible, given the finite number of control actuations. In general, the maximum number of unique boundary points increases as 2^j, up to the number of probing directions selected (360 for this study).

The process of constructing the horizontal displacement for each layer is generally repeated through all M layers up to the release altitude at layer $k = M$. An example of a cross section from an envelope using 360 probe directions for the same wind profile of Fig. 9.76 is shown in Fig. 9.78. Here, the envelope is constructed at the 500th layer ($j = 500$) at an altitude of 9 km (29,500 ft). If the predicted wind data are very accurate, the target location can be reached by any release point that lies within or on the boundary of the performance envelope. Once the envelope has been fully constructed (Fig. 9.79), the ideal CARP can be determined. The aforementioned procedure to construct the performance envelope can be performed beforehand as long as winds are known.

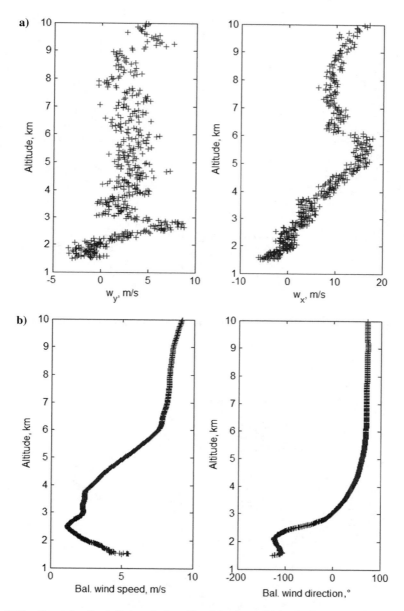

Fig. 9.76   Example of a balloon wind profile a) in the $w_x(h)/w_y(h)$ form and b) in the $W(h)/\chi_W(h)$ form.

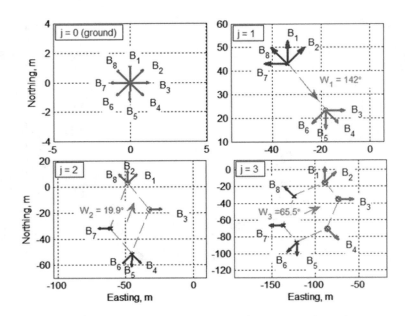

Fig. 9.77   The first four steps of the performance envelope construction process.

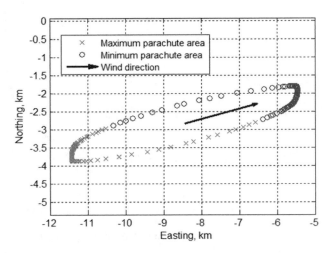

Fig. 9.78   Nine-km-altitude layer from the performance envelope using 360 probe directions.

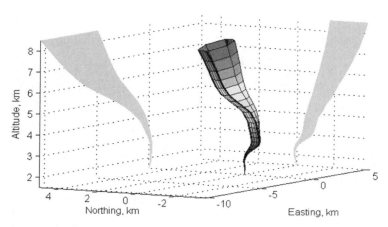

Fig. 9.79 Full performance envelope in 3D with two vertical-plane projections calculated using 360 probe directions.

The control approach for the two-DoF guidance paradigm consists of a series of single–actuation choices. This is because for a given change in altitude, the parachute–payload system can only influence the time spent in that layer and the distance traveled downwind. By making a series of choices (with varying wind directions in each layer), the system can effectively control translation in two dimensions attempting to travel towards the center of the performance envelope, maintaining maximum flexibility for future decisions. (Once at the center, this location provides the maximum degree of control authority to correct for inaccurate wind data.) If the wind data are completely accurate, once at the center of the performance envelope, no further changes to the parachute would be needed to reach IPI. This configuration would result in a descent speed that is the average of the maximum and minimum achievable descent speeds at each altitude. Inaccurate wind data will cause ADS to stray from the center of the performance envelope, prompting the control system to make adjustments to the drag area.

To reach the center of the performance envelope, it might be (mistakenly) postulated that it is desirable to travel towards a line that is both normal to the wind vector and passes through the centroid of the performance envelope (dashed line in Fig. 9.80). Once on the dashed line, the wind can neither help nor hinder progress towards the center within that layer. However, attempting to reach this line can easily result in exiting the performance envelope, rendering the system incapable of reaching the desired target. This is illustrated in Fig. 9.80. Starting at the point labeled "current position," the dashed line lies upwind. To avoid being pushed farther from the dashed line (downwind), the vehicle needs to spend as little time as possible in such unfavorable winds and would configure to minimize the drag area (descending through the unfavorable wind layer quickly).

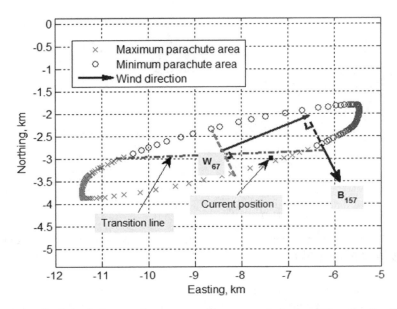

Fig. 9.80 Single layer from performance envelope generated using 360 probe directions with a labeled transition line.

However, as all of the boundary points in the vicinity of the "current position" were generated using a maximum drag area, working towards the dashed line can cause the vehicle to drift towards the boundary (while approaching the dashed line) and actually exit the performance envelope. Once outside of the performance envelope in any wind layer, it is no longer possible to reach the target on the ground.

A more effective algorithm to efficiently reach the center of the envelope begins by defining a transition line that divides the envelope into two decision-making regime locations that require a fast descent rate and locations that require a slow descent rate. Visually, the envelope is bisected by the transition line (Fig. 9.80). The transition line is most easily constructed by identifying and connecting the two points along the perimeter of the envelope whose surface normals are perpendicular to the wind vector. The transition line always passes through the centroid of the performance envelope. This is due to the inherent binary nature of the envelope construction process that produces a closed shape with an apparent 180-deg rotational symmetry. All points upwind of the transition line require canopy configurations that slow the descent in order to spend more time benefiting from the upwind conditions whereas all points downwind of the transition line correspond to canopy configurations that speed the descent to minimize the time spent being carried downwind.

If the parachute-payload system is on the transition line but not at the centroid, the system must wait (as it continues to descend) for a shift in wind direction that permits further progress towards the centroid. Thus, the transition line directly determines the closest achievable distance to the centroid for a given layer. Returning to the preceding example, the transition line lies downwind of the current position (square symbol in Fig. 9.80), and, thus, the vehicle will choose to maximize the drag area, moving the vehicle away from the boundary, closer to the transition line. With this in mind, the goal of the control system is to adjust the drag area so that the vehicle ends up on the transition line. Because the transition line is a linear target, the control methodology already developed in Sec. 9.2.1 can be used at each altitude to approach it.

In each layer during the descent, Eq. (9.56) provides the desired drag area that results in the vehicle reaching the transition line in the minimum amount of time (although it may require more than one layer). The process is then repeated during the descent to continually seek the center point of the performance envelope. During the descent, though, the control algorithm only looks ahead one layer at a time (minimum control horizon). Using the reference trajectory from the performance envelope calculations, this presents a computationally efficient algorithm that can be easily implemented on a small microprocessor included in the flight hardware.

The center of the performance envelope also has the benefit of falling along the minimum actuation path. This is because the center of the performance envelope is the single actuation case using the average descent velocity for the current configuration. If the vehicle is off of this minimum energy (effort) trajectory, the vehicle will navigate towards the center of the envelope. Once at the center, it will require no further actuation (assuming perfectly known wind data).

9.2.4 WIND MODELS AND TWO-DoF GNC SIMULATIONS

To determine the impact of inaccurate wind data on the performance of the suggested GNC paradigm, computer simulations that included three types of errors in the wind data were performed: 1) type A errors were considered to be caused by large spatial and/or temporal differences between the predicted and actually experienced winds; 2) type B errors were caused by measurement precision limitations, inaccuracies, and/or small-scale variations in the wind; and 3) transition wind errors were those in which the predicted wind data exhibit characteristics of both type A and B errors. For all of the results presented in this section, the simulated deployment altitude was 10-km MSL and the IPI altitude was 1.5 km (4,920 ft). Numerical integration was accomplished with $\Delta h = 5$ m (16.4 ft) increments. Other parameters used in these simulations were the same as in Table 9.8.

To investigate these wind errors, simulations were conducted in which the control algorithm made decisions using "inaccurate" wind data for the control

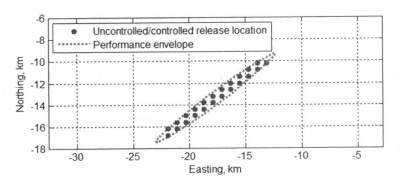

Fig. 9.81   Release locations of simulated uncontrolled and controlled ADS for a typical simulation.

decisions while the ADS' simulated motion was subjected to "true" wind in the kinematic motion simulation. Each controlled descent was paired with an uncontrolled descent vehicle using the nominal f_{nom} drag area configuration. Release locations were selected throughout the performance envelope (at the release altitude) simulating various release location errors. An example of a typical simulation initialization (23 release points) is shown in Fig. 9.81.

In terms of wind data, 14 real wind data sets were collected with seven gathered from balloon ascents, and seven collected from dropsonde descents (following the balloon ascents). These data sets were obtained during two field tests with four balloons launched simultaneously in the first test group, followed by three more in a second test group approximately 1 h later. Dropsondes were released at each balloon burst, which varied in time and location for each balloon. The ascent rates of all balloons were matched to each other nearly identically. Taken together, these 14 wind data sets were mixed and paired to provide a selection of type A, type B, and transition wind error comparisons. Figure 9.82 shows examples of these three types of wind data.

As an outcome from simulations with type A wind errors (Fig. 9.82a), the results were compiled from 54 separate simulations, each considering controlled/uncontrolled releases from 21–43 locations. In total, 1,009 individual descent paths were computed for each of the controlled and uncontrolled descent vehicles. Identical wind data were used for each individual controlled/uncontrolled comparison. For the numerous wind data sets considered, spatial differences varied between 0 and 153 km (95 miles) and temporal differences varied between 63–217 min. The CEP for type A wind errors was found to be 3,083 and 1,811 m (10,115 and 5,940 ft) for the uncontrolled and controlled descent vehicles, respectively. Paired, two-sided t-tests were performed on the simulation results to quantify the effectiveness of the controlled vehicles in comparison with the uncontrolled vehicles. In the 38 completed simulations, the t-test

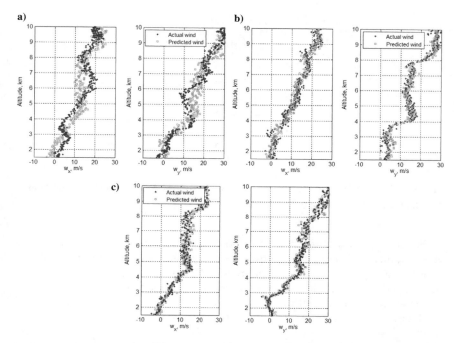

Fig. 9.82   Wind profiles used for predicted and actual wind profiles for a typical a) type A, b) type B, and c) transition wind error simulations.

failed seven times (at a 95% confidence level), with an average p value of 4.45×10^{-2}. That is, the landing location differences were statistically significant in 31 of the 38 cases studied. Landing locations for a typical simulation with type A wind errors are shown in Fig. 9.83.

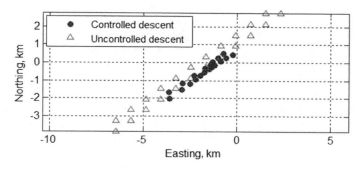

Fig. 9.83   Landing locations of uncontrolled and controlled ADS for a typical simulation using type A wind errors.

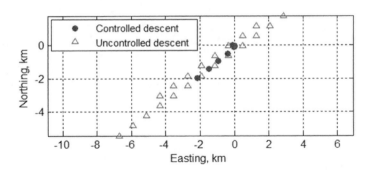

Fig. 9.84   Landing locations of uncontrolled and controlled ADS for a typical simulation containing wind errors of type B.

For the type B wind estimation errors (Fig. 9.82b), nine separate simulations were conducted. These simulations were computed by comparing the balloon-recorded winds for the seven different balloons launched (at two separate times). As discussed earlier, the ascent rates were nearly matched with discrepancies between average ascent rates of approximately 0.15 m/s (0.5 ft/s). The CEP was found to be 2,740 and 295.1 m (9,000 and 970 ft) for the uncontrolled and controlled descent vehicles, respectively. The t-test did not fail for any simulation, with the average p value of 2.86×10^{-7}. Landing locations for a typical simulation with type B wind errors (Fig. 9.82b) are shown in Fig. 9.84.

For the transition-type wind estimation errors (Fig. 9.82c), the results were compiled for seven separate simulations. These results were computed by comparing the dropsonde-descent-recorded winds with other dropsonde descents. These dropsondes varied in time (up to 9-min difference in release time) and location [up to 4.1-km (2.6-miles) difference in release location]. The CEP was found to be 2,614 and 893.2 m (8,580 and 2,930 ft) for the uncontrolled and controlled descent vehicles, respectively. The t-test again did not fail for any simulations, with an average p value of 5.45×10^{-7}. Landing locations for a typical simulation with the transition type wind errors is shown in Fig. 9.85.

An additional simulation was conducted to validate the use of the approach of heading towards the transition line (see Fig. 9.80) as opposed to the center of the performance envelope as discussed in Sec. 9.2.3. This single simulation (using known, real wind data) provides a direct comparison between the two GNC techniques (Fig. 9.86). Conceptually, this technique can be visualized as a virtual target line extended both directions from the center of the performance envelope's cross section perpendicular to the current wind direction. In many situations the control routines will produce the same desired descent speed; however, if the current location is in between the transition line and the virtual target line derived from a wind direction, the algorithm will produce opposing results (one producing a faster descent speed with the other producing a slower

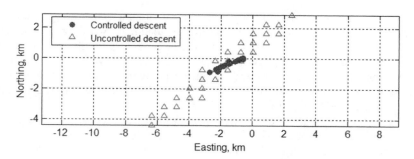

Fig. 9.85   Landing locations of uncontrolled and controlled ADS for a typical simulation containing wind errors in the transition regime.

descent speed). Release locations were identical between the two approaches in much the same manner as the uncontrolled/controlled simulations.

The average error for the transition line approach was 14.4 m (47 ft) with a standard deviation of 23.6 m (77.4 ft) over 27 different simulation paths. The average error for the wind direction approach was much larger, at 92.0 m, with a standard deviation of 84.8 m (278 ft). As mentioned earlier, the guidance technique based on the wind direction does not provide the optimal solution, and although this simulation does not prove optimality in using the transition line approach, it does provide a *more* optimal approach than the wind derived counterpart.

To summarize, although the two-DoF guidance methodology does not produce pin-point accuracy (such as a parafoil is capable of), it is conceptually simple, robust, and can provide greatly improved accuracy over uncontrolled circular parachute descents, provided good wind data are available. (As wind estimates become more accurate, the accuracy of ADS touchdown greatly improves.)

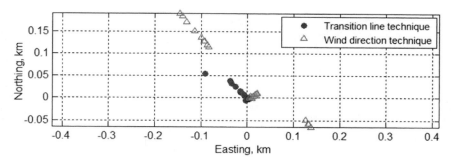

Fig. 9.86   Comparison between transition line and wind direction derived navigation techniques.

Simulations demonstrated the potential for a circular parachute to overcome release location errors (along with parachute/payload mass and drag area estimate errors). For example, if the payload mass is larger than initially estimated, the ADS' controllable descent speed range will be shifted with higher descent speeds for both the maximum and minimum parachute drag areas. As long as the desired descent velocity is not out of this range, ADS will simply reef until the desired descent speed has been reached.

By performing high-altitude, high-opening parachute drops, ADS has a relatively large performance envelope. The typical performance envelope provides more than enough release location error rejection, as many envelopes are 11 km (6.8 miles) in length along the major axis. Even at lower release altitudes, the performance envelope provides adequate size to capture reasonably large release location errors. For example, with a release altitude of 3 km (9,840 ft) AGL, the typical performance envelope will have a major axis length of 2.5 km (1.6 miles) (depending on wind conditions).

In all of the simulations, some controlled descent paths end up significantly farther away from the target, when compared to the other controlled descent paths. This is because the path-planning algorithm uses the predicted winds to create the predicted performance envelope. In reality, the actual performance envelope will differ from the predicted one. This can cause decisions based solely on the predicted envelope to further push the vehicle outside of the actual envelope. The actual envelope cannot be known a priori. For this reason, it is beneficial to release as near to the center of the envelope as possible, thereby providing maximum rejection of inaccurate wind data effects.

CHAPTER 10

Flight-Test Instrumentation

Thomas Jann[*]
Institute of Flight Systems of DLR, Braunschweig, Germany

Oleg Yakimenko[†]
Naval Postgraduate School, Monterey, California

As shown in Chapter 1, all PADS carry AGU that employs different sensors to estimate position and attitude of PADS as a whole. Depending on the location of AGU, it measures parameters of payload (when it sits atop it) or parameters of the confluence point that is more associated with a canopy. For most online applications, having these parameters is enough, but from the standpoint of off-line system identification, especially when considering higher-order-fidelity seven-/eight-/nine-DoF models, more parameters need to be measured and/or estimated. As a result, during developmental tests and evaluations PADS might be equipped with much more sophisticated instrumentation packages. Section 10.1 of this chapter addresses this issue by considering an example of the development, testing, and usage of well-documented instrumentation packages in the past ALEX and FASTWing programs, followed by considering the issue of estimating a relative position of a canopy with respect to the instrumentation package. Section 10.2 looks at how data collected onboard of PADS (for the system identification purposes) can be supplemented or sometimes replaced with data gathered at the ground.

10.1 ONBOARD INSTRUMENTATION

The Institute of Flight Systems of the German Aerospace Center (DLR) has a long-term expertise in the area of the system identification applied to aircraft and general flight vehicles. Since the mid-1990s, the developed identification methods have been applied to different parachute and parafoil systems. Among others, that included two ALEX systems and a FASTWing system. This section is mostly based on [Jann et al. 1999; Jann 2001; Jann and Greiner-Perth 2009] and describes the flight-test instrumentation packages for both systems. Obviously, these instrumentation packages used technology of the 1990s, which partly has become obsolete in the meanwhile. Today, for example,

---

[*]Research Scientist.
[†]Professor.

microcontrollers and sensor systems based on microelectromechanical systems (MEMS) have become more and more efficient and accurate and are now suitable for the application in PADS. Nevertheless, the principles and concepts described in this chapter still remain valid and applicable.

10.1.1 INSTRUMENTATION PACKAGES FOR ALEX AND FASTWING PADS

As mentioned in Sec. 1.2, ALEX PADS (Fig. 1.31) was developed to validate mathematical models of different complexities. The instrumentation package included GPS, inertial and air-data sensors, magnetometer, and actuator position transducers (Fig. 10.1).

The vehicle was equipped with a PC-compatible onboard computer (Fig. 10.2). Some additional modules for analog data acquisition and serial interfacing were connected to its PC/104 interface. Together with a ruggedized hard disc and a power regulating unit, the onboard computer was integrated into a small housing. A 433-MHz radio modem works in semiduplex mode and provided a wireless bidirectional communication link to the ground station. The inertial sensor set consisted of MEMS accelerometers from analog devices and micromechanical rate gyros from Murata, each in a three-axial assembly. Also, a three-axis magnetometer from PNI was used to provide additional attitude information. Two separate GPS receivers were installed, one working in regular GPS-mode (Garmin GPS36) and the other in DGPS-mode (Trimble Lassen SK8). Airflow was measured using a 0.8-m-long noseboom. It was equipped with a pitot tube and two vanes to measure angle of attack and angle of sideslip. Because no practical method was found that would allow protecting the noseboom reliably from damage, it was decided to design the noseboom as simple as possible and to dispose it after the flight. The dynamic pressure was piped through plastic tubes to a differential pressure sensor inside the capsule. The vanes are connected to contactless magnetoresistive potentiometers. A Jenoptik laser altimeter was installed to enable correct flare initiation in autonomous mode. An absolute pressure sensor for barometric altitude, an air temperature sensor, and actuator position transducers were completing the sensor set.

a)   b)

Fig. 10.1   a) ALEX PADS and b) instrumentation package.

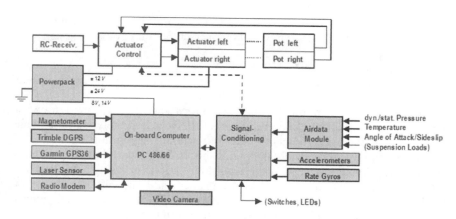

Fig. 10.2   Electrical concept for ALEX system.

An onboard computer, most sensors, radio modem, and other electronic modules were installed on a removable platform that could easily be exchanged between two vehicles. The platform was mounted on shock-absorbing devices and located in the upper part of the vehicle, where it could be accessed through a window in the front. If needed, also other instrumentation packages (e.g., from industry) could be prepared independently on another equally sized platform and be installed in the vehicles. Magnetometer, GPS receiver, and antennas were installed on the tail.

The vehicle was also equipped with an upwards-looking video camcorder that records a video of the apparent motion of the canopy during flight. For postflight analysis, this video was digitized and loaded into an application that was capable of tracking features of the canopy throughout the entire video sequence. Multiple point tracking gave a deep insight into relative positions, velocities, and angles between parafoil and load.

A ground station provided a bidirectional radio modem link to the vehicle. On ground, a first notebook PC collects all of the received telemetry data and is used as terminal for tele commands to the ALEX onboard computer. Some of the telemetry data were transferred to a second notebook that served as "on-line track display" for the ground crew to follow the actual trajectory of ALEX in flight.

The parameter estimation technique developed using ALEX flight tests is described in Secs. 11.1–11.3. The remainder of this section introduces one of the follow-up projects with more detailed description of the instrumentation package. Based on the success with instrumenting two ALEX vehicles, DLR contributed its expertise to the FASTWing project, involving multiple international partners. The main project objectives in FASTWing and follow-on FASTWing CL (delivery of capital loads) comprised the development of a large high-performance parafoil for loads up to 6 tons (Fig. 1.32), the development and application of tools and procedures for the reliable opening of the parachute system, the development of an all-electrical steering box enabling high accuracy and soft landings, and the development and validation of advanced design and simulation tools.

The requirements for the FASTWing CL instrumentation package were that it should 1) measure relevant flight data during the flight tests and 2) log the data on a rugged storage device for postflight analysis.

In addition to that, the instrumentation package was required to be a stand-alone device for test and evaluation of the vehicle prototypes, be capable of tolerating rough environment conditions without significant damage is of crucial importance, and reliably withstand shocks during parachute deployment and landing without interrupting the acquisition process or losing data. By treating the instrumentation packages as independent add-on units, the design of hard- and software for the operational system can focus on the requirements for the end product without considering the specific needs for the evaluation phase. Finally, it was required that the instrumentation package do the following:

- is easy to handle with size and weight being the secondary issues

- has no mechanical and electrical interference with a transport aircraft and PADS itself

- has housing compatible with payload packing/integration method, without sharp edges and delicate parts sticking out (in order to prevent lines or risers from entanglement jeopardizing the parachute deployment and possibly damaging the measurement system)

- possesses interface for data up- and download, monitoring, and programming

- features easy maintenance

The selection of an adequate sensor package or combination was mainly driven by the required performance including accuracy and dynamics, cost, handling including interface to the data logger, available software, and the own experience with a particular device. Criteria for the selection of an appropriate data logger or onboard computer are computing power, data storage capacity, available interfaces, sample rate, shock resistance, and temperature resistance. The power source must fulfill the requirements concerning power consumption of the other components, operation time, and operation altitude (temperature and pressure environment). The design of the housing depends mainly on the requirements regarding the space for components and cabling, the mechanical interface to the payload carrier, robustness issues, and shock resistance.

Because there was no data acquisition system on the market that fulfilled the requirements for the application, a new design approach was required. The following instrumentation packages have been selected within a tradeoff analysis and realized afterwards:

- The Flight Data Acquisition System (FDAS-M) is an independent high-quality flight-test instrumentation package for remotely and autonomously controlled drop tests comprising an extended sensor package, a video camcorder, a data logger, and a power supply integrated into a robust housing.

- The deployable noseboom (DNB) was designed for measuring the airflow affecting the vehicle (airspeed, angle of attack, angle of sideslip).

- A low-cost version of FDAS (FDAS-L) was designed particularly for the high-risk parachute verification tests. It is also a stand-alone device with its own power supply, data logger, and a reduced sensor set. Its main purpose is to acquire adequate flight-test data for qualitative but not for quantitative analysis.

FDAS-M was intended to be used for remotely and autonomously controlled flight tests in order to obtain a suitable database for a quantitative postflight evaluation using system identification methods. As result, a verified and validated flight mechanical model of the parafoil-load system including the parafoil aerodynamics shall be derived. Figure 10.3 shows the layout of the FDAS-M (compare Fig. 10.2). The sensor set selected for FDAS-M is presented in Table 10.1.

In FASTWing CL, FDAS-M operated in combination with an external air data measurement, DNB, and the Guidance, Navigation and Control System (GNCS), which is part of the steering box (Fig. 10.4). The GNCS provided the parachute

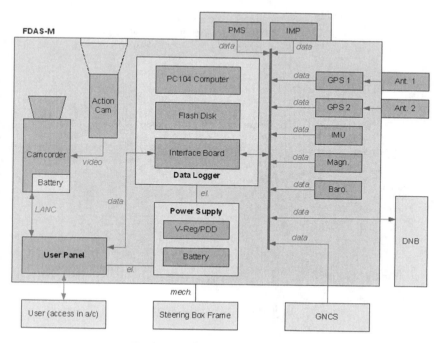

Fig. 10.3   Block diagram of FDAS-M.

TABLE 10.1   FDAS-M SENSORS

Name	Measures	Sensor Type	Range	Update Rate
GPS1	GPS position and speed	Novatel OEM4-G2	Worldwide	20 Hz
GPS2	GPS position and speed	Garmin GPS18x-5Hz	Worldwide	5 Hz
IMU	Accelerations (3-axis)	BAE SiIMU 01	±50 g	50 Hz
IMU	Angular rates (3-axis)	BAE SiIMU 01	±1000 deg/s	50 Hz
Baro.	Static pressure	Honeywell HPA	0–1213 hPa	20 Hz
Magn.	Magnetic field (3-axis)	Honeywell HMR2300	±200 μT	20 Hz
PMS	Temperature	Mela PMS15	−40 to +85°C	0.3 Hz
PMS	Relative humidity	Mela PMS15	0 to 100%	0.3 Hz
IMP	Airspeed	Bräuniger impeller sensor S2	0–40 m/s (0–78 kt)	10 Hz

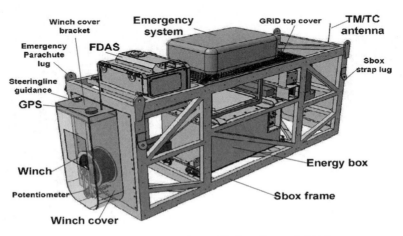

Fig. 10.4   Steering box layout [Hollestelle et al. 2009].

control inputs (i.e., the winch positions for the left and right control lines) as well as ripcord switch status, GPS position, radar altitude, voltages of the power supply of the steering box, and some other status information to FDAS-M. Redundancies in some measurements like GPS, airspeed, static pressure, and others were intentional because of reliability reasons and complementary characteristics of the sensors. The redundant measurement of static pressure, airspeed, and temperature allows using FDAS-M without DNB in such cases where airflow angles are of secondary importance.

In the upper-right corner of Fig. 10.3, PMS stands for "Plug and Measure Serial" PMS15 temperature/humidity sensor, and IMP is an abbreviation for impeller used as an airspeed sensor. The impeller sensor for airspeed measurement was connected to a signal conditioning unit followed by a digital input port of the data logger. All of the other sensor modules (including DNB and GNCS) were connected to the data logger via serial RS-232 interfaces. Because signal conditioning and analog-to-digital conversion was already covered by the sensor modules themselves, using serial interfaces facilitates the layout of the data acquisition system. Only additional power wire routing to each sensor listed in Table 10.1 was required (details on a power supply unit can be found in [Hollestelle et al. 2009]).

The onboard computer executed the data logger software that acquired the data coming from the different sources and stores them on the mass data storage device. The PC104-type boards are used to provide robustness and good electrical connectivity:

- Processor board with low power consumption and fanless operation (AMD Geode GX1)
- Eight-port RS-232/RS-422/RS-485 expansion module
- Sixteen-channel 16-bit analog and digital I/O module
- Rugged DC/DC power supply module

The storage device was a rugged solid state 2.5-in. IDE flash disk with 16-GB memory size to host the data logging software (including operating system) and save flight-test data for 120 min. Each sample of the flight-test data is time stamped. An interface board provides the interfaces between the data logger (i.e., serial expansion module and the analog I/O module) and the sensors or connected modules. The interfaces are configured such that the required power is provided to the sensors. The interface board was designed by DLR specifically for the application in FDAS-M.

A video camera shown in the upper-left corner of Fig. 10.3 looked upwards to the canopy through a window on top of the FDAS-M housing in order to acquire images of the stabilization chute and the parafoil during deployment and flight. The camera assembly comprised a Sony DRC-HC96 digital video camcorder, a robust pen camera with wide angle lens, and Sony's Local Application Control Bus System (LANC) interface. The elements were located in the FDAS-M but were completely independent of the sensor package and the data logger. In this configuration, the video image is provided by the pen camera that is connected via composite video cable to the camcorder that is only used as a recorder. For robustness and survivability reasons, a camcorder using Mini-DV tapes is used. The camcorder is activated and deactivated from outside the FDAS-M using a LANC control device.

On the one hand, the video data allow evaluating the deployment of the parachutes qualitatively; on the other hand, it will be evaluated quantitatively during postflight analysis using image processing techniques for analyzing the relative motion between payload and parafoil.

The housing (Fig. 10.4) was designed in order to accommodate all possible individual components into one box that can be connected to the steering box of the FASTWing CL vehicle. The housing was robust enough to withstand shock loads of up to 20 g without significant damage. The frame is made from welded 25×25 mm (1×1 in.) stainless-steel tubes and covered with 3-mm aluminum sheets. The $L \times W \times H$ size of FDAS-M is $4 \times 3 \times 2.6$ m ($13 \times 10 \times 8.5$ ft). The height increased to 3 m (10.1 ft) when air tunnel and GPS antenna are included. The total weight was 21 kg (46.3 lb). Shock-sensitive parts were mounted on an internal platform that was suspended by helical cable shock isolators.

On top of FDAS-M, a special cover formed an air tunnel wherein airspeed sensor and temperature/humidity sensors are located. The cover protects the sensors from damage and prevents possible entanglement with lines. Distortions of the airspeed measurement are accepted. GPS antennas are mounted on top of this cover to ensure a free and undisturbed view to the sky. FDAS-M is equipped with four shackles for connecting the system to the payload using straps. During the flight tests, the connection must be tight enough to ensure that FDAS-M remains in its position even in presence of significant accelerations during parachute deployment and landing.

As shown in Fig. 10.3, the FDAS-M was equipped with its own power supply that comprised the following:

- Rechargeable batteries (NiCd, 18 V, 7,500 mAh) for more than 2-h operation time
- DC/DC-converter(s) and voltage regulators for the individual voltages of the FDAS-M components
- Power distribution device with connectors

The batteries were rechargeable via the user panel (Fig. 10.5). The panel also provided interfaces to the user for development and testing, flight preparation, and download of the measured data after the flight test. Because these interfaces were not required during the flight, they were covered with a transparent Plexiglas cover.

The flight data acquisition system FDAS-M consisted of two independent subsystems, the video camcorder and the data logger with the connected sensors. The

Fig. 10.5   Integrated FDAS-M with open side cover and without GPS18x antenna.

two subsystems are activated independently from each other. The camera is recording until the end of the tape is reached (typically, after 1 h) or until it is deactivated again. Logging starts immediately after the initialization procedure and is indicated by a red light-emitting diode (LED) flashing. When the GPS receiver has acquired enough satellites for a valid position fix, a second yellow LED is also flashing. The proper reception of GPS satellite signals must already be possible inside the aircraft in order to track the airdrop itself and enable fast GPS satellite reacquisition in the case of signal loss. Hence, a GPS reradiation kit must be installed in the transport aircraft to provide GPS signals in the cargo bay. After both subsystems have been activated and both LEDs indicate proper operation, the drop test can be performed.

The FDAS-M data logger uses the QNX real-time operating system. During the initialization, space is allocated on the flash drive, and log files are created for the complete acquisition time. This is done in order to avoid time-consuming file handling during data acquisition, which otherwise could lead to a loss of data samples, and also ensures that the logged data remain readable even in case of a sudden interruption or loss of power. Finally, this procedure cares for a gentle and predictable usage of the flash memory cells that allow only a limited number of write-and-erase cycles.

Each sensor has its own acquisition and logging routine (thread). These are executed quasi-simultaneously within the real-time software framework. The data acquisition thread is executed with high priority and transfers the data to the logging thread via a random-access-memory (RAM) buffer. Then the logging procedure fills the allocated files with valid sensor data saving each sample with a time stamp for precise postflight synchronization of the data.

After landing, FDAS-M can be disconnected and removed from the steering box and payload. The acquired data remain in the flash drive and can be read out connecting a notebook PC via ethernet cable. For removing and exchanging the mini-DV tape in the camcorder, the housing must be opened. The video is digitized and transferred to the notebook PC, where it can be processed using

standard multimedia tools. For evaluating the relative motion between payload
and canopy, video analyzing and tracking techniques are applied (to be discussed
in Sec. 10.1.4).

Besides the standard laboratory setup that served as basic configuration
for FDAS-M software development, some other ground pretests have been
performed to check out components and software functionality of FDAS-M.

The so-called outdoor walking tests served mainly as battery-powered tests of
the GPS satellite acquisition and performance, which was not possible to assess in
the laboratory setup. On two occasions FDAS-M was dropped from about 1.5 m
(5 ft) height onto ground (grass land) simulating the impact shock at 5 m/s
(9.7 kt) and resulting in an acceleration of 13-g max. FDAS-M survived the
impact without any damage and without any interruption of data acquisition
and logging.

Another test served to test the proper communication and operation between
GNCS, FDAS-M, and DNB. Because the real guidance, navigation, and control
computer of FASTWing CL was not available at DLR, a simulation program
was written and executed on a notebook PC that created a serial output according
to the FDAS-M $\Leftrightarrow$ GNCS communication protocol. While the GNCS simulation program was running, the user could change the ripcord switch parameter
and also simulate the descent prior to landing leading to a reduction of the
GPS and radar altitude in the GNCS data stream. This procedure allowed
testing the proper operation of the deployment and retraction of the noseboom.

Finally, also two climate chamber tests have been performed to ensure
proper operation of FDAS-M under more extreme conditions. During the low
temperature ($-20°C$) and low pressure (500 hPa) test, an atmospheric environment at approximately 4,000 m (13,120 ft) altitude was simulated for 2 h. The
high-temperature test ($+70°C$) served to simulate the possible situation of the
system to be heated by the sun while waiting for recovery after landing. No
problems were encountered in these tests.

Because the vehicle body changes the flowfield of the surrounding air, the
airflow should be measured in a certain distance from an air vehicle. However,
because of the risk of line entanglement jeopardizing parachute and parafoil
deployment, a rigid noseboom that is normally used in aircraft cannot be
applied directly to air-dropped parachute/load systems.

The new design approach used here consists of a noseboom that can be
unfolded after the deployment of the parafoil, and, in order to prevent damage
to noseboom and sensors, it can be folded back again before landing. The
length of the noseboom is about 1 m (3.3 ft) to bring the sensor head into less dis-
turbed air. The remaining measurement errors can be corrected to some extent
during the postflight analysis. A block diagram of the deployable noseboom is
shown in Fig. 10.6.

The mechanism consists of a pivoted boom with a sensor head containing
the airflow sensors (Table 10.2). The boom is actuated by a stepper motor with
a planetary gearbox. The stepper driver generates step impulses for the motor

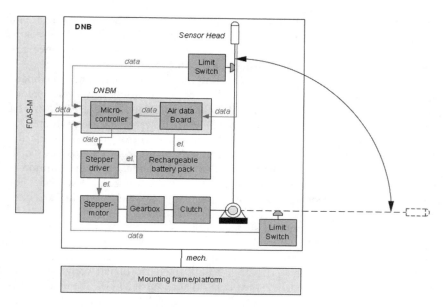

Fig. 10.6   Diagram DNB for air-data measurement.

and is equipped with a serial RS-232 interface for configuration and reception of the control commands coming from the DNB-Module (DNBM).

The sensor head on the noseboom (Fig. 10.6) is equipped with a pitot tube, two vanes for measuring angle of attack and angle of sideslip, and a temperature sensor. The vanes are connected to contactless magnetoresistive potentiometers. From their sinusoidal characteristic the range of ±45 deg with best linearity is used. The air temperature is measured using a silicon-based temperature sensor. The pitot tube is mounted at the nose with some inclination in order to

TABLE 10.2   SENSORS USED IN DNB

Measures	Sensor Type	Range	Update Rate
Dynamic pressure	Honeywell 163PC01D75	±6.23 hPa	25 Hz
Static pressure	Sensortechnics 144SC0811BARO	800 to 1100 hPa	25 Hz
Airflow angles α and β	Midori CP-2UX-SW4	±45 deg	25 Hz
Temperature	Analog devices AD22100	-50 to $+150°C$	25 Hz

achieve a better alignment with the airflow in the expected glide angle range. The total pressure is piped through a plastic tube to a port of the differential pressure sensor inside the DNBM. The second port and the static pressure sensor are measuring the ambient pressure inside the module.

The DNBM consisted of two main components: the microcontroller board and the air-data board. The microcontroller receives continuously a data record including ripcord switch status and GPS and radar altitude from FDAS-M via serial interface. Based on these values, the DNBM software steps through prepro-grammed phases (see next section) and generates control commands for the stepper driver (number of steps, frequency, etc.). The ripcord status is used to trigger the *noseboom-out* command to deploy the boom, and the altitude is used to trigger the *noseboom-in* command for retraction (see next section). For boom end-position feedback and proper alignment, two limit switches are inte-grated into stopper units inside the DNB housing.

The air-data board contains signal conditioning components for the vane signals and sensors for static pressure and differential respectively dynamic pressure. The analog measurements are transferred from the air-data board to the microcontroller, which performs a 10-bit analog-to-digital conversion and forwards the data continuously to the FDAS-M at a rate of 25 Hz. The used micro-controller board is based on the Freescale HCS12 16-bit microcontroller. Except for its software, it is equivalent to the microcontroller board used for FDAS-L.

The DNB is equipped with a rechargeable battery pack (NiCd, 25.2 V, 2,300 mAh) for the power supply of the actuation unit and the DNBM. The housing consists of welded steel pipes forming a frame in which an aluminum plate carries the described components. The frame is covered with an aluminum sheet that provides protection for the noseboom against mechanical damage. The complete assembly (Fig. 10.7) has a size of $1.2 \times 0.4 \times 0.2$ (0.4) m [$4 \times 1.3 \times 0.65$ (1.3) ft], a weight of approximately 37 kg (82 lb), and is mounted in the front of the airdrop platform using bolts.

A general overview of the operation during flight tests is given in Fig. 10.8.

After switching on the main power, the stepper driver and the stepper motor will start up (LEDs on the stepper driver are lighted). Next, the DNBM is acti-vated starting the execu-tion of the data acquisition and control program on the microcontroller.

Fig. 10.7 Assembled DNB during ground testing with FDAS-M in the back.

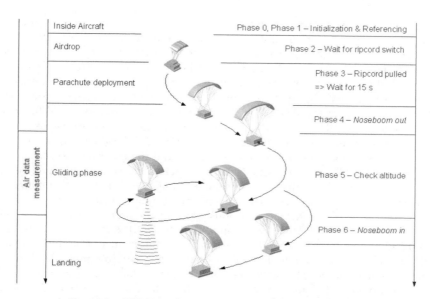

Fig. 10.8   DNB operation concept during a typical flight test.

After initialization, the program enters an endless loop in which air data from the DNB sensors are acquired and transferred to the FDAS-M where they are logged together with the other measurements. At the same time the data coming from the FDAS-M are received and processed. These data essentially include the ripcord switch status and multiple altitude data coming from different sources (GPS, barometric, and radar altimeter). Depending on the information contained therein, the DNBM generates the commands for deploying and retracting the noseboom. For this purpose, the typical flight-test sequence is divided into seven phases:

Phase 0: Initialization and incremental deployment

Phase 1: Referencing (incremental retraction until inner limit switch becomes active)

Phase 2: Wait for ripcord switch

Phase 3: Wait for n seconds after ripcord switch has been pulled (e.g., $n = 15$)

Phase 4: Deploy noseboom $\Longrightarrow$ send *noseboom-out* command

Phase 5: Check altitude $\Longrightarrow$ check radar altitude against GPS height (see the following)

Phase 6: Retract noseboom $\Longrightarrow$ send *noseboom-in* command

The ripcord switch is activated by a static line that is connected to the parafoil and pulled at the beginning of its deployment. After this event, the noseboom extension is delayed for a configurable number of seconds, here 15 s, in order to wait for the system to finish the dereefing process and return to a stable flight state.

The GPS height (above ground) is computed by subtracting the altitude of the foreseen landing point from the GPS altitude, both above MSL. Three configurable decision altitudes, the upper and lower GPS height limits and the radar decision height, are used within the DNBM software to initiate the noseboom retraction. This approach allows validation of the radar altitude and reduction of the risk of an unintentional early noseboom retraction in the case of erroneous radar height measurements. Here, the upper and lower GPS height limits are set to 60 m (196.8 ft) and 30 m (98.4 ft), respectively, and the radar decision height is set to 45 m (148 ft) above ground. If the radar height measurement is erroneous, for example, if it is lower than the decision height of 45 m (148 ft) at that moment when the GPS height passes through the upper height limit at 60 m (196.8 ft), then the retraction will be initiated. Analogously, the noseboom is retracted, when the GPS height passes the lower height limit at 30 m and the radar altitude is still above 45 m (148 ft). For the future, it is planned to implement even more sophisticated algorithms for filtering and cross-checking available multiple altitude information in order to reliably reject faulty values. Phase, limit switch states, and communication status are coded in a status byte that is sent to FDAS-M together with the sensor data.

The large 6-ton (13,228-lb) FASTWing CL payload system consists basically of a so-called superstructure, which is placed on two combined 463L pallets 2.74 × 2.24 m (108 × 88 in.). Superstructure and pallets are connected using straps with honeycomb damping blocks in-between (Fig. 10.9). The superstructure is a 5.5 × 2 m (18 × 6.6 ft) rigid steel platform for the real payload, the parachutes, the steering box, and the flight-test instrumentation devices FDAS-M and DNB. FDAS-M is connected to the steering box on the back of

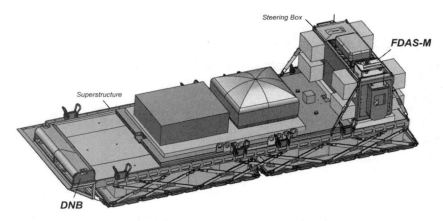

Fig. 10.9   Concept of large FASTWing CL payload (6 tons) with integrated FDAS-M and DNB.

Fig. 10.10   FDAS-M and DNB prepared for the car test.

the system, while the DNB is mounted at the front on a 0.5-m (1.64-ft) skid-like extension of the superstructure. For serial data connection between DNB and FDAS-M, a 7-m (23-ft) long cable is routed along one side of the superstructure. For lighter loads, steering box, parachutes, and instrumentation may also be placed on a smaller pallet.

To test the functionality of FDAS-M and DNB in terms of data acquisition, data consistency, and operation readiness, a so-called car test was carried out. For this test, both devices were installed on a wooden platform that was mounted on the roof of a car (Fig. 10.10). FDAS-M and DNB were connected using a serial cable. To simulate the GNCS computer, which will later be part of the steering box of FASTWing CL, a notebook running the GNCS simulation program is operated from inside the car and also connected to FDAS-M via serial interface.

After activation and initialization of all components, a 30-min drive was performed on public streets near the DLR research center at Braunschweig airport. During the drive, the GNCS program on the notebook simulated an airdrop with the corresponding ripcord switch signal (phase 3) that, as planned, caused 15 s later the deployment of the noseboom (phase 4). The driving speed was varied from 0 to 90 km/h (25 m/s or 49 kt) and corresponds to the expected maximum flight speed of the 6-ton payload below the FASTWing CL parafoil. For noseboom retraction, the GNCS simulation program continuously decreased the radar and GPS altitude values by 5 m/s (phase 5). As expected, when passing the simulated altitude of 45 m (148 ft), the noseboom was retracted (phase 6).

During the car test, the complete set of available FDAS-M, DNB, and GNCS sensor and status data was acquired and logged on the flash drive inside FDAS-M.

In addition to the camcorder inside FDAS-M, a rugged digital video camera was mounted on the DNB capturing the sensor head with the vanes on the noseboom as well as the scene on the street. After the test, the platform was removed from the car, and FDAS-M was connected to the network via ethernet for retrieving the test data.

Clearly, the developed FDAS-M comprised a variety of different sensors that is not cheap and therefore unlikely to be used to test a brand new PADS. Hence, a simpler and cheaper solution was used for the flight data acquisition during the parachute verification tests. The low-cost flight data acquisition system FDAS-L shown in Fig. 10.11a consists basically of a microcontroller-based data logger with reduced sensor package. It is a stand-alone device with its own power supply that can be operated independently from the FASTWing CL vehicle.

Because FDAS-L is designed as a low-cost device, the fulfillment of the original requirements in terms of accuracy and update rate of measurements and logging duration is less restrictive. Hence, the capabilities of FDAS-L are much more limited compared to FDAS-M. Its main purpose consists in acquiring adequate flight-test data for a qualitative but not for a quantitative analysis. No postprocessing of the flight test data is planned.

As shown in Fig. 10.11b, FDAS-L is equipped with only two sensors: 1) Garmin GPS18x-5Hz receiver with antenna and 2) Crossbow CLX25GP3 three-axis acceleration sensors (± 25 g).

The GPS receiver is connected to the microcontroller via a serial RS-232 interface; the three accelerometer channels are connected to three inputs of the 10-bit analog-to-digital converter of the microcontroller. The 10-bit resolution corresponds to 1024 increments for the full measurement range. Accordingly,

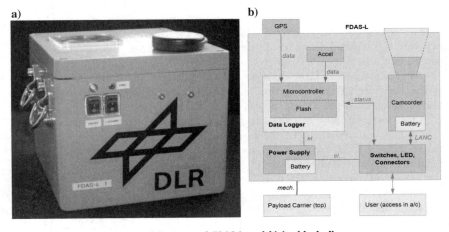

Fig. 10.11   a) Integrated FDAS-L and b) its block diagram.

20 increments correspond approximately to 1 g for the measurement range of ±25 g.

The data logger for FDAS-L consists of a microcontroller board based on the Freescale HCS12 16-bit microcontroller. This board is a multipurpose DLR design, already used for different applications and in the DNB. The microcontroller is the MC9S12DG256 running at a frequency of 24 MHz in the present application as data logger for FDAS-L. To simplify the cabling, an adapter board was designed providing adequate connectors for power supply and data interfaces to sensors and PC. For data read-out and programming, FDAS-L must be opened and connected to a personal computer via null-modem.

FDAS-L is equipped with its own 7.2-V power supply using low self-discharging batteries (Sanyo eneloop). During operation, the average power consumption is less than 100 mA for data logger and connected sensors. Hence, the selected capacity of 2,000 mAh is large enough to provide FDAS-L with power for more than 10 h even after several months of storage. The video camcorder has a separate power supply with a lithium-ion battery.

In addition, a video camcorder is tightly fitted into the damping material inside the FDAS-L housing, looking upwards towards the canopy through a Plexiglas window. The setup here is similar to the one used in FDAS-M, except that no additional pen camera is installed. For robustness and survivability reasons, a Mini-DV camcorder was selected (i.e., the tape shall be readable even after a total crash).

A robust aluminum enclosure was selected in order to accommodate the data logger, batteries, and the acceleration sensors in one box that can be connected to the payload carrier with straps going through the shackles on both sides of the box. The GPS antenna and the camera window on top of the housing must not be covered. Like for FDAS-M, the reception of GPS satellite signals must be possible already inside the aircraft, that is, a GPS reradiation kit must be installed in the cargo bay. The dimensions of the housing are approximately $2.3 \times 2 \times 1.80$ m ($7.5 \times 6.6 \times 5.9$ ft), and the total weight is 5.6 kg (12.3 lb) including all internal components.

The flight data acquisition system FDAS-L comprises two independent subsystems, the video camcorder and the data logger with the connected GPS receiver and acceleration sensors. Both subsystems have their own power supply and are activated independently from each other. Like in the FDAS-M system, the activation of the video camcorder is done here again using a LANC camera control device. The activation of the data logger is done using the on/off switch. Active data logging is indicated by a yellow LED flashing. A fast flashing frequency (2.5 Hz) indicates that the GPS receiver has acquired enough satellites for a valid position fix; a slow flashing frequency indicates that the GPS position is not valid. With the logging switch on, it is guaranteed that the flash memory is not erased accidentally due to shock or loose contact.

Configuration and data read-out is performed using a notebook PC that is connected to the data logger via serial RS-232 interface. The sampling rates can be selected between 0 and 20 Hz for the accelerometer and between 0 and 5 Hz for the GPS receiver (0 Hz means "no acquisition"). For delaying the start of the logging process, a time delay from 0 to 600 s can be introduced.

In the present configuration, the accelerations are sampled at 20 Hz and the GPS positions at 5 Hz, which are the highest possible sampling rates. This leads to a maximum total data acquisition time of 15 min. If the sampling rates are reduced to 5 Hz for the accelerations and 1 Hz for GPS, the total acquisition time increases to 1 h. The limited acquisition time is a result from selecting an available multipurpose microcontroller board with only 256 kB of flash memory, which was originally designed for a completely different application. This was a tradeoff during the concept phase taking also into account the available programming tools and experience with this board as well as the complete access to all resources of the microcontroller. This provided highest flexibility and allowed development of an adequate custom software solution.

With a configured time delay of 180 s, the drop test can be performed after 3 min after both subsystems have been activated. The drop test should be finished before the end of the acquisition time is reached, that is, airdrop and landing must happen within 18 min after activation. Therefore, activation should be scheduled approximately at "T-8 min," that is, 8 min before payload release and airdrop. Under normal circumstances, this also gives the GPS enough time to achieve a valid position fix. After landing, the data logger is switched off again, and the camcorder is deactivated. The acquired data remain in the flash memory of the microcontroller and can be read out using a notebook PC via serial interface.

Having a simplified flight parameters recorder, which does not include a full-fledged IMU, may not be enough to investigate the entire ADS trajectory from the exit from aircraft all of the way to touchdown. Specifically, GPS data might not be available for the first ~30 s of a drop due to limited antenna visibility of the sky upon rotation at exit and the high accelerations at canopy opening that saturate the GPS's oscillators (Fig. 10.12). That is where IMU might be useful. The Payload Derived Position Acquisition System (PDPAS) being developed at YPG aims at closing this first 30-s gap [Tiaden and Yakimenko 2008; Tiaden and Yaki-menko 2009; Tiaden 2011]. With the requirement to have a stand-alone system, not relying on an aircraft 1553 bus data to provide an initial guess on the Euler angles, PDPAS solution intro-duced a secondary Euler angle initializa-tion loop right after GPS acquires a satellite signal. The error between the

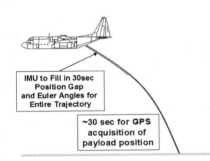

Fig. 10.12   PDPAS position solution throughout ADS trajectory.

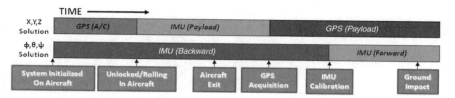

Fig. 10.13   PDPAS position solution source as a function of time [Tiaden 2011].

actual vs IMU-produced position data allows the reinitialization of IMU and integration of Euler angles backward as shown in Fig. 10.13.

10.1.2   INSTRUMENTATION FOR CANOPY MOTION ESTIMATION

Relative motion between parafoil and payload is a phenomenon that is introduced by the usage of textile materials for canopy, harness, and lines. Because of this flexibility, additional degrees of freedom have to be considered in two different domains of system identification (Fig. 10.14). First, relative motion is an unwanted part in all measurements that are addressing the canopy's aerodynamics but taken at the load. Without measuring the relative motion itself, it could be difficult to convert between the two coordinate systems. Second, relative motion possibly affects the flight characteristics of a parafoil-payload system and because of that should be part of the flight-mechanical model itself.

In the past, the problem of relative motion has been addressed by various approaches. The simplest approach is to ignore relative motion. For certain geometries of load, harness, and lines and for particular stages of flight, this is a reasonable assumption. Then, in the analysis the relative motion just appears as measurement noise that is not analyzed in detail. In another approach relative motion is simplified and for example treated as additional pitch and yaw for the payload in the flight mechanical model. This, on one hand, keeps the models straightforward and, on the other hand, introduces basic abilities in the model, for example, the possibility for detailed flare analysis. Finally, the approach for a detailed investigation of relative motion effects is a multibody analysis with all relevant parts incorporated concerning their degrees of freedom, geometry, and mass. For both, transformation of flight-test data and building meaningful models, measurements for the relative motion are essential. For example, Schoenenberger et al. [2005] were able to use a laboratory camera system mounted in a wind tunnel to quantify the trim behavior of parachutes and measure important aerodynamic data such as the pitch damping and pitching-moment coefficients through manual analysis of the video frames.

Parachute and parafoil canopies are made of thin cloth material. This makes it difficult to accurately quantify their behavior. Most conventional means of measuring payload trajectories using onboard instrumentation cannot be used

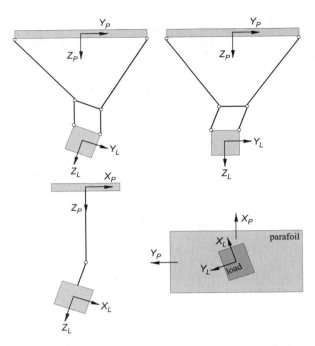

Fig. 10.14   Relative roll, lateral displacement, relative pitch, relative yaw [Jann and Strickert 2005].

in conjunction with a parachute canopy without influencing the shape or behavior of the system. Instead, laboratory experiments have been conducted using near-range optical tracking systems such as the Vicon motion capturing system, which uses a network of grayscale or infrared camera systems to track the 3D motion of reflective markers in real-time [Gonyea et al. 2013; Jones et al. 2007; Tanner et al. 2013]. A problem with this type of analysis is that for parachute drops, only very small segments of flight can be simulated within the test space enclosed by such a system.

Alternatively, miniature IMU could be fixed to desired canopy locations to measure a variety of parameters for multiple canopy parts at the same time. One of such earlier attempts was undertaken by Hur [2005]. Figure 10.15 shows how one single-axis and three tri-axial accelerometers (CXL02LF1 and CXL02LF3 by Crossbow) were sewn into the main chord of the Buckeye PPG (Figs. 1.9a and 5.43a) along with a Crossbow WSC-100 wireless transmitter. The entire setup including cables, connectors, and batteries weighted 1.2 kg (2.7 lb). To minimize the dynamic effect due to weight increase of the parafoil canopy, the heavier parts—transmitter and power supply—were located at

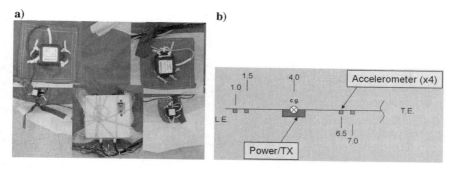

Fig. 10.15 a) CXL02LF1/CXL02LF3 accelerometers along with a wireless transmitter sewn in the canopy and b) their locations along the 3.3 m (11.5 ft) chord of Buckeye PPG [Hur 2005].

canopy's c.g. (Fig. 10.15b). A bluetooth wireless data acquisition system (independent of main acquisition system recording GPS/IMU data) was based on a Valitec ReadyDAQ AD2000 Data logger located in the payload (powered vehicle) and allowed receiving signals within 30.5-m (100-ft) range and stored them at 25-Hz rate [Hur 2005].

Figure 10.16a shows a sample of raw data collected by accelerometers within the main IMU and accelerometers sewn in the canopy. As expected, these sample data demonstrate that the noise level for accelerometers sewn in the canopy is much larger and therefore requires filtering before it can be used in estimation of relative canopy motion. A Butterworth filter with the cutoff frequency of 0.2 Hz allows substantial improvement (Fig. 10.16b). The filtered accelerations were then used to estimate pitch and yaw attitudes and their rates. Sideslip angle was calculated from reduced velocities [Hur 2005].

A decade later, a much more consolidated and powerful system weighed only a few grams. For example, the VN-100 SMD (surface-mount device) miniature high-performance IMU and Attitude Heading Reference System (AHRS) incorporates the latest MEMS sensor technology combining tri-axial accelerometers, tri-axial gyros, tri-axial magnetic sensors, a barometric pressure sensor, and a 32-bit processor into a miniature surface mount module [VN-100 2014]. Along with providing calibrated sensor measurements (at a rate of 100 Hz), this IMU/AHRS also computes and outputs a real-time, drift-free 3D orientation solution that is continuous over the complete 360 deg of motion. A similar system, the Bantam miniature IMU, small enough at 0.001 m^2, 10 g (1.7 in.^{2}, 0.35 oz) to be sewn in the smallest PADS canopy, has been used in the study of comparing different fidelity models of PADS (Fig. 10.17b) [Gorman and Slegers 2011a, 2011b, 2012; Slegers et al. 2015]. IMU data coming out of Bantam IMU were wirelessly transmitted and recorded by a flight computer located atop payload (using a low-power wireless transceiver with a 20-m (66-ft) range).

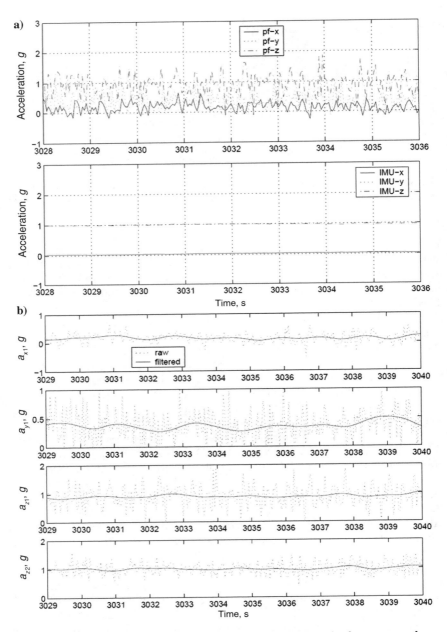

Fig. 10.16   Raw data from main IMU and accelerometers a) sewn in the canopy and b) filtered (vs unfiltered) canopy accelerometers data [Hur 2005].

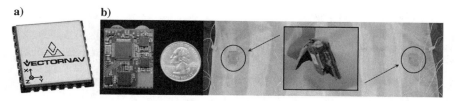

Fig. 10.17   Parachute trajectory characterization: a) VN-100 SMD IMU/AHRS and b) Bantam IMU [VN-100 2015; Gorman and Slegers 2011b].

An example of amplitude spectrum of measured angular velocities for both the payload and parafoil canopy of Snowflake PADS is shown in Figs. 10.18a and 10.18b, respectively. As expected, the frequency spectrums of payload and parafoil canopy have some differences. While the payload yaw rate has two frequency components, a low-frequency component at 0.05 Hz from the control input and a dynamic twisting mode at 1.05 Hz, the canopy yaw-rate spectrum features only one, at 0.05 Hz, coming from the control input, that is, dynamic twisting does not propagate to the canopy. The payload roll and pitch-rate amplitude spectrums have similar components at about 2.2 Hz, but canopy only exhibits this frequency in the pitch channel caused by two-body pitch coupling. The continuation of this effort includes addition of a miniature pressure sensor and installing these IMU/AHRS/pressure sensors in each individual cell of an MC-5 canopy [Slegers et al. 2015].

Another alternative to quantify canopy motion is to use a simple payload-mounted digital camera positioned to look upward and capture video of the canopy while in flight (Fig. 10.19). Using image processing algorithms, both the location and the orientation of the parachute can be determined automatically. As mentioned in Sec. 10.1.1, both ALEX and FASTWing programs used this latter approach (back then it utilized a video camcorder, so that video data were digitized during a postflight analysis).

A variety of tracking algorithms have been developed and can be utilized to identify and track parafoil's canopy throughout the entire video sequence. These include multiple point tracking or color segmentation algorithms. For example, a recent work by Hanke and Schenk [2014] shows a strong potential to implement a photogrammetric approach with exciting bundle adjustment software to find the relative orientation of canopy by tracking a set of markers. Figure 10.20 shows an example of some natural points that could be tracked. These include vertices formed by suspension lines attachment points and ribs endings at leading and trailing edges. Figure 10.21 gives a panoramic view of the entire Ozone Matra R11 canopy featuring 20 artificially introduced tracking targets.

Hanke and Schenk [2014] report that they used a calibrated 12-Megapixel Nikon D200 camera with 23.6 × 15.8 mm CCD (charge-coupled device) taking 3,872 × 2,592 pixel images. Using PhotoModeler software allowing extracting

Fig. 10.18 Sample of angular velocity amplitude spectrum for a) payload and b) canopy [Gorman and Slegers 2011b].

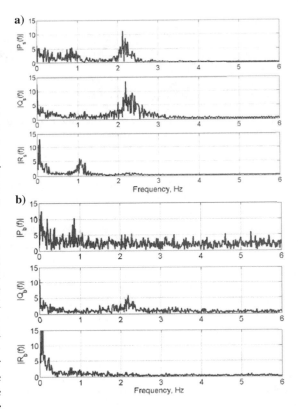

3D measurements and models from photographs taken with an ordinary camera, they were able to reliably track the total of 270 points (Fig. 10.22) with only a few millimeters of error on average.

Figure 10.23 shows the major steps of a color segmentation algorithm that finds and tracks a central red region of a Snowflake PADS canopy as developed by Decker [2013] and discussed in [Decker and Yakimenko 2015]. Beginning with a three-layer RGB digital color image for each frame, a grayscale image is created using the red pixel intensity layer subtracted by half of the sum of the blue and green layers to boost the effect of red pixels while reducing the intensity of green, blue, and consequently white colors. From there, the resulting grayscale image is contrast-stretched and compared to a threshold value to generate a binary

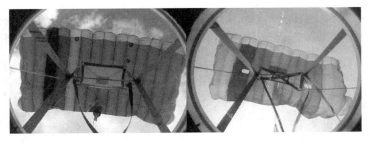

Fig. 10.19   Video camera views of the canopy [Jann and Strickert 2005].

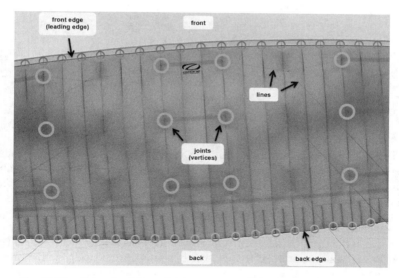

Fig. 10.20   Natural points and features to track [Hanke and Schenk 2014].

image containing mostly red (or colors similar to red) pixels. The binary image then undergoes a sequence of morphological erode and dilation operations until only one large single body is produced.

Once the shape is isolated, the shape center is calculated as the average location of the segmented shape's pixels (depicted in the last image in Fig. 10.23 with a diamond-shaped marker). In addition, the orientation of the shape can be estimated by calculating the inertial moment of the segmented shape's pixels about the shape center. In Fig. 10.24, the axis of the largest moment is depicted with star-shaped markers while the transverse axis is shown with circle-shaped markers.

Figure 10.25 shows an example of data that can be extracted from the looking-upward camera data analysis. Specifically, it shows a relative displacement of the central moment and relative orientation.

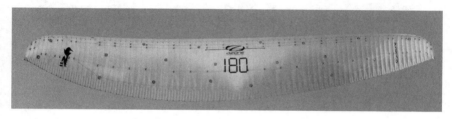

Fig. 10.21   Panoramic image of paraglider canopy [Hanke and Schenk 2014].

Fig. 10.22   Locations of points of interest [Hanke and Schenk 2014].

For system identification, the motion data from the video has to be synchronized to the data coming from the other sources, such as GPS or IMU. This can be done on the basis of events, which are registered in all sensor systems simultaneously such as the parafoil opening or landing event. Another option is the use of a dedicated synchronization code or time code that is generated by the onboard computer and recorded at the video as well.

Because the distance between canopy and payload is known from geometry, the relative displacements can be approximately transformed into relative attitude. Provided that the relative attitude between canopy and payload is precisely synchronized with the rest of data, the attitude and angular rates can be recomputed to give the corresponding values for the canopy (see an example obtained on the ALEX PADS in Fig. 10.26). Using the relative attitude, angles of attack and sideslip, the accelerations and velocities can be realigned to the reference frame $\{b\}$ with the origin at the PADS center of mass. As result of the correction, the

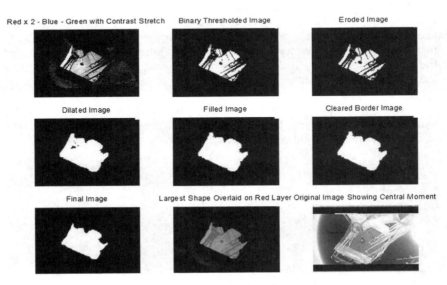

Fig. 10.23   Isolating the red shape of the canopy [Decker 2013].

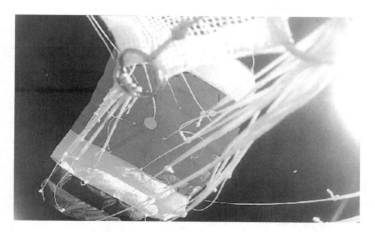

Fig. 10.24   Proper identification of central moment and canopy orientation [Decker 2013].

time histories become smoother because they are not overlaid by the relative motion anymore. This effect is expected because the canopy behaves more sluggish than the suspended payload. Another effect is a short delay of the payload's motion compared to the motion of the canopy. For example, if the right control line is pulled, first the canopy performs the corresponding rolling and yawing motion before it drags the payload after it. Both effects can be observed in the

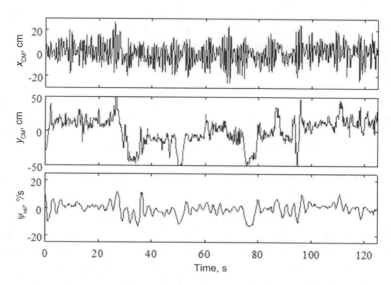

Fig. 10.25   Measurement of relative motion of the canopy [Decker 2013].

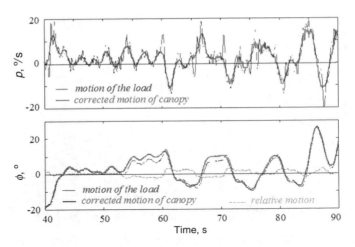

Fig. 10.26   Payload motion, relative motion, and corrected canopy motion [Jann and Strickert 2005].

comparison of between payload and canopy motion in the time histories of Fig. 10.26 and also Fig. 11.6. Because the measurements are corrected for the location of the sensor package, accelerations and velocities resulting from the rotation of the vehicle have to be transformed into $\{b\}$.

Video analysis turned out to be a very valuable tool, which supplements the other measurements. Of course, this method has drawbacks and challenges. The first challenge is to provide an uninterrupted view from the payload to the canopy and to find a suitable location to mount the video camera on the payload. On one hand, the camera must not disturb the parafoil opening; on the other hand, no obstacles such as reefing devices or harness should obstruct the view. For typical ADS, a wide angle lens has to be used to capture the whole canopy in the field of view for accurate determination of canopy center.

Another challenge of video analysis is the sensitivity to lighting conditions. Changing illuminations and hard shadows cause difficulty for simple automated image processing algorithms. This can occur during certain orientations where the sun may appear in a field of view. One solution for this problem is to guide the tracking manually through these difficult scenes. The same approach can be applied when the algorithm loses a tracked marker because of occlusions or because the canopy has traveled out of frame. An additional challenge is that for certain tracking shapes (nearly square) the automated image processing algorithm may struggle to identify the forward direction of the canopy as compared to the lateral direction. As an example, Fig. 10.27 shows a frame where the central red region being segmented by the color-tracking algorithm appears almost square in shape. As a result, the canopy shape was improperly segmented and estimated poorly. Depending on the size of PADS, there might be some additional problems

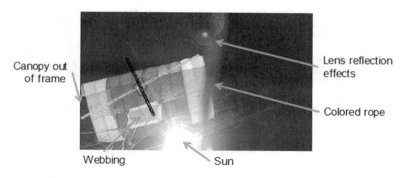

Canopy out of frame

Lens reflection effects

Colored rope

Webbing   Sun

Fig. 10.27   Causes of improper segmentation of canopy [Decker 2013].

related to the suspension lines and slider (see example in Fig. 10.19). In an example of Fig. 10.27, the slider was completely off to the side and did not affect the segmentation process, but it may block the features that the automated image processing algorithm tries to track. The same holds for the suspension lines.

These issues were addressed in [Decker 2013] where it was determined that from the practical standpoint simply changing the canopy color configuration might be used to improve robustness of tracking algorithm. (Red and green colors would be the best choice to take a full advantage of RGB layering.) Tracking multiple rectangular (noncentered) cells would be beneficial as well. If one of these multiple regions could be a different color (Fig. 10.27b), it would alleviate aliasing errors and therefore measure 360 deg of canopy twist. To assess the potential accuracy of the improved color scheme shown in Fig. 10.28, simulations using the canopy 3D model (created using CAD) were conducted (Fig. 10.29). The resolution of the canopy images (600 × 400 pixels) was chosen to replicate the resolution of a typical digital camera used in the canopy relative motion assessment tests. The automated analysis required just a fraction of a second, and in all simulations the orientation angle was correctly determined within 0.24 deg. This number provides a lower bound on a relative orientation of a canopy estimate using the upward-looking cameras (and color segmentation algorithms).

System identification is usually conducted at the earlier stages of PADS development when different color schemes for the canopy could be attempted. For a fielded PADS or PADS with monotonically colored canopies (see examples in

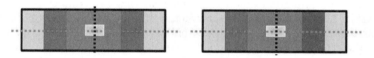

Fig. 10.28   Improved canopy color configurations (light blue/red/dark blue/green/ light blue) [Decker 2013].

Fig. 10.29   Sample analysis of 3D model generated images [Decker 2013].

Figs. 1.10–1.13, 1.17, 1.18, 1.21–1.23, 1.25–1.27) the edge detection algorithm tracking the leading and trailing edges of the canopy should be used.

10.2   GROUND EQUIPMENT

Continuing an image processing theme of Sec. 10.1.4, this section introduces ground-based equipment usually used during ADS testing. First, it demonstrates an example of a fully automated analysis of video data to quantify the motion of ADS or in other words to obtain a so-called time-space-position information (TSPI) solution (usually, it refers to a position of payload centroid). Second, it discusses a potential to obtain a TSPI solution for both canopy and payload and as a result to estimate an attitude of ADS and its components.

10.2.1   GROUND-BASED KTM SYSTEMS

During the airdrop test, multiple ground-based optical systems are positioned at various locations around a test range pointing towards a test article. When the videos from all cameras are correctly synchronized, it is possible to calculate an accurate TSPI history of this test article. In the case of ADS, TSPI data (about a three-DoF translational trajectory of a payload) allow assessing such important parameters as a descent rate history, altitude loss during canopy deployment, canopy opening shock, etc.

For example, kineto-tracking mounts (KTMs) commonly used at the test ranges are sophisticated optical systems capable of capturing a variety of image types including both digital and analog optical video, infrared video, and other information such as radar data (Fig. 10.30). KTM systems are quite large, usually requiring a trailer to position them unless they are permanently located at the range. KTMs feature two-axis rotation capability (in azimuth and elevation), so that the test object can remain in view for as much of the test as practicable. The key performance parameters of KTM are shown in Table 10.3.

The KTM systems are usually operated by a trained technician who sits in a seat in the center of the mount and tries to keep the test article in the center of the image frame. The system shown in Fig. 10.30a is equipped with a variety of optical systems with different focal lengths 0.254 to 6 m (10 to 240 in.), so that the resulting videos can show the test object (descending ADS) at different

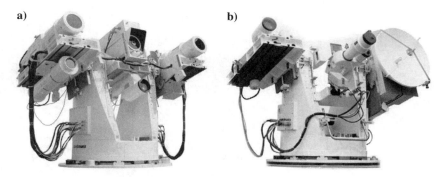

Fig. 10.30   a) Standard KTM and b) radar-enhanced KTM [L-3 Communications 2008].

levels of zoom. Newer models are capable of being controlled remotely and may also use a closed-loop control with a feedback provided by the radar (Fig. 10.30b).

Also, the newest KTM models feature a networking capability, which could allow a rough triangulation of the test object position obtained using multiple KTMs (Fig. 10.31a) in real time. This real-time triangulation solution can then be fed back to KTM to enable automated test object following even without the radar. Otherwise, video scoring is conducted off-line.

TABLE 10.3   KTM PERFORMANCE INDICATORS [L-3 COMMUNICATIONS 2008]

Category	Parameter	Performance Indicator
Range of motion	Elevation	−10 to +190 deg
	Azimuth	−335 to +335 deg
Dynamic performance	Velocity–Az/El	Up to 60 deg/s
	Acceleration–Az/El	Up to 60 deg/s^2
Position encoder	21-bit resolution (standard)	0.6 arc sec (0.00017 deg)
	23-bit resolution (optional)	0.15 arc sec (0.00004 deg)
Geometric errors	Wobble	<3 arc sec (0.00083 deg)
	Orthogonality	<5 arc sec (0.0014 deg)
Performance	LOS pointing accuracy (after calibration)	5 arc sec (0.0014 deg)
	Tracking jitter	<3 arc sec (0.00083 deg)

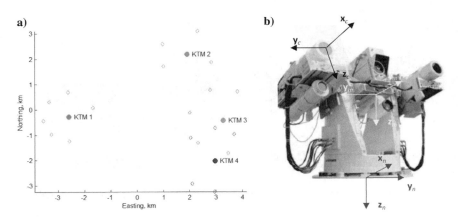

Fig. 10.31   a) Typical test setup with four KTMs and calibration monuments around them and b) three coordinate frames KTM utilizes.

A camera sees a tracking object (canopy center) in the image plane $\{c\}$ with the offsets in pixels

$$u = x_{\text{offset}}, \quad v = y_{\text{offset}} \tag{10.1}$$

Knowing the calibration angle-to-pixel ratios (aka boresight constants) C_{Az} and C_{El}, these correspond to the angular offsets

$$\varepsilon_{Az} = C_{Az}x_{\text{offset}}, \quad \varepsilon_{El} = C_{El}y_{\text{offset}} \tag{10.2}$$

The TSPI solution relies on knowing azimuth and elevation of an optical path of camera as well as horizontal and vertical position of test article in the image frame with respect to the center of the image frame. These quantities are measured in two different coordinate frames. Adding a local tangent plane constitutes a set of three coordinate frames KTM has to deal with as shown in Fig. 10.31b.

The coordinate frame $\{c\}$ associated with each camera has its origin at $\mathbf{r}^m_{\text{arm}} = [x_{\text{arm}}, y_{\text{arm}}, z_{\text{arm}}]^T$ defined in $\{m\}$ with axis $\mathbf{x}_c$ extending along the camera's optical path, axis $\mathbf{y}_c$ aiming along a horizontal KTM gimbal, and axis $\mathbf{z}_c$ perpendicular to $\mathbf{x}_c$ and $\mathbf{y}_c$ and looking downward in the camera plane. The coordinate frame $\{m\}$ has its origin at the surveyed point $\mathbf{r}_{KTM} = [x_{KTM}, y_{KTM}, z_{KTM}]^T$ defined in the Earth-centered Earth-fixed coordinate frame $\{ECEF\}$ with axis $\mathbf{z}_m$ pointing downward along the vertical KTM gimbal, axis $\mathbf{y}_m$ aiming along a horizontal KTM gimbal, and axis $\mathbf{x}_m$ perpendicular to $\mathbf{z}_m$ and $\mathbf{y}_m$ and looking forward. The $\mathbf{x}_n$ axis of the local (at the KTM location) tangent coordinate frame $\{n\}$ points North, $\mathbf{y}_n$ axis points East, and $\mathbf{z}_n$ axis points downward along the plumb line (NED coordinate frame).

The coordinate frame $\{c\}$ is rotated with respect to $\{m\}$ by one angle $El = El_{enc} + bias_{El}$ about axis $\mathbf{y}_m$ ("enc" stands for "as recorded by an encoder") so that for any vector $\mathbf{r}^m$ in $\{m\}$

$$\mathbf{r}^c = {}_m^c\mathbf{R}\mathbf{r}^m = \mathbf{R}_{El}\mathbf{r}^m \qquad (10.3)$$

where the corresponding rotation matrix is

$$
{}_m^c\mathbf{R} = \begin{bmatrix} \cos(El_{enc} + bias_{El}) & 0 & -\sin(El_{enc} + bias_{El}) \\ 0 & 1 & 0 \\ \sin(El_{enc} + bias_{El}) & 0 & \cos(El_{enc} + bias_{El}) \end{bmatrix} \qquad (10.4)
$$

The coordinate frame $\{m\}$ is rotated with respect to $\{n\}$ by three consecutive single rotations $tilt_E$ (about $\mathbf{x}_n$), $tilt_N$ (about $\mathbf{y}_n$), and $Az = Az_{enc} + bias_{Az}$ (about $\mathbf{z}_n$). Hence, for any vector $\mathbf{r}$ in $\{ECEF\}$, we can write

$$\mathbf{r}^m = {}_n^m\mathbf{R}({}_{ECEF}^n\mathbf{R}\mathbf{r}) = \mathbf{R}_{Az}\mathbf{R}_{tilt_N}\mathbf{R}_{tilt_E}({}_{ECEF}^n\mathbf{R}\mathbf{r}) \qquad (10.5)$$

[compare Eq. (5.8)]. Keeping in mind that the tilting or mislevel error defined by $tilt_E$ and $tilt_N$ is very small (less than 0.001 rad), we can write

$$
\mathbf{R}_{tilt_E} = \begin{bmatrix} 1 & 0 & 0 \\ 0 & \cos(tilt_E) & \sin(tilt_E) \\ 0 & -\sin(tilt_E) & \cos(tilt_E) \end{bmatrix} \approx \begin{bmatrix} 1 & 0 & 0 \\ 0 & 1 & tilt_E \\ 0 & -tilt_E & 1 \end{bmatrix} \qquad (10.6)
$$

$$
\mathbf{R}_{tilt_N} = \begin{bmatrix} \cos(tilt_N) & 0 & -\sin(tilt_N) \\ 0 & 1 & 0 \\ \sin(tilt_N) & 0 & \cos(tilt_N) \end{bmatrix} \approx \begin{bmatrix} 1 & 0 & -tilt_N \\ 0 & 1 & 0 \\ tilt_N & 0 & 1 \end{bmatrix} \qquad (10.7)
$$

$$
\mathbf{R}_{Az} = \begin{bmatrix} \cos(Az_{enc} + bias_{Az}) & \sin(Az_{enc} + bias_{Az}) & 0 \\ -\sin(Az_{enc} + bias_{Az}) & \cos(Az_{enc} + bias_{Az}) & 0 \\ 0 & 0 & 1 \end{bmatrix} \qquad (10.8)
$$

$$
{}_{ECEF}^n\mathbf{R} = {}_{UEN}^n\mathbf{R}\mathbf{R}_{lat}\mathbf{R}_{lon} = \begin{bmatrix} 0 & 0 & 1 \\ 0 & 1 & 0 \\ -1 & 0 & 0 \end{bmatrix} \begin{bmatrix} \cos(lat) & 0 & \sin(lat) \\ 0 & 1 & 0 \\ -\sin(lat) & 0 & \cos(lat) \end{bmatrix}
$$

$$
\times \begin{bmatrix} \cos(lon) & \cos(lon) & 0 \\ -\sin(lon) & \cos(lon) & 0 \\ 0 & 0 & 1 \end{bmatrix} \qquad (10.9)
$$

[compare Eq. (5.9)]. The variables $tilt_E$, $tilt_N$, $bias_{Az}$, and $bias_{El}$, along with two boresight constants C_{Az} and C_{El} [defined in Eq. (10.2)] are not known and have to be estimated during camera calibration prior to each test.

The calibration procedure for each camera involves each particular camera looking at multiple calibration monuments with surveyed locations $k = 1, \ldots, K$, $k = 1, \ldots, K$. (For the KTM setup shown in Fig. 10.31a, these

Fig. 10.32   An example of double camera-calibration monument.

calibration monuments surrounding each KTM location
are depicted by a diamond
marker.) An example of a
double calibration monument
with azimuth-elevation data
encoded as a left-edge barcode (and also imprinted by
the upper edge) is shown in
Fig. 10.32a.

Mathematically, given data from observing K calibration monuments by the jth camera (which includes $\mathbf{r}_{KTM}$, $\mathbf{r}_T^k$, $\mathbf{r}_{arm}^{m;j}$, Az_{enc}^k, El_{enc}^k, x_{offset}^{kj}, y_{offset}^{kj}), the calibration procedure consists in finding

$$\min_{\Xi^j} (J^j) \tag{10.10}$$

with a cost function

$$J^j = \sum_{k=1}^{K} \left(\Delta\varepsilon_{Az}^{kj2} + \Delta\varepsilon_{El}^{kj2} \right) \tag{10.11}$$

is a vector of varied parameters

$$\Xi^j = [tilt_E, tilt_N, bias_{Az}, bias_{El}, C_{Az}^j, C_{El}^j] \tag{10.12}$$

In Eq. (10.11),

$$\Delta\varepsilon_{Az}^{kj} = \tan^{-1} \frac{[0, 1, 0]\Delta\mathbf{r}^c(\mathbf{r}_{KTM}, \mathbf{r}_T^k, \mathbf{r}_{arm}^{m;j}, Az_{enc}^k, El_{enc}^k, tilt_E, tilt_N, bias_{Az}, bias_{El})}{[1, 0, 0]\Delta\mathbf{r}^c(\mathbf{r}_{KTM}, \mathbf{r}_T^k, \mathbf{r}_{arm}^{m;j}, Az_{enc}^k, El_{enc}^k, tilt_E, tilt_N, bias_{Az}, bias_{El})}$$
$$- C_{Az}^j x_{offset}^{kj} \tag{10.13}$$

$$\Delta\varepsilon_{El}^{kj} = \tan^{-1} \frac{[0, 0, -1]\Delta\mathbf{r}^c(\mathbf{r}_{KTM}, \mathbf{r}_T^k, \mathbf{r}_{arm}^{m;j}, Az_{enc}^k, El_{enc}^k, tilt_E, tilt_N, bias_{Az}, bias_{El})}{[1, 0, 0]\Delta\mathbf{r}^c(\mathbf{r}_{KTM}, \mathbf{r}_T^k, \mathbf{r}_{arm}^{m;j}, Az_{enc}^k, El_{enc}^k, tilt_E, tilt_N, bias_{Az}, bias_{El})}$$
$$- C_{El}^j y_{offset}^{kj} \tag{10.14}$$

This is based on the fact that ε_{Az} and ε_{El} offsets [see Eq. (10.2)] can be computed using the ratios of components of the vector $\Delta \mathbf{r} = \mathbf{r}_c - \mathbf{r}_T$ connecting the origin of $\{c\}$ and $\mathbf{r}_T$ (translated to $\{c\}$). The latter vector is computed as

$$\Delta \mathbf{r}^c = {}_m^c\mathbf{R}[{}_n^m\mathbf{R}{}_{ECEF}^n\mathbf{R}(\mathbf{r}_T - \mathbf{r}_{KTM}) - \mathbf{r}_{arm}^m]$$
$$= \mathbf{R}_{El}[\mathbf{R}_{Az}\mathbf{R}_{tilt_N}\mathbf{R}_{tilt_E}{}_{ECEF}^n\mathbf{R}(\mathbf{r}_T - \mathbf{r}_{KTM}) - \mathbf{r}_{arm}^m] \qquad (10.15)$$

To optimize the values of six varied parameters of Eq. (10.10), at least three calibration monuments are needed. While spreading these monuments around KTM poses no problem, a certain precaution should be made to ensure some vertical separation between them; otherwise, Eq. (10.14) will have no contribution. That is why one of the calibration monuments (Fig. 10.32) is a double monument reducing a probability of ill-formulated minimization problem.

In practice however, trying to optimize the values of six varied parameters in Eq. (10.12) leads to an erroneous result. The reason is that the values of two boresight constants C_{Az} and C_{El} are connected to each other, meaning they cannot be optimized as two independent variables. Two different technologies for capturing digital images, CCD and CMOS (complementary metal oxide semiconductor), are both based on sensors having a certain aspect ratio (Fig. 10.33). Most image sensors utilize 4:3 aspect ratio (Fig. 10.33b), which corresponds to the 640×480 pixel image and consequently to 4:3 aspect ratio of horizontal and vertical field of view. For instance, the $\frac{1}{2}$-in. CCD/CMOS chip typically used by KTM cameras (nonstandardized "inch" refers to approximately 1.5 length of the actual diagonal of the sensor) features a 4:3 aspect ratio as well. When converting analog video to a standard 480i 4:3 digital format, the resulting

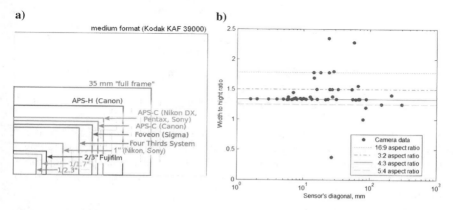

Fig. 10.33   Typical camera sensor sizes compared to a) the "full frame" 35-mm format and b) their aspect ratio [Sensor Sizes 2015].

image of 720×480 pixel happens to be wider than 4:3. The two consequences are that 1) only the center 704 pixels contain actual 4:3 image, and the 8-pixel-wide stripes from either side appear in black (in Fig. 10.32 they are cut off), and 2) the resulting 704×480 pixel image has a pixel aspect ratio of 10:11, meaning that

$$C_{El}/C_{Az} = \lambda = 1.1 \tag{10.16}$$

Hence, rather than addressing a problem represented by Eq. (10.10), it makes sense to eliminate one varied parameter, say C_{El}, and compute it afterwards based on the optimized value of C_{Az} and Eq. (10.16). In practice, a double calibration monument is used to find both boresight constants, and then the least-squares procedure of Eq. (10.10) is run for just four varied parameters

$$\Xi^j = [tilt_E, tilt_N, bias_{Az}, bias_{El}] \tag{10.17}$$

Using a double calibration monument (Fig. 10.32) with a surveyed locations of the centers of lower left and upper right targets $\mathbf{r}_{dTl}$ and $\mathbf{r}_{dTr}$, respectively, boresight constants C_{Az} and C_{El} are estimated as

$$C_{Az} = \frac{\begin{bmatrix} 1 & 0 & 0 \\ 0 & 1 & 0 \end{bmatrix}_{ECEF}^{n}\mathbf{R}||\mathbf{r}_{dTr} - \mathbf{r}_{dTl}||}{x_{offset}^{dR} - x_{offset}^{dL}}, \quad C_{El} = \frac{[0,\ 0,\ 1]_{ECEF}^{n}\mathbf{R}||\mathbf{r}_{dR} - \mathbf{r}_{dL}||}{y_{offset}^{dR} - y_{offset}^{dL}} \tag{10.18}$$

The error in determining these constants is based on a self-explanatory relation

$$
\begin{aligned}
C &= \frac{Angle}{Pix} = \frac{Angle_{true} + \delta_{Ang}}{Pix_{true} + \delta_{Pix}} = \frac{Angle_{true}}{Pix_{true}} \frac{1 + \frac{\delta_{Ang}}{Angle_{true}}}{1 + \frac{\delta_{Pix}}{Pix_{true}}} \\
&\approx \frac{Angle_{true}}{Pix_{true}}\left(1 + \frac{\delta_{Ang}}{Angle_{true}}\right)\left(1 - \frac{\delta_{Pix}}{Pix_{true}}\right) \\
&\approx \frac{Angle_{true}}{Pix_{true}}\left(1 + \frac{\delta_{Ang}}{Angle_{true}} - \frac{\delta_{Pix}}{Pix_{true}}\right)
\end{aligned} \tag{10.19}
$$

Based on Eq. (10.19), the relative error can be estimated as

$$\left|\frac{\delta_C}{C}\right| \leq \left|\frac{\delta_{Ang}}{Angle_{true}}\right| + \left|\frac{\delta_{Pix}}{Pix_{true}}\right| \tag{10.20}$$

Specifically, when observing a double calibration monument (the camera is not moving), the first term on the right-hand side disappears. Then, assuming a 1-pixel error in determining the monument's center and a $\sim$320-pixel horizontal spread between two targets (as we have it for a 60-in. camera of Fig. 10.32), we obtain

$$\left|\frac{\delta_{C_{Az}}}{C_{Az}}\right| \approx \frac{1}{320} = 0.003, \quad \text{i.e., } 0.3\% \tag{10.21}$$

The relative error in determining the boresight constant in the vertical direction (assuming 145 pixel spread as in Fig. 10.31) then becomes

$$\left| \frac{\delta C_{El}}{C_{El}} \right| \approx \frac{1}{145} = 0.007, \quad \text{i.e., } 0.7\% \tag{10.22}$$

When reducing the focal length to say 0.254 m (10 in.), the relative errors of Eqs. (10.21) and (10.22) increase by the factor of six to about 2 and 4%, respectively. To mitigate the effect of having not enough resolution between the two targets of double calibration monument for the short focal-length cameras, an alternative approach can be used. Any (even not surveyed) monument is placed in the four corners of the image frame by rotating KTM/camera and recording azimuth and elevation at all four corners, upper-left (ul), upper-right (ur), lower-right (lr), and lower-left (ll). As a result, almost a full pixel spread of the image is used in both directions, and the boresight constants can be computed as

$$C_{Az} = (\mathbf{A}^T \mathbf{A})^{-1} \mathbf{A}^T \mathbf{b}, \quad C_{El} = \lambda C_{Az} \tag{10.23}$$

where

$$\mathbf{A} = \left[x_{offset}^{ul} - x_{offset}^{ur}, x_{offset}^{ll} - x_{offset}^{lr}, (y_{offset}^{ul} - y_{offset}^{ll})\lambda, (y_{offset}^{ur} - y_{offset}^{lr})\lambda \right]^T$$

$$\mathbf{b} = -\left[Az_{enc}^{ul} - Az_{enc}^{ur}, Az_{enc}^{ll} - Az_{enc}^{lr}, El_{enc}^{ul} - El_{enc}^{ll}, El_{enc}^{ur} - El_{enc}^{lr} \right]^T \tag{10.24}$$

The error estimates of Eqs. (10.21) and (10.22) become of the order of $1/(\frac{3}{4}720) = 0.002$ and $1/(\frac{3}{4}480) = 0.0028$, respectively, regardless of the focal length. The value of a much smaller first term on the right-hand side of Eq. (10.20) can be estimated as the ratio of the 0.0017-deg position encoder error (see Table 10.3) to say $\frac{3}{4}$ of the field of view φ, which is defined by the focal length f and optic sensor size d as

$$\varphi = 2 \tan^{-1}(d/2f) \tag{10.25}$$

Figure 10.34 shows the estimates of the relative errors of boresight constants C_{Az} and C_{El} for the $\frac{1}{2}$-in. CCD/CMOS chip (with width $w = 6.4$ mm and height $h = 4.8$ mm) for four different focal length cameras proving that the usage of a four-corner procedure gives a big advantage.

To conclude, according to Table 10.3 after a proper calibration the pointing error is on the order of 5 arc sec (0.0014 deg). At the typical slant range to the test article of 4 km, it yields about 10-cm (4-in.) error in determining its position, which is five times smaller than that of differential GPS!

To conclude the camera calibration discussion, when deriving Eqs. (10.3) and (10.5) we made a couple of assumptions that might not necessarily be true. We assumed that KTM components, the azimuth bearing surface and the elevation trunnions, are assembled to meet a combination of orthogonal, coaxial, and coplanar requirements. Tolerance violations (including those caused by wear and tear) contribute to the pointing error. The simplest example is a shaft eccentricity error

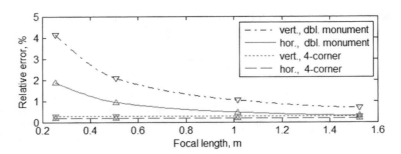

Fig. 10.34   Relative errors in determining boresight constants for 10-, 20-, 40-, and 60-in. focal length cameras.

caused by the axis of rotation being not concentric with the attached digitizer/ encoder. Let us estimate possible contribution of nonorthogonality.

If the horizontal axis of rotation (elevation trunnion) happens to be not orthogonal to the vertical axis of rotation (normal to an azimuthal bearing surface), this introduces the nonorthogonality error. Mathematically, having a small inclination error between the elevation trunnion and azimuthal bearing surface, $\varepsilon_{no} \ll 1$, leads to a necessity to modify the matrix ${}_{n}^{m}\mathbf{R}$ in Eq. (10.5) to

$$
{}_{m}^{c}\mathbf{R}^{*} = \begin{bmatrix} 1 & 0 & 0 \\ 0 & 1 & \varepsilon_{no} \\ 0 & -\varepsilon_{no} & 1 \end{bmatrix} {}_{m}^{c}\mathbf{R} \tag{10.26}
$$

The matrix ${}_{m}^{c}\mathbf{R}$ in Eq. (10.3) might need to be modified to

$$
{}_{m}^{c}\mathbf{R}^{*} = \begin{bmatrix} 1 & \varepsilon_{c} & 0 \\ -\varepsilon_{c} & 1 & 0 \\ 0 & 0 & 1 \end{bmatrix} \begin{bmatrix} 1 & 0 & -\varepsilon_{b} \\ 0 & 1 & 0 \\ \varepsilon_{b} & 0 & 1 \end{bmatrix} {}_{m}^{c}\mathbf{R} \tag{10.27}
$$

to account for the azimuth skew or collimation error $\varepsilon_{c} \ll 1$ due to a possibility of the pointing axis of the lens not being orthogonal to the elevation trunnion axis, and the so-called lens sag or droop or bending moment error $\varepsilon_{b} \ll 1$ in the vertical plane about the elevation trunnion axis. This lens sag error depends on the true elevation angle [McAllister and Gose 1980]

$$
\varepsilon_{b} = \varepsilon_{b}^{\max} \cos \left(El_{enc} + bias_{El} \right) \tag{10.28}
$$

In practice, ε_{no}, ε_{c}, and $\varepsilon_{b}^{\max}$ are negligibly small (less than 0.001 radian) and for low elevation angles are absorbed by $bias_{Az}$ and $bias_{El}$, meaning that they cannot possibly be identified during a calibration routine described in this session. (To determine these errors, the calibration procedure with $El \sim 0$ deg just described should be repeated using a plunged mount configuration with $El \sim 180$ deg.)

Elevation bias $bias_{El}$ also absorbs the average atmospheric refraction error computed for all calibration monuments. Atmospheric refraction is the deviation of light from a straight line as it passes through the atmosphere due to the variation in air density (affected by current temperature, temperature gradient, pressure, and humidity) as a function of altitude, and as such is a varied parameter difficult to estimate. Atmospheric refraction becomes more severe when there are strong temperature gradients and is not uniform when there is turbulence in the air. Hence, it depends on the time of the day. McAllister and Gose [1980] suggested estimating the effect of refraction over a given distance as a correction angle ε_r to be added to the appropriate elevation angle

$$\varepsilon_r \approx -0.00076° \frac{\sqrt{\Delta x^2 + \Delta y^2}}{1 + 0.33\Delta z} \tag{10.29}$$

where $\sqrt{\Delta x^2 + \Delta y^2}$ is the distance from the camera to the projection of the object position onto the horizontal plane and Δz is the distance of the object above that plane, both in kilometers. Even though the empirical coefficients in Eq. (10.29) were obtained for the White Sands Missile Range in New Mexico, this formula enables a ballpark estimate of a possible altitude error due to the fact that calibration monuments are usually located about 1 km away from KTM while the typical slant range to the test article is four times larger. Using Eq. (10.29), it is easy to show that the potential altitude estimate error in this case is bounded by 7.5 cm (3 in.) caused by the unaccounted atmospheric refraction error of 3.7 arc sec (0.001 deg), which is of the order of the pointing error. To assess the change of atmospheric refraction during the tests, error calibration is done prior and after a series of tests.

10.2.2 OBTAINING TSPI SOLUTION

Once all cameras are properly calibrated, the test itself can be conducted. During ADS tracking by multiple cameras, the azimuth/elevation time histories are recorded along with the video. Video data are postprocessed to find x/y offsets of the test article with the camera frame. Figure 10.35 shows a set of sample images starting from extraction of payload from an aircraft with a drogue parachute all of the way down to touchdown. All of these data, conditioned and synchronized, can then be used to find the TSPI solution. Figure 10.36 shows an example of Az/El and offsets data for the payload's centroid obtained from three cameras ready to be processed. Note that KTM operators do a very good job keeping the test article in the center of the frame.

Finding the TSPI solution, a time history of three-dimensional position $\mathbf{P}(t) = \left[x_{pl}(t), y_{pl}(t), z_{pl}(t) \right]^T$ of the test article (e.g., payload's centroid) involves addressing the same problem defined by Eqs. (10.10–10.15) but with a different vector of varied parameters. For each time instance $t = t_i$, we need to find

$$\min_{\Xi^i} (J^i) \tag{10.30}$$

Fig. 10.35   Examples of video images to analyze.

where

$$\Xi^i = \mathbf{P}(t_i) \tag{10.31}$$

$$J^i = \sum_{j=1}^{N_c}\left(\Delta\varepsilon_{Az}^{ij\,2} + \Delta\varepsilon_{El}^{ij\,2}\right) \tag{10.32}$$

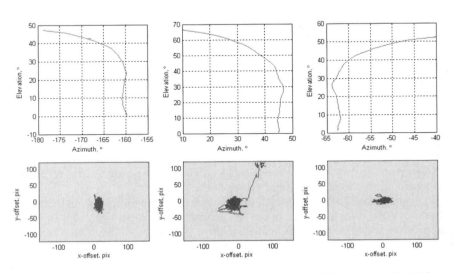

Fig. 10.36   Example of synchronized data for three cameras [Yakimenko et al. 2007].

and

$$\Delta\varepsilon_{Az}^{ij} = \tan^{-1}\left(\frac{[0, 1, 0]\Delta\mathbf{r}^c(\mathbf{P})}{[1, 0, 0]\Delta\mathbf{r}^c(\mathbf{P})}\right) - C_{Az}^{j}x_{\text{offset}}^{ij}$$

$$\Delta\varepsilon_{El}^{ij} = \tan^{-1}\left(\frac{[0, 0, -1]\Delta\mathbf{r}^c(\mathbf{P})}{[1, 0, 0]\Delta\mathbf{r}^c(\mathbf{P})}\right) - C_{El}^{j}y_{\text{offset}}^{ij} \qquad (10.33)$$

Equation (10.15) now takes the form

$$\Delta\mathbf{r}^c = {}_{m}^{c}\mathbf{R}\left[{}_{n}^{m}\mathbf{R}_{ECEF}{}^{n}\mathbf{R}(\mathbf{P} - \mathbf{r}_{KTM}) - \mathbf{r}_{\text{arm}}^{m}\right] = \Delta\mathbf{r}^c$$
$$= \mathbf{R}_{El}\left[\mathbf{R}_{Az}\mathbf{R}_{tilt_N}\mathbf{R}_{tilt_E}{}_{ECEF}^{n}(\mathbf{P} - \mathbf{r}_{KTM}) - \mathbf{r}_{\text{arm}}^{m}\right] \qquad (10.34)$$

Each camera contributes two nonlinear equations of the form of Eq. (10.33). Hence, finding the TSPI solution (three varied parameters as indicated in Eq. (10.31) requires at least two cameras mounted onto KTMs at two different locations ($N_c \geq 2$).

Figure 10.37 shows the TSPI solution for the data set of Fig. 10.36 (with the azimuth and elevation time histories corrected for $bias_{Az}$ and $bias_{El}$). Figure 10.37b also shows the spread of azimuth change while tracking a descending payload, and Fig. 10.38 allows assessing a spread in elevation for each of three cameras used in this high-altitude uncontrolled airdrop. Figure 10.39 shows the estimates of each coordinate (in {NED}) and a descent rate. A simple one-dimensional digital filter with a window size of 80 samples (taken at a 30 frames/s rate) is used to smooth a descent rate estimate. These data were used to prove that a specification limit for the rate of descent was met (which was an objective for this particular test). For this particular drop, the GPS/IMU data

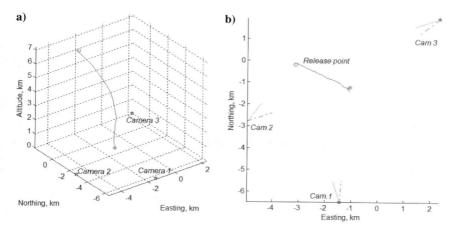

Fig. 10.37   Reconstructed payload trajectory in a) 3D and b) 2D [Yakimenko et al. 2007].

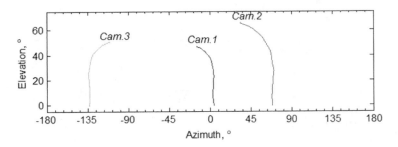

Fig. 10.38   Azimuth and elevation data while tracking a payload by three cameras.

were also available, and comparison with these "true" data demonstrated a fairly good match.

Figures 10.40–10.42 show the similar results for the low-altitude uncontrolled ADS drop that involved six cameras [Yakimenko et al. 2007].

10.2.3   AUTOMATING VIDEO DATA PROCESSING

As mentioned in the preceding section, data from all cameras need to be synchronized. Using a telemetry protocol known as "Chapter 10" [IRIG 2009] allows

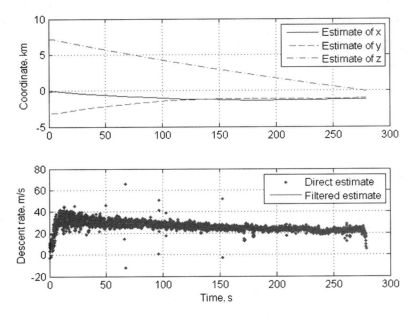

Fig. 10.39   Position and descent rate estimates [Yakimenko et al. 2007].

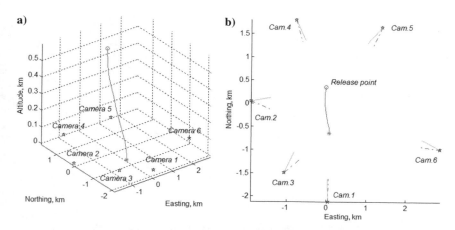

Fig. 10.40   Reconstructed payload trajectory in a) 3D and b) 2D for the six-camera test.

several streams of data to be time-synchronized and saved in one data file, which eases data management and simplifies the postprocessing of test video. A screenshot of the Common Mission Debrief Program (CMDP) [CMDP 2013] that reads Chapter 10 format for a sample parachute drop is shown in Fig. 10.43.

To speed up obtaining the TSPI solution and enhance this capability to track multiple objects (payload and canopy), several software packages could potentially be adopted. The TrackEye motion analysis and TEMA Camera Control are examples of such software packages that can be used to track several specific features on the test object [TrackEye 2015]. These packages, however, still rely on a trained operator to manually enter all auxiliary data and then manually find the object (calibration monument or tracking article), adjust tracking algorithm, initiate object scoring/tracking while monitoring a test video. As an example, Fig. 10.44 features a TrackEye model set to obtain and analyze the TSPI solution

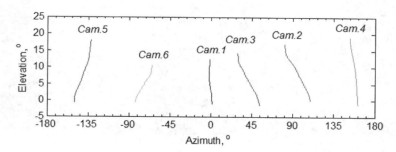

Fig. 10.41   Azimuth and elevation data while tracking a payload by six cameras.

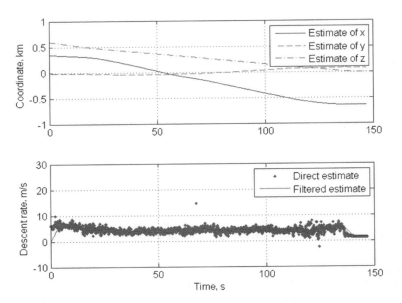

Fig. 10.42   Position and descent rate estimates.

using tracking data from three KTMs. KTM calibration has to be performed for each camera separately prior to that. Figure 10.45 shows an example of scoring one of the calibration monuments. This process involves creating a camera

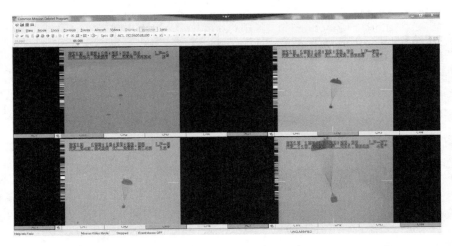

Fig. 10.43   Common Mission Debrief Program displaying Chapter 10 data.

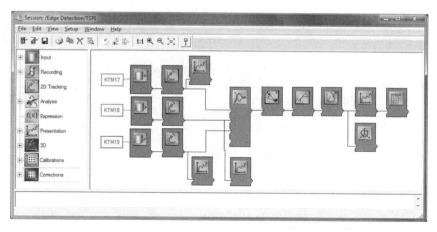

Fig. 10.44   TrackEye model to obtain and analyze the TSPI solution.

model first, so that each particular scoring could be associated with a specific calibration monument. Similarly, Fig. 10.46 demonstrates an example of a test article
tracking (that is where multiple distinctive points within the same test article
could be tracked simultaneously). In both cases decoding left-edge barcode
assumes a separate run of the same video sample and then assembling everything
together in one model (as shown in Fig. 10.44).

The edge-detection-based object segmentation algorithm introduced in
Fig. 10.23 can be easily adjusted to handle automatic data extraction for both
the payload and canopy. As an example, Fig. 10.47 shows both the canopy and
payload being tracked as two separate objects simultaneously.

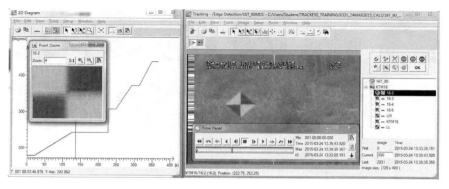

Fig. 10.45   Scoring calibration monuments in TrackEye.

Fig. 10.46   Test item tracking in TrackEye.

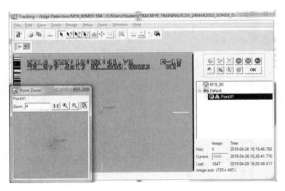

The modified multi-stage algorithm developed by Decker [2013] starts from finding two objects of a sufficient size. To reduce the probability of identifying the risers as pertaining to either the payload or the canopy shapes, the edge detection process starts at a low sensitivity and increases the sensitivity logarithmically until two large objects are found. Then, these objects are analyzed to see if they are parts of ADS. Classification criteria may involve the relative size and position of the objects. For example, if the smaller object (candidate payload) is between one-quarter and one-twentieth the size of the larger object (candidate canopy), the ratio criterion is satisfied. Another criterion can be that the smaller object must appear below the larger object. If both objects are of sufficient size, the ratio is acceptable, and the vertical positioning is correct, the objects are classified as a payload and a canopy. Figure 10.47b shows an example of a successful classification even in the case of a massive airdrop. For parafoils the size-ratio criterion is slightly different because when the canopy is pointed directly at the camera it may appear smaller in area [Decker and Yakimenko 2015]. The vertical position criterion should also be adjusted to account for a drogue parachute if it is identified by accident, or if the payload is wildly swinging around the canopy, as is sometimes the case during the first few moments of a turbulent deployment.

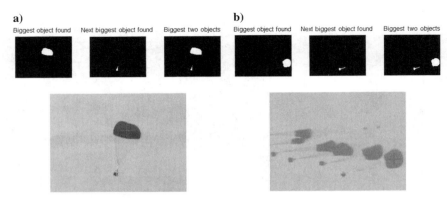

Fig. 10.47   Parachute system dual object tracking [Decker 2013].

Once the system is classified, the entire image is separated onto two parts zoomed in at each of the two objects. A new subimage for the canopy is centered at the previously identified canopy center, and the size of the subimage is chosen to be 150% of the distance between the previously identified canopy center and payload center. The payload subimage is centered at the previously identified payload center, and the size of the window is chosen to be 50% of the distance between the previously identified canopy center and payload center. Each of these two windows is then resized to be 40×40 pixels for a common filtering operation, which simulates a darkest-spot mean-shift tracking algorithm. The resized subimage is inverted in intensity and filtered using a 5×5 pixel Gaussian kernel 20 times in a row until the jagged pixel intensity values resemble a smooth peak. The region surrounding this peak is used to fit a 2D paraboloid surface, and a subpixel estimate of the object center is found as the highest location on this peak. This process is illustrated for finding the payload center in Fig. 10.48. This same process is also conducted to the find the center of the canopy subimage.

The tracking algorithm proceeds going frame by frame unless either object's center has moved to the edge of the original window, or the center has moved more than threshold distance from the previous center location. When this

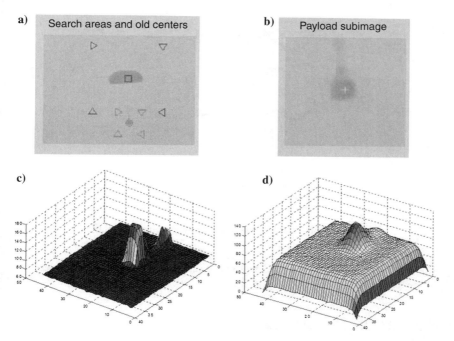

Fig. 10.48   Parachute system tracking: a) original image, b) payload subimage, c) resized and inverted subimage, and d) filtered subimage [Decker 2013].

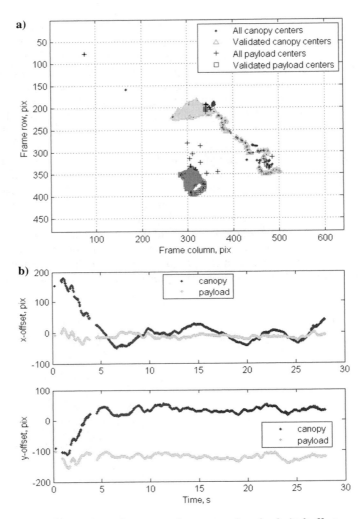

Fig. 10.49   Plots of auto-scored canopy and payload pixel offsets.

happens, the edge detection-based segmentation algorithm is re-initiated, and the process restarts.

10.2.4   ATTITUDE DETERMINATION

Using the automated two-object scoring algorithm presented in the preceding section, it is possible to go beyond a simple TPSI solution described in Sec.

10.1.2 and estimate the motion of parachute-payload system as a whole. As an example, Fig. 10.49a shows the payload and canopy centroids in the image frame {c} detected by the two-object scoring algorithm for one of three cameras used in a low-altitude drop from 300 m (980 ft) AGL. Figure 10.49b presents pixel offset time histories for both objects with several obvious outliers removed.

Figure 10.50 shows the result of a straightforward application of the dual-object tracking analysis to characterize the swinging motion of ADS

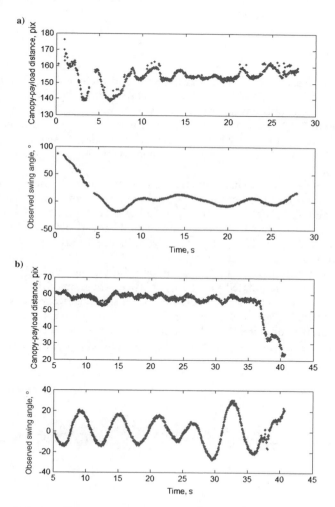

Fig. 10.50   a) Distance between canopy and payload and ADS's attitude derived from Fig. 10.46b data and b) another drop of the same system [Decker 2013].

during its descent. Using differences between the x and y offsets of canopy and payload centroids as seen by just one camera, the swaying motion of the parachute-payload system (swinging of payload relative to a canopy) can easily be retrieved. Obviously, the magnitudes shown in Fig. 10.50 do not necessarily indicate the total swing angle of payload, just the observed angle measurable in the image plane $\{c\}$. A fast Fourier transform (FFT) analysis of data presented in Fig. 10.50 results in a peak oscillation frequency of 0.154 Hz, which corresponds to a median swing period T of about 6.5 s. This matches the vertical dimension of the tested ADS (see notations in Fig. 9.34b)

$$z_1 + z_4 \approx g\left(\frac{T}{2\pi}\right)^2 \tag{10.35}$$

which in this case was about 10.5 m (34.5 ft), and the coupled roll-spiral mode data for a similar-size PADS of Table 6.1. As seen from Fig. 10.50, the distance between two centroids has a periodic behavior as well.

Figure 10.51 shows video data analysis for the same ADS dropped from a higher altitude (with over 200 s of available video data). Again, the swinging motion exhibits very smooth periodic behavior (with only one grossly erroneous raw data point).

Essentially the same characteristics can be obtained by utilizing data from multiple cameras to produce the TSPI solution for both canopy and payload $\mathbf{P}_{canopy}(t)$ and $\mathbf{P}_{payload}(t)$, respectively. Figure 10.52a shows the canopy-payload distance and spatial pitch angle θ_{sp} (aka plumb angle) obtained by analyzing the difference $\Delta\mathbf{p}(t) = \mathbf{P}_{canopy}(t) - \mathbf{P}_{payload}(t)$. Figure 10.52b shows a horizontal projection of this vector (hodograph).

Following notations of Fig. 9.34 for parachutes, the angle θ_{sp} depicted in Fig. 10.52a can be designated as the angle between the z_b axis of canopy-payload system and a plumb line. To compute the spatial angle of attack α_{sp}, which is the angle difference between the longitudinal axis of canopy-payload system and air-speed vector, the components of the airspeed vector need to be computed. To this end, differentiating the components of $\mathbf{P}_{canopy}(t)$ and $\mathbf{P}_{payload}(t)$ allows estimating

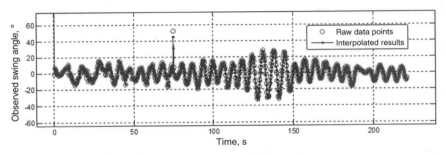

Fig. 10.51   Observed swing angle between payload and canopy in $\{c\}$ [Decker 2013].

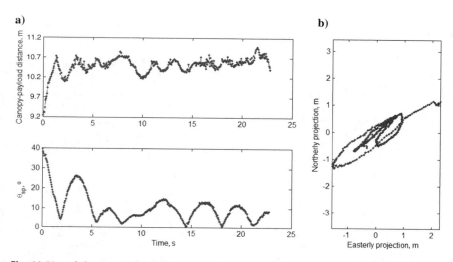

Fig. 10.52 a) Canopy-payload distance and spatial pitch angle and b) horizontal projection of (x–y) position Δp(t) [Decker and Yakimenko 2015].

the components of the ground-speed vector as shown in Fig. 10.53. Additionally, Fig. 10.54a shows the difference in canopy and payload speed vectors.

While data in Fig. 10.54a is free of the influence of winds because the same winds act on both the payload and canopy, data in Fig. 10.53 cannot be used for $\alpha_{sp}(t)$ derivation unless the winds are subtracted. As explained in Sec. 3.1.2 during airdrops, these winds are usually measured and somewhat known. These winds should be subtracted from the canopy and payload ground-speed vector components to yield airspeed vector components.

For the low-altitude descent corresponding to Fig. 10.53, an even easier approach was implemented. It was assumed that during a short descent the uncontrolled ADS essentially goes down, so that with no winds V_N and V_E oscillations in the air-mass-related coordinate frame should have occurred around a zero value. Consequently, the mean values of the V_N and V_E speed components were found and subtracted from Fig. 10.53 data. These mean components (assumed to be due to horizontal winds) were $\overline{V}_N = 3.5\,\text{m/s}$ (6.8 kt) and $\overline{V}_E = 2.1\,\text{m/s}$ (4 kt). Figure 10.54b shows a hodograph $V_N - \overline{V}_N$ vs $V_E - \overline{V}_E$.

Now that an airspeed vector is found, the spatial angle of attack can be computed. For Figs. 10.52–10.54, data $\alpha_{sp}(t)$ both the canopy and payload are presented in Fig. 10.55. For the payload $\alpha_{sp}(t)$ happens to be slightly higher than for the canopy.

These data can be used in modeling and/or verification of ADS dynamics. For example, Table 10.4 shows the results of the FFT analysis run on time histories for different parameters as defined earlier in this section with a goal of determining ADS's fundamental frequencies (harmonics) and time constants. For example,

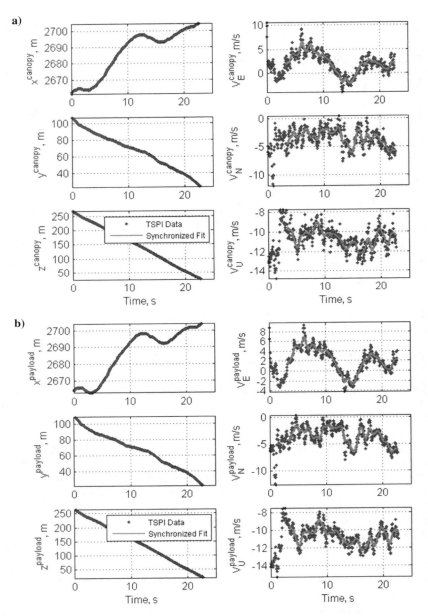

Fig. 10.53   North, East, and Up components of the ground-speed vector for a) canopy and b) payload [Decker 2013].

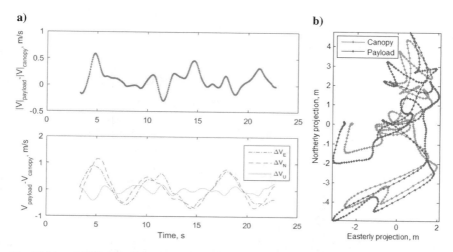

Fig. 10.54   a) Difference between canopy and payload speed vector components and b) horizontal projection of the airspeed vector for canopy and payload [Decker and Yakimenko 2015].

the 6.6-s period seems to correspond to the coupled roll-spiral mode while the 12.8-s period corresponds to the phugoid mode.

Further development could include estimating relative pitch, roll, and yaw for the higher-fidelity models (seven-/eight-/nine-DoF models), that is, considering ADS as a flexible rather than solid system. Towards this goal, Yakimenko et al. [2007] studied several approaches allowing estimating payload's attitude based on tracking multiple distinctive features. These days the Computer Vision System Toolbox of MATLAB offers several methods for selecting regions of an image that have unique locally invariant content, so that you can detect them even in the presence of rotation or scale change. For example, the Features

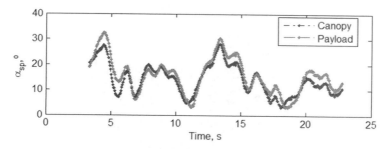

Fig. 10.55   Spatial angle of attack for canopy and payload [Decker 2013].

TABLE 10.4 FUNDAMENTAL FREQUENCIES/HARMONICS OF A SAMPLE ADS

Harmonic, Hz	0.039	0.078	0.1172	0.1563	0.1953	0.2344	0.3125	0.3906
Time constant (period), s	25.6	12.8	8.5	6.6	5.12	4.27	3.2	2.56
System's height	+							
θ_{sp}		+					+	
ΔV_E				+				
ΔV_N		+		+				
ΔV_U		+				+	+	+
$\left|V_{payload}\right| - \left|V_{canopy}\right|$			+		+		+	
α_{sp}			+			+		

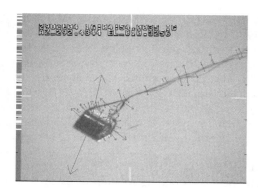

Fig. 10.56 Real drop image featuring 137 scale-invariant key points.

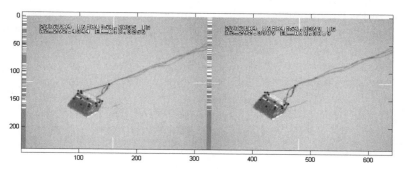

Fig. 10.57 Feature tracking to estimate a payload's rotation in {c} [Yakimenko et al. 2007].

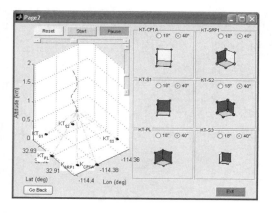

Fig. 10.58 Snapshot of payload's attitude estimate GUI in a six-KTM airdrop.

from Accelerated Segment (FAST), Harris, and Shi/Tomasi methods allow detecting the corner features, and the Speeded Up Robust Features (SURF) and Maximally Stable Extremal Regions (MSER) methods detect the blob features. These features are described by the compact vectors, descriptors, centered around the detected features and relying on local gradient computations. This new representation permits comparison between neighborhoods regardless of changes in scale or orientation. Figure 10.56 features an example of a typical payload tracking video, where an automated Scale-Invariant Feature Transforms (SIFT) [Lowe 2004] technique allowed finding over a hundred descriptive key points in each frame.

Comparing two consecutive frames (1/30 s apart) allows finding several matches that can be used to estimate the corresponding rotation of payload in the camera frame (Fig. 10.57). Combining rotation data from multiple cameras produces the desired estimate of payload's attitude in the inertial coordinate frame. Figure 10.58 features a graphical user interface (GUI) developed and used to visualize the results of payload's attitude estimation using different focal-length cameras in a multicamera experiment setup.

CHAPTER 11

Parametrical Identification of Parachute Systems

Oleg Yakimenko[*]
Naval Postgraduate School, Monterey, California

Thomas Jann[†]
Institute of Flight Systems of DLR, Braunschweig, Germany

Chapter 4 has addressed one of three techniques of determining PADS aerodynamic and control derivatives, specifically using analytical methods including computational fluid dynamics (CFD). The other two methods are wind-tunnel testing and flight testing. As well-known wind-tunnel testing has certain restrictions, it is usually conducted on the scaled models of PADS and in stationary conditions. Moreover, the use of suspension lines and the flexible nature of parachutes and parafoils make it difficult to handle them in the wind tunnel and almost impossible to introduce the sideslip angles into the canopy. In contrast, the flight tests do not suffer from aforementioned limitations allowing exploring a complete flight envelope. Conducting flight tests is the most expensive option and requires appropriate instrumentation of PADS to obtain adequate information for the following system identification (SID). Instrumenting PADS for flight tests was addressed in Chapter 10. Another important piece in SID is a structural identification, that is, defining a number and type of equations of motions parameterized with respect to parameters that need to be determined. This subject was addressed in Chapter 5. Comparing the flight-test results with those generated by the model with the same control inputs, atmospheric parameters, and initial conditions allows adjusting those parameters. This chapter is based on the three well-documented PADS parametric SID efforts undertaken in the past [Jann and Strickert 2005; Jann 2006; Rogers 2002; Rogers 2004; Yakimenko and Statnikov 2005]. These SID efforts were conducted in a systematic manner and involved well-instrumented systems, which resulted in the comprehensive models of high-GR systems. In what follows, Sec. 11.1 starts with introducing the overall methodology of parameter estimation with the differences between three chosen approaches. It is followed by Sec. 11.2 discussing the flight-test data preconditioning, which allows monitoring quality of flight-test data. Next, three different parametric identification techniques are discussed.

---

[*]Professor.
[†]Research Scientist.

Section 11.3 introduces a single-criterion error method used to validate a six-DoF mathematical model of the ALEX PADS featuring 19 varied parameters. Because of the limited accuracy of low-cost sensors and also the need to account for a relative canopy motion, much effort had to be invested into flight-path reconstruction. Comparing differences between the original theoretical and identified models indicated in which direction the aerodynamic model should be extended or changed in order to come closer to reality. Section 11.4 employs multicriteria identification technique to investigate an eight-DoF model of the Pegasus PADS. Being able to utilize 33 varied parameters, mostly coefficients of aerodynamic and control derivatives, and eight adequacy criteria, this approach demonstrated an excellent capability to explore the model itself in more detail performing sensitivity and correlation analysis, studying criterion vs criterion dependencies, and allowing judging on the adequacy of the model structure in general. Section 11.5 demonstrates the extended Kalman filter approach applied in attempt to estimate 55 parameters of the six-DoF model of the X-38 PADS. It discusses the error model and demonstrates the capability to estimate aerodynamic and mass property modeling errors with reasonable certainty when the flight sensor data are of sufficient quality. Finally, for completeness, two more examples involving simplified reduced-order linearized models are presented.

11.1 METHODOLOGY OF PARAMETER ESTIMATION

The objective of SID is to extract a mathematical model that describes the real system behavior from flight-test data, that is, measured inputs and outputs. Assuming that the model structure is known, the model parameters are adjusted by means of an optimization procedure in order to achieve the best match possible between the simulated output $\mathbf{y}$ and the measured output $\mathbf{z}$ for the same input history $\mathbf{u}$.

Key to successful SID is a coordinated approach by carefully and systematically treating the four so-called "Quad-M" requirements (Fig. 11.1) [Hamel and Jategaonkar 1996]:

1. Design the control input shape (*maneuvers*) to excite all interesting modes of the vehicle dynamic motion.

2. Select instrumentation and filters for high-accuracy *measurements*.

3. Define the structure of a possible mathematical *model* regarding the flight vehicle.

4. Select the most suitable identification *method* to ensure the quality of data analysis.

The accuracy and reliability of parameter estimation depends notably on the amount of information available in the vehicles' response to dedicated

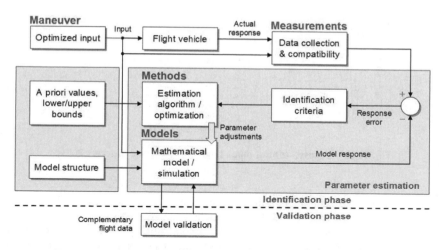

Fig. 11.1 Quad-M basics of system identification [Jategaonkar 2006].

input. This input, or maneuver, on the one hand should excite the frequency range of interest best and on the other hand must be flyable. So, typical control inputs for SID often are a compromise between high-bandwidth sweep inputs and step inputs that are repeatable and simple to fly. Another criterion for the design of an input is its amplitude. If chosen too small, crucial effects such as nonlinearities might not be revealed. On the other side, very big amplitudes can cause the vehicle to leave the scope of the model or even worse to break, collapse, stall, or spiral the vehicle.

For parafoil systems, additional considerations are necessary. First, all of the systems can just be excited in a limited way, using the two control lines. It is not possible to decouple single effects using inputs for dedicated axes like ailerons, rudder, or thrust control. To get most out of the data even with two actuators, it is necessary to drive them symmetrically as well as asymmetrically. Also, different working points should be considered. A turn resulting from a 0/30% (left/ right) deflection can be significantly different to a curve from a 50/80% deflection. To detect potential asymmetric characteristics, some turns should also be flown in the opposite direction. Another problem is the limited power and speed of typical actuators for parafoil systems. The bandwidth they can provide for excitation is also a margin for analysis. Also, for accurate SID it is necessary to measure the control line deflection directly to account for the response of the actuators as well.

Finally, some effort on the flight-test planning should be spent. Normally, both flight time and flight-test area are restricted. A sequence of maneuvers has to be found that results in as much information as possible in the responses and at the same time maintaining the parafoil within the designated test region thereby not violating the safety. To avoid interferences among separate maneuvers, stationary

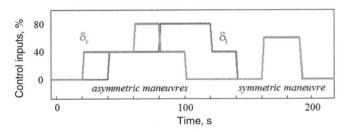

Fig. 11.2   Proposed control input sequence [Jann and Strickert 2005].

phases have to be scheduled. The remote pilot should use a detailed script to perform all desired maneuvers. If the vehicle is capable of autonomous flight, it is a good idea to execute SID maneuvers automatically by the control system for better precision and reproducibility. One example of a suitable maneuver sequence is shown in Fig. 11.2.

Measurements are the basic source for insights into the characteristics of dynamic systems. For aerial vehicles usually very sophisticated sensor systems are necessary to acquire significant data. Each measured quantity passes through a measuring chain, consisting of environmental effects, the sensor itself, signal conditioning electronics, and finally the sampling and analog-digital converting devices, until it is logged digitally on a data carrier. Every single element is subject to errors or inaccuracies. Common systematic errors are, for example, nonlinearities, temperature dependencies, alignment errors, and time delays. Other effects can be caused by electromagnetic influences and noise. Because the accuracy of the parameter estimates is directly dependent on the quality of the measured data, it is crucial to reduce as many sensors errors as possible. This can be done by a detailed sensor chain calibration. Remaining errors and inconsistencies can be cleared out by the so-called flight-path reconstruction. This applies especially for low-cost sensors (with poor accuracy) that are often used in parachute and parafoil systems.

A principal problem in parafoil-load systems is to get in situ measurements from relevant places, for example, at the canopy. It is very difficult to place sensors, cables, data loggers, and batteries on a textile wing that is exposed to packing, deployment, and opening shocks and to guarantee their proper function. Normally, the data are acquired in the payload. If the system was rigid, there would not be a problem, but because it is not, the relative motion falsifies results. This drawback can be eliminated to some extent by measuring and processing the relative motion between canopy and load.

As mentioned already for SID, a mathematical model with a parametric structure is required. The model complexity depends upon the complexity of the problem to solve. For example, if only a trajectory simulation is foreseen, a point-mass model with three or four DoFs might be sufficient (see Sec. 5.1.3). For more detailed investigation of rigid-body dynamics, a six-DoF model is required (Sec. 5.1.5), and, if relative motion is an issue, introduction of additional degrees of freedom or a multibody approach is necessary (Sec. 5.2).

The characteristic motion of the flight vehicle is the result of external forces and moments, like aerodynamic, gravitational, and inertial forces acting on the vehicle. SID normally focuses on the aerodynamic subsystem whereas mass and inertia characteristics are assumed to be well known. Based on Newton's laws and general aerodynamic principles, the mathematical model is derived. The resulting equations of motion can be formulated in state space, which allows a graphical representation as block diagrams and a comfortable treatment with standard methods and many available software tools.

This so-called "white-box" approach requires a priori information about the system in terms of known physical relations among the corresponding quantities. In contrast, the "black box" methods, like neural nets, local model networks, or polynomial classifiers are of more general structure and don't include a priori information. They are better suited for applications where no or only very little information about the real system is available. In exchange, physical interpretability of the parameters is usually lost.

Finally, we came to the Methods block in Fig. 11.1. SID is a postflight analysis methodology, which means that algorithms can be applied to the complete data time history without restrictions in terms of computing time. For flight vehicles, where parametrical models are used, identification methods can be broadly classified into three categories:

1. Equation-error method (e.g., regression)
   - Linear estimation problem
   - Deterministic method

2. Output-error method (e.g., maximum likelihood)
   - Nonlinear estimation problem
   - Deterministic method

3. Filter-error method (e.g., Extended Kalman filter)
   - Nonlinear estimation problem
   - Statistical method

The selection of a particular method depends on the model formulation and assumptions regarding the measurement and process noise [Hamel and Jategaonkar 1996]. If measurement and process noise can be neglected, the faster equation-error method can be used. In case only measurement noise has to be considered, the output-error method is suited. If both measurement and process noise must be taken into account, the computational intensive filter-error method has to be used.

In parachute systems, wind and turbulence have a strong influence on the system (see Sec. 3.1). If present, but not measured, this unknown input must be considered as process noise, requiring the application of the filter-error method.

On the other hand, wind profile and turbulence acting on the flight vehicle can be reconstructed by the so-called flight-path reconstruction, provided that airspeed, angle of attack, and angle of sideslip have been measured during the flight. In this case, also the output-error method can be used.

Software-wise, parametrical identification of such complex systems as six-/seven-/eight-/nine-DoF PADS models relies on self-written software rather than existing packages. The reason is that say well-known System Identification Toolbox of MATLAB only allows identifying parameters of linear or nonlinear models of a very specific class like frequency-response models, impulse-response models, low-order transfer functions (process models), input-output polynomial models, state-space models, transfer function models, nonlinear black-box models, ordinary difference, or differential equations (grey-box models).

The output-error method, widely used for aircraft parametric identification, is often realized as maximum-likelihood cost function in combination with a Gauss–Newton optimization algorithm. Its extension to general nonlinear systems works well in most cases and is also implemented in ESTIMA and FITLAB. Both are software packages that have been developed by DLR Institute of Flight Systems and used for parameter identification [Jategaonkar et al. 2004]. Other optimization techniques, for instance offered by the Optimization Toolbox of MATLAB, can be used as well. Multi-objective optimization can be carried with the Global Optimization Toolbox of MATLAB or other packages.

11.2 FLIGHT-PATH RECONSTRUCTION

The so-called flight-path reconstruction (FPR) is used to identify parameters of a sensor (error) model and check the compatibility of the measured flight-test data, applying the kinematic relations, which are independent from the specific vehicle characteristics. Possible sensor biases, scale factor errors, and unknown time delays along with initial state values can be estimated applying the maximum likelihood output-error algorithm. Also, missing time histories, for example, the Euler angles ϕ, θ, and ψ, which may not be directly available from flight-test data, can be reconstructed.

Because of the low flight velocity of parachute and parafoil systems, their motion is greatly affected by the wind. Especially small and lightweight PADS are subject to gusts and turbulences. The result is a significant difference between inertial speed and airspeed that leads to a shift of the flight trajectory and to dynamic stimulation of the vehicle that has to be considered for SID. One possibility to cope with the wind influence is the precise measurement of the airspeed V_a, the angle of attack α, and the angle of sideslip β during the flight. Provided that GPS and IMU data are also available, a highly reliable wind profile including turbulence can be computed using the flight-path reconstruction and can be used as input for SID. If no air data measurement is available, the wind must be treated as unknown input and turbulence as process noise.

This makes the application of the filter-error method mandatory [Hamel and Jategaonkar 1996; Plaetschke and Saliaris 1995].

To summarize, for the later identification of PADS model parameters a complete set of consistent flight data is required, including the time histories of the following:

- Position components of origin of $\{b\}$ in $\{I\}$: x, y, z ($h = -z$)
- Ground-speed vector components expressed in $\{b\}$: u, v, w
- Accelerations of $\{b\}$ with respect to $\{I\}$ expressed in $\{b\}$: a_x, a_y, a_z
- Attitude Euler angles defining $\{b\}$ with respect to $\{I\}$: ϕ, θ, and ψ
- Angular rates of $\{b\}$ with respect to $\{I\}$ expressed in $\{b\}$: p, q, r
- Airspeed vector components expressed in $\{b\}$: v_x, v_y, v_z (usually measured as the airspeed V_a, angle of attack α, and angle of sideslip β)
- Actuator positions δ_l, δ_r

FPR is used to meet this requirement.

11.2.1 MATHEMATICAL FOUNDATION

Assuming that the sensor package is installed in the payload, the flight-path reconstruction is done for the reference point, which is essentially the payload center of gravity by integrating measured accelerations and angular rates that are computed from the corresponding state equations. Combining Eq. (5.23) with Eq. (5.73) and dividing both sides by PADS' mass yield

$$
\begin{bmatrix} \dot{u} \\ \dot{v} \\ \dot{w} \end{bmatrix} + \begin{bmatrix} p \\ q \\ r \end{bmatrix} \times \begin{bmatrix} u \\ v \\ w \end{bmatrix} = \begin{bmatrix} a_x \\ a_y \\ a_z \end{bmatrix} + g \begin{bmatrix} -\sin(\theta) \\ \cos(\theta)\sin(\phi) \\ \cos(\theta)\cos(\phi) \end{bmatrix} \tag{11.1}
$$

Rearranging terms and accounting for measurement errors result in

$$
\begin{bmatrix} \dot{u} \\ \dot{v} \\ \dot{w} \end{bmatrix} = \begin{bmatrix} a_x^m + \Delta a_x \\ a_y^m + \Delta a_y \\ a_z^m + \Delta a_z \end{bmatrix} - \begin{bmatrix} p_m + \Delta p \\ q_m + \Delta q \\ r_m + \Delta r \end{bmatrix} \times \begin{bmatrix} u \\ v \\ w \end{bmatrix}
$$

$$
+ g \begin{bmatrix} -\sin(\theta) \\ \cos(\theta)\sin(\phi) \\ \cos(\theta)\cos(\phi) \end{bmatrix}, \quad \begin{bmatrix} u(0) \\ v(0) \\ w(0) \end{bmatrix} = \begin{bmatrix} u_0 \\ v_0 \\ w_0 \end{bmatrix} \tag{11.2}
$$

(super- or subscript m stands for measurement). Bias parameters of accelerometers (Δa_x, Δa_y, Δa_z) and gyros (Δp, Δq, Δr) along with the initial values $[u_0, v_0, w_0]^T$ need to be estimated during FPR. (Possible alignment errors of the accelerometer and gyro packages and scale factors of the sensors are assumed to be negligible.)

Similarly, Eq. (5.15) can be represented as

$$
\begin{bmatrix} \dot{\phi} \\ \dot{\theta} \\ \dot{\psi} \end{bmatrix} = \begin{bmatrix} 1 & \sin(\phi)\dfrac{\sin(\theta)}{\cos(\theta)} & \cos(\phi)\dfrac{\sin(\theta)}{\cos(\theta)} \\ 0 & \cos(\phi) & -\sin(\phi) \\ 0 & \sin(\phi)\dfrac{1}{\cos(\theta)} & \cos(\phi)\dfrac{1}{\cos(\theta)} \end{bmatrix} \begin{bmatrix} p_m + \Delta p \\ q_m + \Delta q \\ r_m + \Delta r \end{bmatrix}, \quad \begin{bmatrix} \phi(0) \\ \theta(0) \\ \psi(0) \end{bmatrix} = \begin{bmatrix} \phi_0 \\ \theta_0 \\ \psi_0 \end{bmatrix}
$$

$$(11.3)$$

Integration of Eq. (11.3) (with $[\phi_0, \theta_0, \psi_0]^T$ to be determined during FPR) yields PADS attitude. Magnetometer measurements are used within the FPR framework to support the estimation of the Euler angles that are implicitly contained in the rotation matrix For transforming the constant local magnetic vector, defined in $\{I\}$, into the body-fixed coordinates $\{b\}$ the corresponding rotation [Eq. (5.10)] should be applied as follows:

$$\mathbf{h}^m = {}^b_n\mathbf{R}\mathbf{h}^I - \Delta\mathbf{h} \tag{11.4}$$

Here, magnetometer offsets vector $\Delta\mathbf{h} = [\Delta h_x, \Delta h_y, \Delta h_z]^T$ represent hard iron effects.

Similarly, the airspeed vector expressed in $\{b\}$ can be converted to $\{I\}$

$$
\begin{bmatrix} V_x \\ V_y \\ V_z \end{bmatrix} = {}^b_n\mathbf{R}^T \begin{bmatrix} u \\ v \\ w \end{bmatrix} \tag{11.5}
$$

Their integration provides PADS position [Eq. (5.27)]

$$
\begin{bmatrix} \dot{x} \\ \dot{y} \\ \dot{z} \end{bmatrix} = \begin{bmatrix} V_x \\ V_y \\ V_z \end{bmatrix} \tag{11.6}
$$

Biased parameters V_a^m, α_m, and β_m are measured in a noseboom coordinate frame and need to be transformed into $\{b\}$ using Eq. (5.75)

$$
\begin{bmatrix} v_{xNB}^m \\ v_{yNB}^m \\ v_{zNB}^m \end{bmatrix} = \begin{bmatrix} \cos(\alpha_m + \Delta\alpha)\cos(\beta_m + \Delta\beta) & \cos(\alpha_m + \Delta\alpha)\sin(\beta_m + \Delta\beta) & -\sin(\alpha_m + \Delta\alpha) \\ -\sin(\beta_m + \Delta\beta) & \cos(\beta_m + \Delta\beta) & 0 \\ \sin(\alpha_m + \Delta\alpha)\cos(\beta_m + \Delta\beta) & \sin(\alpha_m + \Delta\alpha)\sin(\beta_m + \Delta\beta) & \cos(\alpha_m + \Delta\alpha) \end{bmatrix}
$$
$$
\times \begin{bmatrix} V_a^m + \Delta V \\ 0 \\ 0 \end{bmatrix} \tag{11.7}
$$

or

$$
\begin{bmatrix} v_{xNB}^m \\ v_{yNB}^m \\ v_{zNB}^m \end{bmatrix} = \begin{bmatrix} \cos(\alpha_m + \Delta\alpha)\cos(\beta_m + \Delta\beta) \\ -\sin(\beta_m + \Delta\beta) \\ \sin(\alpha_m + \Delta\alpha)\cos(\beta_m + \Delta\beta) \end{bmatrix} (V_a^m + \Delta V) \tag{11.8}
$$

The airspeed components at CL (Fig. 4.4) are then computed with account for an additional airflow caused by rotation of PADS, similar to how it was done in Eq. (5.84) for the canopy

$$
\begin{bmatrix} v_x^m \\ v_y^m \\ v_z^m \end{bmatrix} = \begin{bmatrix} v_{xNB}^m \\ v_{yNB}^m \\ v_{zNB}^m \end{bmatrix} - \begin{bmatrix} p \\ q \\ r \end{bmatrix} \times \begin{bmatrix} x_{BN} \\ y_{BN} \\ z_{BN} \end{bmatrix} \tag{11.9}
$$

Here, vector $[x_{BN}, y_{BN}, z_{BN}]^T$ defines the position of noseboom sensor with respect to CL.

Finally, the wind profile $\mathbf{W} = [w_x, w_y, w_z]^T$ is reconstructed using Eq. (5.78) by subtracting the airspeed vector projected onto the $\{I\}$ axes from the ground-speed vector

$$
\mathbf{W} = \begin{bmatrix} V_x \\ V_y \\ V_z \end{bmatrix} - {}_n^b\mathbf{R}^T \begin{bmatrix} v_x^m \\ v_y^m \\ v_z^m \end{bmatrix} \tag{11.10}
$$

The PADS's position is measured by GPS in terms of latitude, longitude (given in degrees), and altitude above MSL. These quantities can be transformed to $\{I\}$ using the following approximate formulas:

$$
\begin{aligned}
x &= (lat - lat_0) \cdot 1{,}852 \text{ m} \cdot 60/\text{deg} \\
y &= (lon - lon_0) \cdot \cos(lat_0) \cdot 1{,}852 \text{ m} \cdot 60/\text{deg} \\
z &= -(alt - alt_0)
\end{aligned} \tag{11.11}
$$

where lat_0, lon_0, alt_0 are the known geodetic coordinates.

Unknown sensor biases along with initial conditions are put together in the extended parameter vector $\boldsymbol{\Xi}$ for the sensor (error) model that is estimated in the FPR

$$
\boldsymbol{\Xi}^{FPR} = \begin{bmatrix} \Delta a_x, \Delta a_y, \Delta a_z, \Delta p, \Delta q, \Delta r, \Delta h_x, \Delta h_y, \Delta h_z, \Delta V, \Delta\alpha, \Delta\beta, \\ \tau_{GPS}, \phi_0, \theta_0, \psi_0, u_0, v_0, w_0, w_{x0}, w_{y0}, w_{z0} \end{bmatrix}^T \tag{11.12}
$$

(The parameter τ_{GPS} is introduced to account for a time delay in the GPS receiver.)

11.2.2   FPR RESULTS

The FPR data presented in this section and used for parameter identification in
Sec. 11.3 use flight data gathered for the ALEX PADS. The flight-path reconstruction is done in two steps before it is completed by the correction for relative
motion. First, the integrated inertial motion measured by accelerometers and
gyros is compared against the GPS measurements, giving the correction parameters (e.g., bias), initial values, and reconstructed time histories for the involved
measurements [Eqs. (11.1–11.7)]. Afterward, the results are incorporated into a
second FPR, where they are checked against the air-data measurements at the
noseboom, in order to estimate the air-data biases and the mean wind profile
[Eqs. (11.8–11.10)]. The initial condition for the position is taken directly from
the GPS measurements.

Usually, not all flight tests are equally suited for evaluation, and so a selection is made. Important is the nominal behavior of the vehicle. Non-nominal
situations, like end cell closure or a stall situation with the canopy collapsing,
must be excluded (at least from the following identification) because they
cannot be represented by the simulation model. A good and suitable dataset
is shown in Fig. 11.3. Along with the measured time histories for horizontal projection of the ground-speed vector V_G, Earth-referenced flight-path angle γ, and
heading angle χ, Fig. 11.3 also shows the reconstructed ones. The match is good;
only at the beginning and the end of the time sections differences in V_G and γ
are visible.

The entire time interval of 205 s was divided in three time sections, for
which initial conditions and bias parameter for accelerometers and gyros

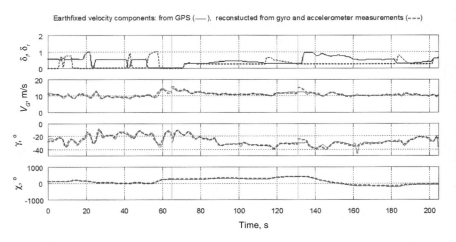

Fig. 11.3   Time histories of measured and reconstructed ground-speed vector
components.

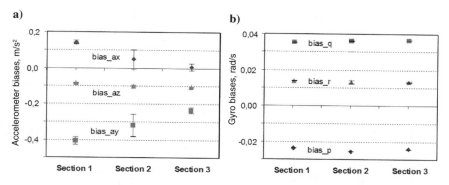

Fig. 11.4 Estimated bias values for used low-cost accelerometers and gyros.

were estimated separately. This was necessary because the biases, particularly from the low-cost accelerometers used here, varied over the flight time (Fig. 11.4). The reason is probably some temperature dependency of the micro-electromechanical system sensors, which is not compensated internally. While the gyros show an offset error (i.e., bias) that can go beyond 2% full scale, the change over the flight time is in the magnitude of 0.1% or below. In contrast, the accelerometers have only an offset error of 1% full scale, but show a possible variation during flight time up to 0.5%, which is worse. The division into time sections was a necessary workaround because no reliable model for the variation of the accelerometer bias was found that could be used in the FPR for compensation.

For the second FPR using the air-data measurements, the noseboom biases are estimated to $\Delta V = -1.95$ m/s $(-3.8$ kt), $\Delta \alpha = -8.3$ deg, and $\Delta \beta = 0.9$ deg. The large offset in the angle-of-attack measurement is a result of an inaccurate manual adjustment of the vane during flight preparation. For the wind, only constant horizontal components are estimated. Because the flight test took place on a hot summer day with calm winds, the results for the horizontal wind are in the magnitude of less than 1 m/s (2 kt). On the other hand, there is vertical wind activity in the order of 1–2 m/s (2–3.9 kt). The differences in the airspeed measurements and the reconstructed time histories from GPS characterize the differences between estimated wind and true wind (Fig. 11.5).

The last step is the correction for relative motion as described in Sec. 10.1.4. Using the relative attitude, which for example is obtained by video analysis, the angles of attack, and sideslip, the accelerations and velocities are transformed into the reference frame {b} with the origin at the PADS center of mass, with elimination of the rotations of the payload relative to the canopy (Fig. 11.6). Finally, the new recomputed measurements are checked for consistency again within the FPR procedure.

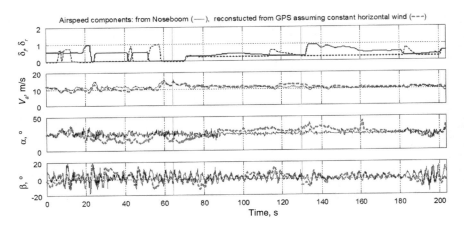

Fig. 11.5   Time histories of measured and reconstructed airspeed components.

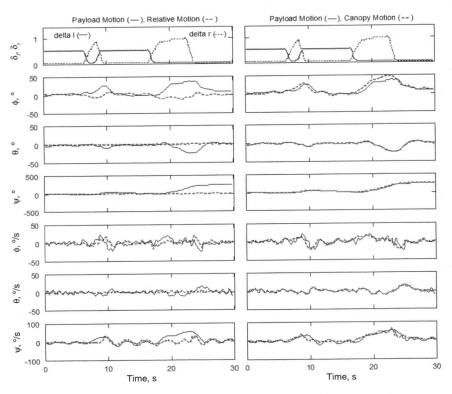

Fig. 11.6   Payload motion, relative motion, and corrected canopy motion.

11.3 PARAMETER ESTIMATION USING OUTPUT-ERROR METHOD

The objective in SID is to extract a mathematical model that describes the real system behavior from flight-test data, that is, measured inputs and outputs. Assuming that the model structure is known, the model parameters Ξ are adjusted by means of an optimization procedure in order to achieve the best match possible between the simulated output $\mathbf{y}$ and the measured output $\mathbf{z}$ with the known input history $\mathbf{u}$. The cost function to be minimized is the product of the output error variances σ of all m output variables,

$$\Phi = \prod_{i=1}^{m} \sigma^2(z_i - y_i) \tag{11.13}$$

with

$$\sigma^2(z_i - y_i) = \frac{1}{N} \sum_{k=0}^{N-1} [z_i(t_k) - y_i(t_k)]^2 \tag{11.14}$$

The cost function in Eq. (11.13) is equal to the maximum-likelihood (ML) cost function for an unknown output-error covariance matrix that is assumed to be diagonal. Starting from an initial guess, the parameters are improved iteratively (using a Gauss–Newton optimization method) solving a set of linear equations in each iteration. The optimization process stops as soon as the relative change either of the cost function or the unknown parameters from one iteration to the next one falls below a specified limit (Fig. 11.7).

In what is presented in Secs. 11.3.1 and 11.3.2, longitudinal and lateral flight dynamics are analyzed separately and then together. For validation, the quality and reliability of the identified parameters are evaluated according to their

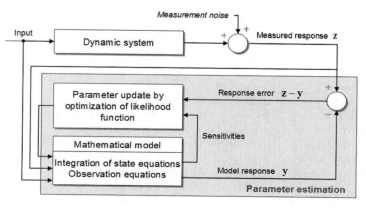

Fig. 11.7   Parameter estimation by the output-error method [Jategaonkar 2006].

standard deviation, correlation with other parameters, and the match of the corresponding simulated and measured time histories. In addition, trim computations are performed, and the identified model is linearized and compared with the theoretical assumptions in the frequency domain.

11.3.1   VALIDATION OF LONGITUDINAL AND LATERAL MOTION OF ALEX PADS

The selection of the parameters to be identified and validated depends on the model characteristics that are considered as uncertain. In the present case this concerns mainly the aerodynamics of the parafoil. Geometry, masses, moment of inertia, and the aerodynamics of load and lines are taken as known. However, because the last parameters are also affected by uncertainties, it is likely that the results of parameter identification not only reproduce the characteristics of the parafoil, but also contain portions of the remaining system.

For getting reliable identification results, it will be intended to analyze long and "interesting" time slices with explicit maneuvers of the flight. However, as already mentioned before, it must be ensured that the model assumptions remain valid all of the time. Especially in the present case of a parafoil-load system, this requirement cannot be always fulfilled because strong maneuvers often lead to heavy deformations of the canopy up to collapsing and also severe relative motions that cannot be corrected. So, usually a representative compromise between a smooth but varied flight is the best option. Unsuitable time slices should be removed from the data set.

The identifiability also depends on the sensitivity matrix, that is, the dependency of the output variables from the parameters, and the correlation among the parameters. For example, the rigging angle μ defining the inclination of the airfoil relative to the suspended payload (Fig. 4.4) has a big influence on the aerodynamics of the complete system. Although this parameter is determined geometrically by the lengths of the suspension lines, there is a remaining uncertainty. It would be highly desirable to validate this parameter in the identification process as well because otherwise it could lead to inaccurate estimation of the parafoil aerodynamic parameters. Unfortunately, the rigging angle cannot be identified for example together with $C_{L\alpha}$ because they are highly correlated. At the end the rigging angle was taken as a given model assumption and set constant at 13 deg for the parafoil (measured against a chord line).

As starting values for the parameter identification, the approximate coefficients derived analytically were taken (Table 4.2). However, the identification has shown that the simultaneous estimation of all derivatives—beside its larger computing time—causes problems. The difficulties result from insufficient information in the acquired flight-test data that prevents distinguishing the effects of all parameters. For the benefit of a good and reliable identification, the number of estimated parameters should be reduced to the most significant derivatives. An effective method to accomplish this consists in separating longitudinal and lateral motion and identifying the corresponding parameters at first

TABLE 11.1   INPUT, OUTPUT, AND VARIED PARAMETERS VECTORS FOR ANALYSIS OF LONGITUDINAL AND LATERAL MOTION

Parameter	Longitudinal Motion	Lateral Motion
Output vector	$\mathbf{z} = [q, \theta, u, w, V, \gamma, V_z, z, V_a, \alpha]^T$	$\mathbf{z} = [p, r, \phi, \psi, v, \chi, x, y, \beta]^T$
Input vector	$\mathbf{u} = [\delta_l, \delta_r, w_x, w_y, w_z]^T$	
Vector of varied parameters	$\Xi = [C_{L\alpha}, C_{L\delta}, C_{D0}, C_{D\delta}, C_{m0},$ $C_{m\alpha}, C_{mq}, C_{m\delta}, K_1, K_2, \mathbf{z}_0]^T$	$\Xi = [C_{Y\beta}, C_{Y\delta}, C_{l\beta}, C_{lp}, C_{l\delta},$ $C_{nr}, C_{n\delta}, \Delta\delta, K_1, K_2, \mathbf{z}_0]^T$

Coefficients K_1, K_2 and actuator offset $\Delta\delta$ are parameters of nonlinear control efficiency described by Eqs. (5.164–5.167).

independently from each other. This is done by reducing the equations of motion to the relevant longitudinal or lateral set and complementing the missing quantities (e.g., angular rates p and r, side speed v for longitudinal motion) for the complete motion by using measurements as so-called pseudocontrol inputs [Plaetschke and Saliaris 1995].

Table 11.1 shows two sets of outputs used for comparing simulated and measured (respectively reconstructed) time histories (vector $\mathbf{z}$) and varied parameters (vector Ξ). These two sets were used for longitudinal and lateral channels separately and then combined. As seen, both sets shared the same input vector $\mathbf{u}$ composed of the left and right control deflections and the reconstructed wind profile. Additionally, for the simulation the complete six-DoF motion, the angular rates p and r, and side speed v were taken from measurements and used as pseudocontrol inputs.

Table 11.2 and Fig. 11.8 show the results of parameter identification for the longitudinal motion. Starting from the nominal values of Table 4.2, the identification process converged quite rapidly in only 10 iterations.

Compared to the original values, the estimated parameters are in about the same magnitude. $C_{L\alpha}$ has increased about 30%, and C_{D0} has almost doubled. Also, the pitching-moment derivatives C_{m0} and $C_{m\alpha}$ have been estimated to a significant value while they were to zero in the original model.

TABLE 11.2   ESTIMATED PARAFOIL AERODYNAMIC DERIVATIVES FOR LONGITUDINAL MOTION

Parameter	Value	Rel. Std. Dev.	Parameter	Value	Rel. Std. Dev.
$C_{L\alpha}$	3.1129	0.9%	C_{m0}	−0.0662	4.5%
$C_{L\delta_s}$	0.1879	7.9%	$C_{m\alpha}$	0.6965	3.7%
C_{D0}	0.1680	3.2%	C_{mq}	0.0025	350%
$C_{D\delta_s}$	0.2477	5.2%	$C_{m\delta_s}$	−0.0624	9.9%
K_1	−0.3375	8.1%	K_2	−0.0719	15.8%

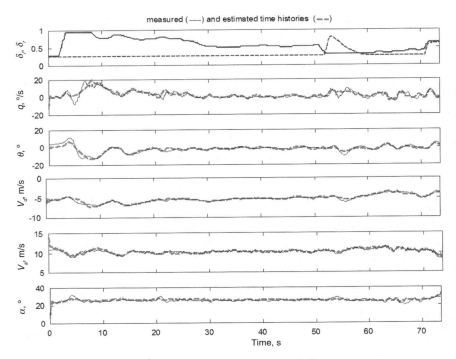

measured (——) and estimated time histories (— —)

Fig. 11.8   Time histories of simulated and measured longitudinal motion after identification.

The corresponding relative standard deviation of the parameters can be taken as a measure for the trustworthiness of the result. Values below 10% can usually be accepted without problems. In the present case the parameters K_2 and particularly C_{mq} exceed this limit; the result for the pitch damping derivative is totally unreliable. In such a case, usually a new identification run under changed conditions must be executed in which, for example, those parameters are kept at a constant value. Here, we continue for the moment and proceed to the validation of lateral motion.

The identification of the parameters concerning the lateral motion was principally done in the same manner using the set of evaluated output variables, inputs, and varied parameters as identified in Table 11.1.

This time, measurements of the pitch rate q and the velocity components u and w were taken as pseudocontrol inputs for the longitudinal motion. As seen from data of Table 11.1, compared to the full set, the vector of lateral aerodynamic parameters was reduced. Among others, $C_{l\beta}$, C_{Yp}, C_{np}, C_{Yr}, and C_{lr} have been set to zero. This was necessary because of high correlations of these parameters with others that heavily reduced the liability of the results.

TABLE 11.3   ESTIMATED PARAFOIL AERODYNAMIC DERIVATIVES FOR LATERAL MOTION

Parameter	Value	Rel. Std. Dev.	Parameter	Value	Rel. Std. Dev.
$C_{Y\beta}$	−0.2088	3%	$C_{Y\delta_a}$	0.1236	1.5%
$C_{l\beta}$	0.1941	3%	C_{nr}	−0.1983	2.2%
C_{lp}	−0.3063	3.8%	$C_{n\delta_a}$	0.1342	1.2%
$C_{l\delta_a}$	−0.1342	2.6%	$\Delta\delta$	−0.0451	0.7%
K_1	−0.3319	0.4%	K_2	0.0373	2.6%

Using the same time slice as in the preceding for identification of lateral motion, the match of simulated and measured time histories was achieved as shown in Table 11.3 and Fig. 11.9.

All of the relative standard deviations are well within the limit of 10%. Compared with the original derivatives, the values on the left side are quite similar. In contrast, the values for $C_{Y\delta_a}$, C_{nr}, and $C_{n\delta_a}$ have reached about the double of the original parameters. Because other derivatives from the original model have been

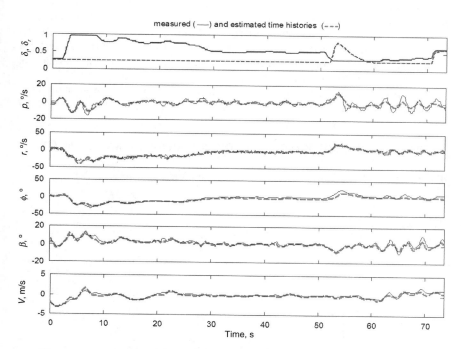

Fig. 11.9   Match of simulated and measured lateral motion after identification.

set to zero, it is likely that the estimated parameters are trying to compensate some portion of the others influence on the flight mechanics.

11.3.2 VALIDATION OF THE SIX-DoF MODEL AND NONLINEAR CONTROL EFFICIENCY

Now, starting with a combination of both parameter sets of Table 11.1, a new identification run is performed using the complete six-DoF model without pseudocontrol inputs. The estimated parameters consist in the combination of both for longitudinal and lateral motion. Table 11.4 shows the results achieved in this case. The match between simulated and measured time histories from the six-DoF identification is demonstrated in Fig. 11.10.

Clearly, six-DoF simulation has a better match compared to those lower-order models of Sec. 11.3.1. Comparing the parameter values with the results of the other two identifications, it becomes obvious that the relative standard deviation has been reduced, especially for the parameters concerning the longitudinal motion. Also, the value for C_{mq} is now much more plausible. The parameters K_1 and K_2 have a correlation of 92%; all others are below 90%. These results give a strong indication that the new parameter set is more realistic and accurate than the former values. Nevertheless, this does not guarantee that the parameters are necessarily true. Because the ML identification method is based on statistical approaches, this only means that—under the given conditions (i.e. given model structure, estimated parameter set, selected flight-test data)—the parameters are the most likely ones for reproducing the measurements with the simulation, that is, the real system with the model.

The new parameter set now allows comparing the original theoretical with the new identified model. Differences between both may give an indication in which

TABLE 11.4   ESTIMATED PARAFOIL AERODYNAMIC DERIVATIVES FOR SIX-DoF MOTION

Parameter	Value	Rel. Std. Dev.	Parameter	Value	Rel. Std. Dev.
$C_{L\alpha}$	3.1700	0.3%	C_{m0}	−0.0529	3.3%
$C_{L\delta_s}$	0.1138	3.3%	$C_{m\alpha}$	0.8292	0.6%
C_{D0}	0.1242	1.9%	C_{mq}	−0.1010	0.5%
$C_{D\delta_s}$	0.3468	1.7%	$C_{m\delta_s}$	−0.1712	3.3%
$C_{Y\beta}$	−0.2466	2.9%	$C_{Y\delta_a}$	0.1256	0.9%
$C_{l\beta}$	0.0926	3.6%	C_{nr}	−0.1301	1.9%
C_{lp}	−0.1708	2.6%	$C_{n\delta_a}$	0.1298	0.9%
$C_{l\delta_a}$	−0.1514	1%	$\Delta\delta$	−0.0352	0.8%
K_1	−0.3111	1.8%	K_2	0.0690	2.8%

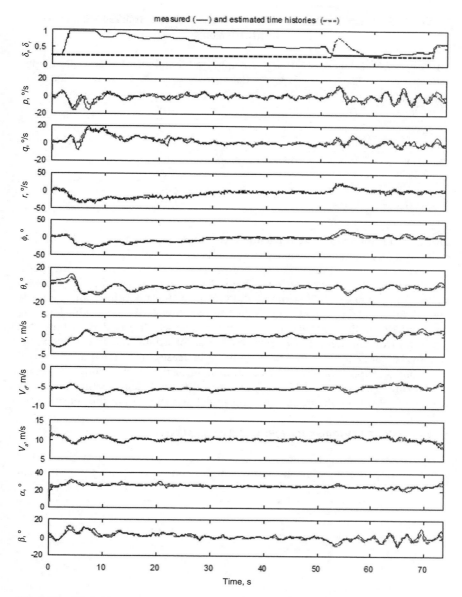

Fig. 11.10 **Time histories of simulated and measured six-DoF motion after identification.**

TABLE 11.5   EIGENVALUES, TIME PERIOD, AND DAMPING OF LINEARIZED ORIGINAL AND
IDENTIFIED MODEL

Motion Mode	Original	Period T, s	Damping Ratio ζ	Identified	Period T, s	Damping Ratio ζ
Short-period mode	$-3.33 \pm i\,4.63$	1.1	0.58	$-2.96 \pm i\,3.05$	1.5	0.7
Phugoid mode	$-0.30 \pm i\,0.69$	8.4	0.4	$-0.31 \pm i\,0.76$	7.7	0.38
Dutch roll mode	$-0.67 \pm i\,3.40$	1.8	0.19	$-2.18 \pm i\,1.46$	2.4	0.83
Coupled roll-spiral mode	$-0.44 \pm i\,0.83$	6.7	0.47	$-0.46 \pm i\,1.24$	4.8	0.35

direction the theory and aerodynamics of gliding parachutes should evolve in order to understand and include the new effects in the model.

The main result of identification is a progressive characteristic for the identified nonlinear control efficiency [Jann 2003; Jann 2004], which was also studied in [Babinsky 1999; Iacomini and Cerimele 1999a; Iacomini and Cerimele 1999b]. Figure 11.11 presents the results from all three trials: longitudinal, lateral, and six-DoF motion identification. All three data sets clearly demonstrate that small control line deflections are probably absorbed by the flexibility of the canopy. Larger deflections lead to a significant increase of the local drag and therefore control effect efficiency. As mentioned in Sec. 5.3, the control efficiency model (the exact values of K_1, K_2 and $\Delta\delta$ in this case) depends on the canopy design and PADS configuration (rigging angle) and the trim of the control lines.

11.3.3   FLIGHT PERFORMANCE AND STABILITY ANALYSIS

A trim computation determines the equilibrated system states for a given set of inputs and parameters. If the trim computation is performed for different

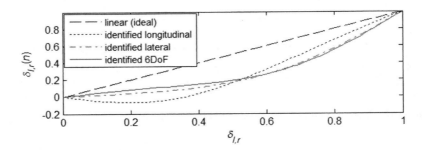

Fig. 11.11   Identified nonlinear control efficiency.

inputs or parameters, the dependency of the flight performance of those variables can be studied. Figure 11.12 shows the influence of a symmetrical control line deflection on flight speed, sink rate, glide ratio, and angle of attack for the original and the new identified six-DoF model.

Compared with each other, the range of the flight performance variables remains about the same; on the other hand, the characteristic of these, particularly the glide ratio, have changed. Instead of decreasing monotonously like in the original model, the glide ratio (L/D) first increases a little before it sinks even below the former value. The reason for the changed characteristics is mainly the combination of the introduction of the nonlinear control efficiency together with a change of the parameters $C_{L\delta_s}$ and $C_{D\delta_s}$.

To analyze the stability characteristics of the model, it was converted into a linear time-invariant system by linearizing it at some trim point. As trim point, the zero control deflection situation (δ_l, $\delta_r = 0$) with the vehicle gliding forward is selected. The eigenvalues of the state matrix in the zero-pole diagram can be used directly for evaluating the stability (Fig. 11.13) (compare Fig. 6.1a). They describe the dynamic behavior of the linearized system and identify frequency and damping of the inherent natural modes as described in detail in Sec. 6.1.

Comparing the eigenvalues of the identified model with those of the original model, phugoid and coupled roll motion have moved towards higher frequencies,

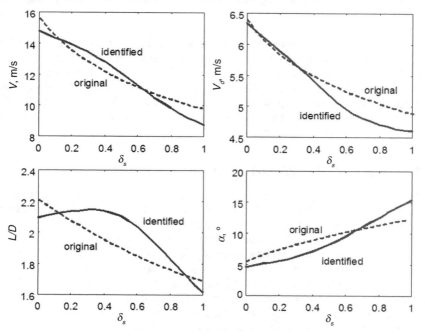

Fig. 11.12   Influence of symmetrical control line deflection on flight performance.

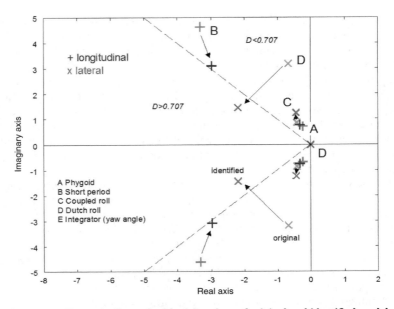

Fig. 11.13   Zero-pole diagram with eigenvalues of original and identified model.

that is, they are a little faster now. Although the damping of the phugoid has almost not changed, the coupled roll seems to be less damped than in the theoretical model. The short-period motion is shifted for some amount towards lower frequency and higher damping. The most significant change affects the Dutch roll, which is much more damped in the identified than in the original model. An explanation could be that the flexibility of the real canopy tends to absorb fast motions, preferably in lateral direction.

11.4   MULTICRITERIA PARAMETRICAL IDENTIFICATION

By their nature, applied identification problems are multicriteria problems. However, as a rule these problems are treated as single-criterion problems. Usually, it is done by using the most important criterion, or by using several criteria, but one at a time. The standard approach however is to develop a single compound criterion that weighs criteria relative to their importance, like the one used in the preceding section. [See how the cost was defined in Eq. (11.3).] In this section the parametric identification is cast as a multicriteria problem involving several dozens of varied parameters defining an eight-DoF model of Pegasus PADS and uses the parameter space investigation (PSI) method and Multicriteria Optimization/Vector Identification (MOVI) software package to produce the Edgeworth-Pareto set of possible solutions (Fig. 11.14).

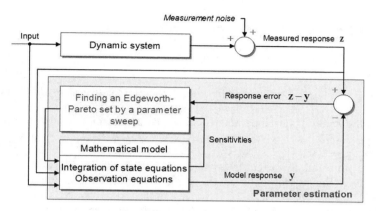

Fig. 11.14 Multicriteria parameter estimation.

11.4.1 IDENTIFICATION PROBLEM FORMULATION FOR PEGASUS PADS

The behavior of the Pegasus PADS was described with the eight-DoF model with kinematic and dynamic equations summarized in Table 5.14 and coefficients of aerodynamic and control derivatives presented in Eqs. (5.96–5.101). This model depends of several dozens of parameters that are not known for sure but rather being approximated using wind-tunnel data for some generic similar-type canopy, approximations of geometry, rigging configuration, and apparent mass and inertia terms. The original model [Eqs. (5.62), (5.96–5.101)] enhances in such a way that each unknown parameter entered equations of motion as a product of some nominal value and scaling multiplier

$$C_D = 0.14k_{CD_0} + (0.25k_{A_2} + 0.2k_{A\delta_s}\delta_s)C_L^2 \tag{11.15}$$

$$C_Y = (-0.005k_{CY\beta} - 0.0001k_{CY\beta}^\alpha\alpha)\beta + (-0.007k_{CY\delta_a} + 0.0012k_{CY\delta_a}^\alpha\alpha)\delta_a \tag{11.16}$$

$$C_L = 0.0375k_{CL\alpha}(\alpha + 10k_{\alpha_0}) + 0.2k_{CL\delta_s}(\delta_s + 0.5|\delta_a|) \tag{11.17}$$

$$C_l = -(0.0014k_{Cl\beta} + 0.001k_{Cl\beta}^\alpha\alpha)\beta - \frac{b}{2V_a}(0.15k_{Clp}p - 0.0775k_{Clr}r)$$
$$- (0.0063k_{Cl\delta_a} + 0.001k_{Cl\delta_a}^\alpha\alpha)\delta_a \tag{11.18}$$

$$C_m = -0.33k_{Cm_0} - 6.39k_{Cmq}\frac{c}{V_a}q + \frac{dm_s}{m_s + m_c}\left(\frac{1}{c} + \frac{\rho C_{B0}}{2W/S}\right)$$
$$\times [C_D \quad C_L]\begin{bmatrix} \cos(\alpha + 8k_{\text{rig}}) \\ -\sin(\alpha + 8k_{\text{rig}}) \end{bmatrix} \tag{11.19}$$

$$C_n = (0.007k_{Cn\beta} - 0.0003k_{Cn\beta}^\alpha \alpha)\beta + \frac{b}{2V_a}(0.023k_{Cnp}p - 0.0936k_{Cnr}r)$$

$$+ (0.03k_{Cn\delta_a} - 0.001k_{Cn\delta_a}^\alpha \alpha)\delta_a \qquad (11.20)$$

$$\mathbf{I}_{a.m.} = \text{diag}([Ak_A, Bk_B, Ck_C]), \quad \mathbf{I}_{a.i.} = \text{diag}([I_A k_A^*, I_B k_B^*, I_C k_C^*]) \qquad (11.21)$$

Hence, the vector of varied parameters Ξ was composed of 31 scaling factors: k_{α_0}, $k_{CL\delta_s}$, $k_{CL\delta_s}$, k_{CD_0}, k_{A_2}, $k_{A\delta_s}$, k_{Cm_0}, k_{rig}, k_{Cmq}, $k_{CY\beta}$, $k_{CY\beta}^\alpha$, $k_{CY\delta_a}$, $k_{CY\delta_a}^\alpha$, $k_{Cl\beta}$, $k_{Cl\beta}^\alpha$, $k_{Cl\delta_a}$, k_{Clp}, k_{Clp}, k_{Clr}, $k_{Cn\beta}$, $k_{Cn\beta}^\alpha$, $k_{Cn\delta_a}$, $k_{Cn\delta_a}^\alpha$, k_{Cnp}, k_{Cnr}, k_A, k_B, k_C, k_A^*, k_B^*, and k_C^*. It also included the unknown controls synchronization error Δt_c and unknown offset along the x_b axis between the parafoil aerodynamic center (c/4 point) and PADS center of gravity x_{CG}.

The problem of Pegasus PADS model identification was in tuning these multipliers (varied parameters) so that the model output matched that observed in the real flight. The six-DoF Simulink model of Pegasus PADS, presented in Fig. 5.9, was enhanced to include two more degrees of freedom and the interactive interface (Fig. 11.15) allowing changing any scaling coefficient(s) to see the impact of the performance of the system. This capability (of running a simulation with an arbitrary set of varied parameters) was then used in parameter identification. Specifically, the Simulink model was called from a MATLAB script to produce a series of outputs to compare them with those obtained in the flight test.

As will be shown later in Sec. 11.4.3, data available from the flight test included the following:

- Local tangent plane coordinates (after processing *lat/lon/alt* information from GPS)

- Components of inertial velocity (V_N, V_E, and V_U, measured by GPS)

- Angular rates (p, q, and r measured by IMU)

Analyzing these available data, one can think of the following adequacy criteria:

- Proximity of simulated trajectory to the real one (that could be further split into the horizontal and vertical components)

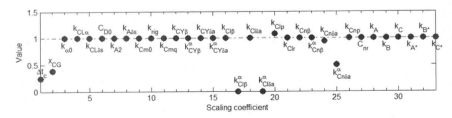

Fig. 11.15   Interactive interface allowing changing the value of any scaling coefficient.

- Closeness of the speed/heading profiles or adequacy of natural eigenvalues for all three translational channels
- Adequacy of natural eigenvalues for all three rotational channels

Ideally, it would be nice to be able to meet all three adequacy criteria simultaneously. In practice however, a single criterion [like the one of Eq. (11.13)] combining multiple outputs is used (see two vectors $\mathbf{z}$ in Table 11.1). The multicriteria optimization (identification) paradigm (Fig. 11.16) allows the investigation of the varied parameters' influence on each adequacy criterion separately.

Three obvious reasons for employing multicriteria (or vector) identification technique are as follows:

1. Although single-criterion identification may result in a relatively good overall match between the model output and real flight data, generally we cannot assert a sufficient correspondence (structurally and parametrically) between the model and real object. In other words, utilizing a single criterion may be to the detriment of the physical essence of a problem.

2. The richness and quality of flight-test data available these days using modern sensors suggests going beyond single compound cost function using several particular criteria to evaluate adequacy of the mathematical model, that is, determining to what extent the mathematical model corresponds to the physical system in principle.

3. Availability of preliminary information about the lower and upper limits on most of the varied parameters would certainly help to speed up a single-criterion identification process. The multicriteria identification allows doing exactly this, that is, exploring the varied parameters' domain, establishing a feasible set of these parameters, and performing sensitivity analysis. This gives an idea about what varied parameters could be dropped off to decrease the problem's dimension, what starting point to use for single-criterion identification, and what lower and upper bounds to use.

The essence of the multicriteria identification (optimization) is presented next.

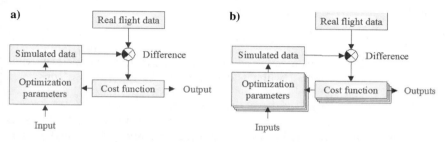

Fig. 11.16   a) Single-criteria identification and b) multicriteria identification.

11.4.2 MATHEMATICAL FOUNDATION AND SOFTWARE

Any nonlinear high-fidelity PADS model developed in Chapter 5 can be represented as

$$\dot{\mathbf{x}} = \mathbf{f}(\mathbf{x}, \mathbf{u}, \boldsymbol{\Xi}, t), \quad t \in [t_0; t_f], \quad \mathbf{x}_{t=t_0} = \mathbf{x}_0 \qquad (11.22)$$

In this model $\mathbf{x} = [x_1, \ldots, x_n]^T$ is the state vector with initial conditions $\mathbf{x}_0$, $\mathbf{u} = [\delta_s, \delta_a, w_x, w_y, w_z]^T$ is the vector of controls, and $\boldsymbol{\Xi} = [\Xi_1, \ldots, \Xi_p]^T$ is a vector of varied parameters to be identified. Three types of constraints can be imposed in the vector $\boldsymbol{\Xi}$. The first group, the parametric constraints, has a simple form

$$\Xi_j^* \leq \Xi_j \leq \Xi_j^{**}, \quad j = 1, \ldots, p \qquad (11.23)$$

establishes the upper and lower bounds on varied parameters (scaling factors in our case) defining a parallelepiped Π in the p-dimensional space.

The second group, the functional constraints, can be written as

$$c_j^* \leq g_j(\boldsymbol{\Xi}) \leq c_j^{**}, \quad j = 1, \ldots, q \quad \text{for } t \in [t_0; t_f] \qquad (11.24)$$

where $g_j(\boldsymbol{\Xi})$ can be a linear or nonlinear algebraic combination of varied parameters themselves or in fact be computed as $g_j(\mathbf{x}(\boldsymbol{\Xi}))$ with $\mathbf{x}(\boldsymbol{\Xi})$ being the state vector integrated according to Eq. (11.22).

The third group of constraints involves adequacy criteria. Let $\Phi_l^m(\boldsymbol{\Xi})$, $l = 1, \ldots, \nu$ denote the outputs of the model described by Eq. (11.22) [it is actually $\Phi_l^m(\mathbf{x}(\boldsymbol{\Xi}))$], and $\Phi_l^{\exp}$, $l = 1, \ldots, \nu$ denote preprocessed experimental data (corrected for biases, drop offs, etc.). Then, the adequacy criteria

$$\Re_l(\Phi_l^m(\boldsymbol{\Xi}), \Phi_l^{\exp}) \qquad (11.25)$$

is what we need to satisfy (minimize or maximize). Here $\Re_l(\Phi_l^m(\boldsymbol{\Xi}), \Phi_l^{\exp})$ denotes some operator applied to simulated and experimental data. [It may have a form of Eq. (11.13), or be their ratio, difference, etc.] Because of the multicriteria nature of a problem, to decrease the total number of reasonable candidate solutions (avoid situations when the values of certain criteria might seem unacceptable), the criterial constraints are introduced as

$$\Re_l(\Phi_l^m(\boldsymbol{\Xi}), \Phi_l^{\exp}) \leq \Re_l^{**}, \quad l = 1, \ldots, \nu \qquad (11.26)$$

Here, $\Re_l^{**}$ is the worst value of a particular criterion that can be tolerated while ameliorating other criteria (without loss of generality hereafter we consider a minimum problem). The values of $\Re_l^{**}$ can depend on experimental data quality and physical sense of the adequacy criteria.

The major difference between criteria (11.25) and hard functional constraints (11.24) is that the values of $\Re_l^{**}$ are not known beforehand and have to be determined while addressing the identification problem. They can be either tightened or loosened. For extra flexibility the functional constraints (11.24) can also be represented in a form of pseudocriteria, especially when they are not firm.

Constraints (11.24) and (11.26) limit the initial space $\mathbf{\Pi}$ to subspace $\mathbf{G}$, $\mathbf{G} \subseteq \mathbf{\Pi}$ and finally to some feasible set $\mathbf{D}$, $\mathbf{D} \subseteq \mathbf{G} \subseteq \mathbf{\Pi}$ as shown in Fig. 11.17. The multi-criteria parameter identification problem is then to find an Edgeworth-pPareto set $\mathbf{EPS}$, $\mathbf{EPS} \subseteq \mathbf{D}$, so that

$$\Re_l(\mathbf{EPS}) = \min_{\Xi \in \mathbf{D}} \Re_l\big(\Phi_l^m(\Xi),\, \Phi_l^{\exp}\big), \quad l = 1, \ldots, \nu \tag{11.27}$$

After finding a set $\mathbf{EPS}$, the most preferable or optimal vector Ξ_0, $\Xi_0 \in \mathbf{EPS}$ can finally be determined (chosen) among several best candidates. However, unlike well-conditioned traditional single-criterion identification, we are not so much interested in finding Ξ_0, but in defining and exploring the feasible and Edgeworth-pPareto sets.

The key feature of any multicriteria identification (optimization) package is to populate the search region $\mathbf{\Pi}$ with the uniformly distributed trial points (variations of vector Ξ) and compute adequacy criteria at them. This capability defines

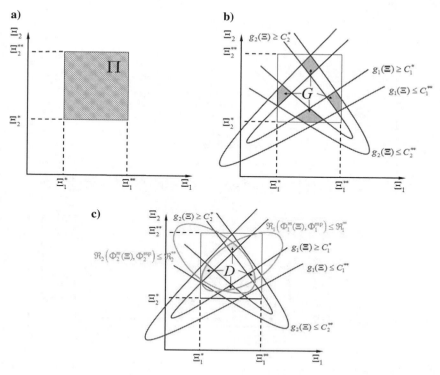

Fig. 11.17 Illustration of subsets a) $\mathbf{\Pi}$, b) G, and c) D in 2D space.

an efficiency of the method. The Pegasus multicriteria parametric identification was carried out using the MOVI software package (ver.1.3) adapted to work with scripts and models created in Mathwork's Matlab/Simulink development environment. This package allows the user to perform feasibility analysis/design in a user-friendly manner. The limit on the total number of varied parameters (they may both be continuous and discrete) exceeds several hundreds, and so this software handled 33 varied parameters for the Pegasus PADS identification with no problem. Criteria can be either minimized or maximized. Some or even all of the criterial constraints (if unknown a priori) can be considered as pseudocriteria and adjusted interactively.

Figures 11.18–11.20 show the test runs of the identification problem where several rather than all varied parameters and four different adequacy criteria were used. Figure 11.18a demonstrates an example of test tables obtained after multiple runs of the model with different parameter vectors. Number N in the

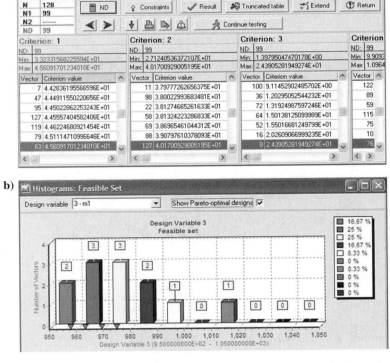

Fig. 11.18 Examples of a) a fully ordered test table and b) a histogram showing a feasible set for one of the parameters.

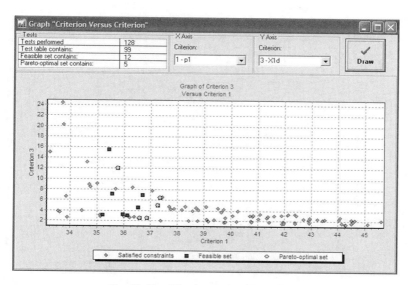

Fig. 11.19   Criterion vs criterion graph.

left-top corner indicates the total number of trials, while ND (ND $\leq$ N)—the number of varied parameter vectors in the feasible set. All functional failures (trials that did not meet the functional constraints) can be considered separately in another window. By softening constraints, part of them may be immediately returned to the feasible set.

The minimum and maximum numbers of each adequacy criterion are presented at the title of each table (vertical column). At this step MOVI allows truncating the whole table to achieve the better results by working with a smaller portion of it at a time, and correcting the value of any criterial constraint to narrow/broaden the feasible set (which can be done for all columns from the left to the right decreasing ND value for every criterion). Eventually, the resulting table can be converted to the simplified form containing information on the subsets of feasible solutions and Pareto-optimal solutions (NP $\leq$ ND).

Further analysis involves graphical representation of data. Figure 11.18a represents a histogram for one specific parameter (how many of the trials fall into the certain range). Figure 11.19 depicts an example of criteria vs criteria graph. Figure 11.20a shows the criteria vs single parameter plot for all trial points (meaning that each point corresponds to a single parameter vector). In addition, the criteria vs single parameter plot can be obtained for any specified parameter by running some additional runs with all other parameters fixed (as it shown in Fig. 11.20b). It is possible to change ranges for the chosen parameter here. Moreover, the values of any other component of the parameter vector can be corrected as well.

a)
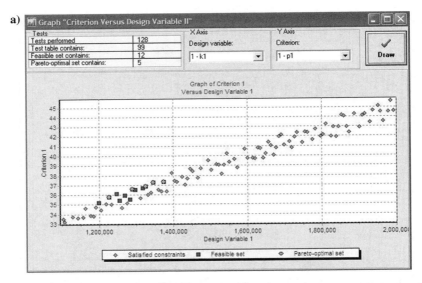

b)
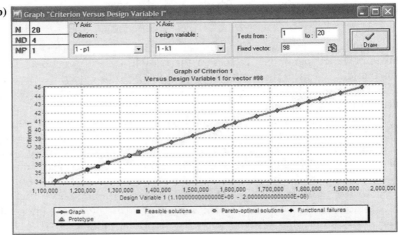

Fig. 11.20   Criterion vs parameter graph constructed for a) all feasible solutions and b) parameter sweep.

MOVI relies on the so-called PSI method to populate the region **II** (p-dimensional cube) with the trial points. Ideally, we would like to have several partitions L for each varied parameter resulting in the total of $M = L^p$ trial points. Figure 11.21a gives an example of having just four partitions for

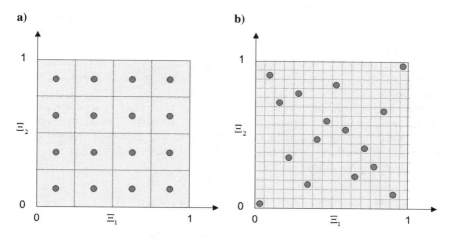

Fig. 11.21   Example of a) straightforward vs b) "wise" Latin-cube sampling.

two (normalized) parameters. Obviously, this straightforward approach is not efficient because it requires a tremendous amount of resources. Table 11.6 shows the
growth of the total number of the trial points with p when you would want to have
just 4 or 10 partitions for varied parameter. It also shows the computer CPU
requirements assuming that each run of the Pegasus PADS Simulink model
roughly requires 1 min.

**TABLE 11.6   NUMBER OF TRIAL POINTS AND CPU RESOURCES REQUIRED IF USING A
STRAIGHTFORWARD APPROACH**

Number of Varied Parameters	5 Sampling Points per Variable		10 Sampling Points per Variable	
	Total Number of Trials	Required CPU Time	Total Number of Trials	Required CPU Time
2	16	25 min	100	1.7 h
3	64	1 h	1,000	17 h
4	256	4.3 h	10,000	7 days
5	1,024	17 h	100,000	2 months
6	4,096	2.8 days	1,000,000	2 years
7	16,384	11.4 days	10,000,000	19 years
8	65,536	1.5 months	100,000,000	1.9 century

The alternative approach would be to use a Latin cube structure. In the example shown in Fig. 11.21b with the same total number of the trial points as in Fig. 11.21a (16), we are able to explore 16 (vs just 4) partitions for each of two varied parameters. Yes, we do not have the orthogonal grid as in the straight-forward approach, but each varied parameter is still tested in every one of its partitions.

Now, to produce a Latin-cube structure for a p-dimensional cube using M trial points, we would use the following approach. First, we partition each (normalized) varied parameter (dimension) into M segments, so that the whole space is partitioned into M^p p-dimensional cells. Second, we choose M cells to contain the trial points by doing the following:

1. Randomly choosing one of the M^p cells as the first point

2. Eliminating all cells that agree with this point on any of its parameters (that is, crossing out all cells in the same row, column, etc.), leaving $(M-1)^p$ candidates

3. Randomly choosing one of these remaining cells and repeating step 2

4. Repeating steps 2 and 3 until only one cell left, which then contains the final trial point

This later approach, however, suffers from trial points not being distributed uniformly along each dimension so that you may end up having "gaps" and not exploring certain combinations of varied parameters. The PSI method employed by MOVI uses specially constructed LP_τ quasi-random sequences [Sobol' 1967; Bratley and Fox 1988; Statnikov and Matusov 1996] so that successively assigned trial points at any stage sort of "know" how to fill in the gaps in the previously generated distribution and keep filling them in, hierarchically (see example in Fig. 11.22). Figure 11.23 shows that the coverage of LP_τ sequence is also much more equally distributed in comparison to the embedded Microsoft Windows random numbers generator (RNG).

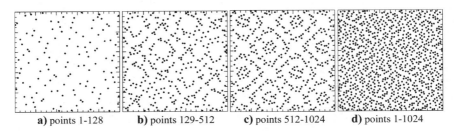

a) points 1-128   **b)** points 129-512   **c)** points 512-1024   **d)** points 1-1024

Fig. 11.22   First 1024 points of the two-dimensional LP$_\tau$ sequence.

a)   b)

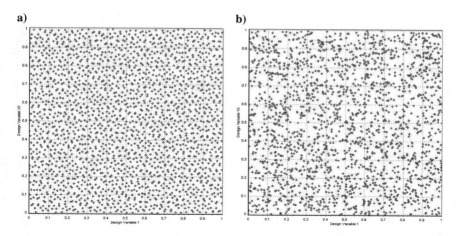

Fig. 11.23   Comparison of a) LP$_\tau$ sequence coverage with b) Windows RNG coverage in the plain of two (first vs tenth) out of 25 parameters for 2048 trials.

Once the trial points are produced, the values of functional dependencies $g_j(\Xi)$ are being computed for these M trial points. If they satisfy corresponding constraints of Eq. (11.24), the quality criteria $\Re_l(\Phi_l^m(\Xi_i), \Phi_l^{\exp})$, $l = 1, \ldots, \nu$ are also being calculated at each trial point $i = 1, \ldots, N$, $N \leq M$.

The varied parameter space is investigated in three stages. First, a table of trials is ascribed to each lth criterion $\Re_l(\Phi_l^m(\Xi), \Phi_l^{\exp})$, and the values of $\Re_l(\Phi_l^m(\Xi_1), \Phi_l^{\exp}), \ldots, \Re_l(\Phi_l^m(\Xi_N), \Phi_l^{\exp})$ are arranged in ascending order (assuming that all the criteria must be minimized) (which is what was shown in Fig. 11.18a). Second, the preliminary criterial constraints $\Re_l^{**}$ are chosen by analyzing the values of $\Re_l(\Phi_l^m(\Xi_i), \Phi_l^{\exp})$ in each ν columns [Eq. (11.26)]. During the third stage, the set of all Ξ_i satisfying all inequalities of Eq. (11.26) simultaneously is determined. A nonempty set of these vectors Ξ_i means existence of the feasible set. Otherwise, the values of $\Re_l^{**}$ need to be loosened, and if this does not help, the number of trials needs to be increased (to have more partitions for each varied parameter) followed by repeating the second stage with a larger table. Alternatively, the bounds in Eqs. (11.23) and (11.24) need to be loosened as well. The procedure is continued until subspace **D** proves to be nonempty and the maximum values of $\Re_l^{**}$ are specified. After that, the Edgeworth-Pareto set **EPS** is constructed and analyzed (see example in Fig. 11.19).

11.4.3   ANALYSIS OF FEASIBLE SOLUTIONS

The SID relied on a rich set of flight-test data collected during four drops for two different configurations of Pegasus PADS. These data included multiple

a)

GPS (10/2*Hz*)	IMU (100/4*Hz*)	Compass (4*Hz*)
Hours	Compass Time UTC (s)	Time UTC (s)
Minutes	Roll (°)	Magnetic Heading (°)
Seconds	Pitch (°)	Pitch (°)
Time UTC (s)	True Heading (°)	Roll (°)
Latitude	Roll Rate (rad/s)	H_x (μT)
Longitude	Pitch Rate (rad/s)	H_y (μT)
Alt AGL (m)	Yaw Rate (r/s)	H_z (μT)
Gnd Spd (m/s)	X accel (ft/s/s)	Temp (°C)
Trk Angle (°)	Y accel (ft/s/s)	
Vel East (m/s)	Z accel (ft/s/s)	
Vel North (m/s)	IMU Time (UTC)	
Vel Up (m/s)		

b)

Dropsonde (10*Hz*)
Time UTC (s)
Latitude (°)
Longitude (°)
Altitude MSL (m)
Q
Std Dev
V_E (m/s)
V_N (m/s)
V_U (m/s)

Commands (4*Hz*)	Motors Position (4*Hz*)
Time UTC (s)	Time UTC (s)
Full Travel Right (%)	Rt Encoder Pos (in)
Full Travel Left (%)	Left Encoder Pos (in)
	Rt Mtr Pos (in)
	Left Mtr Pos (in)
	Rt Mtr Error (in)
	Left Mtr Error (in)

Fig. 11.24   Data collected by the sensor suites aboard a) Pegasus PADS and b) windpack dropsonde.

different-rate sensors as shown in Fig. 11.24a. In addition to that, each drop was preceded by the release of an uncontrolled windpack dropsonde (resembling a small torpedo attached to a circular parachute) used to estimate a current wind profile (as shown in Fig. 11.24b, a GPS suite aboard a dropsonde essentially collects the same data as a GPS suite aboard the Pegasus PADS).

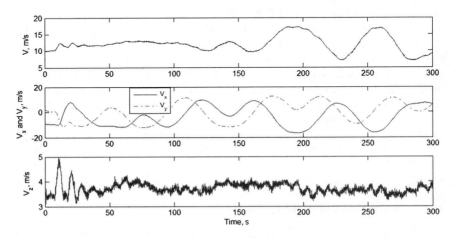

Fig. 11.25   Time histories of ground-speed vector magnitude *V* and its components.

Data preprocessing included the following steps:

- Excluding onboard and on-the-ground portions (if present)
- Excluding discontinuous portions of data (if present)
- Synchronizing data from different sensors by time
- Synchronizing windpack data with GPS data by altitude
- Checking for data consistency
- Fixing bias errors
- Determining initial conditions for simulations

The latter three steps were conducted according to the FPR procedure as was described earlier in Sec. 11.2. Examples of preprocessed conditioned data from each sensor are shown in Figs. 11.25–11.29. Figure 11.30 shows the trajectories of the windpack dropsonde and Pegasus PADS released (a few seconds later).

As outlined in Sec. 11.4.1, Pegasus PADS identification requires estimating 33 varied parameters. Also mentioned are several adequacy criteria to be used. Specifically, eight criteria included the following:

- Two describing the closeness of the horizontal and vertical projections of trajectories, $\int_0^{t_f} \sqrt{[x^m(t) - x^{\exp}(t)]^2 + [y^m(t) - y^{\exp}(t)]^2}\,dt$ and $\int_0^{t_f} |z^m(t) - z^{\exp}(t)|\,dt$, respectively
- Two related to adequacy of simulation and flight-test data in terms of closeness of the power spectrum density for ground-speed components $S_{V_x(t)}$, $S_{V_y(t)}$,

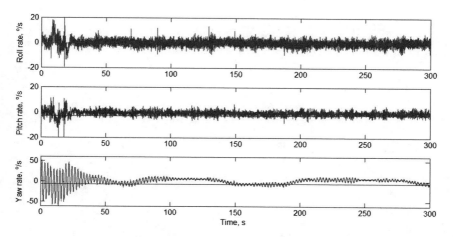

Fig. 11.26   Time histories of angular rate components.

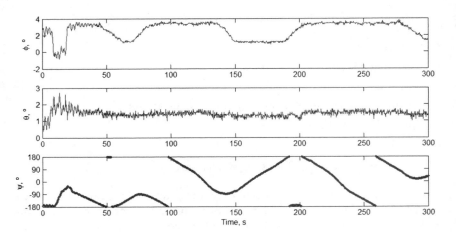

Fig. 11.27   Euler angles' time histories.

and $S_{V_z(t)}$ (to be more precise the norm of the difference in the four lowest frequency components were considered as adequacy criteria)

- Two ascribed to the adequacy of power spectrum density for roll $S_{\phi(t)}$, pitch $S_{\theta(t)}$, and yaw $S_{\psi(t)}$ (similar to the preceding these criteria tried to match one, three, and one lowest frequency components for roll, pitch, and yaw)

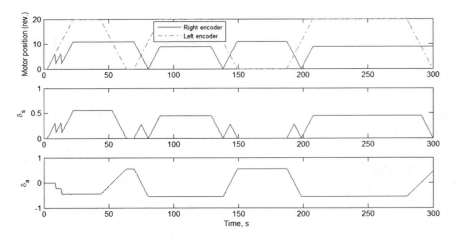

Fig. 11.28   Time histories of right and left motor positions and derived control inputs.

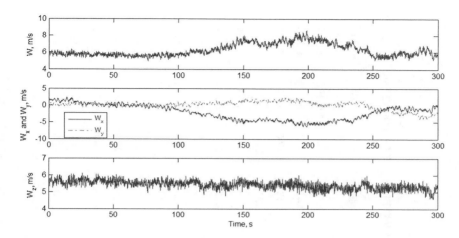

Fig. 11.29   Time histories of wind vector magnitude and its components.

Before we proceed with a discussion of multicriteria identification for the Pegasus PADS, let us think of how the single-criterion approach would be used to identify 33 varied parameters. To this end, Table 11.7 summarizes attempts to find the best values of varied parameters using different gradient-free and

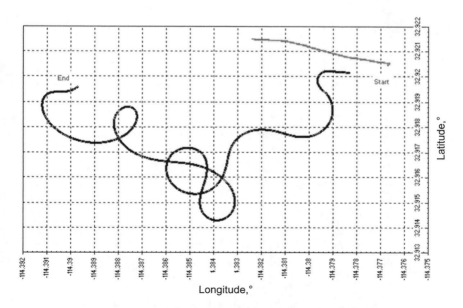

Fig. 11.30   Bird's-eye view of the Pegasus and dropsonde trajectories.

TABLE 11.7　VECTORS OF VARIED PARAMETERS CORRESPONDING TO THE BEST SOLUTION WITH RESPECT TO EACH CRITERION

Variable	Nom. value	Set 0	Set 1	Set 2	Set 3	Set 4	Set 5	Set 6	Set 7	Set 33a	Set 33b
Δt_c	1	1	1	0.240	0.240	0.240	0.240	0.288	0.287	0.495	0.350
x_{CG}	0	0	0	0	0	0.386	0.386	0.463	0.597	0.547	0.535
$k_{\alpha0}$	10	1	1	1	1	1	0.936	1.139	1.174	1.241	1.256
$k_{CL\alpha}$	0.0375	1	1	1	1	1	1.002	0.913	0.967	0.928	0.941
$k_{CL\delta_s}$	0.2	1	1	1	1	1	1.070	1.137	1.271	1.771	1.769
k_{CD_0}	0.14	1	1	1	1	1	1.062	0.867	1.049	0.963	0.961
k_{A_2}	0.25	1	1	1	1	1	0.995	0.801	0.448	0.635	0.630
$k_{A\delta_s}$	0.2	1	1	1	1	1	1.064	1.200	0.759	0.414	0.406
k_{Cm_0}	−0.33	1	1	1	1	1	1.049	0.800	1.111	0.937	0.941
k_{rig}	8	1	1	1	1	1	1	1.052	0.585	0.959	0.957
k_{Cmq}	−6.39	1	1	1	1	1	0.894	0.800	1.226	0.588	0.584
$k_{CY\beta}$	−0.005	1	1	1	1	1	1	0.800	1.180	0.461	0.443
$k_{CY\beta}^{\alpha}$	−0.0001	1	1	1	1	1	1	1.200	0.860	1.441	1.451
$k_{CY\delta_a}$	−0.007	1	1	1	1	1	1	1.200	0.226	1.133	1.282
$k_{CY\delta_a}^{\alpha}$	0.0012	1	1	1	1	1	1	1.200	1.401	1.215	1.341

$k_{Cl\beta}$	−0.0014	1	1	1	1	1	0.957	1.367	0.457	0.463
$k_{Cl\beta}^{\alpha}$	−0.001	0	0	0	0	0	0	0	0	0
$k_{Cl\delta_a}$	−0.0063	1	1	1	1	1	1.200	0.745	0.530	0.622
$k_{Cl\delta_a}^{\alpha}$	−0.001	0	0	0	0	0	0	0	0	0
k_{Clp}	−0.15	1	1	**1.077**	1.077	1.077	**0.861**	1.348	1.061	1.071
k_{Clr}	0.0775	1	1	1	1	1.004	**0.803**	0.688	0.562	0.565
$k_{Cn\beta}$	0.007	1	1	1	1	1	**0.800**	0.575	0.473	0.470
$k_{Cn\beta}^{\alpha}$	−0.0003	1	1	1	1	1	**1.200**	1.089	0.771	0.774
$k_{Cn\delta_a}$	0.03	1	**0.943**	0.943	0.943	0.943	**0.987**	0.892	0.964	0.966
$k_{Cn\delta_a}^{\alpha}$	−0.001	1	**0.501**	0.501	0.501	0.501	**0.401**	0.344	0.644	0.541
k_{Cnp}	0.023	1	1	1	1	0.990	**1.188**	1.519	1.025	1.022
k_{Cnr}	−0.0936	1	1	1	1	**1.001**	**1.201**	1.155	1.106	1.109
k_A	0.899	1	1	1	1	1	1	—	0.539	0.548
k_B	0.339	1	1	1	1	1	1	—	1.383	1.385
k_C	0.783	1	1	1	1	1	1	—	1.259	1.279
k_A^*	0.63	**1.427**	1.427	1.427	1.427	1.427	1.427	1.430	1.799	1.760
k_B^*	0.817	**0.415**	0.415	0.415	0.415	0.415	0.415	**0.410**	0.498	0.525
k_C^*	1	**0.783**	0.783	0.783	0.783	0.783	0.783	**0.780**	0.403	0.202
Method	fminHJ	fminHJ	fminbnd	fminHJ	fminHJ	fmincon [0.8;1.2]		fminsearch		fminHJ
Relative cost	8.843	3.576	3.436	3.342	3.288	3.133	1.534	1.076	1.077	1.000

gradient-based methods available in the MATLAB Optimization toolbox. A single criteria used for this trial was

$$\int_0^{t_f} \sqrt{[x^m(t) - x^{\exp}(t)]^2 + [y^m(t) - y^{\exp}(t)]^2 + [z^m(t) - z^{\exp}(t)]}dt \qquad (11.28)$$

Obviously, with so many varied parameters it was not very easy to set up even initial conditions for all varied parameters, and without it the optimization problem would either diverge or converge to a local minimum. The initial runs not even included all 33 parameters but only considered just a few (shown in bold). Thus, Table 11.7 presents several consecutive quasi-optimal solutions the identification problem converges to. The values of the cost function [adequacy criterion (11.28)] also shown at the bottom of Table 11.7 are normalized to the value of Set 33b.

As seen, the last three sets exhibit about the same cost (the trajectories are indistinguishably closed to each other), but the values of varied parameters are quite different (sometime as much as twofold). That means that having so many varied parameters we are probably overlooking some correlation between them that cannot possibly be established with a single-criterion approach. Figure 11.31 represents the spread of varied parameters [with respect to their nominal (unity) values] for these three sets graphically. After several preliminary runs, the range for all varied parameters for the constrained optimization (the MATLAB function *fmincon* was utilized to obtain Set 5) was established as ([0.2;2]). Even though *fminsearch* and *fminHJ* (another gradient-free function

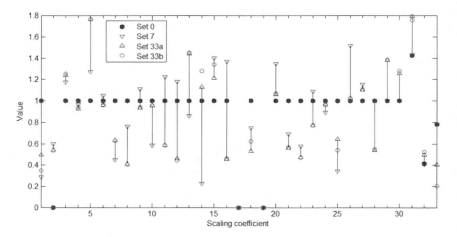

Fig. 11.31   Graphical representation of parameters spread.

described in [Yakimenko 2011]) did not use these constraints, the varied parameters spread seemed to stay within about the same limits ([0.2;1.8]).

All computations were carried on different PCs with the values of parameter and function tolerance of 0.01 and 0.001, respectively. One Simulink model call required 106 s on a 1-GHz PC and 96 s on 1.8-GHz PC. To give a sense of algorithm convergence properties, let us also mention computational performance for just two sets. While Set 5 with only nine varied parameters on a 1-Ghz PC utilizing a preconditioned conjugate gradient method for constrained optimization took 38 h to converge involving 1,276 Simulink model calls (that's where 32% of the time was spent), Set 33a with 33 varied parameters utilizing a nongradient optimization method on a 1.8-GHz PC required 104.4 h and involved 3,935 Simulink model runs (43% of a CPU time).

To summarize, with a tremendous amount of the required CPU resources, the single-criteria identification with dozens of varied parameters can converge to some local minimum, but the solution does not really give an insight of what is going on.

As an alternative to such an approach, Table 11.8 presents 12 feasible solutions (vectors of varied parameters) that were chosen as the best with respect to each criteria during multicriteria identification as described in Sec. 11.4.2 with "just" 1,024 sampling points. [Even though the original number of sample points was 1,024, the set of feasible (and Pareto-optimal) solutions included much fewer trials because of criterial constraints applied to some of the state variables (angle of attack, speed vector components, altitude).] The range for each varied parameter was limited to [0.7;1.3] (except for x_{CG}). Table 11.9 presents the values of all eight criteria for the collection of vectors Ξ shown in Table 11.8. The numbers shown in Table 11.8 represent the values normalized with respect to the minimum value for each corresponding criteria (among all values within each column) so that the value of 1 represents the best solution for each column (these cells are blocked with a heavy border). These vectors of varied parameters can be further used as an initial guess for a single-criterion identification to further tune the scaling factors to satisfy a specific criterion. As clearly seen from Table 11.8 however, optimizing with respect to one criterion does not automatically lead to the best or even acceptable solution with respect to another criterion.

Figure 11.32 shows the real drop trajectory, prototype trajectory with all scaling factors equal to unity ($x_{CG} = 0$), that is, obtained for Ξ_0, and five other trajectories that were found to be the best with respect to the first two and the last three adequacy criteria, that is, obtained for the vectors Ξ_{580}, Ξ_{369}, Ξ_{1005}, Ξ_{623}, and Ξ_{431}.

As seen from Fig. 11.32, all trajectories are somewhat close to each other, which implicitly proves the quality of the developed model. Multicriteria identification allows a deeper understanding of the model. For example, for this particular system the simulated and experimental trajectories' closeness can be achieved for both the six-DoF (Fig. 5.18) model and the eight-DoF model. However, they are the data of Table 11.9 that allow seeing at what cost the trajectory closeness is

TABLE 11.8   COLLECTION OF VECTORS Ξ RESULTING IN A MINIMUM VALUE OF ONE OF EIGHT ADEQUACY CRITERIA

Variable	0	111	140	265	369	431	580	585	623	698	809	1005	Min	Max
Δt_c	1	1.28	0.81	1.04	1.03	1.28	0.78	1.04	1.28	0.92	1.05	1.13	0.78	1.28
x_{CG}	0	-0.62	0.05	0.82	0.79	-0.08	0.38	-1.69	0.14	-1.11	0.96	-1.89	-1.89	0.96
k_{α_0}	1	1.23	0.77	1.21	0.97	1.16	1.19	0.71	0.90	1.05	1.06	1.12	0.71	1.23
$k_{CL\alpha}$	1	0.85	0.78	0.73	1.11	0.75	1.00	0.89	1.16	0.95	0.91	0.88	0.73	1.16
$k_{CL\delta_s}$	1	0.94	0.77	1.24	1.05	0.87	1.24	1.28	1.22	1.27	1.06	0.88	0.77	1.28
k_{CD_0}	1	1.11	0.95	0.81	1.13	0.94	1.16	0.97	1.00	1.29	0.73	0.87	0.73	1.29
k_{A_2}	1	1.13	1.14	1.10	0.86	1.06	0.90	1.17	0.91	0.82	1.29	1.25	0.82	1.29
$k_{A\delta_s}$	1	0.70	1.21	0.93	0.72	0.74	0.80	0.91	0.77	1.18	1.06	1.01	0.70	1.21
k_{Cm_0}	1	0.88	0.83	0.79	1.20	1.00	1.14	0.72	1.09	0.94	0.85	0.76	0.72	1.20
k_{rig}	1	1.03	0.72	0.71	0.83	0.81	1.12	0.78	0.76	0.99	0.88	0.99	0.71	1.12
k_{Cmq}	1	0.92	1.16	0.92	1.06	0.71	0.94	0.90	0.98	1.11	0.83	0.72	0.71	1.16
$k_{CY\beta}$	1	1.00	1.11	1.16	1.03	1.17	1.13	1.24	1.03	0.84	0.70	1.22	0.70	1.24

$k_{CY\beta}^{\alpha}$	1	0.87	1.01	1.26	1.11	0.80	0.81	0.78	1.27	0.85	1.09	0.84	0.78	1.27
$k_{CY\delta_a}$	1	0.79	1.19	1.08	1.22	0.83	1.17	1.28	0.73	1.13	0.77	1.07	0.73	1.28
$k_{CY\delta_a}^{\alpha}$	1	1.27	1.27	0.97	0.85	1.05	1.05	0.78	0.72	1.14	0.71	1.09	0.71	1.27
$k_{Cl\beta}$	1	0.79	1.20	1.13	0.95	1.26	1.03	1.15	1.01	0.75	0.74	0.85	0.74	1.26
$k_{Cl\delta_a}$	1	0.85	1.19	1.04	0.84	1.21	1.24	0.91	1.23	0.74	0.79	1.07	0.74	1.24
k_{Clp}	1	1.12	0.73	0.72	0.96	0.90	1.30	0.96	0.78	1.24	1.14	1.17	0.72	1.30
k_{Clr}	1	0.99	1.21	0.92	1.11	0.95	0.94	1.28	0.71	0.82	1.15	0.93	0.71	1.28
$k_{Cn\beta}$	1	0.98	1.04	1.04	0.74	1.18	1.25	1.14	1.09	1.20	1.14	1.17	0.74	1.25
$k_{Cn\beta}^{\alpha}$	1	0.77	0.82	0.90	0.79	0.77	1.15	1.11	0.84	1.05	0.77	0.84	0.77	1.15
$k_{Cn\delta_a}$	1	1.12	1.23	1.13	1.23	1.25	1.08	0.97	1.05	1.25	1.21	1.26	0.97	1.26
$k_{Cn\delta_a}^{\alpha}$	1	0.85	1.28	0.73	0.99	0.98	0.70	0.74	0.73	0.85	0.71	0.91	0.70	1.28
k_{Cnp}	1	1.19	0.85	0.98	1.20	1.20	1.10	0.99	1.16	1.27	0.82	1.07	0.82	1.27
k_{Cnr}	1	0.94	0.92	1.09	1.05	1.20	1.05	0.94	0.82	1.09	1.02	0.94	0.82	1.20

TABLE 11.9   VALUES OF ALL EIGHT CRITERIA FOR THE COLLECTION OF VECTORS Ξ SHOWN IN TABLE 11.8

Vector ID	Criterion 1 $\int \sqrt{\Delta x^2 + \Delta y^2}\,dt$	Criterion 2 $\int |\Delta z|\,dt$	Criterion 3 $S_{v_x(t)}$	Criterion 4 $S_{v_y(t)}$	Criterion 5 $S_{v_z(t)}$	Criterion 6 $S_{\phi(t)}$	Criterion 7 $S_{\theta(t)}$	Criterion 8 $S_{\psi(t)}$
0	1.51	5.56	1.24	4.33	1.19	64.23	3.18	1.07
111	1.69	7.37	1.51	5.19	1.56	7.33	3.67	1.15
140	2.05	12.98	1.00	3.68	2.11	4.05	3.37	1.00
265	1.44	1.33	1.91	9.44	2.24	42.52	3.69	1.16
369	1.66	1.00	2.96	14.76	4.79	8.97	1.54	1.38
431	1.35	14.07	1.07	1.00	1.24	8.78	3.83	1.00
580	1.00	11.11	1.49	3.54	1.63	63.02	3.42	1.03
585	1.37	4.60	1.61	3.74	1.85	13.32	3.01	1.06
623	1.55	12.98	1.61	6.80	2.18	69.80	1.00	1.23
698	1.38	1.68	1.65	7.93	1.95	1.02	2.98	1.19
809	2.08	1.90	1.45	6.84	1.00	1.96	3.19	1.23
1005	1.77	4.77	1.68	8.80	1.70	1.00	2.98	1.27

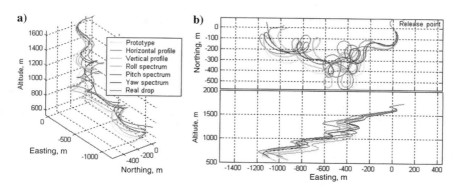

Fig. 11.32 Reference, nominal, and optimal parafoil trajectories with respect to five different criteria.

achieved. Analyze the natural eigenvalues (power spectrum) for Euler angles for both models whose frequencies are missing, thus defining limitations of the six-DoF model vs eight-DoF model.

Another example is a detailed sensitivity analysis. For instance, Fig. 11.33 clearly shows contradictions while trying to minimize different criteria. Figures 11.33a and 11.33b correspond to satisfying the horizontal adequacy criterion while Figs. 11.33c and 11.33d relate to the vertical adequacy criterion (where $k^\alpha_{CY\delta_a}$ is varied in Figs. 11.33a and 11.33c and $k_{Cn\beta}$ in Figs. 11.33a and 11.33c). These plots were obtained for another (larger) set of trial points.

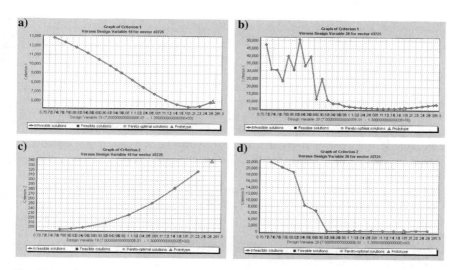

Fig. 11.33 Examples of criteria vs parameter dependencies for a, c) $k^\alpha_{CY\delta_a}$ and b, d) $k_{Cn\beta}$.

TABLE 11.10   SCALING FACTORS WITH A MAGNITUDE OF CORRELATION COEFFICIENT HIGHER THAN 0.51

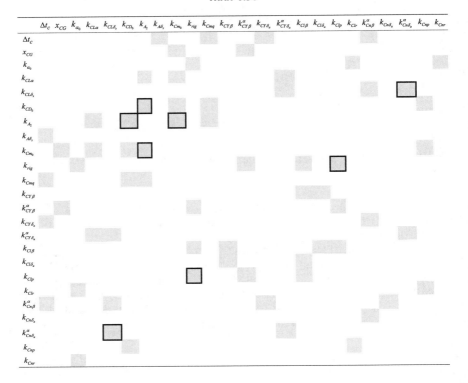

As yet another example, Table 11.10 shows explicit and implicit correlation between some varied parameters revealed while analyzing multicriteria identification data. The shaded cells in Table 11.10 show 0.51 correlation between the corresponding scale factors, and the cells blocked with a heavy border show those with a magnitude of the correlation coefficient exceeding 0.8. A high value of correlation coefficient magnitude means that the relative effect of the systems' performance cannot be distinguished between themselves, that is, they cannot reliably be identified simultaneously. That partially explains the spread of the varied parameters in Fig. 11.31. Fixing some would allow reduction of the original number of varied parameters by about 30% and alternatively (if needed) adding new parameters that were not considered in the original setup without further increase of the identification problem's dimension. Data of Table 11.10 also provide an insight into the set of additional states of the system that might need to be observed and recorded (e.g., relative position of a canopy).

11.5 FILTER-ERROR METHOD

This section presents the filter-error method with the overall computational flow shown in Fig. 11.34. Examples of using this approach include three implementations. First, Secs. 11.5.1–11.5.5 are devoted to the extended Kalman filter (EKF) algorithm that was used to process flight-test data from NASA's X-38 crew return vehicle (CRV) parafoil development program [Rogers 2004]. The lighter version of this approach (involving less error states) has been also implemented on a circular parachute model [Rogers 2002], but in this section, which closely follows [Rogers 2004], the EKF's capabilities are demonstrated for the case when additional sensor data including attitude, angular rate, and air data (angle of attack and sideslip) are available. Last, in Sec. 11.5.6, two more efforts involving applying the recursive weighted least-squares (RWLS) and observer/Kalman filter identification (OKID) algorithms to identify parameters of the linearized reduced-order and linear nonparametric models are presented.

11.5.1 EKF ALGORITHM AND IMPLEMENTATION

The objective of the EKF algorithm implementation was to extract information from X-38 flight data, and then uses these results to modify and improve vehicle flight dynamics simulations that were used to predict system performance. The continuous/discrete form of the Kalman filter algorithm is summarized in Table 11.11 along with the assumptions for this implementation.

The computational flow for this implementation is presented in Fig. 11.35. Inputs to the EKF algorithm are the recorded control deflections and measured atmospheric parameters (air density and winds). The system model described in Table 11.11, as a general nonlinear differential equation, is implemented in the algorithm as mathematical models describing the vehicle six-DoF rigid-body

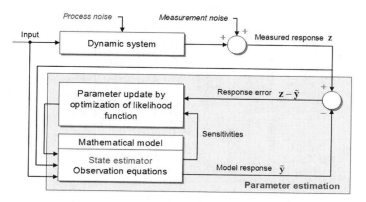

Fig. 11.34 Parameter estimation by the filter-error method.

TABLE 11.11   CONTINUOUS/DISCRETE EXTENDED KALMAN FILTER ALGORITHM

	Equation	Assumptions
System model	$\dot{\mathbf{x}} = \mathbf{f}(\mathbf{x}, \mathbf{u}, t) + \mathbf{w}(t)$	$E[\mathbf{w}(t)\mathbf{w}(\tau)^T] \equiv \mathbf{Q}\delta(t - \tau)$
Observation model	$\mathbf{z}_k = \mathbf{h}_k\mathbf{x}(t_k) + \mathbf{n}_k$	$E[\mathbf{n}_k\mathbf{n}_k^T] \equiv \mathbf{R}_k, \; E[\mathbf{w}\,\mathbf{n}_k^T] \equiv \mathbf{0}$
Initial conditions	$E[\mathbf{x}(t = 0)] = \hat{\mathbf{x}}_0$	$\hat{\mathbf{x}}_0$–usually assumed zero
	$E[(\mathbf{x}_0 - \hat{\mathbf{x}}_0)(\mathbf{x}_0 - \hat{\mathbf{x}}_0)^T] = \hat{\mathbf{P}}_0$	$\hat{\mathbf{P}}_0$–consistent with model
Propagate state	$\dot{\bar{\mathbf{x}}} = \mathbf{f}(\bar{\mathbf{x}}, \mathbf{u}, t)$	Numerically integrated up to observation time instant
Propagate covariance	$\dot{\mathbf{P}}(t) = \mathbf{F}(\mathbf{x}(t), t)\mathbf{P}(t) + \mathbf{P}(t)\mathbf{F}(\mathbf{x}(t), t)^T + \mathbf{Q}(t)$	
Measurement updates	$\hat{\mathbf{x}}_k = \bar{\mathbf{x}}(t_k) + \mathbf{K}_k[\mathbf{z}_k - \mathbf{h}_k(\bar{\mathbf{x}}(t_k))]$	At observation time, updates computed and state/ covariance reinitialized
	$\mathbf{K}_k = \bar{\mathbf{P}}(t_k)\mathbf{H}_k^T[\mathbf{H}_k\bar{\mathbf{P}}(t_k)\mathbf{H}_k^T + \mathbf{R}_k]^{-1}$	
	$\hat{\mathbf{P}}_k = [\mathbf{I} - \mathbf{K}_k\mathbf{H}_k]\bar{\mathbf{P}}(t)_k[\mathbf{I} - \mathbf{K}_k\mathbf{H}_k]^T + \mathbf{K}_k\mathbf{R}_k\mathbf{K}_k^T$	
Linearization		
System	$\mathbf{F}(\bar{\mathbf{x}}(t)) = \dfrac{\partial\mathbf{f}(\mathbf{x}(t), \mathbf{u}, t)}{\partial\mathbf{x}(t)}\bigg|_{\mathbf{x}(t)=\bar{\mathbf{x}}(t)}$	Used in preceding equations
Measurement	$\mathbf{H}_k = \dfrac{\partial\mathbf{h}(\mathbf{x}(t), t)}{\partial\mathbf{x}(t)}\bigg|_{\mathbf{x}(t)=\bar{\mathbf{x}}(t_k)}$	

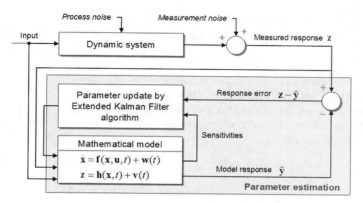

Fig. 11.35   Computational flow of extended Kalman filter algorithm.

dynamics, aerodynamic and mass property parameters. This model represents a reference that is corrected by using estimates generated by the EKF algorithm. (The EKF algorithm tries to estimate mass/inertia properties and aerodynamic/ control coefficients.) Allowances are made in the EKF algorithm to mitigate the effects of payload oscillations present in test data.

The algorithm is implemented for global iteration. In this implementation, observation data are processed in the algorithm as an initial cycle in the iteration procedure. State and variance estimates at the end of this initial cycle are used as initial conditions to reprocess the observation data again in a second cycle. At the end of this second cycle, the EKF estimates are again used to initialize the state and covariance matrix for a third processing cycle. This procedure is repeated until the estimates for the aerodynamic and mass property modeling errors are relatively constant over the data time span for a selected number of processing cycles. In the results to follow, the number of processing cycles is 7.

After the selected number of iteration cycles has been completed, the resulting estimates for the aerodynamic and mass property corrections are used for a final pass through the data. In this cycle, measurement variances are set to unrealistically large values negating the benefits of the measurement updates and effectively producing a simulation of the vehicle dynamics. Results from this simulation pass are compared to the sensor observation data used in the previous filter passes. This comparison is one of the ways the EKF capabilities are assessed. The other assessment is based on an examination of how consistently the EKF estimated the corrections necessary to the reference model.

11.5.2 EQUATIONS OF MOTION AND X-38 PADS AERODYNAMICS MODEL

To simplify the equations of motion and their linearization, the vehicle axis system is assumed to be located at its c.g. (see Fig. 11.36). The following four

Fig. 11.36   Aerodynamics wind and body coordinate frames.

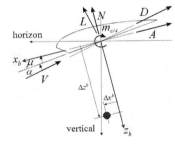

matrix equations are used to describe PADS kinematic and dynamics. The equations describing this motion are vector/matrix differential equations for inertial position, body relative velocity, body-to-inertial direction cosine matrix, and body angular rate:

$$\textit{Position:} \quad \dot{\mathbf{p}} = \mathbf{R}\mathbf{v} \tag{11.29}$$

$$\textit{Velocity:} \quad \dot{\mathbf{v}} = -\mathbf{M}^{-1}\mathbf{S}(\boldsymbol{\omega})\mathbf{M}\mathbf{v} + \mathbf{M}^{-1}[\mathbf{f} + m\mathbf{R}^{T}\mathbf{g}] \tag{11.30}$$

$$\textit{Attitude:} \quad \dot{\mathbf{R}} = \mathbf{R}\mathbf{S}(\boldsymbol{\omega}) \tag{11.31}$$

$$\textit{Angular rate:} \quad \dot{\boldsymbol{\omega}} = -\mathbf{I}^{-1}\mathbf{S}(\boldsymbol{\omega})\mathbf{I}\boldsymbol{\omega} + \mathbf{I}^{-1}\mathbf{T} \tag{11.32}$$

In these equations, $\mathbf{p} = [x, y, z]^{T}$ is the position of origin of $\{b\}$ expressed in $\{I\}$ (for clarity of the following derivations, we dropped all subscripts denoting the reference coordinate frame), $\mathbf{R}$ denotes a transpose of the rotation matrix ${}^{n}_{b}\mathbf{R}$ [introduced in Eq. (5.10)], $\mathbf{v}$ represents the ground-speed vector (as measured by GPS) of origin of $\{b\}$ expressed in $\{b\}$, $\boldsymbol{\omega}$ denotes the vector of the angular rates of $\{b\}$ with respect to $\{I\}$ expressed in $\{b\}$, $\mathbf{S}(\boldsymbol{\omega})$ is the skew-symmetric matrix [see Eq. (5.21)], and $\mathbf{g} = [0, 0, g]^{T}$ is the gravity vector expressed in $\{I\}$. The mass matrix $\mathbf{M}$ and moment of inertia matrix $\mathbf{I}$ are defined as

$$\mathbf{M} = \begin{bmatrix} m_{x} & 0 & 0 \\ 0 & m_{y} & 0 \\ 0 & 0 & m_{z} \end{bmatrix} = \begin{bmatrix} m + \Delta m_{x} & 0 & 0 \\ 0 & m + \Delta m_{y} & 0 \\ 0 & 0 & m + \Delta m_{z} \end{bmatrix}$$

$$\left(\text{so that } \delta\mathbf{M} = \begin{bmatrix} \delta m_{x} & 0 & 0 \\ 0 & \delta m_{y} & 0 \\ 0 & 0 & \delta m_{z} \end{bmatrix} \right) \tag{11.33}$$

and

$$\mathbf{I} = \begin{bmatrix} I_{x} & 0 & J \\ 0 & I_{y} & 0 \\ J & 0 & I_{z} \end{bmatrix} \left(\text{so that } \delta\mathbf{I} = \begin{bmatrix} \delta I_{x} & 0 & \delta J \\ 0 & \delta I_{y} & 0 \\ \delta J & 0 & \delta I_{z} \end{bmatrix} \right) \tag{11.34}$$

The errors Δm_{i} in the mass matrix are attributed to apparent mass. Error matrices $\delta\mathbf{M}$ and $\delta\mathbf{I}$ include nominal mass and mass distribution errors plus those associated with apparent mass (for matrix $\delta\mathbf{M}$ only, because the uncertainties associated with the material mass distribution and resulting moments of inertia do not justify modeling apparent mass for the inertia terms separately).

Note, compared to abbreviations used in Chapter 5, abbreviation **F** now denotes a Jacobian matrix of the system equation as shown in Table 11.11, and **M** denotes the mass matrix [Eq. (11.33)]. That is why we use **f** and **T** to represent the vectors of applied body-referenced aerodynamic force and moment, respectively.

We model the aerodynamic force vector **f** (expressed in $\{b\}$) as

$$\mathbf{f} = 0.5\rho|\Delta\mathbf{v}|^2 S\mathbf{C}_f \tag{11.35}$$

where the airspeed vector $\Delta\mathbf{v} = [v_x, v_y, v_z]^T$ (expressed in $\{b\}$) can be obtained by subtracting the wind vector **W** (expressed in $\{I\}$) converted to $\{b\}$ using the rotation matrix **R** [see Eq. (5.78)]

$$\Delta\mathbf{v} = \mathbf{v} - \mathbf{R}^T\mathbf{W} \tag{11.36}$$

In Eq. (11.35), $\mathbf{C}_f = [C_A, C_Y, C_N]^T$ is the vector of aerodynamic coefficients (expressed in $\{b\}$) computed using angles of attack and sideslip expressed via components of $\Delta\mathbf{v}$ [see Eq. (5.77)] and air density is given by either accurate formula for a nonisothermal troposphere [Eq. (3.12)] or its approximation throughout multiple atmospheric layers [Eq. (3.33)]

$$\rho = \rho_0 e^{-h/H_*} \tag{11.37}$$

Similarly, the aerodynamic force moment $\mathbf{T}_{c/4}$ about the $m_{c/4}$ point (Fig. 11.36) (expressed in $\{b\}$) is given by

$$\mathbf{T}_{c/4} = \frac{1}{2}\rho|\Delta\mathbf{v}|^2 S \begin{bmatrix} b & 0 & 0 \\ 0 & c & 0 \\ 0 & 0 & b \end{bmatrix} \mathbf{C}_m = QS\mathbf{v}\mathbf{C}_m \tag{11.38}$$

where $\mathbf{C}_m = [C_l, C_m, C_n]^T$ denotes the vector of aerodynamic moments (expressed in $\{b\}$). Moment $\mathbf{T}_{c/4}$ needs to be translated to the PADS c.g. to yield

$$\mathbf{T} = \mathbf{T}_{c/4} + \mathbf{S}(\mathbf{f})\Delta\mathbf{r} = \mathbf{T}_{c/4} - \mathbf{S}(\Delta\mathbf{r})\mathbf{f} \tag{11.39}$$

where $\Delta\mathbf{r} = [\Delta x^b, \Delta y^b, \Delta z^b]^T$.

The structure of the reference mathematical model of X-38 parafoil aerodynamics was determined to define the vectors $\mathbf{C}_f$ and $\mathbf{C}_m$ via the following coefficients of aerodynamic and control derivatives (in $\{b\}$):

$$C_A = C_{A0} + C_{A\alpha}\alpha + C_{A\delta_s}\delta_s + C_{A|\delta_a|}|\delta_a| + C_{A\alpha\delta_s}\alpha\delta_s + C_{A\delta_s^2}\delta_s^2 + C_{A\delta_a^2}\delta_a^2 \tag{11.40}$$

$$C_Y = C_{Y\beta}\beta + C_{Y\delta_a}\delta_a + C_{Y\delta_s\delta_a}\delta_s\delta_a + C_{Y|\delta_a|\delta_a}|\delta_a|\delta_a \tag{11.41}$$

$$C_N = C_{N0} + C_{N\alpha}\alpha + C_{N\delta_s}\delta_s + C_{N\alpha\delta_s}\alpha\delta_s + C_{N\delta_s^2}\delta_s^2 + C_{N|\delta_a|}|\delta_a| \quad (11.42)$$

$$C_l = C_{l\beta}\beta + C_{lp}\frac{b}{2V_a}p + C_{lr}\frac{b}{2V_a}r + C_{l\delta_a}\delta_a + C_{l\delta_s\delta_a}\delta_s\delta_a + C_{l|\delta_a|\delta_a}|\delta_a|\delta_a \quad (11.43)$$

$$C_m = C_{m0} + C_{m\alpha}\alpha + C_{mq}\frac{c}{2V_a}q + C_{m\delta_s}\delta_s + C_{m\alpha\delta_s}\alpha\delta_s + C_{m\delta_s^2}\delta_s^2 + C_{m|\delta_a|}|\delta_a|$$

$$(11.44)$$

$$C_n = C_{n\beta}\beta + C_{np}\frac{b}{2V_a}p + C_{nr}\frac{b}{2V_a}r + C_{n\delta_a}\delta_a + C_{n\delta_s\delta_a}\delta_s\delta_a + C_{n|\delta_a|\delta_a}|\delta_a|\delta_a$$

$$+ C_{n\delta_s|\delta_a|\delta_a}\delta_s|\delta_a|\delta_a \quad (11.45)$$

As seen, compared to the traditional model [e.g., Eqs. (5.74) and (5.79)], the aerodynamic force and moment coefficients (11.40–11.45) have additional higher-order terms (HOT) to describe the control effectiveness and interaction of the asymmetric and symmetric flap deflections at high deflection angles. Such a structure was suggested after [Iacomini and Cerimeli 1999a; Iacomini and Cerimeli 1999b], where the control portion of the yawing moment C_n and longitudinal coefficients C_A, C_N, and C_m had even more complex form, specifically

$$\Delta C_n = (c_1 + c_2\delta_s + c_3\delta_s^2)\delta_a + (c_4 + c_5\delta_s + c_6\delta_s^2)|\delta_a|\delta_a \quad (11.46)$$

and

$$\Delta C_{A/N/m} = (c_1 + c_2\delta_s + c_3\delta_s^2 + c_4\delta_s^3)\delta_a + (c_5 + c_6\delta_s + c_7\delta_s^2 + c_8\delta_s^3)\alpha \quad (11.47)$$

These dependences were constructed in such a way because of flight data from [Jessick 1990a; Jessick 1990b] showing little change in coefficients C_A, C_N, C_m, and C_n, with small flap settings, then a rapid buildup for midrange flap values, and diminished effectiveness with large values.

The nominal (reference) values for the coefficients of aerodynamic and control derivatives in Eqs. (11.40–11.45), to be corrected by the EKF algorithm, were taken from [Jessick 1990a; Jessick 1990b; Iacomini and Cerimeli 1999a; Iacomini and Cerimeli 1999b], as shown in Table 11.12.

11.5.3 MODEL LINEARIZATION

According to the algorithm outlined in Table 11.11, we need to linearize Eqs. (11.29–11.32) to establish dynamics of the system errors for the position, velocity, attitude, and angular rate. Let us start from the translational kinematics Eq. (11.29). We can write

$$\delta\dot{\mathbf{p}} = \delta\mathbf{R}\mathbf{v} + \mathbf{R}\delta\mathbf{v} \quad (11.48)$$

TABLE 11.12   REFERENCE VALUES OF AERODYNAMIC AND CONTROL DERIVATIVES

	0	α	β	p	q	r	δ_s	δ_a	$\alpha\delta_s$	δ_s^2	$\delta_s\delta_a$	δ_a^2	$\delta_s\|\delta_a\|\delta_a$
C_A	0.0923	−0.00134	–	–	–	–	0.18	–	−0.007	0.173	–	0	–
C_N	0.194	0.035	–	–	–	–	0.24		0.054	0.153	–	–	–
C_m	−0.00474	−0.00442	–	–	−6.39	–	−0.093	–	0.00021	−0.123	–	–	–
C_Y	0	–	−0.0048	–	–	–	–	0	–	–	0	0	–
C_l	0	–	−0.0001	−0.150	–	0.0775	–	0	–	–	0	0	–
C_n	0	–	0.0010	0.023	–	−0.0936	–	−0.0303	–	–	0.195	0.0758	−0.226

Now, recall Eq. (5.10) for the rotation matrix $_n^b\mathbf{R}$. For the small Euler angles, this equation reduces to

$$_n^b\mathbf{R} = \begin{bmatrix} 1 & \Delta\psi & -\Delta\theta \\ -\Delta\psi & 1 & \Delta\phi \\ \theta & -\Delta\phi & 1 \end{bmatrix} \qquad (11.49)$$

or using the notation of Eq. (5.21) for a skew-symmetric matrix

$$_n^b\mathbf{R} = \mathbf{I} - \mathbf{S}(\boldsymbol{\varphi}) \qquad (11.50)$$

where $\boldsymbol{\varphi}$ represents the vector of attitude errors $\boldsymbol{\varphi} = [\Delta\phi, \Delta\theta, \Delta\psi]^T$.

Hence, the true rotation matrix $\mathbf{R}^{\text{true}} = {}_b^n\mathbf{R}^{\text{true}}$ can be obtained by an additional rotation with respect to the direction cosine matrix $\mathbf{R}$ obtained by integrating Eq. (11.35)

$$\mathbf{R}^{\text{true}} = (\mathbf{I} + \mathbf{S}(\boldsymbol{\varphi}))\mathbf{R} \qquad (11.51)$$

From this equation, we derive an expression for $\delta\mathbf{R}$ as

$$\delta\mathbf{R} = \mathbf{R}^{\text{true}} - \mathbf{R} = \mathbf{S}(\boldsymbol{\varphi})\mathbf{R} \qquad (11.52)$$

Plugging Eq. (11.52) back into Eq. (11.29) finally results in

$$\delta\dot{\mathbf{p}} = \mathbf{S}(\boldsymbol{\varphi})\mathbf{R}\mathbf{v} + \mathbf{R}\delta\mathbf{v} = -\mathbf{S}(\mathbf{R}\mathbf{v})\boldsymbol{\varphi} + \mathbf{R}\delta\mathbf{v} \qquad (11.53)$$

[Equation (11.53) utilizes an equality $\mathbf{S}(\mathbf{a})\mathbf{b} = -\mathbf{S}(\mathbf{b})\mathbf{a}$.]

Let us proceed with rotational kinematics described by Eq. (11.31). Taking a variation of both sides results in

$$\delta\dot{\mathbf{R}} = \delta\mathbf{R}\mathbf{S}(\boldsymbol{\omega}) + \mathbf{R}\delta\mathbf{S}(\boldsymbol{\omega}) \qquad (11.54)$$

Substituting Eq. (11.52) yields

$$\delta\dot{\mathbf{R}} = \mathbf{S}(\boldsymbol{\varphi})\mathbf{R}\mathbf{S}(\boldsymbol{\omega}) + \mathbf{R}\mathbf{S}(\delta\boldsymbol{\omega}) \qquad (11.55)$$

Now, differentiating both parts of equation Eq. (11.52) results in

$$\delta\dot{\mathbf{R}} = \mathbf{S}(\dot{\boldsymbol{\varphi}})\mathbf{R} + \mathbf{S}(\boldsymbol{\varphi})\dot{\mathbf{R}} \qquad (11.56)$$

Equating Eqs. (11.55) and (11.56) and substituting the original attitude kinematics (11.31) for $\dot{\mathbf{R}}$ give

$$\mathbf{S}(\boldsymbol{\varphi})\mathbf{R}\mathbf{S}(\boldsymbol{\omega}) + \mathbf{R}\mathbf{S}(\delta\boldsymbol{\omega}) = \mathbf{S}(\dot{\boldsymbol{\varphi}}_I)\mathbf{R} + \mathbf{S}(\boldsymbol{\varphi})\mathbf{R}\mathbf{S}(\boldsymbol{\omega}) \qquad (11.57)$$

From here we arrive at

$$\mathbf{S}(\dot{\boldsymbol{\varphi}}) = \mathbf{R}\mathbf{S}(\delta\boldsymbol{\omega})\mathbf{R}^{-1} = \mathbf{S}(\delta\boldsymbol{\omega}^I) \qquad (11.58)$$

[Equation (11.58) utilizes an equality $S(\boldsymbol{\omega}^I) = S(^n_b R \boldsymbol{\omega}) = ^n_b R S(\boldsymbol{\omega})^n_b R^T.$]
Finally, we can write Eq. (11.58) in the vector form

$$\dot{\boldsymbol{\varphi}} = R \delta \boldsymbol{\omega} \tag{11.59}$$

Next, let us address the velocity equation (11.30). The error dynamics is expressed by

$$\delta \dot{\mathbf{v}} = -\delta M^{-1} S(\boldsymbol{\omega}) M \mathbf{v} - M^{-1} \delta S(\boldsymbol{\omega}) M \mathbf{v} - M^{-1} S(\boldsymbol{\omega}) \delta M \mathbf{v} - M^{-1} S(\boldsymbol{\omega}) M \delta \mathbf{v}$$

$$+ \delta M^{-1} [\mathbf{f} + m R^T \mathbf{g}] + M^{-1} \delta \mathbf{f} + M^{-1} m \delta R^T \mathbf{g} + M^{-1} m R^T \delta \mathbf{g} \tag{11.60}$$

The sum of the first and third terms on the right-hand side of this equation reduces to

$$-\delta M^{-1} S(\boldsymbol{\omega}) M \mathbf{v} - M^{-1} S(\boldsymbol{\omega}) \delta M \mathbf{v}$$

$$= \begin{bmatrix} (rm_y v - qm_z w)m_x^{-2} & -rvm_x^{-1} & qwm_x^{-1} \\ rvm_y^{-1} & (pm_z w - rm_x u)m_y^{-2} & -pwm_y^{-1} \\ -qum_z^{-1} & pvm_z^{-1} & (qm_x u - pm_y v)m_z^{-2} \end{bmatrix}$$

$$\times \begin{bmatrix} \delta m_x \\ \delta m_y \\ \delta m_z \end{bmatrix} \tag{11.61}$$

The remaining terms are as follows:

$$-M^{-1} \delta S(\boldsymbol{\omega}) M \mathbf{v} = M^{-1} S(M \mathbf{v}) \delta \boldsymbol{\omega}$$

$$= - \begin{bmatrix} 0 & m_z w m_x^{-1} & -m_y v m_x^{-1} \\ -m_z w m_y^{-1} & 0 & m_x u m_y^{-1} \\ m_y v m_z^{-1} & -m_x u m_z^{-1} & 0 \end{bmatrix} \delta \boldsymbol{\omega} \tag{11.62}$$

$$-M^{-1} S(\boldsymbol{\omega}) M \delta \mathbf{v} = \begin{bmatrix} 0 & m_y r m_x^{-1} & -m_z q m_x^{-1} \\ -m_x r m_y^{-1} & 0 & m_z p m_y^{-1} \\ m_x q m_z^{-1} & -m_y p m_z^{-1} & 0 \end{bmatrix} \delta \mathbf{v} \tag{11.63}$$

$$-M^{-1} [\delta \mathbf{f} + m R^T \delta \mathbf{g}] = - \begin{bmatrix} m_x^{-1} & 0 & 0 \\ 0 & m_y^{-1} & 0 \\ 0 & 0 & m_z^{-1} \end{bmatrix} [\delta \mathbf{f} + m R^T \delta \mathbf{g}] \tag{11.64}$$

$$-\delta \mathbf{M}^{-1}\mathbf{a} = \mathbf{M}^{-2}\delta \mathbf{m}\mathbf{a} = \mathbf{M}^{-2}\mathrm{diag}(\mathbf{a})\delta \mathbf{m}$$

$$= \begin{bmatrix} a_1 m_x^{-2} & 0 & 0 \\ 0 & a_2 m_y^{-2} & 0 \\ 0 & 0 & a_3 m_z^{-2} \end{bmatrix} \begin{bmatrix} \delta m_x \\ \delta m_y \\ \delta m_z \end{bmatrix} \qquad (11.65)$$

[in Eq. (11.65), $\mathbf{a} = \mathbf{f} + m\mathbf{R}^T\mathbf{g}$], and

$$-\mathbf{M}^{-1}m\delta \mathbf{R}^T\mathbf{g} = \mathbf{M}^{-1}m\mathbf{R}^T\mathbf{S}(\boldsymbol{\varphi})\mathbf{g} = -\mathbf{M}^{-1}m\mathbf{R}^T\mathbf{S}(\mathbf{g})\boldsymbol{\varphi}$$

$$= mg \begin{bmatrix} m_x^{-1} & 0 & 0 \\ 0 & m_y^{-1} & 0 \\ 0 & 0 & m_z^{-1} \end{bmatrix} \begin{bmatrix} -r_{21} & r_{11} & 0 \\ -r_{22} & r_{12} & 0 \\ -r_{23} & r_{13} & 0 \end{bmatrix} \boldsymbol{\varphi} \qquad (11.66)$$

[Here we used Eq. (11.52) while r_{ij} are the elements of the first two rows of rotation matrix $\mathbf{R}$.]

Now let us find variations of aerodynamic and gravitational forces in Eq. (11.64). The variation of the aerodynamic force vector given by Eq. (11.35) can be expressed as

$$\delta \mathbf{f} \equiv \frac{\partial \mathbf{f}}{\partial \mathbf{p}} \delta \mathbf{p} + \frac{\partial \mathbf{f}}{\partial \mathbf{v}} \delta \mathbf{v} + \frac{\partial \mathbf{f}}{\partial \boldsymbol{\varphi}} \delta \boldsymbol{\varphi} + \frac{\partial \mathbf{f}}{\partial \mathbf{W}} \delta \mathbf{W} + \frac{\partial \mathbf{f}}{\partial \mathbf{C}_f} \delta \mathbf{C}_f \qquad (11.67)$$

In Eq. (11.67) the first two terms can be represented as

$$\frac{\partial \mathbf{f}}{\partial \mathbf{p}} = \frac{\partial \mathbf{f}}{\partial \rho} \frac{\partial \rho}{\partial h} \frac{\partial h}{\partial \mathbf{p}} \qquad (11.68)$$

where

$$\frac{\partial \mathbf{f}}{\partial \rho} = \frac{1}{2} |\Delta \mathbf{v}|^2 S \mathbf{C}_f, \quad \frac{\partial \rho}{\partial h} = -\frac{\rho}{H_s}, \quad \frac{\partial h}{\partial \mathbf{p}} = [0, \ 0, \ -1] \qquad (11.69)$$

and

$$\frac{\partial \mathbf{f}}{\partial \mathbf{v}} = \frac{\partial \mathbf{f}}{\partial \Delta \mathbf{v}} \frac{\partial \Delta \mathbf{v}}{\partial \mathbf{v}} = \frac{\partial \mathbf{f}}{\partial \Delta \mathbf{v}} \mathbf{I}_{3 \times 3} = \frac{\partial \mathbf{f}}{\partial \Delta \mathbf{v}} \qquad (11.70)$$

where

$$\frac{\partial \mathbf{f}}{\partial \Delta \mathbf{v}} = \frac{\partial \mathbf{f}}{\partial Q} \frac{\partial Q}{\partial \Delta \mathbf{v}} + \frac{\partial \mathbf{f}}{\partial \alpha} \frac{\partial \alpha}{\partial \Delta \mathbf{v}} + \frac{\partial \mathbf{f}}{\partial \beta} \frac{\partial \beta}{\partial \Delta \mathbf{v}} \qquad (11.71)$$

In turn, in the last equation,

$$\frac{\partial \mathbf{f}}{\partial Q} = S \mathbf{C}_f, \quad \frac{\partial \mathbf{f}}{\partial \alpha} = S \frac{\partial \mathbf{C}_f}{\partial \alpha}, \quad \frac{\partial \mathbf{f}}{\partial \beta} = S \frac{\partial \mathbf{C}_f}{\partial \beta} \qquad (11.72)$$

and

$$\frac{\partial Q}{\partial \Delta \mathbf{v}} = \rho (\Delta \mathbf{v})^T, \quad \frac{\partial \alpha}{\partial \Delta \mathbf{v}} = \left[\frac{-v_z}{v_x^2 + v_z^2}, 0, \frac{v_x}{v_x^2 + v_z^2} \right]$$

$$\frac{\partial \beta}{\partial \Delta \mathbf{v}} = \frac{1}{|\Delta \mathbf{v}|^2} \left[\frac{-v_x v_y}{\sqrt{v_x^2 + v_z^2}}, \sqrt{v_x^2 + v_z^2}, \frac{-v_y v_z}{\sqrt{v_x^2 + v_z^2}} \right]$$

(11.73)

The third, fourth, and fifth terms in Eq. (11.67) are

$$\frac{\partial \mathbf{f}}{\partial \boldsymbol{\varphi}} = \frac{\partial \mathbf{f}}{\partial \Delta \mathbf{v}} \frac{\partial \Delta \mathbf{v}}{\partial \boldsymbol{\varphi}}, \quad \frac{\partial \mathbf{f}}{\partial \mathbf{W}} = \frac{\partial \mathbf{f}}{\partial \Delta \mathbf{v}} \frac{\partial \Delta \mathbf{v}}{\partial \mathbf{W}}, \quad \frac{\partial \mathbf{f}}{\partial \mathbf{C}_f} = QS\mathbf{I}_{3 \times 3}$$

(11.74)

where accounting for Eqs. (11.36) and (11.54)

$$\frac{\partial \Delta \mathbf{v}}{\partial \boldsymbol{\varphi}} = -\frac{\partial (\mathbf{R}^T \mathbf{W})}{\partial \boldsymbol{\varphi}} = -\mathbf{R}^T \mathbf{S}(\mathbf{W}), \quad \frac{\partial \Delta \mathbf{v}}{\partial \mathbf{W}} = -\mathbf{R}^T$$

(11.75)

and $\partial \mathbf{f} / \partial \Delta \mathbf{v}$ was determined in Eq. (11.71).

Finally, let us linearize the angular rate equation (11.32). The error dynamics is governed by

$$\delta \dot{\boldsymbol{\omega}} = -\delta \mathbf{I}^{-1} \mathbf{S}(\boldsymbol{\omega}) \mathbf{I} \boldsymbol{\omega} - \mathbf{I}^{-1} \delta \mathbf{S}(\boldsymbol{\omega}) \mathbf{I} \boldsymbol{\omega} - \mathbf{I}^{-1} \mathbf{S}(\boldsymbol{\omega}) \delta \mathbf{I} \boldsymbol{\omega} - \mathbf{I}^{-1} \mathbf{S}(\boldsymbol{\omega}) \mathbf{I} \delta \boldsymbol{\omega}$$
$$+ \delta \mathbf{I}^{-1} \mathbf{T} + \mathbf{I}^{-1} \delta \mathbf{T}$$

(11.76)

In Eq. (11.76), the sum of the first, third, and fifth terms

$$-\delta \mathbf{I}^{-1} \mathbf{S}(\boldsymbol{\omega}) \mathbf{I} \boldsymbol{\omega} - \mathbf{I}^{-1} \mathbf{S}(\boldsymbol{\omega}) \delta \mathbf{I} \boldsymbol{\omega} + \delta \mathbf{I}^{-1} \mathbf{T} = \delta \mathbf{I}^{-1} (-\mathbf{S}(\boldsymbol{\omega}) \mathbf{I} \boldsymbol{\omega} + \mathbf{T}) - \mathbf{I}^{-1} \mathbf{S}(\boldsymbol{\omega}) \delta \mathbf{I} \boldsymbol{\omega}$$
$$= -\mathbf{I}^{-2} \delta \mathbf{I} (\mathbf{I} \dot{\boldsymbol{\omega}}) - \mathbf{I}^{-1} \mathbf{S}(\boldsymbol{\omega}) \delta \mathbf{I} \boldsymbol{\omega}$$

(11.77)

and the sum of the second and fourth terms

$$-\mathbf{I}^{-1} \delta \mathbf{S}(\boldsymbol{\omega}) \mathbf{I} \boldsymbol{\omega} - \mathbf{I}^{-1} \mathbf{S}(\boldsymbol{\omega}) \mathbf{I} \delta \boldsymbol{\omega} = \mathbf{I}^{-1} (\mathbf{S}(\mathbf{I} \boldsymbol{\omega}) - \mathbf{S}(\boldsymbol{\omega}) \mathbf{I}) \delta \boldsymbol{\omega}$$

(11.78)

[The scalar form of the sensitivity terms described by Eqs. (11.77) and (11.78) is given in Tables 11.13 and 11.14.]

Similar to Eq. (11.67), recalling Eq. (11.38) for the aerodynamic moments in the angular rate equation (11.32), the torque variation in Eq. (11.76) can be expressed as

$$\delta \mathbf{T} \equiv \frac{\partial \mathbf{T}}{\partial \mathbf{p}} \delta \mathbf{p} + \frac{\partial \mathbf{T}}{\partial \mathbf{v}} \delta \mathbf{v} + \frac{\partial \mathbf{T}}{\partial \boldsymbol{\varphi}} \delta \boldsymbol{\varphi} + \frac{\partial \mathbf{T}}{\partial \boldsymbol{\omega}} \delta \boldsymbol{\omega} + \frac{\partial \mathbf{T}}{\partial \mathbf{W}} \delta \mathbf{W} + \frac{\partial \mathbf{T}}{\partial \mathbf{C}_f} \delta \mathbf{C}_f + \frac{\partial \mathbf{T}}{\partial \mathbf{C}_m} \delta \mathbf{C}_m + \cdots$$

(11.79)

TABLE 11.13　ANGULAR RATE EQUATION LINEARIZATION WITH RESPECT TO THE ANGULAR RATES

	δp	δq	δr
$\delta\dot{p}$	$-\dfrac{J(I_z+I_x-I_y)}{(I_xI_z-J^2)}q$	$-\left[\dfrac{[J^2+I_z(I_z-I_y)]}{(I_xI_z-J^2)}r+\dfrac{J(I_z+I_x-I_y)}{(I_xI_z-J^2)}p\right]$	$-\dfrac{[J^2+I_z(I_z-I_y)]}{(I_xI_z-J^2)}q$
$\delta\dot{q}$	$-\dfrac{(I_x-I_z)}{I_y}r-\dfrac{2J}{I_y}p$	0	$-\dfrac{(I_x-I_z)}{I_y}p-\dfrac{2J}{I_y}r$
$\delta\dot{r}$	$-\dfrac{[I_x(I_y-I_x)-J^2]}{(I_xI_z-J^2)}q$	$-\left[\dfrac{[I_x(I_y-I_x)-J^2]}{(I_xI_z-J^2)}p-\dfrac{J(I_x+I_z-I_y)}{(I_xI_z-J^2)}r\right]$	$\dfrac{J(I_x+I_z-I_y)}{(I_xI_z-J^2)}q$

Here, the first term

$$\frac{\partial \mathbf{T}}{\partial \mathbf{p}}=\frac{\partial \mathbf{T}}{\partial \rho}\frac{\partial \rho}{\partial h}\frac{\partial h}{\partial \mathbf{p}} \tag{11.80}$$

in which

$$\frac{\partial \mathbf{T}}{\partial \rho}=\frac{1}{2}|\Delta\mathbf{v}|^2 S\begin{bmatrix} b & C_l \\ c & C_m \\ b & C_n \end{bmatrix} \tag{11.81}$$

and the remaining derivatives in Eq. (11.80) have been determined in Eq. (11.69) already.

The second term in Eq. (11.79)

$$\frac{\partial \mathbf{T}}{\partial \mathbf{v}}=\frac{\partial \mathbf{T}}{\partial \Delta\mathbf{v}}\frac{\partial \Delta\mathbf{v}}{\partial \mathbf{v}}=\frac{\partial \mathbf{T}}{\partial \Delta\mathbf{v}}\mathbf{I}_{3\times 3}=\frac{\partial \mathbf{T}}{\partial \Delta\mathbf{v}} \tag{11.82}$$

in which

$$\frac{\partial \mathbf{T}}{\partial \Delta\mathbf{v}}=\frac{\partial \mathbf{T}}{\partial Q}\frac{\partial Q}{\partial \Delta\mathbf{v}}+\frac{\partial \mathbf{T}}{\partial \alpha}\frac{\partial \alpha}{\partial \Delta\mathbf{v}}+\frac{\partial \mathbf{T}}{\partial \beta}\frac{\partial \beta}{\partial \Delta\mathbf{v}}+\frac{\partial \mathbf{T}}{\partial |\Delta\mathbf{v}|}\frac{\partial |\Delta\mathbf{v}|}{\partial \Delta\mathbf{v}} \tag{11.83}$$

In Eq. (11.83)

$$\frac{\partial \mathbf{T}}{\partial Q}=S\begin{bmatrix} b & C_l \\ c & C_m \\ b & C_n \end{bmatrix}, \quad \frac{\partial \mathbf{T}}{\partial \alpha}=QS\boldsymbol{v}\frac{\partial \mathbf{C}_m}{\partial \alpha}, \quad \frac{\partial \mathbf{T}}{\partial \beta}=QS\boldsymbol{v}\frac{\partial \mathbf{C}_m}{\partial \beta}$$

TABLE 11.14   ANGULAR RATE EQUATION LINEARIZATION WITH RESPECT TO INERTIA ERRORS

	δJ	δI_x	δI_y	δI_z
$\delta \dot{p}$	$\dfrac{2J(\dot{p} - qr) - (I_z + I_x - I_y)pq + T_z}{I_x I_z - J^2}$	$\dfrac{-I_z \dot{p} - Jpq}{I_x I_z - J^2}$	$\dfrac{I_z qr + Jpq}{I_x I_z - J^2}$	$\dfrac{-I_x \dot{p} - (2I_z - I_y)qr - Jpq + T_x}{I_x I_z - J^2}$
$\delta \dot{q}$	$\dfrac{-r^2 + p^2}{I_y}$	$-\dfrac{pr}{I_y}$	$-\dfrac{\dot{q}}{I_y}$	$\dfrac{pr}{I_y}$
$\delta \dot{r}$	$\dfrac{2J(\dot{r} + pq) + (I_x + I_z - I_y)qr - T_x}{I_x I_z - J^2}$	$-\dfrac{-I_z \dot{r} + (2I_x - I_y)pq + Jqr + T_z}{I_x I_z - J^2}$	$-\dfrac{-I_x pq - Jqr}{I_x I_z - J^2}$	$\dfrac{I_x \dot{r} - Jqr}{I_x I_z - J^2}$

and

$$\frac{\partial \mathbf{T}}{\partial |\Delta \mathbf{v}|} = -\frac{QS}{2|\Delta \mathbf{v}|^2} \begin{bmatrix} b^2 C_{lp} & 0 & b^2 C_{lr} \\ 0 & c^2 C_{mq} & 0 \\ b^2 C_{np} & 0 & b^2 C_{nr} \end{bmatrix} \boldsymbol{\omega} \tag{11.84}$$

The remaining terms in Eq. (11.79) are

$$\frac{\partial \mathbf{T}}{\partial \boldsymbol{\omega}} = \frac{QS}{2|\Delta \mathbf{v}|^2} \begin{bmatrix} b^2 C_{lp} & 0 & b^2 C_{lr} \\ 0 & c^2 C_{mq} & 0 \\ b^2 C_{np} & 0 & b^2 C_{nr} \end{bmatrix}, \quad \frac{\partial \mathbf{T}}{\partial \mathbf{W}} = \frac{\partial \mathbf{T}}{\partial \Delta \mathbf{v}} \frac{\partial \Delta \mathbf{v}}{\partial \mathbf{W}}$$

$$\frac{\partial \mathbf{T}}{\partial \mathbf{C}_f} = -QSS(\Delta \mathbf{r}), \quad \frac{\partial \mathbf{T}}{\partial \mathbf{C}_m} = QS\boldsymbol{v} \tag{11.85}$$

with $\partial \Delta \mathbf{v}/\partial \mathbf{W}$ determined in Eq. (11.75) and $\partial \mathbf{T}/\partial \mathbf{C}_f$ computed using Eq. (11.39).

A summary of linearization of Eqs. (11.29–11.32) is given in Table 11.15 showing how errors associated with the dynamics for position, velocity, attitude, and angular rate equations relate to each other. An abbreviated set of aerodynamic and mass property modeling errors is presented in Table 11.16 as they contribute to the system error equations as well. This includes variations in wind $\delta \mathbf{W}$, aerodynamic force coefficient $\delta \mathbf{C}_f$, aerodynamic moment coefficient $\delta \mathbf{C}_m$, mass $\delta \mathbf{m}$, and inertia $\delta \mathbf{I}$.

As mentioned earlier, the errors in mass $\delta \mathbf{m}$ represent uncertainties in the apparent mass [see Eq. (11.33)]. However, these errors (appearing in the top

TABLE 11.15   SYSTEM-ERROR EQUATION SUMMARY

	$\delta \mathbf{p}$	$\delta \mathbf{v}$	$\boldsymbol{\varphi}$	$\delta \boldsymbol{\omega}$	$\cdots$
$\delta \dot{\mathbf{p}}$	$\mathbf{0}_{3\times3}$	$\mathbf{R}$ Eq. (11.53)	$-\mathbf{S}(\mathbf{R}\mathbf{v})$ Eq. (11.53)	$\mathbf{0}_{3\times3}$	
$\delta \dot{\mathbf{v}}$	$\mathbf{M}^{-1}\dfrac{\partial \mathbf{f}}{\partial \mathbf{p}}$ Eq. (11.67)	$\mathbf{M}^{-1}\mathbf{S}(\boldsymbol{\omega})\mathbf{M} + \mathbf{M}^{-1}\dfrac{\partial \mathbf{f}}{\partial \mathbf{v}}$ Eqs. (11.63), (11.67)	$-\mathbf{M}^{-1}m\mathbf{R}^T\mathbf{S}(\mathbf{g}) + \mathbf{M}^{-1}\dfrac{\partial \mathbf{f}}{\partial \boldsymbol{\varphi}}$ Eqs. (11.66), (11.67)	$\mathbf{M}^{-1}\mathbf{S}(\mathbf{M}\mathbf{v})$ Eq. (11.62)	
$\dot{\boldsymbol{\varphi}}$	$\mathbf{0}_{3\times3}$	$\mathbf{0}_{3\times3}$	$\mathbf{0}_{3\times3}$	$\mathbf{R}$ Eq. (11.59)	
$\delta \dot{\boldsymbol{\omega}}$	$\mathbf{I}^{-1}\dfrac{\partial \mathbf{T}}{\partial \mathbf{p}}$ Eq. (11.79)	$\mathbf{I}^{-1}\dfrac{\partial \mathbf{T}}{\partial \mathbf{v}}$ Eq. (11.79)	$\mathbf{I}^{-1}\dfrac{\partial \mathbf{T}}{\partial \boldsymbol{\varphi}}$ Eq. (11.79)	$\mathbf{I}^{-1}(\mathbf{S}(\mathbf{I}\boldsymbol{\omega}) - \mathbf{S}(\boldsymbol{\omega})\mathbf{I}) + \mathbf{I}^{-1}\dfrac{\partial \mathbf{T}}{\partial \boldsymbol{\omega}}$ Eqs. (11.78), (11.85) and Table 11.12	

TABLE 11.16 ABBREVIATED SET OF PARAMETER-ERROR EQUATIONS

...	$\delta\mathbf{W}$	$\delta\mathbf{C}_f$	$\delta\mathbf{C}_m$	$\delta\mathbf{m}$	$\delta\mathbf{l}$
$\delta\dot{\mathbf{p}}$	$\mathbf{0}_{3\times3}$	$\mathbf{0}_{3\times3}$	$\mathbf{0}_{3\times3}$	$\mathbf{0}_{3\times3}$	$\mathbf{0}_{3\times3}$
$\delta\dot{\mathbf{v}}$	$\mathbf{M}^{-1}\dfrac{\partial\mathbf{f}}{\partial\mathbf{W}}$ Eq. (11.66)	$\mathbf{M}^{-1}\dfrac{\partial\mathbf{f}}{\partial\mathbf{C}_f}$ Eq. (11.66)	$\mathbf{0}_{3\times3}$	See Eq. (11.61)	$\mathbf{0}_{3\times3}$
$\dot{\boldsymbol{\varphi}}$	$\mathbf{0}_{3\times3}$	$\mathbf{0}_{3\times3}$	$\mathbf{0}_{3\times3}$	$\mathbf{0}_{3\times3}$	$\mathbf{0}_{3\times3}$
$\delta\dot{\boldsymbol{\omega}}$	$\mathbf{I}^{-1}\dfrac{\partial\mathbf{T}}{\partial\mathbf{W}}$ Eq. (11.79)	$\mathbf{I}^{-1}\dfrac{\partial\mathbf{T}}{\partial\mathbf{C}_f}$ Eq. (11.79)	$\mathbf{I}^{-1}\dfrac{\partial\mathbf{T}}{\partial\mathbf{C}_m}$ Eq. (11.79)	$\mathbf{0}_{3\times3}$	Eq. (11.77) and Table 11.13

row in Table 11.16) are not varied directly but via an apparent mass coefficient error. Specifically, the apparent mass in Eq. (11.33) is modeled as

$$\Delta\mathbf{m} = \mathbf{k}\rho U \tag{11.86}$$

where the enclosed air mass volume U is computed from the mass property equations developed in Appendix E. The error associated with apparent mass is expressed as

$$\delta\Delta\mathbf{m} = \delta\mathbf{k}\rho U + \mathbf{k}U\frac{\partial\rho}{\partial h}\frac{\partial h}{\partial\mathbf{p}}\delta\mathbf{p} \tag{11.87}$$

where the volume U is assumed known. The EKF implements the apparent mass coefficient error $\delta\mathbf{k}$, and along with the position error this defines the mass error $\delta\mathbf{m}$ [Eq. (11.87)].

Table 11.16 can be extended to include other errors associated with the aerodynamic model, that is, $\delta\mathbf{C}_{f\alpha}$, $\delta\mathbf{C}_{m\delta_a}$, etc., which can be accomplished by including the corresponding terms to Eqs. (11.71) and (11.83). Table 11.17 shows what aerodynamic derivative and control coefficient modeling errors were implemented in this specific SID attempt. Only coefficients that are blocked with a heavy border were selected for implementation in the EKF as the error states (compare them to those that are shaded being part of the reference model presented in Table 11.12).

The rationale for limiting the number of states includes 1) selecting those that are consistently important to each of the flights processed and 2) not including states that cannot be estimated separately. More details are provided in Sec. 11.7.1. For the coefficients that are shaded, the EKF algorithm estimates errors in the reference model values and applies these estimates as corrections to those reference model elements. The blocks that are not shaded are not supported

TABLE 11.17 STATIC AERODYNAMIC MODEL ERROR STATES

	0	α	β	p	q	r	δ_s	δ_a	$\alpha\delta_s$	δ_s^2	$\delta_s\delta_a$	δ_a^2	$\delta_s\|\delta_a\|\delta_a$
C_A	C_{A0}	$C_{A\alpha}$	–	–	–	–	$C_{A\delta_s}$	$C_{A\|\delta_a\|}$	$C_{A\alpha\delta_s}$	$C_{A\delta_s^2}$	–	$C_{A\delta_a^2}$	–
C_N	C_{N0}	$C_{N\alpha}$	–	–	–	–	$C_{N\delta_s}$	$C_{N\|\delta_a\|}$	$C_{N\alpha\delta_s}$	$C_{N\delta_s^2}$	–	–	–
C_m	C_{m0}	$C_{m\alpha}$	–	–	C_{mq}	–	$C_{m\delta_s}$	$C_{m\|\delta_a\|}$	$C_{m\alpha\delta_s}$	$C_{m\delta_s^2}$	–	–	–
C_Y	C_{Y0}	–	$C_{Y\beta}$	–	–	–	–	$C_{Y\delta_a}$	–	–	$C_{Y\,\delta_s\delta_a}$	$C_{Y\|\delta_a\|\delta_a}$	–
C_l	C_{l0}	–	$C_{l\beta}$	C_{lp}	–	C_{lr}	–	$C_{l\delta_a}$	–	–	$C_{l\,\delta_s\delta_a}$	$C_{l\|\delta_a\|\delta_a}$	–
C_n	C_{n0}	–	$C_{n\beta}$	C_{np}	–	C_{nr}	–	$C_{n\delta_a}$	–	–	$C_{n\,\delta_s\delta_a}$	$C_{n\|\delta_a\|\delta_a}$	$C_{n\delta_s\|\delta_a\|\delta_a}$

by existing reference model values, and for these coefficients the EKF algorithm estimates the complete coefficient.

Selecting the aerodynamic model error states as just described allow an evaluation of how well the EKF consistently estimates error in this reference model. If errors are estimated consistently across all flights, then a higher confidence can be associated with the results and that these results reflect the difference between the reference model and the vehicle's actual aerodynamic characteristics. Also, given higher confidence in a reference model modified with these results, a simulation that uses this model to predict system performance will also have a higher confidence associated with their results.

A summary of the errors included in the EKF implementation as error states is presented in Table 11.18. The EKF implementation has been enhanced to include errors for cross product of inertia, velocity apparent mass, and yawing moment and lateral-axis higher-order control effectiveness terms.

11.5.4   MEASUREMENT DATA

Test data available from the X-38 test program for the EKF-based SID presented in Secs. 11.5.1–11.5.3 included four sets. To be more specific, two sets of flight data were available for the instrumented pallet payload (hereinafter referred to as configuration A) and prototype CVR payload (referred to as configuration B), shown in Fig. 1.8. Flights B1, B2, and A1 of these four flights are typical of the profiles flown by a guided parafoil system with guidance provided for lateral vehicle steering. Flight A2 is dissimilar in that large sustained control inputs are used. Shown in Table 11.19 is a summary of the relevant test data available for the various test vehicle designations and data sets.

As seen, configurations A and B used a variety of systems to measure the vehicle's position, velocity, attitude, angular rate, and air data: differential GPS (DGPS), embedded GPS inertial (EGI) navigation system, air-data probe (ADP), and flush air-data system (FADS). The left and right control positions (δ_l and δ_r, respectively) were recorded as well. Winds and air density were provided by weather balloon/dropsonde. Shaded areas indicate data that seem to be common for all flights and therefore were chosen as "measurements" in the filter algorithm are position, angular rate, and air-data system angle-of-attack and sideslip outputs.

Examples of measured data are presented in Figs. 11.37 and 11.38. Shown in Fig. 11.38a are the left and right control positions for Flight B1 (the symmetric and asymmetric control inputs were modeled according to the equation following Fig. 5.10). Shown in Fig. 11.37b are the measured north and east wind components for Flight B2. This figure presents weather balloon and dropsonde measured winds and their average. The EKF includes error states for errors in each of the wind components. Figure 11.38a presents air-data angle of attack and sideslip for Flight A1. These data suggest that this vehicle "flies" at nearly constant angle of attack. Also evident from this figure are oscillations in sideslip angle.

TABLE 11.18   EKF ERROR STATE SUMMARY

State	Description	Math Symbol
1–3	Inertial position	$\delta\mathbf{p}$
4–6	Body velocity	$\delta\mathbf{v}$
7–9	Inertial attitude	φ
10–12	Body rotation	$\delta\omega$
13–15	Inertial wind	$\delta\mathbf{W}$
16–18	Aerodynamic-force coefficient	$\delta\mathbf{C}_{f0}$
19–21	Aerodynamic-moment coefficient	$\delta\mathbf{C}_{m0}$
22–25	Moment of inertia	$\delta\mathbf{I}$
26–30	Damping-moment coefficient with respect to $p/q/r$	$\delta C_{lp},\ \delta C_{np},\ \delta C_{mq},\ \delta C_{lr},\ \delta C_{nr}$
31–33	Force coefficient with respect to α/β	$\delta C_{A\alpha},\ \delta C_{Y\beta},\ \delta C_{N\alpha}$
34–36	Moment coefficient with respect to α/β	$\delta C_{l\beta},\ \delta C_{m\alpha},\ \delta C_{n\beta}$
37–39	Force coefficient with respect to δ_s/δ_a	$\delta C_{A\delta_s},\ \delta C_{Y\delta_s},\ \delta C_{N\delta_s}$
40–42	Moment coefficient with respect to δ_s/δ_a	$\delta C_{l\delta_a},\ \delta C_{m\delta_s},\ \delta C_{n\delta_a}$
43–44	Longitudinal-force coefficient cross axis with respect to δ_a	$\delta C_{A|\delta_a|},\ \delta C_{N|\delta_a|}$
45	Pitch-moment coefficient cross axis with respect to δ_a	$\delta C_{m|\delta_a|}$
46–48	Velocity apparent-mass coefficient	$\delta\mathbf{k}$
49–50	Roll-moment coefficient with respect to δ_s/δ_a	$\delta C_{l|\delta_a|\delta_a},\ \delta C_{l\delta_s\delta_a}$
51–52	Side-force coefficient with respect to δ_s/δ_a	$\delta C_{Y|\delta_a|\delta_a},\ \delta C_{Y\delta_s\delta_a}$
53–55	Yaw-moment coefficient with respect to δ_s/δ_a	$\delta C_{n|\delta_a|\delta_a},\ \delta C_{n\delta_s\delta_a},\ \delta C_{n\delta_s|\delta_a|\delta_a}$

Oscillations also show up in yaw angular rate shown in Fig. 11.38b from Flight A2. The impact of oscillations is minimized by adjusting the uncertainty, measurement variance, assumed for the measurements processed in the algorithm.

Knowing the behavior of the angle of attack allows a better understanding of the choice of the varied aerodynamic derivative coefficients in Table 11.16 (states 16–45, 49–55 in Table 11.17). First, the configuration B series flights are characterized by relatively benign control inputs compared to the configuration A series. For these flights, an aerodynamic model composed of mostly linear

TABLE 11.19   FLIGHT-TEST DATA SUMMARY

	Configuration A		Configuration B
Update rate	5 Hz	100 Hz	5 Hz
Position	*lat*/*lon*/*alt* (DGPS)		*lat*/*lon*/*alt* (EGI)
Velocity	$V_x/V_y/V_z$ (DGPS)		$V_x/V_y/V_z$ (EGI)
Attitude	θ		$\phi/\theta/\psi$ (EGI)
Angular rate	$p/q/r$	$p/q/r$	$p/q/r$ (EGI)
Air data	$\alpha/\beta/Q$ (ADP)	$\alpha/\beta/Q$ (ADP)	$\alpha/\beta/Q$ (FADS)
Controls		δ_l/δ_r	δ_l/δ_r
Winds/density	$w_x/w_y/\rho$		w_x/w_y (weather balloon/ dropsonde)

terms would be adequate. Utilization of additional HOT would result in estimates that vary from flight to flight and would not represent true repeatable error in the reference model. There are some HOT that are important for the configuration A series, and for a common EKF implementation for both flight series, these must be included. Second, as observed from Fig. 11.37a, both configurations fly at nearly constant angles of attack. The terms that contain combinations one coefficient dependency and another with that same dependency plus angle of attack, that is, $(\delta C_{A\delta_s} + \delta C_{A\alpha\delta_s}\alpha)\delta_s$ [Eqs. (11.40–11.45)], enter the error dynamics as linear combinations. With nearly constant angle-of-attack flight characteristics, these errors cannot be estimated separately.

As just mentioned, observations processed in the EKF algorithm as measurement updates are position, angular rates, and air data. These data are independent from the corresponding reference model generated values. The differences between the reference model values and those observed in the flight data are processed in the EKF as measurement updates. The algorithm effectively tries to minimize this difference by adjusting the filter-error states.

Let us next address the data quality. The EGI navigation system was onboard the configuration B flights. This unit provided high-quality vehicle state information for position, velocity, attitude, and angular rate. However, the prototype CRV was suspended at an inclination angle that was not known precisely; therefore, there was some uncertainty in the transformation of the EGI angular rates, referenced to its case axis, into the parafoil body axes. Because the EGI-sensed rates include payload oscillations, there is uncertainty in attributing the sensed rates to vehicle angular rates. The EKF implementation's angular rate measurement variances are selected to reflect these uncertainties. Air data available on the configuration B series flights from FADS were characterized by

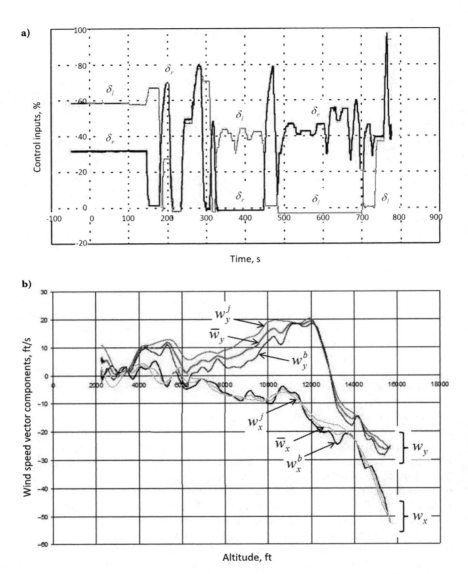

Fig. 11.37   a) Control positions from Flight B1 and b) measured winds from Flight B2 [Rogers 2004].

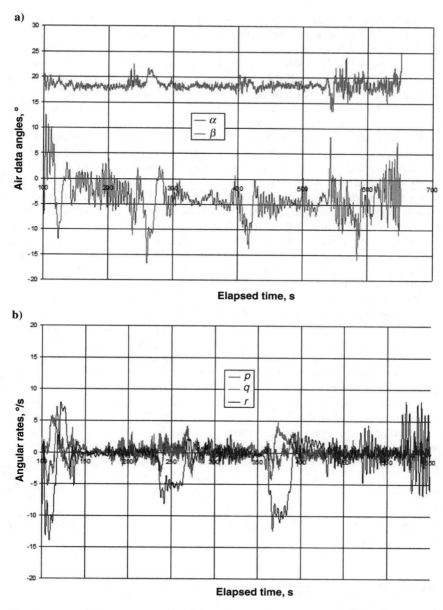

Fig. 11.38 a) Air data from Flight A1 and b) angular rates from flight A2 [Rogers 2004].

large-amplitude (several degrees) noise. The EKF implementation's air-data
measurement variances were selected based on this noise level.

The configuration A flights were instrumented with different angular rate and
air-data sensors (Table 11.19). From the trends in angle-of-attack variation in the
configuration B flights and the lack of the same type variation in the configuration
A flights, the air data for the configuration B flights is treated as suspect. This will
clearly be evident for Flight A2 identification later with the sideslip angle offset
from zero by approximately 20 deg and its trend following the angle-of-attack
data. As a result of the air-data quality for these flights, the EKF implementation's
angular rate and air-data measurement variances were tailored for these flights.

Data availability and quality for the configuration B series flights is such that
the system dynamics equations' initial position, velocity, attitude, and angular
rates can be initialized using data from the EGI unit. However, for the configuration A flights, GPS position and velocity and angular sensor data are used to initialize the corresponding system model states. Attitude is initialized assuming
the vehicle yaw angle corresponds to the vehicle horizontal flight-path angle,
and pitch and roll are assumed initially zero. This initialization approach requires
starting the EKF at a mission time for which the vehicle dynamics are
relatively constant.

As just mentioned, measurements processed in the EKF algorithm include
position, angular rates, and air-data angle of attack and sideslip. As shown in
Table 11.11, the model assumed for these measurements governs the measurement equation matrix. For position and angular rates, the measurement is
formed as the difference between these measurements and the corresponding
values generated by the system model dynamics. For these two measurement
types, the measurement matrix is just an identity matrix.

Air-data measurements are formed as the difference between measured values
and those generated by the EKF reference model. The EKF reference model values
are assumed to contain error, and this error is represented by an overbar in the
following equation:

$$\mathbf{z} = \begin{bmatrix} \alpha_m - \bar{\alpha} \\ \beta_m - \bar{\beta} \end{bmatrix} \tag{11.88}$$

We can write the linearized equations for the angle of attack and sideslip similar
to Eqs. (11.67) and (11.79):

$$\delta\alpha = \frac{\partial\alpha}{\partial\mathbf{v}}\delta\mathbf{v} + \frac{\partial\alpha}{\partial\varphi}\varphi + \frac{\partial\alpha}{\partial\mathbf{W}}\delta\mathbf{W} + \cdots = \frac{\partial\alpha}{\partial\Delta\mathbf{v}}\left(\mathbf{I}_{3\times3}\delta\mathbf{v} + \frac{\partial\Delta\mathbf{v}}{\partial\varphi}\varphi + \frac{\partial\Delta\mathbf{v}}{\partial\mathbf{W}}\delta\mathbf{W} + \cdots\right)$$

$$\delta\beta = \frac{\partial\beta}{\partial\mathbf{v}}\delta\mathbf{v} + \frac{\partial\beta}{\partial\varphi}\varphi + \frac{\partial\beta}{\partial\mathbf{W}}\delta\mathbf{W} + \cdots = \frac{\partial\beta}{\partial\Delta\mathbf{v}}\left(\mathbf{I}_{3\times3}\delta\mathbf{v} + \frac{\partial\Delta\mathbf{v}}{\partial\varphi}\varphi + \frac{\partial\Delta\mathbf{v}}{\partial\mathbf{W}}\delta\mathbf{W} + \cdots\right)$$

$$\tag{11.89}$$

[The partial derivatives in Eq. (11.89) are given in Eqs. (11.73) and (11.75).] The resulting measurement matrix for the air-data observations can be expressed as

$$
z = Hx + n = \begin{bmatrix} 0 & \dfrac{\partial \alpha}{\partial v} & \dfrac{\partial \alpha}{\partial \varphi} & 0 & \dfrac{\partial \alpha}{\partial W} & \cdots \\[2mm] 0 & \dfrac{\partial \beta}{\partial v} & \dfrac{\partial \beta}{\partial \varphi} & 0 & \dfrac{\partial \beta}{\partial W} & \cdots \end{bmatrix} \begin{bmatrix} \delta p \\ \delta v \\ \varphi \\ \delta \omega \\ \delta W \\ \vdots \end{bmatrix} + n \qquad (11.90)
$$

11.5.5 SYSTEM IDENTIFICATION RESULTS

The results of flight data processing are presented to demonstrate the ability of the EKF, in simulation, to reproduce the observed vehicle dynamics using its estimates. Specific results presented for both PADS configurations include plots illustrating the vehicle's short-term dynamics comparing the EKF and corresponding sensor data for angular rates, angle of attack and sideslip, and horizontal and vertical velocities. Plots are also presented illustrating the results for the horizontal plane trajectory and estimated wind modeling error. Following these plots are summary bar charts presenting the EKF estimated aerodynamic and mass property modeling errors, allowing an assessment of the EKF estimates' consistency from flight to flight.

Let us start from configuration B. Shown in Figs. 11.39–11.41 are comparisons between simulation results using EKF estimates and indicated sensor data. Figure 11.39 indicates that the EKF simulation results track the mean roll and yaw angular rates closely for the two configuraiton B flights; however, the pitch rate is not followed as closely. Payload oscillations are evident from the angular rate and air-data plots. Figure 11.40 shows that the EKF simulation results loosely follow the trends in angle of attack, but follow the sideslip more closely. The air-data sensor noise can be seen in these plots. Comparing the angular rate and air-data sensor data for Flight B1, a control input, resulting in an angular rate, is accompanied by a reduction in angle of attack. This is not as evident in Flight B2 results because the control inputs are not sustained over as long a time interval as in Flight B1. Comparisons between EKF simulation results for horizontal and vertical velocity and EGI data are presented in Fig. 11.14. Comparing the angle-of-attack trend in Fig. 11.40 with the horizontal velocity in Fig. 11.41, the reduction in angle of attack is accompanied with an increase in horizontal velocity.

Shown in Figs. 11.42 and 11.43 are the horizontal trajectory and estimated wind modeling error for the configuration B flights. The trajectory plots in Fig. 11.42 compare the EKF simulation results (the dashed lines) and sensor data (the solid lines). The simulation results follow the same general trend as the observed flight, but after some time has elapsed from the initial point, this trajectory begins to depart from the observed trajectory. Even though the EKF

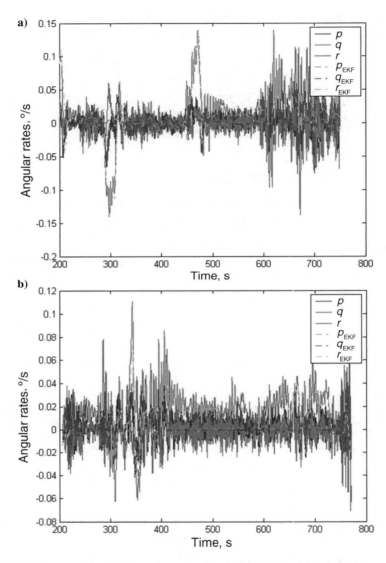

Fig. 11.39   Measured and EKF angular rates for a) Flight B1 and b) Flight B2 [Rogers 2004].

simulated short-term vehicle dynamics tracks the sensor data, small differences in these short-term dynamics can result in larger differences in the longer-term position dynamics.

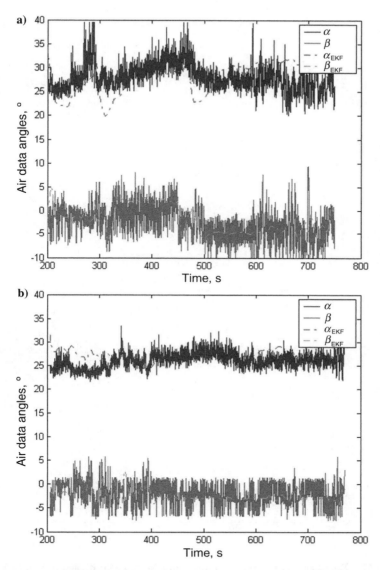

Fig. 11.40   Measured and EKF air data for a) Flight B1 and b) Flight B2 [Rogers 2004].

In general, comparisons just presented for configuration B flights show good agreement in tracking the short-term dynamics between the simulation results, based on EKF estimates, and sensor data.

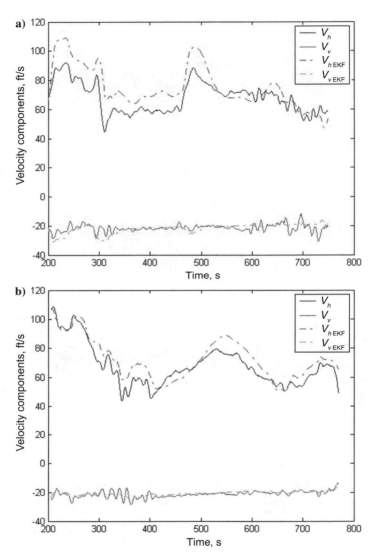

Fig. 11.41   Measured and EKF velocity components for a) Flight B1 and b) Flight B2 [Rogers 2004].

Now, let us address configuration A flights. Shown in Figs. 11.44–11.46 are plots comparing simulated dynamics, based on using EKF estimates, and the indicated sensor data. Figure 11.44a shows that the EKF simulation results track the mean roll and yaw angular rates with payload oscillations being evident again. In

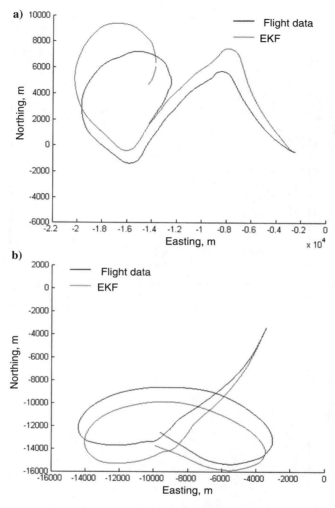

Fig. 11.42   EKF horizontal trajectory for a) Flight B1 and b) Flight B2 [Rogers 2004].

Fig. 11.44b, the yaw rate is tracked closely only for the second of the two maneuvers. Figure 11.45a shows that the EKF simulation results follow the angle-
of-attack trend; however, in Fig. 11.45b, the trend is not followed as closely. In
Fig. 11.45a, Flight A1 sideslip data are processed in the EKF but with increased
measurement variance. Simulated sideslip follows the major changes in the
sensor data, but is offset from the sensor values. Sideslip in Fig. 11.45b for
Flight A2 is the EKF simulation result with a fictional zero beta measurement

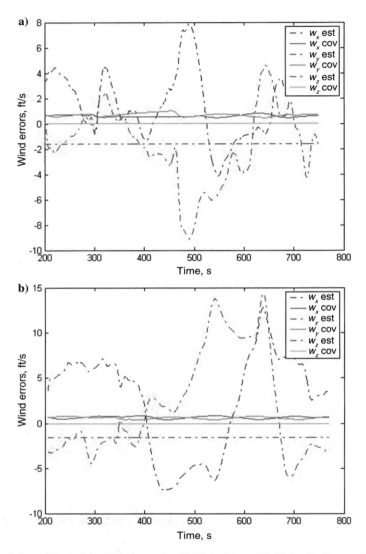

Fig. 11.43   EKF wind modeling error for a) Flight B1 and b) Flight B2 [Rogers 2004].

processed in the EKF algorithm. The assumption is made that the vehicle "flies" with small beta, and to reflect the uncertainty associated with this assumption, an expanded variance is selected for this "measurement."

Shown in Figs. 11.47 and 11.48 are the horizontal trajectory and estimated wind modeling error for the configuration A flights. The trajectory plots in

...

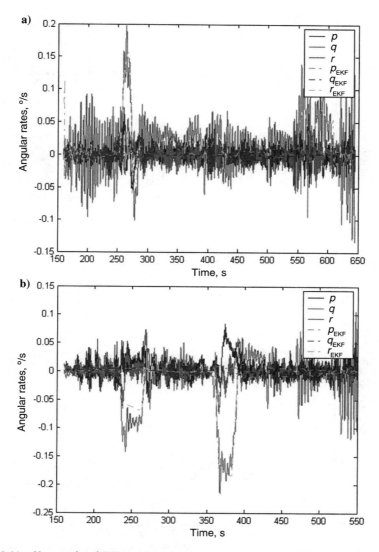

Fig. 11.44 Measured and EKF angular rates for a) Flight A1 and b) Flight A2 [Rogers 2004].

Fig. 11.47 compare the EKF simulation results (the dashed lines) and sensor data (the solid lines). The simulated trajectory follows the same general trend as the observed flight, but, as before, after some time has elapsed from the initial point, the EKF trajectory begins to depart from the observed trajectory.

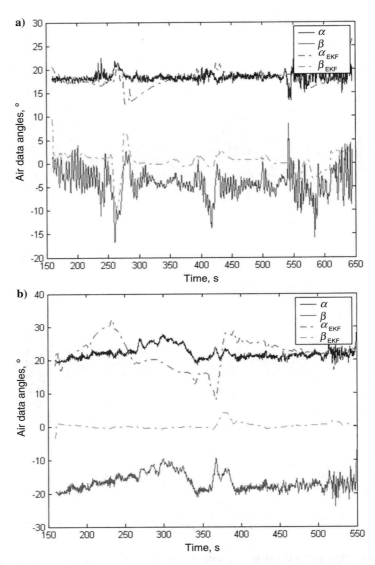

Fig. 11.45   Measured and EKF air data for a) Flight A1 and b) Flight A2 [Rogers 2004].

In general, comparisons presented for the configuration A flights also show agreement in tracking the short-term dynamics between the simulation results, based on EKF estimates, and sensor data. However, sensor quality is definitely a

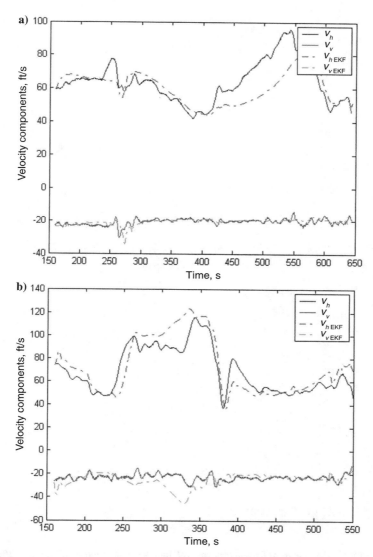

Fig. 11.46   Measured and EKF velocity components for a) Flight A1 and b) Flight A2 [Rogers 2004].

concern for the configuration A series flights. The tailoring just described was required for air-data measurements. Not using this tailoring resulted in less favorable comparisons between simulated and observed sensor data.

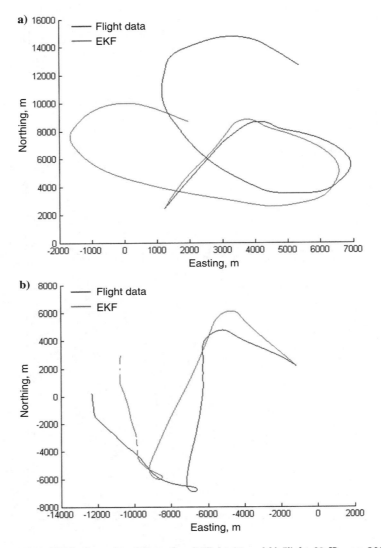

Fig. 11.47   EKF horizontal trajectory for a) Flight A1 and b) Flight A2 [Rogers 2004].

For the configuration B, angles of attack are seen to vary and be in a good agreement with the measured data; however, the results from the configuration A series show less variation in the sensor angles of attack while the EKF simulation results show more variation. Also in configuration A series, the EKF simulated

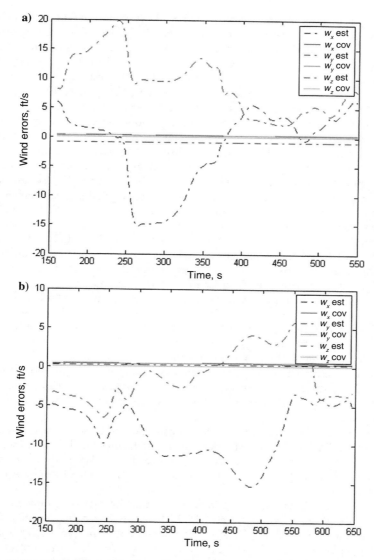

Fig. 11.48   EKF wind modeling error for a) Flight A1 and b) Flight A2 [Rogers 2004].

sideslip sensor data are either offset to unrealistically large values (20 deg in Fig. 11.45b), or required deweighting for the EKF algorithm to produce meaningful results (Fig. 11.45a). These two factors suggest that the air-data quality for these flights is questionable.

Comparing estimates from several flights allows an assessment of reasonableness in assuming that the EKF estimates reflect the actual difference between the reference model and that necessary to reproduce in simulation the vehicle dynamics as observed in flight. Comparisons presented in the following are in the form of summary bar charts presenting the EKF estimated aerodynamic and mass property errors. These charts are presented in Figs. 11.49–11.51.

The results presented in Figs. 11.49 and 11.50 are for the most significant aerodynamic coefficient and inertia property modeling errors as reflected by their consistency from flight to flight. Estimates for most of these error states are consistent for all flights. There is consistency in the zero condition force and moment terms except for Flight A1 roll and yaw moment coefficients and Flight A2 normal-force coefficient (Fig. 11.49). The first of these exceptions can be explained by the deweighting of the air-data sideslip measurement and the second by different aerodynamic "trim" conditions as suggested by a corresponding larger zero condition pitch moment for this flight. There is general consistency in the moment of inertia estimates for the primary axis terms while there is some difference in the cross-product term. Aerodynamic damping coefficient estimates for the configuration B series are in general agreement; however, these are less consistent with the configuration B series, which are less consistent within themselves. The force and moment coefficients due to aerodynamic angles and control effectiveness (Fig. 11.50) are the most consistent of all of the modeling error estimates. Only the results for sideslip angle terms for Flight A2 are different and can be explained from the assumption of the fictional zero sideslip measurement.

Comparing results between the configuration A and configuration B series, there is more consistency between the two configuration B series results. Not only are the signs of the error estimates consistent, but also their magnitudes are consistent.

The results in Fig. 11.51 indicate less consistency for these aerodynamic modeling errors. This lack of consistency can result from several factors including typical variation from vehicle to vehicle, control input magnitude employed in each flight, sensor data quality, etc. Reasons for the lack of consistency are suggested next.

Other analysis has shown a direct relationship between apparent mass and control effectiveness estimates. The EKF results comparing estimates with apparent mass included as part of the error state vector in Table 11.18 and not including apparent mass showed a change most significantly in the control effectiveness estimates. This relationship suggests that for flights with frequent control inputs, apparent mass estimates are difficult to separate from control effectiveness estimates. The first three flights were characterized by frequent control inputs.

The control input magnitudes in the configuration A series were greater than those employed in the configuration B series. For the lower control input levels, the linear control effectiveness terms in Fig. 11.50 are more important than the

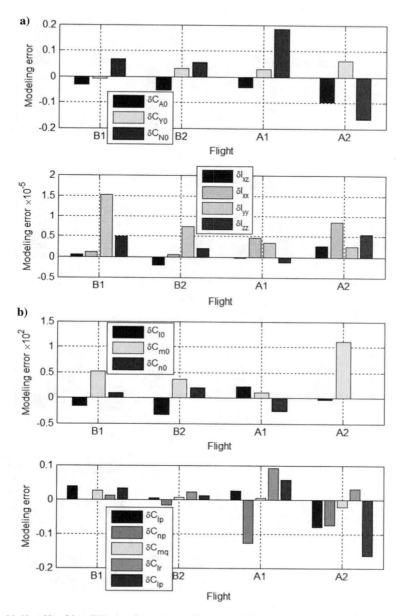

Fig. 11.49 Matching EKF aerodynamic coefficient and inertia property modeling errors (errors states 16–30 in Table 11.17) [Rogers 2004].

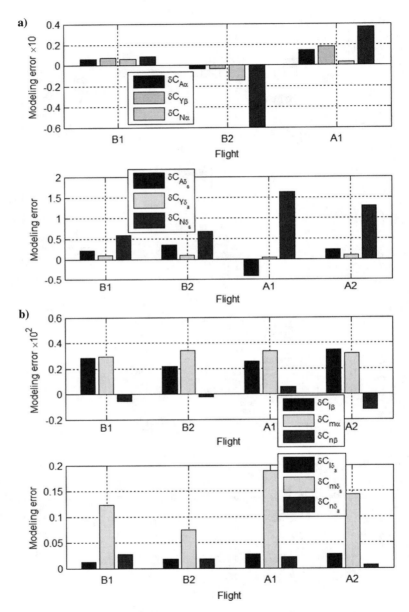

Fig. 11.50   Matching EKF aerodynamic coefficient modeling errors (errors states 31–42 in Table 11.17) [Rogers 2004].

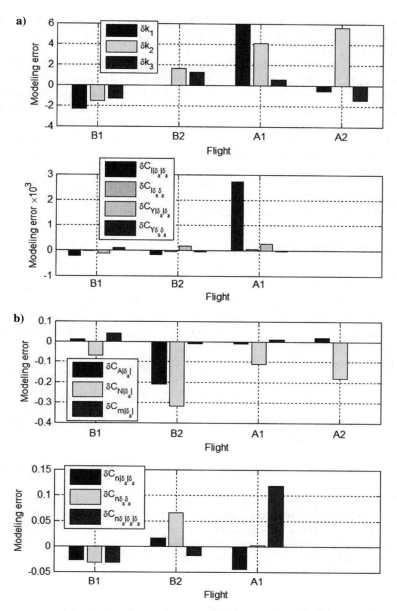

Fig. 11.51   Not matching EKF aerodynamic coefficient and mass property errors (errors states 43–55 in Table 11.17) [Rogers 2004].

higher-order terms presented in Fig. 11.51. With the larger control inputs, these higher-order terms would become more important and larger values estimated for these error states.

Flight A2 was characterized by control inputs at large and sustained levels, and for this flight the EKF algorithm was not able to distinguish between linear and higher-order control effectiveness terms. As a consequence, the lateral higher-order terms were deleted from the error state structure in Table 11.18 for this flight's processing. This approach was also examined in EKF processing for Flight A1. However, for this flight the control input magnitudes varied, and the lateral higher-order terms produced better results compared to results when these terms were excluded.

Comparisons between EKF and observed data presented in Figs. 11.39–11.48 suggest that, in general, the inconsistencies suggested by these last plots do not contribute significantly to the vehicle's flight characteristics.

11.5.6 SIMPLIFIED APPROACHES

While one of the realizations of the filter-error method presented in Secs. 11.5.1–11.5.5 was implemented on a complex PADS model involving several dozen parameters, this and other realizations of this method can also be used on a much smaller scale. For example, let us show how the RWLS estimation algorithm was used to identify several parameters of the linearized reduced-order model of Eq. (5.214).

Having one additional term in Eq. (5.214)

$$[0,\ 0,\ QSbJ_{xz},\ QSbJ_{zz}]^T \delta_{\text{bias}} \qquad (11.91)$$

Slegers and Costello [2005] used RWLS in conjunction with a standard linear measurement equation

$$\mathbf{z}_k = \mathbf{H}_k \mathbf{x} + \mathbf{n}_k \qquad (11.92)$$

In Eq. (11.92), $\mathbf{x} = [C_{l\varphi}, C_{lp}, C_{l\delta_a}, C_{nr}, C_{nr}, C_{n\delta_a}, \delta_{\text{bias}}]^T$ is the vector composed of five constant aerodynamic coefficients and constant bias term, $\mathbf{z}_k = [\delta\dot{p}_k, \delta\dot{q}_k]^T$ is the vector of observed variations of two angular velocities, and $\mathbf{n}_k$ represents zero-mean measurement noise. Matrix $\mathbf{H}_k$ is obtained from a linearized reduced-order model of Eq. (5.214) [with the additional term described by Eq. (11.91)] as follows:

$$\mathbf{H}_k = QSb \begin{bmatrix} J_{xxI}\varphi_k & J_{xxI}\dfrac{b}{2V_a}p_k & J_{xx}\delta_{a;k} & J_{xz}\dfrac{b}{2V_a}r_k & J_{xz}\delta_{a;k} & J_{xz} \\[3mm] J_{xzI}\varphi_k & J_{xz}\dfrac{b}{2V_a}p_k & J_{xz}\delta_{a;k} & J_{zz}\dfrac{b}{2V_a}r_k & J_{zz}\delta_{a;k} & J_{zz} \end{bmatrix} \qquad (11.93)$$

The RWLS estimation to Eq. (11.92) is given by

$$\hat{\mathbf{x}}_k = \hat{\mathbf{x}}_{k-1} + \mathbf{P}_k \mathbf{H}_k^T \mathbf{Q}^{-1}(\mathbf{z}_k - \mathbf{H}_k \hat{\mathbf{x}}_k) \tag{11.94}$$

$$\mathbf{P}_k = \mathbf{P}_{k-1} - \mathbf{P}_{k-1}\mathbf{H}_k^T(\mathbf{Q} + \mathbf{H}_k \mathbf{P}_{k-1} \mathbf{H}_k^T)^{-1} \mathbf{H}_k \mathbf{P}_{k-1} \tag{11.95}$$

where $\mathbf{P}_i$ is the error covariance estimate of the parameters at measurement i and $\mathbf{Q}$ is the measurement noise covariance.

Because the recursive weighted least-squares estimation requires numerical differentiation of measured roll and yaw rates, Slegers and Costello [2005] chose a sinusoidal control input. This way differentiation of roll and yaw rates produced significant signals. Before differentiation, the measured roll and yaw rates were processed with a zero-phase digital filter. The recursive weighted least-squares estimation was initialized with $\mathbf{P}_1 = \text{diag}(0.05[1, 1, 1, 1, 1, 1])$, and the initial guess on the vector of model parameters was $\hat{\mathbf{x}}_1 = [-0.1, -0.5, 0.1, -0.1, 0.1, 0]^T$. The measurement noise covariance matrix was set as $\mathbf{Q} = \text{diag}([0.00475, 0.0005])$.

Another realization was used by Hur [2005] for nonparametrical identification of the elements of the state-space representation of Buckeye PPG (Figs. 1.9 and 5.43a) model in longitudinal and lateral-directional motion. The idea was to substitute matrices $\mathbf{A}$ and $\mathbf{B}$ in Eqs. (5.189) and (5.190) with the same-dimension matrices using data from the flight tests. This effort was based on OKID (Fig. 11.52), an extension of eigensystem realization (ERA), which utilizes the concept of minimum realization. The time-domain ERA algorithm solves for sampled pulse response histories, which is called system Markov parameters [Hur and Valasek 2003]. Based on the concept of stochastic Kalman filter estimation and the techniques of deterministic Markov parameter identification, OKID generates a state-space discrete linear model representation of the nonlinear system in the time domain. The advantages of OKID are that this is a time-domain method allowing handling of nonzero initial conditions and arbitrary multiple inputs and outputs. It works well for lightly damped systems (observer effectively increases damping so that full decay not required) [Hur 2005; Hur and Valasek 2003].

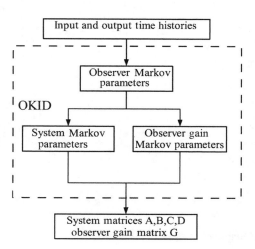

Fig. 11.52 Computational flow in the OKID algorithm [Hur 2005].

Applying OKID algorithm on Buckeye PPG allowed identifying its models in longitudinal and lateral-directional motion as (based on Run 117 and Run 220 data, respectively) [Hur 2005]

$$
\begin{bmatrix} \dot{u} \\ \dot{w} \\ \dot{q} \\ \dot{\theta} \\ \dot{q}_s \\ \ddot{\theta}^{s/p} \end{bmatrix} = \begin{bmatrix} -0.123 & -3.14 & -2.288 & -0.180 & 0.107 & -0.0012 \\ 2.98 & -0.027 & 0.102 & 0.102 & 0.0276 & 0 \\ 0.184 & -0.168 & -0.022 & -0.349 & 0.649 & -0.0011 \\ 0.211 & -0.107 & 0.361 & 0.0044 & 0.627 & 0 \\ 0.286 & 0.182 & -0.483 & -0.652 & -0.292 & 0.0008 \\ -0.209 & -0.231 & -0.093 & 0.0734 & 0.425 & -0.0029 \end{bmatrix} \begin{bmatrix} u \\ w \\ q \\ \theta \\ q_s \\ \theta^{s/p} \end{bmatrix}
$$

$$
+ \begin{bmatrix} 2.51 \\ 1.45 \\ 1.07 \\ -0.22 \\ -3.84 \\ 2.51 \end{bmatrix} \delta_s \qquad (11.96)
$$

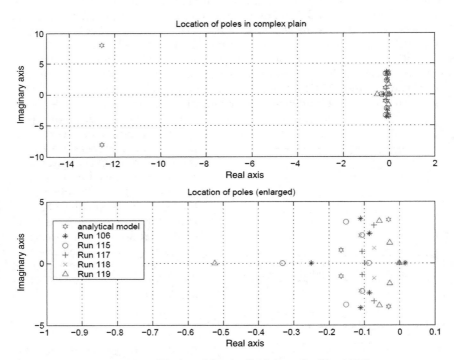

Fig. 11.53 Validation of longitudinal dynamics [Hur 2005].

and

$$
\begin{bmatrix} \dot{v} \\ \dot{p} \\ \dot{r} \\ \dot{\phi} \\ \dot{r}_s \\ \dot{\psi}^{s/p} \end{bmatrix} =
\begin{bmatrix}
-0.17 & -2.79 & 0.0181 & 0.0077 & 0.0038 & -0.00018 \\
2.94 & -0.029 & 0.0094 & -0.00074 & -0.0034 & -0.00019 \\
-0.352 & 0.129 & -0.159 & 0.0353 & 0.0789 & -0.0011 \\
-0.517 & 0.228 & -0.54 & -0.38 & 0.0558 & 0.02 \\
-0.43 & 0.176 & -0.481 & -0.704 & -0.252 & 0.0194 \\
-0.251 & 0.14 & -0.282 & -0.376 & -0.399 & -0.0226
\end{bmatrix}
\begin{bmatrix} v \\ p \\ r \\ \phi \\ r_s \\ \psi^{s/p} \end{bmatrix}
$$

$$
+ \begin{bmatrix} 2.92 \\ -1.27 \\ 2.85 \\ 4.38 \\ 3.63 \\ 2.22 \end{bmatrix} \delta_a \qquad (11.97)
$$

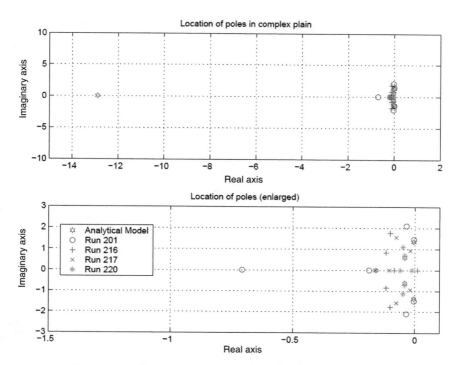

Fig. 11.54 Validation of lateral-directional dynamics [Hur 2005].

These models are the replacements for analytically derived models of Eqs. (5.189) and (5.190). Implementation of OKID on several sets of flight-test data resulted in a slight variation of eigenvalues (Figs. 11.53 and 11.54). Figures 11.53 and 11.54 also show the eigenvalues for the analytically derived models [Eqs. (5.189) and (5.190)] indicating that the usage of the aerodynamic derivatives scaled from a larger system produces the unrealistic results as discussed in Sec. 5.4 [Hur 2005].

Comparing the models given by Eqs. (5.189) and (5.190) with those presented by Eqs. (11.96) and (11.97), it is clear that the latter ones do not match the structure of matrices $\mathbf{A}$ and $\mathbf{B}$ in Eqs. (5.189) and (5.190) (and it was not an intention to begin with) and therefore cannot be used for parametrical identification (identifying aerodynamic derivatives) as it was the case in all other examples of this chapter. These models however can be used in simulations to develop and tune the control algorithms.

AIAA Aerodynamic Decelerator Systems TC Activities

Following the Symposium on Parachute Technology and Evaluation held on September 1964 in El Centro, California, and featuring 39 technical papers, AIAA established the Technical Committee on the subject of Aerodynamic Deceleration, the Aerodynamic Decelerator Systems TC (info.aiaa.org/tac/AASG/ADSTC/default.aspx and www.aerodecelerator.org).

The ADS TC organizes a variety of activities including technical conferences, seminars, and short courses. Table A1 lists all AIAA conferences held since 1966 and dedicated specifically to aerodynamic decelerator systems technology (ADST). In 1993, the ADST confe rence was organized jointly with the British Royal Aeronautical Society (RAeS), and in 1999, with the Confederation of European Aerospace Societies (CEAS). In 2013 and 2015, the ADST conference was held in conjunction with the Parachute Industry Association (PIA) Meeting and Symposium. Also shown in this table is the number of papers presented and archived by AIAA (arc.aiaa.org/series/6.ads). Up to date, this archived collection includes almost 1,500 papers (Fig. A1). Starting from the 12th conference held in 1993, the conference itself has been preceded by a one-day ADST seminar covering some specific area of interest. The information about these seminars is presented in Table A2.

In the even years (between the conferences), the ADS TC organizes the H.G. Heinrich Parachute Systems Short Course. The last short course was held at the National Institute of Aerospace on June 2–6, 2014 (www.nianet.org/parachute-short-course).

To recognize significant contributions to the advancement of aeronautical or aerospace systems through research, development, and application of the art and science of aerodynamic decelerator technology, the ADS TC established the Aerodynamic Decelerator and Balloon Technology Award, which was first given at the 7th ADST conference. In 1986, this award was renamed to the Aerodynamic Decelerator Systems Award, and since 1995 it has been known as the Theodor W. Knacke Aerodynamic Decelerator Systems Award. Information about all awardees is presented in Table A3 (the most up-to-date information can be found among one of the AIAA's Technical Excellence

TABLE A1   AIAA ADST CONFERENCES

Event	Venue	Dates	Year	Number of Papers
23rd AIAA Aerodynamic Decelerator Systems Technology Conference and Seminar (in conjunction with the PIA Meeting and Symposium)	Daytona Beach, FL	March 30–April 2	2015	83
22nd AIAA Aerodynamic Decelerator Systems Technology Conference and Seminar & Co-located Conferences (in conjunction with the PIA Meeting and Symposium)	Daytona Beach, FL	March 25–28	2013	95
21st AIAA Aerodynamic Decelerator Technology Conference and Seminar	Dublin, Ireland	May 23–26	2011	103
20th AIAA Aerodynamic Decelerator Technology Conference and Seminar	Seattle, WA	May 4–7	2009	78
19th AIAA Aerodynamic Decelerator Systems Technology Conference and Seminar & AIAA Balloon Systems Conference and Seminar	Williamsburg, VA	May 21–24	2007	62
18th AIAA Aerodynamic Decelerator Technology Conference and Seminar	Munich, Germany	May 23–26	2005	70
17th AIAA Aerodynamic Decelerator Technology Conference and Seminar	Monterey, CA	May 19–22	2003	64
16th AIAA Aerodynamic Decelerator Technology Conference and Seminar	Boston, MA	May 21–24	2001	61
15th CEAS/AIAA Aerodynamic Decelerator Technology Conference and Seminar	Toulouse, France	June 9–11	1999	59

Conference	Location	Date	Year	
14th AIAA Aerodynamic Decelerator Technology Conference and Seminar	San Francisco, CA	June 3–5	1997	59
13th AIAA Aerodynamic Decelerator Technology Conference and Seminar	Clearwater Beach, FL	May 15–18	1995	59
12th RAeS/AIAA Aerodynamic Decelerator Systems Technology Conference and Seminar	London, England	May 10–13	1993	53
11th AIAA Aerodynamic Decelerator Systems Technology Conference	San Diego, CA	April 9–11	1991	55
10th AIAA Aerodynamic Decelerator Systems Technology Conference	Cocoa Beach, FL	April 18–20	1989	58
9th AIAA Aerodynamic Decelerator & Balloon Technology Conference	Albuquerque, NM	October 7–9	1986	50
8th AIAA Aerodynamic Decelerator & Balloon Technology Conference	Hyannis, MA	April 2–4	1984	44
7th AIAA Aerodynamic Decelerator & Balloon Technology Conference	San Diego, CA	October 21–23	1981	40
6th AIAA Aerodynamic Decelerator & Balloon Technology Conference	Houston, TX	March 5–7	1979	47
5th AIAA Aerodynamic Decelerator Systems Conference	Albuquerque, NM	November 17–19	1975	51
4th AIAA Aerodynamic Decelerator Systems Conference	Palm Springs, CA	May 21–23	1973	45
3rd AIAA Aerodynamic Decelerator Systems Conference	Dayton, OH	September 14–16	1970	49
2nd AIAA Aerodynamic Decelerator Systems Conference	El Centro, CA	September 23–25	1968	38
AIAA Aerodynamic Decelerator Systems Conference	Houston, TX	September 7–9	1966	63

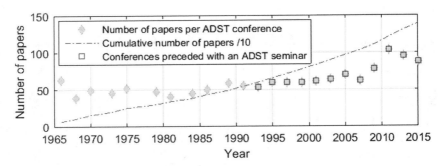

Fig. A1   Number of ADST conference papers archived by AIAA.

TABLE A2   ADST SEMINARS

Event	Year	Seminar Topic	Number of Attendees
12th ADST Seminar	2015	Simulation and Modeling of Modern Aerodynamic Decelerators: A Systems Engineering Perspective	51
11th ADST Seminar	2013	Entry, Descent, and Landing	63
10th ADST Seminar	2011	Systems Engineering in the Real World: ADS Case Studies	49
9th ADST Seminar	2009	Parachute Systems Test and Evaluation: Processes and Technology	76
8th ADST Seminar	2007	Modeling and Simulation of Parachute Systems–Current State of the Art and Look Ahead	80
7th ADST Seminar	2005	A Systems Engineering Approach to Aerodynamic Decelerator Systems	17
6th ADST Seminar	2003	Advances in Modern Fibers, Textiles, and Parachute Material Processes	57
5th ADST Seminar	2001	Advanced Sensor Technology	
4th ADST Seminar	1999	Parachute Materials Technology	
3rd ADST Seminar	1997	Parachute Flight Dynamics and Trajectory Simulation	
2nd ADST Seminar	1995	Ram-Air Parachute Design	
1st ADST Seminar	1993	Computational Fluid Dynamics and Discrete Vortex Methods	

TABLE A3   AIAA THEODOR W. KNACKE AWARD AWARDEES

Year	Awardee	Citation
2015	Robert J. Sinclair	For 30 years of innovation and leadership in the design, testing and fielding of parachutes for military systems, manned spacecraft recovery, and interplanetary probe descent and landing
2013	Gary Thibault	For over 30 years of exceptional commitment to the technical advancement of military airborne equipment and unwavering dedication to the field of aerodynamic deceleration systems
2011	Edwin D. Vickery	For five decades of dedication and innovation in the art of aerodynamic decelerator systems and for exceptional mentorship of the next generation of aerodynamic decelerator engineers
2009	Charles H. "Chuck" Lowry	For over five decades of contributions to significant programs in the field of aerodynamic decelerator systems, and for exhibiting the highest integrity, cooperation, and enthusiasm
2007	Glen J. Brown	For breakthroughs in parafoil flight mechanics and parachute guidance and soft landing as well as advances in supersonic and hypersonic planetary aerobraking and decelerator systems
2005	Roy L. Fox, Jr.	For four decades of outstanding contributions to the developmental and qualification testing of parachute deceleration and landing systems
2003	Vance L. Behr	For his significant contributions to the design of several high-performance parachute systems for NASA and the Department of Energy, and for his ongoing dedication to the education of the next generation of aerodynamic decelerator systems engineer
2001	Phillip R. Delurgio	In recognition of a lifetime of achievement and dedication to the field of aerodynamic decelerator systems
1999	Dean F. Wolf	For sustained excellence in technical contributions to the development of parachutes for NASA programs including Apollo, Pioneer Venus, space shuttle, and space station
1997	J. Stephen Lingard	For breakthroughs in the theory of gliding parachute aerodynamics and advances to personnel safe escape systems and supersonic interplanetary parachute systems

(Continued)

TABLE A3   AIAA THEODOR W. KNACKE AWARD AWARDEES (Continued)

Year	Awardee	Citation
1996	Matts J. Lindgren	For the excellence of his technical contribution, innovative design, and exemplary personal commitment to every project he has come in contact with, especially the CORONA photo reconnaissance reentry vehicle, SR-71, and D-21 programs
1995	Carl W. Peterson	For contributions to the understanding of aerodynamic decelerators and their technical applications, both as an engineer and as an inspiring leader
1994	Robert W. Rodier	For contributions in the development of sophisticated parachute recovery systems used in the U.S. Space Program, military aircraft and U.S. Army Airdrop Systems
1993	Karl-F. Doherr	For demonstrated leadership and experience in modeling flight dynamics of modern parachute systems, and for creativity, technical depth, and other contributions to the science of decelerator systems design
1992	James D. Reuter	For continuing contributions to the art of parachute design, repeatedly pioneering new capabilities beyond the edges of the old performance envelope
1991	Donald W. Johnson	For comprehensive and sustained contributions to parachute technology in the areas of ribbon parachute development, rocket recovery and the applications of Kevlar to high-performance parachute design
1990	Maurice P. Gionfriddo	For over 30 years of dedication, management, and administration that led to a significant amount of advances in the field of aerodynamic decelerator systems
1989	David J. Cockrell	For outstanding sustained and continuing contribution to the science of parachute dynamics and mathematical modeling of parachute motion
1988	John W. Kiker	For outstanding work in the development of parachutes and recovery systems of all types, and in particular for work on airplane and spacecraft parachute systems for the safe recovery of American airmen and astronauts
1986	William B. Pepper	In recognition 27 years of major contributions to parachute technology and design in the areas of weapon parachute systems, reentry vehicle and scientific payload recovery and Kevlar parachute technology
1986	Domina Jalbert	In recognition of a lifetime of innovation in parachutes, balloons and kites, in particular, his invention of the parafoil ram-air wing

(Continued)

TABLE A3   AIAA THEODOR W. KNACKE AWARD AWARDEES (*Continued*)

Year	Awardee	Citation
1984	Herman Engel, Jr.	In recognition of 35 years of outstanding accomplishments and contributions to the design and development of many parachute systems for emergency escape of aircraft crew members and the recovery of unmanned vehicles
1981	Theodor W. Knacke	For the introduction of modern technology and sound engineering principles to the former art of parachute design, thereby enabling aircraft, ordnance, spacecraft, and other advanced system designers to successfully use parachutes to satisfy system requirements
1979	Helmut G. Heinrich	For 40 years of continuous involvement with and significant contributions to the science and technology of aerodynamic deceleration systems and for teaching by doing and doing by teaching

Awards at www.aiaa.org/HonorsAndAwardsRecipientsList.aspx). I am very proud that two of the Knacke Award winners are contributors to this book.

The ADST TC has also made a substantial effort to archive a selection of technical reports written by engineers from 1935 to 2001, which is now available as Parachute History Collection at Linda Hall Library [Gionfriddo 2011]. To search for the papers, one can use lhldigital.lindahall.org/cdm/landingpage/collection/parachute or primo.lindahall.hosted.exlibrisgroup.com/primo_library/libweb/action/search/.do. The download requests of copyrighted material and nonrestricted material should be placed through another link: www.lindahall.org/services/document_delivery/index.shtml (thanks to Storm Dunker for testing and providing the last two links).

Units Conversions

Because this volume is based on papers from the United States and Europe, both international units and imperial units (in the parentheses following the value in international units) are used throughout the text. The figures though are plotted in one system of units or another. To help a reader to convert between these two systems of units, Table B1 presents the conversion coefficients from the imperial units to international units. Table B2 shows the reverse conversion. For temperature conversion, the reader can use the following:

$$T[°C] = \frac{5}{9}(T[°F] - 32), \quad T[°F] = \frac{9}{5}T[°C] + 32 \qquad (B1)$$

$$T[K] = T[°C] + 273.15, \quad T[°C] = T[K] - 273.15 \qquad (B2)$$

For reader's convenience Table B3 lists the MATLAB functions that can be used to do quick conversions between different systems of units. (These functions belong to the Aerospace Toolbox of MATLAB.)
Here are several examples of how these functions could be applied:

convlength(3.3,'ft','m')	converts 3.3 feet to meters (1.0058)
convmass(1,'slug','kg')	converts 1 slug to kilograms (14.5939)
convvel(60,'kts','m/s')	converts 60 knots to meters per second (30.8667)
convforce(1,'lbf','N')	converts 1 pound force to Newtons (4.4482)
convlength(convlength(11,'ft','m'),'ft','m')	converts the area of 11 ft^2 to m^2 (1.0219)

In Microsoft Excel, it is the convert function that converts a number from one measurement system to another. The syntax is

CONVERT(number,from_unit,to_unit)

Table B4 shows the units available for conversion in the categories applicable to this book.

TABLE B1   IMPERIAL UNITS TO METRIC UNITS CONVERSION

Value	Imperial Units	Value	International System of Units (SI)
Length			
1	in.	0.0254	m
1	ft	0.3048	m
1	miles	1609.344	m
1	n miles	1852	m
Area			
1	in.^{2}	0.0006451	m^2
1	ft^2	0.0929030	m^2
Volume			
1	in.^{3}	1.638706e-5	m^3
1	ft^3	0.0283168	m^3
Mass			
1	oz	0.02834951	kg
1	lb	0.45359237	kg
1	slug	14.5939029	kg
Moment of inertia			
1	slug $\cdot$ ft^2	1.3558	kg $\cdot$ m^2
1	lb $\cdot$ ft^2	0.04214	kg $\cdot$ m^2
Speed			
1	ft/s	0.3048	m/s
1	ft/min	0.00508	m/s
1	kt	0.5144444	m/s
1	mph	0.44704	m/s
Force			
1	lb	4.448222	N
Wing loading			
1	lb/ft^2	4.8824	kg/m^2
Density			
1	lb/ft^3	16.018491	kg/m^3

TABLE B2   METRIC UNITS TO IMPERIAL UNITS CONVERSION

Value	Imperial Units	Value	International System of Units
Length			
1	m	39.37007874	in.
1	m	3.280839895	ft
1	m	0.000621371	miles
1	m	0.000539957	n miles
Area			
1	m^2	1550.147264	$in.^2$
1	m^2	10.76391505	ft^2
Volume			
1	m^3	61023.75899	$in.^3$
1	m^3	35.31472483	ft^3
Mass			
1	kg	35.27397828	oz
1	kg	2.204622622	lb
1	kg	0.068521766	slug
Moment of inertia			
1	$kg \cdot m^2$	0.737571913	$slug \cdot ft^2$
1	$kg \cdot m^2$	23.73036	$lb \cdot ft^2$
Speed			
1	m/s	3.280839895	ft/s
1	m/s	196.8503937	ft/min
1	m/s	1.94384466	kt
1	m/s	2.236936292	mph
Force			
1	N	0.224808924	lb
Wing loading			
1	kg/m^2	0.2048	lb/ft^2
Density			
1	kg/m^3	0.062427853	lb/ft^3

TABLE B3   UNIT CONVERSION FUNCTIONS IN MATLAB

Function	Conversion
convlength	from length units to desired length units ('ft', 'm', 'km', 'in', 'mi', 'naut mi')
convmass	from mass units to desired mass units ('lbm', 'kg', 'slugs')
convvel	from velocity units to desired velocity units ('ft/s', 'm/s', 'km/s', 'in/s', 'km/h', 'mph', 'kts', 'ft/min')
convforce	from force units to desired force units ('lbf', 'N')
convdensity	from density units to desired density units ('lbm/ft^3', 'kg/m^3', 'slug/ft^3', 'lbm/in^3')
convpres	from pressure units to desired pressure units ('psi', 'Pa', 'psf', 'atm')
convtemp	from temperature units to desired temperature units ('K', 'F', 'C', 'R')

TABLE B4   UNIT CONVERSION FUNCTIONS IN EXCEL

Function	from_unit or to_unit argument
Distance	Meter "m"
	Statute mile "mi"
	Nautical mile "Nmi"
	Inch "in"
	Foot "ft"
Weight and mass	Gram "g"
	Slug "sg"
	Pound mass (avoirdupois) "lbm"
	Ounce mass (avoirdupois) "ozm"
Force	Newton "N"
	Pound force "lbf"
Pressure	Pascal "Pa" (or "p")
	Atmosphere "atm" (or "at")
	mm of Mercury "mmHg"
Temperature	Degree Celsius "C" (or "cel")
	Degree Fahrenheit "F" (or "fah")
	Kelvin "K" (or "kel")

For example,

= CONVERT(1.0,"lbm","kg")	converts 1 pound mass to kilograms (0.453592)
= CONVERT(68,"F","C")	converts 68 degrees Fahrenheit to Celsius (20)
= CONVERT(CONVERT(100,"ft","m"),"ft","m")	converts 100 sq. feet into square meters (9.290304)

Note, compared to MATLAB Microsoft Excel does not have a conversion for speed and density units.

Canopy Data

Tables C1–C3 show some data on representative ram-air parachutes that could be used for modeling. These data were composed by James Moore during the Orion PADS development. The development of the Orion system was facilitated by a detailed six-degrees-of-freedom canopy dynamics model (CDM), which served as the foundation for all GNC activities. During the GPADS programs for light, medium, and heavy systems as well as the NASA X-38 program (see Sec. 1.2), there was an opportunity to test a succession of cargo parafoils from all of the primary U.S. manufacturers. An interface control document spreadsheet was established to maintain the key dimensions and aerodynamic performance for each canopy. In many cases dimensional data were provided by the canopy manufacturer. In all cases the information was used as inputs to the detailed CDM aeromechanical model where the canopy specific aerodynamic stability derivatives were matched to the available flight-test data.

Some of these canopies were later tested with other PADS and are explicitly mentioned in Chapters 1 and 4. More data including that on the latest canopy designs can be found by looking at the manufacturers' websites, e.g.,

- www.airborne-sys.com/pages/view/cargo-aerial-delivery-systems
- www.cimsa.com/aerial_delivery.aspx
- www.icaruscanopies.aero/index.php?option=com_content&view=article&id=25&Itemid=657
- www.performancedesigns.com/products.asp

TABLE C1   SMALL CANOPY DATA

Parameter	Snowflake (Fig. 1.30)	XP310	MT-1X (Fig. 4.1d)	SET 400-2 (Fig. 4.1d)	BT-80 (Figs. 1.23, 4.1c)	PD500 (Fig. 4.1e)
Manufacturer	RCS	Pioneer	Para-Flite	Strong Enterprise	Parachutes de France	Performance Designs
Airfoil	Clark Y	Clark Y	Clark Y	Clark Y	N/A	Clark Y
Canopy thickness, %	16%	16%	16%	12%	15%	15%
Canopy area, m^2	1	29	34	37	39	46
ft^2	10.8	312	366	398	420	495
Geometric aspect ratio	1.94	2.70	2.19	3.50	2.80	2.73
Canopy span, m	1.4	8.8	8.6	11.4	10.4	11.2
ft	4.6	29.0	28.3	37.3	34.3	36.8
Lower surface chord, m	0.7	3.3	3.9	3.3	3.7	4.1
ft	2.4	10.7	12.9	10.7	12.2	13.5
Rigging angle, deg	6	4	4	7.5	14.6	4

Line area, m²	0.01	0.64	0.63	1.56	0.32	0.69
ft²	0.11	6.89	6.78	16.79	3.44	7.43
Suspension lines length-to-span ratio	0.79	0.75	0.75	0.62	0.59	0.64
Maximum flap deflection in turn, m	0.15	0.81	1.00	0.50	0.44	0.45
ft	0.49	2.66	3.28	1.64	1.44	1.48
L/D at full glide	1.88	2.88	2.53	3.00	3.07	3.11
L/D at maximum brakes	1.31	1.16	1.09	1.00	0.73	1
Canopy weight, kg	0.3	11.3	11.3	6.8	6.8	13.6
lb	0.66	24.9	24.9	15	15	30
Maximum payload mass, kg	6	227	227	680	210	227
lb	13	500	500	1500	462	500
Mass to canopy area ratio, kg/m²	4.9	7.9	6.6	18.2	5.4	4.9
lb/ft²	1	1.6	1.4	3.7	1.1	1.0

TABLE C2   MEDIUM CANOPY DATA

Parameter	C520	MT1-650	GQ ARA	C900	GS750 (Fig. 1.26)	C1200
Manufacturer	Strong Enterprise	Para-Flite	Irvin G.Q.	Strong Enterprise	Pioneer	Strong Enterprise
Airfoil	Clark Y	Clark Y	Clark Y	Clark Y	LS-1	Clark Y
Canopy thickness, %	13%	15.5%	13%	17%	17%	16%
Canopy area, m^2	47	60	65	84	70	112
ft^2	506	650	700	904	753	1206
Geometric aspect ratio	2.8	2.2	3.3	2.7	3	2.5
Canopy span, m	11.5	11.5	14.6	15.1	14.5	16.7
ft	37.6	37.8	48.1	49.4	47.5	54.9
Lower surface chord, m	4.1	5.2	4.4	5.6	4.8	6.7
ft	13.4	17.2	14.6	18.3	15.8	22
Rigging angle, deg	4	4	6	4	10	6

Line area, m²	0.59	0.83	1.43	1.94	1.41	4.41
ft²	6.35	8.93	15.39	20.88	15.18	47.47
Suspension lines length-to-span ratio	0.70	0.75	0.60	0.71	0.72	0.56
Maximum flap deflection in turn, m	0.64	1.4	0.63	1.12	0.61	0.95
ft	2.1	2.7	2.1	3.7	2.0	3.1
L/D at full glide	2.77	2.53	6.03	3.34	4.12	3.33
L/D at maximum brakes	0.73	0.98	1.49	0.73	1.60	2
Canopy weight, kg	12.8	19.5	22.7	26.1	24	27.2
lb	28.2	43	50	57.5	53	60
Maximum payload mass, kg	227	227	544	567	680	998
lb	500	500	1200	1250	1500	2200
Mass to canopy area ratio, kg/m²	4.8	3.8	8.4	6.7	9.8	8.9
lb/ft²	1	0.8	1.7	1.4	2.0	1.8

TABLE C3 LARGE CANOPY DATA

Parameter	PF 5K (Fig. 1.16a)	PF 10K
Manufacturer	Para-Flite	Para-Flite
Airfoil	Modified Clark Y	Modified Clark Y
Canopy thickness, %	16.3%	16.3%
Canopy area, m^2	186	279
ft^2	2,000	3,000
Geometric aspect ratio	3.2	3.2
Canopy span, m	24.4	30
ft	80	98
Lower surface chord, m	7.6	9.3
ft	25	30.6
Rigging angle, deg	3	3
Line area, m^2	4	7
ft^2	43.1	75.4
Suspension lines length-to-span ratio	0.57	0.57
Maximum flap deflection in turn, m	1.28	2.54
ft	4.20	8.33
L/D at full glide	2.88	3.07
L/D at maximum brakes	2	2
Canopy weight, kg	144.7	217.3
lb	319	479
Maximum payload mass, kg	2,268	4,536
lb	5,000	10,000
Mass to canopy area ratio, kg/m^2	12.2	16.3
lb/ft^2	2.5	3.3

Computation of CEP$_{DMPI}$ and CEP$_{MPI}$ for Noncircular Distribution

To compute CEP$_{DMI}$, we need to start with CND that has a bias. (For the purpose of derivation of the correct formula, we may assume that this bias is along one of the axes, say x axis.) Then, Eq. (1.11) becomes

$$p(x, y) = \frac{1}{2\pi\sigma^2} e^{-\frac{(x-\text{Bias})^2+y^2}{2\sigma^2}} \tag{D1}$$

Introducing $x = r\cos(\varphi)$ and $y = r\sin(\varphi)$ allows representing the probability of radius r (CEP$_{DMI}$) being less than some value a as

$$P(r \le a) = \frac{1}{2\pi\sigma^2} \int_0^a \int_0^{2\pi} re^{-\frac{r^2-2r\text{Bias}\cos(\varphi)+\text{Bias}^2}{2\sigma^2}} \, dr \, d\varphi \tag{D2}$$

[which is similar to Eq. (1.12)]. Now, recalling Eq. (1.14) leads to

$$P(r \le a) = \frac{2\ln(2)}{\pi \, \text{CEP}_{MPI}^2} \int_0^a \int_0^{2\pi} re^{-\ln(2)\frac{r^2-2r\text{Bias}\cos(\varphi)+\text{Bias}^2}{\text{CEP}_{MPI}^2}} \, dr \, d\varphi \tag{D3}$$

Finally, introducing $\bar{r} = \dfrac{r}{\text{CEP}_{MPI}}$, $V = \dfrac{\text{Bias}}{\text{CEP}_{MPI}}$ and $k_{\text{Bias}} = \dfrac{\text{CEP}_{DMPI}}{\text{CEP}_{MPI}}$ allows rewriting Eq. (D3) as

$$P(r \le a) = \frac{\ln(2)}{\pi} \int_0^{k_{\text{Bias}}} \int_0^{2\pi} \bar{r}e^{-\ln(2)[\bar{r}^2-2\bar{r}V\cos(\varphi)+V^2]} \, d\bar{r} \, d\varphi \tag{D4}$$

Because we want this probability P to be equal to 0.5 (the definition of CEP$_{DMPI}$), we finally get

$$0.5 = \frac{\ln(2)}{\pi} \int_0^{k_{\text{Bias}}} \int_0^{2\pi} \bar{r}e^{-\ln(2)[\bar{r}^2-2\bar{r}V\cos(\varphi)+V^2]} \, d\bar{r} \, d\varphi \tag{D5}$$

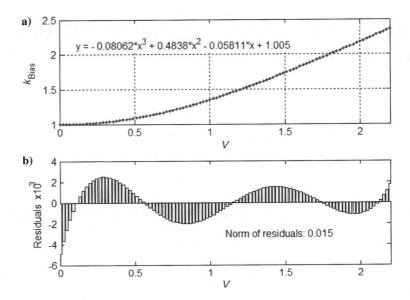

Fig. D1   Numerical solutions of Eq. (D5) and a) cubic regression $k_{Bias} = k_{Bias}(V)$ for $V \in [0; 2.2]$ and b) corresponding residuals.

Unfortunately, Eq. (D5) has no analytical solution (with respect to k_{Bias}), so that we have to solve it numerically for a set of V values. Figure D1 shows such numerical solutions obtained for $V \in [0; 2.2]$ along with a cubic regression with respect to parameter V. This regression, known as RAND 234 formula, is widely used in ballistics. Residuals for this formula are of the order of 10^{-3}.

Because of the wide dispersion of the touchdown points for a self-guided ADS, the V ratio happens to be much smaller compared to that in ballistics. For mature ADS, it is less than 0.1; for new systems it can be around 0.5. Figure D2 shows solutions of Eq. (D5) for another interval, $V \in [0; 0.5]$, along with more accurate cubic regression for this range of V. Figure 1.41 in Sec. 1.3.2 represents the same data but features even simpler, quadratic regression that has the norm of residuals of the order of 2×10^{-4}.

Now let us proceed with the case when $\sigma_x \neq \sigma_y$. Let us assume no bias, $\mu_x = \mu_y = 0$, and no correlation, $\rho = 0$. In this case BND [Eq. (1.8)] reduces to the elliptical distribution

$$p(x, y) = \frac{1}{2\pi\sigma_x\sigma_y} e^{-\frac{x^2}{2\sigma_x^2} - \frac{y^2}{2\sigma_y^2}} \tag{D6}$$

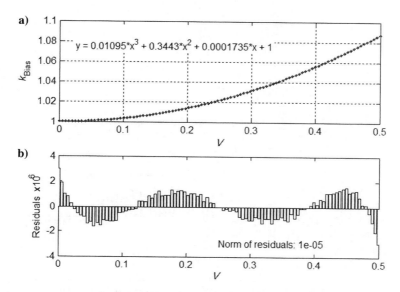

Fig. D2   Numerical solutions of Eq. (D5) and a) a cubic regression $k_{Bias} = k_{Bias}(V)$ for $V \in [0; 0.5]$ b) with the corresponding residuals.

In this case, the probability of radius r being less than some value a is computed as

$$P(r \leq a) = \iint\limits_{S} \frac{1}{2\pi\sigma_x\sigma_y} e^{-\frac{x^2}{2\sigma_x^2} - \frac{y^2}{2\sigma_y^2}} dx\,dy$$

$$= \int\limits_{0}^{r} \int\limits_{0}^{2\pi} \frac{1}{2\pi\sigma_x\sigma_y} e^{-\frac{r\cos(\phi)^2}{2\sigma_x^2} - \frac{r\sin(\phi)^2}{2\sigma_y^2}} r\,d\varphi\,dr \qquad (D7)$$

By equating this expression to 0.5 and solving it numerically, we can establish a dependence of CEP from the ratio $K = \sigma_S/\sigma_L$. These dependences are shown in Fig. D3 for two ranges, $K \in [0.3; 1]$ and $K < 0.3$, along with the linear and quadratic regressions, respectively. As seen from Fig. D3, for these two ranges the best fit yields the following estimates:

$$K \in [0.3; 1]: \quad \text{CEP}_{\text{MPI}} = 0.6183\sigma_S + 0.5619\sigma_L \quad \text{[see Eq.(1.30)]} \qquad (D8)$$

$$K \leq 0.3: \quad \text{CEP}_{\text{MPI}} = (0.9373K - 0.0350)\sigma_S + 0.6757\sigma_L \qquad (D9)$$

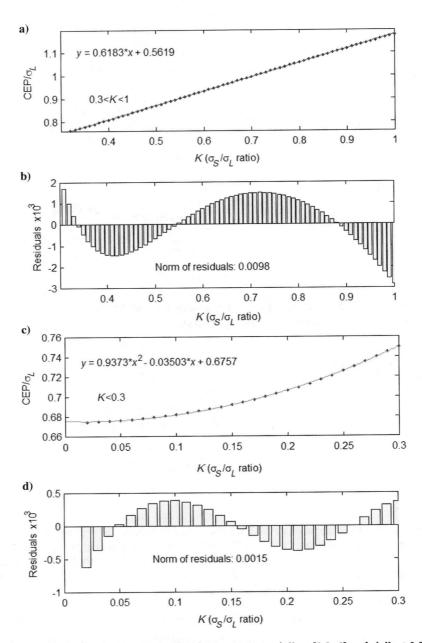

Fig. D3   Numerical solutions of Eq. (D7) for two ranges, a) $K \in [0.3; 1]$ and c) $K < 0.3$, with b) regressions and d) corresponding residuals, respectively.

PADS Geometry and Mass Properties Modeling

For the purpose of system identification, it is quite rare that a full set of required data is readily available. If, however, the set of data includes some imagery of a descending PADS, it may certainly fill the gap or at least serve for the verification purposes. The following presents some analysis performed by Oleg Yakimenko and Robert Rogers.

By analyzing front-view geometry of parafoil system (Fig. E1), the following relationship can be easily established:

$$\bar{R} = (2\varepsilon_b)^{-1} \tag{E1}$$

$$\bar{b}^2 = 8\bar{R}^2[1 - \cos(\varepsilon_b)] - 4\bar{a}^2, \quad \bar{b} = 2(\bar{R} - \bar{a})\tan(\varepsilon_b) \tag{E2}$$

In these equations, a bar denotes relative quantities related to the span of uninflated parafoil b_u. Equations (E2) can be resolved for

$$\bar{a} = \bar{R}\left[1 - \frac{\sqrt{2[1 - \cos(\varepsilon_b)]}}{\text{tg}(\varepsilon_b)}\right], \quad \bar{b} = 2\bar{R}\sqrt{2[1 - \cos(\varepsilon_b)]} \tag{E3}$$

The relative radius $\bar{R}$, span $\bar{b}$, and arch $\bar{a}$ are plotted vs a span angle $2\varepsilon_b$ in Fig. E2.

Figure E2 also shows several data points obtained by processing the Pegasus PADS images and estimating the span angle, relative radius, span, and arch from there. As seen, the centroids of these data points lie exactly on the theoretical curves.

Based on the aforementioned approach, geometric data for the Pegasus PADS (Figs. E1 and 1.11b) were determined as follows:

- Uninflated span, b_u = 11.43 m (37.5 ft)
- Parafoil span angle, $2\varepsilon_b$ = 76.7 $\pm$ 1.5 deg
- Radius, r = 8.55 $\pm$ 0.26 m (28 $\pm$ 0.85 ft)
- Arch, a = 1.74 $\pm$ 0.19 m (5.7 $\pm$ 0.62 ft)
- Span, b = 10.62 $\pm$ 0.23 m (34.8 $\pm$ 0.75 ft)
- Arch-to-span ratio   = 0.164 $\pm$ 0.016
- Front suspension line   = 7.8 $\pm$ 0.26 m (26 $\pm$ 0.85 ft)

a)

b)

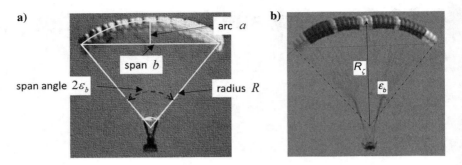

Fig. E1   Front view of the a) Pegasus and b) X-38 PADS.

Given the known chord and airfoil (Clark Y), a few more parameters were estimated as follows:

- Chord, c $= 4.22$ m (13.7 ft)
- Aspect ratio, b/c $= 2.54 \pm 0.04$
- Thickness, t $= 0.63$ m (2 ft)
- Relative thickness, t/c $= 0.15$

Differentiating Eq. (E1) yields

$$\frac{\delta R}{R} = -\frac{\delta \varepsilon_b}{\varepsilon_b} \qquad (E4)$$

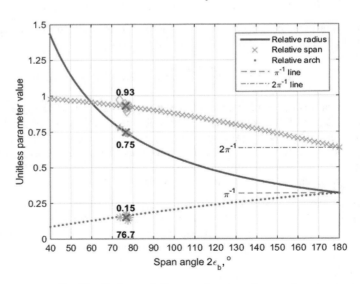

Fig. E2   Basic parafoil parameters vs a flare angle.

Differentiating Eq. (E3) and substituting Eq. (E4) allow tying the errors in span angle to those of parameters R, a, and b.

$$\delta a = -R\left(\frac{1}{\varepsilon_b}\left\{1 - \frac{\sqrt{2[1 - \cos(\varepsilon_b)]}}{\tan(\varepsilon_b)}\right\}\right.$$
$$\left. - \left\{\sqrt{2[1 - \cos(\varepsilon_b)]}\right\}^3 \frac{\cos(\varepsilon_b)[\cos(\varepsilon_b) + 2]}{\sin^4(\varepsilon_b)}\right)\delta\varepsilon_b$$
$$\delta b = b\left\{\frac{\sin(\varepsilon_b)}{\sqrt{2[1 - \cos(\varepsilon_b)]}} - \frac{1}{\varepsilon_b}\right\}\delta\varepsilon_b \qquad \text{(E5)}$$

Figure E3 shows parameters identified from the side-view analysis. Along with chord angle ε_c, the rigging angle μ can be determined (or verified). For Pegasus PADS with a Clark Y airfoil, the rigging angle μ was determined to be -8 deg.

To estimate parafoil's canopy and line mass and inertia properties, the following assumptions have to be made:

- The parafoil is thin ($t \ll c$).

- The parafoil surface is a segment slice of a right circular cone being normal to the cone centerline, the forward slice at the leading edge and the rearward slice at the trailing edge (Fig. E4).

- The cone angle μ is the magnitude of parafoil's rigging angle.

- The radius R_s is normal to the cone centerline and intersects cone surface between the leading and trailing edges at a point $\varsigma \in [0; 1]$ defined as the chord fraction forward of this reference line. (Normally, this is the quarter-chord point, so that $\varsigma = 0.25$.)

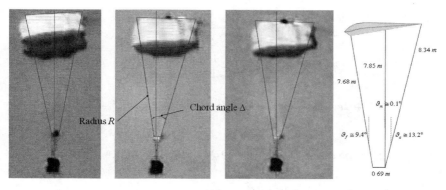

Fig. E3   Estimation of Pegasus PADS geometry from a side-view photo.

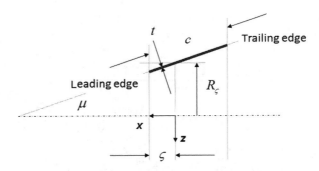

Fig. E4 Longitudinal section of parafoil canopy's conical shape.

The resulting shape of the parafoil is shown in Fig. E5a. In developing expressions for the suspension-line mass and inertia properties, the assumption is that the suspension line geometry forms a pyramid whose density is inversely proportional to the distance from the reference point (Fig. E5b).

With the aforementioned models of the PADS canopy and suspension lines, the canopy moments of inertia are computed by taking the corresponding definite double integrals resulting in

$$
I_{xx}^c = \rho t c \left[\varepsilon_b - \frac{1}{2} \sin(2\varepsilon_b) \right] \left[R_\varsigma^3 + \frac{3}{2} R_\varsigma^2 \mu (1 - 2\varsigma) \xi \right.
$$
$$
+ R_\varsigma \mu^2 (1 - 3\varsigma + 3\varsigma^2) \xi^2 + \frac{1}{4} \mu^3 (1 - 4\varsigma + 4\varsigma^2 - 4\varsigma^3) \xi^3 \right]
$$
$$
+ 2\varepsilon_b \left[\frac{1}{3} R_\varsigma (1 - 3\varsigma + 3\varsigma^2) \xi^2 + \frac{1}{4} \mu (1 - 4\varsigma + 4\varsigma^2 - 4\varsigma^3) \xi^3 \right] \quad \text{(E6)}
$$

$$
I_{yy}^c = \rho t c \left[\varepsilon_b + \frac{1}{2} \sin(2\varepsilon_b) \right] \left[R_\varsigma^3 + \frac{3}{2} R_\varsigma^2 \mu (1 - 2\varsigma) \xi \right.
$$
$$
+ R_\varsigma \mu^2 (1 - 3\varsigma + 3\varsigma^2) \xi^2 + \frac{1}{4} \mu^3 (1 - 4\varsigma + 4\varsigma^2 - 4\varsigma^3) \xi^3 \right]
$$
$$
+ 2\varepsilon_b \left[\frac{1}{3} R_\varsigma (1 - 3\varsigma + 3\varsigma^2) \xi^2 + \frac{1}{4} \mu (1 - 4\varsigma + 4\varsigma^2 - 4\varsigma^3) \xi^3 \right] \quad \text{(E7)}
$$

$$
I_{zz}^c = \rho t c \varepsilon_b \left[2 R_\varsigma^3 + 3 R_\varsigma^2 \mu (1 - 2\varsigma) \xi + 2 R_\varsigma \mu^2 (1 - 3\varsigma + 3\varsigma^2) \xi^2 \right.
$$
$$
+ \frac{1}{2} \mu^3 (1 - 4\varsigma + 4\varsigma^2 - 4\varsigma^3) \xi^3 \right] \quad \text{(E8)}
$$

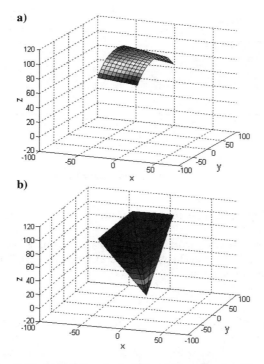

Fig. E5   A 3D view of the a) parafoil and b) suspension lines shape.

$$I_{xz}^c = \rho tc\xi 2 \sin(\varepsilon_b)\left[\frac{1}{2}R_s^2(1-2s) + \frac{2}{3}R_s\mu(1-3s+3s^2)\xi\right.$$

$$\left. + \frac{1}{4}\mu^2(1-4s+4s^2-4s^3)\xi^2\right] \tag{E9}$$

where $\xi = c(1+\mu^2)^{-0.5}$. The canopy mass and center of gravity are given by

$$m^c = \rho tc\varepsilon_b[2R_s + (1-2s)\xi] \tag{E10}$$

$$x_{cg}^c = -\frac{1}{m^c}\rho tc\varepsilon_b\left[R_s^2(1-2s)\xi + \frac{2}{3}\mu(1-3s+3s^2)\xi^2\right] \tag{E11}$$

$$z_{cg}^c = -\frac{1}{m^c}\rho tc\xi 2 \sin(\varepsilon_b)\left[r_\alpha^2 + R_s\mu(1-2s)\xi + \frac{1}{3}\mu^2(1-3s+3s^2)\xi^2\right] \tag{E12}$$

Equations (E6–E12) can also be used for both canopy fabric contributed mass properties, but also the mass of the enclosed air. For the former, the known fabric weight can be used to compute the value of product ρtc; in the latter, atmospheric density ρ is used to compute the enclosed air mass. If

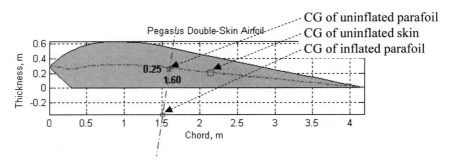

Fig. E6   Determining canopy's c.g.

known, the canopy fabric mass computed from Eq. (E11) can be used as a check of equation formulation.

For the Pegasus PADS, the mass of canopy was estimated as 6.44 kg (14.2 lb) and the mass of trapped air at the sea level as 21.9 kg (48.3 lb). The c.g. of inflated parafoil was determined to lie

$$R\left[1 - (\varepsilon_b)^{-1}\sin(\varepsilon_b)\right] \tag{E13}$$

beneath the c.g. of airfoil along the line tilted clockwise by $-\mu$, that is, 0.62 m (2 ft) (Fig. E6).

For the suspension lines with a known mass m^l, the resulting inertias and c.g. expressions are

$$I^l_{xx} = \frac{1}{3}m^l(R_s^2 + \frac{1}{12}b^2), \quad I^l_{yy} = \frac{1}{3}m^l\left[R_s^2 + \frac{1}{3}c^2 - (1-s)sc^2\right]$$

$$I^l_{zz} = \frac{1}{3}m^l\left[\frac{1}{12}b^2 + \frac{1}{3}c^2 - (1-s)sc^2\right], \quad I^l_{xz} = \frac{1}{6}m^l(1-2s)cR_s \tag{E14}$$

$$x^l_{cg} = -\frac{1}{4}(1-2s)c, \quad z^l_{cg} = -0.5R_s \tag{E15}$$

PADS Penetration Speed

Let us start this appendix from estimating the maximum penetration (forward) speed. Considering a parachute gliding in quiescent air of density ρ, with weight W and wing area S, the true airspeed of the parachute and glide angle γ are determined by Eqs. (2.26) and (2.22). If we rewrite Eq. (2.22) as

$$\cos(\gamma) = \frac{C_L}{\sqrt{C_L^2 + C_D^2}} \tag{F1}$$

and plug the result along with Eq. (2.26) into the first of the two equations of Eq. (2.27), we obtain the expression for the horizontal component of the airspeed vector of parachute V_h (penetration speed). Squaring this resulting equation yields

$$V_h^2 = V^2 \cos^2(\gamma) = \frac{2W}{\rho S} \frac{C_L^2}{\sqrt{C_L^2 + C_D^2}^3} \tag{F2}$$

Noting that $V_h \to 0$ as $C_L \to 0$ or $C_L \to \infty$, V_h has an extremum with respect to C_L. The lift coefficient maximizing V_h can be found from the equation $\partial(\ln(V^2))/\partial C_L = 0$, which yields

$$\frac{2}{C_L} - \frac{3}{C_L^2 + C_D^2}\left(C_L + C_D \frac{\partial C_D}{\partial C_L}\right) = 0 \tag{F3}$$

Assuming a parabolic drag polar

$$C_D = C_{D0} + K C_L^2 \tag{F4}$$

[similar to that of Eq. (6.4) but featuring a different C_{D0}] Eq. (F3) results in

$$4K C_D^2 + C_D(1 - 6K C_{D0}) - C_{D0} = 0 \tag{F5}$$

Equation (F5) has an obvious solution

$$C_D = -\frac{1}{8K}\left(-1 + 6K C_{D0} + \sqrt{1 + 4K C_{D0} + 36K^2 C_{D0}^2}\right) \tag{F6}$$

Being substituted into Eq. (F4), it gives

$$C_L = \frac{1}{2\sqrt{2}\,K}\sqrt{-1 + 2KC_{D0} + \sqrt{1 + 4KC_{D0} + 36K^2C_{D0}^2}} \approx \sqrt{2}C_{D0} \qquad \text{(F7)}$$

At this lift coefficient,

$$\gamma \approx \cos^{-1}\left(\sqrt{2/3}\right) \approx 35\text{ deg}, \quad V \approx \sqrt{\frac{2W}{\rho S C_{D0}\sqrt{3}}}, \quad V_h \approx \sqrt{\frac{4W}{\rho S C_{D0}3\sqrt{3}}} \qquad \text{(F8)}$$

Let us estimate parasitic drag coefficient C_{D0} for the entire ADS. Consider the canopy attached to the payload by the total of 500 m (1,640 ft) of 1.5 mm (0.06 in.) lines of a circular cross section. All lines are assumed approximately perpendicular to the flow, and the aerodynamic interference between them is ignored. Given a typical flight velocity of 10 m/s (19.4 kt), the corresponding crossflow Reynolds number on a single line is of the order of 10^3 [McCormick 1979], and the drag coefficient of all lines, based on their frontal area, should be about 1.1. Accordingly, $C_D^l \approx 0.03$. Assuming a store drag area to be of the order of 1 m^2 (10.8 ft^2), $C_D^s = 1/26 \approx 0.04$ [see Eq. (2.18)]. (For the sake of comparison, for a pilot in a sitting position a store drag area is about 0.5 m^2 (5.4 ft^2) [Hoerner 1965].) Consequently, the parasitic drag coefficient of the entire parachute is $C_{D0} = C_{D0}^c + C_D^l + C_D^s \approx 0.1$.

Assuming $K \approx 0.1$, $C_{D0} \approx 0.1$, the wing loading of 35 N/m^2(0.73 lb/ft^2) and standard sea-level conditions, that is, $\rho \approx 1.225$ kg/m^3 (0.0765 lb/ft^3), the lift coefficient corresponding to the maximum penetration speed happens to be 0.14 [Eq. (F7)]. According to Eq. (F8), this corresponds to $V \approx 18.2$ m/s (35.4 kt) and $V_h \approx 14.8$ m/s (28.8 kt). For comparison, the lift coefficient, maximizing the glide ratio,

$$\sqrt{C_{D0}/K}$$

is approximately 1. The corresponding glide angle and penetrating speed are $\gamma \approx 11.3$ deg, $V \approx 13.4$ m/s (26 kt), and $V_h \approx 7.4$ m/s (14.4 kt).

Next, let us estimate the product $C_{L\alpha}K$ in Eq. (6.14) defining the existence of longitudinal trim conditions for a parafoil-based ADS. For this analysis, we will use the approximation of drag-polar constant K for an arched wing

$$K \approx \frac{1 + \delta}{\pi \text{AR}} + k_s \qquad \text{(F9)}$$

[compare Eq. (2.10)], where δ is a planform efficiency correction factor, and k_s is a separation drag correction factor [McCormick 1979], and AR is the effective aspect ratio of the wing [Iosilevskii 1996].

Using Eqs. (4.38) and (F9), the product $C_{L\alpha}K$ can be written as

$$
\begin{aligned}
C_{L\alpha}K &= \frac{\pi AR C_{L\alpha}^a}{\sqrt{(\pi AR)^2 + C_{L\alpha}^{a\,2}} + C_{L\alpha}^a} \left(\frac{1+\delta}{\pi AR} + k_s\right) \\
&= \frac{C_{L\alpha}^a[(1+\delta) + k_s \pi AR]}{\sqrt{(\pi AR)^2 + C_{L\alpha}^{a\,2}} + C_{L\alpha}^a}
\end{aligned}
\tag{F10}
$$

with all parameters appearing on the right-hand side of this equation being non-negative [Iosilevskii 1995].

Equation (F10) has a maximum

$$
\max_{AR}(C_{L\alpha}K) = \frac{(1+\delta)^2 + k_s^2 C_{L\alpha}^{a\,2}}{2(1+\delta)}
\tag{F11}
$$

at

$$
AR = \frac{2k_s C_{L\alpha}^{a\,2}(1+\delta)}{\pi[(1+\delta)^2 - k_s^2 C_{L\alpha}^{a\,2}]}
\tag{F12}
$$

Therefore, if $k_s C_{L\alpha}^a < 1 + \delta$ (which is usually the case), then $C_{L\alpha}K < 1$. Moreover,

$$
C_{L\alpha}K \leq \frac{(1+\delta)^2 + k_s^2 C_{L\alpha}^{a\,2}}{2(1+\delta)} < 1
\tag{F13}
$$

In the example considered by Iosilevskii [1995], for a gliding parachute with a canopy area of 26 m^2 (280 ft^2), span of 10 m (33 ft) (AR = 3.85), and an LS1-0417 wing section with 8.4% chord air intake, we have $C_{D0}^c \approx 0.03$ and $C_{L\alpha}^a \approx 5$ [Ross 1993]. Using Eq. (4.38), we find $C_{L\alpha} = 3.34$. From [McCormick 1979] and [Ross 1993], we take $\delta \approx 0.1$ and $k_s \approx 0.02$, respectively. Then, Eq. (F9) yields $K = 0.11$. Thus, $C_{L\alpha}K = 0.37$. Substituting the same values into Eq. (F11), we find the maximum value of $C_{L\alpha}K$ to be 0.55. This is illustrated in Fig. F1 where three families of plots for three combinations $[\delta, k_s] = [0.2; 0.05]$, $[\delta, k_s] = [0.1; 0.02]$, and $[\delta, k_s] = [0; 0]$ are plotted for three different values of $C_{L\alpha}^a$. Both $C_{L\alpha}K = 0.37$ and $C_{L\alpha}K = 0.55$ points corresponding to the $C_{L\alpha}^a = 5/\delta$, $k_s = [0.1; 0.02]$ line are shown as well. As seen, the maximum value of $C_{L\alpha}K$ occurs at low aspect ratios, which are not practical. For the AR range of the existing PADS (the shaded area in Fig. F1 corresponds to the actual values of PADS AR as shown in Fig. 1.36), the values of $C_{L\alpha}K$ are indeed smaller than the maximum value defined by Eq. (F11). For sportive parafoils, featuring higher AR compared to cargo parafoils, the $C_{L\alpha}K$ product will be even smaller; hence, Eq. (F13) holds for all parafoils.

Lastly, let us estimate the minimal lift coefficient at which, according to Eq. (6.33), trim is still possible, $C_{L2}(\delta_{\min})$. Toward this end, let us approximate

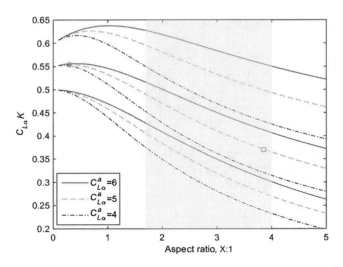

Fig. F1   Dependence of the product $C_{L\alpha}K$ for a parafoil canopy from aspect ratio, $C_{L\alpha}^a$, δ, and k_s.

a canopy as a straight rigid wing equipped with TE full span plain (sealed) flap. Let c_k be the relative chord of the flap and $\delta_s \ll 1$ be the angle of its deflection. Neglecting the dependence of C_{D0}^c on δ_s [McCormick 1979],

$$\partial C_{m0}/\partial \delta_s \approx \partial C_{m0}^c/\partial \delta_s \tag{F14}$$

In the framework of a lifting-line theory, both $\partial C_{m0}/\partial \delta_s$ and $(1/C_{L\alpha})\partial C_{L0}/\partial \delta_s$ are known to be independent of the aspect ratio of the wing [Glauert 1983]. Hence, using results of the thin airfoil theory [Glauert 1983]

$$C_{L2} = \frac{\pi}{z_{cg} - z_c} \frac{\sin(\theta_k)[1 + \cos(\theta_k)]}{2[\theta_k + \sin(\theta_k)]} \tag{F15}$$

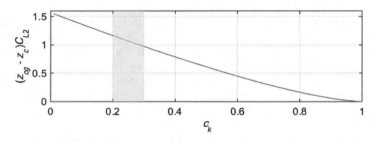

Fig. F2   Minimal lift coefficient as a function of the flaps' relative chord.

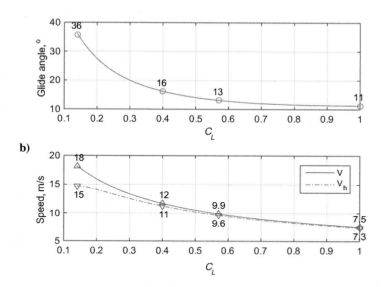

Fig. F3   a) Glide angle and b) total speed and horizontal component of a speed vs the lift coefficient.

where

$$\theta_k = \cos^{-1}(1 - 2c_k) \tag{F16}$$

With $z_{cg} - z_c$ between 2.2 and 2.5, and relative flaps chord between 0.2 and 0.3, Eq. (F15) yields the value of $C_{L2}(\delta_{min})$ between 0.57 and 0.4, about 0.5 on average (see Fig. F2). With a wing loading of 35 N/m² (0.73 lb/ft²), these values correspond to a flight speed between 10 and 12 m/s (19.4 and 23.3 kt) at standard sea-level conditions, in good agreement with the top speed data of current recreational parachute designs.

The exact values of aforementioned glide angles and penetration speeds for the cases of the lift coefficient equal to 0.14, 0.4, 0.57, and 1 are shown in Fig. F3. As seen, lift coefficients providing a larger penetration speed, which is one of the ultimate design goals, happen to be much lower than the typical minimal lift coefficient limit of about 0.5. To fly faster the c.g. position should be moved forward. For a cargo parafoil, this can be done by adjusting a rigging angle as discussed in Secs. 2.5.1 and 8.2 (the human jumper can simply pull the lines attached to the forward half of the canopy).

Geometry and Mass Properties of Different-Weight PADS Models

Table G1 presents data used by Goodrick [1984] in computer simulations to study the scale effects on ram-air ADS performance. These data can be used for modeling and can also serve as a benchmark to compare with the corresponding parameters of the actual PADS. All canopies utilized AR = 2. For convenience, Table G2 presents the same data as in Table G1 but in imperial units.

TABLE G1 C.G. HEIGHT AND MOMENTS OF INERTIA FOR DIFFERENT-SIZE ADS MODELS IN SI

S, m²	m_s, kg	l, m	h_{cg}, m	I_{xx}, kg·m²	I_{yy}, kg·m²	I_{zz}, kg·m²	$\frac{h_{cg}}{l}$, %	$\frac{I_{xx}}{m_s h_{cg}^2}$, %	$\frac{I_{yy}}{m_s h_{cg}^2}$, %	$\frac{I_{zz}}{m_s h_{cg}^2}$, %
3	15	2.7	0.13	5.7	5.4	0.73	4.7	2,315	2,199	296
	30	2.7	0.066	5.9	5.5	0.73	2.4	4,451	4,232	557
	60	2.7	0.038	5.9	5.6	0.73	1.4	6,788	6,459	840
6	30	3.8	0.25	31.4	29.74	4.12	6.6	1,672	1,586	220
	60	3.8	0.13	32.3	30.7	4.12	3.4	3,237	3,075	413
	120	3.8	0.066	32.8	31.2	4.12	1.74	6,280	5,971	788
12	59	5.4	0.47	174.4	165.2	23.3	9.3	1,201	1,138	160
	118	5.4	0.26	182	172.8	23.3	4.8	2,281	2,167	292
	236	5.4	0.14	186	176.9	23.3	2.5	4,140	4,238	558
18.6	165	6	0.5	550	552	70	8	1,333	1,338	170
31.4	165	8	1.3	1,955	1,854	258	16	701	665	98
	250	8	0.9	2,062	1,961	258	11	1,018	968	127
75	250	13	4.3	15,525	14,634	2,276	33	336	316	49
	500	13	2.6	18,276	17,385	2,276	20	541	514	67
150	500	19	7.8	83,070	78,029	12,873	41	273	257	42
	1,000	19	4.9	102,186	97,146	12,873	26	426	405	53
300	1,000	27	13.4	412,524	384,010	72,820	50	230	214	41
	2,000	27	8.9	531,759	503,245	72,820	33	336	318	46
600	2,500	38	20	2,190,032	2,028,730	411,929	53	219	203	41
	5,500	38	12.8	2,956,872	2,795,569	411,929	34	328	310	46

TABLE G2   C.G. HEIGHT AND MOMENTS OF INERTIA FOR DIFFERENT-SIZE ADS MODELS IN IMPERIAL UNITS

S, ft^2	m_s, lb	l, ft	h_{cg} ft	I_{xx}, lb · ft^2	I_{yy}, lb · ft^2	I_{zz}, lb · ft^2
33.2	33.1	0.4	0.4	135	128	17
	66.1	0.2	0.2	138	131	17
	132.3	0.1	0.1	140	133	17
64.6	66.1	0.8	0.8	744	706	98
	132.3	0.4	0.4	767	729	98
	264.6	0.2	0.2	779	741	98
129	130.1	1.6	1.6	4,137	3,921	553
	260.1	0.9	0.9	4,318	4,101	553
	520.3	0.5	0.5	4,415	4,198	553
200	363.8	1.6	1.6	13,052	13,099	1,661
338	363.8	4.3	4.3	46,392	43,995	6,122
	551.2	3.0	3.0	48,931	46,535	6,122
807	551.2	14.1	14.1	368,408	347,265	54,009
	1,102	8.5	8.5	433,689	412,546	54,009
1,615	1,102	25.6	25.6	1,971,251	1,851,628	305,476
	2,205	16.1	16.1	2,424,874	2,305,275	305,476
3,229	2,205	44.0	44.0	9,789,195	9,112,557	1,728,019
	4,409	29.2	29.2	12,618,641	11,942,004	1,728,019
6,458	5,512	65.6	65.6	51,969,459	48,141,763	9,775,075
	12,125	42.0	42.0	70,166,573	66,338,852	9,775,075

Geometry and Mass Properties of the G-12-Based AGAS

Tables H1–H3 contain geometric and mass data used to estimate the position of static mass center, the moments of inertia (Table H4), and arms of the corresponding forces while computing moments acting on a generic G-12-based AGAS.

TABLE H1   DIMENSIONS OF AGAS COMPONENTS

Component	Value
Length of suspension lines l_{SL}	15.55 m (51 ft)
Length of pressurized PMA l_{PMA}	5.80 m (19 ft)
Length of vented PMA l_{PMA}	7.62 m (25 ft)
Radius of uninflated canopy R_0	9.75 m (32 ft)
Radius of inflated canopy $R_p = \dfrac{2}{3}R_0$	6.50 m (21.3 ft)
Canopy shape ratio ε	0.82
Cone half-angle γ	15.31 deg
Dimension of a cubic payload container a_{pl}	1.22 m (4 ft)

TABLE H2   MASSES M_i OF INDIVIDUAL AGAS COMPONENTS

i	Component	Mass
1	Canopy	22.8 kg (50.3 lb)
2	Suspension lines	35.3 kg (77.8 lb)
3	Four PMAs	13.4 kg (29.5 lb)
4	Payload	990 kg (2183 lb)
ADS total		**1061.5 kg (2340 lb)**

TABLE H3   AUXILIARY GEOMETRIC RELATIONS

z-Coordinate of a Specific Point	Value, m
Centroid of hemispheroidal canopy z_1	−2.8 m (9 ft)
Centroid of suspension lines z_2	7.5 m (24.6 ft)
Centroid of PMAs z_3	17.8 m (58.4 ft)
Centroid of payload z_4	21.2 m (69.6 ft)
Static mass center of the whole ADS z_G	20.2 m (66.2 ft)
Center of pressure $z_P = -\dfrac{3}{8}\varepsilon R_p$	−2 m (5.6 ft)

TABLE H4   MOMENTS OF INERTIA OF AGAS COMPONENTS

i	Component	$\bar{I}^i_{aa} = \bar{I}^i_{bb}$	$I^i_{xx} = I^i_{yy}$	$I^i_{zz} = \bar{I}^i_{cc}1$
1	Canopy	365.9 kg · m² (8,683 lb · ft²)	539.2 kg · m² (12,795 lb · ft²)	624 kg · m² (14,808 lb · ft²)
2	Suspension lines	1,047.1 kg · m² (24,848 lb · ft²)	3,032.2 kg · m² (71,956 lb · ft²)	770.8 kg · m² (18,291 lb · ft²)
3	PMAs	53.4 kg · m² (1,267 lb · ft²)	4,296.8 kg · m² (101,965 lb · ft²)	37 kg · m² (878 lb · ft²)
4	Payload	245.6 kg · m² (5,828 lb · ft²)	445,287.9 kg · m² (10,566,842 lb · ft²)	245.6 kg · m² (5,828 lb · ft²)
ADS total			453,156.1 kg · m² (10,753,557 lb · ft²)	1,677.4 kg · m² (39,805 lb · ft²)

References

Abbott, I. H., and von Doenhoff, A. E. (1959), *Theory of Wing Sections*, Dover, New York.

Adkins, C. N. (1982), "The Added Mass and Inertia of Lightly Loaded Aircraft," unpublished report.

Aerolight (2014), "Powered Paragliding," www.aerolight.com

Aguilar, D. M., and Figueiredo, W. (1995), "Feasibility Demonstration of Autonomous Precision Aerial Delivery System (APADS) 'PEGASUS'," *Proceedings of 13th AIAA Aerodynamic Decelerator Systems Technology Conference and Seminar*, AIAA, Washington, D. C.

AID (2014), Airfoil Investigation Base, www.airfoildb.com

Alaibadi, S. K., Garrard, W. L., Karlo, V., Mittal, T. E., Tezduyar, T. E., and Stein, K. R. (1995), "Parallel Finite Element Computations of the Dynamics of Large Ram Air Parachutes," *Proceedings of 13th AIAA Aerodynamic Decelerator Systems Technology Conference and Seminar*, AIAA, Washington, D. C.

Albers, S. (1995), "The LAPS Wind Analysis," *Weather and Forecasting*, Vol. 10, No. 2, pp. 342–352.

Alexander, J., Powell, D. B., and Krainski, W. J. (2007), "Affordable and Lightweight Composite Airdrop Platform," *Proceedings of the 19th AIAA Aerodynamic Decelerator Systems Technology Conference and Seminar*, AIAA, Reston, VA.

Allen, R. F. (1995), "Orion Advanced Precision Airborne Delivery System," *Proceedings of the 13th AIAA Aerodynamic Decelerator Systems Technology Conference and Seminar*, AIAA, Washington, D. C.

Altmann, H. (2009), "Numerical Simulation of Parafoil Aerodynamics and Dynamic Behavior," *Proceedings of the 20th AIAA Aerodynamic Decelerator Systems Technology Conference*, AIAA, Reston, VA.

Altmann, H. (2011), "An Enhanced GNC Functionality Combining Pre-Flight Wind Forecast and in-Flight Identified Wind," *Proceedings of the 21st AIAA Aerodynamic Decelerator Systems Technology Conference*, AIAA, Reston, VA.

Altmann, H. (2013), "Influence of Wind on Terminal Guidance and Landing Precision of Autonomous Parafoil Systems," *Proceedings of the 22nd AIAA Aerodynamic Decelerator Systems Technology Conference*, AIAA, Reston, VA.

Altmann, H., and Windl, J. (2005), "ParaLander: A Medium-Weight Demonstrator for Autonomous, Range-Optimized Aerial Cargo Delivery," *Proceedings of the 18th AIAA Aerodynamic Decelerator Systems Technology Conference and Seminar*, AIAA, Reston, VA.

Babinsky, H. (1999), "The Aerodynamic Performance of Paragliders," *The Aeronautical Journal of the Royal Aeronautical Society*, pp. 421–428.

Baca, K. (1984), "An Experimental Study of the Performance of Clustered Parachutes in a Low Speed Wind Tunnel," *Proceedings of the 8th AIAA Aerodynamic Decelerator Systems Technology Conference*, AIAA, Reston, VA.

Balaji, R., Mittal, S., and Rai, A. K. (2005), "Effect of Leading Edge Cut on the Aerodynamics of Ram-Air Parachutes," *International Journal for Numerical Methods in Fluids*, Vol. 47, pp. 1–17.

Barber, J., Montague, D., and Barello, L. (2011), "Use of Agile Methodologies to Develop Robust and Supportable Parachute Systems for the U.S. DoD," *Proceeding of the 21st AIAA Aerodynamic Decelerator Systems Technology Conference*, AIAA, Reston, VA.

Barrows, T. M. (May 2001), "Apparent Mass of Parafoils with Spanwise Camber," *Journal of Aircraft*, Vol. 39, No. 3, 2002, pp. 445–451; also Barrows, T. M., "Apparent Mass of Parafoils with Spanwise Camber," *Proceeding of the 16th AIAA Aerodynamic Decelerator Systems Technology Conference and Seminar*, AIAA, Reston, VA.

Bashkina, L., Soinov, A., and Tokareva, L. (1989), "An Investigation of Pressure Distribution over Parafoils in Wind Tunnels," *Unsteady Mechanics*, Vol. 22, Physical-Technical Inst., Kazan, Russia.

Bates, J. (1990), *Parachuting, from Student to Skydiver*, Tab Books, Pennsylvania.

Bennett, T., and Fox, R. (2003), "Design, Development and Flight Testing of the NASA X-38 7,500 ft^2 Parafoil Recovery System," *Proceeding of the 17th AIAA Aerodynamic Decelerator Systems Technology Conference and Seminar*, AIAA, Reston, VA.

Bennett, T., and Fox, R. (2005), "Design, Development and Flight Testing of the U.S. Army 4,200 ft^2 Parafoil Recovery System," *Proceedings of the 18th AIAA Aerodynamic Decelerator Systems Technology Conference and Seminar*, AIAA, Reston, VA.

Benney, R., and Accorsi, M. (2001), "Aerodynamic Decelerator Systems – Diverse Challenges and Recent Advances," *Journal of Aircraft*, Vol. 38, No. 5, p. 785.

Benney, R., Barber, J., McGrath, J., McHugh, J., Noetscher, G., and Tavan, S. (2005a), "The Joint Precision Airdrop System Advanced Concept Technology Demonstration," *Proceedings of the 18th AIAA Aerodynamic Decelerator Systems Technology Conference and Seminar*, AIAA, Reston, VA.

Benney, R., Barber, J., McGrath, J., McHugh, J., Noetscher, G., and Tavan, S. (2005b), "The New Military Applications of Precision Airdrop Systems," *Proceedings of the Infotech@Aerospace Conference*, AIAA, Reston, VA.

Benney, R., Brown, G., and Stein, K. (1999), "A New Pneumatic PMA: Its Use in Airdrop Applications," *Proceedings of 15th CAES/AIAA Aerodynamic Decelerator Systems Technology Conference*, AIAA, Reston, VA.

Benney, R., Henry, M., Lafond, K., Meloni, A., and Patel, S. (2009a), "DOD New JPADS Programs & NATO Activities," *Proceedings of the 20th AIAA Aerodynamic Decelerator Systems Technology Conference and Seminar*, AIAA, Reston, VA.

Benney, R., McGrath, J., McHugh, J., Meloni, A., Noetscher, G., Tavan, S., and Patel, S. (2007), "DOD JPADS Programs Overview & NATO Activities," *Proceedings of the 19th AIAA Aerodynamic Decelerator Systems Technology Conference and Seminar*, AIAA, Reston, VA.

Benney, R., Meloni, A., Cronk, A., and Tiaden, R. (2009b), "Precision Airdrop Technology Conference and Demonstration 2007," *Proceedings of the 20th AIAA Aerodynamic Decelerator Systems Technology Conference and Seminar*, AIAA, Reston, VA.

Benney, R., Meloni, A., Henry, M., Lafond, K., Cook, G., Patel, S., and Goodell, L. (2009c), "Joint Medical Distance Support and Evaluation (JMDSE) Joint Capability Technology Demonstration (JCTD) & Joint Precision Air Delivery Systems (JPADS)," *Proceedings of the Special Operations Forces Industry Conference*, Tampa, FL.

Benolol, S., and Zapirain, F. (2005), "The FASTWing Project - Parafoil Development and Manufacturing," *Proceedings of the 18th AIAA Aerodynamic Decelerator Systems Technology Conference and Seminar*, AIAA, Reston, VA.

Benton, J. E., and Yakimenko, O. A. (2013), "On the Development of Autonomous HAHO Parafoil System for Targeted Payload Return," *Proceeding of the 22nd AIAA Aerodynamic Decelerator Systems Technology Conference and Seminar*, AIAA, Reston, VA.

Bergeron, K., Fejzic, A., and Tavan, S. (2011), "AccuGlide 100: Precision Airdrop Guidance and Control via Glide Slope Control," *21st AIAA Aerodynamic Decelerator Systems Technology Conference*, AIAA, Reston, VA.

Bergeron, K., Noetscher, G., Shurtliff, M., Tavan, S., and Deazley, F. (2015), "Longitudinal Control for Ultra Light Weight Guided Parachute Systems," *Proceeding of the 23rd AIAA Aerodynamic Decelerator Systems Technology Conference and Seminar*, AIAA, Reston, VA.

Berland, J., Gargano, W., Bagdonovich, B., and Barber, J. (2009), "Autonomous Precision Delivery of 42,000 Pounds (19,000 kg) Under One Parachute," *Proceedings of the 19th AIAA Aerodynamic Decelerator Systems Technology Conference and Seminar*, AIAA, Reston, VA.

Blevins, R. D. (1990), *Flow-Induced Vibration*, 2nd ed., Krieger Publishing, Malabar, FL.

Boggs, C. (2015), "Computed Air Release Point Data Collection, Post Processing and Applications," *Proceedings of the 23rd AIAA Aerodynamic Decelerator Systems Technology Conference and Seminar*, AIAA, Reston, VA.

Bourakov, E. A., Yakimenko, O. A., and Slegers, N. J. (2009), "Exploiting a GSM Network for Precise Payload Delivery," *Proceedings of the 20th AIAA Aerodynamic Decelerator Systems Technology Conference*, AIAA, Reston, VA.

Bratley, P., and Fox, B. L. (1988), "Algorithm 659: Implementing Sobol's Quasirandom Sequence Generator," *ACM Transactions on Mathematical Software*, Vol. 14, No. 1, pp. 88–100.

Brennen, C. E. (1982), "A Review of Added Masses and Fluid Inertial Forces," Naval Civil Engineering Lab., Rept. CR 82.010, Port Hueneme, CA.

Brocato, B. (2003), "Vertical Wind Analysis," *Proceedings of the 17th AIAA Aerodynamic Decelerator Systems Technology Conference and Seminar*, AIAA, Reston, VA.

Brown, G., and Benney, R. (2005), "Precision Aerial Delivery Systems in a Tactical Environment," *Proceedings of the 18th AIAA Aerodynamic Decelerator Systems Technology Conference and Seminar*, AIAA, Reston, VA.

Brown, G., Haggard, R., Almassy, R., Benney, R., and Dellicker, S. (1999a), "The Affordable Guided Airdrop System (AGAS)," *Proceedings of the 15th AIAA Aerodynamic Decelerator Systems Technology Conference and Seminar*, AIAA, Reston, VA.

Brown, G., Norton, B., and Lingard, J. S. (1999b), "Inertially Controlled Recovery System (ICRS)," *Proceedings of the 15th CEAS/AIAA Aerodynamic Decelerator Systems Technology Conference*, AIAA, Reston, VA.

Brown, G. J. (1993), "Parafoil Steady Turn Response to Control Input," *Proceedings of the 12th RAeS/AIAA Aerodynamic Decelerator Systems Technology Conference and Seminar*, AIAA, Washington, D. C.

Bui, X. N. (1994), "Planifcation de Trajectoires pour un Robot Polygonal Non-Holonome dans un Environement Polygonal," Ph.D. Dissertation, Ecole Nationale Superieure des Mines de Paris, France.

Bui, X. N., Souères, P., and Laumond, J. P. (1994), "The Shortest Path Synthesis for Non-Holonomic Robots Moving Forwards," INRIA Rept. No. 2153.

Burke, B. (1997), *The Canopy Pilot's Handbook*, 1st ed., Skydive Arizona, http://www.axisflightschool.com/pdf/references_1997CanopyPilotHandbook.pdf

Burke, S. M., and Ware, G. M. (1967), *Static Aerodynamic Characteristics of Three Ram-Air Inflated Low Aspect Ratio Fabric Wings*, NASA TN D-4182, Hampton, VA.

Byushgens, A. G., and Shilov, A. A. (1972), "Dinamicheskaja Model' Parachuta i Opredelenie ego Charakteristik (The Dynamic Model of Parachute and Determination of its Characteristics)," *Uchenie Zapiski TsAGI*, Vol. 3, No. 4, pp. 49–58 (in Russian); Translation FTD-MT-24-0551-75.

Cacan, M., Scheuermann, E., Ward, M., and Costello, M. (2005), "Hyperadaptive Control of Heavily Damaged Autonomous Airdrop Systems," *Proceeding of the 23rd AIAA Aerodynamic Decelerator Systems Technology Conference and Seminar*, AIAA, Reston, VA.

Cacan, M. R., Ward, M. B., Scheuermann, E., and Costello, M. (2015), "Human-In-The-Loop Control of Guided Airdrop Systems," *Proceedings of the 23rd AIAA Aerodynamic Decelerator Systems Technology Conference and Seminar*, AIAA, Reston, VA.

Calise, A., and Preston, D. (2006), "Design of a Stability Augmentation System for Airdrop of Autonomous Guided Parafoils," *Proceedings of the AIAA Guidance, Navigation, and Control Conference and Exhibit*, AIAA, Reston, VA.

Calise, A., and Preston, D. (2008), "Swarming/Flocking and Collision Avoidance for Mass Airdrop of Autonomous Guided Parafoils," *Journal of Guidance, Control, and Dynamics*, Vol. 31, No. 4, pp. 1123–1132; also Calise, A., and Preston, D. (2005), "Swarming/Flocking and Collision Avoidance for Mass Airdrop of Autonomous Guided Parafoils," *Proceedings of the AIAA Guidance, Navigation, and Control Conference and Exhibit*, AIAA, Reston, VA.

Calise, A., and Preston, D. (2009), "Approximate Correction of Guidance Commands for Winds," *Proceeding of the 20th AIAA Aerodynamic Decelerator Systems Technology Conference and Seminar*, AIAA, Reston, VA.

Calise, A., Preston, D., and Ludwig, G. (2007), "Modeling for Guidance and Control Design of Autonomous Guided Parafoils," *Proceedings of the 19th AIAA Aerodynamic Decelerator Systems Technology Conference and Seminar*, AIAA, Reston, VA.

Campbell, D., Fill, T., Hattis, P., and Tavan, S. (2005), "An On-Board Mission Planning System to Facilitate Precision Airdrop," *Proceedings of the Infotech@Aerospace Conference*, AIAA, Reston, VA.

Cao, Y., and Zhu, X. (2013), "Effects of Characteristic Geometric Parameters on Parafoil Lift and Drag," *An International Journal of Aircraft Engineering and Aerospace Technology*, Vol. 85, No. 4, pp. 280–292.

Cargo Parachutes (2012), Mills Manufacturing, www.millsmanufacturing.com/products/cargo-parachutes

CARP (2005), "Computed Air Release Point Procedures," U. S. Air Force, AFI-11-231, http://static.e-publishing.af.mil/production/1/af_a3_5/publication/afi11-231/afi11-231.pdf

Carter, D., and Rasmussen, S. (2015), "Games Against the Wind: Guided Parafoil Accuracy," *Proceedings of the 23rd AIAA Aerodynamic Decelerator Systems Technology Conference and Seminar*, AIAA, Reston, VA.

Carter, D., George, S., Hattis, P., McConley, M., Rasmussen, S., Singh, L., and Tavan, S. (2007), "Autonomous Large Parafoil Guidance, Navigation, and Control System Design Status," *Proceedings of the 19th AIAA Aerodynamic Decelerator Systems Technology Conference and Seminar*, AIAA, Reston, VA.

Carter, D., George, S., Hattis, P., Singh, L., and Tavan, S. (2005), "Autonomous Guidance, Navigation, and Control of Large Parafoils," *Proceedings of the 18th AIAA Aerodynamic Decelerator Systems Technology Conference and Seminar*, AIAA, Reston, VA.

Carter, D., Singh, L., Wholey, L., Rasmussen, S., Barrows, T., George, S., McConley, M., Gibson, C., Tavan, S., and Bagdonovich, B. (2009), "Band-Limited Guidance and Control of Large Parafoils," *Proceedings of the 20th AIAA Aerodynamic Decelerator Systems Technology Conference and Seminar*, AIAA, Reston, VA.

Celik, I. B., Ghia, U., Roache, P. J., Freitas, C. J., Coleman, H., and Raad, P. E. (2008), "Procedure for Estimation and Reporting of Uncertainty due to Discretization in CFD Applications," *ASME Journal of Fluids Engineering*, Vol. 130, No. 078001.

Chatzikonstantinou, T. (1989), "Numerical Analysis of Three-Dimensional Non Rigid Wings," *Proceedings of the 10th AIAA Aerodynamic Decelerator Systems Technology Conference*, AIAA, Washington, D. C.

Chatzikonstantinou, T. (1993), "Recent Advances in the Numerical Analysis of Ram Air Wings Using the Three-Dimensional Simulation Code 'PARA3D'," *Proceedings of the 12th RAeS/AIAA Aerodynamic Decelerator Systems Technology Conference and Seminar*, AIAA, Washington, D. C.

Chatzikonstantinou, T. (1999), "Problems in Ram Air Wing Modeling and Their Solution in the Three Dimensional Simulation Code 'PARA3D'," *Proceedings of the 15th CEAS/AIAA Aerodynamic Decelerator Systems Technology Conference*, AIAA, Reston, VA.

Chiel, B. (2015), "Adaptive Control of a 10K Parafoil System," *Proceedings of the 23rd AIAA Aerodynamic Decelerator Systems Technology Conference and Seminar*, AIAA, Reston, VA.

Cleminson, J. R. (2013), "Path Planning for Guided Parafoils: An Alternative Dynamic Programming Formulation," *Proceedings of the 22nd AIAA Aerodynamic Decelerator Systems Technology Conference*, AIAA, Reston, VA.

CMDP (2013), "Common Mission Debrief Program," Eglin AFB, USAF 846TSS/TSI, Ft. Walton Beach, FL.

Cockrell, D. J., and Doherr, K.-F (1981), "Preliminary Consideration of Parameter Identification Analysis from Parachute Aerodynamic Flight Test Data", *Proceedings of the 7th AIAA Aerodynamic Decelerator and Balloon Technology Conference*, AIAA, New York.

Cockrell, D. J., and Haidar, N. I. A. (1993), "Influence of the Canopy-Payload Coupling on the Dynamic Stability in Pitch of a Parachute System," *Proceedings of the 12th RAeS/AIAA Aerodynamic Decelerator Systems Technology Conference and Seminar*, AIAA, Washington, D. C.

Corley, M. S., and Yakimenko, O. A. (2009), "Computation of the Safety Fans for Multi-stage Aerodelivery Systems," *Proceedings of the 20th AIAA Aerodynamic Decelerator Systems Technology Conference and Seminar*, AIAA, Reston, VA.

Crimi, P. (1990), "Lateral Stability of Gliding Parachutes," *Journal of Guidance, Control Dynamics*, Vol. 13, No. 6.

Culpepper, S., Ward, M. B., Costello, M., and Bergeron, K. (2013), "Adaptive Control of Damaged Parafoils," *Proceedings of the 22nd AIAA Aerodynamic Decelerator Systems Technology Conference and Seminar*, AIAA, Reston, VA.

Davydov, Y., and Mosseev, Y. (1990), "Aerodynamic Study of Parachute Systems," Inst. of Parachute Systems, Moscow.

de Lassat de Pressigny, Y., Benney, R., Henry, M., Bechet, R., and Wintgens, J. H. (2009), "PACD 2008: Operational Requirements Fulfilled," *Proceedings of the 20th AIAA Aerodynamic Decelerator Systems Technology Conference and Seminar*, AIAA, Reston, VA.

Decker, R. J. (2013), "A Computer Vision-Based Method for Artillery Launch Characterization," Ph.D. Dissertation, Naval Postgraduate School, Monterey, CA.

Decker, R. J., and Yakimenko, O. A. (2015), "Automated Canopy and Payload Motion Estimation Using Vision Based Methods," *Proceedings of the 23rd AIAA Aerodynamic Decelerator Systems Technology Conference and Seminar*, AIAA, Reston, VA.

Dellicker, S. (1999), "Low Cost Parachute Navigation Guidance and Control," M.S. Thesis, *Naval Postgraduate School*, Monterey, CA.

Dellicker, S., and Bybee, J. (1999), "Low Cost Parachute Guidance, Navigation, and Control," *Proceedings of the AIAA 15th Aerodynamic Decelerator Systems Conference*, AIAA, Reston, VA.

Dellicker, S., Benney, R., and Brown, G. (2001), "Guidance and Control for Flat-Circular Parachutes," *Journal of Aircraft*, Vol. 38, No. 5, pp. 809–817.

Dellicker, S., Benney, R., LeMoine, D., Brown, G., Gilles, B., Howard, R., and Kaminer, I. (2003), "Steering a Flat Circular Parachute – They Said It Couldn't Be Done," *Proceedings of the 17th AIAA Aerodynamic Decelerator Systems Technology Conference and Seminar*, AIAA, Reston, VA.

Dellicker, S., Benney, R., Patel, S., Williams, T., Hewgley, C., Yakimenko, O., Howard, R., and Kaminer, I. (2000), "Performance, Control and Simulation of the Affordable Guided Airdrop System," *AIAA Guidance, Navigation, and Control Conference*, AIAA, Reston, VA.

Delwarde, C., de Lassat de Pressigny, Y., Benney, R., Vallance, M., Norton, B., and Wintgens, J. (2007), "Precision Airdrop Capability Demonstration in France," *Proceedings of the 19th AIAA Aerodynamic Decelerator Systems Technology Conference and Seminar*, AIAA, Reston, VA.

Desabrais, K. (2005), "Aerodynamic Forces on an Airdrop Platform," *Proceedings of the 18th AIAA Aerodynamic Decelerator Systems Technology Conference and Seminar*, AIAA, Reston, VA.

Dietz, A., Sorenson, P., Lafond, K., and Tavan, S. (2007), "A Sodar Height Sensor for Precision Airdrops," *Proceedings of the 19th AIAA Aerodynamic Decelerator Systems Technology Conference and Seminar*, AIAA, Reston, VA.

Dobrokhodov, V., Yakimenko, O., and Junge, C. (2003), "Six-Degree-of-Freedom Model of a Controlled Circular Parachute," *Journal of Aircraft*, Vol. 40, No. 3, pp. 482–493; also Dobrokhodov, V., Yakimenko, O., and Junge, C. (2002), "Six-Degree-of-Freedom

Model of a Controllable Circular Parachute," *Proceedings of the AIAA Atmospheric Flight Mechanics Conference*, AIAA, Reston, VA.

Doherr, K., and Schilling, H. (1992), "Nine-Degree-of-Freedom Simulation of Rotating Parachute Systems," *Journal of Aircraft*, Vol. 29, No. 5, pp. 774–781; also Doherr, K.-F., and Schilling, H. (1991), "9DOF-Simulation of Rotating Parachute Systems," *Proceedings of the 11th AIAA Aerodynamic Decelerator Systems Technology Conference*, AIAA, Washington, D.C.

Doherr, K.-F. (1997), "Parachute Flight Dynamics and Trajectory Simulation," *Proceedings of the 14th AIAA Aerodynamic Decelerator Systems Technology Conference*, AIAA, Reston, VA.

Doherr, K.-F., and Jann, T. (1997), "Test Vehicle ALEX-I for Low-Cost Autonomous Parafoil Landing Experiment," *Proceedings of the 14th AIAA ADS Conference and Seminar*, AIAA, Reston, VA.

Doherr, K.-F., and Saliaris, C. (1981), "On the Influence of Stochastic and Acceleration Dependent Aerodynamic Forces on the Dynamic Stability of Parachutes," *Proceedings of 7th AIAA Aerodynamic Decelerator and Balloon Technology Conference*, AIAA, New York.

Doherr, K.-F., and Saliaris, C. (1987), "Dynamic Stability Analysis of Parachutes" CCG-Univ. of Minnesota, Parachute Systems Technology Short Course, Oberpfaffenhofen.

Doherr, K.-F., deBoer, B., Fenske-Swetlakowa, F., Gockel, W., Saliaris, C., and Stabenau, P. (1994), "Pre-Study on Parachutes for a Crew Transport Vehicle - Capsule Recovery System. Volume 3: Advanced Recovery System Concept," Inst. für Flugmechanik, DLR Technical Rept. HT-TN-E9-3-DLR, Germany.

Dommasch, D. O., Sherby, S. S., and Connolly, T. F. (1951), *Airplane Aerodynamics*, Pitman Aeronautical Publications.

Dribnoy, V., Nisht, M., and Sudakov, A. (1989), "A Numerical Simulation of Ram-Air Airfoil," *Computer-Aided Study of Aircraft Journal*, Vol. 1313, Air Force Engineering Academy, Moscow, Russia.

Driels, M. (2013), *Weaponeering: Conventional Weapon Systems Effectiveness*, 2nd ed., AIAA Educational Series, AIAA, Reston, VA.

Dubins, L. E. (1957), "On Curves of Minimal Length with a Constraint on Average Curvature and with Prescribed Initial and Terminal Positions and Tangents," *American Journal of Mathematics*, Vol. 79, No. 3, pp. 497–516.

Eaton, J. A. (1982), "Added Mass and the Dynamic Stability of Parachutes," *Journal of Aircraft*, Vol. 19, No. 5, pp. 414–416.

Eiff Aerodynamics (2015), "Classic Owners Manuals," Eiff Aerodynamics, Inc., DeLand, FL, http://eiff.com/manuals/HERITAGE.html

Eilertson, W. H. (1969), *Gliding Parachutes for Land Recovery of Space Vehicles*, Bellcomm Inc., Case 730, NASA-CR-108990, Washington. D. C.

Entchev, R. O., and Rubenstein, D. (2001), *Modeling Small Parafoil Dynamics*, Final Rept. 16.622, http://web.mit.edu/rrodin/www/projects/parafoil/#tth_sEcA.

Eslambolchi, A. (2012), "Computation of Flow over a Full-Scale Ram-Air Parachute Canopy," M.S. Thesis, California State Univ., Northridge, CA.

Eslambolchi, A., and Johari, H. (2013), "Simulation of Flowfield Around a Ram-Air Personnel Parachute Canopy," *Journal of Aircraft*, Vol. 50, No. 5, pp. 1628–1636.

Fadali, M. S., and Visioli, A. (2009), *Digital Control Engineering*, Elsevier Inc., Burlington, MA, Chap. 5, pp. 154–160.

Favorin, M. V. (1977), *Momenti Inerzii Tel. Spravochnik (Moments of Inertia of Bodies. Reference Book)*, Mashinostroenie, Moscow (in Russian).

Fields, T. (2013), "A Descent Rate Control Approach to Developing an Autonomous Descent Vehicle," Ph.D. Dissertation, Univ. of Nevada, Reno, NV.

Fields, T., and Basore, N. (2015), "Reversible Control Line Reefing System for Circular Parachutes," *Proceedings of the 23rd AIAA Aerodynamic Decelerator Systems Technology Conference and Seminar*, AIAA, Reston, VA.

Fields, T., LaCombe, J., and Wang, E. (2011), "One Degree of Freedom Approach for an Autonomous Descent Vehicle Using a Variable Drag Parachute," *Proceedings of the 21st AIAA Aerodynamic Decelerator Systems Technology Conference and Seminar*, AIAA, Reston, VA.

Fields, T., LaCombe, J., and Wang, E. (2012), "Autonomous Guidane of a Circular Parachute Using Descent Rate Control," *Journal of Guidance, Control, and Dynamics*, Vol. 35, No. 4, pp. 1367–1370.

Fields, T., LaCombe, J., and Wang, E. (2013), "Flight Testing of a 1-DOF Variable Drag Autonomous Descent Vehicle," *Proceedings of the 22nd AIAA Aerodynamic Decelerator Systems Technology Conference and Seminar*, AIAA, Reston, VA.

Forehand, J. E., and Bair, H. Q. (1968), "Parawing Precision Aerial Delivery System," *Journal of Aircraft*, Vol. 6, No. 5, pp. 463–469.

Forichon, A. (1961), "Parachute," U.S. Patent 2,997,263.

Fraser, R. (2011), "Analysis of Windpack Meteorological Data," *Proceedings of the 21st AIAA Aerodynamic Decelerator Systems Technology Conference*, AIAA, Reston, VA.

Gage, S. (2003), "Creating a Unified Graphical Wind Turbulence Model from Multiple Specifications," *AIAA Modeling and Simulation Technologies Conference and Exhibit*, AIAA, Reston, VA.

Gavrilovski, A., Ward, M., and Costello, M. (2012), "Parafoil Control Authority with Upper-Surface Canopy Spoilers," *Journal of Aircraft*, Vol. 49, No. 5, pp. 1391–1397; also Gavrilovski, A., Ward, M., and Costello, M. (2011), "Parafoil Glide Slope Control Using Canopy Spoilers," *Proceedings of the 21st AIAA Aerodynamic Decelerator Systems Technology Conference and Seminar*, AIAA, Reston, VA.

George, S., Carter, D., Hattis, P., Singh, L., Berland, J. C., Dunker, S. Markle, Lewis, J., Tavan, S., and Barber, J. (2005), "The Dragonfly 4,500 kg Class Guided Airdrop System," *Proceedings of the Infotech@Aerospace Conference*, Arlington, VA.

Giadrosich, D. (1995), *Operations Research Analysis in Test and Evaluation*, AIAA Educational Series, AIAA, Reston, VA.

Gilles, B., Hickey, M., and Krainski, W. (2005), "Flight Testing of a Low-Cost Precision Aerial Delivery System," *Proceedings of the 18th AIAA Aerodynamic Decelerator Systems Technology Conference and Seminar*, AIAA, Reston, VA.

Gimadieva, T. Z. (2001), "Optimal Control of a Gliding Parachute System," *Journal of Mathematical Sciences*, Vol. 103, No. 1, pp. 54–60.

Gimadieva, T. Z. (2006), "Mathematical Modeling and Control Algorithms Design for Parachute Systems," Ph.D. Dissertation, Kazan State Power Engineering Univ., Kazan, Russia (in Russian), www.dissercat.com/content/matematicheskoe-modelirovanie-i-razrabotka-algoritmov- dvizheniya-parashyutnykh-sistem

Gionfriddo, M. (2011), "Aerodynamic Decelerator Systems Technical Committee Effort to Archive Parachute Technology," *Proceedings of the 21st AIAA Aerodynamic Decelerator Systems Technology Conference*, AIAA, Reston, VA.

Glauert, H. (1983), *The Elements of Airfoil and Airscrew Theory*, 2nd ed., Cambridge Univ. Press, Cambridge, England, U.K.

Glider Technical Data (2014), Gradient Gliders, www.gradient.cx/en/gliders

Gockel, W. (1995), *Development of a Mathematical Model of a Large Parafoil-Load System and Flight Mechanic Analysis*, Inst. of Flight Mechanics, Internal Rept. IB 111-95/36, DLR Braunschweig, Germany.

Gockel, W. (1996), *Bewertung der Manovrierbarkeit Geflügelter Raumtransporter im Entwurfsprozess*, Inst. of Flight Mechanics, Research Rept. 96-04, DLR Braunschweig, Germany.

Gockel, W. (1997a), "Concept Studies of an Autonomous GNC System for Gliding Parachute," *Proceedings of the 14th AIAA Aerodynamic Decelerator Systems Technology Conference*, AIAA, Reston, VA.

Gockel, W. (1997b), "Computer Based Modeling and Analysis of a Parafoil-Load Vehicle," Case Study 1, Seminar at the 14th AIAA ADS Conference, San Francisco, CA.

Gockel, W. (1998), "ALEX – Mathematical Vehicle Model and Guidance and Control Function – First Design Steps Prior to Flight Test Analysis," Inst. of Flight Mechanics, Internal Rept. IB 111-98/02, DLR Braunschweig, Germany.

Gockel, W., and Jann, T. (1998), "ALEX-Flugdatenauswertung, Entwicklung Eines Validierten Flugmechanischen Modells," Inst. of Flight Mechanics, Internal Rept. IB 111-98/47, DLR Braunschweig, Germany.

Gonyea, K., Braun, R., Tanner, C. L., Clark, I. G., Kushner, L. K., and Schairer, E. (2013) "Aerodynamic Stability and Performance of Next-Generation Parachutes for Mars Entry, Descent, and Landing," *Proceedings of the 22nd AIAA Aerodynamic Decelerator Systems Technology Conference and Seminar*, AIAA, Reston, VA.

Gonzalez, M. A. (1993), "Prandtl Theory Applied to Paraglider Aerodynamics," *Proceedings of the 12th RAeS/AIAA Aerodynamic Decelerator Systems Technology Conference and Seminar*, AIAA, Washington, D. C.

Goodrick, T. (1969), "Wind Effect on Gliding Parachute Systems with Non-Proportional Automatic Homing Control," Army Natick Labs., Technical Rept. TR 70-28-AD, Natick, MA.

Goodrick, T. (1970), "Estimation of Wind Effect on Gliding Parachute Cargo Systems Using Computer Simulation," *Proceedings of the 3rd AIAA Aerodynamic Decelerator Systems Conference*, AIAA, New York.

Goodrick, T. (1975), "Theoretical Study of the Longitudinal Stability of High Performance Gliding Airdrop Systems," *Proceedings of the 5th AIAA Aerodynamic Decelerator Systems Conference*, AIAA, New York.

Goodrick, T. (1979a), "Hardware Options for Gliding Airdrop Guidance Systems," *Proceedings of the 6th AIAA Aerodynamic Decelerator and Balloon Technology Conference*, AIAA, New York.

Goodrick, T. (1979b), "Simulation Studies of the Flight Dynamics of Gliding Parachute Systems," *Proceedings of the 6th AIAA Aerodynamic Decelerator and Balloon Technology Conference*, AIAA, New York.

Goodrick, T. (1981), "Comparison of Simulation and Experimental Data for a Gliding Parachute in Dynamic Flight," *Proceedings of the 7th AIAA Aerodynamic Decelerator and Balloon Technology Conference*, AIAA, New York.

Goodrick, T. (1984), "Scale Effects on Performance of Ram Air Wings," *Proceedings of the 8th AIAA Aerodynamic Decelerator and Balloon Technology Conference*, AIAA, New York.

Goodrick, T., Pearson, A., and Murphy, A. (1973), "Analysis of Various Automatic Homing Techniques for Gliding Airdrop Systems with Comparative Performance in Adverse Winds," *Proceedings of the 4th AIAA Aerodynamic Decelerator Conference*, AIAA, New York.

Gorman, C. M. (2011), "Modeling, Comparison and Analysis of Multi-Body Parafoil Models with Varying Degrees of Freedom," Ph.D. Disseration, Univ. of Alabama in Huntsville, AL.

Gorman, C. M., and Slegers, N. J. (2011a), "Comparison and Analysis of Multi-Body Parafoil Models with Varying Degrees of Freedom," *Proceedings of the 21st AIAA Aerodynamic Decelerator Systems Technology Conference and Seminar*, AIAA, Reston, VA.

Gorman, C. M., and Slegers, N. J. (2011b), "Modeling of Parafoil-Payload Relative Yawing Motion on Autonomous Parafoils," *Proceeding of the 21st AIAA Aerodynamic Decelerator Systems Technology Conference and Seminar*, AIAA, Reston, VA.

Gorman, C. M., and Slegers, N. J. (2012), "Evaluation of Multi-Body Parafoil Dynamics Using Distributed Miniature Wireless Sensors," *Journal of Aircraft*, Vol. 49, No. 2; also Gorman, C. M., and Slegers, N. J. (2011), "Evaluation of Multi-Body Parafoil Dynamics Using Distributed Miniature Wireless Sensors," *Proceedings of the 21st AIAA Aerodynamic Decelerator Systems Technology Conference and Seminar*, AIAA, Reston, VA.

GPS (2014), Official U.S. Government Information about the Global Positioning System and Related Topics, www.gps.gov

GPS Error Analysis (2014), "Error Analysis for the Global Positioning System," en.wikipedia.org/wiki/Error_analysis_for_the_Global_Positioning_System

Gründer, M. (2001), "Astrium Steuert X-38," *Flug Revue*, No. 1, p. 50.

Guided PADS (2014), "Airborne Systems," www.airborne-sys.com/pages/view/guided-precision-aerial-delivery-systems

Gupta, B., Upadhyaya, S. C., Kumar, V., Krishna, R., and Ghosh, A. K. (2011a), "Trajectory Simulation of Large Ram Air Parachute Using 9-DOF Model," *Proceedings of the 21st AIAA Aerodynamic Decelerator Systems Technology Conference and Seminar*, AIAA, Reston, VA.

Gupta, M., Xu, Z., Zhang, W., Accorsi, M., Leonard, J., and Stein, K. (2001b), "Recent Advances in Structural Modeling of Parachute Dynamics," *Proceedings of the 16th AIAA Aerodynamic Decelerator Systems Technology Conference*, AIAA, Reston, VA.

Hailiang, M., and Zizeng, Q. (1994), "9-DoF Simulation of Controllable Parafoil System for Gliding and Stability," *Journal of National University of Defense Technology*, Vol. 16 No. 2, pp. 49–54.

Hamel, P., and Jategaonkar, R. (1996), "Evolution of Flight Vehicle System Identification," *Journal of Aircraft*, Vol. 33, No. 1, pp .9–28.

Han, Y. H., Wang, Y. W., Yang, C. X., and Xiao, J. (2013), "Numerical Methods for Analyzing the Aerodynamic Characteristics of Cross Parachute with Permeability," *Proceedings of the 22nd AIAA Aerodynamic Decelerator Systems Technology Conference*, AIAA, Reston, VA.

Hanke, K., and Schenk, S. (2014), "Evaluating the Geometric Shape of a Flying Paraglider," *ISPRS Technical Commission V Symposium*, Riva del Garda, Italy.

Harrington, N., and Doucette, E. (1999), "Army After Next and Precision Airdrop," http://www.almc.army.mil/alog/issues/JanFeb99/MS388.htm

Hattis, P. (2007), "Autonomous Large Parafoil Guidance, Navigation, and Control System Design Status," *Proceedings of the 19th AIAA Aerodynamic Decelerator Systems Technology Conference and Seminar*, AIAA, Reston, VA.

Hattis, P., and Benney, R. (1996), "Demonstration of Precision Guided Ram-Air Parafoil Airdrop Using GPS/INS Navigation," *Proceedings of the Institute of Navigation's 52nd Annual Meeting*, Cambridge, MA.

Hattis, P., Angermueller, K., Fill, T., Wright, R., Benney, R., LeMoine, D., and King, D. (2003), "In-Flight Precision Airdrop Planner Follow-on Development Program," *Proceeding of the 17th AIAA Aerodynamic Decelerator Systems Technology Conference and Seminar*, AIAA, Reston, VA.

Hattis, P., Appleby, B., Fill, T., and Benney, R. (1997), "Precision Guided Airdrop System Flight Test Results," *Proceedings of the 14th AIAA Aerodynamic Decelerator Systems Conference*, AIAA, Reston, VA.

Hattis, P., Campbell, D., Carter, D., McConley, M., and Tavan, S. (2006), "Providing Means for Precision Airdrop Delivery from High Altitude," *Proceedings of the AIAA Guidance, Navigation, and Control Conference and Exhibit*, AIAA, Reston, VA.

Hattis, P., Fill, T., Rubenstein, D., Wright, R., and Benney, R. (2000), "An Advanced On-board Airdrop Planner to Facilitate Precision Payload Delivery," *Proceedings of the AIAA Modeling and Simulation Technologies Conference and Exhibit*, AIAA, Reston, VA.

Hattis, P., Fill, T., Rubenstein, D., Wright, R., Benney, R., and LeMoine, D. (2001), "Status of an Onboard PC-Based Airdrop Planner Demonstration," *Proceedings of the 16th AIAA Aerodynamic Decelerator Systems Technology Conference*, AIAA, Reston, VA.

Hattis, P. D., Polutchko, R. J., Appleby, B. D., Barrows, T. M., Fill, T. J., Kachmar, P. M., and McAteer, T. D. (1996), "Final Report: Development and Demonstration Test of a Ram-Air Parafoil Precision Guided Airdrop System," Vols. 1–4 (Rept. CSDL-R-2752) and Addendum (Rept. CSDL-R-2771), Charles Stark Draper Lab., Cambridge, MA.

Hayhurst, J. (1996), "A Brief History of Accuracy Parachutes," Eiff Aerodynamics, Inc., DeLand, FL, http://eiff.com/manuals/HERITAGE.html

Hayter, A. J. (2012), *Probability and Statistics for Engineers and Scientists*, 4th ed., Duxbury Press.

Head, M. R. (1982), "Flow Visualization in Cambridge University Engineering Department," *Proceedings of the 2nd International Symposium on Flow Visualization*, Bochum, West Germany, p. 402; also Van Dyke, M. (1982), *An Album of Fluid Motion*, 14th ed., Parabolic Press, Inc., p. 51.

Henn, H. (1944), "Die Absinkeigenschaften von Fallschirmen (Descent Characteristics of Parachutes)," *Untersuchungen und Mitteilungen*, No. 6202, ZWB, Berlin, Germany (in German); also Royal Aeronautical Establishment Translation No. 223.

Hewgley, C., Yakimenko, O., and Slegers, N. (2011), "Shipboard Landing Challenges for Autonomous Parafoils," *Proceedings of the 21st AIAA Aerodynamic Decelerator Systems Technology Conference*, AIAA, Reston, VA.

Hewgley, C. W. (2014), "Pose and Wind Estimation for Autonomous Parafoils," Ph.D. Dissertation, Naval Postgraduate School, Monterey, CA.

Hewgley, C. W., and Yakimenko, O. A. (2009), "Precision Guided Airdrop for Vertical Replenishment of Naval Vessels," *Proceedings of the 20th AIAA Aerodynamic Decelerator Systems Technology Conference and Seminar*, AIAA, Reston, VA.

Hewgley, C. W., and Yakimenko, O. A. (2011), "Improved Surface Layer Wind Modeling for Autonomous Parafoils in a Maritime Environment," *Proceedings of the 21st AIAA Aerodynamic Decelerator Systems Technology Conference*, AIAA, Reston, VA.

Higgins, M. W. (1979), "Control System for Ram Air Gliding Parachute," U.S. Patent 4,175,722.

Hock, T. F., and Franklin, J. L. (1999), "The NCAR GPS Dropwindsonde," *Bulletin of the American Meteorological Society*, Vol. 80, No. 3, pp. 407–420.

Hoerner, S. F. (1965), *Fluid-Dynamic Drag*, S. F. Hoerner, Midland Park, NJ.

Hoerner, S. F., and Borst, H. V. (1985), *Fluid Dynamic Lift*, 2nd ed., L. A. Hoerner, Brick Town, NJ.

Hogue, J., and Jex, H. (1995), "Applying Parachute Canopy Control and Guidance Methodology to Advanced Precision Airborne Delivery Systems," *Proceedings of the 13th AIAA Aerodynamic Decelerator Systems Technology Conference*, AIAA, Washington, D. C.

Hogue, J., Johnson, W., Allen, R., and Pierce, D. (1993), "Parachute Canopy Control and Guidance Training Requirements and Methodology," *Proceedings of the AIAA/AHS/ASEE Aerospace Design Conference*, AIAA, Washington, D. C.

Hollestelle, P., Jiménez Olazábal, A., and Grunwald, A. (2009), "The Guidance, Navigation and Control System of the Folding, Adaptive, Steerable Textile Wing Structure for Capital Loads," *Proceeding of the 20th AIAA Aerodynamic Decelerator Systems Technology Conference*, AIAA, Reston, VA.

Hur, G. (2005), "Identification of Powered Parafoil-Vehicle Dynamics from Modelling and Flight Test Data," Ph.D. Dissertation, Texas A&M Univ., TX, http://repository.tamu.edu/bitstream/handle/1969.1/3859/etd-tamu-2005A-AERO-Hur.pdf.

Hur, G., and Valasek, J. (2003), "System Identification of Powered Parafoil-Vehicle from Flight Test Data," *Proceedings of the AIAA Atmospheric Flight Mechanics Conference*, AIAA, Reston, VA.

Iacomini, C., and Cerimele, C. (1999a), "Lateral Directional Aerodynamics from a Large Scale Parafoil Test Program," *Proceeding of the 15th CEAS/AIAA Aerodynamic Decelerator Systems Technology Conference*, AIAA, Reston.

Iacomini, C., and Cerimele, C. (1999b), "Longitudinal Aerodynamics from a Large Scale Parafoil Test Program," *Proceeding of the 15th CEAS/AIAA Aerodynamic Decelerator Systems Technology Conference*, AIAA, Reston, VA.

Iacomini, C., and Madsen, C. (1999c), "Investigation of Large Scale Parafoil Rigging Angles: Analytical and Drop Test Results," *Proceeding of the 15th CEAS/AIAA Aerodynamic Decelerator Systems Technology Conference*, AIAA, Reston, VA.

Ibrahim, S. (1965a), "Apparent Mass and Moment of Inertia of Cup-Shaped Bodies in Unsteady Incompressible Flow," Ph.D. Thesis, Univ. of Minnesota, Twin Cities, MI.

Ibrahim, S. (1965b), "Experimental Determination of the Apparent Moment of Inertia of Parachutes," Rept. No. FDL-TDR-64-153, Wright-Patterson AFB, Dayton, OH.

Iosilevskii, G. (1993), "A Paraglider - Performance, Aerodynamics and Flight Mechanics," Technion - Israel Inst. of Technology, TAE-Report No. 692, Haifa, Israel.

Iosilevskii, G. (1995), "Center of Gravity and Minimal Lift Coefficient Limits of a Gliding Parachute," *Journal of Aircraft*, Vol. 32, No. 6, pp. 1297–1302.

Iosilevskii, G. (1996), "Lifting Line Theory of Arched Wings in Asymmetric Flight," *Journal of Aircraft*, Vol. 33, No. 5, pp. 1023–1026.

IRIG 106 (2009), *IRIG 106-07 Chapter 10 Programming Handbook*, Range Commanders Council: Telemetry Group, White Sands Missile Range, NM.

I-View (2004), "Israeli-Weapons," http://www.israeli-weapons.com/weapons/aircraft/uav/i-view/I-View.html

Jackson, G., and Crocker, C. (2014), "The Use of Altimeters in Height Measurement," The Hills Database, http://www.hills-database.co.uk/altim.html

Jalbert, D. (1966), "Multi-Cell Wing Type Aerial Device," U.S. Patent 3,285,546.

Jann, T. (2001), "Aerodynamic Model Identification and GNC Design for the Parafoil-Load System ALEX," *Proceedings of the 16th AIAA Aerodynamic Decelerator Systems Technology Conference*, AIAA, Reston, VA.

Jann, T. (2003), "Aerodynamic Coefficients for a Parafoil Wing with Arc Anhedral - Theoretical and Experimental Results," *Proceedings of the 17th AIAA Aerodynamic Decelerator Systems Technology Conference and Seminar*, AIAA, Reston, VA.

Jann, T. (2004), *Modellierung, Identifizierung und Autonomes Fliegen eines Gleitfallschirm-Last-Systems*, Ph.D. Dissertation, Inst. of Flight Mechanics, DLR-FB-2004-33, DLR Braunschweig, Germany.

Jann, T. (2005), "Advanced Features for Autonomous Parafoil Guidance, Navigation and Control," *Proceedings of the 18th AIAA Aerodynamic Decelerator Systems Technology Conference and Seminar*, AIAA, Reston, VA.

Jann, T. (2006), "Validation of a Gliding Parachute Simulation Model Through System Identification," RTO AVT-133 Specialists Meeting on Fluid Dynamics of Personnel and Equipment Precision Delivery from Military Platforms, Vilnius, Lithuania.

Jann, T. (2011), "Coupled Simulation of Cargo Airdrop from a Generic Military Transport Aircraft," *Proceedings of the 21st AIAA Aerodynamic Decelerator Systems Technology Conference*, AIAA, Reston, VA.

Jann, T. (2015), "Implementation of a Flight Dynamic Simulation for Cargo Airdrop with Complex Parachute Deployment Sequences," *Proceedings of the 23rd AIAA Aerodynamic Decelerator Systems Technology Conference and Seminar*, AIAA, Reston, VA.

Jann, T., and Greiner-Perth, C. (2009), "Flight Test Instrumentation for Evaluation of the FASTWing CL System," *Proceeding of the 20th AIAA Aerodynamic Decelerator Systems Technology Conference*, AIAA, Reston, VA.

Jann, T., and Strickert, G. (2005), "System Identification of a Parafoil-Load Vehicle - Lessons Learned," *Proceeding of the 18th AIAA Aerodynamic Decelerator Systems Technology Conference and Seminar*, AIAA, Reston, VA.

Jann, T., Doherr, K., and Gockel, W. (1999), "Parafoil Test Vehicle ALEX – Further Development and Flight Test Results," *Proceedings of the 15th CEAS/AIAA Aerodynamic Decelerator Systems Technology Conference*, AIAA, Reston, VA.

Jategaonkar, R. (2006), *Flight Vehicle System Identification: A Time Domain Methodology*, AIAA Progress in Aeronautics and Astronautics Series, Vol. 216, AIAA, Reston, VA.

Jategaonkar, R., Fischenberg, D., and Gruenhagen, W. (2004), "Aerodynamic Modeling and System Identification from Flight Data – Recent Applications at DLR," *Journal of Aircraft*, Vol. 41, No. 4, pp. 681–691.

Jessick, M. V. (1990a), "Advanced Recovery System Guidance, Navigation and Controls Model Development and Analysis Program," Boeing, Final Rept. D180-32850-1, App. B, Huntsville, AL.

Jessick, M. V. (1990b), "PARASIM Program Document," Boeing, Final Rept. D180-32851-1, Huntsville, AL.

Johnson, J., Yakimenko, O., Kaminer, I., and Dellicker, S. (2001), "On the Development and Pre-Flight Testing of the Affordable Guided Airdrop System for G-12 Cargo Parachute,"

Proceedings of 16th AIAA Aerodynamic Decelerator Systems Technology Conference and Seminar, AIAA, Reston, VA.

Jones, T., Downey, J., Lunsford, C., Desabrais, K., and Noetscher, G. (2007), "Experimental Methods Using Photogrammetric Techniques for Parachute Canopy Shape Measurements," *Proceedings of the 19th AIAA Aerodynamic Decelerator Systems Technology Conference and Seminar,* AIAA, Reston, VA.

Jorgensen, D. S. (1982), "Cruciform Parachute Aerodynamics," Ph.D. Dissertation, Univ. of Leicester, Leicester, England, U.K.

Jorgensen, D., and Hickey, M. (2005), "The AGAS 2000 Precision Airdrop System," *Proceedings of Infotech@Aerospace Conference,* Arlington, VA; also Jorgenson, D., and Hickey, M. (2005), "The AGAS 2000 Precision Delivery System," *Proceeding of the 18th AIAA Aerodynamic Decelerator Systems Technology Conference and Seminar,* AIAA, Reston, VA.

JPADS (2014), JPADS: Making Precision Airdrop a Reality, Defense Industry Daily, www.defenseindustrydaily.com/jpads-making-precision-airdrop-a-reality-0678/

Kalro, V., Aliabadi, S., Garrard, W., Tezduyar, T. E., Mittal, S., and Stein, K. (1997), "Parallel Finite Element Simulation of Large Ram Air Parachutes," *International Journal for Numerical Methods in Fluids,* Vol. 24, No. 12, pp. 1353–1369.

Kaminer, I., and Yakimenko, O. (2003), "Development of Control Algorithm for the Autonomous Gliding Delivery System," *Proceedings of the 17th AIAA Aerodynamic Decelerator Systems Conference,* Monterey, CA; also Kaminer, I., and Yakimenko, O. (2003), "On the Development of GNC Algorithm for a High-Glide Payload Delivery System," *Proceedings of the IEEE Conference on Decision and Control,* IEEE, Maui, HI, pp. 5438–5443.

Kaminer, I., Yakimenko, O., and Pascoal, A. (2005), "Coordinated Payload Delivery Using High-Glide Ratio Parafoil Systems," *Proceedings of the 18th AIAA Aerodynamic Decelerator Systems Technology Conference,* AIAA, Reston, VA.

Kane, T. R., and Levinson, D. A. (1985), *Dynamics: Theory and Applications,* McGraw-Hill Series in Mechanical Engineering, Mcgraw-Hill, New York.

Ke, P., Yang, C., Sun, X., and Yang, X. (2009), "Novel Algorithm for Simulating the General Parachute-Payload System: Theory and Validation," *Journal of Aircraft,* Vol. 46, No. 1, pp. 189–197.

Kelly, K., and Pena, B. (2001), "Wind Study and GPS Dropsonde Applicability to Airdrop Testing," *Proceedings of 16th AIAA Aerodynamic Decelerator Systems Technology Conference and Seminar,* AIAA, Reston, VA.

Kempel, R. W. (1971), "Analysis of a Coupled Roll-Spiral-Mode, Pilot-Induced Oscillation Experienced with the M2-F2 Lifting Body," Flight Research Center, NASA TN D-6496, Edwards AFB, CA.

Kim, O.-Y., Lu, C., Albers, S., McGinley, J., and Oh, J.-H. (2007), "Improvement of LAPS Wind Analysis by Including Background Error Statistics," *Proceedings of the 22nd Weather Analysis and Forecasting Conference,* Park City, UT.

Klädtke, R., Püttmann, N., and Graf, E. D. (1999), "Europäische Partnerschaft im X-38 Programm," Deutsche Gesellschaft für Luft- und Raumfahrt Jahrestagung.

Klotz, H., Markus, M., Grimm, W., and Strandmoe, S. E. (1999), "Guidance and Control for Autonomous Re-Entry and Precision Landing of a Small Capsule," *Proceedings of the 4th ESA International Conference,* Noordwijk, The Netherlands, http://articles.adsabs.harvard.edu//full/2000ESASP.425..333K/0000333.000.html

Knacke, T. W. (1992), *Parachute Recovery Systems Design Manual*, NWC TP 6575, Para Publishing, Santa Barbara, CA.

Knapp, C., and Barton, W. (1968), "Controlled Recovery of Payloads at Large Glide Distances, Using the Para-Foil," *Journal of Aircraft*, Vol. 5, No. 2, pp. 112–118.

Komerath, N. M., Funk, R., Mahalingam, R. G., and Matos, C. (1998), "Low Speed Aerodynamis of the X-38 CRV: Summary of Research NAG9-927, 5/97-4/98," GITAER-EAG-98-03.

Koopersmith, R. M., and Pearson, A. E. (1975), *Determination of Trajectories for a Gliding Parachute System*, Army Natick Labs., Technical Rept. TR 75-117 AMEL, Natick, MA.

Kothandaraman, G., and Rotea, M. (2005), "Simultaneous-Perturbation-Stochastic-Approximation Algorithm for Parachute Parameter Estimation," *Journal of Aircraft*, Vol. 42, No. 5.

Krenz, H., and Burkhardt, O. (2005), "The FASTWing Project - A Self Navigated Gliding System for Heavy Loads," *Proceedings of the 18th AIAA Aerodynamic Decelerator Systems Technology Conference and Seminar*, AIAA, Reston, VA.

Kroo, I. (1987), "LinAir for the Macintosh," Desktop Aeronautics, Stanford, CA.

Krylov, I. A., and Chernous'ko, F. L. (1963), "On a Method of Successive Approximations for Solving Optimal Control Problems," *USSR Computational Mathematics and Mathematical Physics*, Vol. 2, No. 6, pp. 1371–1382.

Kurashova, M., and Vishnyak, A. (1995), "Identification of a Paraglider Longitudinal Aerodynamic Characteristics," *Proceedings of the 13th AIAA ADST Conference*, AIAA, Washington, D. C.

L-3 Communications (2008), "Kineto-Tracking Mount - MODEL K433," http://www2.l-3com.com/ios/pdf/KTM.pdf

Lee, C. K., and Buckley, J. E. (2004), "Method for Steerable Clustered Round Parachutes," *Journal of Aircraft*, Vol. 41, No. 5, 2004, pp. 1191–1195; also Lee C. K., and Buckley J. E. (2003), "Steerable Round Clustered Parachutes," *Proceeding of the 17th AIAA Aerodynamic Decelerator Systems Technology Conference and Seminar*, AIAA, Reston, VA.

Lee, C. K., Lanza, J., and Buckley, J. (1996), "Apparatus and Method for Measuring Angular Positions of Parachute Canopies," *Journal of Aircraft*, Vol. 33, No. 6, pp. 1197–1199.

Lemko, O. et al. (1983), "An Experimental Study of Aerodynamic Performances of Bodies in Unsteady Flow," KVVAIU, Kiev, Ukraine.

Lemko, O., and Lozhkin, M. et al. (1986), "An Experimental Study of Aerodynamic Performances of Parachutes and Flexible Wings," KVVAIU, Kiev, Ukraine.

Lester, W. G. S. (1962), "A Note on the Theory of Parachute Stability," *Aeronautical Research Council Reports and Memoranda*, No. 3352, London.

Levin, D., and Shpund, Z. (1997), "Canopy Geometry Effect on the Aerodynamic Behavior of Cross-Type Parachutes," *Journal of Aircraft*, Vol. 34, No. 5, 1994, pp. 648–652; also Levin, D., and Shpund, Z. (1997), "Effect of Canopy Geometrical Variation on the Aerodynamic Characteristics of a Cross-Type Parachute," *Proceedings of the 13th AIAA Aerodynamic Decelerator Systems Technology Conference*, AIAA, Washington, D.C.

Li, Y., and Lin, H. (1991), "Theoretical Investigation of Gliding Parachute Trajectory with Deadband Non-Proportional Automatic Homing Control," *Proceedings of the 11th AIAA Aerodynamic Decelerator Systems Technology Conference*, AIAA, Washington, D. C.

Lilliefors, H. (1967), "On the Kolmogorov-Smirnov Test for Normality with Mean and Variance Unknown," *Journal of the American Statistical Association*, Vol. 62, No. 318, pp. 399–402.

Lingard, J. S. (1979), "A Semi-Empirical Theory to Predict the Load Time History of an Inflating Parachute," Royal Aircraft Establishment, RAE Technical Rept. 79141, London.

Lingard, J. S. (1981), "The Performance and Design of Ram-Air Gliding Parachutes," Royal Aircraft Establishment, RAE Technical Rept. TR 81103, London.

Lingard, J. S. (1986), "The Aerodynamics of Gliding Parachutes," *Proceedings of the 9th AIAA Aerodynamic Decelerator and Balloon Technology Conference*, AIAA, New York.

Lingard, J. S. (1995), "Precision Aerial Delivery/Ram-Air Parachute System Design," *Proceedings of the 13th AIAA ADS Conference and Seminar*, AIAA, Washington, D. C.

Lissaman, P. B. S., and Brown, G. J. (1993), "Apparent Mass Effects on Parafoil Dynamics," *Proceedings of the 12th RAeS/AIAA Aerodynamic Decelerator Systems Technology Conference and Seminar*, AIAA, Reston, VA.

Lowe, D. G. (2004), "Distinctive Image Features from Scale-Invariant Keypoints," *International Journal of Computer Vision*, Vol. 60, No. 2, pp. 91–110.

Lozar, N., and Leaman, R. (2009), "Support from Above: The Need for Aerial Delivery Training in Ground Units," The U.S. Marine Corps, Command and Stuff College, Quantico, VA.

Ludtke, W. (1971), "Effects of Canopy Geometry on the Drag Coefficient of a Cross Parachute in the Fully Open and Reefed Conditions for a W/L Ratio of 0.264," Naval Ordnance Lab., NOLTR 71-111.

Ludtke, W. (1972), "Effects of Canopy Geometry on the Spinning Characteristics of a Cross Parachute with a W/L RATIO of 0.264," *Copertina Flessibile*, 1972.

Ludtke, W. (1973), "A Technique for Calculation of the Opening Shock Forces for Several Types of Solid Cloth Parachutes," *Proceedings of the 4th AIAA Aerodynamic Decelerator Conference*, AIAA, New York.

Ludwig, R., and Heins, W. (1963), "Theoretische Untersuchungen zur Dynamischen Stabilität von Fallschirmen (Theoretical Study of the Dynamic Stability of Parachutes)," *Jahrbuch der Wissenschaftlichen Gesellschaft für Luft- und Raumfahrt*, pp. 224–230 (in German; Faraday Translation).

Macdonald, D., Cordell, T., Diefendorf, R. J., Dudis, D., Fraser, H., Hughes, R., Schmitt, R., and Stupp, S. (1997), "Summary Report: New World Vistas, Air and Space Power for the 21st Century," U.S. Air Force Science Advisory Board, http://www.au.af.mil/au/awc/awcgate/vistas/index.htm; also http://www.dtic.mil/dtic/tr/fulltext/u2/a309591.pdf

Machin, R. A., and Ray, E. (2015) "Pendulum Motion in Main Parachute Clusters," *Proceedings of the 23rd AIAA Aerodynamic Decelerator Systems Technology Conference and Seminar*, AIAA, Reston, VA.

Machin, R., Stein, J., and Muratore, J. (1999b), "An Overview of the X-38 Prototype Crew Return Vehicle Development and Test Program," *Proceedings of the 15th CEAS/AIAA Aerodynamic Decelerator Systems Technology Conference*, AIAA, Reston, VA.

Madsen, C., and Cerimele, C. (2000), "Updated Flight Performance and Aerodynamics from a Large Scale Parafoil Test Program," *Proceedings of the AIAA Modeling and Simulation Technologies Conference*, AIAA, Reston, VA.

Madsen, C., and Cerimele, C. (2003), "Flight Performance, Aerodynamics, and Simulation Development for the X-38 Parafoil Test Program," *Proceedings of the 17th CEAS/AIAA Aerodynamic Decelerator Systems Conference*, AIAA, Reston, VA.

Majji, M., Juang, J., and Junkins, J. (2010b), "Time-Varying Eigensystem Realization Algorithm," *Journal of Guidance, Control, and Dynamics*, Vol. 33, No. 1, pp. 13–28.

Manolakis, D., Ingle, V., and Kogon, S. (2005), *Statistical and Adaptive Signal Processing: Spectral Estimation*, Signal Modeling, Adaptive Filtering and Array Processing, Artech House, Inc.

Markus, M. (1999), "Entwicklung und Implementierung eines Algorithmus zur Flugführung und Flugregelung eines autonom Operierenden Parafoilgleiters (Development and Implementation of an Algorithm for Guidance and Flight Control of an Autonomously Operating Paragliders)," M.S. Thesis, RWTH Aachen Univ., Aachen, Germany.

Mårtensson, K. (1972), *New Approach to the Numerical Solution of Optimal Control Problem*, Lund Inst. of Technology, Rept. 7206, Lund, Sweden.

Mårtensson, K. (1973), "A Constraining Hyperplane Technique for State Variable Constrained Optimal Control Problems," *Journal of Dynamic Systems, Measurement, and Control*, Vol. 95, No. 4, doi:10.1115/1.3426739. pp. 380–89.

Matos, C., Mahalingam, R. G., Ottinger, G., Klapper, J., Funk, R., and Komerath, N. M. (1998), "Wind Tunnel Measurements of Parafoil Geometry and Aerodynamics," *Proceedings of the 36th Aerospace Science Meeting and Exhibition*, AIAA, Reston, VA.

MAX Sensors (2014), QinetiQ North America, www.qinetiq-na.com/wp-content/uploads/brochure_meteorological-sensor-suite.pdf

Mayer, R. (1984), "Terminal Descent Controlled Vehicle Recovery," *Proceedings of the 8th AIAA Aerodynamic Decelerator and Balloon Technology Conference*, AIAA, New York.

Mayer, R., Puskas, E., and Lissaman, P. (1986), "Controlled Terminal Descent and Recovery of Large Aerospace Components," *Proceedings of the 9th AIAA Aerodynamic Decelerator and Balloon Technology Conference*, AIAA, New York.

McAllister, S., and Gose, J. B. (1980), "Optical Data Reduction. The White Sands Missile Range Data Systems Manual," Report PD-0089-J, New Mexico State Univ., Las Cruces, NM.

McCormick, B. (1979), *Aerodynamics and Aeronautics and Fight Mechanics*, Wiley, New York.

McGhee, R. J., and Beasley, W. D. (1973), "Low Speed Aerodynamic Characteristics of a 17-Percent Thick Airfoil Section Designed for General Aviation," NASA TN D-7428.

McGrath, J., Strong, T., and Benney, R. (2005), "Status of the Development of an Autonomously Guided Precision Cargo Aerial Delivery System," *Proceedings of the 18th AIAA Aerodynamic Decelerator Systems Technology Conference and Seminar*, AIAA, Reston, VA.

McHugh, J., Benney, R., Miletti, J., and Mortaloni, P. (2005), "Planning, Execution and Results of the Precision Airdrop Technology Conference and Demonstration 2003," *Proceedings of the 18th AIAA Aerodynamic Decelerator Systems Technology Conference and Seminar*, AIAA, Reston, VA.

McQuilling, M., Potvin, J., and Riley, J. (2011), "Simulating the Flows About Cargo Containers Used During Parachute Airdrop Operations," *Journal of Aircraft*, Vol. 48, No. 4, pp. 1405–1411.

Meeks, C. (1991), *Skydiving*, Capstone Press, North Mankato, MN.

Mittal, S., Saxena, P., and Singh, A. (2001), "Computation of Two-Dimensional Flows past Ram-Air Parachutes," *International Journal for Numerical Methods in Fluids*, Vol. 35, pp. 643–667.

Mkrtchyan, H., and Johari, H., "Detailed Aerodynamic Analysis of Ram-Air Parachute Systems in Steady Flight," *Proceedings of the 21st AIAA Aerodynamic Decelerator Systems Technology Conference*, AIAA, Reston, VA.

Mohammadi, M. A., and Johari, H. (2010), "Computation of Flow over a High Performance Parafoil Canopy," *Journal of Aircraft*, Vol. 47, No. 4, pp. 1338–1345.

Mooij, E., Wijnands, Q. G. J., and Schat, B. (2003), "9 DoF Parafoil/Payload Simulator Development and Validation," *Proceedings of the AIAA Modeling and Simulation Technologies Conference*, AIAA, Reston, VA.

Moore, J. (2012), "White Paper: Variable Angle of Attack Control System (VACS) for Gliding Aerial Delivery," Response to Request for Information prepared for InSitech, by JM Technologies, Inc.

Moore, J. (2013), "Drogue Assisted Variable Glide Slope Control," *Proceedings of the 22nd AIAA Aerodynamic Decelerator Systems Technology Conference*, AIAA, Reston, VA.

Mortaloni, P. A. (2001), "Test Methodology and Requirements Interpretation of the Advanced Tactical Parachute System," *Proceedings of the 16th AIAA Aerodynamic Decelerator Systems Technology Conference*, AIAA, Reston, VA.

Mortaloni, P. A. (2003), "On the Development of a Six-Degree-of-Freedom Model of a Low-Aspect-Ratio Parafoil Delivery System," M.S. Thesis, *Naval Postgraduate School*, Monterey, CA.

Mortaloni, P. A., and Stewart, J. (1999), "Aerodynamic Decelerator Testing - U.S. Army History and Capabilities," *Proceedings of the 15th CEAS/AIAA Aerodynamic Decelerator Systems Technology Conference*, AIAA, Reston, VA.

Mortaloni, P. A., Yakimenko, O. A., Dobrokhodov, V. N., and Howard, R. M. (2003), "On the Development of Six-Degree-of-Freedom Model of Low-Aspect-Ratio Parafoil Delivery System," *Proceedings of the 17th AIAA Aerodynamic Decelerator Systems Technology Conference*, AIAA, Reston, VA, pp. 40–49.

Mosseev, Yu. (2001a), "Fluid-Structure Interaction Simulation of the U.S. Army G-12 Parachute," Rept. No. 17-01-RDD Ozon, Moscow, Russia, http://www.mtu-net.ru/mosseev/rd.htm

Mosseev, Yu. (2001b), "Software Tools for the Paraglider Computer-Aided Design Guide," *Proceedings of the 16th AIAA Aerodynamic Decelerator Systems Technology Conference*, AIAA, Reston, VA.

Mosseev, Yu. (2007), "Aerodynamics of Parafoil-Load Systems," unpublished material, Ozon, Moscow, Russia.

Mosseev, Yu., Rysev, O., and Fedorov, V. (1987), "A Mathematical Model of Parafoil Shape," *Fluid-Structure Interaction Journal*, Vol. 20, Physical-Technical Inst., Kazan, Russia.

Moulin, J. (1995), "Wind Influences on Recovery System Performances," *Proceedings of the 13th AIAA ADS Conference and Seminar*, AIAA, Washington, D. C.

Müller, S. (2002), "Modellierung, Stabilität und Dynamik von Gleitschirmsystemen," Ph.D. Dissertation, Herbert Utz Verlag, Germany.

Müller, S., Wagner, O., and Sachs, G. (2003), "A High-Fidelity Nonlinear Multibody Simulation Model for Parafoil Systems," *Proceeding of the 17th AIAA ADST Conference and Seminar*, AIAA, Reston, VA, pp. 149–158.

Mulloy, C. C., Tiaden, R. D, and Yakimenko, O. A. (2011), "The 2nd Generation of Safety Fans GUI," *Proceedings of the 21st AIAA Aerodynamic Decelerator Systems Technology Conference*, AIAA, Reston, VA.

Murphy, A. L. (1971), "Trajectory Analysis of a Radial Homing Gliding Parachute in a Uniform Wind," Army Natick Labs., Technical Rept. TR 73-2-AD, Natick, MA, Sept.

Murray, J. E., Sim, A. G., Neufeld, D. C., Rennich, P. K., Norris, S. R., and Hughes, W. S. (1994), "Further Development and Flight Test of an Autonomous Precision Landing System Using a Parafoil," NASA TM-4599.

Nakajima, T., Hiraki, K., Hinada, M., Yamagami, T., and Ohta, S. (1995), "Guidance Experiment of Gliding Parachute Dropped from Balloon," *Proceedings of the 13th AIAA ADS Conference and Seminar*, AIAA, Washington, D. C.

Nice, R. J., Haak, E. L., and Guteneauf, R. (1965), "Drag and Stability of Cross-Type Parachutes," *Univ. of Minnesota*, FDL-TDR-64-155, Minneapolis, MN.

Nicolaides, J. D. (1965), "On the Discovery and Research of the Para-Foil," *Proceedings of the International Congress on Air Technology*, Little Rock, AR.

Nicolaides, J. D. (1970a), "A Review of Para-Foil Applications," *Journal of Aircraft*, Vol. 7, No. 5, pp. 423–431.

Nicolaides, J. D. (1970b), "Improved Aeronautical Efficiency Through Packable Weightless Wings," *Proceedings of the CASI/AIAA Meeting for Prospects for Improvement in Efficiency of Flight*, AIAA, New York.

Nicolaides, J. D. (1971), "Parafoil Wind Tunnel Tests," U.S. Air Force Flight Dynamics Lab., AFFDL-TR-70-146, Wright-Patterson Air Force Base, OH.

Nicolaides, J. D. (1983), "Free Wing Flyer," U.S. Patent 4,375,280.

Nicolaides, J. D., and Knapp, C. F. (1965), "A Preliminary Study of the Aerodynamic and Flight Performance of the Para-Foil," *Conference on Aerodynamic Deceleration*, Univ. of Minnesota, MN.

Nicolaides, J. D., and Tragarz, M. A. (1970), "Parafoil Flight Performance," *Proceedings of the 3rd AIAA Aerodynamic Decelerator Systems Conference*, AIAA, New York.

Nicolaides, J. D., Speelman, III, and Menard, G. L. C. (1970), "A Review of Para-foil Applications," *Journal of Aircraft*, Vol. 7, No. 5, pp. 423–431.

NOAA (2014), National Oceanic and Atmospheric Administration, www.noaa.gov

Ochi, Y. (2013), "Flight Control of a Powered Paraglider by MIMO PID Control Based on Two-Degree-of-Freedom Integral-Type Optimal Servomechanism," *Proceedings of the 19th IFAC Conference on Automatic Control in Aerospace*, Würzburg, Germany.

Ochi, Y., and Watanabe, M. (2011), "Modelling and Simulation of the Dynamics of a Powered Paraglider," *Proceedings of the Institution of Mechanical Engineers, Part G: Journal of Aerospace Engineering*, Vol. 225, No. 3, pp. 373–386, doi: 10.1177/09544100JAERO888

Ochi, Y., Kondo, H., and Watanabe, M. (2009), "Linear Dynamics and PID Flight Control of a Powered Paraglider," *Proceedings of the AIAA Guidance, Navigation, and Control Conference and Exhibit*, AIAA, Reston, VA.

Onyx PADS (2014), Atair Airspace, www.atair.com/onyx/ul

Oye, I., and Schulte, C. (2009), "Development of Aerodynamic Analysis Software Tools Within the FASTWing CL Project," *Proceedings of the 20th AIAA Aerodynamic Decelerator Systems Technology Conference*, AIAA, Reston, VA.

Pagen, D. (1990), *Walking on Air Paragliding Flight*, Dennis Pagen.

Parachute Military & Sport Systems, www.parachutefactory.com

PATCAD (2001a), "Precision Airdrop Technology Conference and Demonstration," Final Rept. No. 02-019, DTC project No. 8-ES-065-G12-002, Yuma, AZ, http://yuma-notes1.army.mil/mtea/patcadreg.nsf

PATCAD (2001b), "Precision Airdrop Technology Conference and Demonstration," Final Rept., U.S. Army RDECOM, Natick, MA.

PATCAD (2003), "Precision Airdrop Technology Conference and Demonstration," Final Rept., U.S. Army RDECOM, Natick, MA.

PATCAD (2005), "Precision Airdrop Technology Conference and Demonstration," Final Rept., U.S. Army RDECOM, Natick, MA.

PATCAD (2007), "Precision Airdrop Technology Conference and Demonstration," Final Rept., U.S. Army RDECOM, Natick, MA.

PATCAD (2009), "Precision Airdrop Technology Conference and Demonstration," Final Rept., U.S. Army RDECOM, Natick, MA.

Patel, S., Hackett, N. R., and Jorgensen, D. S. (1997), "Qualification of the Guided Parafoil Air Delivery System-Light (GPADS-Light)," *Proceedings of the 14th AIAA Aerodynamic Decelerator Systems Technology Conference*, AIAA, Reston, VA.

Pearson, A. E. (1972), *Optimal Control of a Gliding Parachute System*, Army Natick Labs., Technical Rept. AD-776 335, Natick, MA.

Penwarden, A. D., Grigg, P. F., and Rayment, R. (1978), "Measurements of Wind Drag on People Standing in a Wind Tunnel," *Building and Environment*, Vol. 13, pp. 75–84.

Perkins, C., and Hage, R. (1957), *Airplane Performance, Stability and Control*, Wiley, New York.

Petry, G., Behr, R., and Tscharntke, L. (1999), "The Parafoil Technology Demonstration (PTD) Project: Lessons Learned and Future Visions," *Proceeding of the 15th CEAS/AIAA Aerodynamic Decelerator Systems Technology Conference*, AIAA, Reston, VA.

Petry, G., Hummeltenberg, G., and Tscharntke, L. (1997), "The Parafoil Technology Demonstration Project," *Proceedings of the 14th AIAA Aerodynamic Decelerator Systems Technology Conference*, AIAA, Reston, VA.

Peyada, N. K., Singhal, A, and Ghosh, A. K. (2007), "Trajectory Modeling of a Parafoil in Motion Using Analytically Derived Stability Derivative at High Angle of Attack," *Proceedings of the 19th AIAA Aerodynamic Decelerator Systems Technology Conference and Seminar*, AIAA, Reston, VA.

PGAS (1996a), "Development and Demonstration Test of a Ram-Air Parafoil Precision Guided Airdrop System, Vol. 3, Simulation and Flight Test Results," Draper Lab., Final Rept., CSDL-R-2752, Cambridge, MA.

PGAS (1996b), "Development and Demonstration Test of a Ram-Air Parafoil Precision Guided Airdrop System, Addendum: Flight Test Results and Software Updates after Flight 19," Draper Lab., Final Rept., CSDL-R-2771, Cambridge, MA.

Plaetschke, E. (1997), "Aerodynamic Model Identification of D160-Parafoil from PTD Flight Test Data," DLR Inst. of Flight Mechanics, Internal Rept. No. IB 111-97/40, Germany.

Plaetschke, E., and Saliaris, C. (1995), "Aerodynamic Parameter Estimation of NASA Spacewedge Parafoil-System," DLR Inst. of Flight Mechanics, Internal Rept. No. IB 111-95/39, Germany.

Pontrjagin, L., Boltjanskiy, V., Gamkrelidze, R., and Mishenko, E. (1969), *Mathematical Theory of Optimal Processes*, Nayka, Moscow (in Russian).

Pope, A. (1951), *Basic Wing and Aerofoil Theory*, McGraw-Hill, New York.

Potvin, J. (2003), "Glide Performance Study of Standard and Hybrid Cruciform Parachutes," *Proceedings of the 17th AIAA Aerodynamic Decelerator Systems Technology Conference and Seminar*, AIAA, Reston, VA.

Potvin, J., Charles, R., and Desabrais, K. (2007), "Comparative DSSA Study of Payload-Container Dynamics Before, During and After Parachute Inflation," *Proceedings of the 19th AIAA Aerodynamic Decelerator Systems Technology Conference and Seminar*, AIAA, Reston, VA.

Power Pack (2014), "American Paragliding," www.americanparagliding.com

Poynter, D. (1972), *The Parachute Manual, A Technical Treatise on Aerodynamic Decelerators*, Para Publishing, CA.

Prakash, O. (2006), "Modeling and Simulation of Parafoil-Payload System," Ph.D. Dissertation, Indian Inst. of Technology, Bombay, India.

Prakash, O., and Ananthkrishnan, N. (2006), "Modeling and Simulation of 9-DOF Parafoil-Payload System Flight Dynamics," *Proceedings of the AIAA Atmospheric Flight Mechanics Conference and Exhibit*, AIAA, Reston, VA.

Puranik, A. S. (2011), "Dynamic Modeling, Simulation and Control Design of a Parafoil-Payload System for Ship Launched Aerial Delivery System (SLADS)," Ph.D. Dissertation, Michigan Technological Univ., Houghton, MI.

Puskas, E. (1984), "Ram Air Parachute Design Considerations and Applications," *Proceedings of the 8th AIAA Aerodynamic Decelerator and Balloon Technology Conference*, AIAA, New York.

Puskas, E. (1989), "The Development of a 10,000lb Capacity Ram-Air Parachute," *Proceedings of the 10th AIAA Aerodynamic Decelerator Systems Technology Conference*, AIAA, Washington, D. C.

Quadrelli, M. B., Cameron, J. M., and Kerzhanovich, V. (2001), "Modeling, Simulation, and Control of Parachute/Balloon Flight Systems for Mars Exploration," *Proceedings of the 16th AIAA Aerodynamic Decelerator Systems Technology Conference and Seminar*, AIAA, Reston, VA.

Raab, C. (2009), "Flugmechanisches Gesamtmodell für ein Zukünftiges Militärtransportflugzeug," Ver. 2.0, Inst. of Flight Systems, IB 111-2009/03, DLR Internal Rept., Braunschweig, Germany.

Raab, C., Friehmelt, H., and Spangenberg, H. (2005), "Einsatz von Echtzeitsimulation bei Systemtechnischen Betrachtungen an Militärtransportflugzeugen," *Deutscher Luft- und Raumfahrtkongress*, Friedrichshafen, Germany.

Rademacher, B. J. (2009), "In-Flight Trajectory Planning and Guidance for Autonomous Parafoils," Ph.D. Dissertation, Paper 10597, Iowa State Univ., Ames, IA, www.lib.dr.iastate.edu/etd/10597

Rademacher, B. J., Lu, P., Strahan, A. L., and Cerimele, C. J. (2009), "In-Flight Trajectory Planning and Guidance for Autonomous Parafoils," *Journal of Guidance, Control, and Dynamics*, Vol. 32, No. 6, pp. 1697–1712; also Rademacher, B. J., Lu, P., Strahan, A. L., and Cerimele, C. J. (2008), "Trajectory Design, Guidance and Control for Autonomous Parafoils," *Proceedings of the AIAA Guidance, Navigation, and Control Conference and Exhibit*, AIAA, Reston, VA.

Ray, E. S., and Morris, A. L. (2011), "Measurement of CPAS Main Parachute Rate of Descent," *Proceedings of the 21st AIAA Aerodynamic Decelerator Systems Technology Conference*, AIAA, Reston, VA.

Redelinghuys, R. (2007), "A Flight Simulation Algorithm for a Parafoil Suspending an Air Vehicle," *Journal of Guidance, Control, and Dynamics*, Vol. 30, No. 3, pp. 791–803.

Redelinghuys, R., and Rhodes, S. (2007), "A Graphic Portrayal of Parafoil Trim and Static Stability," *Proceedings of the 19th AIAA Aerodynamic Decelerator Systems Technology Conference and Seminar*, AIAA, Reston, VA.

Render, P. M., and Coulter, P. R. (1993), "The Stability and Aerodynamic Performance of Clusters of Small Cruciform Parachutes," *Proceedings of the 12th RAeS/AIAA Aerodynamic Decelerator Systems Technology Conference and Seminar*, AIAA, Washington, D.C.

Rogers, J. (2009), "Comparative Analysis Involving Wind Profile Data Sources," *Proceedings of the 20th AIAA Aerodynamic Decelerator Systems Technology Conference and Seminar*, AIAA, Reston, VA.

Rogers, R. (2002), "Aerodynamic Parameter Estimation for Controlled Parachutes," *Proceedings of the AIAA Atmospheric Flight Mechanics Conference*, AIAA, Reston, VA.

Rogers, R. (2004), "Continued Testing of the Extended Kalman Filter (EKF) Parameter Estimation Code with X-38 Parafoil Flight Data," Technical Report – Phase II, Yuma Proving Grounds, Yuma, AZ.

Ross, J. C. (1993), "Computational Aerodynamics in the Design and Analysis of Ram - Air Inflated Wings," *Proceedings of the 12th RAeS/AIAA Aerodynamic Decelerator Systems Technology Conference and Seminar*, AIAA, Washington, D. C.

Rotea, M., and Kothandaraman, G. (2001), "Parameter Estimation for Airdrop System: Final Report," Contract No. 00244-00-P-3258, Purdue Univ., West Lafayette, IN.

Rysev, O., Belozerkovsky, S., Nisht, M., and Ponomarev, A. (1987), "Parachutes and Hang-Gliders Computer Investigations," Mashinostroenie, Moscow.

Rysev, O., Ponomarev, A., Vasiljev, M., Vishnjak, A., Dneprov, I., and Mosseev, Yu. (1996), "Parachute Systems," Nauka, Moscow.

Sadeck, J. E., and Lee, C. K. (2009), "Continuous Disreefng Method for Parachute Opening," *Journal of Aircraft*, Vol. 46, No. 2, pp. 501–504.

Scheuermann, E., Ward, M., Cacan, M., and Costello, M. (2015), "Flight Testing of Autonomous Parafoils Using Upper Surface Bleed Air Spoilers," *Proceedings of the 23rd AIAA Aerodynamic Decelerator Systems Technology Conference and Seminar*, AIAA, Reston, VA.

Schlichting, H., and Truckenbrodt, E. (1959), *Aerodynamik des Flugzeuges*, Springer-Verlag, Berlin/Göttingen/Heidelberg.

Schoenenberger, M., Queen, E., and Cruz, J. (2005), "Parachute Aerodynamics from Video Data," *Proceedings of the 18th AIAA Decelerator Systems Technology Conference and Seminar*, AIAA, Reston, VA.

Sego, K., Jr., (2001), "Development of a High Glide, Autonomous Aerial Delivery System – Pegasus 500 (APADS)," *Proceedings of the 16th AIAA Aerodynamic Decelerator Systems Technology Conference*, AIAA, Reston, VA.

Seher-Weiß, S. (2007), "FITLAB User's Guide – Parameter Estimation Using Matlab®, Version 2.0," Inst. of Flight Mechanics, Internal Rept. IB 111-2007/27, DLR Braunschweig, Germany.

Sensor Sizes (2015), Image Sensor Format, www.wikipedia.org/wiki/Image_sensor_format.

Shaw, M. A. (1994), "Autonomous Survival Kit Delivery by Parafoil," M.S. Thesis, Cranfield Univ., Cranfield, England, U.K.

Shen, C. O., and Cockrell, D. J. (1988), "Aerodynamic Characteristics and Flow Round Cross Parachutes in Steady Motion," *Journal of Aircraft*, Vol. 25, No. 4, pp. 317–323;

also Shen, C. O., and Cockrell, D. J. (1986), "Aerodynamic Characteristics and Flow Round Cross Parachutes in Steady Motion," *Proceedings of the 9th AIAA Aerodynamic Decelerator and Balloon Technology Conference*, AIAA, Reston, VA.

Sherpa PADS (2014), Mist Mobility Integrated Systems Technology Inc., http://www.mmist.ca/Media/Docs/Brochures/Sherpa_Ranger_Brochure.pdf

Shilov, A. A. (1971), "Ob Ustoychivosti Dvizhenija Parachuta na Rezhime Ustanovivshegosja Snizhenija (The Stability of Motion of Parachute in the Steady-State Reduction Mode)," *Uchenie Zapiski TsAGI*, Vol. 2, No. 4, pp. 76–83 (in Russian); Translation FTD-MT-24-0465-75.

Shilov, A. A. (1973), "Analiz Ploskikh Slabo-Dempfirovannikh Oszilljaziy Parachuta v Svobodnom Ustanovivshemsja Snizhenii (An Analysis of the Plane Slightly Damped Oscillation of a Parachute in Free Steady Descent)," *Uchenie Zapiski TsAGI*, Vol. 4, No. l, pp. 137–143 (in Russian); Translation FTD-MT-24-0510-75.

Shpund, Z., and Levin, D. (1997), "Dynamic Investigation of Asymmetric Cross-Type Parachutes," *Proceedings of the 14th AIAA Aerodynamic Decelerator Systems Technology Conference*, AIAA, Reston, VA.

Sim, A. G., Murray, J. E., Neufeld, D. C., and Reed, R. D. (1994), "Development and Flight Testing of a Deployable Precision Landing System," *Journal of Aircraft*, Vol. 31, No. 5, pp. 1101–1108; also Sim, A. G., Murray, J. E., Neufeld, D. C., and Reed, R. D. (1993), "The Development and Flight Test of a Deployable Precision Landing System for Spacecraft Recovery," NASA-TM-4525.

Siouris, G. (1993), *Aerospace Avionics Systems: A Modern Synthesis*, Academic Press, San Diego, CA.

Skywalk (2014), Glider Technical Data, www.skywalk.info

Slegers, N. (2004), "Dynamic Modeling, Control Aspects and Model Predictive Control of a Parafoil and Payload System," Ph.D. Dissertation, Oregon State Univ., Corvallis, OR.

Slegers, N. (2010), "Effects of Canopy-Payload Relative Motion on Control of Autonomous Parafoil," *Journal of Guidance, Control, and Dynamics*, Vol. 33, No. 1, pp. 116–125.

Slegers, N., and Costello, M. (2003), "Aspects of Control for a Parafoil and Payload System," *Journal of Guidance, Control, and Dynamics*, Vol. 26, No. 6, doi:10.2514/2.6933, pp. 898–905.

Slegers, N., and Costello, M. (2005), "Model Predictive Control of a Parafoil and Payload System," *Journal of Guidance, Control, and Dynamics*, Vol. 28, No. 4, pp. 816–821; also Slegers, N., and Costello, M. (2004), "Model Predictive Control of a Parafoil and Payload System," *Proceedings of the AIAA Atmospheric Flight Mechanics Conference*, AIAA, Reston, VA.

Slegers, N., and Yakimenko, O. (2009), "Optimal Control for Terminal Guidance of Autonomous Parafoils," *Proceedings of the 20th AIAA Aerodynamic Decelerator Systems Technology Conference and Seminar*, AIAA, Reston, VA.

Slegers, N., and Yakimenko, O. (2011), "Terminal Guidance of Autonomous Parafoils in High Wind to Airspeed Ratios," *Proceedings of the Institution of Mechanical Engineers, Part G: Journal of Aerospace Engineering*, Vol. 225, No. 3, pp. 336–346, doi:10.1243/09544100JAERO749

Slegers, N., Beyer, E., and Costello, M. (2008), "Use of Variable Incidence Angle for Glide Slope Control of Autonomous Parafoils," *Journal of Guidance, Control, and Dynamics*, Vol. 31, No. 3, pp. 585–596.

Slegers, N., Kyle, J., and Costello, M. (2006), "A Nonlinear Model Predictive Control Technique for Unmanned for Air Vehicles," *Journal of Guidance, Control, and Dynamics*, Vol. 29, No. 5, pp. 1179–1188.

Slegers, N., Scheuermann, E., and Bergeron, K. (2015), "High Fidelity in-Flight Pressure and Inertial Canopy Sensing," *Proceedings of the 23rd AIAA Aerodynamic Decelerator Systems Technology Conference and Seminar*, AIAA, Reston, VA.

Smith, B. A. (2000), "Large X-38 Parafoil Passes First Flight Test," *Aviation Week and Space Technology*, Vol. 152, No. 5, pp. 40–44.

Smith, B. D. (1995), "Steering Algorithms for GPS Guidance of RAM-AIR Parachutes," *Proceedings of the 8th International Technical Meeting of the Institute of Navigation*, Palm Springs, CA.

Smith, J., and Bennett, T. (2000), "Design and Testing of the X-38 Spacecraft Primary Parafoil," *Proceedings of the AIAA Modeling and Simulation Technologies Conference*, AIAA, Reston, VA.

Smith, J., Bennett, T., and Fox, R. (1999), "Development of the NASA X-38 Parafoil Landing System," *Proceedings of the 15th AIAA Aerodynamic Decelerator Systems Technology Conference*, AIAA, Reston, VA.

Smith, J., Witkowski, A., and Woodruff, P. (2001), "Parafoil Recovery Subsystem for the Genesis Space Return Capsule," *Proceedings of the 16th AIAA Aerodynamic Decelerator Systems Technology Conference*, AIAA, Reston, VA.

Snook, J., Cram, J. M., and Schmidt, J. (1995), "LAPS/RAMS. A Nonhydrostatic Mesoscale Numerical Modeling System Configured for Operational Use," *Tellus Series A*, Vol. 47, No. 5, pp. 864–875.

SnowGoose UAV (2014), Mist Mobility Integrated Systems Technology, Inc., http://www.mmist.ca/pg_ProductsSnowGooseOverview.php

Sobieski, J. (1994), "The Aerodynamics and Piloting of High Performance Ram-Air Parachute," http://www.afn.org/skydive/sta/highperf.pdf.

Sobol', I. M. (1967), "On the Distribution of Points in a Cube and the Approximate Evaluation of Integrals," *USSR Computational Mathematics and Mathematical Physics*, Vol. 7, No. 4, pp. 784–802.

SOP (2013), "Pilot Proficiency System, Standard Operating Procedure 12-02," The United States Hang Gliding and Paragliding Association, Inc., https://ushpa.aero/policy/SOP-12-02.pdf.

Soppa, U., Görlach, T., and Roenneke, A. J. (2005), "German Contribution to the X-38 CRV Demonstrator in the Field of Guidance, Navigation and Control (GNC)," *Acta Astronautica*, Vol. 56, No. 8, pp. 737–749.

Soppa, U., Strauch, H., Goerig, L., Belmont, J. P., and Cantinaud, O. (1997), "GNC Concept for Automated Landing of a Large Parafoil," *Proceedings of the 14th AIAA Aerodynamic Decelerator Systems Technology Conference*, AIAA, Reston, VA.

Spalart, P. R., and Allmaras, S. R. (1992), "A One-Equation Turbulence Model for Aerodynamic Flows," *Proceedings of the 30th Aerospace Sciences Meeting and Exhibit*, AIAA, Reston, VA.

Spearman, C. (1904), "The Proof and Measurement of Association Between Two Things," *American Journal of Psychology*, Vol. 15, No. 1, pp. 72–101.

Speelman, R. J., Berndt, R. J., and Babish, C. A. (1972), "Parafoil Steerable Parachute, Exploratory Development for Airdrop System Application," USAF Flight Dynamics Lab., AFFDL-TR-71-37, OH.

STARA Airdrop (2014), STARA Technologies, Inc., http://www.staratechnologies.com/airdrop.html

Statnikov, R. B., and Matusov, J. B. (1996), "Use of P_τ Nets for the Approximation of the Edgeworth-Pareto Set in Multicriteria Optimization," *Journal of Optimization: Theory and Applications*, Vol. 91, No. 3, pp. 543–560.

Statnikov, R. B., and Matusov, J. B. (2002), *Multicriteria Analysis in Engineering Using the PSI Method with MOVI 1.0*, Kluwer Academic, Dordrecht, The Netherlands.

Stein, J., Machin, R., and Muratore, J. (1999), "An Overview of the X-38 Prototype Crew Return Vehicle Development and Test Program," *Proceedings of the 15th AIAA Aerodynamic Decelerator Systems Technology Conference*, AIAA, Reston, VA.

Stein, J., Madsen, C., and Strahan, A. (2005), "An Overview of the Guided Parafoil System Derived from X-38 Experience," *Proceedings of the 18th AIAA Aerodynamic Decelerator Systems Technology Conference and Seminar*, AIAA, Reston, VA.

Strahan, A. (2003), "Testing of Parafoil Autonomous Guidance, Navigation & Control for X-38," *Proceedings of the 17th AIAA Aerodynamic Decelerator Systems Technology Conference and Seminar*, AIAA, Reston, VA.

Strickert, G. (2004), "Study on the Relative Motion of Parafoil-Load-Systems," *Aerospace Science and Technology*, Vol. 8, No. 6, pp. 479–488.

Strickert, G., and Jann, T. (1999), "Determination of the Relative Motion Between Parafoil Canopy and Load Using Advanced Video-Image Processing Techniques," *Proceedings of the 15th AIAA Aerodynamic Decelerator Systems Technology Conference*, AIAA, Reston, VA.

Strickert, G., and Witte, L. (2001), "Analysis of the Relative Motion in a Parafoil-Load System," *Proceedings of the 16th AIAA Aerodynamic Decelerator Systems Technology Conference*, AIAA, Reston, VA.

Strong, T., McGrath, J., and Tavan, S. (2005), "The Screamer Airdrop System for Precision Airdrop of 1,000 and 4,000 kg Class Payloads," *Proceedings of the Infotech@Aerospace Conference*, Arlington, VA.

Tanner, C. L., Clark, I. G., Gallon, J. C., Rivellini, T. R., and Witkowski, A. (2013) "Aerodynamic Characterization of New Parachute Configurations for Low-Density Deceleration," *Proceedings of the 22nd AIAA Aerodynamic Decelerator Systems Technology Conference and Seminar*, AIAA, Reston, VA.

Taranenko, V. T. (1968), "Experience of Employing Ritz's, Poincaré's, and Lyapunov's Methods for Solving Flight Dynamics Problems," Air Force Engineering Academy, Moscow, (in Russian).

TASK Sensors (2014), QinetiQ North America, www.qinetiq-na.com/products/survivability/task

Tavan, S. (2006), "Status and Context of High Altitude Precision Aerial Delivery Systems," *Proceedings of the AIAA Guidance, Navigation, and Control Conference*, AIAA, Reston, VA.

Taylor, A. P. (2003), "An Investigation of the Apparent Mass of Parachutes Under Postinflation Dynamic Loading Through the Use of Fluid Structure Interaction Simulations," *Proceeding of the 17th AIAA Aerodynamic Decelerator Systems Technology Conference and Seminar*, AIAA, Reston, VA.

TEMP (2012), "Raytheon's Unmanned Technology Transforms Commercial Cargo Ships into Aid Delivery Platforms," *Unmanned Systems Technology Magazine*, No. 1, http://www.unmannedsystemstechnology.com/2012/07/raytheons-unmanned-technology-transforms-commercial-cargo-ships-into-aid-delivery-platforms/#sthash. QJ2XqKS5.dpuf

Tezduyar, T., Karlo, V., and Garrad, W. (1999) "Advanced Computational Methods for 3D Simulation of Parafoils," *Proceedings of the 15th CEAS/AIAA Aerodynamic Decelerator Systems Technology Conference*, AIAA, Reston, VA.

Tiaden, R. D (2011), "Payload Derived Position Acquisition System (PDPAS) Algorithm Implementation with the Common Range Integrated Instrumentation System Rapid Prototype Initiative (CRIIS-RPI)," *Proceedings of the 21st AIAA Aerodynamic Decelerator Systems Technology Conference and Seminar*, AIAA, Reston, VA.

Tiaden, R. D, and Yakimenko, O. A. (2008), "Development of a Payload Derived Position Acquisition System for Parachute Recovery Systems," *Proceedings of the AIAA Guidance, Navigation and Control Conference and Exhibit*, AIAA, Reston, VA.

Tiaden, R. D, and Yakimenko, O. A. (2009), "Concept Refinement of a Payload Derived Position Acquisition System for Parachute Recovery Systems," *Proceedings of the 20th AIAA Aerodynamic Decelerator Systems Technology Conference and Seminar*, AIAA, Reston, VA.

Toglia, C., and Vendittelli, M. (2010), "Modeling and Motion Analysis of Autonomous Paragliders," Technical Rept. No. 5, Dept. of Computer, Control and Management Engineering, Univ. degli Studi di Roma "La Sapienza," Rome Italy, http://www.dis. uniroma1.it/~labrob/pub/papers/TRDIS510_ParaMod.pdf

Toglia, C., Vendittelli, M., and Lanari, L. (2010), "Path Following for an Autonomous Paraglider," *Proceedings of the 49th IEEE Conference on Decision and Control*, IEEE, pp. 4869–4874.

Toohey, D. (2005), "Development of a Small Parafoil Vehicle for Precision Delivery," Massachusetts Inst. of Technology, Cambridge, MA.

Tory, C., and Ayres, R. (1977), "Computer Model of a Fully Deployed Parachute," *Journal of Aircraft*, Vol. 14, No. 7, pp. 675–679.

TrackEye (2015), Flight Test and Test Ranges, www.imagesystems.se/image-systems-motion-analysis.

Tribot, J.-P., Rapuc, M., and Durand, G. (1997), "Large Gliding Parachute Experimental and Theoretical Approaches," *Proceedings of the 14th AIAA Aerodynamic Decelerator Systems Technology Conference*, AIAA, Reston, VA.

T-Series (2013), T-Series Hardware, Informational Brochure, Vicon, Inc., www.vicon.com/ Content/PDFs/Systems/T-series.pdf

Tuma, J. J. (1987), *Engineering Mathematics Handbook*, McGraw-Hill, New York.

Tutt, B. A., and Taylor, A. P. (2005), "The Use of LS-DYNA to Simulate the Inflation of a Parachute Canopy," *Proceedings of the 18th AIAA Aerodynamic Decelerator Systems Technology Conference and Seminar*, AIAA, Reston, VA.

Tutt, B. A., Taylor, A. P., Berland, J., and Gargano, B. (2005), "The Use of LS-DYNA to Access the Performance of Airborne Systems North America Candidate ATPS Main Parachutes," *Proceedings of the 18th AIAA Aerodynamic Decelerator Systems Technology Conference and Seminar*, AIAA, Reston, VA.

Tweddle, B. E. (2006), *Simulation and Control of Guided Ram Air Parafoils*, Univ. of Waterloo, Canada.

Uhl, E., and Krenz, H. (2009), "Overview About the FASTWing CL Project, a Self-Navigated Gliding System for Capital Loads," *Proceedings of the 20th AIAA Aerodynamic Decelerator Systems Technology Conference*, AIAA, Reston, VA.

Ulich, B., Steele, K., McMillin, P., Benney, R., and Bagdonovich, B. (2003), "A Sodar Height Sensor for RRDAS," *Proceedings of the 7th AIAA Aerodynamic Decelerator Systems Technology Conference and Seminar*, AIAA, Reston, VA.

Visioli, A., and Zhong, Q.-C. (2011), *Control of Integral Processes with Dead Time*, Springer-Verlag, New York, Chap. 2, pp. 24–34.

VN-100 (2015), "IMU/AHRS Surface-Mount Device," Vectornav, www.vectornav.com/products

Wailes, W. (1993), "Development Testing of Large Ram Air Inflated Wings," *Proceedings of the 12th RAeS/AIAA Aerodynamic Decelerator Systems Technology Conference and Seminar*, AIAA, Washington, D. C.

Wailes, W. (1998), "Advanced Recovery Systems for Advanced Launch Vehicles (ARS). Phase 1 Study Results," *Proceedings of the 10th AIAA Aerodynamic Decelerator Systems Technology Conference*, AIAA, Reston, VA.

Wailes, W., and Hairington, N. (1995), "The Guided Parafoil Airborne Delivery System Program," *Proceedings of the 13th AIAA Aerodynamic Decelerator Systems Technology Conference*, AIAA, Washington, D. C.

Ward, M. (2012), "Adaptive Glide Slope Control for Parafoil and Payload Aircraft," Ph.D. Dissertation, School of Aerospace Engineering, Georgia Inst. of Technology, Atlanta, GA.

Ward, M., and Costello, M. (2013), "Autonomous Control of Parafoils Using Upper Surface Spoilers," *Proceedings of the 22nd AIAA Aerodynamic Decelerator Systems Technology Conference*, AIAA, Reston, VA.

Ward, M., Gavrilovski, A., and Costello, M. (2011a), "Flight Test Results for Glide Slope Control of Parafoil Canopies of Various Aspect Ratios," *Proceedings of the 21st AIAA Aerodynamic Decelerator Systems Technology Conference and Seminar*, AIAA, Reston, VA.

Ward, M., Gavrilovski, A., and Costello, M. (2011b), "Performance of an Autonomous Guided Airdrop System with Glide Slope Control," *Proceedings of the 21st AIAA Aerodynamic Decelerator Systems Technology Conference*, AIAA, Reston, VA.

Ward, M., Gavrilovski, A., and Costello, M. (2013), "Glide Slope Control Authority for Parafoil Canopies with Variable Incidence Angle," *Journal of Aircraft*, Vol. 50, No. 5, pp. 1505–1513.

Ward, M., Montalvo, C., and Costello, M. (2010a), "Performance Characteristics of an Autonomous Airdrop System in Realistic Wind Environments," *Proceedings of the AIAA Atmospheric Flight Mechanics Conference*, AIAA, Reston, VA.

Ward, M., Slegers, N., and Costello, M. (2010b), "On the Benefits of in-Flight System Identification for Autonomous Airdrop Systems," *Journal of Guidance, Control, and Dynamics*, Vol. 33, No. 5, pp. 1313–1326.

Ward, M., Slegers, N., and Costello, M. (2012), "Specialized System Identification for Parafoil and Payload Systems," *Journal of Guidance, Control, and Dynamics*, Vol. 35, No. 2, pp. 588–597.

Ware, G. M., and Hassel, J. L., Jr. (1969), "Wind Tunnel Investigation of Ram Air Inflated All-Flexible Wings of Aspect Ratio 1.0 to 3.0," NASA TM SX-1923.

Watanabe, M., and Ochi, Y. (2007), "Modeling and Motion Analysis for a Powered Paraglider (PPG)," *Proceedings of the SICE Annual Conference*, Takamatsu City, Japan, Sept., pp. 3007–3012.

Watanabe, M., and Ochi, Y. (2008), "Modeling and Simulation of Nonlinear Dynamics of a Powered Paraglider," *Proceedings of the AIAA Guidance, Navigation, and Control Conference and Exhibit*, AIAA, Reston, VA.

Watkins, J. W. (2011), "Effects of Non-Standard Atmospheric Temperature on Barometric Altitude Measurements and Calculated Aerodynamic Coefficients," *Proceedings of the 21st AIAA Aerodynamic Decelerator Systems Technology Conference*, AIAA, Reston, VA.

Wegereef, J., and Jentink, H. (2003), "SPADES: A Parafoil Delivery System for Payloads Until 200 Kg," *Proceedings of the 17th AIAA Aerodynamic Decelerator Systems Technology Conference and Seminar*, AIAA, Reston, VA.

Wegereef, J., and Jentink, H. (2005), "Parafoil Characterisation Tests with SPADES," *Proceedings of the 18th AIAA Aerodynamic Decelerator Systems Technology Conference and Seminar*, AIAA, Reston, VA.

Wegereef, J., Benolol, S., and Krenz, H. (2009), "A High-Glide Ram-Air Parachute for 6,000kg Payloads, Tested with the FASTWing CL Test Vehicle," *Proceedings of the 20th AIAA Aerodynamic Decelerator Systems Technology Conference*, AIAA, Reston, VA.

Wegereef, J., Doejaaren, F., Benolol, S., and Zapirain, J. F. (2011), "FASTWing CL Flight Tests with a High-Glide Ram-Air Parachute for 6000kg Payloads," *Proceeding of the 21st AIAA Aerodynamic Decelerator Systems Technology Conference and Seminar*, AIAA, Reston, VA.

Wegereef, J., Leiden, B. V., and Jentink, H. (2007), "Modular Approach of Precision Airdrop System SPADES," *Proceedings of the 19th AIAA Aerodynamic Decelerator Systems Technology Conference and Seminar*, AIAA, Reston, VA.

Wei, K.-C., and Pearson, A. E. (1974), *Numerical Solution to the Optimal Control of a Gliding Parachute System*, Army Natick Labs., Technical Rept. AD-A013 009, Natick, MA.

White, F. M. (2003), *Fluid Mechanics*, 5th ed., McGraw-Hill, New York.

White, F. M., and Wolf, D. F. (1968), "A Theory of Three-Dimensional Parachute Dynamic Stability," *Journal of Aircraft*, Vol. 5, No. 1, pp. 86–92.

Wilcoxon, F. (1945), "Individual Comparisons by Ranking Methods," *Biometrics Bulletin*, Vol. 1, No. 7, pp. 80–83.

Wise, K. A. (2006), "Dynamics of a UAV with Parafoil Under Powered Flight," *Proceedings of the AIAA Guidance, Navigation, and Control Conference and Exhibit*, AIAA, Reston, VA.

Wolf, D. F. (1968), "The Dynamic Stability of a Non-Rigid Parachute and Payload System," Ph.D. Dissertation, Univ. of Rhode Island, Kingston, RI.

Wolf, D. F. (1971), "Dynamic Stability of Nonrigid Parachute and Payload System," *Journal of Aircraft*, Vol. 8, No. 8, pp. 603–609.

WRF Model (2014), "The Weather Research and Forecasting Model," http://www.wrf-model.org/index.php

Wright, R., Benney, R., and McHugh, J. (2005a), "Precision Airdrop System," *Proceedings of the 18th AIAA Aerodynamic Decelerator Systems Technology Conference and Seminar*, AIAA, Reston, VA.

Wright, R., Benney, R., and McHugh, J. (2005b), "An On-Board 4D Atmospheric Modeling System to Support Precision Airdrop," *Proceedings of the Infotech@Aerospace Conference*, Arlington, VA.

Wyllie, T. (2001), "Parachute Recovery for UAV Systems," *Aircraft Engineering and Aerospace Technology*, Vol. 73, No. 6, pp. 542–551.

Wytlie, T., and Downs, P. (1997), "Precision Parafoil Recovery - Providing Flexibility for Battlefield UAV Systems?," *Proceedings of the 14th AIAA Aerodynamic Decelerator Systems Technology Conference*, AIAA, Reston, VA.

X-38 (2014), NASA Armstrong Flight Research Center, http://www.nasa.gov/centers/armstrong/multimedia/imagegallery/X-38/index.html

Yakimenko, O. (2000), "Direct Method for Rapid Prototyping of Near-Optimal Aircraft Trajectories," *Journal of Guidance, Control, and Dynamics*, Vol. 23, No. 5, pp. 865–875.

Yakimenko, O. (2005), "On the Development of a Scalable 8-DoF Model of a Generic Parafoil-Based Delivery System," *Proceedings of the 18th AIAA Aerodynamic Decelerator Systems Technology Conference and Seminar*, AIAA, Reston, VA.

Yakimenko, O. (2011), *Engineering Computations and Modeling in MATLAB/Simulink*, AIAA Education Series, AIAA, Reston, VA.

Yakimenko, O.A. (2013), "Statistical Analysis of Touchdown Error for Self-Guided Aerial Payload Delivery Systems," *Proceedings of the 22nd AIAA Aerodynamic Decelerator Systems Technology Conference*, AIAA, Reston, VA.

Yakimenko, O., and Slegers, N. (2009), "Using Direct Methods for Terminal Guidance of Autonomous Aerial Delivery Systems," *Proceedings of the European Control Conference*, Budapest, Hungary.

Yakimenko, O., and Slegers, N. (2011), "Optimization of the ADS Final Turn Maneuver in 2D and 3D," *Proceedings of the 21st AIAA Aerodynamic Decelerator Systems Technology Conference and Seminar*, AIAA, Reston, VA.

Yakimenko, O., and Statnikov, R. (2005), "Multicriteria Parametrical Identification of the Parafoil-Load Delivery System," *Proceedings of the 18th AIAA Aerodynamic Decelerator Systems Technology Conference and Seminar*, AIAA, Reston, VA.

Yakimenko, O., Berlind, R., and Albright, C. (2007), "Status on Video Data Reduction and Air Delivery Payload Pose Estimation," *Proceedings of the 19th AIAA Decelerator Systems Technology Conference and Seminar*, AIAA, Reston, VA.

Yakimenko, O., Dobrokhodov, V., and Kaminer, I. (2004), "Synthesis of Optimal Control and Flight Testing of Autonomous Circular Parachute," *Journal of Guidance, Control, and Dynamics*, Vol. 27, No. 1, pp. 29–40.

Yakimenko, O., Dobrokhodov, V., Johnson, J., Kaminer, I., Dellicker, S., and Benney, R. (2002), "On Control of Autonomous Circular Parachute," *Proceedings of AIAA Guidance, Navigation, and Control Conference*, AIAA, Reston, VA.

Yakimenko, O., Dobrokhodov, V., Kaminer, I., and Berlind, R. (2005), "Autonomous Video Scoring and Dynamic Attitude Measurement," *Proceedings of the 18th AIAA Aerodynamic Decelerator Systems Technology Conference and Seminar*, AIAA, Reston, VA.

Yakimenko, O., Slegers, N., and Tiaden, R. (2009), "Development and Testing of the Miniature Aerial Delivery System Snowflake," *Proceedings of the 20th AIAA Aerodynamic Decelerator Systems Technology Conference and Seminar*, AIAA, Reston, VA.

Yakimenko, O., Slegers, N., Bourakov, E., Hewgley, C., Jensen, R., Robinson, A., Malone, J., and Heidt, P. (2011), "Autonomous Aerial Payload Delivery System 'Blizzard'," *Proceedings of the 21st AIAA Aerodynamic Decelerator Systems Technology Conference*, AIAA, Reston, VA.

Yavus, T., and Cockrell, D. (1981), "Experimental Determination of Parachute Apparent Mass and its Significance in Predicting Dynamic Stability," *Proceedings of 7th AIAA Aerodynamic Decelerator and Balloon Technology Conference*, AIAA, Reston, VA.

Yavuz, T. (1989), "Determining and Accounting for a Parachute Virtual Mass," *Journal of Aircraft*, Vol. 26, No. 5, pp. 432–437.

Yingling, A. J., Hewgley, C. W., Seigenthaler, T. A., and Yakimenko, O. A. (2011), "Miniature Autonomous Rocket Recovery System (MARRS)," *Proceedings of the 21st AIAA Aerodynamic Decelerator Systems Technology Conference*, AIAA, Reston, VA.

Zhu, Y., Moreau, M., Accorsi, M., Leonard, J., and Smith, J. (2001), "Computer Simulation of Parafoil Dynamics," *Proceedings of the 16th AIAA Aerodynamic Decelerator Systems Technology Conference*, AIAA, Reston, VA.

Zimmerman, C. H. (1932), "Characteristics of Clark Y Aerofoils at Small Aspect Ratio," NACA Rept. 431.

Zimmerman, C. H. (1937), "An Analysis of Lateral Stability in Power-Off Flight with Charts for Use in Design," NACA Rept. 589.

CONTRIBUTING AUTHORS

Oleg Yakimenko, Ph.D, D.Sc.
Professor, Department of Systems Engineering/
Department of Mechanical and Aerospace Engineering
Naval Postgraduate School
Monterey, California

Chapters 1, 3, 4, 5, 6, 7, 8, 9, 10, 11

Steve Lingard, Ph.D.
Technical Director, Vorticity Ltd.
Chalgrove, Oxfordshire, United Kingdom

Chapter 2

Horst Altmann, Ph.D.
Chief Engineer, Parafoil Systems
Airbus Defence & Space
Manching, Germany

Chapter 3

Hamid Johari, Ph.D.
Professor, Mechanical Engineering Department
California State University
Northridge, California

Chapter 4

Thomas Jann, Ph.D.
Research Scientist, German Aerospace Center (DLR)
Institute of Flight Systems, Braunschweig, Germany

Chapters 4, 6, 7, 10, 11

Nathan Slegers, Ph.D.
Associate Professor, Mechanical Engineering
George Fox University
Newberg, Oregon

Chapter 5

Gil Iosilevskii, Ph.D.
Associate Professor, Department of Aerospace Engineering
Technion — Israel Institute of Technology
Haifa, Israel

Chapter 6

Peter Crimi
President, Andover Applied Sciences, Inc. (retired)
West Boxford, Massachusetts

Chapter 6

Glen Brown
President, Vorticity US
Santa Cruz, California

Chapter 6

James Moore
President, JM Technologies LLC
Tucson, Arizona

Chapter 8

Travis Fields, Ph.D.
Assistant Professor, Civil and Mechanical Engineering
University of Missouri-Kansas City
Kansas City, Missouri

Chapter 9

INDEX

Note: Page numbers followed by *f* or **t** (indicating figures or tables).

A-22 cargo bag, 239*f*
A-22 Container Delivery System
(A-22 CDS), 236
AALCT. *See* Advanced Airdrop for Land
Combat' Technology
Above ground level (AGL), 3, 128
ACTD. *See* Advance Concept Technology
Demonstration
AD. *See* Attitude determination
ADP. *See* Air-data probe
ADS. *See* Aerodynamic decelerator
system
ADS TC. *See* Aerodynamic Decelerator
Systems TC
ADST. *See* Aerodynamic decelerator
systems technology
Advance Concept Technology
Demonstration (ACTD), 12
Advanced Airdrop for Land Combat'
Technology (AALCT), 21
Advanced Precision Air Delivery System
(APADS), 21
Advanced recovery system (ARS),
121–122
Aerial guidance unit (AGU), 391
Aerodynamic coefficients, 200
aerodynamic and control coefficients,
206, 213
for asymmetric control line
deflection, 210
geometry and mass properties of
ALEX PADS, 212**t**
lateral aerodynamic derivatives, 209
local anhedral angle, 207
parafoil-payload model, 211*f*
pitching moment, 208, 210
stability and control derivatives for
ALEX PADS, 212**t**
yaw damping, 209
arc-anhedral geometry and force
distribution, 200
ALEX PADS and canopy shape
model, 201*f*

anhedral angle, 201
forces and moments on canopy,
202, 204
Fourier coefficients, 205
Kutta–Joukowski theorem, 204
lift, drag, and side-force
distributions, 203
local side-force distribution, 206
MT-4 canopy, 201*f*
normalized lift and side-force
distribution for wing, 206*f*
dependence on angle of attack, 213
comparison of lift, drag coefficients
and lift-drag ratios, 214*f*
lift and drag, 213
rolling and yawing moments,
217–218
schematic of forces acting on
parafoil-payload system,
215–216*f*
Aerodynamic decelerator system (ADS), 2
quasi-steady-state descent of, 153
Aerodynamic Decelerator Systems TC
(ADS TC), 829
ADST seminars, 832**t**
AIAA Theodor W. Knacke Award
Awardees, 833–835**t**
conferences, 830–831**t**
selection of technical reports, 835
Aerodynamic decelerator systems
technology (ADST), 829
Aerodynamic forces and moments
identification, 637
effect of changing riser length, 639*f*
dependence, 640*f*
descent rate comparison, 644*f*
eigenfrequencies and corresponding
periods, 645**t**
Euler angles PSD analysis, 645*f*
Flight angles determination, 638*f*
G-12 parachute aerodynamics, 638*f*
horizontal and vertical plane projections
of trajectory, 642*f*

Aerodynamic forces and moments
    identification (*Continued*)
    measured wind magnitude *vs.* altitude
        profile, 643*f*
    objective of parameter identification,
        641
    optimization parameters values, 646**t**
    PSD, 643
    scattergram of wind vectors, 643*f*
    standard nonlinear system identification
        technique, 641
    three-dimensional projections, 642*f*
    x and *y* components of ground-speed
        vector comparison, 644*f*
    yaw rotation scheme, 639*f*
Affordable Guided Airdrop System
    (AGAS), 24, 601, 604
AFWA. *See* Air Force Weather Agency
AGAS. *See* Affordable Guided Airdrop
    System
AGL. *See* Above ground level
AGU. *See* Aerial guidance unit; Airborne
    guidance unit
AHRS. *See* Attitude Heading Reference
    System
Air Force, 10
Air Force Weather Agency
    (AFWA), 169
Air-data probe (ADP), 803
Air-referenced flight path.
        See Flight path angle
Airborne guidance unit (AGU),
    1, 15
Airborne systems, 13
Aircraft approach, 532
Airspeed estimation, 398
    algorithms, 410
    dilution of precision, 403*f*
    errors in measuring PADS
        states, 409**t**
    landing accuracy, 407
    linear algebraic equations, 402
    Monte Carlo simulations, 406
    NEF for wind and airspeed, 405*f*
    Newton–Raphson formula, 408
    onboard wind estimation, 407*f*
    PADS trajectories, 399–400*f*
    Taylor-series expansion, 401

ALE template. *See* Arbitrary Lagrangean
    Eulerian template
ALEX. *See* Small Autonomous Parafoil
    Landing Experiment
Along-track errors (ATRK errors), 534
Altitude margin, 425
Angle of attack (AoA), 203, 534
    dependence of aerodynamic
        coefficients on, 213
    comparison of lift, drag coefficients
        and lift-drag ratios, 214*f*
    lift and drag, 213
    rolling and yawing moments, 217–218
    schematic of forces acting on
        parafoil-payload system, 215–216*f*
Angle to wind (ATW), 538, 544
Anhedral angle, 84*f*, 201
ANSYS Fluent, 222
AoA. *See* Angle of attack
APADS. *See* Advanced Precision Air
    Delivery System
Apparent mass pressures, 277
    fluids kinetic energy, 277
    PADS models, 281
    parafoil coordinate frame, 278
    parafoil geometry, 278*f*
    volumetric representation, 279*f*
    VSAERO panel code, 280
APRS. *See* Automatic Packet Reporting
    System
AR. *See* Aspect ratios
Arbitrary Lagrangean Eulerian template
    (ALE template), 249
ARS. *See* Advanced recovery system
Aspect ratios (AR), 43, 200
    canopies, 555
Assumed *vs.* real surface-layer profiles, 128
    causes for over-/undershooting
        target, 139**t**
    data from Windpack release, 135*f*
    downwind and crosswind components,
        137–138
    ground wind offsets and crosswind
        component, 131*f*
    Snowflake PADS drops, 130*f*
    Snowflake performances, 131–133
    Snowflake test site, 128*f*
    Snowflake trajectories, 129, 131

terminal guidance, 137, 139
Windpack data *vs.* Snowflake
  estimation, 136*f*
Windpack parachutes and geodetic
  survey, 129*f*
Windpack release, 131, 134*f*
Asymmetric cross-type parachutes, 648*f*
Asymmetric spoiler configurations, 590.
  See also Symmetric spoiler
  configurations
  configurations, 590
  lateral control authority *vs.* asymmetric
    spoiler configuration, 591*f*
  maximum turn rate, 590
  parafoil and payload system, 590
  turn rate *vs.* asymmetric spoiler
    deflection, 591*f*, 592*f*
Asymmetric upper-surface spoiler
  configurations, 591
ATRK errors. *See* Along-track errors
Attitude determination (AD), 732–733
  airspeed vector, 735
  canopy-payload distance and spatial
    pitch angle, 735*f*
  distance between canopy and payload,
    733*f*
  feature tracking, 739*f*
  features, 740
  FFT analysis, 734
  frequencies/harmonics, 738**t**
  ground-speed vector components,
    736*f*
  payload's attitude, 739*f*
  real drop image, 739*f*
  relative pitch, roll, and yaw
    estimation, 737
  spatial angle of attack for canopy and
    payload, 737*f*
  swing angle between payload and
    canopy, 734*f*
Attitude Heading Reference System
  (AHRS), 705
ATW. *See* Angle to wind
Autoflight mode, 448
Automatic Packet Reporting System
  (APRS), 164
Automating video data processing,
  726–727

auto-scored canopy and payload pixel
  offsets, 732*f*
edge-detection-based object
  segmentation algorithm, 729
KTM calibration, 728–729
modified multistage algorithm, 730–761
parachute system
  dual object tracking, 730*f*
  tracking, 731*f*
TEMA Camera Control, 727–728
TrackEye motion analysis, 727, 729*f*
  scoring calibration monuments, 729*f*
  test item tracking, 730*f*
tracking algorithm, 731–732
Autonomous Parafoil Landing
  Experiment PADS (ALEX PADS),
  38, 686
and canopy shape model, 201*f*
eigenvalues of linearized model of,
  354–355
geometry and mass properties, 212**t**
instrumentation packages, 686
  bidirectional radio modem link, 687
  car test, 699–700
  DNB, 694–697, 699*f*
  DNBM, 695–696
  electrical concept, 687*f*
  FDAS-M, 688–689, 692–693, 699*f*
  flight-test sequence, 697
  GNCS, 689, 691, 694
  GPS height, 698
  LANC interface, 691–692
  outdoor walking tests, 694
  PC104-type boards, 691
  PMS, 691
  ripcord switch, 697–698
  steering box layout, 690*f*
  within tradeoff analysis, 688–689
longitudinal and lateral motion
  validation, 754
  approximate coefficients, 754–755
  estimated parafoil aerodynamic
    derivatives, 755**t**
  estimated parafoil aerodynamic
    derivatives, 757**t**
  identifiability, 754
  input, output, and varied parameters
    vectors, 755**t**

Autonomous Parafoil Landing
    Experiment PADS (ALEX PADS)
    (*Continued*)
    match of simulated and measured
       lateral motion, 757*f*
    time histories, 756*f*
    model root locus, 358–359*f*
    stability and control derivatives, 212**t**

Ballistic winds, 149–150
    computation, 155*f*
    hourly wind data conversion, 154*f*
    predicated impact point
       calculation, 153*f*
    raw *vs.* ballistic winds, 156*f*
    sample winds aloft data file, 152**t**
    for weather balloons, 153
    wind profiles, 150
    winds' measurements, 151
Band-limited guidance (BLG), 449, 459, 483
    computer simulation, 462*f*
    flight path of MegaFly PADS, 464*f*
    IDVD *vs.*, 486**t**
    MegaFly PADS, 463
    size of mesh grid, 461**t**
    trajectory optimization, 463
Bang–bang control, 420, 612
Bank of trajectories, 461
    computer simulation, 463*f*
    flight path of MegaFly PADS, 465*f*
    MegaFly PADS, 463–464
    size of mesh grid, 462**t**
    trajectory optimization, 464
Baseline 2K PADS Monte Carlo simulation
    results, 543*f*
BC. *See* Boundary conditions
Bivariate normal distribution (BND), 46
BLG. *See* Band-limited guidance
BND. *See* Bivariate normal distribution
Body-fixed [*b*] frame, 264
Body-fixed coordinate system, 370*f*
Boundary conditions (BC), 420
British Royal Aeronautical Society
    (RAeS), 829

c. p. *See* center of pressure
CADS. *See* Controlled Aerial Delivery
    System

Calculated aerial release point
    (CARP), 668
Canopy aerodynamic coefficients, 223
    computed lift characteristics, 225
    flowfield characteristics, 223
    higher-velocity regions, 228
    lift, drag, and side-force
       coefficients, 229
    lift coefficient, drag coefficient,
       lift-to-drag ratio, and
       pitching-moment coefficient, 226*f*
    pitching-, rolling-, and yawing-moment
       coefficients, 230
    pressure coefficient distribution in
       spanwise plane, 225*f*, 228*f*
    pressure coefficient distribution in
       streamwise plane, 224*f*, 227*f*
    velocity magnitude distribution in
       streamwise and spanwise planes,
       224*f*, 227*f*
Canopy data, 843
    large, 848**t**
    medium, 846–847**t**
    small, 844–845**t**
Canopy dynamics model (CDM), 540
Canopy modeling, 219
    computational domain, 221*f*
    MC-4 parachute in flight, 219*f*
    mesh refinement factor, 223
    RANS equations, 222
    robust (Octree) tetrahedral/mixed
       unstructured grid, 221
    views, 220*f*
Canopy motion estimation
    instrumentation, 703
    angular velocity amplitude spectrum,
       708*f*
    causes of improper segmentation, 713*f*
    CXL02LF1/CXL02LF3 accelerometers,
       705*f*
    distance between canopy and
       payload, 710
    identification of central moment and
       canopy orientation, 711*f*
    improved canopy color
       configurations, 713*f*
    isolating red shape of canopy, 710*f*
    locations of points of interest, 710*f*

miniature IMU, 704–705
natural points and features, 709*f*
parachute trajectory characterization,
    707*f*
paraglider canopy, 709*f*
payload motion, relative motion,
    and corrected canopy
    motion, 712*f*
raw data from main IMU and
    accelerometers, 706*f*
relative motion measurement, 711*f*
relative motion problem, 703
relative roll, lateral displacement,
    relative pitch, relative yaw, 704*f*
sample analysis of 3D model generated
    images, 714*f*
short delay of the payload's motion, 711
system identification, 713–714
tracking algorithms, 707
video analysis, 712–713
views of canopy, 708*f*
VN-100 SMD, 705
Canopy rigging, 93
drag coefficient for ram-air wing
    without lines, 96*f*
lift coefficient for ram-air wing without
    lines, 96*f*
pitching-moment coefficient for ram-air
    wing
    without lines, 97*f*
    with rigging angles, 97–98*f*
rigging angle, 94
static stability analysis, 95*f*
Capewell Components Company, 629
Capital Loads (CL), 41
Capsule recovery simulation (CARESI),
    442–443
CARESI. *See* Capsule recovery simulation
CARP. *See* Calculated aerial release point;
    Computed air release point
Cartesian coordinate systems, 264
CCD. *See* Charge-coupled device
CDM. *See* Canopy dynamics model
CDS. *See* Container delivery system
CEA. *See* Circular error average
CEAS. *See* Confederation of European
    Aerospace Societies
center of pressure (c.p.), 204

CEP. *See* Circular error probability
CEP_{DMPI} computation, 849–852
CEP_{MPI} computation, 849–852
CFD. *See* Computational-fluid-dynamics
Changing surface-layer wind influence,
    176–177
change in wind speed, 179*f*
control authority, 179–180
dying wind profiles, 178*f*
landing precision with a glide slope
    angle control, 181*f*
Monte Carlo simulation, 180
PADS aerodynamics modeling, 180*f*
wind profiles, 178
Charge-coupled device (CCD), 707
CI. *See* Confidence interval
CIM-2016 ram-air airfoils, 251*f*
Circle of equal probability. *See* Circular
    error probability (CEP)
Circular error average (CEA), 55–56
Circular error probability (CEP), 13,
    46, 407
bivariate normal distribution, 46*f*
features, 49
parametric and nonparametric
    tests, 48**t**
probability density function, 47
sampling from CND, 47–48
Circular normal distribution (CND), 46
CL. *See* Capital Loads
Closed-form solution, 464–465
errors in measuring PADS states, 469**t**
final turn trajectories, 470*f*
reference turn function, 465–466
sensor errors, 467
simulated guidance in ideal
    conditions, 468*f*
terminal guidance maneuver, 465*f*
CMDP. *See* Common Mission Debrief
    Program
CMDS. *See* Countermeasures dispenser
    system
CMOS. *See* Complementary metal oxide
    semiconductor
CND. *See* Circular normal distribution
COI. *See* Critical operational issues
Common Mission Debrief Program
    (CMDP), 727

Complementary metal oxide
   semiconductor (CMOS), 719
Complete pads aerodynamics, 242
   canopy L/D ratio, 243–244
   estimate of drag contribution of ADS
      components, 248*f*
   glide ratio of model ram-air parachute
      system, 247*f*
   map of line segments in lock-in and
      stationary regimes, 246*f*
   model canopy lift-to-drag ratio, 245*f*
   relative contribution of component of
      model ram-air parachute, 247**t**
   stationary and vibrating suspension
      lines, 246*f*
   suspension line vibration, 244
Computational domain, 221*f*
Computational-fluid-dynamics (CFD),
      76, 200, 213, 218, 741
   canopy aerodynamic coefficients, 223
      flowfield characteristics, 223
      higher-velocity regions, 228
      lift characteristics, 225
      lift coefficient, drag coefficient,
         lift-to-drag ratio, and
         pitching-moment coefficient,
         226*f*
      lift, drag, and side-force
         coefficients, 229
      pitching-, rolling-, and
         yawing-moment coefficients, 230
      pressure coefficient distribution in
         spanwise plane, 225*f*, 228*f*
      pressure coefficient distribution in
         streamwise plane, 224*f*, 227*f*
      velocity magnitude distribution in
         streamwise and spanwise planes,
         224*f*, 227*f*
   canopy modeling, 219
      computational domain, 221*f*
      MC-4 parachute in flight, 219*f*
      mesh refinement factor, 223
      RANS equations, 222
      robust (Octree) tetrahedral/mixed
         unstructured grid, 221
      views, 220*f*
   comparison with lifting-line
      theory, 230

lift and drag coefficients, 231*f*
nominal AR for MC-4 canopy, 231
2D data, 233
3D model with lifting-line
   prediction, 232*f*
complete pads aerodynamics, 242
   canopy L/D ratio, 243–244
   estimate of drag contribution of ADS
      components, 248*f*
   glide ratio of model ram-air
      parachute system, 247*f*
   map of line segments in lock-in and
      stationary regimes, 246*f*
   model canopy lift-to-drag ratio, 245*f*
   relative contribution of component of
      model ram-air parachute, 247**t**
   stationary and vibrating suspension
      lines, 246*f*
   suspension line vibration, 244
suspension lines, payload, and slider
   aerodynamics, 233
   A-22 cargo bag, 239*f*
   classification of line segments, 238*f*
   drag coefficient of flag as function of
      aspect ratio, 243*f*
   forces on rigid cylinder, 234
   lift, drag, and pitch-moment
      coefficients, 242*f*
   map of lock-in regime, 237*f*
   natural frequency of line
      segment, 235
   outline of procedure for
      determination of line
      vibration, 236*f*
   PADS examples, 233*f*
   payload extraction, 238*f*
   Strouhal number, 234
   type-V platform with nose
      bumper, 241*f*
   vibration amplitude as function of
      damping parameter, 237*f*
   x-velocity distribution and
      aerodynamic coefficients, 240*f*
Computed air release point (CARP), 10,
      181–182, 602, 606
   bird's-eye view, 187*f*
   LAPS, 185
   locations, 182*f*

Monte Carlo analysis tool, 187
PADS architecture and top-level
    functions, 183*f*
PADS wireless communication
    GUI, 187*f*
PAPS component of PADS, 186–187
sample CARP diagram, 603*f*
solution tab, 184*f*
top-level GUI, 183–184
WindPADS
    component, 183, 185
    weather GUI, 186*f*
Cones, 621*f*
Confederation of European Aerospace
    Societies (CEAS), 829
Confidence interval (CI), 55
Configuration A flights, 803, 808
    EKF horizontal trajectory, 818*f*
    EKF wind modeling error, 819*f*
    measured and EKF air data, 816*f*
    measured and EKF angular rates,
        815*f*
    measured and EKF velocity
        components, 817*f*
Configuration B flights, 803, 809
    EKF horizontal trajectory, 813*f*
    EKF wind modeling error, 814*f*
    measured and EKF air data, 811*f*
    measured and EKF angular rates,
        809–810
    measured and EKF velocity
        components, 812*f*
Confluence point, 24
Coning motion, 614
Container delivery system (CDS),
    21, 602
Continuously operating reference stations
    (CORS), 411
Control derivatives, 366
    expression, 366
    inequality of equation, 367–368
    predictable longitudinal control, 367
Control efficiency models, 327
    asymmetric TE deflection
        GR variation, 332*f*
        horizontal speed variation, 333*f*
        turn rate against, 335*f*
        on turn rate for PADS, 337**t**

vertical speed variation, 334*f*
    nonlinear control efficiency, 331*f*
    regression analysis, 331–332
    semi-empiric approach, 328
    SPADeS PADS, 336
    symmetric TE deflection
        GR variation, 332*f*
        horizontal speed variation, 333*f*
        on longitudinal motion parameters
            for PADS, 336**t**
        vertical speed variation, 334*f*
Control rate influence and actuator
    dynamics, 169, 171. *See also*
    Unknown surface-layer wind
    effect; Winds-aloft data utilization
    changing surface-layer wind influence,
        176–181
    modeling realistic controller, actuator,
        and atmosphere effect,
        174–176
    simulation setup and reference
        simulations, 171–174
Control-activation rule, 614*f*
Controlled Aerial Delivery System
    (CADS), 32
Coordinated Universal Time (UTC), 197
CORS. *See* Continuously operating
    reference stations
Countermeasures dispenser system
    (CMDS), 141
Coupled FSI–CFD analysis, 245
    ram-air airfoils and wings
        aerodynamics, 250
        FASTWing parafoil in flight, 255*f*
        load or control line action, 253
        LS(1)-0417 and CIM-2016 ram-air
            airfoils, 251*f*
        objectives, 251
        semirigid parafoil model, 255*f*
        visualization of flow over airfoil with
            inlet in water tunnel, 252*f*
        wind-tunnel testing of semirigid
            parafoil models, 254*f*
    shape-related factors influencing
        parafoil aerodynamics, 253
        anhedral-arc shape effect of
            parafoil, 255
        cell shape for rib-rib distance, 257*f*

Coupled FSI–CFD analysis (*Continued*)
  contours of load-bearing ribs, 260*f*
  flow direction over top and bottom
    surfaces of wing load-bearing
    rib, 256*f*
  ideal parafoil shape *vs.* inflated
    one, 256*f*
  influence of control lines
    deflection, 260*f*
  large-span rectangular and elliptical
    parafoils predicted shape, 258*f*
  negative wing twisting effect, 261
  parawing shape in brake regime,
    259*f*
  predicted shape 10-triple-cells
    rectangular parafoil, 258*f*
  predicted shape for 14-cell
    rectangular parafoil, 257*f*
  predicted shape for FASTWing
    tapered parafoil, 257*f*
  relative shrinkage for 10-triple-cells
    rectangular parafoil, 259**t**
  relative shrinkage for
    large-aspect-ratio elliptical
    parafoil, 259**t**
  relative shrinkage for
    large-aspect-ratio rectangular
    parafoil, 259**t**
  relative shrinkage for low-aspect-ratio
    rectangular parafoil, 258**t**
  relative shrinkage for tapered
    fastwing parafoil, 259**t**
  wind-tunnel test of FASTWing
    parafoil, 256*f*
  3D effects, 253
  simulation tools for shape and
    aerodynamics analysis, 247
    chordwise pressure coefficient
      distribution, 250*f*
    DVM2, 249
    MONSTR INTEGRATED software
      package, 248
    PARAD, 249
Coupled roll-spiral mode, 356*f*
Crab angle, 267
Crew return vehicle (CRV), 21
Critical operational issues (COI), 12
Cross-track (XTRK), 538

Cross-type parachutes control, 646
  aerodynamic characteristics, 647
  AGAS, 648
  asymmetric cross-type parachutes, 648*f*
  cruciform parachute cluster
    design, 651*f*
  modifications of geometry of, 647*f*
  nominal locations of parachutes in
    cluster, 650*f*
  specifications of cruciform parachute,
    647*f*
  steerable cluster designs, 650*f*
  steerable cluster of two parachutes, 651*f*
  suspended weight in pounds of cargo
    parachutes, 649**t**
Crosswinds and vertical wind disturbances
    accommodation, 495
  inverse kinematics, 496
  prototype system, 501*f*
  reference trajectory, 499
  Snowflake PADS, 50*f*
  UL-and ML-weight PADS, 500–501
  vertical component of winds, 497–498
Crown rigging, 84
Cruciform parachute cluster design, 651*f*
CRV. *See* Crew return vehicle

Daimler-Benz Aerospace group
    (DASA group), 439
DASA group. *See* Daimler-Benz
    Aerospace group
Degree-of-freedom model
    (DoF model), 115
Deployable noseboom (DNB), 689,
    694–695
  assembled, 696*f*
  for car test, 699*f*
  operation concept, 697*f*
  sensors, 695**t**
Deployed parachute, 658*f*
Deployed parafoil test system, 547*f*
Descent and time rate to touchdown,
    153–154
  air density models, 162*f*
  APRS, 164–165
  using ballistic winds, 159
  correction coefficient, 159*f*
  descent time, 158*f*, 164*f*

geometric and geopotential altitude difference, 157*f*
geopotential altitude, 157
gravitational acceleration, 154
PADS cutdown and looking-up camera snapshot, 165*f*
parameters of air density regression models, 161*f*
parameters of exponential approximations of air density, 162**t**
parameters of regression errors, 161*f*
partial pressure of water vapor, 164
single exponential regression, 160, 162
standard atmosphere model *vs.* single exponential model, 163*f*
for stratospheric drops, 160
time histories of altitudes for balloon and Snowflake PADS, 165*f*
time histories of vertical speed of Snowflake PADS, 166*f*
using Snowflake PADS balloon drops, 164
vertical speed *vs.* altitude for balloon, 166–167*f*
Desired MPI (DMPI), 49
DGPS. *See* Differential GPS
Different-weight PADS models, 865
height and moments of inertia, 866–867**t**
Differential AoA turn control, 573*f*
Differential GPS (DGPS), 803
Direct-method approach, 449
DMPI. *See* Desired MPI
DNB. *See* Deployable noseboom
DNB-Module (DNBM), 695
DNBM. *See* DNB-Module
DoF model. *See* Degree-of-freedom model
"Dog curve," 437
Downrange miss distance histograms, 581*f*
Downwind direction, 398
Drag coefficient, 226*f*
aerodynamic of ram-air parachute, 85–86
aerodynamics of ram-air wings, 80–81
contributory elements, 81
experimental and theoretical drag coefficient for ram-air wing, 82*f*
parameter for rectangular planform wings *vs.* aspect ratio, 81*f*

theoretical drag coefficient for ram-air wings, 82*f*
Drogue add-on 2K PADS Monte Carlo simulation results, 543*f*
Drogue GA augmentation, 536
Drogue-assisted glide-angle control, 530
drogue deployment strategy, 539, 541*f*
baseline 2K PADS Monte Carlo simulation results, 543*f*
and computer simulations, 539
drogue add-on 2K PADS Monte Carlo simulation results, 543*f*
GNC activities, 539
GPADS, 540
PADS free-body diagram with parafoil and drogue, 541*f*
PADS GNC, 540
PADS Monte Carlo tool, 542
three-dimensional visualization of drogue-assisted precision approach, 542*f*
limitation of PADS precision landing strategies, 530, 534
aircraft approach, 532
GS, 531
hang-gliding landing proficiency ratings system, 533**t**
high-glide PADS, 530–531
JPADS-XL Screamer, 531
traditional paraglider landing approaches, 532*f*
parafoil glide augmentation, motivation and performance, 534
drogue bridle, 539
Drogue GA augmentation, 536
final approach error contributors, 535*f*
stall turn, 534
subscale drogue deployment, 540*f*
subscale test parameters, 539**t**
subscale testing, 538
TE brakes application, 535
2K PADS add-on, 536*f*
2K PADS drag polar with simulated steady-state trim conditions, 537*f*
2K parafoil glide augmentation, 538*f*

Drogue-assisted precision approach, three-dimensional visualization of, 542*f*

Drop zone (DZ), 2, 604

Dubins car model, 420

Dubins curves, 420

DuBois area (A_{Du}), 241

Dutch-roll mode, 356*f*

DVM2 aerodynamic code, 249

Dynamic stability model, 102

DZ. *See* Drop zone

Earth frame, 265

Earth-referenced flight-path angle, 565

East-North-Up coordinate frame (ENU coordinate frame), 264

Edge-detection-based object segmentation algorithm, 729

EGI navigation system. *See* Embedded GPS inertial navigation system

Eigensystem realization (ERA), 825

Eight-degree-of-freedom (Eight-DoF), 23

Eight-DoF. *See* Eight-degree-of-freedom

Eight-DoF models, 316

    moment equations, 319

    parameters of Buckeye PPG, 330**t**

    set of equations, 318

    summary of equations for, 350**t**

EKF algorithm. *See* Extended Kalman filter algorithm

Electronic speed controller (ESC), 665

Elevation bias, 723

EM. *See* Energy management

Embedded GPS inertial navigation system (EGI navigation system), 803, 805

EMC. *See* Energy management circle

Energy management (EM), 392

Energy management circle (EMC), 440

ENU coordinate frame. *See* East-North-Up coordinate frame

Equations of motion, 264. *See also* Rigid-body models

    asymmetric TE deflection

        GR variation, 332*f*

        horizontal speed variation, 333*f*

        turn rate against, 335*f*

        on turn rate for PADS, 337**t**

        vertical speed variation, 334*f*

filter-error method, 789–792

higher-fidelity models, 26–327

linearized models, 338

    aerodynamic derivatives, 341

    aerodynamic moment vector, 347

    lateral-directional motion, 339

    linearized eight-DoF model, 340

    longitudinal and lateral-directional channels, 343–344

    longitudinal motion, 344

    MATLAB function, 338

    pitch motion, 345, 349

    seven-DoF model, 339

    two-body PPG dynamics, 341

models of control efficiency, 327

nonlinear control efficiency, 331*f*

powered systems, extension to, 347

    examples, 350*f*

    incorporating complex-shape payload, 348–350

    summary of equations for three-DOF model, 348**t**

regression analysis, 331–332

semi-empiric approach, 328

SPADeS PADS, 336

symmetric TE deflection

    GR variation, 332*f*

    horizontal speed variation, 333*f*

    on longitudinal motion parameters for PADS, 336**t**

    vertical speed variation, 334*f*

ERA. *See* Eigensystem realization

ESA. *See* European Space Agency

ESC. *See* Electronic speed controller

European Space Agency (ESA), 22, 439

Extended Kalman filter algorithm (EKF algorithm), 787–789

FA maneuver. *See* Final approach maneuver

FAA. *See* Federal Aviation Administration

FADS. *See* Flush air-data system

FAST. *See* Features from Accelerated Segment

Fast Fourier transform analysis (FFT analysis), 734

FAST-Wing. *See* Folding, Adaptive, Steerable Textile Wing Structure

FASTWing PADS instrumentation
    packages, 686, 688
  configuration and data read-out,
    701–702
  drop test, 702
  FASTWing CL payload system, 698
  FDAS-L, 700–701
  GPS receiver, 700
  PDPAS, 702–703
FASTWing parafoil
  sections in flight, 255f
  wind-tunnel test, 256f
FDAS-L. See Low-cost version of FDAS
FDAS-M. See Flight Data Acquisition
    System
FEA. See Finite element analysis
Feasible solution analysis, 773–774
  bird's-eye view of Pegasus and
    dropsonde trajectories, 777f
  criteria vs. parameter dependencies,
    785f
  data preprocessing, 775
  fminsearch and fminHJ function,
    780–781
  Pegasus PADS identification, 775
  reference, nominal, and optimal parafoil
    trajectories, 785f
  representation of parameters
    spread, 780f
  scaling factors with magnitude of
    correlation coefficient, 786t
  sensitivity analysis, 785
  time histories
    of angular rate components, 775f
    Euler angles, 776f
    of ground-speed vector magnitude,
      774f
    of right and left motor positions, 776f
    of wind vector magnitude, 777f
  vectors of varied parameters, 778–779t,
    781–783
  values, 784t
Features from Accelerated Segment
    (FAST), 738, 740
Federal Aviation Administration (FAA),
    169, 410
FFT analysis. See Fast Fourier transform
    analysis

Filter-error method, 787. See also
    Output-error method
  EKF algorithm and implementation,
    787–789
  equations of motion, 789–792
  measurement data, 803
    air data and angular rates, 807f
    air-data measurements, 808–809
    angle of attack, 804
    configuration A flights, 803, 808
    configuration B flights, 803, 808
    control positions and measured
      winds, 806f
    EGI navigation system, 805
    flight-test data summary, 805t
  model linearization, 792, 794–803
  reference values of aerodynamic and
    control derivatives, 793t
  simplified approaches, 824
    ERA, 825
    OKID algorithm, 825–826, 828
    RWLS, 824–825
    validation of lateral-directional
      dynamics, 827f
    validation of longitudinal
      dynamics, 826f
  system identification results, 809
    control input magnitudes, 820, 824
    EKF horizontal trajectory, 813f, 818f
    EKF wind modeling error, 814f, 819f
    matching EKF aerodynamic
      coefficient and inertia property
      modeling errors, 821f
    matching EKF aerodynamic
      coefficient modeling errors, 822f
    measured and EKF air data, 811f, 816f
    measured and EKF angular rates,
      809–810, 815f
    measured and EKF velocity
      components, 812f, 817f
    not matching EKF aerodynamic
      coefficient and mass property
      errors, 823f
    result comparison, 820
  X-38 PADS aerodynamics model,
    789–792
Final approach maneuver
    (FA maneuver), 418

Finite element analysis (FEA), 246
Flares, 74
Flight Data Acquisition System
    (FDAS-M), 688–689, 691
  for car test, 699f
  data logger, 693
  sensors, 690t
  subsystems, 692–693
Flight management system (FMS), 539
Flight-path angle, 266
Flight-path reconstruction (FPR),
    746–747
  mathematical foundation, 747–749
  payload motion, relative motion, and
    corrected canopy motion, 752
  results, 750–751
Flight-test instrumentation
  ground equipment, 714
    attitude determination, 732–740
    automating video data processing,
      726–732
    ground-based KTM systems,
      714–723
    TSPI solution, 723–726
  onboard instrumentation, 685–686
    instrumentation for canopy motion
      estimation, 703–714
    instrumentation packages for
      ALEX and Fastwing PADS,
      686–703
Flight-test sequence, 697
Fluid-structure interaction (FSI), 200
Flush air-data system (FADS), 803
FMS. See Flight management system
Folding, Adaptive, Steerable Textile Wing
    Structure (FAST-Wing), 41
Forecast Systems Laboratory (FSL), 169
Forward center-of-gravity limit, 364
  negative lift coefficients, 364
  TE deflections, 365–366
Four-DoF models, 271
  dynamic effects, 276–277
  Earth-referenced flight-path angle, 271
  GNC algorithms, 273
  L/D ratio, 274
  lift and drag coefficients, 276
  parameters of ALEX PADS, 275t
  summary of equations for, 348t

Fourier coefficients, 205
FPR. See Flight-path reconstruction
Froude number (F_r), 105
FSI. See Fluid-structure interaction
FSL. See Forecast Systems Laboratory

G-12-based AGAS
  auxiliary geometric relations, 870t
  dimensions of AGAS components,
    869t
  masses of individual AGAS
    components, 571t
  moments of inertia of AGAS
    components, 870t
GA. See Glide angle
Garmin GPS36, 686
GC. See Glide coefficient
GDGPS. See Global differential GPS
    system
GDS. See Generic Delivery System
Generic Delivery System (GDS), 27
GigaFly PADS, 37
Glide angle (GA), 430, 529
Glide coefficient (GC), 544
Glide path (GP), 531
Glide ratio (GR), 3, 296, 409, 530, 612
Glide slope (GS), 418, 531
Glide slope angle (GSA), 430, 531
Glide slope intercept (GSI), 531
Glide-angle control, 529
  drogue-assisted glide-angle control, 530
    drogue deployment strategy and
      computer simulations, 539–546
    limitation of PADS precision landing
      strategies, 530–534
    parafoil glide augmentation,
      motivation and performance of,
      534–539
  using upper-surface spoilers, 582
    approach path profiles for example
      trajectories, 596f
    asymmetric spoiler configurations,
      590–592
    asymmetric symmetric control
      inputs, 597f
    computer simulations, 593
    ground tracks for example approach
      trajectories, 596f

landing dispersions from 250 landing simulations, 598*f*
miss distance, 598*f*
Monte Carlo simulations, 597
response to asymmetric spoiler deflection, 594*f*
response to symmetric spoiler deflection, 593*f*
simulation results, 592–599
spoiler control configuration, 593*f*
effect of spoiler on GC over ground, 595*f*
symmetric spoiler configurations, 583–590
wind profile for example cases, 595*f*
with variable rigging angle, 544
of attack, pitch, and flight-path angles, 545*f*
canopy trim conditions, 546*f*
experiments with larger PADS, 555–567
experiments with ULW ADS, 546–555
larger pads mechanization, 568–574
quarter-chord, 544
relationship of GC to flight-path angle, 545*f*
terminal rigging-angle control, 574–582
varying effects, 544–546
Gliding parachute performance improvement, 88
drag contributions, 90*f*
large ram-air parachute, 89**t**
small ram-air parachute, 89**t**
inlet height, 90
large-scale ram-air parachutes, 91
lift-to-drag ratio *vs.* wing area, 91*f*
Global differential GPS system (GDGPS), 411
Global positioning system (GPS), 6
GNC approach. *See* Guidance, navigation, and control approach
GNCS. *See* Guidance, navigation and control system
GP. *See* Glide path
GPADS. *See* Guided Parafoil Air Delivery System

GPS. *See* Global positioning system
GR. *See* Glide ratio
Graphical user interface (GUI), 15, 181, 740
Ground equipment, 714. *See also* Onboard instrumentation
attitude determination, 732–740
automating video data processing, 726–732
ground-based KTM systems, 714–723
TSPI solution, 723–726
GS. *See* Glide slope
GSA. *See* Glide slope angle
GSI. *See* Glide slope intercept
GUI. *See* Graphical user interface
Guidance, navigation, and control approach (GNC approach), 391. *See also* Surface-layer wind modeling; T-approach guidance
activities, 539
for autonomous PADS, 395
band-limited guidance, 459–464
bank of trajectories, 459–464
earlier maneuver-based strategies, 429
CARESI, 442–443
cone of silence, 433
Draper Laboratory, 438
Draper Laboratory PGAS, 438*f*
EM phase, 434
emergency landing pattern and standard visual traffic pattern, 432*f*
geometric relations for radial homing guidance, 434*f*
for NASA Spacewedge, 436
PTD guidance strategy, 439–441
reference trajectory for X-38 PADS, 441*f*
trajectory of 50%-wind case, 431*f*
trajectory of zero-wind case, 431*f*
UAV, 430, 433*f*
wind penetration parameter, 435
functional representation, 393*f*
navigation solutions, 395
determination of altitude above surface, 410–415
miss distance, 410*f*
schemes, 394*f*
wind accommodation, 415–418

Guidance, navigation, and control
    approach (GNC approach)
    (*Continued*)
    winds and airspeed estimation,
        398–410
    yaw/heading estimation, 396–398
    optimal guidance, 418–429
    practical approaches, 418
    precision placement guidance
        algorithms, 443–450
    quasi-optimal terminal guidance, 464
    closed-form solution, 464–469
    two-DoF terminal guidance,
        469–487
    sequence of events from PADS
        deployment, 392*f*
Guidance, navigation and control system
    (GNCS), 689
Guided Parafoil Air Delivery System
    (GPADS), 21, 540
Gusts effects, 111

HAHO technique. *See* High-altitude
    high-opening technique
HALO technique. *See* High-altitude
    low-opening technique
Hamiltonian isolines for time-optimal
    control problem, 610*f*
Hang-gliding landing proficiency ratings
    system, 533**t**
High Altitude Release Point (HARP), 602
High-altitude balloon payload train,
    658*f*
High-altitude high-opening technique
    (HAHO technique), 2
High-altitude low-opening technique
    (HALO technique), 2
High-mobility, multipurpose, wheeled
    vehicle (HMMWV), 13
Higher-fidelity models, 296–297
    eight-DoF models, 316–319
    model comparison, 319, 322
        ground track comparison, 323*f*
        parafoil and payload pitch, 325*f*
        parafoil and payload pitch rate, 324*f*
        parafoil and payload roll, 326*f*
        parafoil and payload roll rate,
            324*f*, 328*f*

parafoil and payload yaw, 325*f*
parafoil and payload yaw rate, 23*f*,
    327*f*
nine-DoF model, 305–311
parameters of seven-/eight-/nine-DoF
    models of Snowflake PADS,
    320–321**t**
payload constrained rotational
    kinematics, 311–315
rigging schemes, 297, 299–302
seven-DoF models, 316–319
six-DoF model rewritten for connection
    point, 302–305
Higher-order terms (HOT), 792
HMMWV. *See* High-mobility,
    multipurpose, wheeled vehicle
Holding pattern, 437
Homing, 391, 431, 437
HOT. *See* Higher-order terms

IDVD method. *See* Inverse dynamics in
    virtual domain method
IGS. *See* International GNSS service
IMP. *See* Impeller
Impact points (IPs), 619
Impeller (IMP), 691
IMU. *See* Inertial measurement unit
Inboard spoilers, 591
Inertia tensors, 277
    fluids kinetic energy, 277
    PADS models, 281
    parafoil coordinate frame, 278
    parafoil geometry, 278*f*
    volumetric representation, 279*f*
    VSAERO panel code, 280
Inertial frame $[I]$, 269
Inertial measurement unit (IMU), 24, 395
Inflating parachute, 123*f*
Inflation of ram-air parachutes, 121
    analysis, 122
        apparent mass, 123
        equations of motion of system, 122
        inflating parachute, 123*f*
        simulation of inflation load for
            midspan reefed parachute, 125*f*
        simulation of inflation load of slider
            reefed ram-air parachute, 124*f*
    reefing techniques, 121–122

Institute of Flight Systems of German
Aerospace Center, 685
Instrumented pallet payload. *See*
Configuration A flights
Intended point of impact (IPI), 2,
264, 392
International GNSS service (IGS), 411
International Space Station (ISS), 21
International Standard Atmosphere
(ISA), 143
International Union of Geodesy and
Geophysics (IUGG), 157
Inverse dynamics in virtual domain
method (IDVD method), 449
BLG *vs.*, 486t
IPI. *See* Intended point of impact
IPs. *See* Impact points
ISA. *See* International Standard
Atmosphere
ISS. *See* International Space Station
IUGG. *See* International Union of Geodesy
and Geophysics

JMUA. *See* Joint Military Utility
Assessment
Joint Military Utility Assessment
(JMUA), 12
Joint Precision Air Drop System
(JPADS), 10
AGU software, 15–16
categories, 12
failure modes affecting MOE, 14
JMUA COI 1, 13–14
JPADS KPPs, 15
JPADS-MP development, 14–15
program and key requirements, 11
2K and 10K systems, 13
Joint Precision Airdrop Capability
Demonstration (JPACD), 16
JPACD. *See* Joint Precision Airdrop
Capability Demonstration
JPADS. *See* Joint Precision Air Drop
System
JPADS-XL Screamer, 531

Key performance parameters (KPP), 13
Kineto-tracking mounts (KTMs), 714
calibration procedure, 717–719

camera sensor size comparison, 719*f*
CCD/CMOS chip, 719, 721
coordinate frame, 716–717
double camera-calibration monument,
718*f*, 720
elevation bias, 723
performance indicators, 715t
relative error, 721–722
standard and radar-enhanced, 715*f*
synchronized data for three
cameras, 724*f*
test setup and calibration monuments,
716*f*
TSPI solution, 716
video images to analysis, 724*f*
KPP. *See* Key performance parameters
KTMs. *See* Kineto-tracking mounts
Kutta–Joukowski theorem, 204

L/D ratio. *See* Lift-to-drag ratio
Lagrange approach, 631
LANC interface. *See* Local Application
Control Bus System interface
Landing phase, 437–438
LAPS. *See* Local Analysis and Prediction
System
Larger PADS mechanization, 568
airspeed and wind vector, 557–558
airspeed *vs.* rigging angle for low AR
canopy, 565
AR canopies, 555
behavior of GC over ground *vs.* rigging
angle, 566*f*
canopy, rigging, and payload parameters
for flight-test vehicle, 559t
comparison of low- and medium-AR
canopies in flight, 559*f*
descent rate estimate from constant
control segment, 561*f*
differential AoA turn control, 573*f*
estimating lift and drag, 561*f*
experiments with, 555
extracting forward airspeed from GPS
ground speed, 560*f*
flight tests, 559
GA control, 568
GC and airspeed envelope for low AR
canopy, 565*f*

Larger PADS mechanization (*Continued*)
   GC *vs.* rigging angle for low and
      medium AR canopies, 563*f*
   GC *vs.* rigging for low AR canopy, 564*f*
   glide slope angle, 566*f*
   GPS track for constant control segment,
      560*f*
   lift and drag behavior *vs.* AoA and
      symmetric brake, 563*f*
   lift and drag characteristics for low AR
      canopy, 564**t**
   lift and drag coefficients *vs.* AoA for low-
      and medium-AR canopies, 562*f*
   lift and drag parameters with zero brake
      deflection, 562**t**
   parafoil VACS with A-B-C-D line
      groups, 571*f*
   paraglider speed system, 572*f*
   payload CG, 570
   pilot-operated speed system, 571–572
   range of GC over ground
      light wind, 567*f*
      no wind, 567*f*
      strong wind, 568*f*
   range of glide slope over ground in
      moderate wind, 568*f*
   rotation axis tradeoffs, 573*f*
   speed systems, 571
   symmetric braking, 564
   TE control, 569
   use of symmetric TE brake deflection,
      567
   VACS, 570–571, 574
Lateral instabilities, 120
Lateral motion, 115
   added mass, 115
   lateral instabilities, 120
   ram-air parachute axes, 117*f*
   turn maneuver, 117
      sine of angle of bank, 120
      yaw and roll axes in steady turning
         flight, 118
      yawing moment on ram-air wing,
         119*f*
Lateral-directional motion, 369
   lateral-directional stability in steady
      glide, 369
      body-fixed coordinate system, 370*f*

      coordinates, 371
      dihedral angle, 370
      effect of glide slope on oscillatory
         response, 375*f*
      effect of glide slope on spiral
         divergence, 375*f*
      effect of glide slope on stability
         boundaries, 376*f*
      location of spiral divergence
         boundary, 372
      nominal PADS, 373
      effect of suspension line length,
         373–374*f*
   steady turn response to control
      input, 376
      aerodynamic and control derivatives,
         380**t**
      assumptions, 376
      components of apparent mass
         tensor, 381
      control deflection, 383
      glide ratio and path angle in spiraling
         turn, 379
      mass ratio, 377
      parafoil geometry, 377*f*
      turn performance of PADS, 380
      turn response and scale effect for four
         weight PADS, 382*f*
      yaw equilibrium, 378
LCF. *See* Linear complementary filter
Leading edge (LE), 535
LIDAR. *See* Light detecting and ranging
Lift coefficient, 76, 226*f*
   aerodynamic of ram-air parachute,
      84–85
   aerodynamics of ram-air wings, 76
      for Clark Y ram-air parachute
         airfoil, 77*f*
      experimental and theoretical lift
         coefficients for ram-air wing, 79*f*
      lifting-line theory, 77
      parameter for rectangular planform
         wings *vs.* aspect ratio, 77–78
      for ram-air wing for different aspect
         ratios, 80*f*
      sheet strength, 76
      slope of lift curve for low-aspect-ratio
         wing *vs.* aspect ratio, 78*f*

Lift-to-drag ratio (L/D ratio), 226f, 243,
        274, 534
    aerodynamic of ram-air parachute, 86
        for aspect ratios and line lengths,
            86–87
        with aspect ratios and
            line-length-to-span ratios, 87f
        basic parachute design parameters,
            86t
        difference in performance, 88
    aerodynamics of ram-air wings, 82–83
Lifting-line theory, 77, 230
    lift and drag coefficients, 231f
    nominal AR for MC-4 canopy, 231
    2D data, 233
    3D model with lifting-line
        prediction, 232f
Light detecting and ranging (LIDAR), 395
Line length effect, 111, 115
Linear and logarithmic surface-layer
            wind models accommodation, 490
    altitude budget equation, 494
    averaged winds, 495
    inverse kinematics, 492
    new terminal point, 491–492
    wind magnitude, 493
Linear complementary filter (LCF), 397
Linearized models, 338
    aerodynamic derivatives, 341
    aerodynamic moment vector, 347
    lateral-directional motion, 339
    linearized eight-DoF model, 340
    longitudinal and lateral-directional
        channels, 343–344
    longitudinal motion, 344
    MATLAB function, 338
    pitch motion, 345
    seven-DoF model, 339
    two-body PPG dynamics, 341
Local Analysis and Prediction System
        (LAPS), 185
Local anhedral angle, 207
Local Application Control Bus System
        interface (LANC interface),
        691–692
Longitudinal dynamics, 101. See also
        Ram-air parachute
    dynamic stability model, 102

    gusts effects, 111
    line length effect, 111, 115
    rate of retraction effect of trailing edge
        on flare, 115
    scaling effects, 105
        design parameters for large ram-air
            parachute, 106t
        dimensionless time of pull, 106
        mass ratio, 105
        response of large ram-air parachute,
            108–110f, 112–114f, 116f
        test flight, 107
    of small ram-air parachutes, 103
        design parameters, 103t
        highly damped phugoid motion, 105
        response of small ram-air parachute
            to symmetric TE deflection, 104f
    trailing-edge release effects, 111
Longitudinal static stability
    analysis, 361
        control derivatives, 366–368
        forward center-of-gravity limit,
            364–366
        minimal lift coefficient at trim and aft
            center-of-gravity limit, 368–369
        notation and reference frame, 362f
        pitching-moment coefficient of
            parachute, 362
        quadratic equation, 363
        trim and stability, 361–364
    canopy rigging, 93
        drag coefficient for ram-air wing
            without lines, 96f
        lift coefficient for ram-air wing
            without lines, 96f
        pitching-moment coefficient for
            ram-air parachute with rigging
            angles, 97– 98f
        pitching-moment coefficient for
            ram-air wing without lines, 97f
        rigging angle, 94
        static stability analysis, 95f
    trailing-edge deflection effect, 98
        flight parameters, 98
        influence of symmetric TE deflection
            on flight velocities, 101f
        influence of symmetric TE deflection
            on glide angle, 102f

Longitudinal static stability (*Continued*)
   influence of symmetric TE
      deflection on lift and drag
      coefficients, 100*f*
   influence of symmetric TE deflection
      on trim values, 100*f*
   pitching-moment coefficient for
      ram-air parachute with rigging
      angles, 99*f*, 101–102*f*
Low Velocity Airdrop Delivery System
   (LVADS), 238
Low-cost version of FDAS (FDAS-L),
   689, 700–701
LS(1)-0417 ram-air airfoils, 251*f*
LVADS. *See* Low Velocity Airdrop
   Delivery System

M-code, 15
Marine Corps Warfighting Lab
   (MCWL), 12
Mass ratio (M_r), 105, 377
Mathematical foundation and
   software, 766
   criterial constraints, 766
   Edgeworth-pPareto set, 767
   functional constraints, 766
   MOVI, 769–770
   multicriteria identification package
      feature, 767–768
   parameter space, 773
   PSI method, 772
   test runs of identification problem,
      768–770
   trial points and CPU resources,
      771–772
Maximally Stable Extremal Regions
   (MSER), 740
MAXMS dropsonde. *See* Micro
   Air-Launched Expendable
   Meteorological Sensor dropsonde
MC-4 parachute in flight, 219*f*
MCWL. *See* Marine Corps Warfighting
   Lab
Mean point of impact (MPI), 48
Mean sea level (MSL), 131
Measures of effectiveness (MOEs), 13
Measures of performance (MOPs), 14
Median radial error (MRE), 46, 56

MegaFly PADS, 36–37
MEMS. *See* Microelectromechanical
   systems
Mesh refinement factor, 223
Micro Air-Launched Expendable
   Meteorological Sensor dropsonde
   (MAXMS dropsonde), 140
Microelectromechanical systems
   (MEMS), 686
MicroFly PADS, 36–37
Mid-Span Reefing, 122
Minimum Principle, 420
Model linearization, 792, 794–796
   abbreviated set of parameter-error
      equations, 801**t**
   angular rate equation linearization,
      797–800
   EKF error state summary, 804**t**
   static aerodynamic model error states,
      802–803
   system-error equation summary, 800**t**
   variations of aerodynamic and
      gravitational forces, 796–797
Model predictive control (MPC), 478
Model predictive controllers, 522
   discretized system, 524
   performance index, 522
   yaw profile, 525
Modified multistage algorithm, 730–761
MOEs. *See* Measures of effectiveness
Monkey Cortex Navigation Platform,
   664–665
MONSTR INTEGRATED software
   package, 248
Monte Carlo simulations, 553–554, 581,
   597
MOPs. *See* Measures of performance
MOVI software. *See* Multicriteria
   Optimization/Vector
   Identification software
MPC. *See* Model predictive control
MPI. *See* Mean point of impact
MPSS. *See* Orion Mission Planner and
   Simulation Software
MRE. *See* Median radial error
MSER. *See* Maximally Stable Extremal
   Regions
MSL. *See* Mean sea level

Multicriteria Optimization/Vector
    Identification software (MOVI
    software), 762, 769–770
Multicriteria parametrical identification,
    762. *See also* Filter-error method;
    Parameter estimation
    feasible solution analysis, 773–786
    identification problem formulation for
        Pegasus PADS, 763–765
    mathematical foundation and software,
        766–773
    multicriteria parameter estimation, 763*f*

NASA X-38 program, 21–22
National Oceanic and Atmospheric
    Administration (NOAA), 169
Nationwide GPS system (NDGPS), 411
Navigational frame [*n*], 265
NCF. *See* Nonlinear complementary filter
NDGPS. *See* Nationwide GPS system
NDI controller. *See* Nonlinear dynamic
    inversion controller
NED frame. *See* North-East-Down frame
NEF. *See* Nonlinear estimation filter
Newton's second law, 269
Nine-DoF model, 301*f*, 305
    aerodynamic force, 310
    parafoil dynamics, 307
    parameters of PADS, 329**t**
    payload dynamics, 308
    payload subsystem, 311
    rotation matrix, 308
    single matrix equation, 309
    six-DoF model *vs.*, 305–306
    summary of equations for, 349**t**
    zero-moment connection, 306
NOAA. *See* National Oceanic and
    Atmospheric Administration
Non-gliding parachute systems control
    guidance using canopy distortion, 601
    aerodynamic forces and moments
        identification, 637–646
    AGAS, 604, 629
    AGAS model *vs.* real AGAS drop,
        619*f*
    AGAS-2 control profile *vs.* altitude,
        628*f*
    Capewell Components Company, 629

CARP, 602, 607
    computer simulation and flight-test
        results, 619–629
    concept of operations, 602–607
    cones, 621*f*
    control of cross-type parachutes
        and parachute clusters,
        646–652
    delay and hysteresis to fight yaw
        oscillations, 621*f*
    electromechanically driven AGU,
        629*f*
    number of PMA activations decrease,
        622*f*
    optimal control synthesis, 607–614
    PATCAD 2001 results, 626**t**
    payload suspension scheme, 605*f*
    practical control architecture,
        614–618
    prediction term, 621
    results AGAS drops, 623–624
    rigged AGAS and two AGAS steering,
        625*f*
    RT-tracking *vs.* target-seek
        trajectories, 625*f*
    sample CARP diagram, 603*f*
    simplified *vs.* more advanced
        algorithm with prediction term,
        624*f*
    simulations with operating angles,
        620*f*
    six-DoF model of controllable
        circular parachute, 629–637
    target-seek control strategy, 623
    touchdown errors, 628*f*
    uncontrolled trajectories with
        different wind profiles, 606*f*
    guidance using reefing control, 652
    one-DoF GNC simulations and flight
        test, 663–672
    one-DoF guidance, 652–663
    performance envelope for 2-DoF
        guidance, 673–679
    wind models and 2-DoF GNC
        simulations, 679–684
Nonlinear complementary filter
    (NCF), 397
Nonlinear control efficiency, 758–760

Nonlinear dynamic inversion controller
     (NDI controller), 525. *See also*
     Proportional, derivative, and
     integral controllers
     (PID controllers)
  inner-loop inversion, 525–526
Nonlinear estimation filter (NEF), 404
  for wind and airspeed, 405*f*
Nonlinear proportional GC controller, 582
North-East-Down frame
     (NED frame), 264
Northwest direction (NW direction), 53
NRDEC. *See* U.S. Army Research,
     Development, and Engineering
     Center
NW direction. *See* Northwest direction

OA. *See* Operating angle
Observer/Kalman filter identification
     algorithm (OKID algorithm),
     787, 825–826, 828
Octree, 221
OKID algorithm. *See* Observer/Kalman
     filter identification algorithm
Onboard instrumentation, 685–686.
     See also Ground equipment
  for canopy motion estimation,
     703–714
  packages for ALEX and Fastwing
     PADS, 686–703
One-DoF guidance, 652. *See also* 2-DoF
     guidance
  ADS, 655
  average coefficient of drag, 659*f*
  components, 655
  descent rate, 656*f*
  four reversible reefing techniques with
     solid lines, 653*f*
  GNC algorithm, 658
  GNC simulations and flight test, 663
     ADS and control parameters, 669**t**
     assembled hardware with mounted
        electronics, 665*f*
     communication between electrical
        components of ADS, 665*f*
     controlled and uncontrolled descent
        paths, 670*f*, 672*f*
     drop testing, 672

     empirically tuned PI and PID
        controller gains, 668**t**
     ESC, 665
     fully operational ADS in flight, 665*f*
     ground control station GUI, 666*f*
     Monkey Cortex Navigation Platform,
        664–665
     predicted and actual winds, 671*f*
     software control routine, 667*f*
     step response of ADS, 668*f*
     touchdown errors distribution in
        Monte Carlo simulation, 664*f*
     Ziegler–Nichols tuning rules,
        666–667
  GPS loggers, 656–657
  ground projection of simulated descent
     profile, 657*f*
  high-altitude balloon payload train and
     deployed parachute, 658*f*
  parachute without and with
     reefing, 652*f*
  predictive improvements, 661*f*
  projection of descent path, 654*f*
  projection of descent path prediction,
     662*f*
  reversible reefing system for circular
     canopies, 652
  sequence of descent profile simulations,
     660*f*
  two loop skirt reefing drop-test data,
     654*f*
Operating angle (OA), 609
Optimal control synthesis, 607
  flight path computed with usage of real
     yaw-angle profile, 613*f*
  fuel-minimum problem, 609
  generalized case, 611*f*
  Hamiltonian isolines
     fuel-minimum problem, 611*f*
     time-optimal control problem, 610*f*
  influence of constant yaw rate, 612*f*
  influence of OA's magnitude, 612*f*
  OA, 609
  Pontrjagin's maximum principle,
     607–609
  projection of optimization problem onto
     horizontal plane, 607*f*
  second cost function, 608

TA, 611
time-optimal control, 609*f*
time-optimal trajectory and
    time-optimal controls, 609
Optimal guidance, 418–419
    in AGU, 427
    altitude margin, 425
    combined minimum-time and
        minimum control-effort
        trajectory, 428*f*
    cost function, 429
    EM maneuver, 428*f*
    guidance triplet partitions, 423*f*
    Hamiltonian for system, 419
    minimum control-effort trajectories,
        426*f*
    Minimum Principle, 420
    time-minimum optimal trajectories,
        421*f*
    TPBV problem, 424
Optimal trajectory, 574–575
Orion Mission Planner and Simulation
    Software (MPSS), 17
Orion™ PADS, 536
"Outboard" spoilers, 591–592
Outdoor walking tests, 694
Output-error method, 746. *See also*
    Filter-error method
    parameter estimation using, 753–754
        flight performance and stability
            analysis, 760–762
        longitudinal and lateral motion
            validation of ALEX PADS,
            754–758
        six-DoF model validation and
            nonlinear control efficiency,
            758–760

PACD. *See* Precision Airdrop Capability
    Demonstration
PADS. *See* Parafoil aerial delivery systems;
    Precision aerial delivery systems
PADS control schemes, 501–502
    model predictive controllers, 522–525
    NDI controller, 525–527
    PID controllers, 502
        advanced controllers, 509–521
        AGU for DragonFly PADS, 505

        bode plot of tuned MegaFly
            closed-loop transfer
            function, 509*f*
        closed-loop reference model, 508
        DragonFly heading rate controller,
            506*f*
        heading controller for ALEX PADS,
            503*f*
        inner-loop PID controller, 506–507
        measuring DragonFly response,
            504*f*
        MegaFly controller, 506*f*
        standard PID controller in Simulink,
            507*f*
PADS landing precision affecting
    factors, 127
    control rate influence and actuator
        dynamics, 169, 171
    changing surface-layer wind
        influence, 176–181
    modeling realistic controller,
        actuator, and atmosphere effect,
        174–176
    simulation setup and reference
        simulations, 171–174
    unknown surface-layer wind
        effect, 128
        assumed *vs.* real surface-layer profiles,
            128–139
        ballistic winds, 149–155
        descent and time rate to touchdown,
            153–167
        surface-layer wind measuring,
            139–149
    winds-aloft data utilization, 181
        CARP and Mission Planner,
            181–187
        safety fans tool, 187–197
Para-Point cargo delivery system, 6
Parachute Industry Association
    (PIA), 829
Parachute system, 2
PARAD code. *See* Pressure from
    aerodynamic code
Parafoil aerial delivery systems
    (PADS), 264
Parafoil dynamics simulator (PDS), 23
Parafoil geometry, 377*f*

Parafoil glide augmentation, 534
  drogue bridle, 538
  Drogue GA augmentation, 536
  final approach error contributors, 535f
  stall turn, 534
  subscale drogue deployment, 540f
  subscale test parameters, 539t
  subscale testing, 538
  TE brakes application, 535
  2K PADS add-on, 536f
  2K PADS drag polar with simulated
      steady-state trim conditions, 537f
  2K parafoil glide augmentation, 538f
Parafoil technology demonstrator
    project (PTD project), 439
Parafoil-payload model, 211f
Parafoils, aerodynamic characterization
    of, 199
  aerodynamic coefficients, theoretical
      derivation of, 200
    aerodynamic and control coefficients,
        206–213
    arc-anhedral geometry and force
        distribution, 200–206
    dependence on angle of attack,
        213–218
  CFD analysis, 218
    canopy aerodynamic coefficients,
        223–230
    canopy modeling, 219–223
    comparison with lifting-line theory,
        230–233
    complete pads aerodynamics,
        242–245
    suspension lines, payload, and slider
        aerodynamics, 233–241
  coupled FSI–CFD analysis, 245
    ram-air airfoils and wings
        aerodynamics, 250–253
    shape-related factors influencing
        parafoil aerodynamics,
        253–261
    simulation tools for shape and
        aerodynamics analysis,
        247–250
Paraglider (PPG), 23–24, 327
  speed system, 572f
ParaLander PADS, 34

Parameter estimation. *See also* Filter-error
    method; Multicriteria parametrical
    identification
  methodology, 742
    effort on flight-test planning, 743–744
    mathematical model, 744–745
    measurements, 744
    for parafoil systems, 743
    principal problem in parafoil-load
        systems, 744
    proposed control input sequence,
        744f
    Quad-M basics, 743f
    SID, 745–746
    six-/seven-/eight-/nine-DoF PADS
        models, 746
    "white-box" approach, 745
  using output-error method, 746,
      753–762
Parameter space investigation method
    (PSI method), 762, 772
Parametrical identification of parachute
    systems
  filter-error method, 787
    EKF algorithm and implementation,
        787–789
    equations of motion, 789–792
    measurement data, 803–809
    model linearization, 792, 794–803
    reference values of aerodynamic and
        control derivatives, 793t
    simplified approaches, 824–828
    system identification results,
        809–824
    X-38 PADS aerodynamics model,
        789–792
  FPR, 746–752
  multicriteria parametrical identification,
      762
    feasible solution analysis, 773–786
    identification problem formulation
        for Pegasus PADS, 763–765
    mathematical foundation and
        software, 766–773
    multicriteria parameter estimation,
        763f
  parameter estimation methodology,
      742–746

parameter estimation using output-error method, 753
flight performance and stability analysis, 760–762
longitudinal and lateral motion validation of ALEX PADS, 754–758
six-DoF model validation and nonlinear control efficiency, 758–760
PATCAD. *See* Precision Airdrop Technology Conference and Demonstration
Path-planning algorithm, 684
Payload constrained rotational kinematics, 311
angular velocity of payload, 312
eight-DoF model, 314–315
payload yaw angle, 313
seven-and eight-DoF models, 311–312
seven-DoF model, 312
Payload Derived Position Acquisition System (PDPAS), 702–703
Payload extraction, 238*f*
Payload suspension scheme, 605*f*
PC104-type boards, 691
PDPAS. *See* Payload Derived Position Acquisition System
PDS. *See* Parafoil dynamics simulator
Pegasus PADS, 23, 438, 774*f*, 853
configurations, 24*f*
eight-DoF model, 762
estimation from side-view photo, 855*f*
geometry and mass properties, 25**t**
GNC algorithm, 509
identification, 775, 777
problem formulation, 763–765
PID controller implementation, 509*f*
PGAS. *See* Precision guided airdrop system
PI. *See* Point of impact
PIA. *See* Parachute Industry Association
PID controllers. *See* Proportional, derivative, and integral controllers
Pilot-operated speed system, 571–572
Pitching-moment coefficient, 226*f*
Plug and Measure Serial (PMS), 691
PMAs. *See* Pneumatic muscle actuators
PMS. *See* Plug and Measure Serial

Pneumatic muscle actuators (PMAs), 25, 605–606
Point of impact (PI), 2
Point-of-use delivery, 8–9
Pontrjagin's maximum principle, 607–609
Position of payload centroid. *See* Time-space-position information solution (TSPI solution)
Power spectrum density (PSD), 643
Powered systems, extension to, 38
examples, 350*f*
incorporating complex-shape payload, 348–352
summary of equations for three-DOF model, 348**t**
PPG. *See* Paraglider
Practical control architecture, 612
AGAS deployment and risers untwisting sequence, 618*f*
bang-bang, 612
coning motion, 614
control actions history right after release, 618*f*
control-activation rule, 614*f*
effect of PMA transition moment, 616*f*
GNC system, 618
outer and inner cones, 615*f*
PMA, 614
ways of decreasing influence of yaw oscillations, 616*f*
Precision aerial delivery systems (PADS), 1–2
accuracy, 44, 46
analysis of outliers, 63–71
CEP, 46–48
touchdown accuracy analysis, 48–58
trends in precision airdrop, 58–63
components and deployment sequence, 7*f*
at demonstration events, 18–19**t**
gain, 517
geometry and mass properties modeling, 853–858
JPADS program and key requirements, 11–16
by JPADS weight categories, 20**t**
Mission Planner, 147
Monte Carlo tool, 542

Precision aerial delivery systems (PADS)
    (*Continued*)
  operations, 2
    airdrop, 4
    earlier steerable parachute, 3*f*
    flight pattern for sport skydiving, 8*f*
    gliding parachutes, 3
    military applications, 11
    mission parameters, 4*f*
    PADS mission, 10*f*
    PADS release procedure, 6
    point-of-use delivery, 8–9
    ram-air gliding parachutes, 8
    U.S. Army/U.S. Air Force
      program, 10
  Orion PADS, 20–21
  penetration speed, 859–863
  performance envelope, 45*f*
  precision aerial delivery, 2
    earlier steerable parachute, 3*f*
    flight pattern for sport skydiving, 8*f*
    JPADS program and key
      requirements, 11–16
    mission parameters, 4*f*
    operations, 2–11
    PADS mission, 10*f*
Precision airdrop, trends in, 58
  M PADS and L PADS performance, 64*f*
  miss distance distribution, 62*f*
  MRE estimation, 61
  PATCAD 2001–2009 performance, 61*f*
  PATCAD 2003 and PATCAD 2005
    performance, 59*f*
  PATCAD 2007 and PATCAD 2009
    performance, 60*f*
  touchdown accuracy, 63
  UL PADS performance, 65–66*f*
  XL PADS performance, 65*f*
Precision Airdrop Capability
  Demonstration (PACD), 16
Precision Airdrop Technology
  Conference and Demonstration
  (PATCAD), 16
Precision airdrop technology
  demonstrations, 16
  AGU configurations for SPADeS
    PADS, 28**t**
  ALEX PADS flight testing, 39*f*

CADS, 32
CADS PADS, 33*f*
configurations of Pegasus
  PADS, 24–25
descent rate *vs.* forward speed, 45*f*
DragonFly AGU, 41–42
DragonFly PADS deployment, 35*f*
dual-antenna GPS, 35
FireFly PADS, 36*f*
G-11 parachutes, 30
GDS, 27
GigaFly PADS, 37
glide ratio *vs.* aspect ratio, 44*f*
GPAS-L qualification tests results, 21*f*
MegaFly PADS, 36–37
Mosquito PADS, 28–29
Onyx 500, 31*f*
Onyx ML, 32*f*
Onyx UL, 31*f*
Orion development, 20
parafoil-based PADS, 23–24
ParaLander PADS, 34*f*
PATCAD 2007, 31–32
precision airdrop demonstration
  events, 17**t**
properties of different-weight PADS,
  40**t**, 43**t**
Snowflake PADS, 38
SnowGoose, 24*f*
SPADeS PADS, 26
vehicle configurations, 22*f*
X-38 AGU logic, 23
Precision guided airdrop system
  (PGAS), 438
Precision placement guidance
  algorithms, 443
  examples of flight path, 444–445*f*
  GNC algorithms, 443
  guidance strategy for Snowflake
    PADS, 450*f*
  high-wing-loading
    parafoils, 445–446
  system, 446
  T-approach, 447
  tracking points, 450
Pressure altimeter, 411
Pressure from aerodynamic code
  (PARAD code), 248–249

Proportional, derivative, and integral
     controllers (PID controllers), 502.
     See also Nonlinear dynamic
     inversion controller (NDI
     controller)
  advanced controllers, 509
     adaptive heading rate controller, 517*f*
     AGU oscillations, 516
     airspeed–ground-speed diagram, 519*f*
     error dynamics, 513
     genesis of dependence, 521*f*
     integrated guidance and control
        algorithm, 509
     kinematics for Euler angles, 512
     PADS kinematic equations, 510
     PID controller with bank angle
        feedback, 516*f*
     PID controller with yaw rate
        feedback, 515*f*
     problem geometry, 510*f*
     servo model, 515
     simulation results of reference
        trajectory tracking, 514*f*
     tracking performance, 518*f*
     turn-around-a-point techniques,
        519*f*
  AGU for DragonFly PADS, 505
  bode plot of tuned MegaFly closed-loop
     transfer function, 509*f*
  closed-loop reference model, 508*f*
  heading controller for ALEX PADS,
     503*f*
  inner-loop PID controller, 506–507
  measuring DragonFly response, 504*f*
  MegaFly controller, 506*f*
  standard PID controller in Simulink,
     507*f*
Prototype CVR payload. *See* Configuration
     B flights
PSD. *See* Power spectrum density
PSI method. *See* Parameter space
     investigation method
PTD project. *See* Parafoil technology
     demonstrator project

QinetiQ North America (QNA),
     139–140, 153
QNA. *See* QinetiQ North America

"Quad-M" requirements, 742
Quarter-chord, 544
Quasi-optimal terminal guidance, 464
  closed-form solution, 464–469
  two-DoF terminal guidance, 469
     bird's-eye view of two typical
        Snowflake PADS trajectories, 485*f*
     BLG approach, 483–484
     duration of maneuver, 469–470
     error statistics, 484**t**
     IDVD *vs.* BLG, 486**t**
     IDVD-based approach, 475
     Monte Carlo simulation dispersion,
        483*f*
     MPC, 478
     numerical algorithm, 475–476
     optimal guidance with wind
        disturbance, 477*f*
     optimal real-dynamics-augmented
        guidance, 479*l*–480*f*
     optimal terminal guidance algorithm,
        476, 478
     precision placement trajectory,
        481*f*
     solution of TPBV problem, 471
     speed factor, 473
     terminal guidance decision variables,
        482*f*
     two-point backward difference
        approximation, 474

Radio frequency (RF), 32
Radiosonde Wind Sounding balloon
     (RAWIN balloon), 139–140
RAeS. *See* British Royal Aeronautical
     Society
RAM. *See* Random-access-memory
Ram-air parachute, 73–74
  aerodynamic, 83
     anhedral angle, 84*f*
     crown rigging, 84
     drag coefficient, 85–86
     gliding parachute performance
        improvement, 88–91
     influence of line length on lift
        coefficient, 85*f*
     lift coefficient, 84–85
     lift-to-drag ratio, 86–88

Ram-air parachute (*Continued*)
  airfoil sections, 75
  axes, 117*f*
  extension of ram-air technology, 76
  flight performance, 92–93
  flight velocities *vs.* wing loading, 94*f*
  inflation, 121
    analysis, 122–125
    reefing techniques, 121–122
  lateral motion, 115
    added mass, 115
    lateral instabilities, 120
    ram-air parachute axes, 117*f*
    turn maneuver, 117–120
  longitudinal static stability
    canopy rigging, 93–98
    trailing-edge deflection effect, 98–101
  in steady gliding flight, 92*f*
  suspension lines, 74
Ram-air wing aerodynamics, 76
  drag coefficient, 80
    contributory elements, 81
    experimental and theoretical drag
      coefficient for ram-air wing, 82*f*
    parameter for rectangular planform
      wings *vs.* aspect ratio, 81*f*
    theoretical drag coefficient for
      ram-air wings, 82*f*
  lift coefficient, 756
    for Clark Y ram-air parachute
      airfoil, 77*f*
    experimental and theoretical lift
      coefficients for ram-air wing, 79*f*
    lifting-line theory, 77
    parameter for rectangular planform
      wings *vs.* aspect ratio, 77–78
    for ram-air wing for different aspect
      ratios, 80*f*
    sheet strength, 76
    slope of lift curve for low-aspect-ratio
      wing *vs.* aspect ratio, 78*f*
  lift-to-drag ratio, 82–83
Random numbers generator (RNG), 772
Random-access-memory (RAM), 693
RANS equations. *See* Reynolds-averaged
    Navier–Stokes equations
RAWIN balloon. *See* Radiosonde Wind
    Sounding balloon

Recursive least-squares (RLS), 417
Recursive weighted least-squares
    algorithm (RWLS algorithm), 787,
    824–825
Reefing techniques, 121–122
Reference trajectory (RT), 604
Regression analysis, 331–332
Reynolds-averaged Navier–Stokes
    equations (RANS equations), 222
RF. *See* Radio frequency
Rigging angle, 94
  control, 547*f*
Rigging schemes, 300
  eight-DoF model, 301*f*
  nine-DoF model, 301*f*
  payload, 302*f*
  seven-DoF model, 300–301
Rigid-body models. *See also* Six-DoF model
  apparent mass pressures, 277–281
  inertia tensors, 277–281
  rotation matrices, 264
    crab angle, 267
    Earth frame, 265
    inertial frame, 269
    rotation matrix between NED and
      ENU, 268
    views of PADS, 265*f*
  rotational dynamics, 269–270
  three-and four-DoF models, 271–277
  translational dynamics, 269–270
Ripcord switch, 697–698
RLS. *See* Recursive least-squares
RMS. *See* Root-mean-square
RMSE. *See* Root mean squared error
RNG. *See* Random numbers generator
Robust (Octree) tetrahedral/mixed
    unstructured grid, 221
Root mean squared error (RMSE), 397, 660
Root-mean-square (RMS), 535
Ropes and rings technique, 121
Rotation axis tradeoffs, 573*f*
Rotation matrices, 264
  crab angle, 267
  Earth frame, 265
  inertial frame, 269
  rotation matrix between NED and
    ENU, 268
  views of PADS, 265*f*

Rotational dynamics, 269–270
RT. *See* Reference trajectory
RWLS algorithm. *See* Recursive weighted least-squares algorithm

S-patterns, 452
S-turn pattern, guidance strategies utilizing, 456
    ideal landing trajectory, 458*f*
    ParaLander PADS, 459
    for surface-layer wind identification, 459
    two-stage FA examples, 459*f*
SA. *See* Selective availability
SAASM. *See* Selective Availability Anti-Spoofing Module
Safety fans tool, 187–188
    CARP computation, 195, 197
    entered ADS data correctness check, 196*f*
    GUI, 181, 194–197
    PADS input data, 195*f*
    parameters of generic two-stage pads 192**t**
    stuck in fixed position, 193*f*
    three-stage system, 188–192
SBIR program. *See* Small Business Innovation Research program
Scale-Invariant Feature Transforms (SIFT), 740
Scaling effects, 105, 383
    anhedral angle, 383
    control authority per 1-deg tilt *vs.* normalized roll moment of inertia, 388*f*
    control authority *vs.* canopy area, 387*f*
    design parameters for large ram-air parachute, 106**t**
    dimensionless time of pull, 106
    mass ratio, 105
    modeled ADS payload mass *vs.* canopy area, 384*f*
    modeled PADS height *vs.* canopy area, 384*f*
    normalized moments of inertia *vs.* canopy area, 387*f*
    PADS c. g. height above payload c. g. *vs.* canopy area, 385*f*

PADS moments of inertia *vs.* canopy area, 386*f*
phugoid period *vs.* PDS canopy area and length, 388*f*
pitch amplitude decay ratio *vs.* PADS canopy area and length, 389*f*
response of large ram-air parachute, 108–110*f*
test flight, 107
Second cost function, 608
Selective availability (SA), 6
    errors, 410
Selective Availability Anti-Spoofing Module (SAASM), 15
Semirigid parafoil model, 255*f*
    wind-tunnel testing, 254*f*
Sensitivity analysis, 785
Seven-DoF models, 316
    moment equations, 319
    set of equations, 317
    single matrix equation, 309
    summary of equations for, 350**t**
Sherpa 2200 PADS examples, 233*f*
SID. *See* System identification
SIFT. *See* Scale-Invariant Feature Transforms
Simplex search Nelder–Mead algorithm, 464
Simulink model, 764
Single-criteria identification, 765*f*
Single-input single-output discrete system (SISO discrete system), 522
SISO discrete system. *See* Single-input single-output discrete system
Six-DoF model, 281
    aerodynamic force, 283
    apparent mass, 284
        center, 285
    of controllable circular parachute, 629
        apparent mass coefficients values, 635–636
        assumptions, 630
        coordinate frames and ADS geometry, 630*f*
        distortion of canopy's shape, 631
        geometry-mass properties of AGAS components, 633
        Lagrange approach, 631

Six-DoF model (*Continued*)
  relevant formulas of moments of
    inertia for parachute components,
    634t
  static mass center and moments of
    inertia, 633
  dynamic equations, 282, 285–286
  implementations, 288
    aerodynamic coefficients, 290, 292
    comparison of flight-test data, 298f
    GR, 296
    longitudinal response, 295f
    modeling of coefficient, 291–293f
    nonstandard nonlinear system
      identification technique, 296
    roll and yaw rate responses, 296f
    in Simulink developmental
      environment, 290f
    vertical profiles *vs.* time, 297f
    WindPack dropsonde, 295–296
  lift and drag coefficients, 288f
  nonlinear differential equations, 288
  parafoil-payload system, 282f
  parameters of six-DoF model of
    Snowflake PADS, 289t
  rewritten for connection point,
    302–305
  summary of equations for, 349t
  system dynamics, 287
  validation
    eigenvalues, time period, and
      damping of linearized original and
      identified model, 760t
    estimated parafoil aerodynamic
      derivatives, 758t
    identified nonlinear control
      efficiency, 760f
    and nonlinear control efficiency, 758
    result of identification, 760
    time histories, 759f
Small Autonomous Parafoil Landing
  Experiment (ALEX), 38
Small Business Innovation Research
  program (SBIR program), 31
Small ram-air parachutes, longitudinal
  dynamics of, 103
  design parameters, 103t
  highly damped phugoid motion, 105

response of small ram-air parachute to
  symmetric TE deflection, 104f
Smart Parafoil Autonomous Delivery
  System (SPADeS), 25–26
  with Para-Flite FireFly canopies, 27f
SMD. *See* Surface-mount device
SNCA. *See* Système de Navigation pour
  Charge Accompagnée
Snowflake PADS, 38–39
SODAR. *See* Sonic detection and ranging
Software control routine, 667f
Sonic detection and ranging
  (SODAR), 395
SPADeS. *See* Smart Parafoil Autonomous
  Delivery System
SPADeS 1000 PADS examples, 233f
Spalart–Allmaras turbulence model, 222
Speed systems, 571
Speeded Up Robust Features (SURF), 740
Spiral divergence, 120
Spoiler control configuration, 593f
Stability and steady-state performance, 353
  lateral-directional motion, 369
    lateral-directional stability in steady
      glide, 369–376
    steady turn response to control input,
      376–383
  longitudinal static stability analysis, 361
    control derivatives, 366–368
    forward center-of-gravity limit,
      364–366
    minimal lift coefficient at trim and aft
      center-of-gravity limit, 368–369
    notation and reference frame, 362f
    pitching-moment coefficient of
      parachute, 362
    quadratic equation, 363
    trim and stability, 361–364
  scaling effects, 383
    anhedral angle, 383
    control authority per 1-deg tilt *vs.*
      normalized roll moment of inertia,
      388f
    control authority *vs.* canopy area, 387f
    modeled ADS payload mass *vs.*
      canopy area, 384f
    modeled PADS height *vs.* canopy
      area, 384f

normalized moments of inertia *vs.* canopy area, 387*f*

PADS c.g. height above payload c. g. *vs.* canopy area, 385*f*

PADS moments of inertia *vs.* canopy area, 386*f*

phugoid period *vs.* PDS canopy area and length, 388*f*

pitch amplitude decay ratio *vs.* PADS canopy area and length, 389*f*

stability analysis of linearized model, 354

ALEX PADS model root locus, 358–359*f*

amplitude and phase angle, 357**t**

coupled roll-spiral mode, 356*f*

Dutch-roll mode, 356*f*

eigenvalues, 354–355, 361

eigenvalues of seven-DOF model of Snowflake PADS, 360**t**

magnitudes of eigenvectors, 358*f*

short-period and phugoid modes, 355*f*

Stall turn, 534

Static stability analysis, 95*f*

Steerable cluster designs, 650*f*

Strouhal number, 234

Subscale drogue deployment, 540*f*

Subscale test parameters, 539**t**

Superstructure, 698

SURF. *See* Speeded Up Robust Features

Surface-layer wind measuring

accommodating MAXMS dropsonde on UAV, 148–149

closeness of proximity and time, 143

MAXMS dropsonde, 140–141

miniature weather station, 147–148

PADS Mission Planner, 147

RAWIN balloon, 139–140

sample formats of atmospheric data, 142–143*f*

sounding balloon data, 146*f*

TASK, 140–141

technology improvement and miniaturization, 140

wind data, 145*f*, 147*f*

wind profiles, 143–144

WindPack and QNA dropsonde, 140*f*

Surface-layer wind modeling, 166–167, 487

accommodating crosswinds and vertical wind disturbances, 495

inverse kinematics, 496

prototype system, 501*f*

reference trajectory, 499

Snowflake PADS, 500*f*

UL-and ML-weight PADS, 500–501

vertical component of winds, 497–498

accommodating linear and logarithmic, 490

altitude budget equation, 494

averaged winds, 495

inverse kinematics, 492

new terminal point, 491–492

wind magnitude, 493

altitude dependences of PADS estimation, 489*f*, 491*f*

comparison of wind models, 171*f*

four-dimensional simulation grid spanning, 169

Gaussian white noise, 169

landing precision for Monte Carlo simulations, 171*f*

near-surface wind models, 167–168

effect of unsettled, 487–490

x, y, and *z* components, 170*f*

Surface-mount device (SMD), 705

Symmetric braking, 564

Symmetric spoiler configurations, 583. *See also* Asymmetric spoiler configurations

addition of sealing flaps to simple slits, 584

effects of varying spoiler construction, 585*f*

flight-test results, 584, 585

glide rate control authority *vs.* spoiler span, 587*f*

glide ratio control authority *vs.* chordwise location, 590*f*

LE *vs.* TE deflection of slit, 584

effect of moving spoilers aft of nominal location, 588*f*

effect of moving spoilers forward of nominal location, 589*f*

Symmetric spoiler configurations
  (*Continued*)
  effect of spoiler span, 586
  unactuated and actuated upper-surface
    slit spoiler on test canopy, 583*f*
  effect of varying spoiler span, 586*f*
System identification (SID), 741, 745–746
  objective in, 753
  Quad-M basics of, 743*f*
System Markov parameters, 825
Système de Navigation pour Charge
  Accompagneè (SNCA), 32

T-approach guidance, 447, 450–451
  computer simulations, 453–454
    GNC algorithm, 456
    Monte Carlo simulations, 455
    steep-descent Monte Carlo analysis,
      455*f*
    steep-descent trajectory, 455*f*
  guidance output, 453
  guidance strategies utilizing S-turn
    pattern, 456–459
  for routes, 453*f*
  S-patterns, 452
  T-distribution, 451
TA. *See* Target area
Tactical Atmospheric Sounding Kit
  (TASK), 140
Target area (TA), 611
Target-seek control strategy, 623
TASK. *See* Tactical Atmospheric
  Sounding Kit
Taylor-series expansion, 401
TBL. *See* Turbulent boundary layer
TE. *See* Trailing edges
TE deflection. *See* Trailing-edge deflection
TEMA Camera Control, 727–728
Terminal rigging-angle control, 574
  CEP/average miss distance, 581**t**
  control inputs during example
    approach, 580*f*
  downrange miss distance histograms,
    580*f*
  final approach trajectories with and
    without GC control, 579*f*
  glide slope over ground *vs.* rigging angle
    and brake, 577*f*

Monte Carlo simulations, 581
  nominal forward speed and GC, 575
  nonlinear proportional GC
    controller, 582
  normalized error between nominal and
    required GC, 576
  optimal trajectory, 574–575
  optimal set of control inputs, 578
  PID controller in discrete form, 574
  range of "optimal" control inputs at
    increasing wind levels, 578*f*
  simulated landing accuracy *vs.* mean
    wind speed, 582*f*
  visualization of commanded
    GC logic, 577*f*
  wind profiles for example approach,
    582*f*
Three-degree-of-freedom model
  (Three-DoF model), 263, 271
  dynamic effects, 276–277
  Earth-referenced flight-path angle, 271
  GNC algorithms, 273
  L/D ratio, 274
  lift and drag coefficients, 276
  parameters of ALEX PADS, 272–273
  summary of equations for, 348**t**
Three-dimensional model (3D model), 219
  projections, 642*f*
Three-DoF model. *See*
  Three-degree-of-freedom model
Time-domain ERA algorithm, 825
Time-optimal control, 609*f*
Time-space-position information solution
  (TSPI solution), 714, 716, 723
  azimuth and elevation data, 726–727*f*
  position and descent rate estimates,
    726*f*, 728*f*
  reconstructed payload trajectory,
    725*f*, 727*f*
  time history of 3D position, 723, 725
TIP. *See* Turn initiation point
Touchdown accuracy analysis, 49
  CEP estimation, 50
  data rotation and CEP_{MPI}
    calculation, 54*f*
  effectiveness *vs.* suitability, 57*f*
  F statistic, 57
  nonparametric data conditioning, 51*f*

normality test, 52*f*
self-guided PADS, 58
t test, 53
UCL, 55
Touchdown error, 1, 13, 46, 127, 131, 173,
    178, 424, 483, 628*f*, 664*f*
TPBV problem. *See* Two-point
    boundary-value problem
TR. *See* Turn rate
TrackEye motion analysis, 727, 729*f*
    scoring calibration monuments, 729*f*
    test item tracking, 730*f*
Tracking algorithm, 731–732
Trailing edges (TE), 3, 75
Trailing-edge deflection (TE deflection),
    200, 272, 529
    deflection effect, 98
    flight parameters, 98
    influence
        flight velocities, 101*f*
        on glide angle, 102*f*
        lift and drag coefficients, 100*f*
        trim values, 100*f*
        pitching-moment coefficient for ram-air
            parachute with rigging angles,
            99*f*, 101–102*f*
Trailing-edge release effects, 111
Translational dynamics, 269–270
Trim mode, 448
Trimble Lassen SK8, 686
TSPI solution. *See* Time-space-position
    information solution
Turbulent boundary layer (TBL), 251
Turn initiation point (TIP), 465
Turn maneuver, 117
    sine of angle of bank, 120
    yaw and roll axes in steady turning
        flight, 118
    yawing moment on ram-air wing, 119*f*
Turn rate (TR), 43
Two-DoF guidance, 673. *See also* One-DoF
    guidance
    performance envelope, 673
        balloon wind profile, 675*f*
        boundary envelope construction for
            winds, 674
        construction process, 676*f*
        control approach, 677

full performance envelope in 3D, 677*f*
        minimum actuation path, 679
        Nine-km-altitude layer, 676*f*
        parachute-payload system, 679
        single layer from performance
            envelope, 678*f*
    wind models, 679
        CEP, 680
        landing locations of uncontrolled and
            controlled ADS, 681–682*f*
        path-planning algorithm, 684
        release locations of simulated
            uncontrolled and controlled ADS,
            680*f*
        transition line and wind direction
            comparison, 683*f*
        types of errors, 679
        wind profiles used for predicted and
            actual wind profiles, 681*f*
Two-DoF terminal guidance, 469
    bird's-eye view of two typical Snowflake
        PADS trajectories, 485*f*
    BLG approach, 483
    duration of maneuver, 469–470
    error statistics, 484**t**
    IDVD *vs.* BLG, 486**t**
    IDVD-based approach, 475
    Monte Carlo simulation
        dispersion, 483*f*
    MPC, 478
    numerical algorithm, 475–476
    optimal guidance with wind
        disturbance, 477*f*
    optimal real-dynamics-augmented
        guidance, 479–480*f*
    optimal terminal guidance algorithm,
        476, 478
    precision placement trajectory, 481*f*
    solution of TPBV problem, 471
    speed factor, 473
    terminal guidance decision variables,
        482*f*
    two-point backward difference
        approximation, 474
Two-point boundary-value problem
    (TPBV problem), 422
U.S. Army Research, Development, and
    Engineering Center (NRDEC), 20

U.S. Army Yuma Proving Ground
    (YPG), 128
UAV. *See* Unmanned aerial vehicle
UCL. *See* Upper confidence limit
UH-1A helicopter, 128
UHF radiosonde. *See* Ultra-high-
    frequency radiosonde
Ultra-high-frequency radiosonde
    (UHF radiosonde), 140
ULW ADS, experiments with,
    546–547
    altitude *vs.* distance traveled for four
        flight tests, 520*f*
    CEP, 554–555
    comparison of simulated and measured
        GA, 551*f*
    deployed and undeployed parafoil test
        system, 547*f*
    dispersion for sensor errors and no
        wind, 555*f*
    dispersion for sensor errors and
        varying wind, 559*f*
    error statistics, 554*t*
    estimated lift coefficient, drag
        coefficient, and C_L/C_D, 520*f*
    flight tests, 548
    flight-test summary, 549*t*
    glide-slope guidance
        geometry, 553*f*
    histogram for sensor errors
        and no-wind condition, 556*f*
        and varying winds, 558*f*
    Monte Carlo simulations, 553–554
    parafoil rigging angle change and
        canopy geometry, 548*f*
    rigging angle control, 547*f*
    simulated AoA and velocities varying
        rigging μ, 553*f*
    simulated GC varying rigging μ, 552*f*
    system characteristics, 549*t*
Undeployed parafoil test
    system, 547*f*
Units conversions
    functions in Excel, 840*t*
    functions in MATLAB, 840*t*
    imperial units to metric, 838*t*
    metric units to imperial, 839*t*
    for temperature conversion, 837

Unknown surface-layer wind effect, 128.
    See also Control rate influence and
    actuator dynamics; Winds-aloft
    data utilization
    assumed *vs.* real surface-layer profiles,
        128–139
    ballistic winds, 149–153, 155–156
    descent and time rate to touchdown,
        153–167
    surface-layer wind measuring, 139–149
    surface-layer wind modeling, 166–171
Unmanned aerial vehicle (UAV), 23, 430
Upper confidence limit (UCL), 55
Upper-surface spoilers, glide-angle control
    with, 582
    approach path profiles for example
        trajectories, 596*f*
    asymmetric spoiler configurations,
        590–592
    asymmetric symmetric control inputs,
        597*f*
    computer simulations, 593
    effect of spoiler on GC over ground, 595*f*
    ground tracks for example approach
        trajectories, 596*f*
    landing dispersions from 250 landing
        simulations, 598*f*
    miss distance, 598*f*
    Monte Carlo simulations, 597
    response to asymmetric spoiler
        deflection, 594*f*
    response to symmetric spoiler
        deflection, 593*f*
    simulation results, 592–599
    spoiler control configuration, 593*f*
    symmetric spoiler configurations,
        583–590
    wind profile for example cases, 595*f*
UTC. *See* Coordinated Universal Time

Variable AoA Control System (VACS),
    570–571, 574
Variable rigging angle, glide-angle control
    with, 544
    of attack, pitch, and flight-path angles,
        545*f*
    canopy trim conditions, 546*f*
    experiments with larger PADS, 555–567

experiments with ULW ADS, 546–555
larger pads mechanization, 568–574
quarter-chord, 544
relationship of GC to flight-path
    angle, 545f
terminal rigging-angle control,
    574–582
varying effects, 544–546

WAAS. *See* Wide-area augmentation
    system
Waypoints (WPs), 393
WCA. *See* Wind correction angle
Weather Research and Forecasting Model
    (WRF Model), 169
"White-box" approach, 745
Wide-area augmentation system
    (WAAS), 411
Wind accommodation, 415
    comparison of logarithmic wind
        model, 418f
    correcting IPI position, 416f
    weighting coefficient for wind
        corrections, 417f
Wind correction angle (WCA), 519
Wind frame [w], 264–265
Wind-to-track angle (WTA), 519
WindPack radiosonde, 139–140
Winds estimation, 398
    algorithms, 410
    dilution of precision, 403f
    errors in measuring PADS states, 409t
    landing accuracy, 407
    linear algebraic equations, 402
    Monte Carlo simulations, 406
    NEF for wind and airspeed, 405f

Newton–Raphson formula, 408
    onboard wind estimation, 407f
    PADS trajectories, 399–400f
    Taylor-series expansion, 401
Winds-aloft data utilization, 181. *See also*
    Control rate influence and actuator
    dynamics; Unknown surface-layer
    wind effect
    CARP and Mission Planner, 181–187
    safety fans tool, 187–197
WPs. *See* Waypoints
WRF Model. *See* Weather Research and
    Forecasting Model
WTA. *See* Wind-to-track angle

X-38 PADS aerodynamics model,
    789–792
XTRK. *See* Cross-track

Yaw damping, 209
Yaw dynamics, 272
Yaw rotation scheme, 639f
Yaw/heading estimation, 396
    heading reconstruction with LCF and
        NCF, 398f
    LCF and NCF, 397
    three-point backward difference
        approximation, 397
YPG. *See* U.S. Army Yuma Proving
    Ground; Yuma Proving Ground
Yuma Proving Ground (YPG), 21

Ziegler–Nichols tuning rules, 666–667

2K PADS add-on, 536f
2K parafoil glide augmentation, 538f

SUPPORTING MATERIALS

A complete listing of titles in the Progress in Astronautics and Aeronautics series is available from AIAA's electronic library, Aerospace Research Central (ARC) at arc.aiaa.org. Visit ARC frequently to stay abreast of product changes, corrections, special offers, and new publications.

AIAA is committed to devoting resources to the education of both practicing and future aerospace professionals. In 1996, the AIAA Foundation was founded. Its programs enhance scientific literacy and advance the arts and sciences of aerospace. For more information, please visit www.aiaafoundation.org.

a) 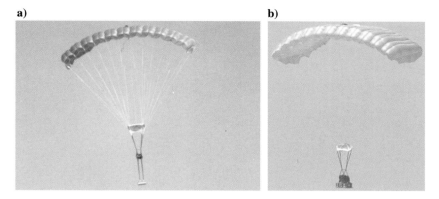 b)

Fig. 1.11 Two configurations of Pegasus PADS [Sego 2001].

a) 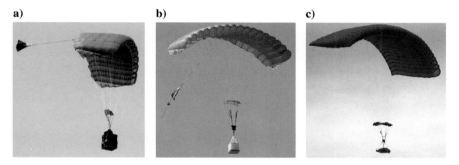 b) c)

Fig. 1.13 Sherpa Provider: a, b) 2K, and c) 10K [PATCAD 2007; PATCAD 2009].

a) 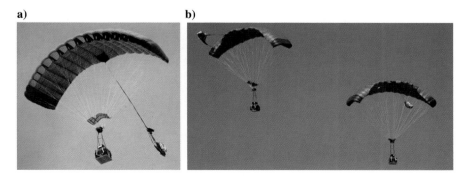 b)

Fig. 1.15 SPADeS with Para-Flite FireFly canopies [PATCAD 2009].

a)   b)

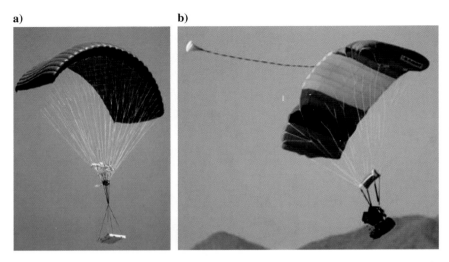

Fig. 1.16   a) Para-Flite 5K and b) Snowbird PADS [PATCAD 2003; PATCAD 2005].

a)   b)

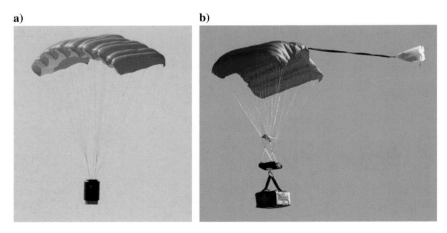

Fig. 1.17   a) STARA's 1 stage GNAT and b) 18 kg (40 lb) Mosquito PADS [PATCAD 2005; PATCAD 2007].

a)   b)

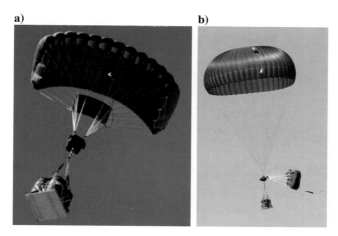

Fig. 1.18   The a) first and b) second stages of the Screamer 2K PADS [PATCAD 2005].

a)   b)

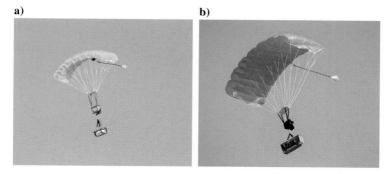

Fig. 1.22   Onyx ML under a) 4.7 m^2 (50 ft^2) and b) 11 m^2 (120 ft^2) canopies [PATCAD 2009].

a)   b)

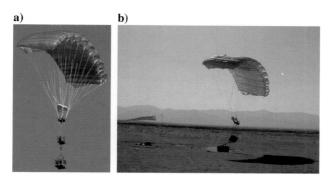

Fig. 1.23   CADS PADS [PATCAD 2005].

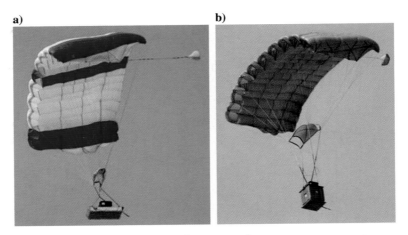

Fig. 1.24   Panther 500 under a 34-m^2, 5.4-kg (365-ft^2, 12-lb) parafoil [PATCAD 2005].

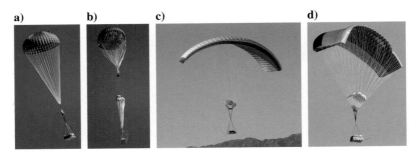

Fig. 1.26   DragonFly PADS deployment, glide, and flaring [George et al. 2005; PATCAD 2005].

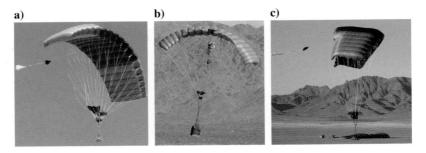

Fig. 1.27   FireFly PADS [Guided PADS 2014; PATCAD 2005].